Petroleum Well Construction

Petroleum Well Construction

Michael J. Economides
Texas A & M University

Larry T. Watters
Halliburton Energy Services

Shari Dunn-Norman
University of Missouri-Rolla

JOHN WILEY & SONS
Chichester • Weinheim • New York • Brisbane • Singapore • Toronto

Other Wiley Editorial Offices

John Wiley & Sons, Inc., 605 Third Avenue,
New York, NY 10158-0012, USA

Wiley-VCH Verlag GmbH, Pappelallee 3,
D-69469 Weinheim, Germany

Jacaranda Wiley Ltd, 33 Park Road, Milton,
Queensland 4064, Australia

John Wiley & Sons (Asia) Pte Ltd, 2 Clementi Loop #02-01,
Jin Xing Distripark, Singapore 0512

John Wiley & Sons (Canada) Ltd, 22 Worcester Road,
Rexdale, Ontario M9W 1LI, Canada

Library of Congress Cataloging-in-Publication Data

Petroleum well construction / edited by M.J. Economides, L.T. Watters, and
S. Dunn-Norman.
    p.   cm.
   Includes bibliographical references (p.   ) and index.
   ISBN 0-471-96938-9
   1. Oil wells — Design and construction.   I. Economides, Michael J.
II. Watters, L.T., Dunn-Norman, S.
TN871.2.P487   1988
622′.338 — dc21   97-31698
                                                   CIP

British Library Cataloguing in Publication Data

A catalogue record for this book is available from the British Library

ISBN 0 471 96938 9

Typeset by Keyword Typesetting Services Ltd.
Printed and bound in Great Britain by Bookcraft (Bath) Ltd
This book is printed on acid-free paper responsibly manufactured from sustainable
forestation, for which at least two trees are planted for each one used for paper production.

Contents

6 Rock Mechanics In Wellbore Construction 143

7 Casing And Tubing Design 175

15 Inflow Performance/Tubing Performance 403

16 Artificial-Lift Completions 433

17 Well Stimulation 471

18 Sand Stabilization And Exclusion 509

Preface

Halliburton is pleased to contribute to this important volume addressing vital technologies of the petroleum industry. Well construction represents a formidable investment for our customers and we are committed to providing high-value solutions encompassing the entire spectrum of the field.

This work represents joint authorship by some of the finest experts in our organization and recognized experts from our client's organizations and other institutions. I am particularly grateful for the contributions of the editors, Profs. Michael J. Economides of Texas A&M University, Shari Dunn-Norman of the University of Missouri-Rolla and Larry T. Watters from Halliburton's Duncan Technology Center.

The transfer of knowledge is essential to the health and success of an industry. It is our hope that this book will contribute to that success.

Dick Cheney
CEO and Chairman of Halliburton Company

Acknowledgments

In a multi-author book such as this, there are always many people to thank, the first of which are our contributors, who have invested substantial amounts of their time to produce a textbook for an important technical subject that did not exist before.

We certainly wish to express our gratitude to the managements of the various companies and organizations for allowing their employees to contribute to this professional activity. The management of Halliburton Energy Services deserves special recognition for both encouraging the project and bearing the bulk of the expenses for the preparation of the manuscript.

Certain people deserve thanks because without their work and perseverance in dealing with us, this book would never have come to fruition. Kristen Ashford, Angela Bailey, Laura Smith, Juleigh Giddens, and Darla-Jean Weatherford performed an admirable job in trying to bring some editorial consistency and performed an almost impossible task, considering the 40+ authors involved in this book. Jim Giddens, Kathy Mead, and Shanon Garvin prepared excellent manuscript illustrations and kept the flow of documents unimpeded. Keith Decker is acknowledged for his work on the cover and other artistic advice. Finally, our deep thanks go to Michele Moren who has kept this project going, often in spite of ourselves. Clearly, the completion of this book is to a large extent because of her work.

The Editors
Duncan, Oklahoma
July 2, 1997

List of Contributors

Editors

Michael J. Economides, *Texas A&M University*
Larry T. Watters, *Halliburton Energy Services*
Shari Dunn-Norman, *University of Missouri-Rolla*

Contributors

Hazim Abass, *Halliburton Energy Services*
Morris Baldridge, *Halliburton Energy Services*
James Barker, *Halliburton Energy Services*
Robert Beirute, *Amoco Production Company*
R. Clay Cole, *Halliburton Energy Services*
Dan R. Collins, *Halliburton Energy Services*
Ali A. Daneshy, *Halliburton Energy Services*
Iain Dowell, *Halliburton Energy Services*
Shari Dunn-Norman, *University of Missouri-Rolla*
Michael J. Economides, *Texas A&M University*
K. Joe Goodwin, *Mobil Exploration and Production*
Mary Hardy, *Halliburton Energy Services*
W. E. (Bill) Hottman, *Halliburton Energy Services*
Richard Jones, *ARCO Exploration and Production Technology*
Hon Chung Lau, *Shell E&P Technology Company*
James F. Lea, *Amoco Production Company*
Thomas Lockhart, *Eniricerche*
James R. Longbottom, *Halliburton Energy Services*
Wade Meaders, *Halliburton Energy Services*
Andrew A. Mills, *Esso Australia Ltd.*

John W. Minear, *Halliburton Energy Services*
Stefan Miska, *University of Tulsa*
Robert F. Mitchell, *Enertech*
Larry Moran, *Conoco*
Justo Neda, *Intevep*
Lewis Norman, *Halliburton Energy Services*
Ronald E. Oligney, *Texas A&M University*
Michael L. Payne, *ARCO Exploration & Production Technology*
Kris Ravi, *Halliburton Energy Services*
Clark E. Robison, *Halliburton Energy Services*
Colby Ross, *Halliburton Energy Services*
Phil Snider, *Marathon Oil Company*
Mohamed Soliman, *Halliburton Energy Services*
Mike Stephens, *M-I Drilling Fluids L.L.C.*
Lance D. Underwood, *Halliburton Energy Services*
Peter Valko, *Texas A&M University*
Sanjay Vitthal, *Halliburton Energy Services*
Randolf R. Wagner, *Enertech*
Larry T. Watters, *Halliburton Energy Services*
Charlie Williams, *Halliburton Energy Services*
Mario Zamora, *M-I. Drilling Fluids L.L.C.*

Nomenclature

A	area	k	reservoir permeability
A_0	fracture area at tip screen out	k_1	permeability in the equivalent radial damaged zone
A_c	cross-sectional area		
A_{cem}	cross-sectional area that is cemented	k_2	permeability in the zone of fracture-face invasion (outside the radial damage zone), md
A_h	surface area of hole for heat transfer		
A_i	surface area of inner casing for heat transfer	k_3	permeability in the zone of fracture-face invasion (inside the radial damage zone), md
B	formation volume factor		
b_1	radial damage radius	k_f	proppant-pack permeability
b_2	fracture-face damage depth	k_{for}	thermal conductivity of a formation
c	specific heat capacity	k_g	effective permeability to gas
c	compressibility	K	fluid consistency index
c_b	bulk compressibility	K_{IC}	critical stress intensity factor (fracture toughness)
c_{ma}	matrix compressibility		
c_t	total reservoir compressibility	k_o	effective permeability to oil
C	cohesive strength	k_r	reservoir permeability
C_{fD}	dimensionless fracture conductivity	k_w	effective permeability to water
C_j	cohesion for jointed rocks	L	length
C_L	Carter leakoff coefficient	L_D	dimensionless horizontal well length
$C_{L,p}$	Carter leakoff coefficient with respect to permeable layer	L_f	fracture half-length, ft
		m	mass flow rate
C_o	uniaxial compressive strength	$m(p)$	gas pseudopressure
C_s	concentration of solids by volume in the fluid	M	mass
		M_{tso}	total sand weight
C_w	wall component of leakoff coefficient	n	flow behavior index
D	diameter	N_{re}	Reynolds number
D_i	inner diameter of pipe	N_{ReBP}	Reynolds number of a Bingham plastic fluid
D_o	outer or cut diameter of pipe	p	pressure
D_w	hole diameter	p_{BHP}	bottomhole pressure
e	eccentricity	p_C	closure pressure
E	Young's modulus	p_c	confining pressure
E'	plane strain modulus	p_D	dimensionless pressure
f	friction factor	p_e	outer boundary constant pressure
g	acceleration of gravity, ft/sec^2	p_f	fracture propagation pressure for a given rate
g_c	units conversion factor, 32.17 lbm ft/(lbfsec^2)	p_i	initial pressure
h	formation thickness, ft	p_m	mud pressure required to prevent plastic zone generation
h_f	fracture height, m		
h_p	permeable height, m	p_{nw}	non-wet fracture pressure
H	depth, ft	p_r	reservoir pressure
$ISIP$	instantaneous shut-in pressure	p_w	wellbore pressure
J	Productivity index		

p_{wf}	flowing bottomhole pressure		V	volume of one fracture wing
Δp	friction pressure		$V_F(t_0)$	fracture volume at tip screen out
Δp	pressure change		V_i	volume of injected fluid into 1 fracture wing
Δp	drawdown pressure		V_u	pipe capacity
Δp_n	net pressure during fracturing		w	fracture width
Δp_{nw}	pressure drawdown in the fracture caused by nonwetting tip		w_e	average fracture width at end of pumping
$\Delta p(t_0)$	net pressure at tip screen out		w_{nw}	fracture width at the beginning of the nonwetting tip
q	volumetric flow rate		w_p	average propped width
q_L	fluid leakoff rate		Δx	effective formation thickness for heat transfer
r	distance from fracture tip		x_f	fracture half-length
r_e	drainage radius		Z	gas compressibility factor
R_f	radius of a radial fracture			
r_i	inner radius of casing			
r_o	outer radius of casing			
R_o	filter-cake resistance			
r_p	ratio of permeable to fracture area		α	poroelastic constant (Chapters 6-7)
r_w	well radius		α	exponent of fracture area growth, dimensionless (Chapter 17)
s	skin factor			
s_d	effective skin factor caused by radial damage and fracture-face damage		α	parameter of the Fan-Economides model (Chapter 19)
s_f	fracture pseudo-skin		β	index of anisotropy
s_{ND}	skin due to non-Darcy flow		γ	ratio of average to maximum width, dimensionless (Chapter 17)
s_t	total composite skin			
S_f	fracture stiffness		γ	gas gravity (Chapter 15)
S_p	spurt-loss coefficient		$\dot{\gamma}$	shear rate
$S_{p,p}$	spurt loss coefficient		ε	proppant schedule exponent and also pad fraction, dimensionless
t	time			
t_a	pseudotime		ϕ	effective porosity, fraction
T	tensile strength		ϕ_j	friction angle for jointed rock
T	torque		ϕ_p	proppant pack porosity, dimensionless
T	temperature		μ	viscosity
T_o	temperature of outgoing fluid		μ_a	apparent viscosity
T_a	temperature in annulus		μ_e	equivalent Newtonian viscosity
T_c	temperature in casing		μ_p	plastic viscosity of a Bingham plastic fluid
t_D	dimensionless time (with respect to well radius)		μ_r	viscosity of reservoir fluid, Pas
t_{Dxf}	dimensionless time (with respect to fracture half-length)		ν	Poisson's ratio, dimensionless
			η	displacement efficiency
t_e	equivalent pseudotime		η	hydraulic diffusivity
t_p	production time, hr		η	fluid efficiency, dimensionless
T_{for}	temperature in formation		η_e	fluid efficiency at end of pumping, dimensionless
T_i	temperature of incoming fluid		ρ	density, lb/ft^3
U_h	overall heat transfer coefficient based on hole area		σ	stress, psi
			σ'	effective stress, psi
U_i	overall heat transfer coefficient based on inside area		σ_n	stress normal to a failure plane
			τ	shear stress, Pa
U_o	overall heat transfer coefficient based on outside area		τ_f	shear stress exerted by fluid
			τ_o	yield stress of a Bingham plastic fluid
V	volume of one fracture wing, m^3		τ_y	resistance of gelled drilling-fluid downhole
v	velocity		τ_o	inherent shear strength
v_d	deposition velocity		Ω	rotational speed of a concentric viscometer

1 Introduction to Drilling and Well Completions

Michael J. Economides
Texas A&M University

Shari Dunn-Norman
University of Missouri-Rolla

Larry T. Watters
Halliburton Energy Services

1-1 IMPORTANCE OF OIL AND GAS WELLS

Few industries and certainly no other materials have played such a profound role in modern world history and economic development as petroleum.

Yet deliberate access to geologic formations bearing petroleum through drilled wells is relatively recent. The "Drake well," drilled in the United States by Colonel Edwin L. Drake in 1859, is considered by many to be the first commercial well drilled and completed. It heralded the creation of an industry whose history is replete with international adventure, color, frequent intrigue, and extraordinary characters. Many believe that the majority of twentieth century social and political events, including two World Wars, a Cold War, and many regional conflicts are intimately connected to petroleum.

Until the late 1950s, much petroleum activity was originated and based in the U.S. Amyx, *et al.* (1960) reported that through 1956, the cumulative world crude-oil production was 95 billion bbl, of which 55 billion had been produced in the U.S.

Beginning tenuously in the early 1900s, speeding up in the period between the World Wars, and accelerating in the 1960s, petroleum exploration and production became a widely international activity. In the late 1990s, the U.S. is still the world's largest oil consumer both in terms of shear volume (18.2 million bbl/d) and, overwhelmingly, per capita (28 bbl/person/year compared to 1 bbl/person/year in China). The U.S. is also the largest petroleum importer (9.5 million bbl/d, representing over 50% of consumption); worldwide production is about 62 million bbl/d. The bulk of petroleum reserves is clearly outside the industrialized world of North America and Western Europe (combined 57 billion bbl vs. 1.1 trillion bbl worldwide). The majority of petroleum is found in the Middle East, where 600 billion bbl are produced, 260 billion of which are from Saudi Arabia alone.

Drilling activity is reflected by the geographical shifting of petroleum operations. The numbers of drilling rigs are now roughly equally distributed between North America and the remainder of the world, although this statistic is somewhat misleading. Wells drilled in mature petroleum environments, such as the continental U.S., are far less expensive, and drilling prices rely on mass utilization but, of course, production rewards are lackluster. In the U.S. and Canada, approximately 34,000 wells were drilled during 1995 and 1996, representing almost 60% of all wells drilled worldwide (about 58,000). Yet the United States and Canada, combined, account for only 13% of the world's petroleum production.

On the contrary, offshore drilling from either platforms or drill ships, drilling in remote locations, or drilling in industrially and developmentally deficient countries is far more expensive and involved.

Maturity in petroleum production is characterized by a marked decrease in both the total production rate and the petroleum rate per well, in addition to an increase in the water-oil ratio. "Stripper wells," representing the vast majority of wells in the USA, imply a production of less than 20 bbl/d of petroleum and a total production rate where water constitutes more than 90%.

Darcy's law, the most fundamentally basic petroleum engineering relationship, suggests that the production rate is proportional to the pressure driving force (drawdown) and the reservoir permeability:

$$q \propto k\left(\bar{p} - p_{wf}\right) \qquad (1\text{-}1)$$

This law can readily explain current worldwide petroleum activities and the petroleum industry's shifting focus. Mature petroleum provinces are characterized by depletion in the reservoir pressure or by the necessity to exploit less attractive geologic structures with lower permeability, k.

The permeability in Equation 1-1 is effective; that is, it is the product of the *absolute* permeability and the *relative* permeability of a flowing fluid competing with other fluids for the same flow paths. The relative permeability is a function of saturation. Thus, water influx from an underlying aquifer not only results in an increase in water production (which is a nuisance in itself) but an associated decrease in petroleum relative permeability and the petroleum portion of the total production rate.

Although such problems do not burden newer reservoirs to the same extent, it must be emphasized that all petroleum reservoirs will follow essentially the same fate.

The two extreme fields of operation, mature reservoirs on land in developed nations and newer discoveries either offshore or in developing countries, result in very different well construction costs.

These costs range from a few hundred thousand dollars to several million dollars. (Or tens of millions if ancillary costs such as the extraordinary testing or the building of an artificial island in the Arctic are considered.)

The total annual worldwide expenditure for petroleum well construction is estimated at over \$100 billion. To give a relative measure for this figure (and to avoid a misunderstanding from a reader) only a handful of nations have national budgets of larger magnitude. This book will provide a comprehensive and integrated treatment of today's technology for the substantial and profoundly international industrial activity of constructing oil and gas wells.

1-2 PETROLEUM FORMATIONS

Because the development of well construction technology has had a rather fragmented past and many practitioners are not trained as either reservoir or production engi-

neers, it is worthwhile here to provide an elementary description of the targets that the drilling of a well is supposed to reach.

1-2.1 Petroleum Fluids

Petroleum is a mixture of hydrocarbons consisting of about 11 to 13% (by weight) hydrogen and 84 to 87% carbon. Chemically, "crude" petroleum may include several hundred compounds, encompassing practically all open-chain and cyclic hydrocarbons of single, double, and triple bonds.

A description of these mixtures by composition was abandoned early in industrial history with the exception of very generic divisions that denote important distinguishing content (such as, paraffinic or asphaltenic crudes). Instead, bulk physical properties such as density and viscosity have been used to describe crude behavior.

Specifically, the phase and thermodynamic behavior has been reduced to the simplifying division of crude petroleum into (liquid) oil and (natural) gas. While such a description is apparent and relatively easy to comprehend given a temperature and pressure, crude petroleum content is generally referred to as volumes at some standard conditions (for example, 60°F and atmospheric pressure). With the definition of pressure and temperature, a volume unit also clearly denotes mass.

Oil, then, consists of higher-order hydrocarbons such as C6+ with much smaller and decreasing quantities of lower-order hydrocarbons, while gas consists of lower-order hydrocarbons—primarily methane and some ethane—with much smaller amounts of higher-order hydrocarbons.

An important variable is the *bubblepoint pressure* which, for a given temperature, denotes the onset of free-gas appearance. At lower pressures, oil and gas coexist.

Petroleum found at conditions above the bubblepoint pressure is all liquid and is referred to as *undersaturated*. Below the bubble point, the petroleum is referred to as *two-phased* or *saturated*. At considerably lower pressure and below the *dew point* pressure, hydrocarbons are all in the gaseous state.

In all natural petroleum accumulations, water is always present either as interstitial, cohabiting with the hydrocarbons, or underlying, in the form of (at times very large) aquifers.

1-2.2 The Geology of Petroleum Accumulation

Petroleum is found chiefly in sedimentary basins, and although fanciful theories of inorganic origin have surfaced in the past, it is almost universally accepted that petroleum has its organic origin in a *source* rock.

Decay of organic remains under pressure and temperature and under conditions preventing oxidation and evaporation has been the most likely process in the formation of petroleum. Associated saline water suggests environments near ancient seas, and thus, a plausible and often repeated scenario is one of ancient rivers carrying organic matter along with sediments and depositing successive layers, eventually buried by substantial overburden.

The formation of petroleum was followed by accumulation. The gravity contrast between hydrocarbons and water, along with capillary effects, would force oil and gas to migrate upward through rock pores. Connected pores provide permeability, and the ratio of pore volume to the bulk volume, the *porosity*, is one of the most important variables characterizing a petroleum reservoir.

The natural tendency of hydrocarbons to migrate upward would continue to the surface unless a trapping mechanism intercedes. This is precisely what happened.

At depths as shallow as a few tens of feet to over 30,000 ft, natural *traps*, which are special geological formations, allowed the accumulation of the migrating hydrocarbons. Common to all cases is an overlain impermeable layer forming a *caprock*.

Figure 1-1 (after Wilhelm, 1945) sketches some of the most common petroleum traps. *Convex traps,* either by simple folding (Figure 1-1A) or because of differences in reservoir thickness (Figure 1-1B) and overlain by an impermeable layer are the easiest to intersect with drilling.

A *permeability trap* (Figure 1-1C) and a *pinchout* trap (Figure 1-1D) denote that laterally and upward the

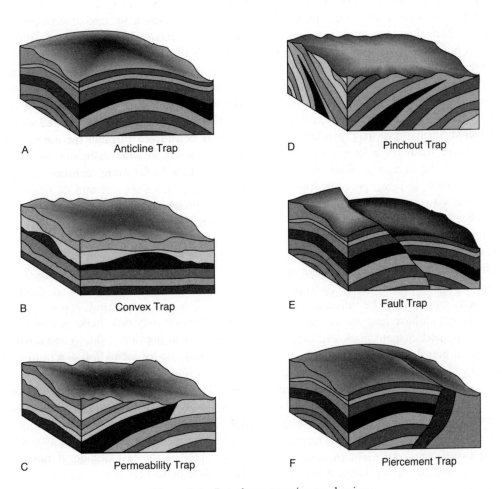

A Anticline Trap	D Pinchout Trap
B Convex Trap	E Fault Trap
C Permeability Trap	F Piercement Trap

Figure 1-1 Petroleum trapping mechanisms

permeable rock vanishes. Such traps may have been created by the rotation of layers, and they can sometimes be associated with continental rift and subsequent drift.

An interesting trapping mechanism is provided by the movement of faults (Figure 1-1E). Upward or downward motion of the layers on one side of the fault may bring an impermeable layer against a permeable one, and this interface can form a very effective trap. At times, the "sister" formation of a structure can be found several hundred feet above or below, and it may also contain attractive quantities of petroleum.

Finally, *piercement traps* (Figure 1-1F), formed by the intrusion of a material of different lithological composition, may form an effective seal to a petroleum trap.

While a trap may contain a petroleum reservoir (defined as a structure in hydraulic communication), oil may coexist with overlain gas, gas may be the only hydrocarbon and, in all cases, water is likely to underlie the hydrocarbons.

An oil field (or a gas field) may contain many reservoirs distributed either laterally or in layers, often separated by nonhydrocarbon formations that may be considerably thicker than the reservoirs themselves. Furthermore, the contained hydrocarbons, reflecting geological eras that may be separated by millions of years, may have considerably differing make-ups. Coupled with different lithological properties and reservoir pressures, more often than not, petroleum production from multilayered formations may preclude the commingling of produced fluids for a variety of operational reasons (including the danger of fluid crossflow through the well from higher to lower pressure zones).

A reservoir itself may be separated into different *geological flow units*, reflecting the varying concentration of heterogeneities, anisotropies, and reservoir quality, such as thickness, porosity, and lithological content.

The era of finding petroleum reservoirs through surface indicators (such as outcrops), conjecture, and intelligent guesses has been replaced by the introduction of seismic measurements, which have had one of the most profound influences on modern petroleum exploration and, in recent years, on petroleum production.

Artificially created seismic events (air bubbles offshore, large vibrators on land) send seismic waves downward. Reflected and refracted through formations, these vibrations are detected back on the surface. Processing of the signals results in the construction of seismic response images that can be two-dimensional (2D), three-dimensional (3D), or even four-dimensional (4D), if taken at different time intervals.

Seismic measurements are then processed and can be represented by a 3D visualization (Figure 1-2). For such an image to be constructed, massive amounts of data are collected and processed through very powerful computers that use sophisticated algorithms.

Figure 1-2 displays a typical, processed seismic 3D volume of amplitude vs. time. Much more data is collected than is displayed in Figure 1-2. Seismic attributes such as reflection, strength, phase frequency, and others may be correlated with several reservoir properties such as porosity, net pay, fluid saturation, and lithological content.

Modern formation characterization is the integration of many measurements that allow for a more appropriate reservoir description and improved reservoir exploitation strategies. Formation characterization involves the combination of various modeling approaches, including geological descriptions and pore volumes, and it is often combined with production history matching.

With powerful visualization and interpretation technologies, as shown in Figure 1-3, geoscientists and engineers can examine a seismic or geological data volume and identify and isolate significant features in ways not possible before. This new means of geological visualization is the basis of modern formation characterization; it is rapidly forcing the abandonment of the traditional, yet simplistic reservoir approximations of parallelepiped boxes or cylinders. While single-well drainages can be tolerably considered through the use of simple approximations, reservoir-wide estimates of hydrocarbons-in-place can now be far more realistic and inclusive of heterogeneities.

The expression for oil (or gas) in place is provided in Equation 1-2.

$$N = Ah\phi(1 - S_w) \qquad (1\text{-}2)$$

This expression may now make use of seismic measurements that can provide A (area) and h (thickness).

Obviously, better formation description can allow for *targeted drilling*. The fraction of dry holes is likely to be reduced and optimum reservoir exploitation can be envisioned, especially with the emergence of horizontal and multilateral/multibranched wells. Along with seismic images, these wells constitute the two most important technologies of the last decade, if not the entire post-World War petroleum era.

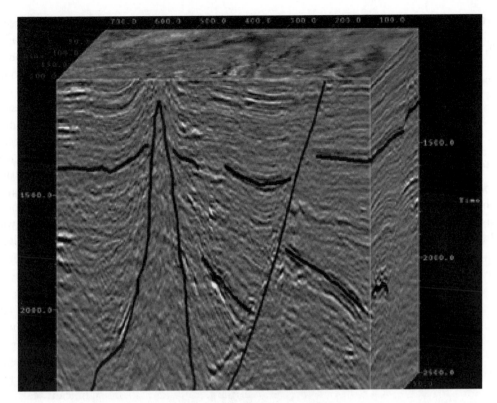

Figure 1-2   3D seismic volume

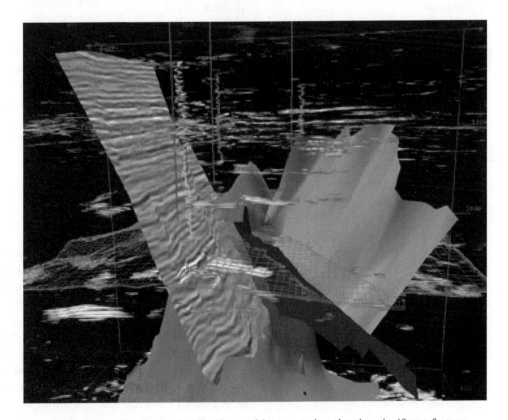

Figure 1-3   3D seismic visualization and interpretation showing significant features

1-3 DRILLING FUNDAMENTALS

The basic principles and technology of drilling an oil or gas well are established and are described in texts (Bourgoyne *et al.*, 1991; Mitchell, 1993; Gatlin, 1960). The following is a brief overview of drilling fundamentals.

1-3.1 Equipment

Drilling a petroleum well is a complex process that requires large, heavy-duty equipment. A conventional *drilling rig* consists first of a structure that can support several hundred tons. A "million-pound" rig is routinely supposed to support 10,000 ft and, in some cases, as much as 30,000 ft of drillpipe and additional equipment.

A *drill bit* (Figure 1-4) is attached to the bottom of the drillpipe by one or more *drill collars*. The entire assembly ends at the floor of a drilling rig and is connected to a rotary table. This table, along with a special joint called the *kelly,* provides rotational motion to the drilling assembly.

While rotary drilling has dominated the petroleum industry in the last 50 years, cable-tool drilling preceded it and was the mainstay of early drilling. In some rare cases, it is still used today. For cable-tool drilling, the drilling assembly is suspended from a wire rope. The assembly is then reciprocated, striking blows to the for-

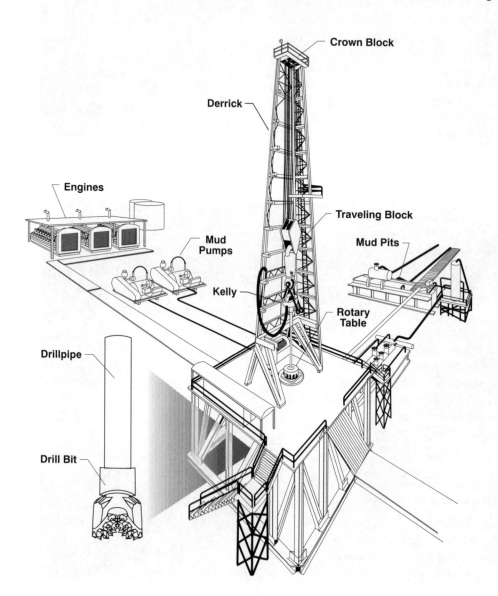

Figure 1-4   Rotary drilling rig with the important components

mation, which becomes fragmented. The drilling assembly is retrieved, and cuttings are brought to the surface with a lowered *bailer*.

Rotary drilling can continue uninterrupted unless a worn-out drill bit must be replaced. Manufacturers have conducted extensive research to improve the durability of drill bits so that the number of trips (pulling the drilling assembly out and then running it in the hole) can be reduced, which results in reduced drilling time.

1-3.2 Drilling Fluids

A critical component of drilling is the drilling fluid, which is also widely referred to in the industry as *drilling mud*. One of the main roles of drilling fluid is to lift the drilling-rock cuttings to the surface and to lubricate the bit in its grinding, rotary action against the rock.

The drilling fluid has other important functions. The weight of the drilling fluid (the fluid density) and the resulting hydrostatic pressure at the drilling point are supposed to impart a positive pressure into the formation. Otherwise, formation fluids under pressure may cause a *kick*, which is an involuntary influx of fluids into the well. Under extreme circumstances, a kick may cause a catastrophic *blowout*.

To provide drilling fluids with the appropriate density for the pressure ranges that will likely be encountered, drilling operators must select the appropriate weighting agent. Drilling fluid weights have ranged from about 8.5 lb/gal (almost neat water) to as much as 15 lb/gal for highly overpressured and deep reservoirs.

Although bentonite clay has been widely used as the main constituent in water-based drilling fluids, other drilling fluid formulations have been used. General families include oil-based and gas-liquid-based fluids. These fluids are supposed to reduce the formation damage caused by water-based fluids and their contained solids when they penetrate the porous medium. One mechanism of controlling formation damage is the formation of a filter cake, which coats the walls of the well, thus reducing fluid leak-off.

1-3.3 Vertical, Deviated, and Horizontal Wells

Through the mid-1980s, vertical wells were drilled almost exclusively. Earlier, deviated wells were introduced, which allowed for the use of surface drilling sites that could be a considerable distance from the targeted formation. This type of well became particularly useful both offshore, where drilling from platforms is necessary, and in the Arctic and other environmentally sensitive areas, where drilling pads can be used.

Although Soviet engineers had drilled several horizontal wells in the 1950s, such activity was limited until the early 1980s, when two western companies, Agip and Elf, reported some impressive results with horizontal wells in an offshore Adriatic oil field. Not only was oil production from the horizontal well several times greater than that of vertical wells in the same field, but the water-oil ratio, a considerable problem with vertical wells, was significantly reduced.

This success literally ushered a new era in the petroleum industry, and although horizontal wells today account for perhaps 10% of all wells drilled, their share is steadily increasing. More importantly, their share in new hydrocarbons produced is disproportionately favorable. Estimates suggest that by the year 2000, perhaps 50% of all new hydrocarbons will come from horizontal and multilateral wells.

The following three categories of horizontal wells are based on the rate of angle build-up in the well trajectory from vertical to horizontal:

- *Long-radius* wells may turn the angle at a rate of 2° to 8° 100 ft; thus, they require a vertical entry point about 1500 ft away from the desired reservoir target. For these wells, conventional drilling assemblies can be used, and conventional well sizes can be constructed. The horizontal lengths of such wells can be considerable; records have been established at over 10,000 ft, but typical horizontal lengths range between 3000 and 4000 ft.

- *Medium-radius* wells require approximately 300 ft to complete a turn from vertical to horizontal. Medium-radius wells use directional control equipment similar to that used in long-radius wells, but drilling practices for such wells are somewhat different.

- *Short-radius* wells can go from vertical to horizontal in 50 ft or less. Specialized, articulated drilling assemblies are needed, and typical well diameters are generally smaller than for conventional wells. For these wells, coiled tubing drilling is often used. Ultrashort-radius drilling technology is available, which allows a well to run from vertical to horizontal within a few feet.

A good driller, aided by modern measurement-while-drilling (MWD) equipment and an appropriate reservoir description, can maintain a well trajectory within ±2 ft from the target. Therefore, if the well is intended to be perfectly horizontal or slightly dipping to reflect reservoir dipping, the departure from the well trajectory can be controlled and minimized.

Once a well is drilled, it must be completed. Section 1-4 provides an overview of well completions.

1-4 WELL COMPLETIONS

The purpose of drilling oil and gas wells is to produce hydrocarbons from, or to inject fluids into, hydrocarbon-bearing formations beneath the earth's surface. The borehole, described previously, provides a conduit for the flow of fluids either to or from the surface. Certain equipment must be placed in the wellbore, and various other items and procedures must also be used to sustain or control the fluid flow. This equipment, and any procedures or items necessary to install it, are collectively referred to as a *well completion*.

In the early twentieth century, oil and gas wells were commonly completed with only a single string of *casing*. The casing was a large diameter (e.g. 7-in.) string of steel pipe, consisting of threaded sections. Initially, casing was set with drilling fluid only.

A casing string in a well extends from the surface to some setting depth. If the top of a casing string is set at a depth below the surface, it is referred to as a *liner*. Liners are commonly found in wells completed during the early part of the twentieth century.

Cementing technology evolved in the 1920s, and by the 1930s most casing strings were set with some cement. Cementing a well is an essential step in almost all well completions, irrespective of whether a perfect bond is achieved between the reservoir and the casing. Currently, most wells are cemented at least some distance above the target reservoir.

In early completions, casing was either set at the top of the producing zone as an *openhole completion* (Figure 1-5) or set through the producing reservoir.

Openhole completions minimize expenses and allow for flexible treatment options if the well is deepened later, but such completions limit the control of well fluids. Phillips and Whitt (1986) show that openhole completions can also reduce sand and water production. Although many wells completed in this manner are still operating today, this method of completion has been superseded by cased completions (Figure 1-6).

In a *cased completion*, casing is set through the producing reservoir and cemented in place. Fluid flow is established by the creation of holes or *perforations* that extend beyond the casing and cement sheath, thereby connecting

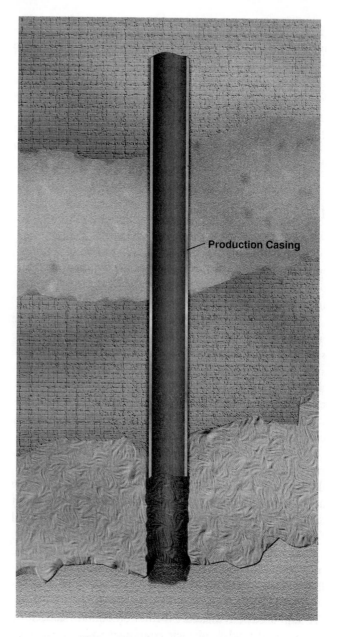

Production Casing

Figure 1-5 Openhole completions

and opening the reservoir to the wellbore (Figure 1-6). Wells that are cased through the producing reservoir provide greater control of reservoir fluids because some or all of the perforations can be cemented off, or downhole devices can be used to shut off bottom perforations. However, openhole wireline logs must be run before the casing is set so that the exact perforation interval is known.

Cased-hole completions are more susceptible to *formation damage* than openhole completions. Formation damage refers to a loss in reservoir productivity, nor-

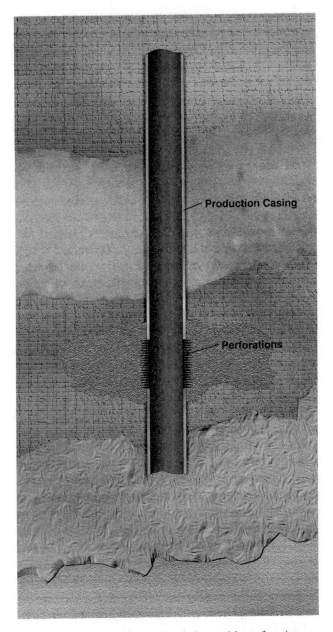

Figure 1-6   Cased-hole completions with perforations

from the wellbore, the well may be *acidized* to dissolve or remove the damage. Matrix acidizing is used to restore initial productivity. *Hydraulic fracturing* is a stimulation technique that creates a fracture that is intended to extend beyond the damage area. Significant advances in well stimulation have been made since the early part of this century, when openhole completions were stimulated by jars of nitroglycerin that were placed and detonated downhole.

As noted previously, reservoir pressure will decline as hydrocarbons are produced. Many of the wells completed in the early part of the twentieth century were produced through the casing, and the reservoirs had sufficient pressure for the hydrocarbons to flow to the surface. With declining reservoir pressures and producing volumes, production through smaller-diameter *tubing* became necessary, since the velocity through the casing could not sufficiently sustain natural flow.

Figure 1-7 shows a simple cased-hole completion with tubing.

These early completion techniques proved adequate in relatively shallow wells. However, as deeper, multiple, and higher-pressure reservoirs were encountered, it was recognized that the completions imposed limitations on well servicing and control, and designs would require improvement to meet increasing requirements for wellbore re-entry and workover operations.

A wide range of downhole equipment has been designed and manufactured to meet the needs of more complex well completions. In situations where multiple reservoirs cannot be commingled, the zones are separated with a *production packer*. Packers are devices that are run on, or in conjunction with, a string of tubing. The packer has a rubber element that is extruded by compression to form a seal between the tubing and the casing (Figure 1-8). Packers are used for a variety of reasons in well completions.

Another component that has become an integral part of well completions is the *sliding sleeve*. The sliding sleeve provides annular access between the tubing and the casing. It is used to produce a reservoir isolated between two production packers and for circulating a well above the uppermost packer. The sleeve is opened or closed through the use of *wireline servicing methods*. Many other functions can be performed with wireline devices set in *landing nipples*.

The evolution of offshore drilling in the 1930s (ETA, 1976) and the production of wells from offshore platforms in the 1940s (Graf, 1981) demanded methods of well shut-in for safety and environmental concerns. *Storm chokes* were pressure-controlled devices set inside

mally associated with fluid invasion, fines migration, precipitates, or the formation of emulsions in the reservoir. Loss of productivity is expressed as a *skin factor*, *s*, in Darcy's equation as follows:

$$q \propto kh\Delta p \div (\ln r_e/r_w + s) \qquad (1\text{-}3)$$

A positive skin value indicates that a well is damaged.

Formation damage can be removed or bypassed through the use of *stimulation* techniques. In instances where the formation damage extends only a few feet

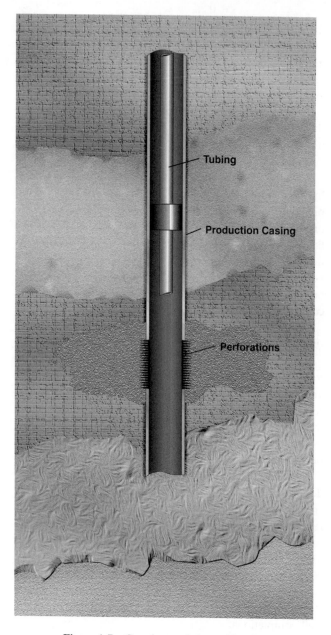

Figure 1-7   Cased completion with tubing

Many reservoirs contain sediments that are so poorly consolidated that sand will be produced along with the reservoir fluids unless the production rate is severely restricted. Sand production may erode the tubing or surface valves and flowlines. In addition, sand could accumulate in downhole equipment and create problems in wireline servicing. *Gravel-packing* was devised as a means of eliminating sand production without greatly restricting production rates. In a gravel-pack completion (Figure 1-9), sand with a grain size larger than the average formation sand grain is placed between the formation and a screen or slotted liner (Economides *et al.,* 1994). More recently, *high-permeability fracturing (frac-packing)* has been proven as a technique for sand control.

Engineers designing well completions must consider that the wells will eventually be unable to flow naturally to the surface. The loss of natural flow occurs because the reservoir pressure declines with production and reservoirs produce increasing amounts of water with time, which increases the density of the flowing fluid. Various techniques of artificially lifting fluids from the wellbore have been developed. *Artificial lift* techniques include sucker rod pumping, electrical submersible pumps, gas lift, and other types of hydraulic lift. Each method of artificial lift requires unique downhole and surface equipment that must be considered during the design of the well completion.

Well stimulation techniques introduced in the early part of the twentieth century have been improved through a more complete understanding of the processes involved. Acidizing models have been developed to describe the use of various types of acids in a range of lithologies. Hydraulic fracturing has experienced even more dramatic improvements since the introduction of crosslinked polymer fluids, high-strength proppants, and analytical techniques, such as the net pressure plot. Such techniques have enabled engineers to substantially improve the flow from both low-permeability and high-permeability reservoirs.

Another notable advance in well completion design is the evolution of *coiled tubing* for servicing and completing wells. Coiled tubing servicing involves the deployment of a continuous string of small-diameter tubing into the wellbore. This coiled tubing is run concentric to existing tubulars, used for the required service, and then removed without damaging the existing completion. Coiled tubing servicing is of increasing importance in highly deviated and horizontal wells, since wireline servicing poses problems at angles greater than 50°.

the tubing string. These devices were intended to shut in wellflow during storms or a major platform catastrophe. At today's offshore locations, these direct-control devices have been supplanted by *surface-controlled subsurface safety valves (SCSSVs)*.

Equipment such as packers, sleeves, landing nipples, and safety valves provide various functions for well control. These devices are only a few examples of an extensive range of equipment that enables engineers to control fluid flow selectively and to stimulate producing reservoirs.

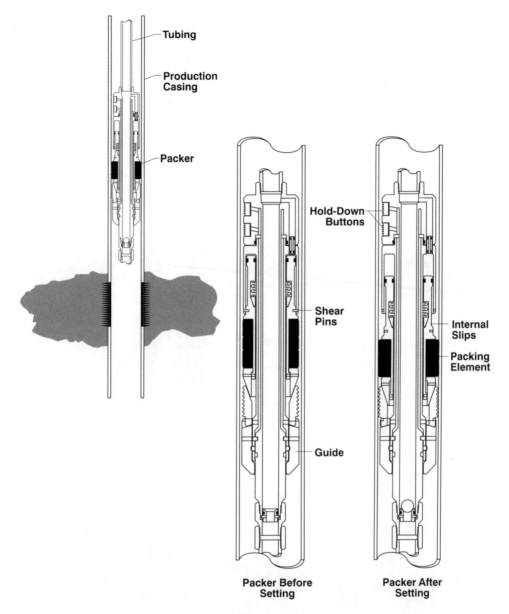

Figure 1-8 Production packer with detail of annular seal

Completion methods such as gravel-packing and stimulation, a variety of downhole equipment, and enhancements to servicing methods, have enabled engineers to design more complex well completions which offer greater fluid flow control, stimulation alternatives, and operational flexibility. An extensive range of downhole designs has been implemented to meet a number of producing requirements. Example designs include dual completions, slimhole and monobore completions (Ross *et al.,* 1992; Robison, 1994), completions for high-pressure, high-temperature (HPHT) reservoirs (Schulz *et al.,* 1988), subsea completions (Cooke and

Cain, 1992) whose wellheads are located on the seafloor, and waterflood or CO_2 injection applications (Stone *et al.,* 1989). Two examples, a dual completion (Figure 1-10) and a subsea completion with gravel-packing and artificial lift (Figure 1-11), illustrate the wide range of well completion designs available today.

Figure 1-10 depicts a *dual completion.* Dual completions are used when multiple reservoirs will be produced. Two tubing strings and at least two production packers are included. The packers may separate two or more producing reservoirs. A sliding sleeve can be included between or above packers so that one or more reservoirs

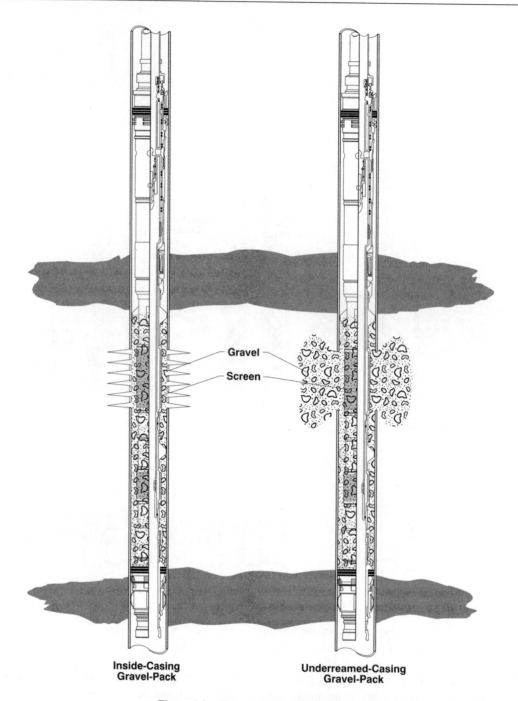

Figure 1-9   Gravel-pack completions

can be selectively produced at any time. Other downhole equipment, such as landing nipples, safety valves, or side-pocket mandrels (for gas lift) may be included in a dual completion. Sanku *et al.* (1990) show the use of a dual completion with gas lift in the Sockeye Field, offshore California (Figure 1-10A). Farid *et al.* (1989) show the application of a dual completion for gas injection in a three-layered reservoir in Abu Dhabi (Figure 1-10B).

Figure 1-11 depicts a *single-string subsea completion.* This completion has been run in the Balmoral Field in the North Sea (Shepherd, 1987). Initial test data indicated that the Balmoral wells would produce significant amounts of sand, and it was decided to gravel-pack the wells to control sand production. The gravel screen is set across the producing zone and a packer is set above the gravel pack. A tubing expansion joint, run above

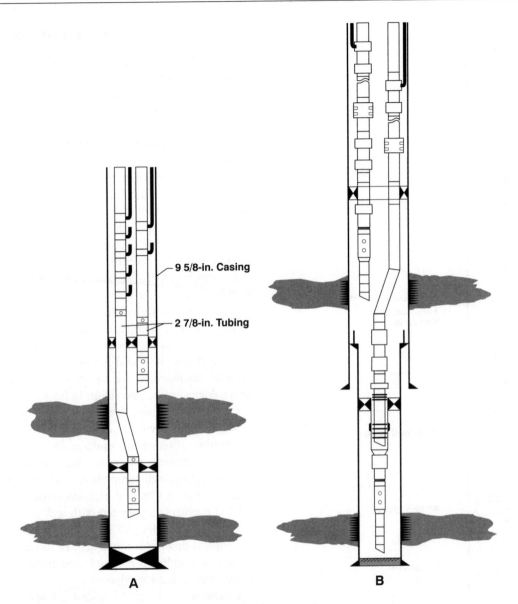

Figure 1-10   Dual completions

the packer, allows the tubing to expand or contract with changes in downhole pressure or temperature. A sleeve is run above the expansion joint to circulate the well, and a number of gas-lift mandrels, equipped with dummy valves, are included in the tubing above the sleeve. The gas-lift mandrels were included in the Balmoral design to provide for future gas lift, since reservoir models predicted a rapid onset of water production and the need for artificial lift. This type of forward planning is crucial in subsea wells, where the cost of mobilizing offshore rigs is substantial.

The evolution and growing application of horizontal drilling techniques has provided additional challenges in well completion design. At present, most horizontal wells are either completed with an openhole horizontal section, with a slotted liner laid in the openhole section (Cooper and Troncoso, 1988; Lessi and Spreux, 1988), or with a gravel-pack screen (McLarty *et al.,* 1993). To date, the use of casing, production packers, sleeves, and other downhole devices has been limited because they cannot provide a mechanical/hydraulic seal at the junction between the vertical wellbore and the horizontal hole. Completion technology in this area is evolving rapidly, and such capabilities will likely be available in the near future, enabling the use of downhole devices and techniques that will provide greater control of fluid flow and stimulation in horizontal and multilateral wells.

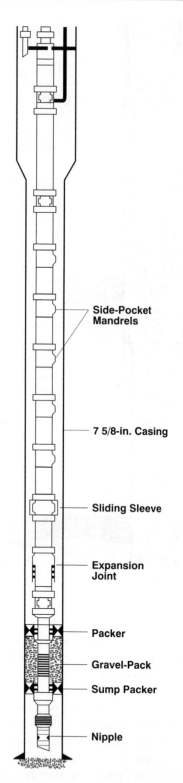

- Side-Pocket Mandrels
- 7 5/8-in. Casing
- Sliding Sleeve
- Expansion Joint
- Packer
- Gravel-Pack
- Sump Packer
- Nipple

Figure 1-11 Balmoral subsea completion (Shepherd, 1987)

1-5 ORGANIZATION OF THIS BOOK

The following chapters were written by experts from various Halliburton companies, operating companies, and from academia. The chapters included in this book are divided into two major categories: drilling (Chapters 2 to 7) and well completions (Chapters 8 to 21). A brief overview of each chapter is provided in the following paragraphs.

Chapter 2 describes technologies for drilling-trajectory monitoring and control, drilling assemblies, and requirements for the complicated well systems of today.

Chapter 3 describes the wide variety of well configurations possible with drilling systems today. Horizontal, multilateral, and multibranch wells are explained, and their applications for a variety of reservoir management problems are outlined.

Measurement-while-drilling (MWD) and logging-while-drilling (LWD) equipment and techniques are discussed in Chapter 4. The formation characterization that these methods allow and the ability to drill complex well systems intelligently are the two most important new technologies in the petroleum industry.

Chapter 5 discusses desirable drilling fluid characteristics, the various types of damage caused by drilling, and the minimization of this damage.

Rock mechanics principles have several applications during drilling, well completion, and subsequent production. Chapter 6 lists and explains most rock mechanics-related problems associated with well construction.

Chapter 7 is a detailed review of casing design and the subsequent downhole movement of tubulars subjected to their own weight, to pressure, and to thermally induced forces during their service.

Chapters 8 through 11 describe, respectively, primary cementing, gas-migration problems, cement sheath evaluation, and in the case of detected problems, remedial cementing.

Chapter 12 describes completion fluids, their properties, and the effect of fluid selection on such completion processes as gravel-packing, setting hardware, and perforating.

All cased-hole completions must be perforated, and these perforations can affect the well's capacity to produce reservoir fluids. Chapter 13 describes perforating technology, which has evolved considerably in recent years in terms of diameter, density (shots per foot), tunnel length of the perforation, and orientation.

A wide variety of completion hardware is available for controlling fluid flow or providing operational flexibility in a well. Chapter 14 describes attributes of basic downhole equipment, such as packers, sliding side-doors, landing nipples, side-pocket mandrels, and SCSSVs.

Well performance is dominated by two main components: reservoir deliverability, described traditionally by the well inflow performance relationship (IPR), and flow in the production string. The combination of these two components is well deliverability. Chapter 15 provides a comprehensive description of these concepts for oil, gas, and two-phase reservoirs fitted with vertical, deviated, or horizontal wells.

Chapter 16 describes fundamental methods of artificial lift. Artificial lift must be included in completions in which the reservoirs no longer have sufficient pressure to sustain natural flow. Artificial lift can also be used to enhance production in flowing wells by augmenting the natural reservoir pressure. Challenges of applying artificial lift systems in horizontal and multilateral wells are also discussed in Chapter 16.

Stimulation can also improve well productivity. Stimulation includes both acidizing and fracturing a reservoir. Stimulation improves a well's ability to flow by providing a more direct flowpath, which results in more production per unit of drawdown. The main concepts of stimulation are discussed in Chapter 17.

Sand production is a significant problem in many wells, particularly wells producing from incompetent reservoirs or reservoirs that have experienced pressure depletion. Sand production is a serious concern because it affects the completion and surface facilities and may also deconsolidate the formation with a potentially catastrophic collapse.

Two general approaches to solving sand production problems are available: sand exclusion techniques, which require the use of filters, and sand-production control techniques, which affect the sand production mechanisms within the pore structures. The prominent sand exclusion technique is gravel-packing augmented by screens, as described in Chapter 18. Sand-production control has been demonstrated by high-permeability fracturing or *fracpack* processes (Chapter 19).

Water production and its control is the other great, persistent problem in reservoir management. Chapter 20 describes the mechanisms of water production and a holistic approach to water-production management.

Finally, proper engineering suggests that well construction must be based on the entire life of the well. Thus, planning completions for such well management is essential. Chapter 21 provides such an approach and combines several of the concepts previously presented in this book.

REFERENCES

Amyx, J.W., Bass Jr., D.M., and Whiting, R.L.: *Petroleum Reservoir Engineering*, McGraw-Hill Book Company, New York (1960) 1–20.

Bourgoyne Jr., A.T., Millheim, K.K., Chenevert, M.E., and Young Jr., F.S.: *Applied Drilling Engineering*, SPE, Richardson, TX (1991).

Cooke, J.C., and Cain, R.E.: "Development of Conventionally Uneconomic Reserves Using Subsea Completion Technology: Garden Banks Block 224, Gulf of Mexico," paper OTC 7004, 1992.

Cooper, R.E., and Troncoso, J.C.: "Overview of Horizontal Well Completion Technology," paper SPE 17582, 1988.

Economides, M.J., Hill, D.A., and Ehlig-Economides, C.A.: *Petroleum Production Systems*, Prentice Hall, Englewood Cliffs, NJ (1994) 119.

ETA Offshore Seminars: *The Technology of Offshore Drilling, Completion and Production*, PennWell Books, Tulsa, OK (1976) 3–32.

Farid, E.A., Al-Khaffaji, N.H.J., and Ryan, M.A.: "Drilling and Completion of Dual Gas Injection Wells in Multilayered Reservoirs," paper SPE 17982, 1989.

Gatlin, C.: *Petroleum Engineering—Drilling and Well Completions*, Prentice-Hall, Inc., Englewood Cliffs, NJ (1960).

Graf, W.J.: *Introduction to Offshore Structures—Design, Fabrication, Installation*, Gulf Publishing Company, Houston (1981) 4–19.

Lessi, J., and Spreux, A.: "Completion of Horizontal Drainholes," paper SPE 17572, 1988.

McLarty, J.M., Dobson, J.W., and Dick, M.A.: "Overview of Offshore Horizontal Drilling/Completion Projects in the Unconsolidated Sandstones in the Gulf of Mexico," paper OTC 7352, 1993.

Mitchell, B.: *Advanced Oilwell Drilling Engineering Handbook*, Mitchell Engineering, Houston (1993).

Phillips, F.L., and Whitt, S.R.: "Success of Openhole Completions in the Northeast Buttery Field, Southern Oklahoma," *SPEPE* (May 1986) 113–119.

Robison, C.E.: "Monobore Completion for Slimhole Wells," paper OTC 7551, 1994.

Ross, B.R., Faure, A.M., Kitsios, E.E., Oosterling, P., and Zettle, R.S.: "Innovative Slim-Hole Completion," paper SPE 24981, 1992.

Sanku, V., Weber, L.S., and Masoner, L.O.: "Development of Sockeye Field in Offshore California," paper SPE 20047, 1990.

Schulz, R.R., Stehle, D.E., and Murali, J.: "Completion of a Deep, Hot, Corrosive East Texas Gas Well," *SPEPE* (May 1988) 153–157.

Shepherd, C.E.: "Subsea Completions in the Balmoral Field," paper OTC 5433, 1987.

Stone, P.C., Steinberg, B.G., and Goodson, J.E: "Completion Design for Waterfloods and CO$_2$ Flood," *SPEPE* (Nov. 1989) 365-370.

Wilhelm, O.: "Classification of Petroleum Reservoirs," *Bull. of Am. Assoc. Petr. Geol.* (1945) **29**.

2 Directional Drilling

Lance D. Underwood
Halliburton Energy Services

Michael L. Payne
Arco Exploration and Production Technology

2-1 INTRODUCTION

Directional drilling began with the use of devices such as whipstocks or techniques such as jetting to kick off, rotary assemblies to control inclination in tangent sections, and wireline steering tools to orient and survey. These tools possessed limited directional control capabilities, required frequent tripping of the drillstring, and made directional drilling an expensive, difficult, and sometimes risky proposition. Directional well planning was more an art than a science, and capabilities and boundaries were based largely on empirical observations and historical tool performance.

Recently, technological advances have contributed to a significant increase in the use and scale of directional drilling. Perhaps the technologies with the highest impact have been steerable mud motors, measurement-while-drilling (MWD) tools, and logging-while-drilling (LWD) tools. These tools in combination have provided the ability to follow complex, 3D well profiles without changing bottomhole assemblies (BHAs), and to measure where the bit has drilled without having to run a wireline to survey or log. Equally important, engineering models have provided the fundamental tools for evaluating drillstrings, hydraulics, BHAs, and the drilled formations themselves. These advances have enabled the drilling of extended-reach, horizontal, and multiple-target well profiles once thought impractical, uneconomical, or impossible.

Most books currently available either discuss deviation control—the attempt to keep vertical wells truly vertical—or discuss older directional-drilling practices, such as whipstocks, jetting, or the use of straight mud motors with bent subs. These practices, while still in use, are now the exception rather than the rule. This chapter, then, primarily discusses directional drilling as it is performed today. Emphasis will be placed on identifying the principles and mechanics that define the capabilities and limitations of current directional-drilling technology.

2-2 WELL PLANNING

Planning even the simplest vertical well is a task that involves multiple disciplines. A casual observer might think that planning a directional well would require only a few geometry calculations in addition to the usual tasks. On the contrary, almost every aspect of well planning is affected when a directional well is planned. Various software systems are available to assist in these engineering efforts, but effective application of such software requires a good understanding of the underlying engineering principles. The fundamental variables that dictate the planned wellpath are the surface location for the rig and wellhead and the location(s) of the target(s) downhole. However, many other variables also impact the final wellpath chosen.

2-2.1 Well Profiles and Terminology

A simple build/hold/drop well profile, known as an "S" well, is shown in Figure 2-1. The *kickoff point* (KOP) is the beginning of the build section. A build section is frequently designed at a constant *buildup rate* (BUR)

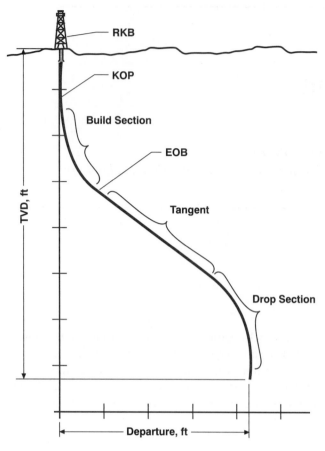

Figure 2-1 Well profile terminology

until the desired hole angle or *end-of-build* (EOB) target location is achieved. BUR is normally expressed in terms of degrees per hundred feet ($°/100\,\text{ft}$), which is simply the measured change in angle divided by the measured depth (MD) drilled. Hole angle, or *inclination*, is always expressed in terms of the angle of the wellbore from vertical. The direction, or *azimuth* of the well is expressed with respect to some reference plane, usually true north. The location of a point in the well is generally expressed in Cartesian coordinates with the wellhead or the rig's *rotary kelly bushing* (RKB) as the reference location. *True vertical depth* (TVD) is usually expressed as the vertical distance below RKB. *Departure* is the distance between two survey points as projected onto the horizontal plane. The EOB is defined in terms of its location in space as expressed by coordinates and TVD. The EOB specification also contains another important requirement, which is the angle and direction of the well at that point. The correct angle and direction are critical in allowing the next target to be achieved; also, it may be necessary to penetrate the payzone at some optimum angle for production purposes.

A tangent section is shown after the build section. The purpose of the tangent is to maintain angle and direction until the next target is reached. In the example well, a drop section is shown at the end of the tangent. The purpose of a drop is usually to place the wellbore in the reservoir in the optimum orientation with respect to formation permeability or in-situ formation stress; alternatively, a horizontal extension may be the preferred orientation in the case of a payzone that contains multiple vertical fractures or that has potential for gas or water coning.

A general classification of build rates is shown in Figure 2-2. See Bourgoyne *et al.* (1991) and Le Peuvedic *et al.* (1990) for further information concerning well profiles and the various reasons to drill directionally.

2-2.2 Factors in Wellpath Design

Completion and reservoir drainage considerations are key factors in wellpath design. For fracturing, gravel-packing, completion in weak formations, or depletion-induced compaction, it may be desirable to limit the inclination of the well through the reservoir or even to require a vertical or near-vertical trajectory. These conditions are also true in laminated or layered reservoirs. Often, it may be desirable for the wellpath in the reservoir to be horizontal to provide as much reservoir drainage and production rate as possible. In horizontal wells, correct TVD placement will minimize gas coning or water production. In vertically fractured formations in which the fractures may aid in the flow of hydrocarbons, the direction of the wellpath in the reservoir may be chosen to intersect multiple fractures. Alternatively, it may be desirable to place the wellbore in a given direction to avoid faults that are expected to allow water migration. Optimal placement of the wellbore in the reservoir will result in maximum production and should actually be the starting point for wellpath design. These issues are addressed in Chapters 3 and 15.

Additional considerations will influence the design of the trajectory from the surface location to the reservoir-target entry point. Some shallow formations in sedimentary geologies are weak and, as a result, building inclination is difficult because of the lack of reactive forces against the BHA. If this condition is anticipated, the KOP should be designed deeper, where formations are more competent.

The interrelationship of the wellpath design and the casing/hole program must also be recognized. The casing/hole program for the well is generally designed on the

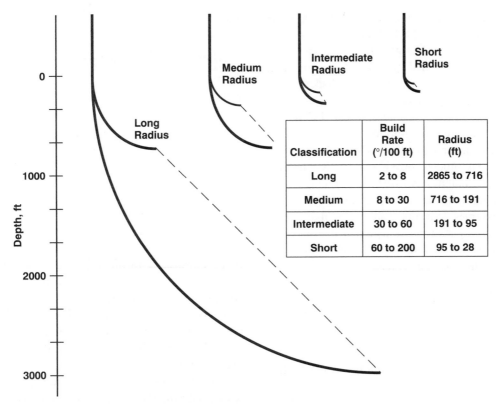

Figure 2-2 Build-rate classification

Classification	Build Rate (°/100 ft)	Radius (ft)
Long	2 to 8	2865 to 716
Medium	8 to 30	716 to 191
Intermediate	30 to 60	191 to 95
Short	60 to 200	95 to 28

basis of the desired completion, the pore pressure regimes for the well, the presence of trouble zones, and regulatory requirements. The casing program influences the planned trajectory in several ways. For a given casing design, the trajectory plan should be optimized for operational efficiency. For example, required builds and turns should be executed fully within a single hole section. When this method is used, the well will be "lined out" towards the reservoir target, and the remaining hole section can be drilled as a straight tangent section without additional directional work. Likewise, in troublesome zones, such as underpressured sands or reactive shales that can increase the risk of stuck pipe, it may be desirable to avoid directional work that requires *sliding* (drilling without rotating the drillstring). Thus, the design of a build section may need to include a short tangent through the troublesome section that will allow it to be rotary-drilled as rapidly as possible.

When the various constraints are considered, a feasible and optimized directional trajectory plan should result. An optimized wellpath often cannot be described by the simplest geometry that can be conceived to connect a series of targets. Even with a simple build-and-hold profile, additional wellpath optimization is possible. For

example, drilling experience in an area should allow for the definition of the typical walk rate (the tendency of the BHA to turn slightly in the azimuthal direction) for certain BHAs in that area. With walk rates defined, the well can and should be "led," or initially directed away from the target in the direction opposite to the anticipated walk. If the well is properly led, steering will not be required in this interval, since the natural walk tendency will gradually bring the well into the target. If the wellpath had been designed as a straight line from one target to the next, frequent steering would be required throughout the interval to counteract the natural walk tendency.

2-2.3 Modeling the Wellpath

In addition to refining the trajectory plan to account for drilling tendencies such as walk, trajectory planning in development projects must also account for the location of existing wells and the requirement that the planned well safely bypasses all existing wellbores. This aspect of planning, known as "collision avoidance," must account for the uncertainty associated with the ability to survey the well.

Wellbore trajectory calculation methods use data sets called *survey stations*, each of which consists of inclination, azimuth, and measured depth. Directional measurements are normally provided by MWD sensors, and measured depth is provided by traveling block sensors or by pipe tally. Many survey models are available (Bourgoyne *et al.*, 1991; Craig and Randall, 1976), and each is based on different assumptions on the shape of the wellbore between survey stations. Except for the tangent method, most models provide virtually identical results. The most commonly used survey calculation method is the minimum curvature method (Figure 2-3). This method assumes that the wellbore is a constant curve between survey stations, and that it is tangent to the measured angle at each station. The minimum curvature survey calculation is as follows:

$$\Delta North = \frac{\Delta MD}{2}[\sin(\phi_1) \cdot \cos(\theta_1) + \sin(\theta_2) \cdot \cos(\phi_2)] \cdot RF \tag{2-1}$$

$$\Delta East = \frac{\Delta MD}{2}[\sin(\theta_1) \cdot \sin(\phi_1) + \sin(\theta_2) \cdot \sin(\phi_2)] \cdot RF \tag{2-2}$$

and

$$\Delta Vert = \frac{\Delta MD}{2}[\sin(\phi_1) + \cos(\theta_2) \cdot \cos(\phi_2)] \cdot RF \tag{2-3}$$

where

$$RF = \frac{360}{\pi DL}\left[\tan\frac{DL}{2}\right] \tag{2-4}$$

and

$$\cos DL = \cos(\theta_2 - \theta_1) - \sin\theta_1 \cdot \sin\theta_2[1 - \cos(\phi_2 - \phi_1)] \tag{2-5}$$

In Equations 2-1 through 2-5, $\Delta North$ is the change in the north coordinate (ft), $\Delta East$ is the change in the east coordinate (ft), $\Delta Vert$ is the change in the TVD coordinate (ft), ΔMD is the change in measured depth (ft), θ is the inclination angle, ϕ is the azimuth angle, RF is the ratio factor, and DL is the dogleg (the total angle change over the interval).

While various calculation methods give near-identical results with the same raw survey data, the accuracy of the

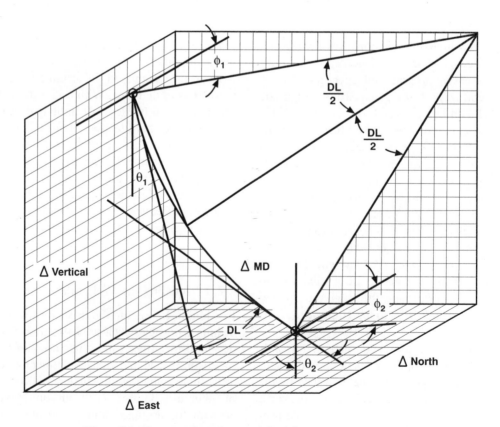

Figure 2-3   Survey calculation model-minimum curvature method

data itself can cause uncertainty as to the precise location of a wellbore. All surveying instruments (MWD, single- and multi-shots, and gyros) have accuracy tolerances and biases. The subject of surveying uncertainty is complex, and readers are encouraged to review Wolff and de Wardt (1981) and Thorogood (1990). In this text, the issue is considered only qualitatively in terms of its impact on trajectory design. To account for uncertainty, the trajectory path is plotted in terms of its possible location by expanding the nominal (planned) trajectory by the tolerances associated with azimuthal and inclination surveying measurements. Azimuthal survey uncertainty is greater than inclination survey uncertainty. As a result, the possible wellpath at any depth can be plotted as an elliptical envelope, often called the *ellipse of uncertainty*, (Figure 2-4). Cumulative uncertainty causes the size of the ellipse to increase with drilled depth. Although the most likely location of the well is near the center of the ellipse, it could be located anywhere within the area of the ellipse. Thus, the ellipse must be calculated and used for evaluating whether the planned well might intersect any existing wells.

The surveying uncertainty that impacts the ability to locate the planned well precisely also impacts the ability to describe the precise location of existing wells. As a result, collision avoidance planning on large development projects, such as offshore platforms and onshore drillpads, is complex. Special computing and plotting techniques assist in such efforts. Traveling cylinder plots (Thorogood and Sawaryn, 1991) serve to describe how close the planned wellpath will come to existing wells. As shown in Figure 2-5, the plot is generated based on the wellpath of interest being the centerline of the cylinder. Offsets of increasing distances are plotted radially around this centerline as a measure of clearance from existing wells. The wellpaths of the existing wells are then plotted with this radial grid as a function of the drilled depth of the current well. The plot allows engineers and operators to monitor the proximity to other wells while drilling and to exercise additional caution when drilling by close approaches. Traveling cylinder plots can be generated on the basis of nominal well locations, but they are usually generated in a manner that already accounts for the surveying uncertainty of the planned and existing wells. These uncertainties are determined by statistical survey tool variation and each operating company's policy with regard to risk management.

Individual wells are usually described in two plots: the profile view and the plan view. Each view is two-dimensional. *Spider plots* and 3D views (Figure 2-6) provide an additional tool to assist in planning and visualizing the trajectories of multiple wells in a development project. These views show the paths of existing wells and how the planned trajectory will be drilled through them.

2-2.4 Torque and Drag

Torque and drag may be critical factors in determining whether the desired wellpath can actually be drilled and cased. Torque/drag models consider well trajectory, drillstring configuration, doglegs, friction factors, and casing depth to predict torque and drag in the well. Torque-and-drag modeling is used for various purposes, including:

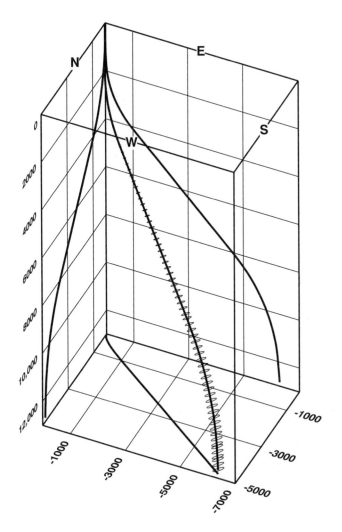

Figure 2-4 Wellpath with superimposed ellipses of uncertainy

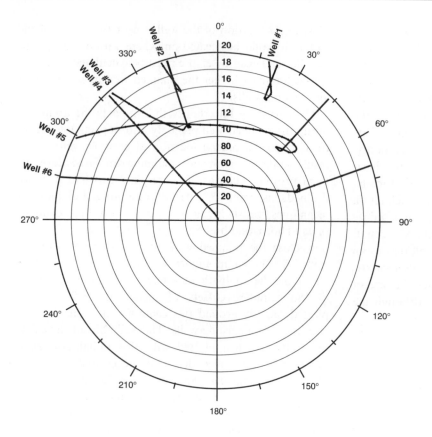

Figure 2-5 Traveling cylinder plot

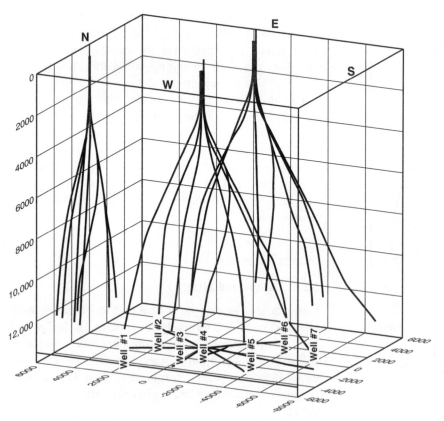

Figure 2-6 3D view-multiwell platform

- evaluating and optimizing wellpaths to minimize torque and drag

- fine-tuning wellpaths to minimize local effects, such as excessive normal loads

- providing normal force loads for inputs into other programs, such as casing wear models

- identifying depth or reach capabilities or limitations, both for drilling and running casing/tubing

- matching the strength of drillstring components to the loads (axial, torsional, or lateral) in the wellbore

- identifying the hoisting and torque requirements of the drilling rig

The most commonly used torque/drag models are based on the "soft-string" model developed by Johancsik *et al.* (1983). The drillstring is modeled as a string or cable that is capable of carrying axial loads but not bending moments. Friction is the product of normal forces and a coefficient of friction. The normal force at each calculation node has two components: (1) the buoyed weight of the pipe in drilling fluid, and (2) the lateral reaction force resulting from drillstring tension through curved sections of the wellbore. A simplified drillstring element, shown in Figure 2-7, has net axial forces and normal forces acting upon it. The equations for these forces are

$$F_N = [(T \Delta\phi \sin\theta_{\text{avg}})^2 + (T \Delta\phi + W \sin\theta_{\text{avg}})^2]^{1/2} \quad (2\text{-}6)$$

$$\Delta T = W \cos\theta_{\text{avg}} \pm f F_N \quad (2\text{-}7)$$

$$\Delta M = f F_N R \quad (2\text{-}8)$$

and

$$F_F = f F_N \quad (2\text{-}9)$$

where F_N is the net normal force, T is the axial tension at the lower end of the element, W is the buoyed weight of drillstring element, F_F is the sliding friction force acting on the element, R is the characteristic radius of element, M is the torsion at the lower end of element, θ is the inclination angle at lower end of element, ϕ is the azimuth angle at lower end of element, f is the coefficient of friction, and $\Delta(T, M, \phi, \theta)$ is the change in those values over the length of the element.

In Equation 2-7, the product $f F_N$ can be positive or negative, depending on whether the drillstring is advancing into the hole or being pulled out of the hole.

If accurate friction factors are derived from existing field data, soft-string models yield reasonably accurate results for most sizes of drillpipe and hole curvatures. However, since the soft-string model does not consider the stiffness of the drillstring, its accuracy will degrade as drillpipe diameter increases and as hole curvature increases. Both of these increases result in high normal forces and increased torque/drag. Finite-element models that incorporate drillstring properties are available for such applications, and they may be necessary for modeling casing. Neither type of model can accommodate localized hole and drillstring/BHA mechanical interactions (for example, a stabilizer hanging up on a ledge or a dogleg). Such a model would require much more information about actual hole geometry than is available. However, empirically derived macrolevel friction factors are adequate for predictive analysis, since these factors are calculated from wells that include similar localized geometry.

Friction factors should be derived from analogous case histories. The properties of the drilling fluid used in the baseline wells and the planned well should be similar. However, the ranges in Table 2-1 can be used as starting points if prior experience is unavailable (Johancsik *et al.,* 1983; Rasmussen *et al.,* 1991).

Field experience has shown that axial drillstring drag is reduced when the drillstring is rotated. Torque-and-drag

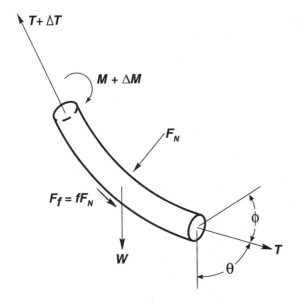

Figure 2-7 Drillstring element for "softstring" torque-and-drag model (from Johancsik *et al.,* 1983)

Table 2-1 Ranges of friction factors in casing and in formation (from Johancsik *et al.*, 1983 and Rasmussen *et al.*, 1991)

Drilling fluid	f in casing	f in formation
Oil-based	0.16 to 0.20	0.17 to 0.25
Water-based	0.25 to 0.35	0.25 to 0.40
Brine	0.30 to 0.40	0.30 to 0.40

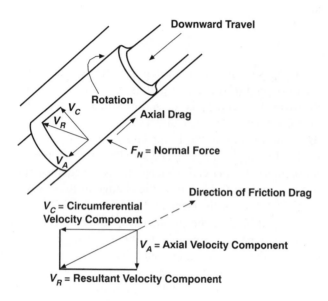

Figure 2-8 Effect of drillstring rotation on axial friction (from Dellinger *et al.*, 1980)

models account for this mathematically by the use of velocity vectors (Dellinger *et al.*, 1980) (Figure 2-8). The resultant velocity V_R of a contact point on the drillstring is the vector sum of two components, *circumferential velocity* V_C (caused by rotation) and *axial velocity* V_A (affected by drilling rate or tripping speed). The *direction* of the resultant frictional force is assumed to act in the direction opposite to that of resultant velocity V_R; therefore, its vector components will be in proportion to those of resultant velocity. The *magnitude* of resultant frictional force is simply the product of normal force F and friction coefficient f, and it does not vary with velocity. Since the magnitude of the vector sum of these components is a fixed quantity, as the circumferential component increases, the axial component must decrease. Logically, as drillstring rotation speed increases, it increases the circumferential component, which decreases axial friction.

Well planning should include torque/drag modeling with worst-case friction factors to ensure that the drillstring can be advanced, rotated, slid if oriented drilling is necessary, and pulled out of the hole. Similar modeling should be used to ensure that friction will not prevent the casing from being run, and that the casing can be pulled if necessary. The torque and tension/compression at any point in the drillstring must be compared to the torsional, tension, and buckling capabilities of the drillstring

Table 2-2 Properties of Range 2 drillpipe (after API Publication RP7-G)

Size OD	New wt. nom. W/ thds & couplings	Torsional yield strength based on uniform wear, ft-lb				Tensile data based on uniform wear load at minimum yield strength, lb.			
in.	lb/ft	E	95	105	135	E	95	105	135
$3\frac{1}{2}$	9.50	9612	12,176	13,457	17,302	132,793	168,204	185,910	239,027
	13.30	12,365	15,663	17,312	22,258	183,398	232,304	256,757	330,116
	15.50	13,828	17,515	19,359	24,890	215,967	273,558	302,354	388,741
4	11.85	13,281	16,823	18,594	23,907	158,132	200,301	221,385	284,638
	14.00	15,738	19,935	22,034	28,329	194,363	246,193	272,108	349,852
	15.70	17,315	21,932	24,241	31,166	219,738	278,335	307,633	395,528
$4\frac{1}{2}$	13.75	17,715	22,439	24,801	31,887	185,389	234,827	259,545	333,701
	16.60	20,908	26,483	29,271	37,637	225,771	285,977	316,080	406,388
	20.00	24,747	31,346	34,645	44,544	279,502	354,035	391,302	503,103
	22.82	27,161	34,404	38,026	48,890	317,497	402,163	444,496	571,495
5	16.25	23,974	30,368	33,564	43,154	225,316	285,400	315,442	405,568
	19.50	27,976	35,436	39,166	50,356	270,432	342,548	378,605	486,778
	25.60	34,947	44,267	48,926	62,905	358,731	454,392	502,223	645,715

and tool joints. Table 2-2 contains properties for Range 2 drillpipe.

The results from torque/drag analysis are usually expressed graphically with torque and/or drillstring tension on one axis and measured depth on the other (Figure 2-9). The subject well is a build/hold/drop profile similar to that of Figure 2-1, with a 5000-ft tangent sec-

tion at a 60° angle. A friction factor of 0.2 was used for cased holes and a factor of 0.3 was used for open holes. The torque/drag analysis qualifies this well as drillable with the operating parameters used, and allows the selection of drillstring components with a reasonable safety margin as compared to loads. Axial load values never drop into the buckling region, even during sliding.

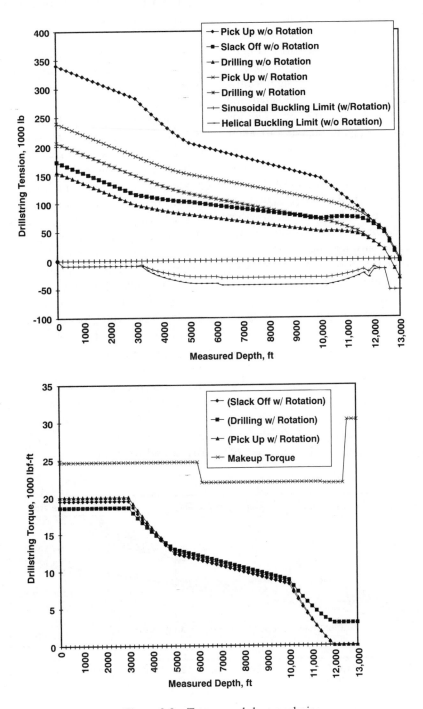

Figure 2-9   Torque-and-drag analysis

Because of the relatively high hole angle, torque is nearly 20,000 ft-lb, dictating the use of 5-in., high-strength (S-135) drillpipe in the upper 6000 ft of the well. Maximum tension in the drillstring is about 340,000 lb, which leaves a reasonable safety margin below the yield strength of 487,000 lb. On the basis of torque/drag, this well is drillable with standard drillstring components and practices if the friction factors used in modeling are accurate and unusual hole problems do not occur.

During drilling, a log of predicted and actual torque and/or drag vs. depth should be maintained (Figure 2-10). Such a log enables friction factors to be updated and verified; if actual friction factors are significantly different than planned, problems can be predicted and prevented instead of dealt with after the fact. The log will also show the effect of changing bit types or altering operational parameters. Note the increase in overall torque when PDC bits are run rather than roller-cone rock bits. The log can also reveal deteriorating hole conditions, such as the buildup of cuttings and local hole features like doglegs. Remedial actions, whether preventive or after the damage occurs, could include (1) enhancing hole cleaning through higher flow rates, rotation, or modified drilling-fluid rheology, (2) making short trips to condition the hole, (3) reaming out ledges, key seats, or doglegs, (4) changing mud type, or even (5) altering the well profile or changing the casing or hole program.

2-2.5 BHA Modeling

BHA modeling is another key component of well planning. A good well plan provides rigsite personnel with the predicted capabilities and tendencies of each planned BHA. BHA modeling should identify the response of each BHA to variations in operating parameters, such as weight-on-bit (WOB), hole angle, overgauge or undergauge hole, stabilizer wear, and formation tendencies. Modeling should also identify the directional response to BHA design parameters, such as stabilizer diameter, drill collar length, or motor bend angle.

BHA modeling can also be used for calculating bending moments and stresses. In some applications, bending stress caused by hole curvature may be high enough to cause additional problems of fatigue or overload. Side forces on bits and stabilizers can also be calculated. These values are useful for motor component design work and as inputs into torque/drag and casing-wear programs.

Various types of directional prediction models exist, but all are based on the principle that directional control is accomplished when forces are applied to the bit that will cause it to drill in the desired direction. Two kinds of models are commonly used: equilibrium (constant hole curvature) models and drill-ahead models.

Equilibrium models are typically static beam models that solve for the hole curvature in which all bending moments and forces on the beams and BHA components are in equilibrium. A typical 2D model (Williamson and Lubinski, 1986) applies known loads (including weight-on-bit, buoyancy, and the weight of the BHA itself) and derived loads (bit side-loads resulting from formation anisotropy) to the BHA elements. Effects of rotation or dynamics are not considered. A single, empirically derived, bit-formation interaction factor is normally used. The premise of the model is that while both the bit and formation may have anisotropic properties, only the net effect can be measured. These models have proved to be reasonably accurate for a wide range of BHAs and formations.

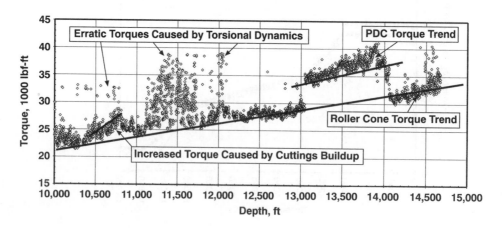

Figure 2-10   Log of actual vs. predicted drillstring torque

Drill-ahead models (Larson and Azar, 1992) create a constantly changing wellbore path, the instantaneous direction of which is based on the combination of force vectors on the bit and anisotropy factors. Anisotropic cutting properties may be assigned to the bit, meaning that its ability to drill sideways in response to a given force is not the same as its ability to drill forward in response to the same force. Formation anisotropy properties may also be assigned, meaning that the formation is more easily drilled in one direction than another. Formation anisotropy properties are oriented with respect to the dip angle and direction of bedding planes. Three-dimensional forces resulting from rotation or reactive torque can also be applied in drill-ahead models. Such models are useful for analysis, but since the shape of the wellbore is generated in increments of inches rather than dozens of feet, these models are typically too computationally intensive for everyday well planning. Some drill-ahead models can be used to characterize the well post-mortem, thus optimizing future well plans.

Formation anisotropy is a variable that can significantly affect the directional characteristics of a BHA. The exact mechanism of formation anisotropy is not known. Lubinski (1953) theorized that formations have a higher drillability perpendicular to the bedding plane than parallel to it (Figure 2-11). This theory would seem to be supported by the science of rock mechanics, since it is known that the compressive strength of many rocks is anisotropic. Rollins (1959) proposed that thinly laminated formations fracture perpendicular to the bedding plane, creating miniature whipstocks that force the bit updip. Murphey and Cheatham (1965) proposed the drill collar moment theory, suggesting that when a bit drills from a soft formation into a hard formation, the hard formation supports most of the bit load, causing a bending moment to be applied to the drill collar. The collar would then bow to the opposite side of the hole and point the bit updip.

Although the formation anisotropy mechanism may not be fully understood, the manner in which BHAs respond to formation anisotropy and dip can be observed. For well-planning purposes, detailed information from offset wells should be compiled. This information should consist of data from intervals in which consistent operating parameters have been used. Formation dip and direction may be obtained either from seismic information or by correlations of logs from several offset wells. BHA design, operating parameters, and hole curvature can be found in daily reports and well surveys. With these as input parameters, the

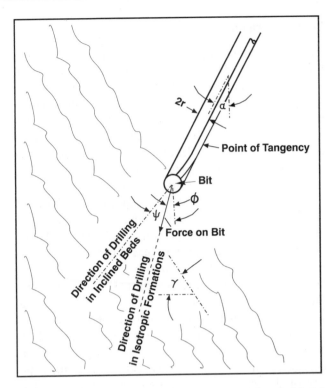

Figure 2-11 Formation drillability theory (after Lubinski, 1953)

BHA modeling program can solve for the remaining variable, which is the formation anisotropy factor. As field development progresses, factors should continually be evaluated, updated, and then applied to the planning of future wells. When BHAs do not exhibit the predicted directional characteristics, parametric studies must be performed to isolate whether variation from plan resulted from formation anisotropy or some other effect, such as hole erosion or stabilizer wear. In nearly all cases, predictive BHA modeling requires disciplined integration of mechanical models with empirical field experience.

To capture the full value of BHA modeling, it must be used for more than simply specifying the BHA configuration that is sent out to the rigsite. BHA modeling should be used to make rig personnel aware of BHA responses to design and operating parameters. Knowing the response of BHAs to these parameters will enable the directional driller to vary BHA configuration or operating parameters to control directional tendencies, as well as to have BHAs available to respond to contingencies. Finally, modeling should be used to establish operating guidelines. In particular, the maximum dogleg in which each BHA can be rotated should be identified.

2-2.6 Hydraulics

Although annulus hydraulic pressure-loss calculations for directional wells are similar to calculations in vertical wells, the mechanism of cuttings transport is different. The physical model itself differs in that the drillstring can be assumed to be offset to the low side of the hole for almost the entire length of the well. Drillstring eccentricity affects velocity profiles along the cross section of the annulus; a low-velocity zone occurs in the vicinity of the least annular clearance. Cuttings transport is also significantly affected by hole angle because the gravitational velocity component acts radially on solids instead of axially. Both drilled cuttings and weighting solids are affected. The low-velocity zone and gravitational component are both factors that contribute to the buildup of cuttings on the low side of the hole (Thomas *et al.*, 1982; Slavomir and Azar, 1986). For these reasons, directional wells require higher-than-normal circulation rates to facilitate cuttings removal. Flow rates increased by 50%, as compared to vertical wells, are not uncommon, depending on hole angle and the cuttings-carrying capacity of the mud. Drillstring rotation is also critical to cuttings removal. Pipe rotation mechanically agitates the cuttings, lifting them up off bottom into the high-velocity flowstream (Lockett *et al.*, 1993).

Throughout most of the well, the drillstring is lying on one side of the wellbore; thus, differential sticking is of greater concern than in vertical wells. Stuck pipe is generally the most common form of trouble in directional wells, and prevention and remediation of stuck pipe should be addressed by training all rig personnel in proper operating practices and optimum drilling-fluid properties. During drilling, the presence of cuttings beds must be monitored by the tracking of changes in circulating pressure and torque and drag; remedial action must be taken when predetermined levels are reached.

Equivalent circulating density (ECD) is also of greater concern in directional wells. ECD is defined as the sum of hydrostatic pressure resulting from the column of mud (and cuttings) in the annulus, plus the pressure drop in the annulus during circulating. The higher flow rates required in directional wells result in a high circulating-pressure drop in the annulus, and the angle of the wellbore with respect to *in-situ* formation stress will generally result in a formation fracture at a lower ECD than in a vertical hole. Both of these factors narrow the range of safe drilling fluid weight. In fact, as illustrated in Figure 2-12, it is possible for the pore pressure and fracture gradient to be such that some high-angle wells cannot be safely drilled (Guild *et al.*, 1994). The relationship

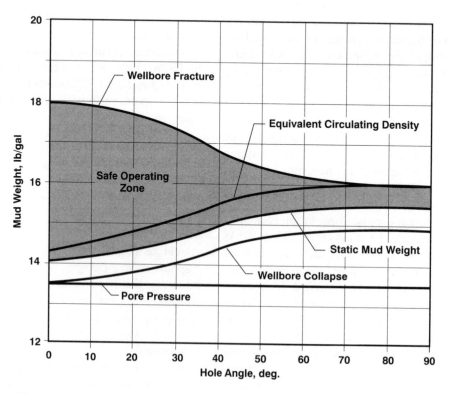

Figure 2-12   Safe drilling-fluid weight range decreases as hole angle increases

between mud weight, ECD, fracture gradient, and hole angle is a key screening criterion when evaluating the feasibility of a directional well and a key factor in planning casing design and hole intervals.

Large-diameter drillpipe is commonly used in directional wells. A good starting point is to specify one size larger than would be used in vertical wells. This practice provides two benefits relative to hydraulics: (1) the larger inside diameter (ID) greatly reduces the pressure drop through the bore, allowing higher flow rates to be used without an increase in surface pressure, and (2) the larger outside diameter (OD) increases annular velocity at a given flow rate, which improves hole cleaning. An additional benefit is the larger pipe's higher torsional capacity which, as seen in the next section, can be necessary in high-angle wells.

The ideal drilling fluid for directional wells should exhibit good lubricity and cuttings-carrying capacity, minimum solids and ECD, and maximum formation inhibition. Since some of these properties may be mutually exclusive, iterative analysis with the appropriate models may be necessary for determining which of the many parameters (hole cleaning, formation stability, torque/drag, etc.) must be given priority and which can be successfully de-emphasized. Some drilling fluids are engineered specifically for directional wells, and drilling-fluid companies should be consulted for specific recommendations regarding flow rates and optimum rheology. Chapter 5 provides additional information regarding drilling-fluid properties.

2-2.7 Rig Capabilities

In addition to geological, geometric, hydraulic, and mechanical considerations, directional planning must also consider the capabilities of the rig and associated drilling equipment. The "severity" of the directional well being planned can be limited by a number of rig aspects. (In this chapter, the term "severity" refers to a qualitative measure of the extent of doglegs, turns, angle, reach, depth, and other factors that influence hoisting, torsional, and hydraulic capabilities of the drilling rig.) Hoisting capability may impose directional limitations on the well, depending on the TVD of the target, the amount of frictional drag associated with lifting the drillstring or deep casing strings, and the amount of mechanical drag induced by doglegs, keyseats, etc. Directional wells induce greater rotary torque than vertical wells. In severe directional wells, torque demands may exceed the capacity of the rotary table, top-drive system, or drill-

string members. These issues should be evaluated through the use of torque-and-drag models (Section 2-2.4) before the rig is specified. Top drives are becoming standard equipment on rigs for drilling directional wells. The ability to circulate and rotate while tripping is an absolute necessity in some directional wells, and always a welcomed capability in others.

Rigs for drilling directional wells may need upgrades in depth-tracking, torque-sensing, and pressure-sensing equipment. Because measured depth is one of the main inputs for calculating well trajectory, knowing the exact measured depth is critical in directional wells. Accurate torque sensing is necessary because of the higher torque demands of directional wells; in fact, torque-limiting or even active torsional control systems with feedback loops may be necessary (Sananikone *et al.*, 1992). High-resolution standpipe pressure instrumentation that can detect motor operating pressure fluctuations as low as 50 psi may be needed even when standpipe pressure is 3000 to 5000 psi.

Crossover subs, whether supplied by the drilling contractor or directional company, should be subjected to a higher level of design, quality, and inspection scrutiny than those used in vertical wells. Crossovers are subjected to high stresses in directional wells and are frequently weak points in the drillstring or BHA. Crossovers should have stress-relief features, be made from high-quality material, and be properly inspected at regular and frequent intervals.

Because inclined wells require higher flow rates to clean the cuttings from the well, the pressure and/or flow rate capacity of the pumps and surface plumbing should be evaluated. The rig must generate enough power to run the pumps, the rotary or top drive, and the draw-works simultaneously at elevated operating parameters. Similarly, the rig's solids-control system must have a volumetric capacity that matches the flow rate requirements, or elevated flow rates will not be sustainable. Because of the various interrelationships between drilling mechanics and the directional trajectory, an integrated and comprehensive approach to rig selection must be taken during the planning and evaluation of directional wells.

2-3 DIRECTIONAL-DRILLING TOOLS

Directional control in most controlled-trajectory drilling, whether long-, medium-, or short-radius, is provided by two basic types of BHAs: drilling motors and rotary assemblies. Techniques such as jetting and whipstocks, and directional measurement tools such as steering tools, are still used occasionally, but they are covered sufficiently in Bourgoyne *et al.* (1991). MWD/LWD tools are found in Chapter 4. A discussion of future directional tools is found in Section 2-7.4. The following section will be confined to the directional control tools most commonly in use today.

2-3.1 Drilling Motors

The positive-displacement motor (PDM) has evolved into the primary method of directional control. PDMs are fluid-driven drilling tools that turn the drill bit independent of drillstring rotation. PDMs are referred to as "mud motors," derived from the fact that drilling fluid is the driving fluid. The power of a PDM is generated by a rotor and stator based on geometry described by Moineau (1932). Both the rotor and stator have helical lobes that mesh to form sealed helical cavities (Figure 2-13). The flow of drilling fluid through these cavities forces the rotor to rotate. The stator profile, which always has one more lobe than the rotor, is molded of rubber inside the motor housing. The rotor, which travels in an orbiting motion around the axis of the tool, is connected to a flexible or articulated coupling that transmits torque while eliminating the orbital motion. The coupling enables the motor housing to feature a bend, usually from 0° to 3°. The coupling transmits torque to a

drive shaft, which is housed in bearings to enable it to transmit both axial ("bit weight") and lateral loads from the drillstring to the bit. PDMs typically have output speeds of from 100 to 300 rpm, while delivering enough torque to power even aggressive polycrystalline diamond compact (PDC) bits.

PDMs are unique in that motor rate is nearly linearly proportional to flow rate, and that torque is proportional to the pressure drop generated, as shown by the power curve in Figure 2-14. This relationship of input pressure to output torque allows the driller to easily detect a stall condition by watching for an increase in standpipe pressure.

Before the introduction of the steerable motor in the late 1980s, the typical drilling motor configuration was a straight motor about 20 ft long with a bent sub above the motor. These tools were capable of build rates up to approximately 5°/100 ft, but they were subject to severe build-rate variations in anisotropic formations and had difficulty kicking off in firm formations. The greatest limitation of the motor/bent-sub combination was that it was only capable of one build rate for a given bent-sub angle and could not be rotated to vary the build rate.

A *steerable* motor (Figure 2-15), typically configured with a bend in the external housing and two or more stabilizers, is a PDM configured to operate as a two-mode system. The two modes of operation are the "sliding," or oriented mode, and the "rotary" mode. In the

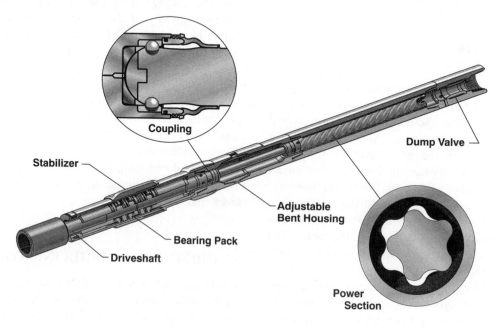

Figure 2-13   Positive-displacement motor (PDM) components

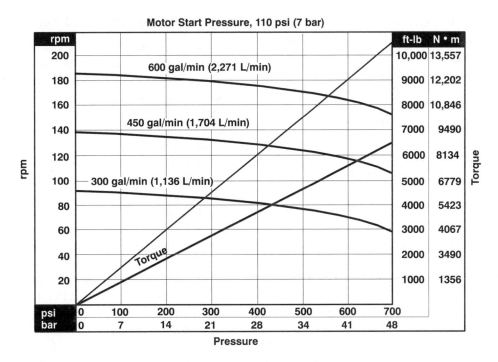

Figure 2-14 Typical PDM power curve

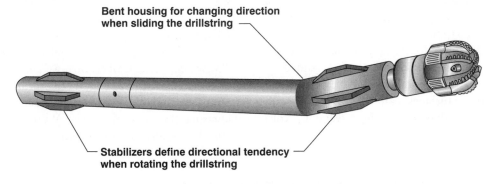

Figure 2-15 Typical steerable motor configuration

sliding mode, the steerable motor is oriented by slowly rotating the drillstring, using MWD signals to determine toolface or bend orientation. Once the desired downhole toolface orientation is achieved, the drillstring is then slid (i.e., advanced without rotating), maintaining the desired toolface. The rotation required to drive the bit is generated entirely by the PDM. The combination of stabilizers and bent housing generates a side load on the bit, causing it to drill in the direction of the toolface. The build or turning capability of steerable motors, referred to as dogleg capability, is typically from 1 to 10°/100 ft.

Placement of the bend in the connecting rod housing instead of above the motor reduces stress and makes rotation possible. In the rotary mode, the drillstring is rotated and the effect of the bend is negated, at least as far as changing direction is concerned. When rotated, the steerable motor behaves directionally like a rotary assembly (Section 2-3.2), in which the directional tendency of the motor is determined by the diameter and placement of its stabilizers and by its stiffness. Steerable motors are normally set up to drill straight ahead in the rotating mode, although they can be configured to build or drop angle while rotating.

This two-mode system gives only two build rates, but since the system can be manipulated from the surface by varying the percentages of sliding and rotating, those modes can be varied so that they provide a number anywhere between the two build rates. In the build section of a well, the sliding mode will be used the majority of the time—perhaps 60%—with the rotating mode used to reduce the build rate as needed. In a tangent or a horizontal section, the rotary mode will be used the majority of the time—on the order of 60% to 95%—with the sliding mode being used only to make correction runs when the rotary mode varies from plan.

Like PDMs, *turbine drilling motors* are powered by drilling fluid, and they rotate the bit. Unlike PDMs, which are powered by an elastomeric stator, a turbine motor uses metal blades that provide greater resistance to chemically aggressive drilling fluids and extreme bottomhole temperatures. Unlike PDMs, however, bit speed decreases as the torque demand of the bit increases, and an increase in torque does not result in a pressure increase that can be seen at the surface. Therefore, detecting a stalled turbine is more difficult than detecting a stalled PDM. Turbines tend to be longer than PDMs and their speed is much higher, typically about 1000 rpm. The high speed of turbines makes them best suited to natural diamond bits or high-speed PDC bits, while PDMs are best suited to three-cone rock bits or medium-speed PDC bits. Much like PDMs, modern turbines can be configured with bends and stabilizers that give them "steerable" capability.

2-3.2 Rotary Assemblies and Adjustable Stabilizers

Rotary assemblies are still occasionally preferred over steerable systems, usually in tangent sections where the directional objective is to drill straight ahead. Rotary assemblies are most commonly used where formation tendencies are predictable and rig economics are not conducive to the use of steerable motors. In a rotary assembly, the weight of the collar gives it a tendency to sag or flex to the low side of the hole; collar stiffness and length and stabilizer diameter and placement are engineered as a means of controlling the amount of flexure to give the desired hold, drop, or build tendency (Figure 2-16). Fixed rotary assemblies have a limited ability to adjust for variation from plan, but they may be practical in some intervals of multiwell developments, where the characteristics of formations can be identified and the assemblies can be optimized. For further discussion of fixed rotary assemblies, see Jogi *et al.* (1990) and Bourgoyne *et al.* (1991).

The limited ability to vary the directional tendency of a fixed rotary assembly comes primarily from varying WOB. Since fixed rotary assemblies have only a single directional tendency, increasing or decreasing WOB will not generally change the directional tendency (i.e., from a build to a drop) but it may be used to "tune" that tendency. For example, the angle-building assembly of Figure 2-16 consists of a near-bit stabilizer and a drill collar which, because of its own weight, flexes to the low side of the hole. Increasing WOB will buckle the collar, shifting the point of tangency closer to the bit and increasing the build rate. The WOB that is optimum from the standpoint of directional control, however,

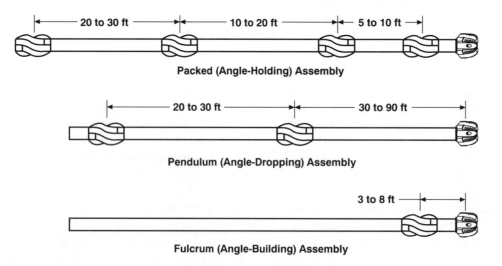

Figure 2-16 Commonly used rotary assemblies

may be sub-optimum from the standpoint of the rate of penetration (ROP) or bit life. The economics of using WOB to control inclination must be weighed against the economics of running a steerable motor or an adjustable stabilizer.

Adjustable-diameter stabilizers (Figure 2-17) are an improvement over fixed rotary assemblies because the diameter of the stabilizer can be adjusted downhole to accommodate variations from plan. Using only the diameter change of the adjustable stabilizer, a properly designed BHA can produce an inclination tendency that ranges from a build to a drop. Adjustable stabilizers control only the inclination of the well, but in many cases, that is sufficient; as wellbore angle increases, the tendency of the bit to "walk" (deviate from plan in the azimuth plane) decreases. Controlling inclination during rotation (as opposed to sliding a motor) has many advantages, including improved ROP, improved hole cleaning through continuous agitation of cuttings beds, smoother wellbores, less chance of stuck pipe, and improved weight transfer to the bit. Adjustable stabilizers are actuated with either WOB or flow, and their position is communicated to the surface through either flow restriction or mud pulses.

Adjustable stabilizers can be used in either rotary-only BHAs or run in conjunction with steerable motors for optimizing the motor's rotary-mode directional tendency. For more information on adjustable stabilizers, see Eddison and Symons (1990), Underwood and Odell (1994), and Odell and Payne (1995).

2-4 CONVENTIONAL (LONG-RADIUS) WELLS

Conventional, or long-radius, wells are typically defined as those with build rates from 1 to approximately 8°/100 ft. However, the definition can vary with hole size. A better definition may be that the steerable motor that is used in the sliding mode to drill the build section may also be safely rotated in that section, and that the hole curvature in the build section is not high enough to cause drillpipe failure from fatigue. This definition suggests that the maximum build rate for a large hole size will be lower than the maximum build rate for a smaller hole size. Equation 2-10 relates the stress in a tubular member that is bent through a given curvature.

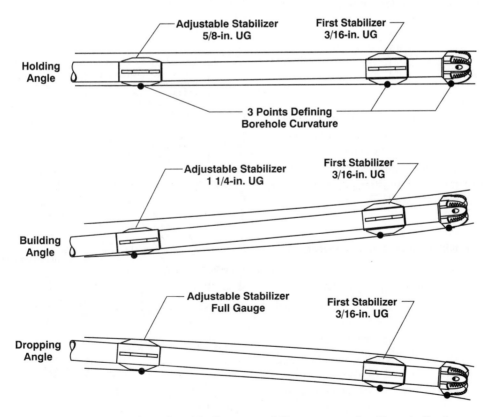

Figure 2-17   Using adjustable-diameter stabilizers to control wellbore inclination

Clearly, for a given acceptable stress level, the smaller the tubular diameter, the higher the tolerable build rate

$$BUR = \frac{137,510\sigma}{ED} \qquad (2\text{-}10)$$

where *BUR* is the build rate(°/100 ft), σ is the allowable stress (psi), *E* is the Young's modulus (psi), and *D* is the pipe diameter (in.).

Equation 2-10 shows the relationship between pipe diameter, curvature, and stress. It does not include the effects of stress risers or of pressure, tension, or torsion, which must be combined to calculate the resultant stresses.

Another characteristic of conventional wells is that the maximum angle of any section of the well can allow the drillstring and wireline tools to advance through that section by the force of gravity alone—that is, without being pushed or pumped down. This criterion would normally limit the maximum angle to about 65° to 80°, depending on friction factors. Friction factors are primarily a function of formation type and mud type, but overall friction factors include the effect of ledges and the effect of the geometry of stabilizers and other points at which the drillstring or BHA contacts the formation. The critical wellbore angle, above which the drillstring will no longer advance down the hole under the force of gravity alone, is θ_c, (Mueller *et al.*, 1991):

$$\theta_c = \tan^{-1}\left[\frac{1}{f}\right] \qquad (2\text{-}11)$$

where *f* is a coefficient of friction.

Although by definition, the maximum angle of conventional wells will allow them to be surveyed and logged by wireline, advances in the accuracy, features, and resolution of MWD and LWD have allowed these tools to become the definitive surveying and logging method for many conventional directional wells.

2-4.1 Drillstring Considerations For Long-Radius Wells

In conventional directional wells, because the angle of the well is low enough to allow gravity to be used, drill collars are commonly used in the BHA to apply weight-on-bit. Unlike drillpipe, the collar's weakest part is the connection; therefore, rotating collars through high doglegs should be avoided. The American Petroleum Institute (API) has published guidelines for rotating drillpipe through doglegs, but not for rotating drill collars through doglegs. Most service companies have generated

internal guidelines either for collars or their own tools, and some of these have been published (Cheatham *et al.*, 1992). Although MWD/LWD tools and motors normally have weaker connections than drill collars, they are generally more limber and can generally tolerate the same build rates as collars of equivalent diameter.

If the maximum angle of conventional wells is low enough that the drillstring will advance under its own weight, then tension in the drillstring will always exist in the build section. Lubinski (1973) and Hansford and Lubinski (1973) showed the relationship of stress in drillpipe as a function of both the dogleg severity to which it is subjected and the amount of tension in the drillpipe while in that dogleg. Abrupt doglegs cause higher stress in drillpipe than gradual doglegs. As shown in Figure 2-18, with $4\frac{1}{2}$-in. S-135 drillpipe in a 6°/100 ft dogleg, tension in the dogleg portion of the drillstring must be limited to about 80,000 lb to avoid fatigue damage. On the other hand, if dogleg severity is limited to 3°/100 ft, then 200,000 lb of drillstring tension in the dogleg can be tolerated. Lubinski's guidelines for both Grade E and S-135 drillpipe are published in API Recommended Practice 7G and should be one of the screening criteria for evaluating the feasibility of any directional well.

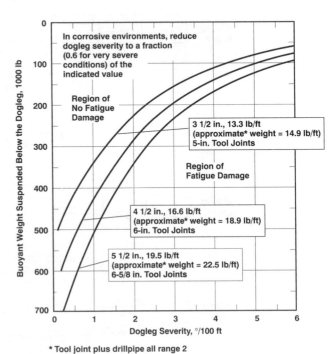

Figure 2-18 Dogleg severity limits for S-135 drillpipe (after Lubinski, 1953)

Particularly in deep wells, unplanned doglegs in the build section should be minimized to reduce both drillstring fatigue and drag. In extremely deep directional wells with a high KOP, the planned build rate itself may have to be minimized. Conversely, if for a given well depth, the KOP can be designed as deep as possible, higher build rates can be safely used. Higher build rates are possible with a deep KOP because less pipe is present below the KOP and because a higher percentage of its weight is supported by the borehole. The result of these factors is less tension in the drillpipe that is in the build section of the hole.

2-4.2 Drilling the Build Section

Steerable motors are normally run in the build section because they can compensate for almost any variation from plan. General well planning guidelines are to design the steerable motor with a dogleg capability of about 25% to 50% more than the planned build rate; the motor is then rotated as required to eliminate the excess build rate. In the build section, however, the greatest fear is falling "behind the curve." To ensure that this situation will not occur, the tendency is to design the steerable motor for a greater build rate than needed. Cases have

been seen where the actual dogleg capability of a steerable motor is two to three times the planned build rate. Such a selection can result in the creation of excessive doglegs in the build section as the assembly repeatedly builds more than is needed; then, the motor must be rotated to reduce the overall build rate. When the motor is first rotated after each interval of oriented drilling, it will also encounter high bit side-loads that may reduce its life.

Figure 2-19 shows build-rate capabilities and bit sideloads for a $7\frac{3}{4}$-in. steerable motor for various bend angles. The highest bend angle shown (1.5°) can build about 10°/100 ft. However, when the motor is rotated in this curvature, the resulting bit side-load is about 19,000 lb. If the desired build rate is 4°/100 ft, the use of a 1° bend should result in a 6°/100 ft dogleg capability. The build-rate safety factor is 50%, but bit side-loads will have been reduced to about 12,000 lb, a reduction of 32%. In addition, since the actual build rate is closer to the necessary build rate, fewer slide/rotate sets will have to be used, resulting in a smoother wellbore.

High bit side loads can cause damage to the gauge or bearings of the bit and limit motor life by causing driveshaft fatigue, radial bearing wear, and stator damage. Stabilizer loads and associated wear also increase. The

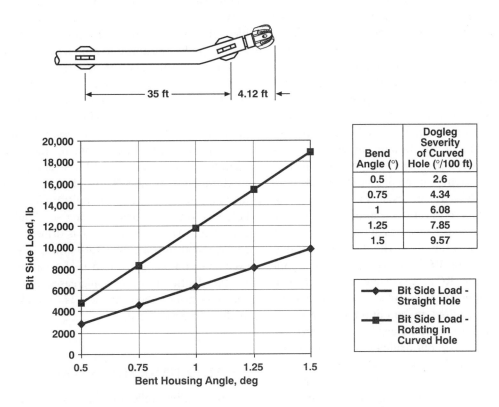

Bend Angle (°)	Dogleg Severity of Curved Hole (°/100 ft)
0.5	2.6
0.75	4.34
1	6.08
1.25	7.85
1.5	9.57

Figure 2-19   Effect of bent housing angle on build rates and bit side-loads

high doglegs resulting from excessive bend angles can cause drillpipe fatigue and difficulty in sliding the drillstring or running logging tools, casing strings, or completion equipment in the well in the future. The preferred well planning strategy for most wells is to ensure that the build assembly will have a sufficient build-rate safety margin to cover any contingency while maximizing the life of other BHA components and minimizing doglegs. This strategy is especially true if the steerable motor is scheduled to drill a long tangent section after drilling the build section, which often occurs. As a rule of thumb, a build assembly with a dogleg capability of 25% greater than the desired build rate is adequate. As experience is gained in a given field, this safety factor can be reviewed and modified. If build rates are predictable, the safety factor could be reduced; if build rates are unpredictable, the safety factor may have to be increased.

2-4.3 Drilling the Tangent Section

In a tangent section or lateral, the goal is normally to drill straight in the direction established at the end of the build. Tangents of 10,000 ft and laterals of 5000 ft are not uncommon. Steerable motors, of course, are designed to drill straight ahead in the rotating mode. Some variation from plan is acceptable, but the ability of the steerable motor to accommodate variation from plan can lull the directional driller, well planner, or drilling engineer into uneconomical drilling practices. Therefore, the inherent advantage of steerable motors—that when the rotary-mode directional tendency varies from plan, it is possible to correct by sliding—can also be a trap. Since corrections are relatively easy to make, there may be less effort to optimize the rotary-mode directional characteristics during the planning phase, especially if the focus is on merely hitting targets, or "staying on the line" geometrically.

However, compelling reasons exist for optimizing the rotary-mode tendency of steerable motors. One simple reason is economics; experience has shown that even in conventional directional wells without severe torque/drag problems, sliding ROP is typically only about 60% of rotary-mode ROP. In severe wells, effective ROP can be as low as 5 to 10%. (This percentage includes the time required to condition the hole and pick up extra collars to apply weight before attempts to slide, orient, and circulate out drilled cuttings following the slide). Secondly, the doglegs created by excessive sliding can cause problems with drillpipe fatigue or with further attempts at sliding later in the well; doglegs may also limit the drillable depth of the well. See Hogg

and Thorogood (1990) and Banks *et al.* (1992) for in-depth analyses of steerable motor optimization and the effects of doglegs.

2-4.4 Hole Erosion

Conventional or long-radius drilling is common on offshore platforms. Offshore geology often consists of relatively soft formations, especially at shallow depths, in which hole erosion can reduce build rate significantly. Hole erosion may be caused by chemical, hydraulic, or mechanical effects. When angle is being built in sloughing clays, drilling should be as fast as possible, since clay hydration is time-dependent. Mud filter-cake properties should be optimized to control clay hydration and sloughing. In unconsolidated sands, such as those found in the Gulf of Mexico, fluid velocities can cause hydraulic erosion. During drilling in unconsolidated sands, one technique that combats hole erosion is to reduce the flow rate while drilling, then pull a joint of pipe off bottom and circulate at a higher flow rate to remove the drilled cuttings out of the hole without eroding the hole in the vicinity of the bit and first stabilizer. Drillstring rotation can damage the filter cake and thereby promote sloughing; therefore, if hole erosion is a concern, rotation should be limited in the build section.

When severe hole erosion becomes a significant problem affecting the predictability of steerable motors, tool design can be optimized to minimize its impact. Steerable motors typically have the first stabilizer on the bearing pack, approximately 2 to 3 ft from the bit. Moving the stabilizer further up the motor will reduce the system's sensitivity to hole erosion (Figure 2-20). Rotary assemblies can also be optimized to lessen the impact of hole erosion through the use of some of the same design and operational techniques used for steerable systems.

Although increasing the bend angle of the motor (and therefore its build-rate capability) may compensate for hole erosion, the effect of doing so should be carefully considered. When the motor has drilled past the area of concern and into a firmer formation, side loads may be excessive and cause reduced performance or motor life. As the depth of the well increases and the formation becomes firmer, hole erosion is less likely, and bend angles can be reduced.

2-5 MEDIUM-RADIUS WELLS

Medium-radius wells use many of the same BHA components and well planning tools used in long-radius

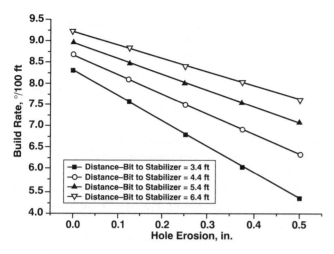

Figure 2-20   Effect of hole erosion on build-rate capability

wells. The main difference is that medium-radius build rates place some limitations on the ability to rotate, and that these limitations can affect the well profile. Medium-radius wells can be broadly characterized by the following:

- The BHA used for drilling the build section cannot be rotated in that section because of stresses in motor and MWD housings and connections. At best, limited rotation is allowed.

- Because of the hole curvature in the build section, the component of drillpipe stress caused by bending is high enough that either drillstring rotation must be limited while in tension, or the stress component resulting from tension must be limited by well profile design.

- To eliminate any tension in the drillstring in the build section, any footage drilled beyond the build must be at or above the critical angle referred to in Section 2-4. As a general rule, medium-radius build rates are used only for wells with high-angle or horizontal laterals.

The definition of medium-radius wells, like that of long-radius wells, will vary with hole size. Table 2-3 shows approximate guidelines for medium-radius wells.

Table 2-3   Medium-radius well guidelines

Hole size (in.)	Build Rate (°/100 ft)	Radius (ft)
6 to 6¾	12 to 25	478 to 229
8½	10 to 18	573 to 318
12¼	8 to 14	716 to 409

2-5.1 Medium-Radius Well Profile Considerations

If build rates never varied, and the precise location of the formation was always known, most well profiles would probably feature a continuous build rate to EOB. Since such conditions do not exist, various well profiles (as shown in Figure 2-21) have been used that provide contingencies for build rate variability and geological uncertainty. The build/tangent/build profile is less preferred because it requires a trip to change the BHA at the end of each of the three sections. A build/tangent/build profile also sacrifices some potential payzone exposure. The dual build rate, as compared to the build/tangent/build rate, saves a trip or, if the well planner is fortunate, two trips if the first build rate is on target and the second BHA can be avoided. The best well profile is usually the "soft landing," in which a medium-radius BHA is used to drill most of the build section, and the remainder of the build section is then drilled, mostly in the sliding mode, with the same steerable assembly that will be used for drilling the lateral. The length of this final section should be at least as long as that of the BHA, so that all BHA components will be through the high build-rate section before they encounter significant rotation. This well profile minimizes the number of trips, applies the

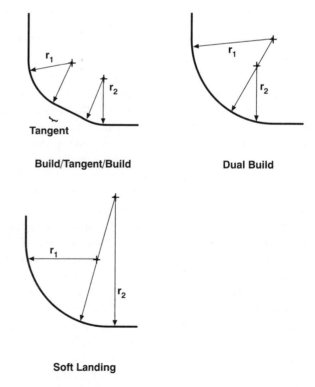

Figure 2-21   Medium-radius well profiles

least stress on motors, and allows precise TVD control at the EOB.

Variations of medium-radius profiles are dual opposing laterals (which eliminate the need to drill a single lateral of twice the length), stacked laterals, or even stacked opposing laterals. For these profiles, the upper laterals are drilled first, then the lower laterals are drilled. After kickoff, both laterals are completed openhole. Multiple lateral wells with full wellbore isolation are discussed in detail in Chapter 3.

Horizontal reach capability may be limited by friction, drillpipe buckling, available pump pressure, or the weight available for pushing the string into the lateral. Torque-and-drag models that incorporate buckling should be used, but local experience should also be thoroughly considered. Horizontal lateral lengths of over 10,000 ft have been achieved, but in many areas, 4000 ft to 6000 ft is a practical limit, after which drilling becomes increasingly difficult and expensive. Often, reservoir considerations suggest much shorter wells. However, when longer wells are drilled with water-based drilling fluids, oriented drilling with steerable motors normally becomes difficult at a lateral length of about 3000 ft; drilling becomes impossible somewhere between 3000 ft and about 5000 ft. The use of oil-based drilling fluid may extend these lengths by 40% to 50%.

2-5.2 Drillstring Design for Medium-Radius Wells

In vertical wells and conventional directional wells, WOB is provided by the weight of the collars directly above the bit, and most of the drillstring is in tension. In the lateral section of medium-radius/horizontal wells, however, the weight of the drillstring is supported by the side of the hole, and it cannot contribute to advancing the drillstring. The force to advance (push) the drillstring in the lateral must come from the vertical portion of the well, or at least from some section of the well that is above the critical angle. To provide this force, drill collars or heavy-wall drillpipe (HWDP), which can be run in compression, are usually run in the vertical portion of the well. Some well profiles require a KOP so close to the surface that the limited amount of vertical hole will not allow the use of enough drill collars or HWDP to provide the necessary force. In such a case, either traveling block weight or a hydraulic pull-down system must be used for applying the force needed to push the drillstring into the hole.

Drill collars serve no purpose in the horizontal lateral. In fact, the weight of collars in the lateral only increases torque and drag and hinders the drillstring from advancing. Dawson and Paslay (1984) showed that conventional drillpipe will buckle with low axial loads in near-vertical holes, but drillpipe can transmit substantial compressive loads without buckling in high-angle wellbores. Figure 2-22 shows the relationship between hole angle and compressive load capability for several commonly used sizes of drillpipe. Common practice is to run conventional drillpipe in compression in the lateral of horizontal wells to transmit axial loads to the bit. Two or three joints of nonmagnetic compressive-service drillpipe are frequently used above the MWD/LWD tools for magnetic spacing.

As mentioned in Section 2-2.4, the torque required to rotate the drillstring in deviated holes is a function of drillstring weight, hole angle, the diameter of tool joints and tubulars, and the coefficient of friction. In high-angle holes, the torque required for rotating the drillstring can be substantial. Torque/drag modeling should be performed to ensure that the torsional strength of tool joints and tubulars exceeds the maximum anticipated drilling torque. The ability to rotate is extremely important because in some high-angle wells, the drillstring cannot be advanced—or even withdrawn—without rotation. This condition is especially true when poor hole cleaning or differential sticking becomes a problem. Torque/drag modeling should also be used during drillstring design to ensure that the drillstring can withstand the tension that occurs during pick-up and the buckling that occurs during slack-off. Maximum loads, both tensile and compressive, should be plotted along the length of the drillstring, and sections of the drillstring should be optimized based on those loads.

The profile and drillstring of a sample medium-radius well is shown in Figure 2-23. The drillstring sections have been sized to withstand the torsional and tensile loads derived by torque-and-drag modeling (Figure 2-24). From the surface to about 8500 ft, the drillstring is in tension, so drillpipe is used. Note, however, that the maximum drillstring tension at the surface is 446,000 lb. According to Table 2-2, S-135 drillpipe is required to provide a reasonable safety margin.

In the vertical part of the subject well, HWDP is used instead of drill collars to apply weight. HWDP is easier to handle on the rig floor than drill collars and, unlike collars, it can withstand passing through a medium-radius build section. If collars were used, an extra trip would be necessary so that the collars could be removed from the string before they reached the curve. For modeling purposes, HWDP is used from 7500 ft to total depth (TD) based on the assumption that a single PDC bit will

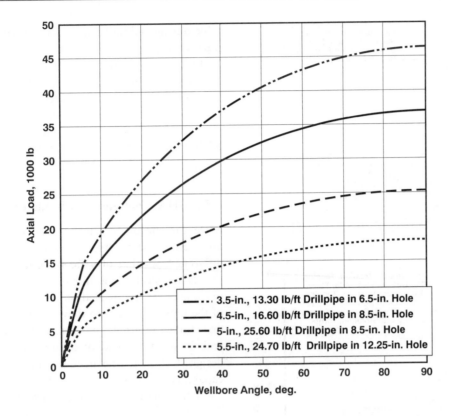

Figure 2-22   Compressive load capability of drillpipe that increases with wellbore angle

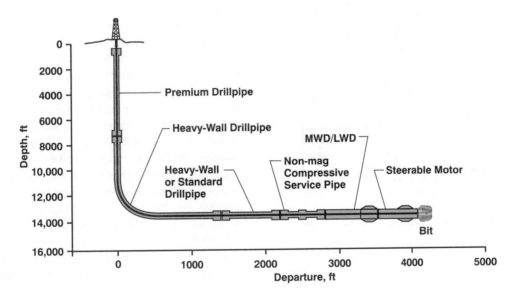

Figure 2-23   Typical drillstring for medium-radius well

drill the entire lateral section, and that the HWDP used to drill the build section will advance into the lateral. If a trip for a worn bit were necessary in the lateral, torque and drag could be reduced with the substitution of standard drillpipe for some of the HWDP in the lateral.

According to Figure 2-22, 5-in. drillpipe in an $8\frac{1}{2}$-in. hole can transmit approximately 45,000 lb without buckling. As shown in Figure 2-24, during sliding with 30,000-lb WOB, the drillstring compression reaches 45,000 lb approximately 1000 ft from the bit. Thus, the last

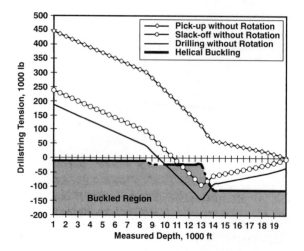

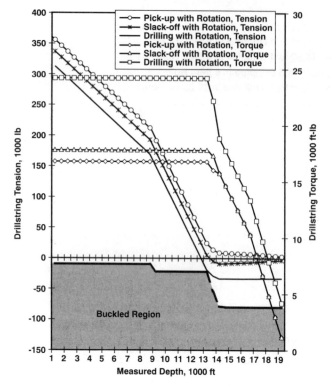

Figure 2-24   Drillstring loads for Figure 2-23 well

1000 ft of HWDP could be replaced by drillpipe without buckling.

Note that during drilling without rotation (i.e., sliding), drag is 126,000 lb greater (surface-indicated weight is 126,000 lb less) than drilling with rotation. In this well, the HWDP in the vertical section of the hole provides sufficient axial force to overcome drag when sliding. To reduce drillstring cost, drillpipe could be run in place of the top 2000 ft or so of HWDP, although some of this pipe would buckle when sliding. Some buckling is accep-

table during sliding because fatigue will not occur if the pipe is not rotating. However, severe buckling can create a "lock-up" condition in which additional weight only serves to increase buckling and side forces. When lock-up does occur, it may be alleviated by the use of stiffer pipe or by a reduction of drag in the lateral. Near the TD of such a well, a lighter pipe may have to be substituted for HWDP in the lateral so that drilling can progress.

2-5.3 BHA Configuration in the Build

Since the motor for drilling the build section is not intended to be rotated, its configuration is somewhat different from that of a steerable motor. The stabilizers that give a steerable motor its rotating-mode directional tendencies are not needed; in fact, they reduce the ability of the motor to slide, so they are typically not used. The first contact point of a medium-radius BHA is generally a pad or sleeve instead of a stabilizer, and it is usually designed close to the bend to maximize the build-rate capability.

Eliminating the possibility of using drillstring rotation to vary build rate means that more emphasis must be placed on accurate trajectory prediction. The effects of variation of downhole parameters, such as hole erosion or formation anisotropy, must be more carefully evaluated than in long-radius wells. Moving the first contact point further from the bit has a beneficial effect if hole erosion is an issue, as mentioned in Section 2-4.4, but it can also make the assembly more sensitive to formation anisotropy. If unpredictable build rates caused by formation anisotropy are a problem, the best approach may be to reduce the planned build rate enough that limited rotation can be used so that some degree of steerability is allowed. If casing shoe depth and geology dictate build rates that are too high for rotation to be feasible, the best solution may be the use of a motor with a lower contact point, such as a stabilized bearing pack. This assembly will exhibit less sensitivity to formation anisotropy.

A second bend at the top of the motor can also be used as a means of improving build-rate capability and predictability. The second bend establishes a definitive contact point and ensures that the top of the motor will stay on the low side of the hole in a near-vertical wellbore. These factors make the build rate more predictable at kickoff and reduce the sensitivity of the assembly to variations in downhole parameters. However, the second bend (1) greatly reduces the chances of rotating the BHA, (2) reduces the ability to configure the motor at the rigsite, and (3) makes it more difficult for the motor to pass through the blowout preventer (BOP) stack. For these reasons, the use of double-bend motors is declining, but they should still be considered where high build rates are required in anisotropic or unconsolidated formations and when a predictable build rate is critical.

2-5.4 Medium-Radius Drilling Practices

Despite efforts to design BHAs that will produce consistent build rates, variation from plan sometimes occurs. Depending on hole curvature and BHA design, minor corrections to the build rate may be possible during drilling in the build section without tripping for a BHA change. As in long-radius drilling, the BHA may be designed to build angle at a slightly higher rate than necessary, so that variations of steerable motor techniques can be used to reduce the build rate. One technique, known as *rocking* or *wagging* the toolface, consists of orienting the motor left for some interval, then right for an equal interval. Another method of reducing the build rate is to rotate the drillstring very slowly, on the order of 1 to 10 rpm. This technique is referred to as *pigtailing* because of the corkscrewed hole it would seem to produce. Both techniques can make an aggressive angle-build BHA drill a tangent-like trajectory when viewed in the vertical plane, allowing precise TVD control. However, these practices (1) produce doglegs and ledges in the wellbore, (2) create offset wellbore intervals through which stabilizers and motors, MWD collars, or other tubulars may have difficulty traversing, and (3) may cause excessive stress in motor or MWD housings when the BHA passes through the high doglegs created. These practices should therefore be used with extreme caution and a recognition of the possible consequences. BHA modeling should be used for analyzing forces on bits and stabilizers and for analyzing bending stresses at connections and critical cross-section changes, with assemblies oriented in the model both highside and lowside.

In the build section, a lack of rotation to agitate cuttings beds requires higher flow rates and/or low-then-high viscosity sweeps to remove cuttings. If cuttings are allowed to build up, the combination of cuttings beds and lack of rotation may increase the possibility of differential sticking.

Time drilling is a technique used especially for both kicking off out of an open hole and creating high doglegs. For kicking off in an open hole, the motor is positioned off-bottom and oriented at the proper location in the wellbore; then, drilling fluid is circulated to rotate the motor for a period. It should be possible to see pump pressure gradually decrease as the bit drills a ledge in the borehole sidewall, although the reduction in pressure will not be as great when compared to on- to off-bottom pressure. When pressure stops dropping, or after a specified period, which depends on formation hardness, the motor is advanced another inch or two and allowed to side-cut for another period. This procedure is repeated for several feet until a sufficient ledge has been created that will allow the bit to begin to take drillstring weight. Time drilling may also be used to ensure high doglegs in hard formation when on-bottom orientations are made. The motor is oriented just off bottom, initially with no WOB, and the same procedure is followed. Time drilling is useful if the rate of build, turn, or drop must be maximized to reach a given directional objective with a given BHA. However, time drilling creates ledges on which stabilizers or tool joints can hang; applying normal weight-on-bit while making orientations will result in a smoother wellbore, and this practice should be used where possible.

The lateral section is usually drilled with conventional tools and practices used in long-radius wells (steerable motors and, less frequently, rotary assemblies). Doglegs that occur early during the drilling of the lateral can reduce the ability to slide later, although they seem to have less influence on the torque required to rotate the drillstring (Hogg and Thorogood, 1990). As in conventional wells, the importance of minimizing doglegs and optimizing the rotary mode directional tendency must be emphasized. When oriented drilling with steerable motors becomes impossible, drilling can continue in the rotary mode if some drift is acceptable. Alternatively, inclination can be controlled with the use of an adjustable stabilizer.

2-6 SHORT-RADIUS WELLS

Issues of drillstring and BHA bending become even more critical for short-radius wells. Short-radius wells have the following characteristics:

- Hole curvature is so high that the BHA must be articulated so it can pass through the build section.

- Drillpipe in the build section is stressed beyond the endurance limit—and in some cases beyond the yield strength—of the material, so that even in the lateral section, the allowable rotation ranges from limited to almost zero.

- Horizontal lateral length capability is reduced by eliminating the possibility of drillstring rotation.

These conditions dictate a challenging set of design constraints and result in unusual well profiles, BHA components, and operational practices. As in other applications, hole size will affect the radius that can be drilled. Table 2-4 contains guidelines for short-radius wells.

Table 2-4 Short-radius well guidelines

Hole size (in.)	Build rate (°/100 ft)	Radius (ft)
$8\frac{1}{2}$	48 to 88	120 to 65
6 to $6\frac{3}{4}$	57 to 115	100 to 50
$4\frac{3}{4}$	64 to 143	90 to 40
$3\frac{3}{4}$	72 to 191	80 to 30

The decision to drill a short-radius well may be driven by the need to (1) set casing very close to the payzone, (2) place artificial lift as close as possible to the payzone, (3) maximize the length of the horizontal lateral in the payzone in fields with close well spacing, or (4) minimize the amount of directionally drilled footage. Frequently,

short-radius wells are drilled by re-entering existing wells, and the radius of the well is simply defined by the distance from the existing casing seat to the payzone.

2-6.1 Well Profiles and Drilling Efficiency

Although the reasons for drilling short-radius wells are usually based on a physical constraint of some sort, there is frequently some choice about just how short the radius is. This decision is critical, not just because it affects how the build section is drilled, but because the radius affects how the rest of the well can be designed and drilled. Drilling efficiency in the lateral and lateral-reach capability are the factors that are most affected by the radius of curvature in the build section.

The lateral section of a short-radius well is drilled either in the sliding mode or with very slow drillstring rotation. If build rate and the resulting pipe stresses preclude rotation, the lateral must be drilled entirely in the oriented mode. Though the motor used to drill the lateral has a fixed build rate, the well can be kept within a small TVD plane by alternating orientations to the right and left, resulting in a sinusoidal profile when viewed in the horizontal plane (Figure 2-25). This procedure is acceptable in many cases, but the increased drillstring drag

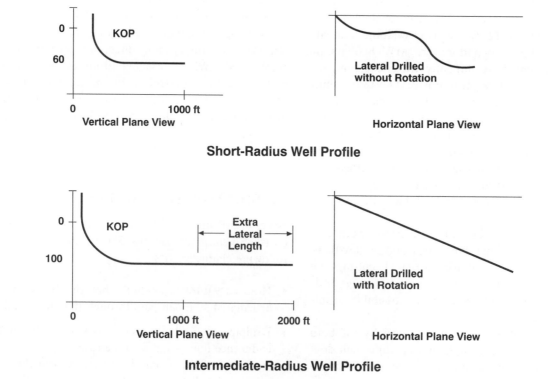

Figure 2-25 Drillstring rotation enables smoother well profiles, further extension

resulting from this technique will eventually limit the length of lateral that can be drilled. Oriented drilling will also progress at a much slower ROP than when the drillstring is rotated.

If the build rate and the resulting drillpipe stresses are low enough to allow limited rotation, drilling the lateral with rotation is certainly preferable to sliding. In addition to improved ROP and reach capability, and the ability to drill a straight hole, rotating results in better hole cleaning and a reduced chance of stuck pipe. However, depending on drillpipe size and the radius of the curvature, the pipe is stressed somewhere between its endurance limit and its yield point. Since damage from fatigue is a function of the stress level and the accumulated number of cycles, at

some number of cycles, the drillstring will fail. Figure 2-26 provides guidelines for limited rotation for various radii of curvature and drillpipe sizes and grades, based on both calculations and historical data. However, cycles-to-failure are difficult to quantify because of variations in material strength, previous stress history, corrosion, surface finish, and wear.

While cycles-to-failure cannot be predicted precisely, one thing is certain: the longer the radius of the well, the lower the chance of failure, and the more contingency options that are available. The radius of the build section directly affects the ability to rotate, and thereby affects drilling efficiency in the lateral. Since the lateral comprises 80% to 90% of footage drilled in short-radius

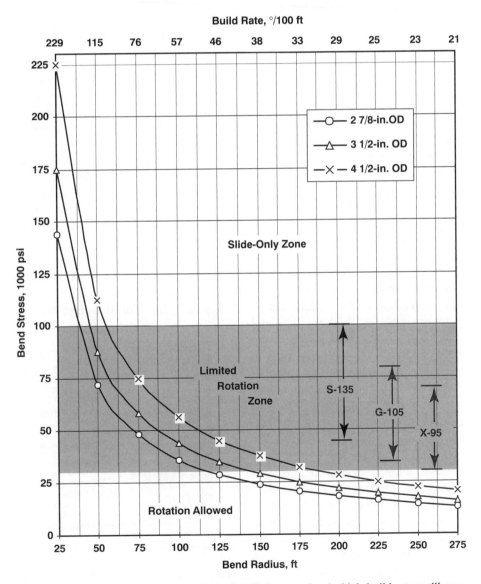

Figure 2-26   General guidelines for limited drillpipe rotation in high-build-rate wellbores

wells, it is the main factor in drilling costs. Drillpipe damage, ROP, and trouble time all factor heavily into short-radius economics. Some operators have settled on an established radius of about 75 ft or greater as optimum from the viewpoint of overall economics. At or above a 75-ft radius, drillstring damage is normally minimal, and limited rotation can be used to maximize ROP and minimize hole problems. Also, above a 75-ft radius, some completion equipment, such as screens or packers, can also be used.

Even if the build rate allows limited rotation, the length of the lateral that can be drilled will eventually be limited by drillstring drag, which is normally exacerbated by buckling. In the oriented mode, laterals of up to 1000 ft have been achieved, although 500 ft may be a more realistic expectation. With rotation, laterals of over 2000 ft have been achieved, with 1000 ft as a realistic expectation.

2-6.2 Milling

When the short-radius well is a re-entry of an existing well, a milling operation of some type is required. The choice of section milling or window milling will depend on several considerations.

For section milling, a packer is typically set below the kickoff point, and an under-reaming casing mill is used to remove existing casing and cement. It is usually desirable to mill enough casing (normally about 60 ft) to eliminate magnetic interference and allow the directional assembly to be oriented with MWD or a steering tool instead of a gyro. If the wellbore is angled enough to allow gravity toolface orientation, the milled section can be shorter. Optimum circulating fluid rheology and very high annular velocities are required for removing milled cuttings. Screens and magnets should be used in the surface equipment for removing all cuttings from the drilling fluid. Metal shavings can damage drilling motors if they are recirculated into the system or are allowed to enter from reverse flow into the string during tripping. After milling, a hard cement plug is set from the packer to a distance of a hundred feet or so above the kickoff point. After the cement is allowed to harden, it is then dressed with a mill or rock bit down to the kickoff point. The short-radius BHA is then tripped in, and the kickoff is initiated. To prevent the bit from preferentially drilling into the cement instead of the formation, the cement plug must be at least as hard as the formation.

When casing is window-milled, the whipstock is oriented and set through the use of a gyro. The whipstock is carried in on the starter mill, but one to two extra mill runs may be necessary for enlarging and dressing the window. After milling, the short-radius BHA is run into the hole and oriented with a gyro, and drilling begins. A whipstock set in a milled window has the advantage of providing a means of positive curve initiation and easily allows tools to re-enter the new wellbore as long as the whipstock is in place. A flow-through whipstock allows continued communication with the casing below the kickoff point to be maintained in the event that the lower wellbore is still producing. However, whipstocks usually require more milling trips and the expense of gyro orientation. Whipstocks occasionally slip after being set, causing errors in the heading of the new well or re-entry problems in the milled window.

2-6.3 Drilling the Short-Radius Build Section

The build rate of short-radius wells requires that large-diameter tubulars (motors or survey collars) must be articulated to pass through the build section. Articulations are knuckle joints or hinge points that transmit axial loads and torque, but not bending moment. BHA components are shortened into lengths that will traverse through the build section without interference (Figure 2-27). Without articulations, excessive bending stress and high side-loads would result.

Since the articulated joints decouple the bending moment from one section of the BHA to another, the build rate of the steering section is unaffected by the stiffness or weight of the sections above it. Build rate is completely defined by three contact points: the bit, the first stabilizer or pad, and the first articulation point (Figure 2-28). For the same reasons that other BHA

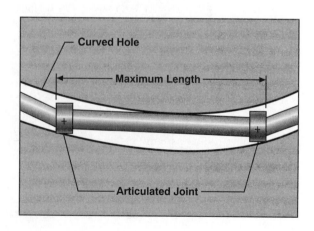

Figure 2-27   Short-radius BHA components must be short to prevent binding in the wellbore

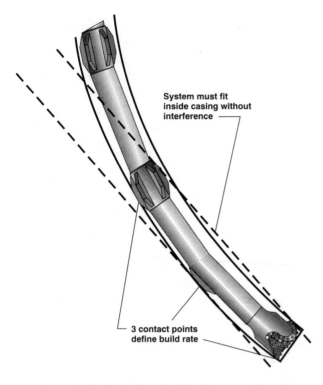

System must fit inside casing without interference

3 contact points define build rate

Figure 2-28 Articulated BHA design considerations

components must pass through the build section without interference, the overall length and offset of these components must allow the assembly to pass through the casing without interference. For this reason, a bit at least 1/8 in. smaller than the casing drift diameter is normally specified.

Because of the relatively short length of short-radius BHA components, slight interference with the borehole will cause high contact loads. The geometry of some short-radius BHAs may allow them to be rotated in the intermediate-radius hole curvature range without exceeding critical loads or stresses. However, the manufacturer's engineering guidelines should be strictly followed; if short-radius in-hole curvatures are rotated in excess of design guidelines, they can bind in the wellbore or break immediately.

The effect of hole erosion on steerable motors was discussed in Section 2-4.4. Hole erosion has even greater effects on short-radius assemblies because of the short spacing of the components that define build rate. Before drilling begins, the characteristics of the formation and its interaction with the drilling-fluid system should be evaluated to avoid build-rate problems caused by erosion or hydration. Care should be taken during milling to avoid damaging the formation in the vicinity of the window or milled section, or build rate may suffer.

In unconsolidated formations, flow rate should be minimized through the build section to minimize hole erosion. If the formation in the build section is known to be prone to hydration, an inhibitive drilling fluid should be used for both milling and drilling the build section.

Like hole erosion, variations in BHA geometry are more significant in short-radius systems than in long or medium-radius systems. Minor variations in stabilizer diameter or bend angle —caused by abrasive wear or even simply resulting from manufacturing tolerances— can cause large variations in build rate. Critical dimensions of each short-radius BHA (including bit diameter) should be measured accurately before and after every run. These measured dimensions, instead of nominal design dimensions, should be used in modeling to ensure that planned build rates are achieved.

Because of their lower torque requirements, roller-cone bits are preferred over PDC bits for the short-radius build section. Toolface orientation is easier to maintain with roller-cone bits, which is important when considering the limber tubulars used in short-radius work; toolface orientation is particularly important when gyros are used. Insert bits may be preferred over milled-tooth bits because of their ability to resist gauge diameter wear. Bit life is seldom a concern in the build section, since typically less than 100 ft of formation must be drilled before tripping for a lateral assembly.

In many areas, short-radius capabilities are well known and yield predictable build rates, allowing TVD tolerances of a few feet to be accomplished with a single BHA. When short-radius capabilities are not well known, the build section of the well can be designed with a dual build rate. The last 10 or 20 ft of inclination in a dual build section are built by a steerable motor that can continue to drill the lateral. In this scenario, the initial build is designed to come in high. As much as a third of the build section may be drilled before useful surveys are acquired, but frequent surveying thereafter should establish the build-rate trend and allow the steerable tool to be run in at the correct time. This method increases the probability of hitting the target, but it will result in slightly less usable hole in the payzone. A gyro may be necessary for orienting the steerable motor so that it trips through the build section without binding.

2-6.4 Orientation and Surveying

Since short-radius wells are usually drilled either out of or immediately below casing, magnetic interference from the casing must be considered. Wells with sufficient initial angle can use MWD or steering-tool tool-face measurements to orient the toolface, if the heading of the original wellbore is known. If enough TVD is available between the casing seat and the payzone, the practical approach is to deepen the existing wellbore and drill a shorter radius, which allows magnetic toolface measurements to be used. If the initial wellbore is near-vertical, and the distance to the payzone is not sufficient to allow a large section to be milled or the well to be deepened, a gyro will be required for the initial orientation. Once consistent gravity toolface measurements are either received (in the case of MWD) or can be expected based on projected inclination (in the case of a wireline steering tool), the gyro can be pulled, and gravity tool-face measurements can be used until valid magnetic azimuth measurements are received.

Either MWD or wireline steering tools are used in short-radius work. MWD tools have the advantage of easier rig-up and no wireline. Wireline steering tools allow for instantaneous toolface updates, and circulation is not required to get surveys. Regardless of the type of survey tool used, the tool must be configured to place the inclination sensor as close to the bit as possible. Most tools can be configured to place the sensor within about 15 to 25 ft from the bit, which is adequate for most build rates.

With sensors 15 to 25 ft from the bit, it is important to track the build-rate trend to ensure that the well is landed at the correct TVD. Surveys should be taken with greater frequency than in conventional directional drilling, on the order of every 10 ft in the build section. If the TVD target tolerance is tight at or near the EOB, it may be advisable to run a survey tool at the end of the drillpipe with a bullnose or used bit. Knowing the exact inclination at this point will aid in determining the configuration and initial orientation of the next BHA to ensure that the lateral lands at the desired TVD.

2-6.5 Drilling the Short-Radius Lateral

Although articulated motors may be run in the lateral, nonarticulated tools can be expected to provide more predictable control. Motors with flexible housings, as mentioned in Section 2-6.7, may be required for passing through the build section without yielding or getting stuck. At the beginning of the build section, the lateral motor may need to be oriented so that it can be tripped through that section without binding. To minimize stress cycles, top drives or power swivels should be used to turn the drillstring as slowly as possible, preferably at 1 to

10 rpm. Drillstring rotation should be avoided when picking up the drillstring or tripping out of the hole, since the combined stress in the build section will be maximum in this condition. Drillpipe fatigue cracks can usually be detected as washouts before the cracks part, so the drillstring should be tripped at the first sign of a loss in pump pressure.

Because most short-radius footage is drilled oriented or at a low rate, the drillstring cannot be relied upon to agitate the cuttings beds and lift cuttings into the annular flowstream. High annular velocities and drilling-fluid rheology must remove cuttings; sliding wiper trips may also be necessary.

2-6.6 Drillstring Design and Maintenance for Short-Radius Wells

Drillstring design for short-radius wells is similar to that of medium-radius wells in that drillpipe is run in compression through the build section and in the lateral. In short-radius wells, drillpipe run in compression can be expected to buckle, often inelastically. The length of drillpipe to be run in compression should be no more than the combined length of the lateral and build section so that the amount of damaged pipe section is minimized. If available, heavy-wall drillpipe can be run above the build so that the rest of the string is in tension.

Drillpipe size and grade should be selected relative to stresses in the build section. Generally, if there is uncertainty regarding which of two grades to use, the stronger drillpipe grade should be chosen. However, because high-strength steel is more prone to attack by H_2S, the lower strength material may be necessary in such an environment. Torque-and-drag modeling should be used for calculating drillstring loads during sliding and rotating. If any rotation of the drillstring is intended, stress should be limited to 75% of yield. The strongest tool joint for a given pipe grade should be selected relative to torsional capacity as given in API RP 7G (1995). Torsional capacity can be assumed to be an approximate indicator of bending strength. If retrievable wireline steering tools will be used, the tool joint ID must be larger than the steering tool OD, which is typically $1\frac{3}{4}$ in. In this section, the term *drillpipe* refers to either drillpipe or tubing. For example, $2\frac{7}{8}$-in. P110 tubing with premium connections has been used as the drillstring in a $4\frac{3}{4}$-in. hole size. Dimensions and torsional specifications of tubing connections are usually proprietary and must be provided by the manufacturer.

The drillstring used in short-radius work should be in new or good condition. After each well, all drillpipe that has been run in compression should be subjected to a comprehensive program of inspection, stress relieving and, if necessary, straightening. Threaded connections should be inspected frequently, through the use of wet magnetic particle or dye penetrant inspection. A system of tracking stress cycles should be implemented, and high-risk pipe should be set aside. Depending on the extent of damage or rework cost, drillpipe that has been run in compression in short-radius wells should either be considered for retirement after use in a few wells, or it should be used as line pipe or production tubing. Pipe run in tension in the vertical part of the well should also be inspected at regular intervals like any drillstring, even though it will avoid the damage of pipe run in a buckled state.

2-6.7 Intermediate-Radius Build Rates and Flexible Motors

As stated previously, the one factor that most clearly differentiates short-radius drilling from medium-radius drilling is that short-radius BHA components must be articulated to pass through casing and the build section without interference. Articulated BHAs are significantly different from long- and medium-radius BHAs both in design and operation. Another factor is that drillpipe is stressed so highly that extreme rotation limitations must be observed. These factors define the lower and upper limits of short radius for each hole size.

A gap exists between the lower limit (in terms of buildup rate) of short radius drilling and the upper limit of medium-radius drilling. In this range, drillpipe rotation is acceptable, but conventional medium-radius motors cannot achieve the necessary build rates.

Articulated motors have suffered from unpredictable build rates in this range, especially in unconsolidated formations. For a 6-in. to $6\frac{3}{4}$-in. hole size, for example, this range would be about 25° to 57°/100 ft. For conventional medium-radius motors to achieve build rates in this range, extremely high bent-housing angles would have to be used. Such high bend angles would produce so much bit interference with casing that it would be difficult or impossible to force the BHA through the casing without damaging the bit. At kickoff, the bit side-load would be so high that the motor drive shaft would be in danger of breaking. If the kickoff can be initiated successfully, the motors tend to bind in the curve and have difficulty sliding.

The gap between medium- and short-radius build rates has become known as the intermediate radius. PDMs with specially engineered flexible sections (Figure 2-29) have successfully achieved build rates in this range. Compared to a standard medium-radius motor, build-rate capability is greater while the bit side-load at kickoff is less. The desired flexibility is achieved through a reduced-diameter housing and/or use of a material with a lower modulus of elasticity. If properly engineered, the flexible section (1) minimizes the forces required to push the assembly through casing to get to the KOP, (2) reduces bit side-loads to a manageable level at kickoff, (3) enhances the effect of WOB on build rate, and (4) extends the allowable hole curvature in which the motor can be rotated. Build rates in unconsolidated formations are more predictable than with articulated motors, and the toolface is easier to maintain. The flexible section results in an increase of about 5 ft in tool length (and therefore bit-to-sensor distance) and increased sensitivity to some downhole parameters. These assemblies should

	Maximum Build Rate	Allowable DLS for Rotation	Bit Side-Load at Kickoff (lb)
4 3/4-in. Std Motor	25	15	3500
4 3/4-in. Flex Motor	60	25	2000

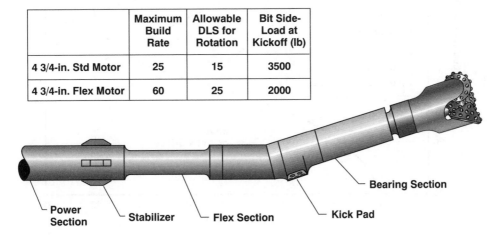

Figure 2-29 Flexible PDM for intermediate-radius build rates

be modeled extensively to evaluate their sensitivity to WOB, formation anisotropy, and hole curvature. Before using these assemblies, directional drillers should be aware of the modeling results.

2-7 EXTREME DIRECTIONAL WELLS

Directional-drilling limitations result from the interaction of the wellbore with the capacities of the drilling rig, drilling fluid, and drilling equipment. One limitation is how rapidly angle can be built or direction can be changed. This boundary is now being expanded with the increased application and advancement of short- and intermediate-radius drilling technology previously described. Another limitation is the maximum drilling depth of directional wells. This limitation is linked with the tension and torque loads that deep directional drilling imposes above the loads in deep vertical wells. Deep directional drilling has limits defined by the tension and torque capacities of the drillpipe and rig equipment.

Analogous to short-radius and deep directional drilling, there are also limitations as to how long or complex a departure can be drilled. This section will discuss these additional boundaries of directional-drilling technology. Technologies for and operations of extended-reach drilling and so-called "designer-well drilling" are rapidly

expanding, and both will become more significant in the future.

2-7.1 Extended-Reach Drilling (ERD)

Various definitions exist for extended-reach drilling (ERD)-class wells. One common definition is a departure of twice or more of the TVD of the well, (a reach-to-TVD ratio of two or more). This definition separates conventional directional wells from wells that could require ERD technology. However, since state-of-the-art ERD can involve wells with reach-to-TVD ratios of four to five or even higher, it should be recognized that distinct classes of ERD wells do exist, as shown in Figure 2-30. These definitions are listed as follows:

- Reach-to-TVD Ratio < 2 → Conventional Directional Drilling (Non-ERD)

- Reach-to-TVD Ratio 2 to 3 → Extended-Reach Drilling

- Reach-to-TVD Ratio > 3 → Severe Extended-Reach Drilling

These classifications are somewhat arbitrary, and drilling a well with a reach-to-TVD less than two can be quite challenging, depending on the formations involved and

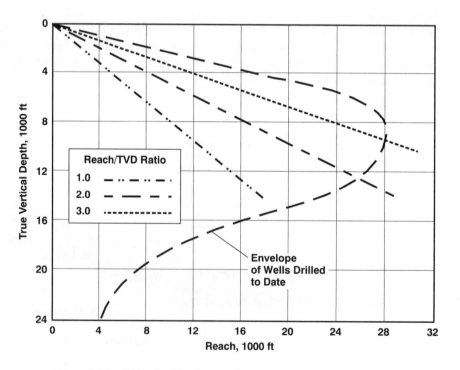

Figure 2-30 ERD classifications and currently achievable envelope

the rig available. Engineers must understand the technical interactions between the rig and equipment and the well being attempted so that constraints and risks can be identified and resolutions sought as early as possible. Various technologies must be applied for the successful planning and execution of an ERD well. These technologies include

- Optimized directional trajectory design

- Refined mechanical wellbore stability estimates for defining mud-weight windows

- Properly engineered drilling fluids for optimization of chemical wellbore stability, cased and openhole lubricity, and cuttings transport and suspension

- Calibrated torque-and-drag models to allow projection of field loads and diagnosis of deviations from predictions during field execution

- Sizing of rig components to provide adequate rotary, hydraulic, hoisting, solids control, and power capabilities

- Properly designed drillstring that minimizes drillpipe pressure loss (allows higher flow rates for given surface pressure limitations)

- Other technologies such as (1) soft-torque top drives, (2) special running systems and techniques for casings, liners, and completion strings, and (3) rotary-steerable directional-drilling systems

The unique issues associated with ERD wells largely stem from the high inclination of the well that is required to reach the objective departures. Tangent "sail" angles in some ERD wells have been 80° or higher. "Average" angles from surface to TD have been 70° or higher. Some "ERD" wells have in fact been drilled as extremely long horizontal wells where the bulk of the well has been drilled horizontally (defined as an inclination of 88° or more).

At such high inclinations, the transport of cuttings from the well is more difficult than in vertical wells. As a result, higher flow rates, tighter control of drilling-fluid rheology, and the use of nonconventional mechanical means to assist hole cleaning must be used. Such mechanical means might include high drillstring rotation (off-bottom) or the use of special, bladed drillpipe to stir cuttings beds mechanically. High rotary speeds and backreaming help clean the hole but can increase drilling shocks and cause fatigue or backoff of motor housings. Therefore, such practices should be viewed as secondary

hole-cleaning methods and used only if primary hole cleaning is inadequate. A common method of removing cuttings beds is the use of low-viscosity sweeps that scour the cuttings off the low side of the hole; high-viscosity sweeps are then used to carry the dislodged cuttings to the surface. The fundamental driver of hole cleaning, however, is flow rate, and high flow rates are strongly recommended throughout high-inclination sections if ECD allows. Flow rates of 1000 to 1100 gal/min in $12\frac{1}{4}$-in. hole and 500 to 600 gal/min in $8\frac{1}{2}$-in. hole are not uncommon. For this reason, the rig's pump and piping capacities with regard to flow rate, pressure, and the drillstring design are critical.

Other implications of high inclination include the greater likelihood of mechanical wellbore instability and hence the need for careful mud-weight planning. Since more formation is exposed for longer periods, chemical stability also becomes more critical.

Aside from hydraulics and formation stability, another potential ERD constraint is the ability to sustain drilling torque and run tubulars in the well. Both of these processes are impacted by optimizing the well trajectory and field control of variations from this directional plan. In terms of design optimization, the ERD trajectory will be subject to similar geologic constraints as discussed in Section 2-2.2. Additionally, the trajectory should be designed to minimize the induced torque during drilling and to maximize the available weight while casing is run. Satisfaction of these objectives varies depending on (1) the TVD and departure of the target and (2) the frictional behaviors of the various hole sections and the location of those sections. One observation from several studies (Banks *et al.*, 1992; Dawson and Paslay, 1984), however, is that multiple build rates should be used to initiate inclination into the well gradually. Increasing the build rates in several steps minimizes near-surface doglegs and the associated torque and drag.

The final tangent angle and shape of the trajectory through the reservoir can be determined on the basis of several considerations. If the well design is close to the limit of wireline intervention capabilities (65 to 70°), it may be best to keep the KOP high and limit the inclination so that wireline operations are feasible. If the well angle is more severe and the inclination will already exceed wireline operating limits, it may be appropriate to slightly deepen the KOP to place the build section in more-competent formation. In such cases, although a higher final inclination angle will result, the higher inclination may in fact be optimal since it inhibits buckling of the drillstring and coiled tubing, which could be controlling factors in severe ERD wells.

In addition to an optimized trajectory plan suited for the specific well application, directional drilling must be performed expertly in the field to minimize unexpected doglegs and variations from plan. Such doglegs will increase drilling torque, make tubular running more difficult, and provide potential trouble spots for excessive casing wear, which can lead to failures. These serious detriments have been recognized previously (Sheppard *et al.*, 1987; Mueller *et al.*, 1991), and a specific measure is available for defining how closely the well is being or has been drilled according to plan. This measure is known as *tortuosity*, which is the sum of all the increments of curvature along the section of interest, subtracted from the planned curvature, and divided by the footage drilled. In the well-planning phase, specific tortuosity guidelines should be generated for each section of the well, based on torque-and-drag modeling. Tortuousity limitations should generally be the most stringent in the upper portion of the well and in the build section, where high drillstring tension causes high side-loading, which results in torque and drag, keyseating, and worn tool joints and casing. Reducing tortuosity requires an integrated engineering approach as described:

- The rotary drilling directional tendency of BHAs should be tracked, quantified, and refined to yield rotary behaviors that are as close as possible to planned build rates. This process minimizes the need for oriented drilling as a means of inclination control.

- When steering is required, it should be executed in a controlled manner. For example, doglegs can be minimized by sliding for only part of a joint; then, rotary drilling can be used for the remaining footage. This process can then be repeated as required to offset the build, drop, or turn behavior of the rotary BHA that is causing the problem.

- Sliding repeatedly in short intervals is greatly preferred over executing one continuous, long, sliding section, which will cause a significant and sharp dogleg.

Systematic analysis and planning of BHAs and careful field execution will allow good control of well tortuosity and will ensure that predrill torque-and-drag projections are not exceeded. Projecting requirements for ERD wells, properly sizing the drilling rig and equipment, and optimizing the trajectory are all subject to detailed, site-specific engineering. The key to success is to recognize those

technologies that are critical and those trade-offs that must be examined to engineer the best possible approach for a given well.

2-7.2 Designer Wells

"Designer wells" are similar to ERD wells, but they may be even more complex, because they involve multiple-target trajectories with substantial and potentially difficult azimuthal steering. Therefore, designer wells may require inclination control combined with strong azimuthal curvature to allow intersection of these offset targets (Figure 2-31). Although the industry has no widely accepted definitions for what constitutes a designer well, some definitions have been proposed based on MD-to-TVD ratios. Such a distinct definition from the reach-to-TVD definition for ERD is necessary because designer wells involve substantial drilled footages that may not necessarily increase well departure.

Because designer wells are frequently being drilled in development environments and to optimize reservoir penetration, the uncertainty of wellbore placement in the reservoir is frequently dealt with by drilling and logging pilot holes before drilling the reservoir section of the ongoing well. Figure 2-32 shows the process whereby, before being plugged, a pilot hole is drilled through a reservoir section and logged (usually with LWD equipment) to obtain the best possible placement for the main wellbore. Examples of such procedures exist where multiple pilot holes and resulting sidetracks have been drilled to optimize wellbore placement, resulting in a well in which the drilled depth was almost twice the final measured depth (Justad *et al.*, 1994).

2-7.3 Drillstring Dynamics

Drillstring dynamics are not unique to designer and ERD wells; drillstring dynamics occur in all well types, and in fact are sometimes severe in vertical wells. Nevertheless, the extreme torque-and-drag levels of designer and ERD wells do result in more stored energy in the drillstring than in other wells. High torque-and-drag levels also cause drillstring components to be stressed near their limits, so that additional loading from dynamic forces can cause failure.

As discussed in Section 2-2.5, BHA prediction software is fundamentally based on static force considerations. Since static force models usually predict BHA tendencies adequately, dynamic steady-state and transient behaviors are usually not an issue in terms of direc-

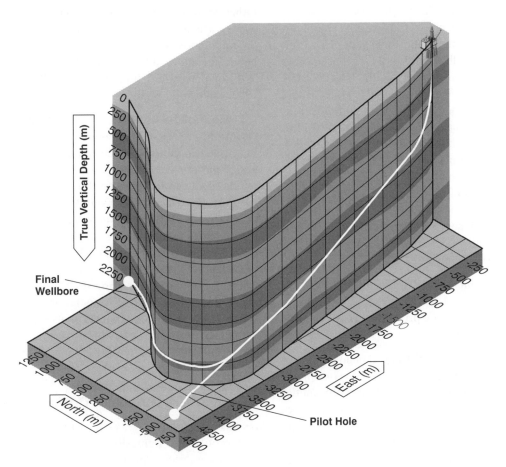

Figure 2-31   "Designer" well (after Justad *et al.*, 1994)

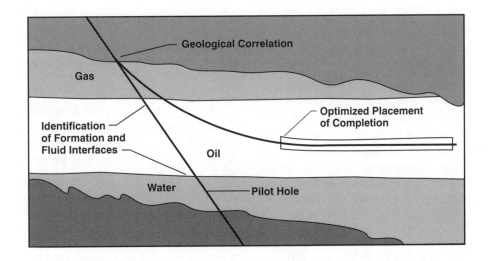

Figure 2-32   Using a pilot hole to optimize wellbore placement

tional tendency prediction. This implies that the average behavior of a dynamically active BHA can be characterized by its static displacement and force state. However, dynamic behaviors still significantly impact other aspects, such as structural integrity of the bit and BHA and equipment fatigue (Fear *et al.*, 1997; Dufeyte and Henneuse, 1991). Our experience has been that most fatigue failures, connection backoffs, and MWD/LWD electronics failures are associated with drillstring dynamics. Moreover, in some drilling situations, the dynamic behaviors can be quite dominant and severely impact the directional-drilling operation (Birades, 1986; Brakel, 1986; Brett *et al.*, 1989).

Various modes of dynamic behaviors can be active in the "structure" composed of the drillstring, BHA, and bit. These include the following:

Axial vibrations or "bit bounce" These vibrations are most commonly associated with three-cone bits drilling hard rock and are usually more prone to occur at shallow depths and in vertical wells.

Torsional vibrations of the drillstring Torsional vibration occurs when the drillstring "twists" according to the frequency of the primary torsional vibration mode. Torsional vibration can be induced purely from the frictional loads along the drillstring. It can also be exacerbated by the cutting behavior of PDC bits, which exhibit nonlinear-force rate behaviors. In extreme cases, torsional vibrations can become severe enough to cause "stick-slip" problems. Under stick-slip conditions, the bit exhibits an erratic speed profile whereby it rapidly decelerates and accelerates as the torsional vibration arrives at and reflects from the bit. This erratic motion can rapidly damage the bit and cause very poor ROP performance.

Lateral vibrations of the BHA Harmonic instability or contact-induced motions, such as stick-slip, can cause BHAs and drillstrings to undergo severe lateral vibrations. These vibrations are most likely to occur when motor or drillstring stabilizers hang and then break free when rotating through formation stringers. The resulting lateral shocks and vibrations can cause fatigue failures or backoffs of BHA connections and rapid wear of tool joints and stabilizers.

Bit whirl PDC bits can exhibit a chaotic behavior whereby they gyrate off-center and induce severe dynamic motion into the BHA. PDC bit whirl can cause cutter damage or failure, which rapidly results in failure of the bit itself. ROP is usually poor when bit whirl becomes active.

Other vibration modes A number of other vibration modes are associated with drillstring motion or with the interaction of some of the vibration modes discussed here. For more information about drilling dynamics, additional technical literature is available (Brett *et al.*, 1989; Payne and Spanos, 1992).

Because of the likelihood for problems when severe vibrations are active, it is important to provide an ability to sense such vibrations and modify equipment or operating parameters to alleviate such dynamics. This requirement can be achieved through the use of dynamic monitoring surface sensors and downhole accelerometers in MWD/LWD. Remedial action may include such manual activities as varying the WOB or drillstring rotation rate (Macpherson and Mason, 1993; Payne *et al.*, 1995), or such automated activities, as in the case of closed-loop feedback systems for top drives (Fear and Abbassian, 1994; Nicholson, 1994; Sananikone, *et al.*, 1992). Closed-loop feedback systems can reduce potentially damaging torsional vibrations to manageable levels, and should be considered for extreme wells.

2-7.4 Emerging Directional-Drilling Systems

Because the industry seeks to drill increasingly difficult well profiles economically, better means of directional control are needed. In recent years, significant advances have been made with regard to adjustable-diameter stabilizers. Initially, such tools were mechanically set and capable of two size settings over a limited size range. These tools now provide more capability. One new tool, which is still mechanically set, allows three different gauge settings. Another advancement has been a hydraulically instructed adjustable-diameter stabilizer with many size settings over a wide size range. These new tools provide greater inclination control while maximizing rotary-mode drilling.

Further advances beyond adjustable gauge systems are needed to impact the drilling of ERD and designer wells. Currently, four major systems are in various stages of development and field testing. These *rotary steerable systems* are designed to steer in both inclination and azimuth planes while rotating the drillstring. One common component of all rotary-steerable drilling systems currently under development is a nonrotating section in an otherwise rotating tool. The nonrotating section provides a stationary mechanical and/or electronic reference plane that is used for controlling orientation, while the drillstring continues to rotate to reduce torque and drag. The various rotary-steerable drilling systems use different mechanisms of operation. In all cases, the mechanisms of steering induce either force or BHA curvature to build, hold, drop, or turn as required. Rotary-steerable systems will allow full 3D control of the wellpath while providing for rotary-mode drilling, thereby yielding optimal ROP and eliminating the drilling performance penalty currently inherent with oriented drilling. Just as the steerable mud motor changed directional-drilling technology in the 1980s, rotary-steerable drilling systems should prove to be a major advancement in directional-drilling capabilities in the near future.

REFERENCES

API RP 7G, "Recommended Practice for Drill Stem Design and Operating Limits," 15th Edition, API (Jan. 1, 1995).

Banks, S.M., Hogg, T.W., and Thorogood, J.L., "Increasing Extended-Reach Capabilities Through Wellbore Profile Optimization," paper SPE 23850, 1992.

Birades, M., "ORPHEE 3D: Static and Dynamic Tridimensional BHA Computer Models," paper SPE 15466, 1986.

Bourgoyne Jr., A.T., Millheim, K.K., Chenevert, M.E., and Young Jr., F.S.: *Applied Drilling Engineering*, SPE Monograph Series, Richardson, TX (1991).

Brakel, J.D.: "Prediction of Wellbore Trajectory Considering Bottomhole Assembly and Drillbit Dynamics," PhD Dissertation, U. of Tulsa, Tulsa, OK (1986).

Brett, J.F., Warren, T.M., and Behr, S.M.: "Bit Whirl: A New Theory of PDC Bit Failure," paper SPE 19571, 1989.

Cheatham, C.A., Comeaux, B.C., and Martin, C.J.: "General Guidelines for Predicting Fatigue Life of MWD Tools," paper SPE 23906, 1992.

Craig Jr., J.T., and Randall, B.V.: "Directional Survey Calculation," *Pet. Engr.* (Mar. 1976) 38–54.

Dawson, R., and Paslay, P.R.: "Drillpipe Buckling in Inclined Holes," *JPT* (Oct. 1984) 1734–1738.

Dellinger, T., Gravley, W., and Tolle, G.C.: "Directional Technology Will Extend Drilling Reach," *O&GJ* (Sept. 15, 1980) 153–169.

Dufeyte, M.P., and Henneuse, H.: "Detection and Monitoring of the Stick-Slip Motion: Field Experiments," paper SPE 21945, 1991.

Eddison, A., and Symons, J.: "Downhole Adjustable Gauge Stabilizer Improves Drilling Efficiency in Directional Wells," paper SPE 20454, 1990.

Fear, M.J., and Abbassian, F., "Experience in the Detection and Suppression of Torsional Vibration From Mud Logging Data," paper SPE 28908, 1994.

Fear, M.J., Abbassian, F., and Parfitt, S.H.L.: "The Destruction of PDC Bits by Severe Slip-Stick Vibration," paper SPE 37639, 1997.

Guild, G.J., Jeffrey, J.T., and Carter, J.A.: "Drilling Extended-Reach/High-Angle Wells Through Overpressured Shale Formation," paper SPE 25749, 1994.

Hansford, J.E., and Lubinski, A.: "Cumulative Fatigue Damage of Drillpipe in Doglegs," *SPE Reprint Series* (1973 Rev.) **No. 6a**, 281–285.

Hogg, W.T., and Thorogood, J.L.: "Performance Optimization of Steerable Systems," *ASME PD* (1990) **27**, 49–58.

Jogi, P.N., Burgess, T.M., and Bowling, J.P.: "Predicting the Build/Drop Tendency of Rotary Drilling Assemblies," *Directional Drilling, SPE Reprint Series* (1990) **No. 30**.

Johancsik, C.A., Friesen, D.B., and Dawson, R.: "Torque and Drag in Directional Wells—Prediction and Measurement," paper SPE 11380, 1983.

Justad, T., Jacobsen, B., Blikra, H., Gaskin, G., Clarke, C., and Ritchie, A.: "Extending Barriers to Develop a Marginal Satellite Field From an Existing Platform," paper SPE 28294, 1994.

Larson, P.A., and Azar, J.J.: "Three-Dimensional, Quasi-Static, Drill Ahead BHA Model for Wellbore Trajectory Prediction and Control," PED, Drilling Technology, *ASME* (1992) **40**.

Le Peuvedic, J.P., Astier, B., Baron, G., Boe, J.C., Dumas-Planeix, M., Koenig, L., Mabile, C., Prin, R., and Toutain, P.: "Directional Surveying (Ch. 3)," and "Directional Drilling Practice (Ch. 4)," *Directional Drilling and Deviation Control Tech.* (1990) 47–137.

Lockett, T.J., Richardson, S.M., and Worraker, W.J.: "The Importance of Rotation Effects for Efficient Cuttings Removal During Drilling," paper SPE 25768, 1993.

Lubinski, A.: "Factors Affecting the Angle of Inclination and Doglegging in Rotary Boreholes," *API Drill. and Prod. Prac.* (1953) 222.

Lubinski, A.: "Maximum Permissible Doglegs in Rotary Boreholes," *SPE Reprint Series* (1973 Rev.) **No. 6a**, 256–275.

Macpherson, J.D., and Mason, J.S.: "Surface Measurement and Analysis of Drillstring Vibrations While Drilling," paper SPE 25777, 1993.

Moineau, J.D.: "Gear Mechanism," US patent 1,892,217 (1932).

Mueller, M.D., Quintana, J.M., and Bunyak, M.J.: "Extended Reach Drilling From Platform Irene," *SPEDE* (June 1991) 138.

Murphey, C.E., and Cheatham Jr., J.B.: "Hole Deviation and Drillstring Behavior," paper SPE 1259, 1965.

Nicholson, J.W.: "An Integrated Approach to Drilling Dynamics Planning Identification and Control," paper SPE 27537, 1994.

Odell, A., and Payne, M.: "Application of a Highly Variable Gauge Stabilizer at Wytch Farm to Extend the ERD Envelope," paper SPE 30462, 1995.

Payne, M.L., and Spanos, P.D.: "Advances in Dynamic Bottomhole Assembly Modeling and Dynamic Response Determination," paper SPE 23905, 1992.

Payne, M.L., Abbassian, F.A., and Hatch, A.J., "Drillstring Dynamic Problems and Solutions for Extended-Reach Drilling Operations," paper presented at the 1995 ASME Energy-Sources Technology Conference, Houston, Jan. 29–Feb. 1.

Rasmussen, B., Sorheim, E., Seiffert, O., Angeltvadt, O., and Gjedrem, T.: "World Record in Extended Reach Drilling, Well 33/9-C10, Statfjord Field, Norway," paper SPE 21984, 1991.

Rollins, H.M.: "Are 3° and 5° 'Straight Holes' Worth Their Cost?" *O&GJ* (1959).

Sananikone, P., Kamoshima, O., and White, D. B.: "A Field Method for Controlling Drillstring Torsional Vibrations," paper SPE 23891, 1992.

Sheppard, M.C., Wick, C., and Burgess, T.: "Designing Well Paths To Reduce Drag and Torque," SPEDE (Dec. 1987) 344–350.

Slavomir, S.O., and Azar, J.J.: "The Effects of Mud Rheology on Annular Hole Cleaning in Directional Wells," *SPEDE* (Aug. 1986) 297.

Thomas, R.P., Azar, J.J., and Becker, T.E.: "Drillpipe Eccentricity Effect on Drilled Cuttings Behavior in Vertical Wellbores," *JPT* (Sept. 1982) 1929.

Thorogood, J.L.: "Instrument Performance Models and Their Practical Application to Directional Surveying Operations," *SPEDE* (Dec. 1990) 294–298.

Thorogood, J.L., and Sawaryn, S.J.: "The Traveling CylinderA Practical Tool for Collision Avoidance," *SPEDE* (Mar. 1991).

Underwood, L., and Odell, A.: "A Systems Approach to Downhole Adjustable Stabilizer Design and Application," paper SPE 27484, 1994.

Williamson, J.S., and Lubinski, A.: "Predicting Bottomhole Assembly Performance," paper SPE 14764, 1986.

Wolff, C.J.M., and de Wardt, J.P.: "Borehole Position Uncertainty—Analysis of Measuring Methods and Derivation of Systematic Error Model," *JPT* (Dec. 1981) 2339–2350.

3 Horizontal, Multilateral, and Multibranch Wells in Petroleum Production Engineering

Michael J. Economides
Texas A&M University

Dan R. Collins
Halliburton Energy Services

W.E. (Bill) Hottman
Halliburton Energy Services

James R. Longbottom
Halliburton Energy Services

3-1 INTRODUCTION

Rather than attempting to move hydrocarbons into vertical wellbores that may not be well positioned, the industry is now resorting to the use of horizontal, multilateral, and multibranch wells that move the wellbore closer to the hydrocarbons in place.

Multilateral well systems allow multiple producing wellbores to be drilled radially from a single section of a "parent" wellbore. A major difference between this method and conventional sidetracking is that both the parent wellbore and the lateral extensions produce hydrocarbons.

Because only a single vertical wellbore is required, multilateral well designs require less drilling time, often have fewer equipment and material requirements, and increase hydrocarbon production. Typical multilateral applications include

- Improving productivity from thin reservoirs

- Draining multiple, closely spaced target zones with horizontal exposure of each zone

- Improving recovery in tight, low-permeability zones (increasing the drainage radius of a given well)

- Preventing water and/or gas coning

- Controlling sand production through lower drawdown at the sand face

- Improving the usability of slot-constrained platform structures

- Improving waterflood and enhanced oil recovery efficiency

- Intersecting vertical fractures

3-2 BACKGROUND

Multilateral drilling and completion methods have been practiced since the mid-1940s. The first applications were developed for mining, where multiple bores were drilled from the parent shaft. These short, directional displacement bores were achieved with bent subs and the conventional rotary drilling technology of the time. Several patents were issued covering multilateral or multibore tools and methods for use in mining (Gilbert, Rehm), but the technology was not initially used in the oil field.

For years, hydraulic fracturing (although not really a competitor of modern multilateral drilling) provided large areal exposure between the well and the reservoir. However, with the significant advancements in horizontal drilling technology in the mid-1980s and its evolution into multilateral drilling in the mid-1990s, the performance of a vertical well with a hydraulic fracture can now be readily surpassed by a properly oriented horizontal or multilateral well in an areally anisotropic reservoir. Furthermore, horizontal wells provide better results in reservoirs with large gas caps or water aquifers.

The first extensive modern application of multilateral drilling was in the Austin Chalk formations in Texas during the late 1980s. High initial production rates and high decline rates required increased reservoir face exposure for the achievement of maximum production in the shortest possible time.

Austin Chalk and Buda Chalk reservoirs are conducive to multilateral applications because of the deposition geometry and integrity of the reservoirs. Chalk formation depositions are wide-ranging and highly fractured. Horizontal drilling tools and techniques intercepted and connected many of these natural fractures. The inherent stability of the formations allowed extended horizontal sections to be drilled without the threat of formation collapse and subsequent loss of the producing wellbore.

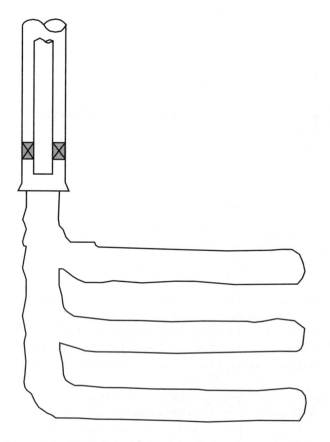

Figure 3-1 Multiple drainholes as applied to the Austin Chalk

Accurately drilling through and connecting or isolating fractured sections of the reservoirs required several technologies, including horizontal drilling, directional drilling, and measurement-while-drilling (MWD) tools and processes (Figure 3-1).

3-3 DESIGN CONSIDERATIONS

When designed properly, multilateral wells can provide considerably higher productivity or injectivity indices and increased reservoir recovery at relatively low incremental cost. Before deciding to drill such a well, engineers must carefully evaluate the expected well performance, operational and economic risks, possible production scenarios, and most importantly, the wellbore management and maintenance of the individual drainholes within the overall reservoir. With advancements in coiled-tubing technology, short-radius, lateral well branches can be drilled, as detailed in Chapter 2. However, selecting the right candidates on the basis of the geological flow units, their configuration, and production and completion technologies is critical. Thus, several new developments and improvements of existing concepts are expected.

3-4 CANDIDATE SELECTION

Clearly, several logistical and operational issues must be considered before certain well systems can be completed. Such issues may be dictated by obvious reservoir exploitation strategies and schemes. The depictions in this chapter represent current activities, but are by no means complete; new configurations are emerging.

Selecting the most beneficial well system for a given reservoir is a challenge. For reservoir engineering, the degree of communication among the drainage areas of individual branches is probably the most substantial issue. The following three major drainage categories are likely, and combinations of these are possible:

- Drainage of a single layer in which areal permeability anisotropy is critical

- Drainage of several ("stacked") layers, which may or may not communicate

- Drainage of several compartments, which may or may not communicate

The second option favors a vertical parent hole, whereas the first and third options favor a horizontal parent hole, with the exception of multibranch wells (including the limiting case of dual opposing laterals).

A more important aspect, however, is the degree of selective wellbore management that the well completion type supports. Thus, the following three configurations are possible:

- Commingled production

- Commingled production with individual branches that can be shut off by gates and re-entered easily. (Underground gathering systems, replete with appropriately sized chokes, can be installed)

- Individual production tubings tied back to the surface

These production options correspond to the previously mentioned reservoir management issues because the need for selective wellbore control increases as the communication of the drained portions of the reservoir increases. For example, if pore pressures and layer and/or fluid properties in the individual layers differ widely, a well system draining different layers will require selective management of individual layers or branches. Numerous completion options are available. Currently, multilaterals with connectivity, isolation, and access are installed (Longbottom, 1996). When the completion is selected, appropriate wellbore management should be assessed, first according to the reservoir type and planned production scenario. The remaining options should then be ranked in order of economics and operational and economic risks.

Once modern reservoir characterization is used, the traditional drainage cylinder of only vertical wells, or the "box" of more recent numerical simulation schemes, are both almost forcibly abandoned. Far from being homogeneous, reservoirs are revealed to be compartmentalized and layered, with irregularly shaped geological flow units. In this environment, multilateral wells drilled from the same vertical parent are essentially independent horizontal wells sidetracked at different points from the vertical well (Figure 3-2). These laterals are positioned with reservoir descriptions that include 3D seismics, petrophysical analysis, and real-time payzone steering. Real-time payzone steering involves an array of tools and measurements that are intended to maintain the well trajectory within the target zone.

A frequent motivation for multilateral or horizontal wells is the lack of available drilling slots in offshore drilling pads. Figure 3-3 shows a typical configuration where few or no drilling slots are available, a problem that is particularly common in the Gulf of Mexico. Thin-bed reservoirs are good candidates for horizontal wells.

Figure 3-2 Multilateral wells targeting discontinuous geological flow units

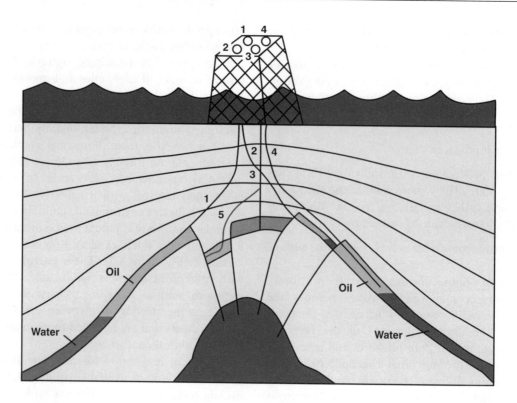

Figure 3-3   Drilling of a horizontal well segment in a location with limited or no drilling slots

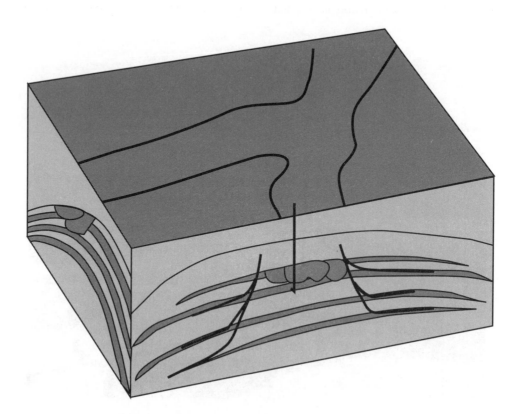

Figure 3-4   Thin-bedded reservoirs with multilateral wells

Layered formations, consisting of several thin beds separated by impermeable layers, can be exploited as shown in Figure 3-4. Several North Sea reservoirs fit this description.

Waterflooding with traditional injector/producer configurations can result in two problems: (1) inefficient sweep in a specific layer or reservoir and (2) water breakthrough. Figure 3-5 provides an example solution of inefficient sweep in specific layers; this solution is used in some west Texas reservoirs. Horizontal laterals drilled off vertical well injectors augment the sweep efficiency in targeted layers.

Water-injection wells are positioned where the water front is not aided by the decidedly anisotropic natural fractures (Figure 3-6). This positioning prevents rapid water breakthrough. Multibranch wells drilled from a parent horizontal well are then used as producers. Figures 3-7 and 3-8 demonstrate how robust reservoir description can influence well configurations. Figure 3-7 shows the use of multilateral wells targeting braided channels, which are characteristic of many deposits worldwide, especially in the north coast of South America (Venezuela and Colombia). The top schematic

in Figure 3-8 shows a horizontal well drilled on the basis of an insufficient reservoir description; several reservoir flow units are not penetrated. The bottom schematic shows guided multilaterals from a parent horizontal wellbore that were designed from an accurate reservoir description from the parent well to point to the position of the geological units. Clearly, the second configuration is more desirable.

If the pressure drop becomes detrimental while fluids flow in a horizontal well, dual opposing laterals (Figure 3-9) can be drilled, as they have been in several Austin Chalk formations and throughout western Canada. In thick formations with low-mobility crudes (such as heavy oils), stacked multilaterals can be constructed (Figure 3-10). These configurations and variants have been drilled or planned in Venezuela, California, and Alberta. Finally, an innovative technique suggested by Ehlig-Economides *et al.* (1996) is to drill multilateral vertical wells from a horizontal trunk and hydraulically fracture the vertical branches (Figure 3-11). The parent well can be drilled in a competent, overlain formation, which would prevent wellbore stability problems.

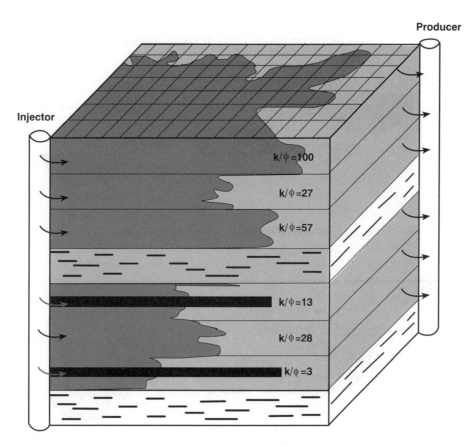

Figure 3-5 Horizontal sidetracks to augment sweep efficiency of specific layers during waterflood

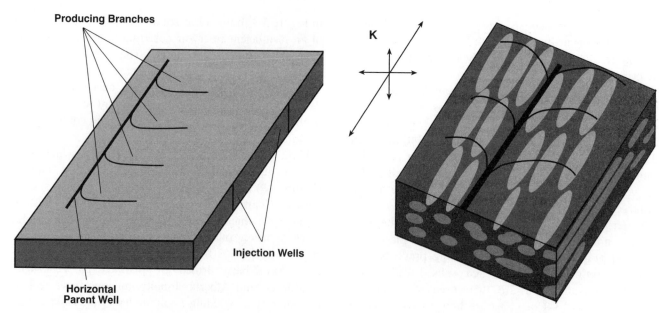

Figure 3-6   Multibranch producers and water injectors in an anisotropic reservoir under waterflood (after Ehlig-Economides *et al.*, 1996)

Figure 3-7   Multibranch well exploiting several braided channels with limited individual capacity

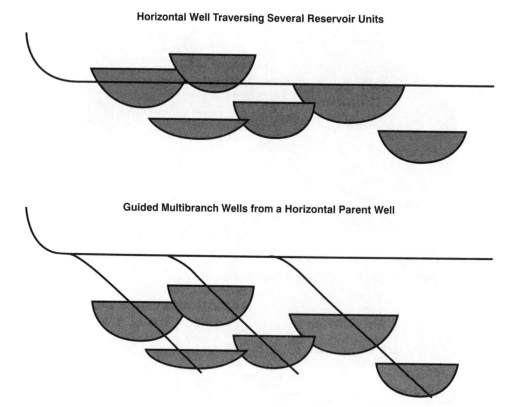

Figure 3-8   Horizontal well in a geostatistically defined reservoir (Top). Guided multibranch wells with appropriate formation characterization from a parent well (Bottom) (after Ehlig-Economides *et al.*, 1996)

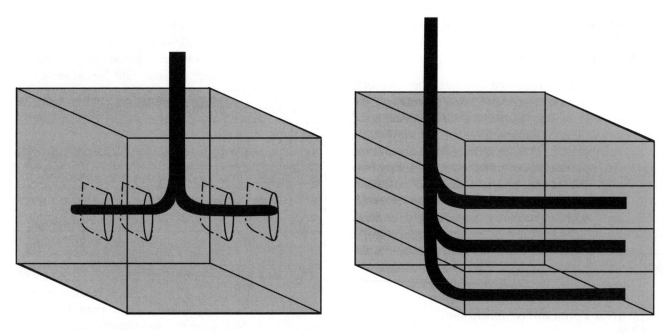

Figure 3-9   Dual opposing laterals

Figure 3-10   Multilateral wells for heavy crude reservoirs

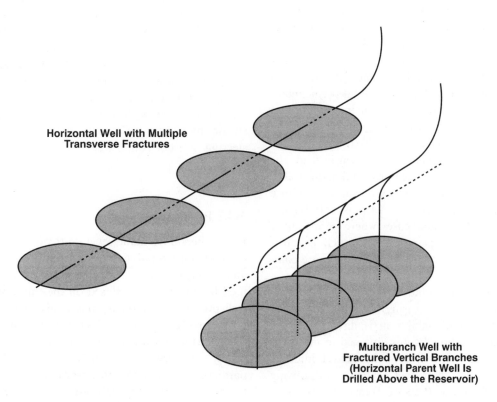

Horizontal Well with Multiple
Transverse Fractures

Multibranch Well with
Fractured Vertical Branches
(Horizontal Parent Well Is
Drilled Above the Reservoir)

Figure 3-11   Fractured vertical multibranch wells vs. a transversely fractured horizontal well (after Ehlig-Economides *et al.*, 1996)

3-5 PRODUCTION FROM HORIZONTAL, MULTILATERAL, AND MULTIBRANCH WELLS

To select the appropriate well configuration, engineers must not only have a good reservoir description, but they must also be able to predict well performance and reserves as well as optimize well systems (Economides *et al.*, 1994; Salas *et al.*, 1996). A comprehensive multi- and single-well productivity or injectivity model has been introduced that allows arbitrary positioning of the well(s) in anisotropic formations (Economides *et al.*, 1994). This flexible, generalized model can be used for the study of several plausible scenarios, especially the economic attractiveness of drilling horizontal and multilateral wells.

3-5.1 Areal Permeability Anisotropy

Traditionally, petroleum engineers have not been too concerned about horizontal permeability anisotropy. When vertical wells are considered, permeability anisotropy is not very important, because in cylindrical flow, the average permeability, k_H, is in the horizontal plane and is simply $\sqrt{k_X k_Y}$. Several studies have shown that large permeability anisotropies in the horizontal plane are common in many reservoirs (Warpinski, 1991; Deimbacher *et al.*, 1992). Naturally fractured formations, which are generally excellent candidates for horizontal wells, are likely to exhibit horizontal permeability anisotropy. In this situation, two principal horizontal permeabilities can be identified: $k_{H\max}$ and $k_{H\min}$. The direction to drill a horizontal well is often based on the shape of the presumed drainage area, when instead, the deciding parameter should be the horizontal permeability anisotropy, particularly in the cases of natural fracture orientation or depositional trends. Warpinski (1991) and Buchsteiner *et al.* (1993) note cases where permeability ratios in the horizontal plane are as much as 50:1, although ratios of 3:1 or 4:1 are considerably more common.

Permeability anisotropy and direction can be determined either with stress measurements in a vertical pilot well before a horizontal well is directed, or by experiments with directional cores obtained in the vertical pilot well. Horizontal well tests or multiwell interference tests are the best permeability anisotropy measurement techniques. Pressure transient tests at a well are most commonly used for measures of the magnitude and direction of permeability, while interference tests are seldom used in the field.

Before a horizontal well is drilled, a vertical pilot well should first be drilled, followed by a partial-penetration drillstem test. This test can be performed in two different ways: (1) by drilling only partially into the net pay or (2) by drilling through the net pay and then packing off only a small portion of the interval. The spherical flow regime, the negative one-half slope straight line on the early-time pressure derivative curve, provides the spherical permeability, $\sqrt{k_H k_V}$. A second test conducted with the entire pay thickness open to flow should provide the horizontal permeability, k_H. The vertical and horizontal permeabilities and the reservoir thickness can indicate the feasibility of drilling a horizontal well in the tested formation. If a horizontal well will be drilled, the proper well azimuth should then be determined (Beliveau, 1995).

The stress field in a reservoir can be described with three principal stresses: a vertical stress, σ_V, a minimum horizontal stress, $\sigma_{H\min}$, and a maximum horizontal stress, $\sigma_{H\max}$. Measurements can identify the maximum and minimum horizontal stress directions, usually coinciding with the maximum and minimum horizontal permeability directions. Figure 3-12 indicates how the horizontal stress components affect the state of fissures in the formation. Fissures perpendicular to the maximum stress direction are compacted, while fissures perpendicular to the minimum stress are relatively open. Large fissure widths imply larger permeability along those fissures. This idea illustrates why the maximum stress components normally coincide with maximum permeability and the minimum stress components usually coincide with minimum permeabilities. The measurement of stresses in a vertical pilot well are valuable for the proper steering of a horizontal well.

3-5.2 Productivity Model

Researchers have made several attempts to describe and estimate horizontal and multilateral well productivity and/or injectivity indices. Several models have been used for this purpose. Based on the tradition of vertical well productivity models, analogous well and reservoir geometries have been considered in the pursuit of simple but elegant analytical or semi-analytical models. A widely used approximation for the well drainage is a parallelepiped model with no-flow or constant-pressure boundaries at the top or bottom and either no-flow or infinite-acting boundaries at the sides. Numerical simulation can be used for incorporating heterogeneities and other reservoir complexities. However, the use of analy-

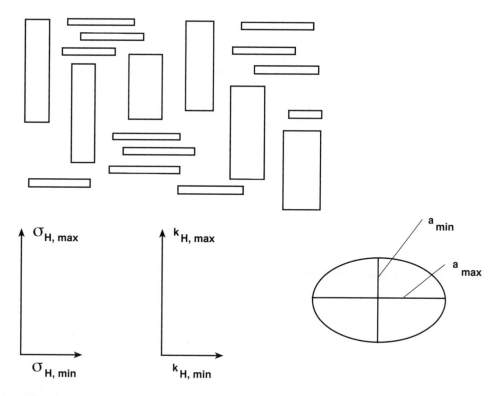

Figure 3-12   Conceptual relationship between stress and permeability components

tical models for productivity index calculations is both attractive and instructive.

Borisov (1964) introduced one of the earliest models, which assumed a constant pressure drainage ellipse in which the dimensions depended on the well length. This configuration evolved into Joshi's (1988) widely used equation, which accounted for vertical-to-horizontal permeability anisotropy. It was adjusted by Economides *et al.* (1991) for a wellbore in elliptical coordinates. This model, while useful for first approximations and comparisons with vertical well productivity indices, does not account for either early-time or late-time phenomena nor, more importantly, realistic well and reservoir configurations.

Babu and Odeh (1989) used an expression for the pressure drop at any point by integrating appropriate point source (Green's) functions in space and time. Their solutions for various no-flow boundary positions include infinite-sum expressions, which account for individual pseudosteady-state pressure drops. These forms are rather complicated and difficult to calculate. Using vertical well analogs, Babu and Odeh (1989) grouped their solution into reservoir/well configuration shape factors and a (horizontal) partial-penetration skin effect. This work agrees with the classic Dietz (1965) shape factors that were incorporated for vertical well performance by

Matthews, Brons, and Hazebroek (1955). In all cases, the Babu and Odeh (1989) work assumes that the well is parallel to the *y*-axis of the parallelepiped model.

For horizontal well pressure-transient response, Goode and Thambynayagam (1987) solved a model in Laplace space and inverted the solution using a numerical inverter. Their solution is very useful for pressure-transient test diagnosis because it identifies limiting flow regimes. Previous work in the pressure-transient response of horizontal wells was presented by Daviau *et al.* (1985) and Clonts and Ramey (1986). Kuchuk *et al.* (1988) extended the Goode and Thambynayagam (1987) approach to include constant pressure (top and/or bottom) boundaries.

While other boundary effects have led to several occasionally controversial results, the inner boundary condition (flow into the well) is even more interesting. In the past, the simplest flow model used was the model of uniform flux, in which the flow rate per unit of well length is constant for each section of the well. A more realistic inner boundary condition would be one of uniform pressure (the infinite conductivity of the wellbore). Uniform flux solutions predict a well-flowing pressure drop that, in general, varies along the well. It is highest toward the center and decreases considerably near the ends of the well. To obtain a well bottomhole pressure, the Kuchuk *et al.* (1988) approach averages the pressure along the well length. Babu and Odeh (1989) record the pressure at the well midpoint, and their model predicts a slightly smaller productivity index than the one by Kuchuk *et al.* (1988). The difference between these two approaches is negligible for fully penetrating wells, but the difference becomes increasingly obvious for smaller penetration. None of the previous models can adequately represent such well configurations as either multiple horizontal wells of arbitrary direction emanating from the same vertical well or as open and closed segments within the same horizontal wells.

The Economides *et al.* (1994) solution obtains dimensionless pressures for a point source of unit length in a no-flow boundary "box" (Figure 3-13). Using a line source with uniform flux, it integrates the solution for the point source along any arbitrary well trajectory. Careful switching of early- and late-time semianalytical solutions allows very accurate calculations of the composite dimensionless pressure of any well configuration.

The productivity index, J, is related to the dimensionless pressure under transient conditions (in oilfield units):

$$J = \frac{q}{\bar{p} - p_{wf}} = \frac{\bar{k} x_e}{887.22 B \mu \left(p_D + \frac{x_e}{2\pi L} \sum s \right)} \quad (3\text{-}1)$$

where $\bar{p}$ is the reservoir pressure, p_{wf} is the flowing bottomhole pressure (psi), μ is the viscosity (cp), B is the formation volume factor, p_D is the calculated dimensionless pressure, and $\bar{k}$ is the average reservoir permeability $\left(\sqrt[3]{k_x k_y k_z} \right)$. $\sum s$ is the sum of all damage and pseudoskin factors. Dimensioned calculations are based on the reservoir length, x_e; L is the horizontal well length.

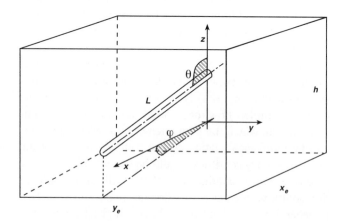

Figure 3-13 Basic parallelepiped model with appropriate coordinates (after Besson, 1990)

The generalized solution to the dimensionless pressure, p_D, starts with early-time transient behavior and ends with pseudosteady state if all drainage boundaries are felt. At that moment, the three-dimensional (3D) p_D is decomposed into one two-dimensional (2D) and one one-dimensional (1D) part,

$$p_D = \frac{x_e C_H}{4\pi h} + \frac{x_e}{2\pi L} s_x \quad (3\text{-}2)$$

where is C_H a "shape" factor, characteristic of well and reservoir configurations in the horizontal plane, and is the skin accounting for vertical effects.

The expression for this skin effect (after Kuchuk *et al.*, 1988) is

$$s_x = \ln \left(\frac{h}{2\pi r_w} \right) + \frac{h}{6L} + s_e \quad (3\text{-}3)$$

and s_e, describing eccentricity effects in the vertical direction, is

$$s_e = \frac{h}{L} \left[\frac{2z_w}{h} - \frac{1}{2} \left(\frac{2z_w}{h} \right)^2 - \frac{1}{2} \right] - \ln \left[\sin \left(\frac{\pi z_w}{h} \right) \right] \quad (3\text{-}4)$$

which is negligible if the well is placed near the vertical middle of the reservoir.

Shape factors for various reservoir and well configurations, including multilateral systems, are given in Table 3-1.

3-5.3 Performance of a Single Horizontal Well: The Effect of Areal Anisotropy

Well orientation is critical to horizontal well performance in areal anisotropic reservoirs. A horizontal well drilled normal to the direction of maximum permeability will lead to higher productivity than one drilled in any other arbitrary direction. This observation corresponds with large reported variations of horizontal well productivity as compared with expectations (Beliveau, 1995).

Here, a study is presented based on the Economides *et al.* (1994) model, showing the horizontal well productivity index vs. time as a function of well orientation. Pertinent variables include the following: $k_x = 10 \, \text{md}$, $k_y = 2 \, \text{md}$, $k_z = 1 \, \text{md}$, $x_e = y_e = 2000 \, \text{ft}$, $h = 50 \, \text{ft}$, and $L = 1200 \, \text{ft}$. The well is rotated at the center of the drainage area and is at the vertical middle of the reservoir. Figure 3-14 presents the results, demonstrating the

Table 3-1 Shape factors for various single, multilateral, and multibranch well configurations (from Economides *et al.*, 1994; Retnanto *et al.*, 1996)

Configuration	L/x_e	C_H
$x_e = 4 y_e$	0.25	3.77
	0.5	2.09
	0.75	1.00
	1	0.26
$x_e = 2 y_e$	0.25	3.19
	0.5	1.80
	0.75	1.02
	1	0.52
$x_e = y_e$	0.25	3.55
	0.4	2.64
	0.5	2.21
	0.75	1.49
	1	1.04
$2 x_e = y_e$	0.25	4.59
	0.5	3.26
	0.75	2.53
	1	2.09
$4 x_e = y_e$	0.25	6.69
	0.5	5.35
	0.75	4.63
	1	4.18
$x_e = y_e$ (cross)	0.25	2.77
	0.5	1.47
	0.75	0.81
	1	0.46
$x_e = y_e$ (star)	0.25	2.66
	0.5	1.36
	0.75	0.69
	1	0.32

Configuration	φ	C_H
$x_e = y_e$, $L/x_e = 0.75$	0	1.49
	30	1.48
	45	1.48
	75	1.49
	90	1.49

Configuration		C_H
$x_e = y_e$, $L_x/x_e = 0.4$	$L_y = 2L_x$	1.10
	$L_y = L_x$	1.88
	$L_y = 0.5L_x$	2.52
$x_e = y_e$, $L_x/x_e = 0.4$	$L_y = 2L_x$	0.79
	$L_y = L_x$	1.51
	$L_y = 0.5L_x$	2.04
$x_e = y_e$, $L_x/x_e = 0.4$	$L_y = 2L_x$	0.66
	$L_y = L_x$	1.33
	$L_y = 0.5L_x$	1.89
$x_e = y_e$, $L_x/x_e = 0.4$	$L_y = 2L_x$	0.59
	$L_y = L_x$	1.22
	$L_y = 0.5L_x$	1.79

importance of proper well orientation. For this case study, the ratio between the best productivity index (normal to k_x) and the worst index (normal to k_y) is 1.6 at 2 months. Permeability anisotropies in the horizontal plane of 3:1 or larger are common; therefore, proper horizontal well orientation can mean very large differences in early-time well performance. Obviously, after pseudosteady-state conditions emerge, these productivity index differences diminish. However, transient productivity indices are a valuable tool for differentiating between an economically attractive well and one that is far less attractive, especially when large drainage areas are involved.

When 2D coordinate transformations are used, the ratio of isotropic to anisotropic productivity indices would indicate the risk of misorienting a horizontal well, which is expressed by the following equation (Retnanto *et al.*, 1996):

$$\frac{J_{\text{aniso}}}{J_{\text{iso}}} = \sqrt{\frac{1}{I_{\text{ani,ar}}} \cos^2 \varphi + I_{\text{ani,ar}} \sin^2 \varphi} \qquad (3\text{-}5)$$

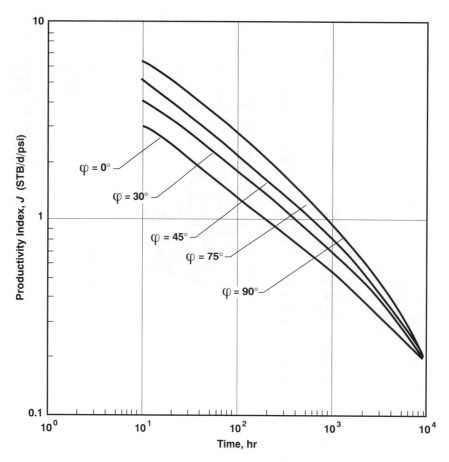

Figure 3-14   Productivity index vs. rotation of a horizontal well in a permeability-anisotropic reservoir (from Economides *et al.*, 1994)

where $I_{ani,ar}$ is the permeability anisotropy in the horizontal plane, defined as the square root of the ratio of the permeabilities in the x and y directions, and φ is the angle of the well with the x direction. The difference can be significant, as shown in Figure 3-15. A permeability ratio of 16 ($I_{ani,ar} = 4$) would result in a PI reduction of 50% in the case of the worst angle ($\varphi = 0°$) but an increase of 100% in the case of the optimum angle ($\varphi = 90°$) when it is compared with the isotropic case.

Changes in the areal anisotropy ratio and well orientation considerably impact the cumulative production. For example, a 2000-ft, single horizontal well in a 100-ft thick formation has a horizontal-to-vertical index of permeability anisotropy, I_{ani}, equal to 3. The average horizontal permeability is 10 md, but the horizontal permeability magnitudes and direction are uncertain. Four horizontal permeability anisotropy ratios are examined: 1:1 (isotropic), 5:1, 10:1, and 50:1. Figure 3-16 shows that for a single horizontal well drilled in the optimal direction, cumulative production accelerates in early time as the areal anisotropy increases. For a given average horizontal permeability as the areal anisotropy ratio increases, the reservoir drainage process accelerates, which makes higher anisotropic cases more attractive than isotropic cases. However, as shown in Figure 3-15, as the well deviates from the optimal direction, the productivity decreases, which slows cumulative production.

Changes in the areal anisotropy ratio and well orientation have an expected effect on the calculated net present value (NPV). Figure 3-17 shows the incremental NPV with increasing areal anisotropy and various well orientations over the single horizontal-well base case. The base case is a 2000-ft, single horizontal well in an isotropic reservoir. The incremental NPVs (above 1, which is for the base case) are moderate when the well is drilled in the optimal direction, while NPVs below 1 pose a large relative departure from the base case, which reflects the effect of well misorientation. Poor reservoir characterization (unknown directional permeabilities) results in a high risk of uncertainty.

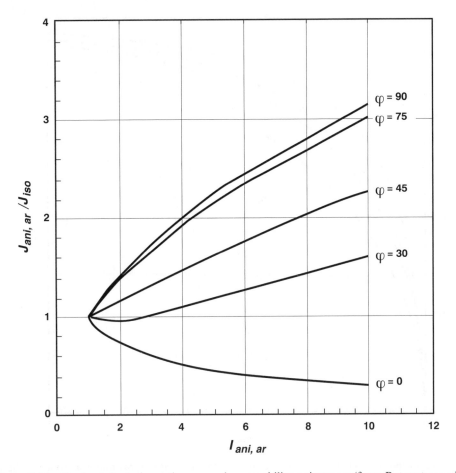

Figure 3-15   Productivity index ratio vs. areal permeability anisotropy (from Retnanto *et al.*, 1996)

3-5.3.1 Example application for a horizontal well in an anisotropic formation

A reservoir has the following dimensions: $x_e = 3000$ ft, $y_e = 3000$ ft, and $h = 100$ ft. First assume that $k_x = k_y = k_z = 10$ md, the well is in the vertical middle (i.e., $z_w = 50$ ft) and that $r_w = 0.328$ ft, $B = 1.15$ res bbl/STB, and $\mu = 0.8$ cp. A horizontal well with length $L = 2250$ ft is drilled in the x-direction.

1. Calculate the productivity index, J.
2. If $k_y = 0.1k_x$, calculate the early-time productivity index ratio between an anisotropic case and an isotropic case.
3. If $k_y = 0.1k_x$ but $\varphi = 45°$, repeat the calculation.

Solution

1. Since the well is in the vertical middle, $s_e = 0$ (otherwise Equation 3-4 should have been used, which accounts for eccentricity effects). From Equation 3-3,

$$s_x = \ln\left(\frac{100}{(2)(3.14)(0.328)}\right) + \frac{100}{(6)(2250)} = 3.89$$

From Table 3-1, noting that $x_e = y_e$ and $L/x_e = 2250/3000 = 0.75$, the shape factor C_H is obtained. It is equal to 1.49. Then from Equation 3-2,

$$p_D = \frac{(3000)(1.49)}{(4)(3.14)(100)} + \frac{3000}{(2)(3.14)(2250)}3.87 =$$
$$3.56 + 0.82 = 4.38$$

Finally, from Equation 3-1 based on the assumption of no skin damage, the productivity index, J, is

$$J = \frac{(10)(3000)}{(887.22)(1.15)(0.8)(4.38)} = 8.4 \, \text{STB/d/psi}$$

[Note: For vertical-to-horizontal permeability anisotropy ($k_z = 1$ md), the average permeability $\bar{k} = \sqrt[3]{(10)(10)(1)} = 4.6$ md, and the reservoir dimensions

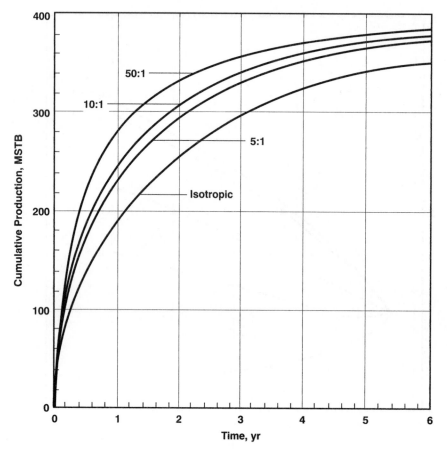

Figure 3-16   Increasing area anisotropy increases reservoir recovery rate for a single horizontal well drilled in the optimal direction (from Smith *et al.*, 1995)

x_e, y_e, h, and the well length must be adjusted accordingly as shown in Economides *et al.*, 1994. The productivity index calculated in this exercise would be reduced by more than 30% for such permeability anisotropy.]

2. Since $k_y = 0.1 \, k_x$, the areal index of anisotropy is

$$I_{\text{ani,ar}} = \sqrt{k_x/k_y} = 3.16$$

From Equation 3-5

$$\frac{J_{\text{aniso}}}{J_{\text{iso}}} = \sqrt{\frac{1}{3.16}} = 0.56$$

3. For $\varphi = 45°$

$$\frac{J_{\text{aniso}}}{J_{\text{iso}}} = \sqrt{\frac{1}{3.16}\cos^2 45 + 3.16\sin^2 45}$$
$$= \sqrt{0.16 + 1.58} = 1.32$$

3-6 MULTILATERAL TECHNOLOGY

Because the drilling, casing/tubing design, completion equipment re-entry processes, and abandonment requirements are interdependent, multilateral wells require more planning than conventional vertical or horizontal wells.

3-6.1 Directional Drilling

The steerable drilling motors and MWD and LWD technology developed in the 1980s ushered a new era of horizontal and extended-reach well scenarios. This technology allowed access to reservoirs outside the vertical plane beneath the wellhead structure. Horizontal well drilling exposes more of the formation face as the wellbore tracks laterally through the structure. This increased formation exposure allows increased production, lower formation drawdown, or a combination of both.

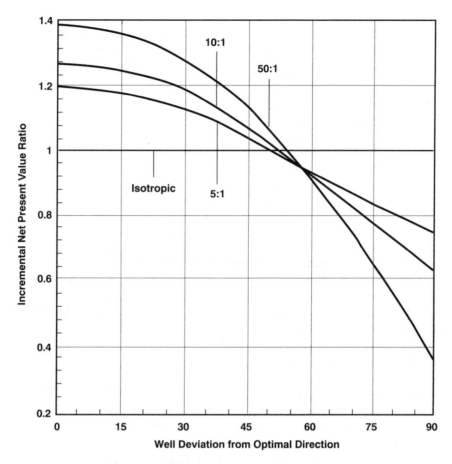

Figure 3-17 Incremental net present value ratio of single horizontal well compared to the single horizontal well base case (from Smith *et al.*, 1995)

Proper planning and path determination for directional drilling are essential for a successful multilateral project. One critical area of concern is the build rate of the lateral borehole, which is the angle in degrees per 100 ft required to steer the wellbore from a vertical plane to a horizontal plane. The lower the build rate, the more gentle the resulting curve. Build rates and radius classifications commonly associated with horizontal drilling are shown in Table 3-2.

The resulting build rate and angle affect the ultimate completion design and the later servicing of the wellbore. For example, the build rate directly affects whether casing can be installed in the lateral wellbore. A short-radius build rate may be required to hit the target from the parent bore casing exit point; however, casing-bending properties prevent normal installation in a bore with a build rate greater than 30°/100 ft.

Another consideration is the bending ability of the completion equipment for installation inside the lateral casing string. Most completion packers, gravel-pack screens, and other large outside-diameter (OD) service tools will only pass through casing with a maximum horizontal build rate/bend radius of 22°/100 ft. If casing and internal completion equipment will be required for managing the lateral wellbore, long- or medium-radius horizontal wellpath planning procedures should be used. If the rock properties of the formation are stable and do not require casing, short-radius or intermediate-radius directional drilling tools and techniques can be considered. During the well designing/planning phase, engineers must balance this consideration against the effects of lateral wellbore re-entry, production logging, stimulation, flow control, and future abandonment procedures.

3-6.2 Completion Technology

For the best results to be achieved from a multilateral well, compatible completion tools must be properly selected. Many technologies and processes affect the design and installation of the final completion suite of tools. Prior planning must be made if lateral wellbore re-

Table 3-2 Directional drilling-radius classifications

Hole Size	2 to 8°/100 ft	9 to 30°/100 ft	31 to 60°/100 ft	61 to 180°/100 ft	Tool Size
12.25 in.	Long Radius	Medium Radius			7.75 in. to 8 in.
9.875 in.	Long Radius	Medium Radius			6.75 in. to 7.75 in.
8.5 in.	Long Radius	Medium Radius	Intermediate Radius		6.5 in. to 6.75 in.
6 to 6.75 in.	Long Radius	Medium Radius	Intermediate Radius	Short Radius	4.75 in.
4.75 in.		Medium Radius	Intermediate Radius	Short Radius	3.5 in. to 3.75 in.
3.75 in. to 4.5 in.			Intermediate Radius	Short Radius	2.875 in.

entry will be required. Possible re-entry objectives such as tool manipulation, zonal isolation, recompletion, zonal stimulation, and data collection should be determined based on experience with other wells in similar reservoir structures. These factors should be communicated as part of the total multilateral system requirement.

Since most lateral wells will be in a horizontal configuration, job designers must ensure compatibility with coiled tubing or jointed pipe operations. These operational requirements will determine whether rig intervention is mandated—another factor in total well economic viability.

Subsea well completions represent another area of special challenge where intervention plans should be avoided because of the major costs of mobilizing a subsea rig or work platform. When subsea intervention is required to bring the well back on production, the subsystems must already be in place. Multilateral well completions are driven by the following criteria:

- Pressure, temperature, and chemical properties of produced fluids and gas
- Zonal segregation and isolation requirements
- Lateral wellbore rock properties and stability
- Regulatory requirements for zone production and management
- Workover/re-entry options and methods
- Sand and water production potential
- Abandonment requirements
- Equipment performance records and risk analysis

3-7 MULTILATERAL WELL DESIGNS

Three major categories of multilateral well designs are now available: openhole multilateral wells, limited-isolation/access multilateral wells, and complete multilateral wells. All three types are discussed in detail in the following sections.

3-7.1 Openhole Multilateral Wells

Although openhole multilateral well designs are relatively simple, detailed reservoir information must be available, and target selection must be performed and coordinated with the directional drilling plan. As shown in Figure 3-18, openhole multilateral wells begin with the drilling of the primary wellbore to a depth above the producing intervals and lateral kickoff points. A surface or intermediate casing string is then set and cemented in place. Drilling continues out through the bottom of the casing, with steerable drilling assemblies that guide the drilling assembly laterally from the parent borehole to create a multiple-lateral hole. Other holes are drilled from the parent bore until all drainage portals for the design have been created (Figure 3-19).

3-7.1.1 Applications

Openhole multilaterals can improve the areal drainage of a common reservoir. When directional drilling tools are used, the direction and inclination of the drill bit is steered and monitored so that drainage holes are placed quickly and accurately. Current directional drilling techniques are accurate to about ±3 ft at distances of up to

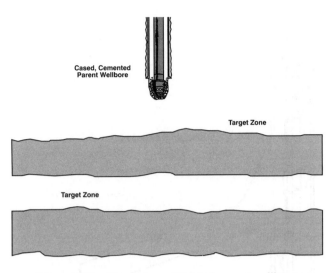

Figure 3-18   Openhole multilateral parent wellbore

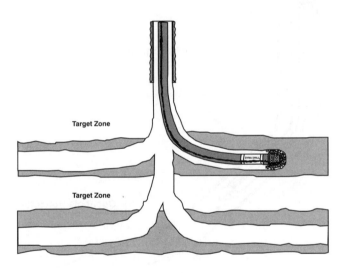

Figure 3-19   Openhole multilateral drilling

6000 ft of horizontal displacement. Most openhole applications have involved relatively short horizontal sections, rarely exceeding 1200 ft of horizontal displacement. Because lateral-bore casing installation is not a consideration in these wells, short-radius drilling techniques can be used to reach the target zones.

3-7.1.2 Reservoir Considerations

Stable, hard-rock formations, such as those found in the Austin Chalk, are appropriate for this type of well design. Many of these formations have steep decline rates in which high initial production is followed by rapid decline within a few months.

The drainage objective for such a formation is to achieve the highest amount of initial production and to drain the reservoir as rapidly as possible, until either water coning or natural rate decline occurs. Any well expected to exhibit a high decline rate will be planned with quicker payout deadlines, thereby increasing the economic constraints on the project's viability.

Since this type of well design requires the commingling of production from each producing bore, the pore pressures of each target zone must be analyzed. This analysis ensures compatibility and minimizes crossflow or interference. Logging data, well-test information, and field experience in this type of reservoir are the primary tools used for analysis. Mud weight changes occurring as each lateral is drilled are also key indicators as to zone segment compatibility for commingled flow.

Lateral-bore plugback and drilling redirection may occur "on-the-fly" during this type of operation to access areas of common reservoir characteristics. Core samples should be obtained, and rock stress properties should be tested for the determination of the relative stress planes and rock stability parameters.

For openhole completions, stable, nonsloughing, heterogeneous hard rock reservoirs should be targeted. Otherwise, wellbore collapse and loss of production could result. Previous instances of wellbore collapse in openhole designs have occurred at the heel of the lateral (where the lateral bore initially departs the parent bore). The selected reservoirs should not require continuous intervention or selective isolation since casing strings will not be used. Re-entry access to any specific lateral bore will be a difficult trial-and-error process and cannot be relied upon as an effective option.

3-7.1.3 Installation Considerations

Completion equipment installation and design requirements are minimal with this type of well program. Typically, a single string of production tubing is anchored to a packer that is set near the base of the parent casing. As required, landing nipples for plugging and safety valves can also be added. The production tubing string should allow maximum flow rates with minimal pressure drop to match the well's production parameters.

3-7.1.4 Operational Considerations

If planned properly, openhole multilaterals pose minimal risk. The major consideration in risk assessment is avoiding unreal expectations that the well will perform in a manner for which it was not designed. Water-coning tendencies should be studied, and risk should be assigned. This completion design allows no contingency for shutting off unanticipated water production, which could lead to complete loss of revenue-generating production. Wellbore collapse potential should also be reviewed, and risk should be carefully considered. Any known risk in a single horizontal well will be multiplied in multilateral wells. The openhole multilateral well should present no more manageable risk implications than other types of openhole completions.

3-7.2 Limited-Isolation/Access Multilateral Systems

3-7.2.1 Applications

As the reservoir requirements for zonal isolation and re-entry access become prerequisite for a successful long-term project, multilateral systems are available to address these needs. Figure 3-20 shows a system that isolates the flow entering the primary bore from the flow entering the lateral bore. This type of installation is appropriate if one of the reservoirs is likely to produce unwanted water or gas during the life of the well. This undesirable production can then be shut off from the producing bore. Expected pressure changes between the lateral-bore reservoir targets can also be managed through this "on/off" arrangement.

3-7.2.2 Reservoir Considerations

Proper reservoir analysis and target selection are required as with any multilateral well design. However, the reservoir management capability offered through this type of completion system allows some control over reservoir incompatibility or nonperformance. As a result, zonal homogeneity becomes less critical.

3-7.2.3 Installation Considerations

As with openhole designs, this limited-isolation/access system does not allow casing to be set in the lateral bore and mechanically reconnected to the parent bore. This system does, however, allow re-entry into the lateral borehole during the drilling and initial completion phases. A stable, nonsloughing, hard-rock formation is the best candidate for this well design.

When installed in the system, slotted liners hung on openhole inflatable packer anchors provide a means of

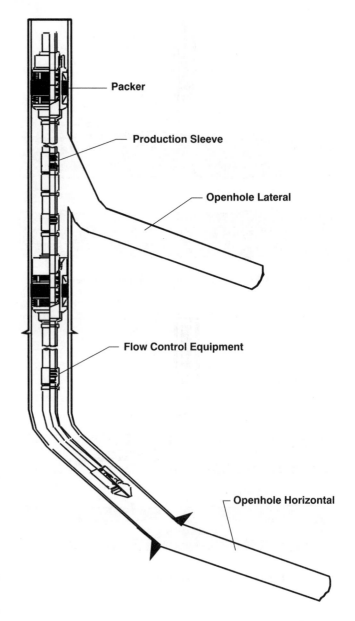

Figure 3-20 Limited-isolation multilateral completion

hole-size maintenance and stability. If sand control problems are anticipated, prepacked screens can be installed in a similar manner.

Since insertable openhole liners and tools cannot be mechanically reconnected to the parent casing, a non-permeable section of stable shale or hard rock should be selected for the lateral exit. Failures in this type of installation tend to occur at the unprotected, openhole section adjacent to the parent-bore casing (the heel of the lateral).

Before the system is installed (Figure 3-20), the primary borehole should be drilled and cased below the point of the planned lateral exits. The primary bore may also be directionally drilled to penetrate lower target sands, serving as a producing bore. An alternative configuration involves placing the primary bore vertically through lower target sands for logging that will determine the exact sand location and thickness for enhanced accuracy of uphole lateral departures.

Once the borehole is cased and logged, a retrievable whipstock is installed at the required lateral exit point. This wedge-shaped diverter is oriented for both depth and azimuth, with the sloped face of the tool aimed in the general direction of the planned lateral bore path. Sometimes, the design includes an integral anchor or separate packer that provides a stable, nonrotating base for the whipstock to engage, which helps maintain placement accuracy during drilling. Pilot mills, window mills, and watermelon mills are run against the face of the whipstock to cut away the section of casing opposite the diverter face. When this process is used, a full-gauge window opening is cut and sized in the casing. This opening allows required directional drilling assemblies to pass.

For best results, the formation immediately adjacent to the exit window should be stable, nonsloughing, hard rock or shale. The window exit point should be selected while the directional drilling path is being planned, so that the created bore track will allow for both target achievement and future liner or tool placement.

Directional drilling assemblies and MWD and/or LWD tools drill the lateral borehole through the desired targets. After completion of the drilling phase, liner, screen, or other tools may be installed, with the face of the whipstock once again being used as a diverter into the openhole lateral bore. Once completed, the whipstock is retrieved from the parent bore, which allows access to lower lateral or parent bore zones.

Additional lateral exits are created in a similar manner, with the process normally starting at the bottom of the parent wellbore and working up the hole. When lateral exits are started at the bottom, the previously created laterals can be temporarily plugged with retrievable bridge plugs as a means of preventing formation contamination from drilling or stimulation fluids used on subsequent laterals. Underbalanced drilling or other drilling and fluid systems can then be used on each lateral while well control is maintained, and unwanted zonal flow is prevented. Once all laterals have been drilled and stabilized, the plugs are removed, and the final phase of casing-completion equipment is installed.

Final completion scenarios for this type of well are numerous and offer more production management options than openhole multilateral wells. Typically, production packers are installed in the primary wellbore casing above and below each lateral exit point. Single or dual strings of production tubing could also be installed to commingle or segregate production as required.

Sometimes, sliding sleeves are placed at the lateral exit point to provide on/off production control from each lateral bore. Through-tubing access sleeves could also be used, adjacent to the lateral exit. These access sleeves contain a precut window that aims toward the lateral borehole. The window may be shifted open through the use of through-tubing slickline or coiled tubing-conveyed tools. Once open, an internal deflector is installed that diverts strings of tools into the openhole section for future logging, tool actuation, or borehole stimulation. Figures 3-21 and 3-22 show some of the completion options available with this type of installation.

When properly planned, limited-isolation/access designs pose fewer risks than openhole multilateral wells because additional production control options are available to adapt to changes in zonal inflow.

3-7.3 Complete Multilateral System

3-7.3.1 Applications

A complete multilateral system provides two to five laterals from one new or existing wellbore. Applications for this type of multilateral system are similar to the limited-isolation/access design, but a complete multilateral system allows project designs for deepwater or subsea environments. In new wells, the exit portal must be capable of being placed in the vertical, inclined, or horizontal plane, and the window exit azimuth must be capable of being oriented after the primary casing string has been installed. The system must also be compatible with cementing operations for liners and/or slotted liners and prepacked screens for sand control.

When a complete multilateral system is used, the lateral wellbore is cased back to the primary bore exit, and the liner casing string is mechanically connected to the primary bore casing; the lateral- to main-wellbore junction must be hydraulically sealed. Any complete lateral bore or portions of any lateral can be isolated as needed to control the production inflow profile. Each lateral must also be accessible for re-entry without rig intervention. The system should allow for washdown methods that help transport long liners to the bottom, and it should accommodate high build rates (45 to 60°/100 ft) after tools exit the lateral window junction.

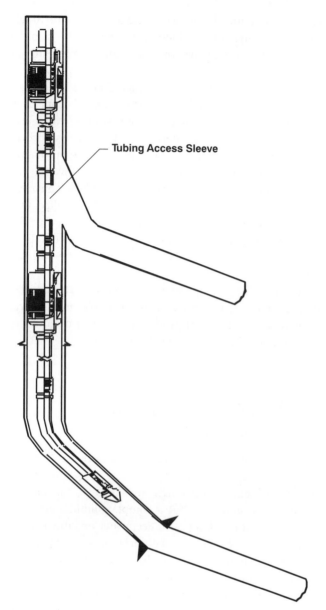

Figure 3-21 Limited-isolation/access multilateral completion (production is commingled)

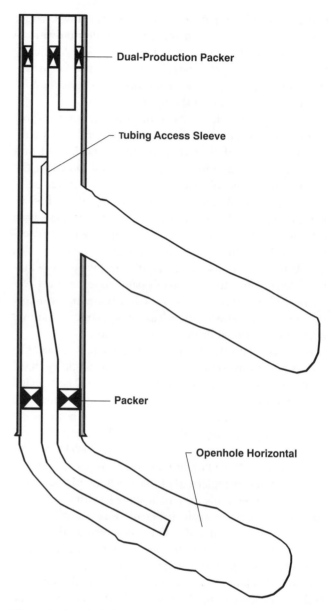

Figure 3-22 Limited-isolation/access multilateral completion (production is segregated through dual tubing strings)

3-7.3.2 Reservoir Considerations

Multilateral systems capable of delivering additional reservoir management options increase the choice of target zones. Again, proper reservoir modeling and target selection must occur during the project planning phase, and a stable, nonsloughing, impermeable shale or hardrock formation is desirable at the exit site. If, however, target selection requires exit in unconsolidated sands or in the producing interval itself, the unconsolidated sand can be stabilized with cement or plasticized material. Since low to medium build rates will be used to simplify

casing installation, engineers must select targets and plan drill paths with such considerations in mind.

3-7.3.3 Installation Considerations

During installation of a complete multilateral system, the primary bore is drilled, and the primary production casing string is cemented in place across all anticipated lateral-bore exit points. The primary bore is normally drilled into a producing zone and completed for final production.

Temporary plugs are normally set above this lower zone to isolate it from fluids used to drill the upper lateral bores. If required, the lateral exit area can be pretreated through the use of oriented perforating and squeeze techniques, without damaging the lower primary zone.

As shown in Figure 3-23, a combination assembly consisting of a whipstock packer anchor, a hollow whipstock temporarily filled with a composite plugging material, and a starter mill are run to depth, oriented to the correct azimuth, and set hydraulically. The starter mill is sheared loose from the now-anchored whipstock, and the first section of exit window is rotary-milled. Additional window and watermelon mills are run to open the exit portal to a full-gauge dimension (Figure 3-24).

Oil-based or special salt-sized drilling fluids are circulated to the surface in preparation for directional drilling of the lateral bore. Lower-bore isolation plugs and the internal plug contained in the whipstock anchor packer prevent contamination of the lower, completed producing zone.

Directional-drilling bottomhole assemblies (BHAs) consisting of steerable drilling motors, MWD tools, and LWD tools are used to drill the lateral bore to casing setting depth (Figure 3-25). Drilling-fluid additives and maintenance of mud weight are the primary concerns in regards to maintaining a stable borehole and preventing uncontrolled flow as pressured zones are encountered.

The hole size must be consistent so that the casing can be properly set and cemented. Therefore, when drilling is complete, openhole caliper logs are run and analyzed. If necessary, hole-opening and reaming steps are then performed.

A liner casing string assembly is made up and run to depth. This string consists of (1) float equipment for cement circulation, (2) applicable mechanical centralizers, (3) sufficient casing to extend 150 to 300 ft back into the primary bore, and (4) a hydraulically set liner hanger with a cement-compatible running tool (Figure 3-26). Typical primary casing/lateral liner sizes are listed in Table 3-3.

Unlike slimhole well designs, multilateral applications require larger casing and tubing diameters. For maximum benefit, the optimum primary casing size should be $9\frac{5}{8}$ in. or larger, allowing for the installation of a 7-in. liner. This selection allows (1) the use of widely available drilling, MWD, and LWD tool sets, (2) the installation of large-diameter production packers, and (3) the use of tubing strings that are $3\frac{1}{2}$ in. and larger.

When the liner string has been positioned, circulation is established and fluids are changed as required for upcom-

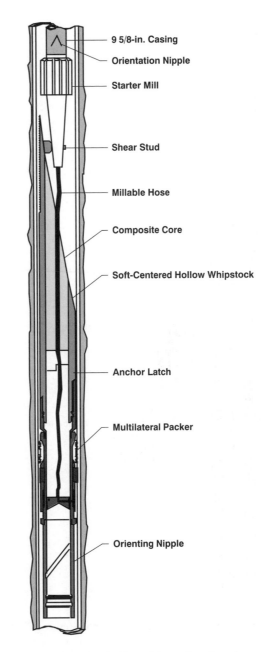

Figure 3-23 Running hollow whipstock and packer; orienting, and setting packer

Table 3-3 Typical primary casing/lateral liner sizes

Primary Casing (in.)	Lateral Liner (in.)
$13\frac{3}{8}$	$9\frac{5}{8}$
$10\frac{3}{4}$	7
$9\frac{5}{8}$	7
$7\frac{5}{8}$	5
7	$4\frac{1}{2}$
$5\frac{1}{2}$	$3\frac{1}{2}$
$4\frac{1}{2}$	$2\frac{7}{8}$

The following labels appear in Figure 3-23:
- 9 5/8-in. Casing
- Orientation Nipple
- Starter Mill
- Shear Stud
- Millable Hose
- Composite Core
- Soft-Centered Hollow Whipstock
- Anchor Latch
- Multilateral Packer
- Orienting Nipple

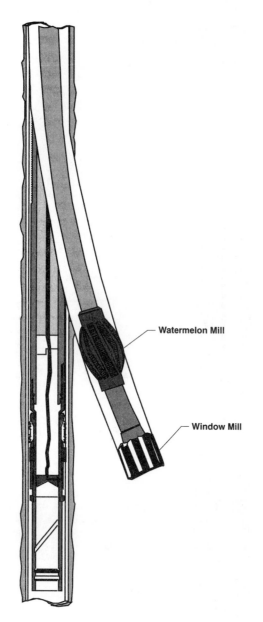

Figure 3-24 Window and watermelon mills used to ream/feather window

ing cementing operations. A special blend of cement and additives is then mixed and pumped down and around the liner string. If long liners are used, multistage cement processes can be used to place general-purpose cement for the lower liner section and to place special sealant in the junction area. During cementing, casing rotation and reciprocation will help ensure complete cement coverage. Once proper displacement is achieved, wiper plugs are released from the surface, wiping the liner string of excess cement and providing a pressure block for setting the hydraulic liner hanger (Figure 3-27).

The hydraulically sealed, cement-anchored lateral liner can now be perforated and completed with production packers, sliding sleeves, sand-control tools, or other completion tools that are required to manage and produce the target reservoir. Any completion design used in single-bore vertical or horizontal wells can be installed in the lateral liner string (Figure 3-28). Additional openhole drilling out the lower end of the lateral liner may also occur to meet completion objectives requiring openhole prepacked screens or slotted liners (Figure 3-29).

At this point, access to the lower producing bore is achieved. A milling assembly consisting of a retrievable downhole mill guide with an internal pilot mill is run into the well and positioned in the lateral liner across from the face of the plugged, hollow whipstock below the liner. The mill guide is oriented and depth-positioned by radioactive tags in the whipstock and conventional gamma ray and collar-locator electric-line logging practices.

When positioned properly, the mill guide is hydraulically set, and the pilot mill is released. Timed, limited weight milling is used to cut a pilot hole through the side of the lateral liner in the vertical plane of the primary casing alignment (Figure 3-30). Large-diameter window and watermelon mills, used in follow-up runs, open the vertical access hole in the liner string and mill out the temporary plugging material in the hollow whipstock. This process creates a full-gauge hole that is dimensionally compatible with the drift diameter of the liner string (Figure 3-31). Completion fluids are circulated through the well, and the final plugs are milled in the base of the whipstock anchor packer. Temporary isolation plugs can now be retrieved from the lower primary bore, and the lower zones are open for completion and production. Other lateral wellbores are then drilled and completed farther up the primary wellbore.

When all installed lateral bores have vertical access to the primary bore, uphole primary bore completion equipment is installed. Completion options range from single-string, commingled flow designs to multistring, segregated production installations as shown in Figures 3-32 and 3-33.

Each lateral wellbore can be re-entered through the production tubing string with size-compatible tools or with a rig, if the tools required are larger than the tubing ID. Since all lateral exit bores are created from the "high" side of the primary bore, tools will track the low side of the primary bore, travel past the open lateral liners and remain in the primary bore unless they are mechanically forced into a selected lateral.

As a means of tool deflection, a diverter tool is installed in the orienting/anchor nipple profile in each primary-

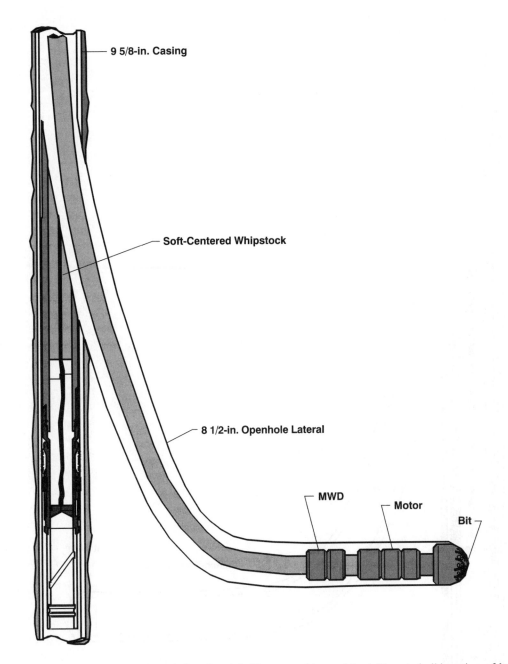

Figure 3-25   Formation bit and directional drilling assembly used to drill angle-build section of hole

bore whipstock anchor packer. As the diverter is set and anchored in the profile, the orienting track rotates the diverter face to align in the direction of the lateral bore. The diverter fills in the milled-out section of the lateral liner and serves as a temporary base to the liner; tools strike the deflector face and are guided into the lateral casing string (Figure 3-34). Slickline, coiled tubing, or jointed pipe-conveyed toolstrings can then be run into any selected lateral bore to shift sliding sleeves, set or retrieve plugs, perform production logging services, or pump stimulation chemicals or other chemicals into the lateral producing zone.

The same diverter tool is used at each lateral. Plugs and downhole chokes can also be set to isolate any portion of any lateral or primary bore to shut off unwanted water or gas production, allowing production control over each zone or segment.

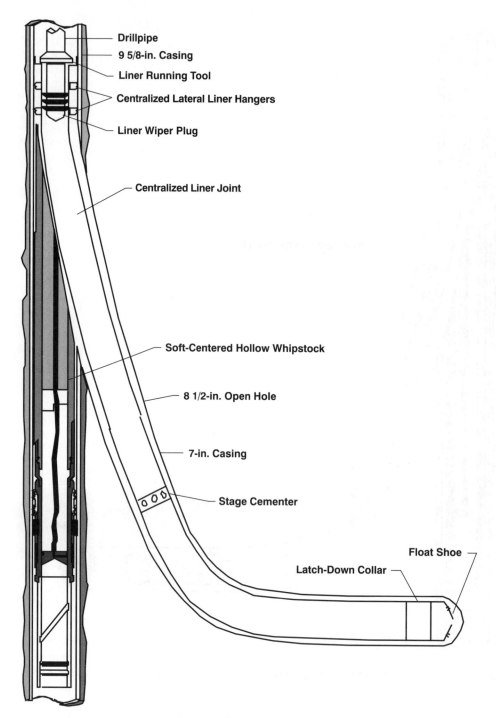

Figure 3-26   Run lateral liner and cement in hole

3-8 PERFORMANCE OF MULTILATERAL AND MULTIBRANCH WELLS

3-8.1 Advantages and Drawbacks

Multilateral wells have many of the same advantages and disadvantages as horizontal wells, only to a greater degree. The advantages generally include higher productivity indices, the possibility of draining relatively thin layers, decreased water and gas coning, increased exposure to natural fracture systems, better sweep efficiencies, and specific EOR applications such as steam-assisted gravity drainage. Drawbacks of multilateral

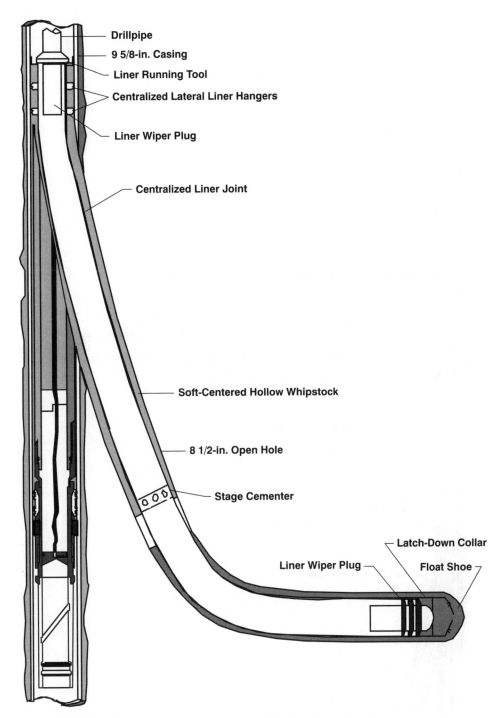

Figure 3-27   Placing specialized sealant at junction

wells include (1) higher initial costs, (2) increased sensitivity to heterogeneities and anisotropies (both stress and permeability), (3) sensitivity to poor (effective) vertical permeability, (4) complicated drilling, completion, and production technologies, (5) complicated and expensive stimulation, (6) often slower and less effective cleanup, and (7) cumbersome wellbore management during production. Disadvantages specific to multilateral wells include difficult selection of appropriate candidates, interference of well branches (all of which should be optimized), crossflow, and difficult production allocation. In addition, if damage occurs during

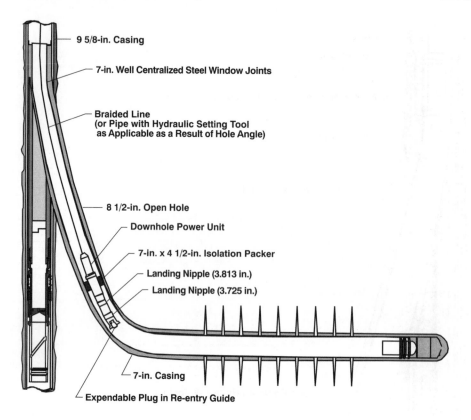

Figure 3-28   Setting packer in lateral

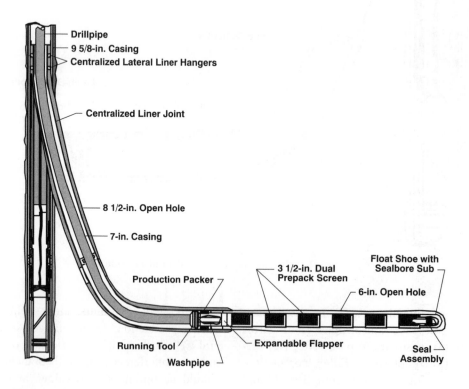

Figure 3-29   Washdown system for screens run in open hole (sized-salt mud displaced)

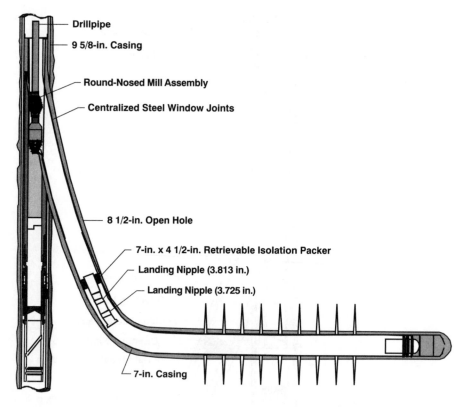

Figure 3-30   Run round-nosed mill sssembly to complete pilot hole through hollow whipstock

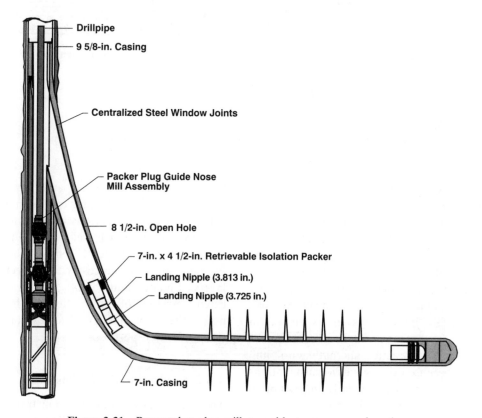

Figure 3-31   Run packer plug mill assembly to remove packer plug

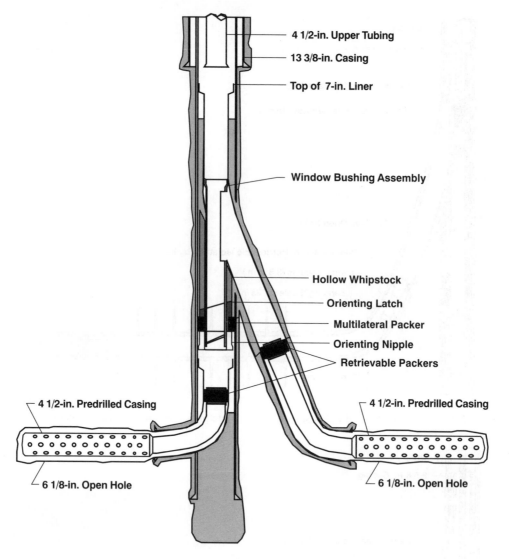

Figure 3-32   Advanced multilateral completion (production commingled)

drilling, stimulation will be more difficult and far more costly than for vertical wells.

Additional motivations to drill multilaterals include increased reserves per well because of increased drainage volume and better vertical and areal sweep, especially for irregular or odd-shaped drainage volumes. Reservoirs can be drained at enhanced economic efficiency, particularly when a limited number of slots is available (offshore) or when surface installations are expensive (in environmentally sensitive areas).

Multilaterals are also attractive alternatives to infill drilling operations because existing surface installations can be used. This feature is particularly important in tight or heavy oil formations. In heterogeneous reservoirs (layered, compartmentalized, arbitrarily oriented

fissure swarms), more oil and gas pockets can be exploited, and more fissures can be intersected.

In anisotropic formations with unknown permeability directions, multibranch wells can reduce economic risk. Furthermore, multilateral wells can be used to balance the nonuniform productivity and injectivity of different layers. Finally, multilateral well systems provide an extensive source of information on the reservoir, which makes them a valuable tool for exploration and formation evaluation purposes.

3-8.2 Production Predictions

The following case studies demonstrate the attractiveness of multilateral horizontal wells.

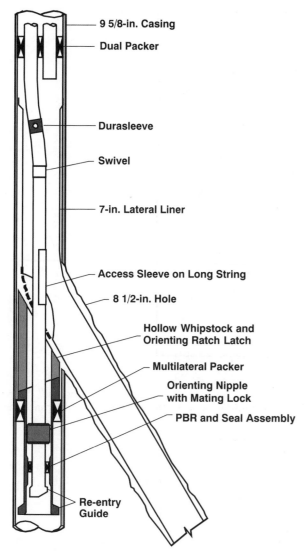

- 9 5/8-in. Casing
- Dual Packer
- Durasleeve
- Swivel
- 7-in. Lateral Liner
- Access Sleeve on Long String
- 8 1/2-in. Hole
- Hollow Whipstock and Orienting Ratch Latch
- Multilateral Packer
- Orienting Nipple with Mating Lock
- PBR and Seal Assembly
- Re-entry Guide

Figure 3-33 Advanced multilateral completion (production segregated)

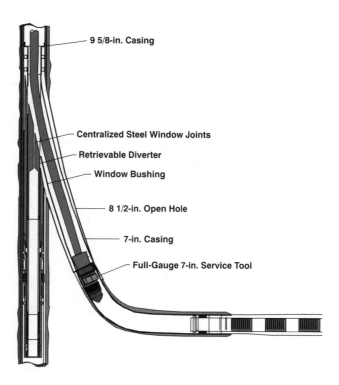

- 9 5/8-in. Casing
- Centralized Steel Window Joints
- Retrievable Diverter
- Window Bushing
- 8 1/2-in. Open Hole
- 7-in. Casing
- Full-Gauge 7-in. Service Tool

Figure 3-34 Re-entry access for full-gauge tools

3-8.2.1 Multibranch Wells

Retnanto and Economides (1996) examined four well configurations: (1) a 2000-ft horizontal well along k_y, (2) a 2000-ft horizontal well along k_x, (3) a "cross" configuration of four laterals, which formed the equivalent of two 2000-ft wells intersecting orthogonally at the reservoir middle, and (4) a "star" configuration with eight laterals, which formed the equivalent of four 2000-ft wells intersecting at 45° angles at the reservoir center. For these studies, three permeabilities were used $k_x = 0.1$, 1, and 10 md). Figures 3-35 and 3-36 show the major impact of permeability on the cumulative production ratio for the "cross" and "star" configurations vs.

the base case of a single 2000-ft horizontal well. For low reservoir permeability, this ratio is maintained at a high level for a considerable time. Multilaterals in low- to moderate-permeability reservoirs can produce with higher production rates, and therefore, with higher cumulative productions at early times compared to single horizontal wells.

Both high- and low-permeability reservoirs exhibit productivity improvements with multilaterals during transient conditions. However, once pseudosteady state is reached, the increase in the productivity index is diminished. Multiple wells in the same drainage area will interfere with each other. Although high-permeability formations experience larger absolute productivity increases, they also experience interference effects earlier than other formations. Therefore, the relative economic benefit of multiwell configurations in high-permeability reservoirs begins to diminish quickly, while low-permeability formations exhibit interference effects much later, and the relative economic benefits are both higher and prolonged.

In high-permeability reservoirs, an additional consideration emerges. Because of the high flow rates, the tubing in the parent wellbore could hinder production, resulting in limited economic gain. Low-permeability formations are rarely tubing-constrained.

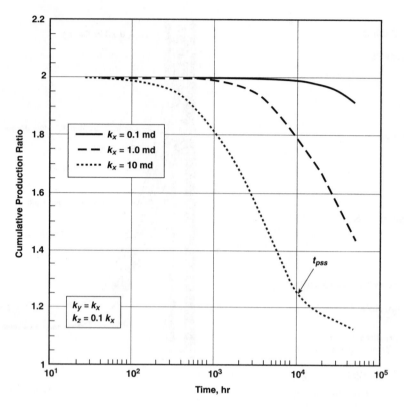

Figure 3-35   Cumulative production ratio for real time for the "cross" configuration, $k_y = k_x$ (from Retnanto and Economides, 1996)

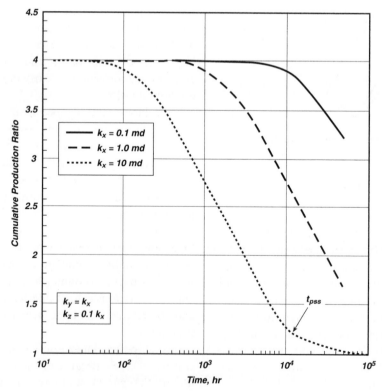

Figure 3-36   Cumulative production ratio for real time for the "star" configuration, $k_y = k_x$ (from Retnanto and Economides, 1996)

A new reason to adopt multilateral horizontal wells is that they reduce the economic risk associated with poor reservoir characterization in areally anisotropic formations and increase the incremental net present value (NPV) more than single horizontal well designs. The following case study compares single horizontal completions to a multilateral "cross" configuration in areally anisotropic reservoirs. The average horizontal permeability is 10 md, but the magnitudes and the direction of the horizontal permeability are uncertain. The study involves four horizontal permeability anisotropy ratios: 1:1 (isotropic), 5:1, 10:1 and 50:1, and it shows the effect of well orientation by rotating both configurations within the horizontal plane. The economic base case is a 2000-ft, single horizontal well that is oriented in the optimal direction, perpendicular to the direction of maximum horizontal permeability. The index of anisotropy is equal to 3 and the thickness is 100 ft. Figure 3-37 shows the incremental NPV for the "cross" configuration vs. the single-well base case. The increases in incre-

mental NPV are moderate; however, regardless of well orientation and areal anisotropy ratio, the incremental NPV is positive for the cross configuration. Compare this with Figure 3-17, where much lower NPVs are shown for a misoriented horizontal well. Multilateral wells drilled at 90° to each branch can alleviate the effects and the risks of areal permeability anisotropy.

3-8.2.2 Multilateral Wells

The following case study demonstrates the attractiveness of multilateral horizontal wells with branches emanating from a horizontal parent wellbore. Figure 3-38 contains the normalized productivity index vs. t_D for $L_y = L_x$ compared to the base case. L_x is the horizontal parent hole and L_y is the tip-to-tip length of the branches. For this study, the base case is a 2000-ft long single horizontal well that is draining an isotropic reservoir. At early time, Branches 2, 4, 6, and 8 show about 2, 3, 4, and 5 times, respectively, the base productivity index.

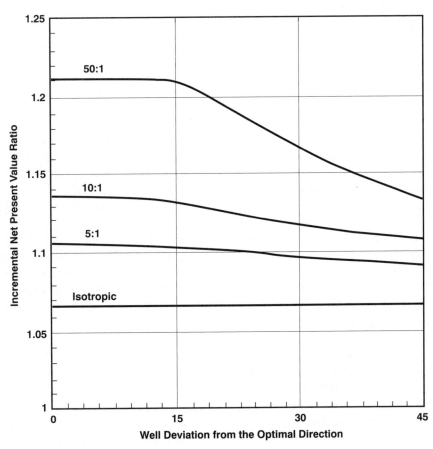

Figure 3-37   Incremental net present value ratio for the "cross" configuration compared to the single horizontal well base case (from Smith *et al.*, 1995)

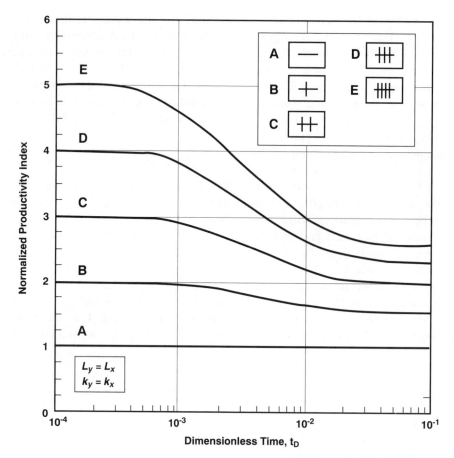

Figure 3-38   Normalized productivity index for the $L_y = L_x$ Configuration ($k_y = k_x$) (from Retnanto *et al.*, 1996)

However, at the onset of pseudosteady state, these normalized productivity indices begin to decline in late time because of interbranch interference for both the isotropic and two anisotropic cases. Figure 3-39 presents the normalized productivity index and the number of branches for $L_y = L_x$. Clearly, for the areally isotropic case, the slope of the normalized productivity index becomes steeper as the number of branches increases. For the anisotropic cases, the slope of the normalized productivity index vs. branches is flatter. However, a smaller number of branches can provide a substantially higher productivity index when compared to a single horizontal well. Of course, in all cases, absolute incremental benefits must be assessed only with incremental discounted economics.

Assuming $10/STB oil and a 25% time-value of money, Figure 3-40 shows the 3-year NPV. The revenue NPV of a single horizontal well in a permeability-isotropic reservoir is $2.5 million, whereas for the 5:1 and 50:1 horizontal permeability ratios, the revenue NPVs are $3.5 million and $5.8 million, respectively. Figure 3-40 also contains the revenue NPVs for multibranch config-

urations. The relative attractiveness of these branched wells and the optimum number of branches will depend on the net NPV, which must incorporate the cost of drilling and well completion.

For example, an eight-branch configuration with a 5:1 areal permeability anisotropy would have an incremental revenue NPV of $2.8 million ($6.3 to $3.5 million). If the drilling and completion of the eight branches is less than this amount, they are financially attractive. Similar conclusions can be drawn for all other permeability anisotropies and corresponding numbers of branches.

3-8.2.3 Incremental Reserves

Additional incremental benefits from horizontal and multilateral wells can be claimed from increases in the probable reserves. Increased production and probable reserves are sufficient motivations for investigating new and more innovative projects. For this study, the term "probable reserves" refers to the amount of reservoir oil contained in a box surrounding a shell that has a pres-

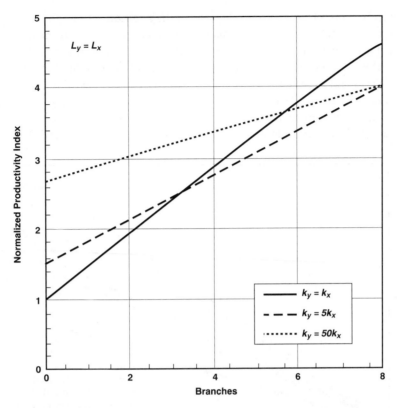

Figure 3-39 Normalized productivity index vs. number of branches ($L_y = L_x$) (from Retnanto *et al.*, 1996)

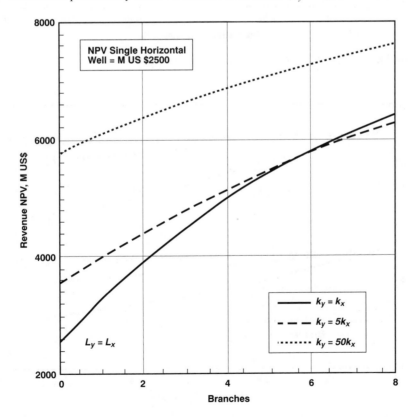

Figure 3-40 The 3-year net present value (NPV) vs. number of branches ($L_y = L_x$) (from Retnanto *et al.*, 1996)

sure equal to the initial reservoir pressure minus 1 psi after 1 year of production. Although this criterion may be appear to be somewhat arbitrary, it is a consistent and reliable measure of reserves. Figure 3-41 presents the calculated probable reserves for vertical, horizontal, and multilateral wells that are 20 to 500 ft thick. For all reservoir thicknesses, the increases in reserves from horizontal and multilateral wells are considerable.

In addition to the already considerable improvement in reserves realized from horizontal wells, the folds of increase in these reserves show a substantial preference for multilateral wells (Figure 3-42). For increasing reservoir thickness, the folds decrease. For example, in a 50-ft thick formation and $I_{ani} = 1$ (Figure 3-42), the folds of increase in reserves from a horizontal well and a multilateral well are 1.64 and 2.33 respectively, when com-

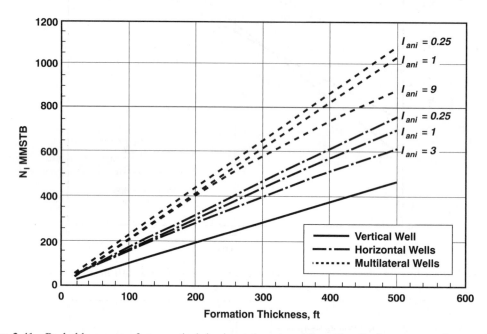

Figure 3-41   Probable reserves from vertical, horizontal, and multilateral wells for a range of reservoir heights

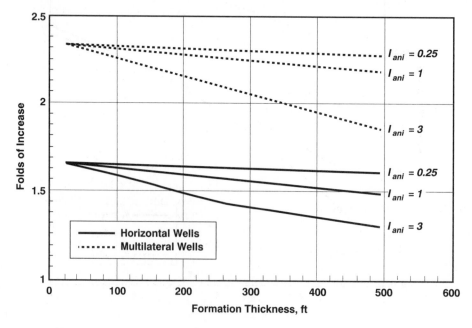

Figure 3-42   Folds of increase in probable reserves over a vertical well

pared to a vertical well. The multilateral well represents an additional 42% increase. However, in a 500-ft thick formation with the same index of anisotropy, these folds of increase are 1.49 and 2.59 respectively.

3-9 FUTURE MULTILATERAL APPLICATIONS/ADVANCEMENTS

As multilateral wells become more common within the industry, applications will likely expand, just as they did when single horizontal technology become widely accepted. Following are some of the most likely future multilateral-related developments.

3-9.1 Instrumented Wellbores

As wells become more complex, engineers will need to monitor various downhole conditions to ensure optimal reservoir management. The following wellbore instruments should provide improved reservoir management.

- Real-time temperature and pressure survey tools will be installed in each lateral wellbore.

- Downhole flowmeters, developed from surface-based technology, will record and report the flow rates at multiple locations within the well.

- Various instruments will allow the monitoring of interference at preset intervals within the laterals. Tools can then be actuated that will reduce or eliminate the interference and maintain or increase total production.

- Equipment will be available that can monitor and report casing and tubing corrosion, allowing the treatment, repair, or replacement of affected casing or tubing.

3-9.2 Interactive Reservoir Management

Interactive reservoir management (IRM) refers to tools and processes that allow the well engineer to observe real-time events in each of the well laterals. When well conditions require modification, downhole tools can be remotely actuated for optimal well production. Processes such as subsurface fluid and gas separation and reinjection will be possible and are under development.

3-10 CONCLUSIONS

When proper formation evaluation techniques are used, multilateral and multibranch wells can solve many current petroleum reservoir management problems. Not only are the productivity or injectivity indices generally superior to those of vertical wells, but when used with other advancements in 3D seismic and reservoir modeling, multilateral wells allow re-entry into existing wells. As a result, the significant amounts of hydrocarbons left in place can be accessed and produced. The overall cost savings delivered by this single technology will allow increased exploration and production activity to occur.

REFERENCES

Babu, D.K., and Odeh, A.S.: "Productivity of a Horizontal Well," *SPERE* (Nov. 1989) 417–421.

Beliveau, D.: "Heterogeneity, Geostatistics, Horizontal Wells and Blackjack Poker," *JPT* (Dec. 1995) 1068–1074.

Besson, J.: "Performance of Slanted and Horizontal Wells on an Anisotropic Medium," paper SPE 20965, 1990.

Borisov, J.P.: *Oil Production Using Horizontal and Multiple Deviation Wells*, Nedra, Moscow (1964). Translated by J. Strauss, S.D. Joshi (eds.), Phillips Petroleum Co., the R&D Library Translation, Bartlesville, OK (1984).

Buchsteiner, H., Warpinski, N.R., and Economides, M.J.: "Stress-Induced Permeability Reduction in Fissured Reservoirs," paper SPE 26513, 1993.

Clonts, M.D., and Ramey Jr., H.J.: "Pressure Transient Analysis for Wells with Horizontal Drainholes," paper SPE 15116, 1986.

Daviau, F., Mouronval, G., Bourdarot, G., and Curutchect, P.: "Pressure Analysis for Horizontal Wells," paper SPE 14251, 1985.

Deimbacher, F.X., Economides, M.J., Heinemann, Z.E., and Brown, J.E.: "Comparison of Methane Production from Coalbeds Using Vertical or Horizontal Fractured Wells," *JPT* (Aug. 1992) 930–935.

Dietz, D.N.: "Determination of Average Reservoir Pressure From Build-Up Surveys," *SPEJ* (June 1965) 117–122; *Trans.*, AIME, **261**.

Economides, M.J., Brand, C.W., and Frick, T.P.: "Well Configurations in Anisotropic Reservoirs," paper SPE 27980, 1994.

Economides, M.J., Deimbacher, F.X., Brand, C.W., and Heinemann, Z.E.: "Comprehensive Simulation of Horizontal Well Performance," *SPEFE* (Dec. 1991) 418–426.

Ehlig-Economides, C.A., Mowat, G.R., and Corbett, C.: "Techniques for Multibranch Well Trajectory Design in the Context of a Three-Dimensional Reservoir Model," paper SPE 35505, 1996.

Gilbert, W.E.: US patent no. 2,452,920.

Goode, P.A., and Thambynayagam, R.K.M.: "Pressure Drawdown and Buildup Analysis of Horizontal Wells in

Anisotropic Media," *SPEFE* (Dec. 1987) 683–697; *Trans.*, AIME, **283**.

Joshi, S.D.: "Augmentation of Well Productivity With Slant and Horizontal Wells," *JPT* (June 1988) 729–739.

Kuchuk, F.J., Goode, P.A., Brice, B.W., Sherrard, D.W., and Thambynayagam, R.K.M.: "Pressure Transient Analysis and Inflow Performance for Horizontal Wells," paper SPE 18300, 1988.

Longbottom, J.R.: "Development and Testing of a Multi-Lateral System," paper SPE 35545, 1996.

Matthews, C.S., Brons, F., and Hazebroek, P.: "A Method for Determination of Average Pressure in a Bounded Reservoirs," *Trans.*, AIME (1955) **204**.

Rehm, W.A.: US patent no. 4,415,205.

Retnanto, A. and Economides, M.J.: "Performance of Multiple Horizontal Well Laterals in Low-To Medium-Permeability Reservoirs," *SPERE* (May 1996) 73–77.

Retnanto, A., Frick, T.P., Brand, C.W., and Economides, M.J.: "Optimal Configurations of Multiple-Lateral Horizontal Wells," paper SPE 35712, 1996.

Salas, J.R., Clifford, P.J., and Jenkins, D.P.: "Multilateral Well Performance Prediction," paper SPE 35711, 1996.

Smith, J., Economides, M.J., and Frick, T.P.: "Reducing Economic Risk in Areally Anisotropic Formations with Multiple-Lateral Horizontal Wells," paper SPE 30647, 1995.

Warpinski, N.R.: "Hydraulic Fracturing in Tight and Fissured Media," *JPT* (Feb. 1991) 146–152.

4 Measurement-While-Drilling (MWD), Logging-While-Drilling (LWD), and Geosteering

Iain Dowell
Halliburton Energy Services

Andrew A. Mills
Esso Australia Ltd.

4-1 INTRODUCTION

No other technology used in petroleum well construction has evolved more rapidly than measurement-while-drilling (MWD), logging-while-drilling (LWD), and geosteering. Early in the history of the oilfield, drillers and geologists often debated environmental and mechanical conditions at the drill bit. It was not until advances in electronic components, materials science, and battery technology made it technically feasible to make measurements at the bit and transmit them to the surface that the questions posed by pioneering drillers and geologists began to be answered.

The first measurements to be introduced commercially were directional, and almost all the applications took place in offshore, directionally drilled wells. It was easy to demonstrate the savings in rig time that could be achieved by substituting measurements taken while drilling and transmitted over the technology of the day. Single shots (downhole orientation taken by an instrument that measures azimuth or inclination at only one point) often took many hours of rig time since they were run to bottom on a slick line and then retrieved. As long as MWD achieved certain minimum reliability targets, it was less costly than single shots, and it gained popularity accordingly. Achieving those reliability targets in the harsh downhole environment is one of the dual challenges of MWD and LWD technology. The other challenge is to provide wireline-quality measurements.

In the early 1980s, simple qualitative measurements of formation parameters were introduced, often based upon methods proven by early wireline technology. Geologists and drilling staff used short, normal-resistivity and natural gamma ray measurements to select coring points and casing points. However, limitations in these measurements restricted them from replacing wireline for quantitative formation evaluation.

Late in the 1980s, the first rigorously quantitative measurements of formation parameters were made. Initially, the measurements were stored in tool memory, but soon the 2-MHz resistivity, neutron porosity, and gamma density measurements were transmitted to the surface in real time. In parallel with qualitative measurements and telemetry, widespread use of MWD systems (combined with the development of steerable mud motors) made horizontal drilling more feasible and, therefore, more common.

Soon, planning and steering horizontal wells on the basis of a geological model became inadequate. Even with known lithology of offset wells and well-defined seismic data, the geology of a directional well often varied so significantly over the horizontal interval that steering geometrically (by using directional measurements) was quickly observed to be inaccurate and ineffective. In response to these poor results from geometric steering, the first instrumented motors were designed and deployed in the early 1990s. Recent developments in MWD and LWD technology include sensors that measure the formation acoustic velocity and provide electrical images of dipping formations.

Before this discussion of MWD and LWD technology, it should be understood that the terms *MWD* and *LWD* are not used consistently throughout the industry. Within the context of this chapter, the term *MWD* refers to directional and drilling measurements. *LWD* refers to wireline-quality formation measurements made while drilling.

The purpose of this chapter is to describe the rationale behind the design of current MWD, LWD, and geosteering systems, and to provide insight into the effective use of these systems at the wellsite.

4-2 MEASUREMENT-WHILE-DRILLING

Although many measurements are taken while drilling, the term *MWD* is more commonly used to refer to measurements taken downhole with an electromechanical device located in the bottomhole assembly (BHA). Normally, the capability of sending the acquired information to the surface while drilling continues is included in the broad definition of MWD. Telemetry methods had difficulty in coping with the large volumes of downhole data, so the definition of MWD was again broadened to include data that was stored in tool memory and recovered when the tool was returned to the surface. All MWD systems typically have three major subcomponents of varying configurations: a power system, a directional sensor, and a telemetry system.

4-2.1 Power Systems

Power systems in MWD may be divided into two general classifications: battery and turbine. Both types of power systems have inherent advantages and liabilities. In many MWD systems, a combination of these two types of power systems is used to provide power to the toolstring with or without drilling fluid flow, or during intermittent drilling fluid flow conditions.

Batteries can provide tool power without drilling-fluid circulation, and they are necessary if logging will occur during tripping in or out of the hole.

Lithium-thionyl chloride batteries are commonly used in MWD systems because of their excellent combination of high-energy density and superior performance at LWD service temperatures. They provide a stable voltage source until very near the end of their service life, and they do not require complex electronics to condition the supply. These batteries, however, have limited instantaneous energy output, and they may be unsuitable for applications that require a high current drain. Although these batteries are safe at lower temperatures, if heated

above 180°C they can undergo a violent, accelerated reaction, and explode with a significant force. As a result, there are restrictions on shipping lithium-thionyl chloride batteries in passenger aircraft. Even though these batteries are very efficient over their service life, they are not rechargeable, and their disposal is subject to strict environmental regulations.

The second source of abundant power generation, turbine power, uses what is available in the rig's drilling-fluid flow. A rotor is placed in the fluid stream, and circulating drilling fluid is directed onto the rotor blades by a stator. Rotational force is transmitted from the rotor to an alternator through a common shaft. The power generated by the alternator is not normally in an immediately usable form, since it is a three-phase alternating current of variable frequency. Electronic circuitry is required to rectify the alternating current (AC) to usable direct current (DC). Turbine rotors for this equipment must accept a wide range of flow rates so that multiple sets of equipment will not be required to accommodate all possible mud pumping conditions. Similarly, rotors must be capable of tolerating considerable debris and lost-circulation material (LCM) entrained in the drilling fluid. Surface screens are often recommended to filter the incoming fluid.

4-2.2 Telemetry Systems

Although several different approaches have been taken to transmit data to the surface, mud-pulse telemetry is the standard method in commercial MWD and LWD systems. Acoustic systems that transmit up the drillpipe suffer an attenuation of approximately 150 dB per 1000 m in drilling fluid (Spinnler and Stone, 1978). Advances in coiled tubing promise new development opportunities for acoustic or electric-line telemetry. Several attempts have been made to construct special drillpipe with an integral hardwire. Although it offers exceptionally high data rates, the integral hardwire telemetry method requires expensive special drillpipe, special handling, and hundreds of electrical connections that must all remain reliable in harsh conditions.

Low-frequency electromagnetic transmission is in limited commercial use in MWD and LWD systems. It is sometimes used when air or foam are used as drilling fluid. The depth from which electromagnetic telemetry can be transmitted is limited by the conductivity and thickness of the overlying formations. Some authorities suggest that repeaters or signal boosters positioned in the drillstring extend the depth from which electromagnetic systems can reliably transmit.

Three mud-pulse telemetry systems are available: positive-pulse, negative-pulse, and continuous-wave systems. These systems are named for the way their pulse is propagated in the mud volume.

Negative-pulse systems create a pressure pulse lower than that of the mud volume by venting a small amount of high-pressure drillstring mud from the drillpipe to the annulus. *Positive-pulse systems* create a momentary flow restriction (higher pressure than the drilling mud volume) in the drillpipe. *Continuous-wave* systems create a carrier frequency that is transmitted through the mud and encode data using phase shifts of the carrier. Positive-pulse systems are more commonly used in current MWD and LWD systems. This may be because the generation of a significant-sized negative pulse requires a significant pressure drop across the BHA, which reduces the hole-cleaning capacity of the drilling fluid system. Drilling engineers can find this pressure drop difficult to deliver, particularly in the extended-reach wells for which the technology is best suited. Many different data coding systems are used, which are often designed to optimize the life and reliability of the pulser, since it must survive direct contact from the abrasive, high-pressure mud flow.

Telemetry signal detection is performed by one or more transducers located on the rig standpipe, and data is extracted from the signals by surface computer equipment housed either in a skid unit or on the drill floor. Real-time detection of data while drilling is crucial to the successful application of MWD in most circumstances. Successful data decoding is highly dependent on the signal-to-noise ratio.

A close correlation exists between the signal size and telemetry data rate; the higher the data rate, the smaller the pulse size becomes. Most modern systems have the ability to reprogram the tool's telemetry parameters and slow down data transmission speed without tripping out of the hole; however, slowing data rate adversely affects log-data density.

The sources of noise in the drilling-fluid pressure trace are numerous. Most notable are the mud pumps, which often create a relatively high-frequency noise. Interference among pump frequencies leads to harmonics, but these background noises can be filtered out using analog techniques. Pump speed sensors can be a very effective method of identifying and removing pump noise from the raw telemetry signal.

Lower-frequency noise in the mud volume is often generated by drilling motors. As the driller applies weight to the bit, standpipe pressure increases; as the weight is drilled off, standpipe pressure is reduced. The problem is exacerbated when a polycrystalline diamond-compact (PDC) bit is being used. Sometimes, the noise becomes so great that even at the lowest data rates, successful transmission can only occur when bit contact is halted and mud flow is circulated off-bottom. Well depth and mud type also affect the received signal amplitude and width. In general, oil-based muds (OBMs) and pseudo-oil-based muds (POBM) are more compressible than water-based muds; therefore, they result in the greatest signal losses. This effect can be particularly severe in long-reach wells where OBM and POBM are commonly used for their improved lubricity. Nevertheless, signals have been retrieved without significant problems from depths of almost 9144 m (30,000 ft) in compressible fluids.

4-2.3 Directional Sensors

The state of the art in directional sensor technology for several years has been an array of three orthogonal flux-gate magnetometers and three accelerometers. Although in normal circumstances, standard directional sensors provide acceptable surveys, any application where uncertainty in the bottomhole location exists can be troublesome.

Extended-reach wells, by nature of their measured depth, can suffer significant errors in bottomhole location. Geographical locations where the horizontal component of the earth's magnetic field is small affect directional sensor accuracy the most. Typical worst-case scenarios are seen when drilling east to west in the North Sea or Alaska. Sag in BHA components in high-angle or horizontal wells can also cause a systematic directional error. Finally, diurnal variations in the earth's magnetic field, and local magnetic interference from BHA components can induce directional errors. Existing models for the prediction of errors in directional surveys were not designed for some of the extreme conditions encountered in today's drilling methods and well conditions.

Numerous methods of varying effectiveness are available to help correct raw magnetic readings for interference. Early corrections assumed that all interference was axial (along the drillstring's axis); more recent methods analyze for both permanent and induced interference on three axes. If magnetic readings can be corrected for variations in the diurnal field, even greater confidence can be placed in the accuracy of bottomhole location.

Along with uncertainties in the measured depth, bottomhole location uncertainties are one contributor to errors in the absolute depth. Note that all methods of real-time azimuth correction require raw data to be transmitted to the surface, which imposes load on the telemetry channel.

Gyroscope (gyro)-navigated MWD offers significant benefits over existing navigation sensors. In addition to greater accuracy, gyros are not susceptible to interference from magnetic fields. Current problems with gyro technology center upon incorporating mechanical robustness, minimizing external diameter, and overcoming temperature sensitivity.

4-2.4 MWD and LWD System Architecture

As MWD and LWD systems have evolved, the importance of customized measurement solutions has increased. The ability to add and remove measurement sections of the logging assembly as wellsite needs change is valuable, thus prompting the design of modular MWD/LWD systems. Methods for such design and operational issues as fault tolerance, power sharing, data sharing across tool joints, and memory management have become increasingly important in LWD systems.

In most cases, a natural division in system architecture exists when tool (drilling collar) ODs are $4\frac{3}{4}$ in. or less. Smaller-diameter tool systems tend to use positive-pulse telemetry systems and battery power systems, and are encased in a probe-type pressure housing. The pressure housing and internal components are centered on rubber standoffs and mounted inside a drill collar. Some MWD/LWD systems, although located in the drill-collar ID, are retrievable and replaceable, in case tool failure or tool sticking occurs. Retrievability from the drill collar in the hole often compromises the system's mounting scheme, and therefore, these types of systems are typically less reliable. Since the MWD string can be changed without tripping the entire drillstring, retrievable systems can be less reliable but cost-effective solutions.

In tool ODs of $6\frac{3}{4}$ in. or more, LWD systems are often turbine-powered since larger collar ODs enable optimal mud flow. When used with other modules, interchangeable power systems and measurement modules must both supply power and transmit data across tool joints. Often, a central stinger assembly protrudes from the lower collar joint and mates with an upward-looking electrical connection as the collar-joint threads are made up on the drillfloor. These electrical and telemetry connections can be compromised by factors such as high build rates

in the drillstring and electrically conductive muds. Recent MWD/LWD designs ensure that each module contains an independent battery and memory so that logging can continue even if central power and telemetry are interrupted. Stand-alone module battery power and memory also enable logging to be performed while tripping out of the hole.

4-2.5 Drilling Dynamics

The aim of drilling dynamics measurement is to make drilling the well more efficient and to prevent unscheduled events. Approximately 75% of all lost-time incidents over 6 hours are caused by drilling mechanics failures (Burgess and Martin, 1995). Therefore, extensive effort is made to ensure that both the drilling mechanics information acquired is converted to a format usable by the driller, and that usable data are provided to the rig floor.

The downhole drilling mechanics parameters most frequently measured are weight on bit (WOB), torque on bit (TOB), shock, and temperature. Downhole pressures, ultrasonic caliper data, and turbine speed can also be acquired. The data that are provided by these measurements are intended to enable informed, timely decisions by the drilling staff and to thereby improve drilling efficiency.

Stuck drillpipe is one of the major causes of lost time on the drilling rig. This condition is normally caused either by differential sticking of the drillstring to the borehole wall or by mechanical problems such as the drillstring becoming keyseated into the formation. Excessive friction applied to the string by a swelling formation can also cause downtime. Models have been developed that can calculate and predict an axial or torsional friction factor. These models are normalized to account for hole inclination and BHA configuration. A buildup in these friction factors usually suggests that preventive action (such as a wiper trip) is necessary. By directly comparing this information with the drilling history of adjacent offset wells, professionals can often determine whether problems are caused by formation characteristics or other mechanical factors such as bit dulling.

To have a positive effect on drilling efficiency, drilling dynamics must have a quick feedback loop to the drilling decision-makers on the rig site. An example of quick feedback loop benefits is found in the identification of bit bounce. As downhole processing power has increased, recent advances have made it possible to observe the cyclic oscillations in downhole weight-on-bit (Hutchinson *et al.,* 1995). If the oscillations exceed a predetermined threshold, they can be diagnosed as bit bounce and a warning is transmitted to the surface. The driller can then take corrective action (such as altering weight on the bit) and observe whether the bit has stopped bouncing on the next data transmission (Figure 4-1). Other conditions, such as "stick-slip" (intermittent sticking of the bit and drillstring with rig torque applied, followed by damaging release or slip) and torsional shocks can also be diagnosed and corrected.

Another means of acquiring drilling dynamics data through the use of downhole shock sensors, has become increasingly popular in recent years. Typically, these sensors count the number of shocks that exceed a preset force threshold over a specific period. This number of occurrences is then transmitted to the surface. Downhole shock levels can be correlated with the design specification of the MWD tool. If the tool is operated over design thresholds for a period, the likelihood of tool failure increases proportionally. A strong correlation, of course, exists between continuous shocking of the BHA

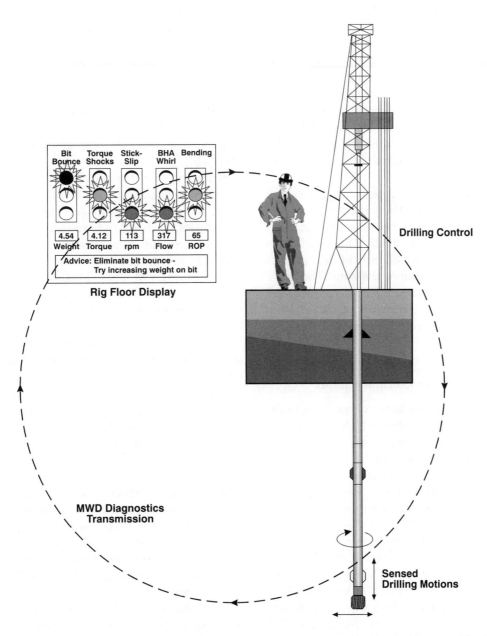

Figure 4-1 Downhole sensors provide useful drilling measurements when combined with a user-friendly display

and the mechanical failure that causes the drillstring to part. In most cases, lateral shock readings have been observed at significantly higher levels than axial shock (along the tool axis), except when jarring the drillstring. Multiaxis accelerometers are available and enable a more detailed analysis of shock forces.

Downhole pressure-measurement-while-drilling is an often misunderstood concept. Conventional formation testers, which isolate a section of formation from the borehole, are not currently available in "while-drilling" form. Pressure-measurement-while-drilling has proved valuable in extended-reach wells where long tangent sections may have been drilled. Studies performed on such wells have shown that hole cleaning can be difficult and that cuttings can build up on the lower side of the borehole. If this buildup is not identified early enough, loss of ROP and sticking problems can result. A downhole annulus pressure measurement can monitor backpressure while circulating the mud volume and, assuming that flow rates are unchanged, can precisely identify when a wiper trip should be performed to clean the hole. Pressure measurements can also help monitor and alter mud weight and optimize equivalent circulation density (ECD).

Historically, drilling dynamics measurements have not been commercially successful . Many rigsite staff choose to rely on experience gained rather than measurements made. A major reason for discounting the integrity of these systems is that the intelligence of the systems needs to be improved to prevent false alarms. The future of drilling dynamics measurements most probably lies with the MWD companies themselves as they demonstrate their product's effectiveness in integrated contracts.

4-2.6 Reliability and Environmental Factors

MWD systems are used in the harshest operating environments. Obvious conditions such as high pressure and temperature are all too familiar to engineers and designers. The wireline industry has a long history of successfully overcoming these conditions.

Most MWD tools can continue operating at tool temperatures up to 150°C. A limited selection of sensors is available with ratings up to 175°C. MWD tool temperatures may be 20°C lower than formation temperatures measured by wireline logs; this trend is caused by the cooling effect of mud circulation. The highest temperatures commonly encountered by MWD tools are those measured while running into a hole where the drilling fluid volume has not been circulated for an extended period. In cases such as this, it is advisable to break

circulation periodically while running in the hole. Using a Dewar flask to protect sensors and electronics from high temperatures is common in wireline, where downhole exposure times are usually short. Using flasks for temperature protection is not practical in MWD because of the long exposure times at high temperatures that must be endured.

The development of a broader range of high-temperature sensors in MWD is not governed by technical issues but rather by economic ones. Electronic component life above 150°C is significantly shortened, imposing high maintenance and repair costs of tools. System reliability in the 150°C to 175°C range can naturally be expected to suffer. Furthermore, battery-powered systems, such as Lithium alloy cells, have a lower energy density and hence shorter (but still adequate) life in this temperature range.

Downhole pressure is less a problem than temperature for MWD systems. Most toolstrings are designed to withstand up to 20,000 psi. The combination of hydrostatic pressure and system backpressure rarely approach this limit.

Shock and vibration present MWD systems their most severe challenges. Contrary to expectations, early tests using instrumented downhole systems showed that the magnitudes of lateral (side-to-side) shocks are dramatically greater than axial shocks during normal drilling. The exception to this rule is found when using jars on the drillstring. Modern MWD tools are generally designed to withstand shocks of approximately 500 G for 0.5 msec over a life of 100,000 cycles.

A more subtle, but no less destructive, force is caused by torsional shock. The mechanism that induces torsional shock, *stick-slip*, is caused by the tendency of certain bits in rare circumstances to dig into the formation and stop. The rotary table continues to wind up the drillstring until the torque in the string becomes great enough to free the bit. When the bit and drillstring break free, severe instantaneous torsional accelerations and forces are applied. If subjected to repeated stick-slip, tools can be expected to fail.

Many modern MWD devices are equipped with accelerometers that provide real-time measurements of the shock levels encountered and transmit these data to the surface. Drilling staff can then take remedial action to prevent either drillstring failure or MWD failure. No matter what preventive actions are taken however, some failures occur during drilling. A very high proportion of failures take place in the 5% of wells with the toughest environmental conditions. In these severe-condition wells, shock levels may exceed tool design specifications.

Early work done to standardize the measurement and reporting of MWD tool reliability statistics focused on defining a failure and dividing the aggregate number of successful circulating hours by the aggregate number of failures. This work resulted in mean-time between-failure (MTBF) number. If the data were accumulated over a statistically significant period, typically 2,000 hours, meaningful failure analysis trends could be derived. As downhole tools became more complex, however, the IADC published recommendations on the acquisition and calculation of MTBF statistics (Ng, 1989).

A common misunderstanding exists between MTBF (which may be quoted as 250 hours for a triple combo system) and the service interval (which may be quoted as 200 hours). The service interval refers to the point at which the device might normally be expected to fail because of battery-life exhaustion or seal failure if it is not replaced by a serviced tool. The MTBF, as its name implies, is a mean. Statistically, no more probability of a failure exists in Hour 250 than in Hour 25, if the system has been properly serviced and is running within design specifications.

Drawing a parallel between automobiles and MWD systems, one commonly cited example illustrates the reliability challenge faced by MWD. For example, an automobile has an MTBF of 350 hours and it is being driven at 60 mph. On average, it will travel 21,000 miles before breaking down, roughly the circumference of the earth at the equator. To be considered economically viable, MWD tools are expected to perform the equivalent of these automotive service figures without the benefit of the maintenance or refueling that is an everyday requirement in automobile operation.

4-3 LOGGING-WHILE-DRILLING

Perhaps more than any other service, the use of LWD and geosteering technology demands teamwork. Successful operations can often be traced back to good planning and communications among geology, drilling, and contractor staffs. If drilling staff members manage the contract, they may choose a device that is not well suited to the replacement of wireline in a particular environment. If geology staff fail to communicate well with the drilling staff, then the tool may be run in the BHA at a point that renders the data unusable. Important issues that must be discussed cross-functionally include the location of tools within the assembly, flow rates, mud types, and stabilization.

4-3.1 Resistivity Logging

The electromagnetic wave resistivity tool has become the standard of the LWD environment. The nature of the electromagnetic measurement requires that the tool be typically equipped with a loop antenna that fits around the OD of the drill collar and emits electromagnetic waves. The waves travel through the immediate wellbore environment and are detected by a pair of receivers. Two types of wave measurements are performed at the receivers. The attenuation of the wave amplitude as it arrives at the two receivers yields the attenuation ratio. The phase difference in the wave between the two receivers is measured, yielding the phase-difference measurement. Typically, these measurements are then converted back to resistivity values through the use of a conversion derived from computer modeling or test-tank data.

The primary purpose of resistivity measurement systems is to obtain a value of true formation resistivity (R_t) and to quantify the depth of invasion of the drilling fluid filtrate into the formation. A critical parameter in MWD measurements is formation exposure time (FET), the difference between the time the drill bit disturbs in-situ conditions and the time the sensors measure the formation. MWD systems have the advantage of measuring R_t after a relatively short formation exposure time, typically 30 to 300 minutes. Interpretation difficulties can sometimes be caused by variable formation exposure time, and logs should always contain at least one formation-exposure curve.

A knowledge of formation exposure time does not, however, rule out other effects. Figure 4-2 shows a comparison between phase and attenuation resistivity with an FET of less than 15 minutes and a wireline laterolog run several days later. Even the attenuation resistivity has been dramatically affected by invasion, reading about 10 ohm-m, whereas the true resistivity is in the region of 200 ohm-m.

Another example, shown in Figure 4-3, illustrates invasion effects in the interval from 2995 to 3027 m. Very deep invasion by conductive muds in the reservoir has caused the 2-MHz tool to read less than 10 ohm-m in a 200-ohm-m zone. Between 3058 and 3067 m, the deep invasion has caused the hydrocarbon-bearing zone to be almost completely obscured. Only by comparison with the overlying, deeply invaded zone from 2995 to 3027 m was this productive interval identified.

Similarly, LWD data density is dependent upon ROP. Good-quality logs typically have graduations or "tick" marks in each track to give a quick-look indication of measurement density with respect to depth.

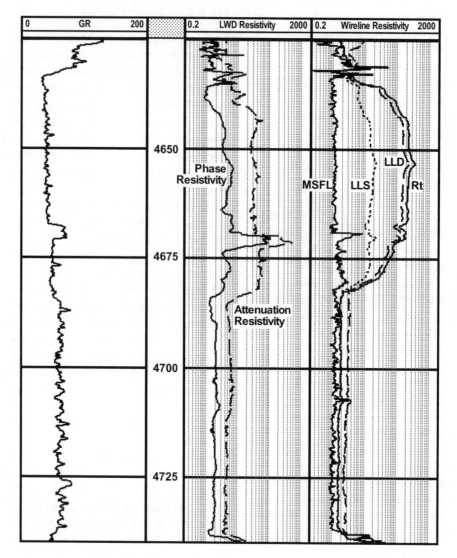

Figure 4-2 Comparison of EWR and wireline resistivity in a deeply invaded, high-permeability sandstone reservoir (EWR formation exposure time was 15 minutes compared with 3 days for the wireline laterolog.)

Early resistivity systems emphasized the difference between the phase and attenuation curves and suggested that one curve was a "deep" (radius of investigation) curve and another a "medium" curve. Difficulties with this interpretation in practice (Shen, 1991) led to the development of a generation of tools that derive their differences in investigation depth from additional physical spacings. A model of the measurement proportions from different areas around the borehole is shown in Figure 4-4. Note that the attenuation measurement often reads deeper into the formation but has less vertical resolution than the phase measurement. Figure 4-5 shows a "tornado" chart that can be used to evaluate the invasion of saline or fresh mud filtrates.

Identification and presentation of invasion profiles, particularly in horizontal holes, can lead to a greater understanding of reservoir mechanisms.

Many of the applications in which LWD logs have replaced wireline logs occur in high-angle wells. This trend leads to an emphasis in LWD on certain specialist-interpretation issues.

The depth of investigation of 2-MHz wave resistivity devices is dependent on the resistivity of the formation being investigated. Measurement response of a device (both phase and attenuation) with four different receiver spacings is shown in Figure 4-6. The region measured by the 25-in. (R25P) is based on a 25-in. diameter of investigation in a formation known to have a resis-

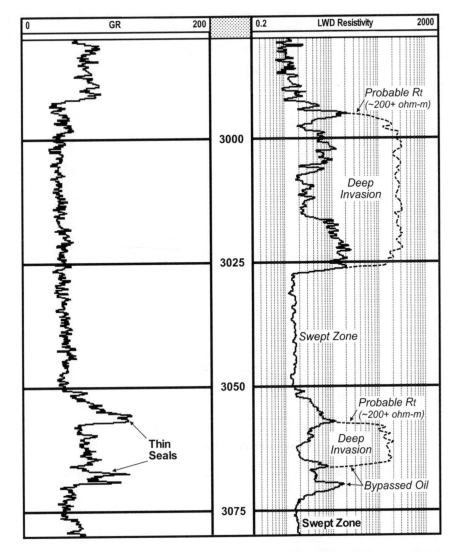

Figure 4-3 Effects of very deep invasion by conductive muds. The oil zone between 3058 and 3067 m was almost overlooked because its EWR resistivity response was masked by deep invasion

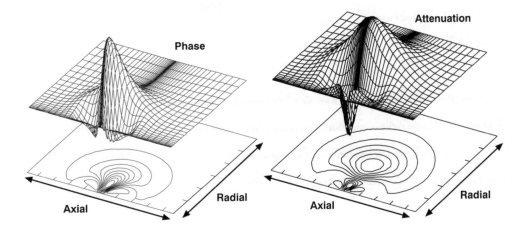

Figure 4-4 Contribution of zones around the borehole to the total measurement for phase and attenuation resistivities

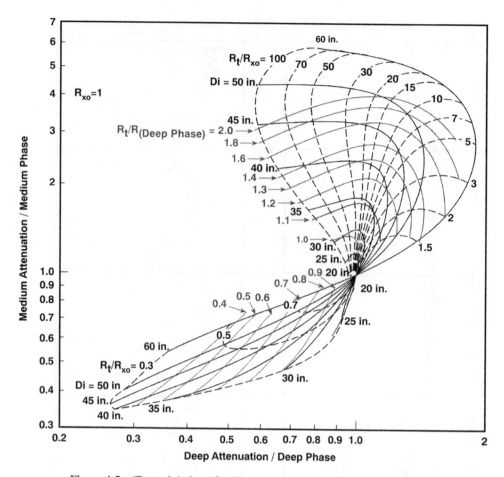

Figure 4-5 'Tornado' chart for the evaluation of saline or freshwater filtrates

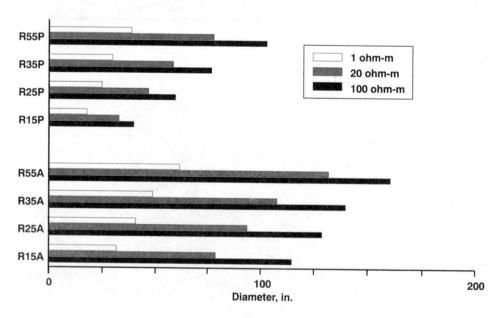

Figure 4-6   Depths of investigation for various spacings in formations of differing resistivity

tivity of 1 ohm-m. The phase measurement looks deeper (away from the borehole) and loses vertical resolution as the charts progress to greater resistivities. In contrast, the amplitude ratio at first looks deeper than the phase measurement, and the expected penalty of poorer vertical resolution is paid. In the most resistive case, the attenuation measurement shows a 129-in. diameter of investigation.

Dielectric effects are responsible for some discrepancies between phase and attenuation resistivity measurements. In defining the transform from the raw measurement to the calculated resistivities, certain assumptions are made about the formation dielectric constant. Dielectric constants are believed to be from 1 to 10. In shales and shaly formations, however, this assumption about dielectric constant is false; as a result, phase readings will read too low and attenuation readings will read too high. Errors are greatest in the most resistive formations.

Further discrepancies between phase and attenuation resistivity measurements may also be attributed to the effects of formation anisotropy. Anisotropy may also be responsible for the separation of measurements taken at different spacings or at different frequencies. Anisotropy effects are caused by differences in the resistance of formation when measured across bedding planes (R_v) or along bedding planes (R_h). An assumption is generally made that R_h is independent of orientation. As borehole inclination increases, the angle between the borehole and formation dip typically increases. When this relative angle exceeds about $40°$, resultant effects become significant. Anisotropy has the effect of increasing the observed resistivity over R_h. Effects are greater on the phase measurements than the attenuation measurements and greater on longer receiver spacings than short ones.

Wave resistivity tools are run in most instances where LWD systems are used, but toroidal resistivity measurements are desirable under some circumstances (Gianzero, 1985). Toroidal resistivity tools typically consist of a transmitter that is excited by an alternating current, which induces a current in the BHA. Two receivers are placed below the transmitter, and the amount of current measured exiting the tool to the formation between the receivers is the lateral (or ring) resistivity. The amount of current passing through the lower measuring point is the bit resistivity (Figure 4-7). Because of the large number of variables involved, bit resistivity measurements have been difficult to quantify although there are signs that this is changing.

In formations with high resistivities (greater than 50 ohm-m) or where thin-bed identification is important, measurements with a toroidal resistivity tool may be more appropriate than measurements with other tool types.

The log example in Figure 4-8 shows a case where 2-MHz measurements have saturated because of the high salinity of the mud. If the drilling fluid is conductive or if conductive invasion is expected, then a toroidal resistivity measurement is preferred. If early identification of a coring or casing point is crucial, then bit resistivity measurements give a good first look. In geosteering applications, toroidal bit resistivity measurements are an immediate indicator of a fault crossing.

The first formation images while drilling were acquired through the use of toroidal resistivity tools. When a small-button electrode is placed on the OD of a stabilizer, the current flowing through that electrode can be monitored. The current is proportional to the formation resistivity in the immediate proximity. Effective measurements are best taken in salty muds with resistive formations. Vertical resolution is 2 to 3 in. and azimuthal resolution is less than 1 in. (Rosthal *et al.*, 1996). With the tool rotating at least 30 rpm, internal magnetometer readings are taken and resistivity values are scanned and stored appropriately. The tool memory capacity is adequate for 150 hours of operation. At the surface, tool memory is dumped and the data are related to the correct depth. Quality checks are made to ensure that poor micro-depth measurements are not affecting the reading.

Imaging while drilling can provide a picture of formation structure, nonconformities, large fractures, and other visible formation features. Azimuthal density devices may also be processed to provide dip information. Research into acquiring and transmitting stratigraphic azimuth and dip while drilling is in progress.

4-3.2 Nuclear Logging

Gamma ray measurements have been made while drilling since the late 1970s. These measurements are relatively inexpensive although they require a more sophisticated surface system than is needed for directional measurements. Log plotting requires a depth-tracking system and additional surface computer hardware.

Applications have been made in both reconnaissance mode, where qualitative readings are used to locate a casing or coring point, and in evaluation mode. Verification of proper MWD gamma ray detector func-

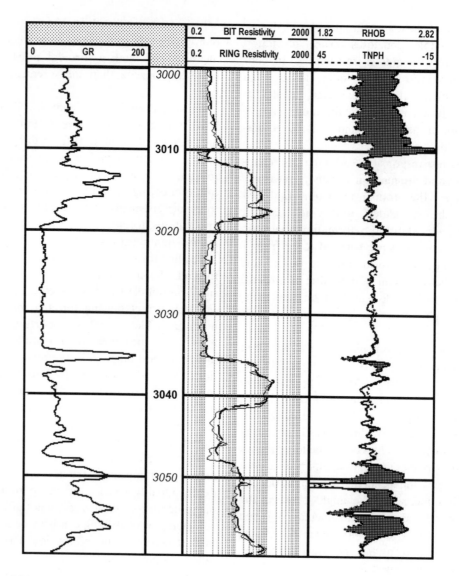

Figure 4-7 Ring and bit resistivity measurements showing good corroboration. Bit resistivity measurements have only recently become truly quantitative

tion is normally performed in the field with a thorium blanket or an annular calibrator (Brami, 1991).

The main differences between MWD and wireline gamma ray curves are caused by spectral biasing of the formation gamma rays and logging speeds (Coope, 1983).

Neutron porosity (ϕ_n) and bulk density (ρ_b) measurements in LWD tools are typically combined in one sub or measurement module, and they are generally run together. Reproducing wireline density accuracy has proved to be one of the most difficult challenges facing LWD tool designers. Tool geometry typically consists of a cesium gamma ray source located in the drill collar and two detectors, one at a short spacing from the source and one at a long spacing from the source. Gamma counts

arriving at each of the detectors are measured. Count rates at the receivers depend upon the density of the media between them. Density measurements are severely affected by the presence of drilling mud between the detectors and the formation. If more than 1 in. of standoff exists, the tendency of the gamma rays to travel the typically less dense mud path, thereby "short circuiting" the formation measurement path, becomes overwhelming. Solving the gamma ray short-circuit problem is accomplished by placing the gamma detectors behind a drilling stabilizer. With the detector mounted in the stabilizer in gauge holes, the maximum mud thickness is 0.25 in. and the mean mud thickness is 0.125 in. Response of the tool is characterized for various stand-

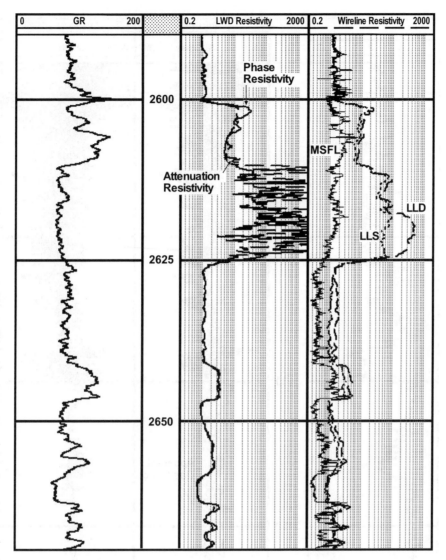

Figure 4-8 EWR phase and attenuation resistivity measurements saturate when run in high-salinity, water-based muds and formation resistivities exceed 50 to 100 ohm.

offs in various mud weights, and various formations and corrections are applied.

Placing the gamma detector in the stabilizer does have some drawbacks. Detector placement can affect the directional tendency of the BHA. In horizontal and high-angle wells in which the density measurement is most frequently run, the stabilizer can sometimes hang up and prevent weight from being properly transferred to the bit. It is important to note that in enlarged boreholes, gamma detectors deployed in the drilling stabilizer do not accurately measure density.

Assuming an $8\frac{1}{2}$-in. bit and an $8\frac{1}{4}$-in. density sleeve are used and the tool is rotating slowly in the hole, the average standoff is 0.125 in. and the maximum standoff is

0.25 in. If, however, the borehole enlarges to 10 in., the average standoff increases to 0.92 in., and the maximum standoff increases to 1.75 in. In big-hole conditions, very large corrections are required to obtain an accurate density reading. An example of an erroneous gas effect using older-generation density neutron devices in an enlarged $9\frac{7}{8}$-in. hole is shown in Figure 4-9.

Two approaches have been developed to obtain accurate density measurements in enlarged boreholes: an azimuthal density method and a constant standoff method. Azimuthal density links the counts to an orientation of the borehole by taking regular readings from a magnetometer (Holenka *et al.*, 1995). When this method is used, the wellbore (which is generally inclined) is divided into

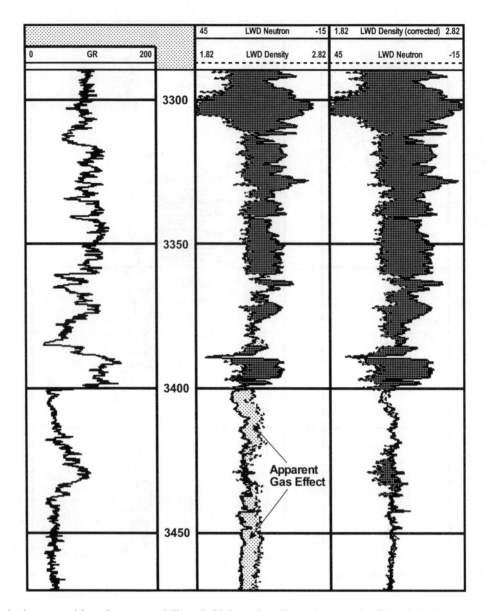

Figure 4-9   Density logs run with undergauge stabilizers in high-angle wells can be severely affected. Quality control checks must be run on each log to ensure correct measurement of formation properties.

four quadrants: bottom, left, right, and top. Incoming gamma counts are placed into one of the four bins. From this, four quadrant densities and an average density are obtained. A coarse "image" of the borehole can be obtained when beds of varying density arrive in one quadrant before another. Azimuthal density can be run without stabilization, but it relies on the assumption that standoff is minimal in the bottom quadrant of the wellbore.

The other method of obtaining density in enlarged boreholes relies on constant measurement of standoff using a series of ultrasonic calipers (Moake *et al.*,

1996). A standoff measurement is made at frequent intervals and a weighted average is calculated. High weight is given to gamma rays arriving at the detector when the standoff is low, and low weight is given to those gamma rays that arrive when the standoff is high (Figure 4-10). This method attempts to replicate the wireline technique of dragging a tool pad up the side of the borehole. The constant standoff method can also be applied to neutron porosity tools.

All density measurements suffer if the drillstring is sliding in a horizontal borehole with the gamma detectors pointing up (away from the bottom of the wellbore).

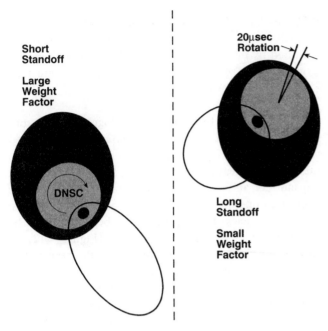

Figure 4-10 Use of ultrasonic measurements to compensate for mud effects

required compared to wireline (Figure 4-11). Standoff between the tool and the formation requires corrections of about 5 to 7 porosity units (p.u.) per inch. Borehole diameter corrections can range from 1 to 7 p.u./in., depending on tool design. Neutron porosity measurements are also affected by mud salinity, hydrogen index, formation salinity, temperature, and pressure. However, these effects are generally much smaller, requiring corrections of about 0.5 to 2 p.u.

Statistical effects are quite significant to nuclear measurements. Uncertainties increase as ROP increases. LWD nuclear measurements can be performed either while drilling or while tripping. Logging-while-drilling rates vary because of ROP changes, but they typically range from 15 to 200 ft/hr, whereas instantaneous logging rates can be significantly higher. Tripping rates can range from 1500 to 3000 ft/hr. Typical wireline rates are about 1800 ft/hr and constant. Statistical uncertainty in LWD nuclear logging also varies with formation type. In general, log quality begins to suffer increased statistical uncertainties at logging rates above

To overcome this problem, orientation devices are often inserted in the toolstring. As the BHA is being made up, the offset between the density sleeve and the tool face is measured. Adjusting the location of the orientation device allows the density measurement to be set to the desired offset. While the drillstring is sliding to build angle, the density detectors can be oriented downward by setting the offset to 180°.

LWD porosity measurements use a source (typically americium beryllium) that emits neutrons into the formation. Neutrons arrive at the two detectors (near and far) in proportion to the amount that they are moderated and captured by the media between the source and detectors. The best natural capture medium is hydrogen, generally found in the water, oil, and gas in the pore spaces of the formation. The ratio of neutron counts arriving at the detectors is calculated and stored in memory or transmitted to the surface. A high near/far ratio implies a high concentration of hydrogen in the formation and hence high porosity.

Neutron measurements are susceptible to a large number of environmental effects. Unlike wireline or LWD density measurements, the neutron measurement has minimal protection from mud effects. Neutron source/detector arrays are often built into a section of the tool that has a slightly larger OD than the rest of the string. The effect of centering the tool has been shown (Allen *et al.*, 1993) to have a dramatic influence on corrections

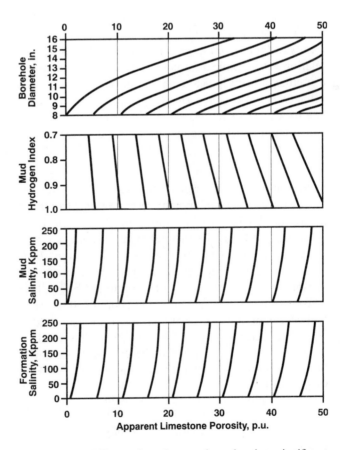

Figure 4-11 Effects of tool centering showing significant effects of corrections compared to wireline

100 ft/hr. This limits the value of logging while tripping to repeating formation intervals of particular interest.

4-3.3 Acoustic Logging

Ultrasonic caliper measurements while drilling have been introduced principally for improving neutron and density measurements. Caliper transducers consist of two or more piezoelectric crystal stacks placed in the wall of the drill collar. These transducers generate a high-frequency acoustic signal, which is reflected by a nearby surface, ideally the borehole wall. The quality of the reflection is determined by the acoustic impedance mismatch between the original and reflected signal. Often, there are restrictions in the quality of the caliper measurement in wells with high drilling fluid weights. The ultrasonic caliper measurement's sensitivity to gas has led some to suggest its use as a downhole gas influx detector (Orban *et al.*, 1991). Compared to the wireline mechanical caliper, the ultrasonic caliper provides readings with much higher resolution.

The major wireline measurement missing from the LWD suite until recently has been acoustic velocity. Acoustic data are important in many lithologies for correlation with seismic information. These data can also be a useful porosity indicator in certain areas. Shear-wave velocity can also be measured and can be used to calculate rock mechanical properties. Four main challenges in constructing a LWD acoustic tool are described below (Aron *et al.*, 1994):

- Preventing the compressional wave from traveling down the drill collar and obscuring the formation arrival. Unlike wireline tools, the bodies of LWD tools must be rigid structural members that can withstand and transmit drilling forces down the BHA. Therefore, it is impractical to adopt the wireline solution of cutting intricate patterns into the body of the tool to delay the arrival of the compressional wave. Isolator design is crucial and is still implemented to enable successful signal processing in a wide variety of formations, particularly the slower ones (those having a compressional delta (Δt_C) slower than approximately 100 μsec).

- Mounting transmitters and receivers on the OD of the drill collar without compromising their reliability

- Eliminating the effect of drilling noise from the measurement

- Processing the data so that it can be synthesized into a single (Δt_C) and so that this data point can be transmitted by mud-pulse. This is particularly challenging, given the large quantity of raw data that must be acquired and processed.

In its most basic form, an acoustic logging device consists of a transmitter with at least two receivers mounted several feet away. Additional receivers and transmitters enhance the measurement quality and reliability. The transmitters and receivers are piston-type piezoelectric stacks that operate in the 20-kHz range, far from drilling noise frequencies. Drilling noise has been shown to be concentrated in the lower frequencies (Figure 4-12). A data acquisition cycle is performed as the transmitter fires and the waveforms are measured and stored. Arrival time is measured from the time the transmitter fires until the wave arrives at each receiver. From this acoustic velocity information, the tool's downhole data processing electronics calculate the formation slowness, or Δt_C, using digital signal processing techniques. This value is the reciprocal of velocity and is expressed in units of μsec/ft. Waveforms are also stored in tool memory for later processing at the surface when the memory is dumped.

A development of the basic configuration is the compensated measurement (Minear *et al.*, 1995). In this transmitter/receiver array, an additional transmitter is mounted an equal distance on the other side of the receivers and a standoff transducer is added. The classical wireline advantages gained by compensating acoustic measurements are that the effects of sonde tilt and borehole washouts are virtually eliminated from the log. Even more significant in LWD than wireline is the fact that compensation provides redundancy in the measurement. An upper and a lower Δt_C are calculated. These two Δt_C values provide a good preliminary indication of the quality of the downhole processing. If these values are relatively equal, processing is more likely to be correct. Memory size is very important in LWD acoustic tools. Typically, LWD acoustic tools require 10 to 20 times the memory capacity of other LWD devices to accommodate waveform storage.

The log in Figure 4-13 shows an example of a log processed at the surface from waveforms stored downhole. Here, the Δt_C values have been reprocessed from the stored waveforms. When compared with a wireline log, this log is clearly less affected by the washout below the shoe and in the shale at X235 MD. LWD acoustic devices, by nature of their size, fill a much larger portion of the borehole than wireline devices and are less suscep-

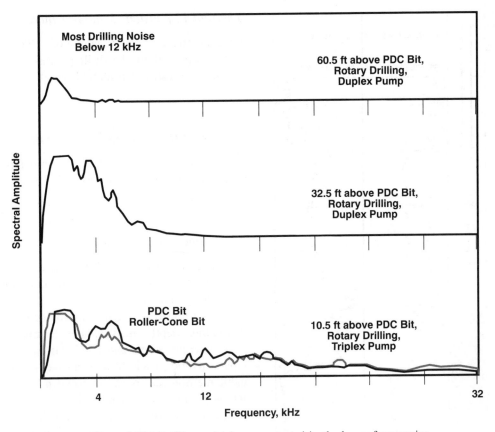

Figure 4-12 Drilling noise is concentrated in the lower frequencies

tible to the effects of borehole washout. The standoff measurement added to LWD acoustic tools can provide a useful indication of borehole conditions.

Synthetic seismograms can be produced when acoustic and density data are combined, which yield valuable correlations with seismic information. In certain acoustic applications, the shear-wave component can be extracted from the waveform and can be used to compute rock mechanical properties.

4-3.4 Depth Measurement

Good, consistent estimates of the absolute depth of critical bed boundaries are important for geological models. A knowledge of the relative depth from the top of a reservoir to the oil/water contact is vital for reserve estimates. Nevertheless, of all the measurements made by wireline and LWD, depth is one of the most critical, yet it is the one most taken for granted. Depth discrepancies between LWD and wireline have plagued the industry.

LWD depth measurements have evolved from mudlogging methods. Depth readings are tied, on a daily basis, to the driller's depth. Driller's depths are based on measurements of the length of drillpipe going in the hole and are referenced to a device for measuring the height of the kelly or top drive with respect to a fixed point. These instantaneous measurements of depth are stored with respect to time for later merging with LWD downhole memory data. The final log is constructed from this depth merge.

On fixed installations, such as land rigs or jackup rigs, a number of well-documented sources exist that describe environmental error being introduced in the driller's depth method. Floating rigs can introduce additional errors. One study suggested the following environmental errors would be introduced in a 3000 m well (Kirkman and Seim, 1989).

Drillpipe Stretch: 5- to 6-m Increase

The weight of the string itself causes the bit to be significantly deeper than it was measured on surface. An additional effect can be noted when the driller allows the

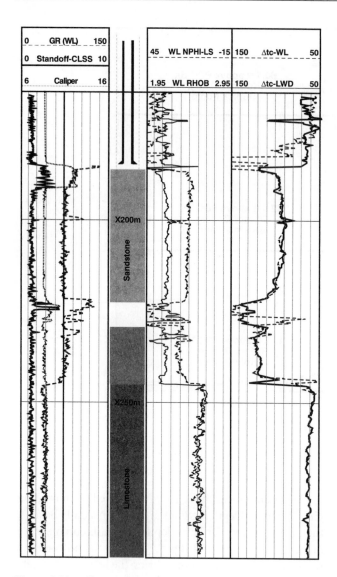

Figure 4-13   Comparison of wireline and LWD acoustic measurements showing that acoustical size minimizes washout problems

Thermal Expansion: 3- to 4-m Increase

Thermal effects, over the length of a 3000-m wellbore at 100°C higher than the drillpipe was measured on the surface, can cause considerable axial expansion.

Pressure Effects: 1- to 2-m Increase

Circulating pressures exerted on the drillpipe can cause collective axial length increase.

Ballooning: 2-m Decrease

The collective outward radial forces exerted on the ID of the drillpipe cause an overall contraction along the drillstring longitudinal axis.

In addition to the actual environmental effects on the drillstring, depth measurement techniques themselves have inherent errors associated with them. The two principal methods each have drawbacks. The measuring-cable method (geolograph wire) can be affected by high winds pulling more of the cable from the drum than is necessary. A second method, which uses an encoding device on the shaft of the draw-works, induces error because it cannot compensate for the number of cable wraps on the drum. Outer wraps have more depth associated with a revolution of the drum than inner wraps.

Floating rigs have special problems associated with depth measurements. Errors are principally caused by rig heave and by tidal action. In LWD, these effects are sufficiently overcome by the placement of compensation transducers on the relatively fixed rucker line.

Wireline measurements are also significantly affected by depth errors, as shown by the amount of depth shifting required between logging runs, which are often performed only hours apart. Given the errors inherent to depth measurement, if wireline and LWD ever tagged a marker bed at the same depth, it would be sheer coincidence.

Environmentally corrected depth would be a relatively simple measure to implement in LWD. Although this measure would certainly reduce depth errors, it most probably would not eliminate them. The "cost" of corrected depth is an additional depth measurement that must be monitored. The industry has yet to indicate that this additional measurement is merited. Running a cased-hole gamma ray during completion operations is a practice adopted by many operators as a check against LWD depth errors and lost data zones.

weight on bit to be drilled off. In this case, the bit may drill up to 2 m of formation without any measurable increase in depth at the surface. All new data recovered will be logged at the same depth.

Conversely, applying weight to the bit can lead to an apparent depth increase of up to 2 m as the drillpipe "squats" inside the hole. Pipe-squatting effects can cause a boundary to be logged at one depth by a density sensor; then, by the time the neutron porosity log is run over the same interval, additional weight may be applied to the bit, causing pipe squat to occur and the boundary bed to appear (erroneously) to have shifted downhole by 1 m.

4-4 GEOSTEERING

Although horizontal wells were occasionally drilled before the advent of MWD, the early 1990s saw a dramatic increase in horizontal activity. This drilling anomaly was a result of a combination of factors. Offshore, many of the structures installed during the late 1970s and early 1980s needed to tap fresh reserves to remain commercial. Previously bypassed formations began to look accessible and appealing with the introduction of horizontal completion techniques. The more widespread use of 3D seismic techniques identified multiple small targets that showed economic potential if produced with horizontal technology. Several authorities suggested that during the planning of a well, the question should not be, "Shall we drill a horizontal well?" but, "Why should we drill a vertical well?"

4-4.1 Geosteering Tools

Most early horizontal wells were drilled using MWD and steerable systems in the traditional way with measurement arrays located up to 80 ft behind the bit. Geologists created a prognosis and the well planner would provide a trigonometric well path. The directional driller would follow the planner's path and hope that it intersected the payzone. In thicker zones, trigonometric steering is still practiced successfully. However, in thinner zones (less than 15 ft thick) it was soon recognized that MWD geological measurements could help steer the wellbore to and through the payzone, thus maximizing well efficiency. *Efficiency* is defined as the percentage of the well path passing through the payzone divided by the well's total horizontal length. Efficiency is closely related to productivity, and one study in the North Sea suggested that the effect of reducing a net horizontal hole from 2000 ft to 1000 ft was a 30% productivity loss (Peach and Kloss, 1994).

A good definition of geosteering is "the drilling of a horizontal or other deviated well, where decisions on well path adjustments are made based on real-time geological and reservoir data." Biostratigraphy and analysis of relative hydrocarbons in the drilling-mud gases can also be used effectively where ROPs are low. If the ROP is sufficiently low and the sensors are located an excessive distance behind the bit, biostratigraphy and relative hydrocarbon data may arrive before real-time MWD.

To avoid the multiple difficulties of trying to steer the well from far behind the bit, instrumented motors were developed. These drilling motors have onboard sensors to measure resistivity, gamma, and inclination. The data

are sent back to the main mud pulser by a telemetry link, and the data are subsequently sent to the surface. Typical telemetry links currently used include a hardwire routed through the motor or electromagnetic transmission. Although acquiring data close to the bit is important, designers must be careful not to compromise either the predictability of the motor or its ability to change path. A BHA used for geosteering is shown in Figure 4-14.

Downhole adjustable stabilizers are often run in combination with extended-reach geosteering applications. The blade diameter of the adjustable stabilizer is addressable from the surface. Thus, inclination may be controlled without resorting to sliding the drill motor. Resultant lower friction gives the BHA the advantage of a greater reach.

Many geosteering authorities believe that the most important sensor in an instrumented motor is not a formation sensor, but the inclinometer. This belief in inclinometer importance is even more true now that data can be acquired and transmitted during rotation. Any deviation from the planned TVD can be instantly observed and corrected. Early deviation recognition reduces the

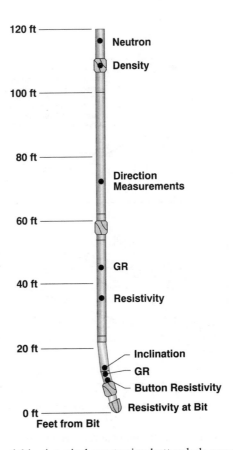

Figure 4-14 A typical geosteering bottomhole assembly

tortuosity of the wellbore and enables extended reach. Having inclination data at the bit immediately confirms that the corrective action was successful. It has been suggested that every foot the sensor is back from the bit leads to 2 ft in the recovery distance (Kashikar and Lesso, 1996). Other significant factors that affect the recovery distance are angle of incidence, reaction time, correction curve rate, hold distance, changes in structure, and curve distance to recovery (Figure 4-15).

The relative merits of the various formation measurements are application-dependent. All rely on a good contrast between different marker beds or fluids. In most wells, either gamma or resistivity can provide a good indication. The best type of resistivity measurement (toroidal or wave) may vary, although resistivity data should always be available at the bit. In some geologic areas, neutron porosity and density measurements are the primary steering tools.

Two different approaches are currently being taken to geosteer with resistivity. The first approach includes a shallow-reading measurement at the bit. The second integrates an electromagnetic wave reading into the motor, farther behind the bit.

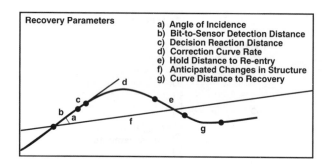

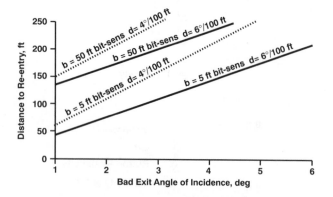

Figure 4-15 Factors affecting the re-entry of a payzone exited by mistake

The resistivity-at-bit method is an extension of the toroidal method described in Section 4-3.1. In water-based mud, the electrical current passes down the body of the mud motor and exits into the formation. In oil-based mud, current flow relies on direct contact with the formation, achieved through the bit teeth contacting the formation. When logging occurs with a toroidal tool in oil-based drilling fluid, an electromagnetic device is usually run farther up the hole for formation evaluation purposes. Often, difficulties arise in resolving differences between the two resistivity measurements. Toroidal measurements can detect an approaching conductive bed more readily than an approaching resistive boundary. A further refinement, applicable in water-based drilling fluids, is a small button electrode. The electrode is linked to the high side of the motor, and in water-based muds, it indicates whether the approaching bed is above or below the motor.

The second approach to geosteering with resistivity involves repackaging a standard wave resistivity measurement around an extended joint in the middle of the drilling motor. Very deep measurements can be made by altering the frequency of the transmission (from 2 MHz to 400 kHz). As the wellbore approaches a boundary, the resistivity reading will begin to deviate from its previous value. In most cases, it is unlikely that a change of less than 20% will be significant. Although this method does sense approaching beds from farther away, the depth of investigation may be counterproductive in thinner reservoirs. The electrode is necessarily 15 ft back from the bit and will not detect faults as quickly as a true at-bit measurement (Figure 4-16). Currently, no azimuthal measurements of wave resistivity are available. In practice, given the relatively high percentage of geosteered wells that are drilled with oil-based muds, azimuthal resistivity has a narrow application.

In the geosteering environment, measurement issues such as formation anisotropy, shoulder-bed effects, and polarization horns become concerns. These formation characteristics must be modeled for a variety of well trigonometries. These models are used later as an aid to real-time interpretation at the rigsite.

The response of gamma ray devices can be made azimuthal. Indeed, the off-center packaging of most instrumented motors dictates that the detectors will be more sensitive to an approaching bed on one side of the motor than the other. This effect can be exploited by shielding the detector and linking it to magnetometers so that either up, down, or quadrant information can be obtained. Azimuthal gamma ray measurements are prin-

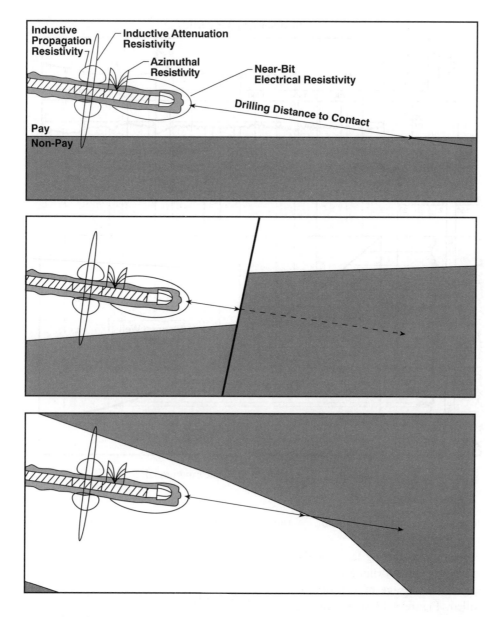

Figure 4-16 The advantages of having a deep measurement compared to a shallower measurement at the bit

cipally useful for indicating whether the drillstring has exited the top or the bottom of a bed.

4-4.2 Geosteering Methods

Cross-functional teamwork and planning are the keys to success in geosteering wells. Typical geosteering team participants include geoscientists, drilling staff, the geosteering coordinator, and the directional driller.

Given the input data, the responsibility of the geoscientist is to provide an expected geological structure for the well.

The geosteering coordinator constructs a series of contingency-forward models. These models predict the response of the formation sensors to a different well path throughout the payzone. Factors that affect the path include formation thickness, shoulder bed effects, and the size of polarization horns. Figure 4-17 shows an example of geosteering in which an offset induction-gamma ray log from a vertical well is used as input. The expected geosteering response at a given depth is predicted and displayed in the horizontal plane above the planned well trajectory. Two thin beds, evident in the wireline log above the main payzone, appear thicker

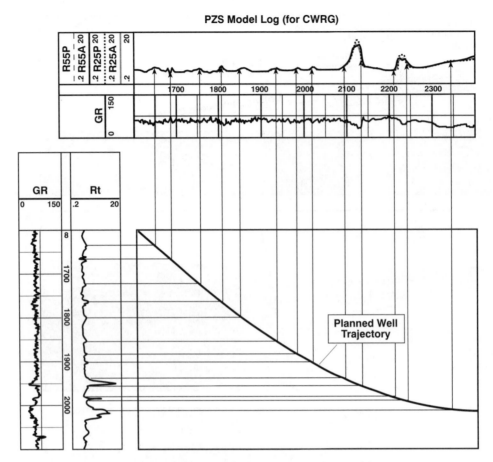

Figure 4-17   A typical payzone steering model

on the geosteered log because of the inclination of the well.

Time is a luxury that the geosteering team rarely has for making decisions. At a typical drilling rate of 30 ft/hr, the engineer receives an average of two datapoints per foot of hole drilled. Figure 4-18 shows the actual log transmitted, alongside the model. All the information correlates until the upper sand is exited. At this point, a rapid decision must be made about how to steer the well. Azimuthal testing can help determine which direction to drill. This testing involves placing azimuthal sensors above, at, and below the point in the wellbore where the unexpected event occurred. After taking a series of azimuthal measurements, the geosteering team can use this information to take the best remedial action to correct the geological model with known data. In practice, azimuthal data must be acquired and transmitted carefully since there is a possibility of washing out the hole or creating an inadvertent sidetrack in the wellbore. The effective presentation of data to make the decision-making process more

intuitive is one of the primary challenges facing geosteering development in the years to come.

4-5 USES OF LOG DATA

Log data, whether acquired by wireline or LWD technology, have a wide range of applications; the most common of which relate to evaluation of the properties of formations penetrated and the fluids they contain. Examples of these applications include

- Reservoir characterization (lithology, mineralogy, producibility)

- Identification of hydrocarbons

- Discrimination of reservoir fluid types (gas, liquid)

- Quantification of hydrocarbon volumes (porosity, saturation)

- Structural and stratigraphic studies

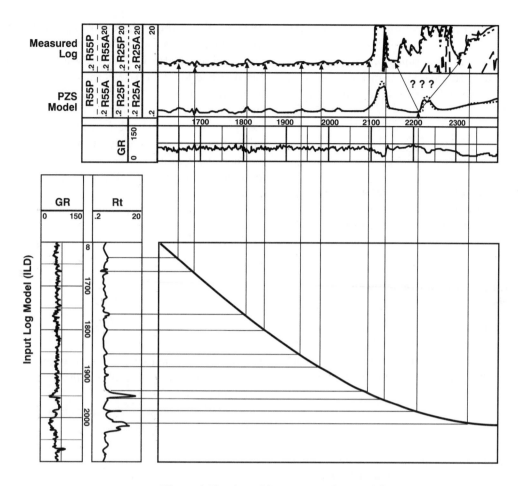

Figure 4-18   Actual log compared to model

● Evaluation of source rock and seal

All of these applications (and many more) fulfill critical roles in the task of finding and producing oil and gas. This section summarizes key aspects of the application of log data to the characterization of reservoirs and the identification and quantification of the hydrocarbons they contain. Although many different wireline logging tools have been designed to answer these questions, the following discussion will deal only with those tools that are currently in common LWD use and those that form a basic logging suite: gamma ray, resistivity, density, and neutron (Jackson and Heysse, 1994; Tait and Hamlin, 1996).

4-5.1 Reservoir Characterization

Perhaps the most fundamental of all issues addressed by log data is the discrimination of reservoir and nonreservoir rocks. In clastic rock sequences, this practice commonly entails using the gamma ray curve to separate sandstones (reservoir) from shales (nonreservoir). Shales contain high percentages of clays with associated radioactive potassium and thorium, giving rise to high gamma ray (GR) count rates. Sandstones, however, are commonly quartz-rich and result in relatively lower gamma ray counts (Figure 4-7). In certain environments, sandstones may contain high levels of other minerals, such as potassium feldspar, and mica which, because of their associated radioactive potassium, result in high gamma ray count rates. These "hot" sandstones may be mistaken for shales and require density and neutron logs to characterize correctly. (See Figure 4-7 between 3912 and 3919 m.)

Between these two end members is a continuum of shaly sandstones and siltstones that can contain abundant clay but can still contain hydrocarbons. In these reservoirs, it is imperative to quantify accurately the percentage of clay that occupies porespace, which reduces the porosity or storage capacity of the reservoir and dramatically reduces the permeability or flow capacity of the reservoir.

The percentage of clay in a sandstone is commonly referred to as the shale volume, (V_{sh}). In the absence of "hot" sandstones, shale volume may be calculated by using Equation 4-1:

$$V_{sh} = \frac{GR - GR_{min}}{GR_{max} - GR_{min}} \qquad (4\text{-}1)$$

where GR is the gamma ray log measurement, GR_{min} is the gamma ray measurement in clean (no clay) sandstone, and GR_{max} is the gamma ray measurement in shale.

4-5.2 Identification and Quantification of Hydrocarbons

All fluids in sedimentary rocks occur within the void space between the mineral grains, or matrix. Before the absolute volume of such fluids, whether water, oil, or gas can be quantified, this void space or porosity must first be quantified.

4-5.2.1 Porosity

The density-neutron tool combination is the primary source of porosity information used in LWD logging today because of its versatility as a porosity, gas, and shale indicator and the difficulty of making a sonic transit time measurement in the harsh drilling environment.

Before any quantitative work is done it is imperative to check that the logs are measuring correctly. To ensure accuracy, the logging contractor must compute neutron porosity in units that are compatible with the formation lithology limestone or sandstone. The density log should then be scaled so that it overlies the neutron in a clean (no clay), water-saturated reservoir. Typical scales appear in Table 4-1.

Where the lithology is sandstone but the neutron has been calculated in limestone units, the simple expedient of plotting density on a scale of 1.82 to 2.82 gm/cm^3 will achieve a similar overlay (Figure 4-7 between 3920 and

Table 4-1   Typical Scales

Lithology	Limestone		Sandstone	
Neutron (p.u.)	45	-15	60	0
Density (gm/cm^3)	1.95	2.95	1.65	2.65

3933 m). When this technique was used (Figure 4-9), the apparent gas effect in the known water-saturated sandstone below 3400 m in Track 2 was immediately recognized as bad log data. Use of the original data resulted in an overcalculation of porosity by 6.7 p.u., or a 29% error in volume. After the density log was corrected, the expected response was observed (Track 3), and the calculated porosity was consistent with surrounding wells.

Once the quality of the logs has been checked, the porosity can be calculated. If calculated with the appropriate matrix, neutron porosity (ϕ_n) can be read directly off the log, although care should be taken to check which corrections have been made to this log.

Density porosity (ϕ_d) can be calculated from the following equation:

$$\phi_d = \frac{\rho_{ma} - \rho_b}{\rho_{ma} - \rho_f} \qquad (4\text{-}2)$$

where ρ_{ma} is the matrix grain density (gm/cm^3), ρ_b is the bulk density (gm/cm^3), and ρ_f is the formation fluid density (gm/cm^3).

Formation porosity is then estimated by averaging density and neutron porosities:

$$\phi_t = \frac{\phi_d + \phi_n}{2} \qquad (4\text{-}3)$$

The porosity calculated is the total porosity (ϕ_t) available for fluid storage. In clastic rocks, total porosity can be broadly subdivided into macro- and micro-porosity.

Macro-porosity is commonly associated with the term "effective" porosity (ϕ_e), which has many definitions, all of which attempt to define the percentage of the total porosity that is available to the storage of moveable fluids; therefore, this pore volume is capable of storing producible hydrocarbons.

Micro-porosity, often termed "ineffective" porosity, is that percentage of the total porosity that is filled with immovable formation water, bound in place by a range of physical and chemical processes. This pore volume (ϕ_{sh}) is commonly associated with clay and other fine-grained particles; this volume is unavailable to hydrocarbon storage.

Effective porosity may be calculated using the following equation:

$$\phi_e = \phi_t - (V_{sh} \times \phi_{sh}) \qquad (4\text{-}4)$$

Where ϕ_t is the total porosity (fraction), ϕ_e is the effective (macro-) porosity (fraction), ϕ_{sh} is the shale (micro-) porosity (fraction), and V_{sh} is the shale volume (fraction).

4-5.2.2 Water Saturation

In clean formations, (total) water saturation can be calculated using the Archie equation:

$$S_{wt}^n = \frac{R_w a}{R_t \phi_t^m} \qquad (4\text{-}5)$$

where S_{wt} is the percentage of the porosity filled with water (fraction), R_w is the formation water resistivity (ohm-m), R_t is the true formation resistivity (invasion corrected, ohm-m), a is the tortuosity ($\sim$1), m is the cementation factor ($\sim$2), and n is the saturation exponent ($\sim$2).

Equation 4-5 breaks down in formations of shaly or complex lithologies. To calculate total water saturation in these reservoirs, more complex equations such as dual water equation or the Waxman-Smits equation are required to account for excess conductivity related to clays or other conductive minerals (Dewan, 1983; Serra, 1984). Once total water saturation has been calculated in these formations, it is commonly divided into effective and ineffective water saturation, on the assumption that hydrocarbon can be reservoired only in effective (macro-) porosity, whereas water occurs in both (Figure 4-19). This saturation can be calculated by using the equation

$$\phi_e(1 - S_{we}) = \phi_t(1 - S_{wt}) \qquad (4\text{-}6)$$

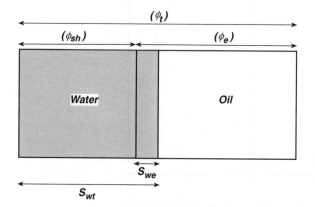

Figure 4-19 Hydrocarbon saturation in complex lithologies is assumed to occur only in the effective porosity first

which can be rewritten as

$$S_{we} = 1 - \left((1 - S_{wt}) \times \frac{\phi_t}{\phi_e} \right) \qquad (4\text{-}7)$$

where ϕ_t is the total porosity (fraction), ϕ_e is the effective (macro-) porosity fraction, S_{wt} is water saturation in total porosity (fraction), and S_{we} is water saturation in effective porosity (fraction).

4-5.3 Formation Evaluation with LWD Instead of Wireline

Intensive efforts have been made by LWD contractors to design and manufacture reliable tools that will provide measurements that are representative of formation properties. In addition, operators have gone to great expense running LWD/wireline comparisons to ensure consistency of these measurements. Despite these efforts, differences still remain because of the differing formation exposure times and logging environment confronted by LWD and wireline tools, and the differing technologies that have been used. In addition, the industry is now using LWD to drill and log high-angle, extended-reach wells never before contemplated with wireline technology.

The following is a summary of the observed differences between LWD and wireline data that could lead to errors in quantifying formation properties.

4-5.3.1 Depth Control

As discussed in Section 4-3.4, a difference of 5 to 10 ft between LWD and wireline depth systems is common and may take the form of incremental stretch or squeeze or the loss or gain of entire sections of data. Before any quantitative work can be attempted, it is essential that all logs be exactly on depth. Therefore, a cased-hole gamma ray log should be run immediately after the casing so that any significant depth discrepancies can be identified. These depth discrepancies will affect each detector at different depths because of their different locations on the BHA relative to the GR which is used for checking with the cased-hole GR.

A further complication to LWD depth control arises when mixing circumferential and azimuthal measurements in formations at high apparent dip to the wellbore. This complication occurs when water saturation is calculated with a circumferential EWR resistivity measurement and azimuthal, down-quadrant, density-neutron data. The azimuthal tools will "see" bed boundaries sooner (and sharper) than the circumferential tool that averages the resistivity measurement from both above and below the bed boundary (Figure 4-20). This effect becomes progressively worse at higher apparent dips; as the wellbore becomes horizontal, the data may become unusable for quantitative analysis.

4-5.3.2 Gamma Ray

Gamma ray repeatability is a function of count rates that are related to detector volume and logging speed. The statistical repeatability of the LWD GR measurement is similar to that of the wireline tool, since the much lower count rates of LWD GR detectors (because of

their reduced size) are largely compensated for by the much slower logging speed.

LWD GR detectors are more sensitive to gamma rays from potassium since the drill collars attenuate uranium and thorium gamma rays more than potassium gamma rays. While this effect, called spectral biasing, is generally less than 25%, it is most pronounced in shales and potassium-rich intervals. Although this effect is likely to be small when GR is used quantitatively, it may result in errors in calculated volumes of clay or potassium feldspar, which will affect porosity and saturation calculations.

Other sources of apparent difference between the wireline and LWD GR measurements are borehole effects. The hole diameter is commonly different between LWD

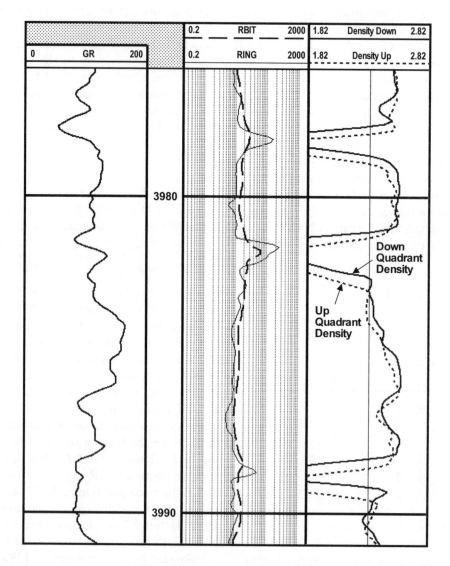

Figure 4-20 Azimuthal tools run in highly deviated wells suffer depth shift, which must be corrected before quantitative analysis

and wireline logging, and corrections are made to different standard conditions.

4-5.3.3 Resistivity Logs

The resistivity measurement is affected by a complex array of physical properties of the formation, the wellbore, and the measurement system as discussed in Section 4-3.1. It is often difficult to resolve which effect is causing observed differences between LWD and wireline measurements, since all effects may be contributing to differing degrees. In these situations, resistivity may have to be computer-modeled to quantify and correct for these effects.

Frequency

Resistivity is sensitive to the frequency at which it is measured. Because wireline tools operate at lower frequencies than LWD EWR devices, they tend to measure higher resistivity. Apparent formation water resistivity calculated from the LWD resistivity log is often inconsistent with that from wireline logs. At these higher operating frequencies (2 MHz), LWD tools are also more affected by dielectric effects than wireline tools. Above 20 ohm-m, dielectric properties become significant.

Anisotropy

Archie's equation (Equation 4-5) requires that resistivity be measured parallel to the formation bedding (R_h). At dips as low as 45°, LWD resistivities can read 10 to 20% higher than R_h in shales and other anisotropic formations. Quantitative formation evaluation with LWD resistivity at relative dips greater than 60° will certainly require anisotropy correction. This effect is much smaller on wireline than LWD because of the lower operating frequency, and it can often be ignored.

Invasion

LWD data are commonly acquired within 1 to 4 hours of drilling, whereas wireline logging commonly occurs 1 to 4 days after drilling. Although LWD resistivity tools are relatively shallow reading devices, the shallower depth of invasion at the time of LWD logging usually enables these devices to measure true formation resistivities. However, examples in high-permeability reservoirs as illustrated in Figures 4-2 and 4-3 demonstrate that this is not always true, and the logs must be carefully scrutinized to ensure that no hydrocarbon zones have been overlooked. Figure 4-3 is an excellent example of this

problem. The lower, deeply invaded zone (3060 m) was identified after careful comparison with the overlying known oil zone. An updip well was drilled to target bypassed oil trapped between the thin seals and discovered several million barrels that would not have been produced by existing wells. Invasion correction charts are only now becoming available to address this issue.

4-5.3.4 Density-Neutron

Standoff

Standoff is a common area of discrepancy between LWD and wireline porosity data. Standoff is commonly estimated from bit size, ultrasonic caliper, or a caliper derived from near and far detector count rates. In deviated wells where a change in wellbore trajectory may require sliding rather than rotating, detectors may be facing away from the formation, and standoff corrections may be incorrectly applied. All density-neutron logs should have the zones of sliding and rotating drillstring clearly marked and the method of standoff correction should be annotated.

Clay Hydration

Extended periods of exposure to water-based drilling fluid systems before wireline logging can cause shale hydration. Clays within the shales absorb water; depending on their mineralogy, they can swell between 2 and 50 times their original volume. This phenomenon causes wireline tools to record lower formation densities in shales than the densities recorded by LWD tools.

If hydration continues, it can cause sloughing and borehole stability problems. Washouts and hole rugosity can then severely degrade wireline porosity data, especially density data, which relies on pad contact with the borehole wall.

Invasion

Invasion also affects the gas effect shown on density-neutron logs, and it may either enhance or reduce that effect depending on the depth of invasion at the time of logging and the relative depth of investigation of the particular tools.

REFERENCES

Allen, D.F., Best, D.L., Evans, M., and Holenka, J.M.: "The Effect of Wellbore Condition on Wireline and MWD Neutron Density Logs," *SPEFE* (March 1993), 50–58.

Aron, J., Chang, S.K., Dworak, R., Hsu, K., Lau, T., Masson, J.P., Mayes, J., McDaniel, G., Randall, C., Kostek, S., and Pluna, T.J.: "Sonic Compressional Measurements While Drilling," *Trans.,* 1994 SPWLA Annual Logging Symposium, paper SS.

Brami, J.B.: "Current Calibration and Quality Control Practices for Selected Measurement While Drilling Tools," paper SPE 22540, 1991.

Burgess, T.M., and Martin, C.A.: "Wellsite Action on Drilling Mechanics Information Improves Economics," paper SPE 29431, 1995.

Coope, D.F.: "Gamma Ray Measurements While Drilling," *The Log Analyst* (Jan-Feb 1983) 39.

Dewan, J.T.: *Essentials of Modern Open-Hole Log Interpretation,* Pennwell, Tulsa, OK (1983).

Gianzero, S. *et al.:* "A New Resistivity Tool for Measurement While Drilling," *Trans.,* 1985 SPWLA Annual Logging Symposium, 3–22.

Holenka, J., Best, D., Evans, M., Kurkoski, P., and Sloan, W.: "Azimuthal Porosity While Drilling," *Trans.,* 1995 SPWLA Annual Logging Symposium, paper BB.

Hutchinson, M., Dubinski, V., and Henneuse, H.: "An MWD Assistant Driller," paper SPE 30523, 1995.

Jackson, C.E., and Heysse, D.R.: "Improving Formation Evaluation by Resolving Differences Between LWD and Wireline Log Data," paper SPE 28428, 1994.

Kashikar, S.V., and Lesso, W.G.: "Principles and Procedures of Geosteering," paper SPE 35051, 1996.

Kirkman, M., and Seim, P.: "Depth Measurement with Wireline and MWD Logs," paper presented at 1989 SPWLA meeting Bergen, Norway, reprinted in SPE Reprint No. 40, "Measurement While Drilling" (1989) 27–33.

Minear, J., Birchak, R., Robbins, C., Linyaev, E., Mackie, B., Young, D., and Malloy, R.: "Compressional Slowness Measurements While Drilling," *Trans.,* 1995 SPWLA Annual Logging Symposium, paper VV.

Moake, G., Beals, R., and Schultz, W.: Reduction of Standoff Effects on LWD Density and Neutron Measurements," *Trans.,* 1996 SPWLA Annual Logging Symposium, paper V.

Ng, F.: "Recommendations for MWD Tool Reliability Statistics," paper SPE 19862, 1989.

Orban, J.J., Dennison, M.S., Jorion, B.M., and Mayes, J.C.: "New Ultrasonic Caliper for MWD Operations," paper SPE 21947, 1991.

Peach, S.R., and Kloss, P.J.L.: "A New Generation of Instrumented Steerable Motors Improves Geosteering in North Sea Horizontal Wells," paper SPE 27482, 1994.

Rosthal, R.A., Young, R.A., Lovell, J.R., Buffington, L., and Arceneaux, C.L.: "Formation Evaluation and Geological Interpretation From the Resistivity at Bit Tool," paper SPE 30550, 1996.

Serra, O.: *Fundamentals of Well Log Interpretation: Vol. 1—The Acquisition of Logging Data,* Elsevier, Amsterdam (1984).

Shen, L.C.: "Theory of a Coil-Type Resistivity Sensor for MWD Application," *Log Analyst* (Sept.-Oct. 1991) 603–611.

Spinnler, R.F., and Stone, F.A.: "Mud Pulse Logging While Drilling Telemetry System Design, Development and Demonstrations," paper presented at the 1978 IADC Drilling Technology Conference.

Tait, C.A., and Hamlin, K.H.: "Solution to Depth-Measurement Problems in LWD," paper presented at the 1996 Energy Week Conference and Exhibition, Houston, Jan. 29–Feb. 2.

5 Drilling Fluids

Mario Zamora and Mike Stephens,
M-I L.L.C.

5-1 INTRODUCTION

Drilling fluid is the lifeblood of modern drilling operations. An oil or gas well simply cannot be drilled without continuous circulation of the drilling fluid to facilitate drilling the hole and to preserve the hole until it can be protected by casing. Although drilling-fluid expenditures average less than 8% of tangible costs, the drilling fluid is vital to the overall success of any well-construction project.

Drilling-fluid technology is dominated by three factors: performance, economics, and environmental concerns (Figure 5-1). Underlying these factors is the element of risk, an integral part of every drilling operation. Compromises are often required. Clearly, the principal task of the drilling fluid is to contribute to the overall success of the well-construction process, with particular emphasis on reservoir evaluation, well productivity, and ultimate hydrocarbon recovery. However, drilling fluids must also be affordable and must not harm the natural environment nor the health and safety of rig workers.

Finding the right balance among these factors is the great challenge both for the drilling-fluids industry, which provides this technology, and for the drillers who use it. The focus of this chapter is to assist in these endeavors. Basic drilling-fluid concepts are included, particularly when they form the foundation for current practices and/or help establish future technical and market trends.

5-2 HISTORICAL PERSPECTIVE

Drilling fluids have evolved with the petroleum industry. In 1901, the discovery well of the Spindletop field on the Texas Gulf Coast marked the first use of the rotary drilling method still in use today. The Spindletop discovery well also introduced the use of drilling fluids, planting the seed for what has grown into the drilling-fluids industry of today. The first modern drilling fluid was created by running a herd of cattle through an earthen slush pit. Drillers used the resulting muddy water to control a potentially catastrophic quicksand problem (Clark and Halbouty, 1980). This unusual event gave drilling fluids their well-known nickname, "drilling mud," or more simply, "mud."

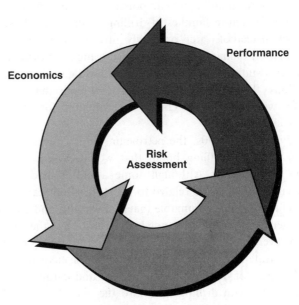

Figure 5-1 Major factors of drilling-fluid selection

The new-found capability of drilling soft and unconsolidated formations rapidly helped spread the use of rotary drilling (Darley and Gray, 1988). Hole stability was achieved by plastering the hole walls with clay materials contained in formation cuttings or added to the mud at the surface. Generally overlooked until years later, however, was the contribution of mud weight (density) to hole stability and, more importantly, to subsurface pressure control. In the early 1920s, barite (ground barium sulfate ore) was selected from a group of high-specific-gravity minerals as the best mud-weight additive for use in pressured formations. Barite was inert in water, not too abrasive, and readily available. Other important mud additives followed. Bentonite clay, mined in the state of Wyoming, was recognized for its superior mud-making properties and became the primary additive for viscosity, solids suspension, and filtration control. On the basis of tonnage, barite and bentonite remain the two most common mud additives in use today (Bacho, 1994).

Increasing demands for petroleum promoted new drilling challenges. Drilling-fluids technology kept pace. Considerable resources were allocated to improve the understanding of mud chemistry, to apply new testing methods, to refine field procedures, and to develop special additives to prevent and correct mud-related problems. Innovative mud systems, some radically different, emerged to satisfy ever-expanding technical requirements and lower costs. Air and natural gas were exploited to increase drilling rates in hard, dry formations. Natural and synthetic polymers became the foundation for entire families of drilling fluids. Muds based on oil instead of water, originally introduced to minimize formation damage, later evolved into high-performance systems suitable for use in the most hostile drilling conditions. Technology had advanced enough that cost-effective mud systems could be formulated for almost any application.

By the mid-1970s, the petroleum industry was in an extraordinary drilling boom that would peak early in the next decade. Although some attention had been given to safety issues related to the storage and handling of corrosive and flammable materials, health and environmental concerns were not high priorities at the time. However, in 1978, both U.S. federal and state agencies, in separate moves, resolved to minimize the environmental impact of discharging whole mud and cuttings contaminated with drilling mud into the sea (Bleier *et al.*, 1993). In addition to performance and cost, drilling operators now had to consider environmental concerns when selecting a drilling fluid.

High-performance, environmentally friendly, synthetic-based muds, introduced in the early 1990s, have arguably made the biggest impact. Despite unit costs two to eight times higher than conventional fluids, impressive performance/cost ratios and environmental acceptability helped establish synthetic-based fluids as the best choice for some critical wells.

Today's rapidly changing technical and business climates have further intensified the focus on drilling fluids and their role in well construction. Currently, the performance of nonconventional wells, including horizontal, multilateral, extended-reach, deepwater, slimhole, and designer wells depends, more than ever, on the overall effectiveness and efficiency of the drilling fluid. Special alliances involving operators, rig contractors, mud companies, and chemical suppliers continue to open new avenues for cooperative research.

5-3 BASIC FUNCTIONS OF A DRILLING FLUID

The three primary drilling-fluid functions are to (1) transport drilled cuttings and cavings to the surface, (2) control subsurface pressures, and (3) support, preserve, and protect the open hole until casing is run and cemented. Additionally, the mud should promote high penetration rates, cool and lubricate the bit and drillstring, buoy the weight of the drillstring and casing, and help obtain information on subsurface formations.

While providing these functions, the drilling fluid should not create side effects that could compromise the well-construction process. Specifically, the drilling fluid must not damage productive formations, endanger the health and safety of personnel, contaminate the environment, nor corrode or otherwise harm the integrity of the drilling equipment. Often, concerns for these side effects become dominant issues in mud-related decisions.

Drilling mud cannot perform its functions unless it remains fluid, stable, and usable, despite continual exposure to subsurface contaminants and hostile conditions. Highly reactive drill solids (especially colloidal-sized particles), corrosive acid gases, saltwater flows, evaporites, and cement are common contaminants. Hostile conditions include high temperatures and pressures.

All basic drilling-fluid functions must essentially be satisfied in each well, regardless of whether the well is a shallow land well or an ultra-extended-reach offshore well. In some cases, such as the control of subsurface pressures, compromise is not an option. Other priorities, however, may change, depending on the goals and conditions of the particular drilling operation. For example, hole cleaning and formation damage are especially critical for horizontal wells. Mud lubricity (for minimum torque and drag), barite sag, penetration rates, wellbore stability, and drilling-fluid hydraulics are among the key concerns for extended-reach drilling. Low mud toxicity is paramount for wells in environmentally sensitive land and offshore areas.

5-4 FORMATION DAMAGE AND WELL PRODUCTIVITY

The fluid used to drill the production zone can have an important impact on well productivity. A reduction in the productive capacity of the well by drilling, completion, and/or production activities is termed *formation damage*. Issues of formation damage can be quite complex because they involve elements of the reservoir formation, the completion techniques that are used, and the composition and properties of the drilling and completion fluids. Filtration control, bridging, and filtrate chemistry are among the most important drilling-fluid properties for minimizing or limiting formation damage. Additionally, drilling fractured formations or drilling horizontal wells may provide special challenges to obtain undamaged wells.

5-4.1 Filtration Control

The abilities of the drilling fluid to bridge pore openings, form a suitable filter cake, and limit the invasion of filtrate are important for limiting the amount of formation damage that occurs. Drilling fluids are commonly slurries consisting of fine solids suspended in a liquid. When a permeable rock is drilled, a phenomenon known as *filtration* occurs. The first step in the filtration process is the bridging of pore openings in the formation by particles in the fluid that cannot enter the pore structure of the rock because they are too large to pass into the pores. The bridging process causes a filter cake to form, which traps the solid particles from the drilling fluid and prevents them from entering the pore network of the rock. As filtration occurs, solids collect in the filter cake, and filtrate enters the pore structure of the rock.

Completion methods are important to any discussion of formation damage because wells completed with casing, cement, and perforations are much less impaired by damage from drilling fluids than wells that are completed openhole, with slotted liners, or with prepacked screens.

Most vertical wells are completed by perforations in clear brine completion fluids. The perforation tunnel typically penetrates the solids damage from the drilling fluid and may even penetrate any damage caused by mud filtrate. If the perforations penetrate the damaged region caused by drilling fluid, then the drilling fluid will not significantly impact well productivity.

Several methods are available for estimating how deeply filtrate invades the reservoir formation. One method uses resistivity or induction logging tools that measure properties of the formation at various depths of investigation. In an oil or gas reservoir, the electrical conductivity of the filtrate-invaded zone will differ from the conductivity deeper into the reservoir. From these electric logs, a general estimate can be made of how deeply mud or filtrate has penetrated into the rock. A second method of estimating the depth of penetration uses the fluid-loss properties of the mud under downhole conditions to make a calculation of filtrate invasion depth (Stephens, 1993). Using geometric considerations and the assumption that the dynamic fluid-loss rate over time is equal to the rate of fluid loss measured in a static 30-min fluid-loss test, the invasion radius, r_s (cm), is

$$r_s = \sqrt{\left(\frac{4.13 \times 10^{-2} DFt}{\phi} + \frac{D^2}{4}\right)} - \frac{D}{2} \qquad (5\text{-}1)$$

where D is well diameter in cm, F is the 30-min fluid loss in cm^3 when standard fluid-loss equipment is used at reservoir temperature and downhole pressure differential, t is exposure time in hours, and ϕ is porosity as a fraction. The invasion depths in Table 5-1 were calculated for a 22.2-cm diameter well into a reservoir with 0.25 porosity and a fluid loss of 5 cm^3/30 min.

While in many cases perforations penetrate 30 cm or more into the reservoir, the depth of penetration depends on mechanical properties of the reservoir rock and the type of perforating device used. Thus, reducing filtration is one of the most important methods of limiting formation damage caused by drilling fluids. Because the amount of filtration that occurs increases with increased

Table 5-1 Filtrate invasion vs. time ($D = 22.2$ cm, $\phi = 0.25$, $F = 5$ cm^3/30 min)

t	24 hr	48 hr	72 hr	96 hr
r_s	12.6 cm	20.6 cm	26.9 cm	32.3 cm

time of exposure of the formation to the fluid, it is generally wise to minimize this exposure.

Filtration control is achieved in drilling fluids by the addition of bentonite clay, fluid-loss control polymers, asphalts, resins, lignite, or other additives that help build a thin, low-permeability filter cake. The additives depend on the mud composition and specific well requirements. A drilling fluid's filtration properties are routinely measured while the fluid is being used to drill the well. It is common practice to reduce the filtration rate of the drilling fluid before the reservoir formation is drilled.

5-4.2 Bridging, Leakoff, and Spurt Loss

While electric-log estimates of invasion depth and calculated estimates from geometrical considerations are often in fairly good agreement, sometimes the inferred invasion depth from logs is much deeper than could be caused by filtration alone. Leakoff of whole mud can occur in certain zones. In fact, some leakoff loss occurs in most wells before the filter cake forms; this loss is called *spurt* loss. A spurt loss is often recorded when the filtration properties of drilling fluid are measured. Both leakoff and spurt losses from drilling fluids can add to the invasion depth and, more importantly, transport fine particles from the drilling fluid into the formation. These fine particles can reduce permeability in the near-wellbore region.

Many cases of extreme formation damage observed during well testing of high-permeability formations are caused by the leakoff of whole mud into the reservoir or by extremely large spurt losses. The fundamental reason that leakoff occurs is that particles in the drilling fluid are too small to bridge the pore openings in the formation. Leakoff should be expected in formations with extremely large openings, such as fractures or vugs. If the drilling fluid retains a low viscosity and does not gel into a solid or semisolid, the drilling fluid can be produced from the vugs or fractures as the well is brought into production. Generally, low-solids fluids are preferred for drilling fractured or vuggy formations.

Most reservoir formations are characterized by matrix permeability instead of fractures or vugs. The leakoff of drilling fluid to formations with matrix permeability occurs when particles in the drilling fluid are too small to bridge the pore openings in the formation. When leakoff occurs, whole drilling fluid is injected into the pore space. The pore space in these formations is not homogeneous in size, so fluid leaks off into the larger pores.

The drilling fluid can then undergo filtration into the smaller pores and solidify, blocking the larger pore network in the formation. This process can significantly reduce the permeability of the reservoir immediately surrounding the wellbore and can severely reduce production rates. To some degree, spurt losses act similarly to leakoff, but the depth of invasion of spurt losses from the mud is limited.

One method to reduce leakoff and minimize the spurt losses from drilling fluids is to ensure that the drilling fluid contains particles of sufficient size to bridge the pore openings in the formations being drilled. The proper particle size for the drilling fluid is a function of the pore-opening diameter in the formation. Pore-opening size can be measured using standard petrographic thin sections from core samples and a microscope. Alternative means of pore-size measurement (such as mercury-injection porosimetry) are likely to miss the largest pore diameters in some formations because of the extremely low threshold pressure for injection into large-diameter pores. When specific measurements of pore diameter are not available for sands, the bridging particle size in microns can be estimated based on the square root of the permeability in millidarcies.

For bridging the pore openings at the rock surface, particle diameters larger than one-third to one-half the pore diameter are typically needed. For most sandstone reservoirs, this particle size is in the silt size range (5 to 74 μm). Bridging particles included in this size range include barite, calcium carbonate, sized salt, and oil-soluble resins. Except for barite, these materials are available in custom grind sizes that can be closely matched to reservoir needs. In addition, fibrous cellulose additives can be used as bridging additives to reduce leakoff. The minimum amount of bridging particles needed to prevent leakoff or excessive spurt losses is generally 1 to 3% by volume. If large particles are present in the fluid, then a distribution of finer particle grades is needed to help form an impermeable filter cake that prevents fine particles from entering the formation with the filtrate.

Reducing leakoff and minimizing spurt losses is important in preventing formation damage from drilling fluids in most drilling operations. Suitable particle-size distributions in drilling fluids can be achieved by the ordinary addition of clay and barite and the incorporation of drilled solids into the fluid. As the well is drilled, solids-control equipment removes solids from the drilling fluid. This equipment includes screens, hydroclones, and centrifuges, and tends to remove larger particle sizes from the drilling fluid. Intense processing by solids-control equipment, especially centrifuges, removes the bridging silt-sized particles from fluid. Drilling the reservoir formation with a highly processed drilling fluid can result in excessive spurt losses or leakoff because the fluid is depleted in the particle size range needed for bridging. This problem can be solved by adding 1 to 3% by volume of a suitable bridging agent immediately before the production zone is drilled.

5-4.3 Filtrate Chemistry

In most drilling operations, drilling-fluid filtrate penetrates beyond the depth of the perforations. The chemical makeup of the filtrate can interact with reservoir components to reduce the permeability of the affected zone. Stephens (1991a) has highlighted the need to characterize the reservoir fluid and mineral components so that filtrate chemistry can be designed to prevent formation damage.

Gas, oil, and brine comprise the reservoir fluids. Gas frequently contains carbon dioxide, and in reservoirs with high carbon dioxide content, calcium in drilling-fluid filtrate may precipitate as calcium carbonate scale, reducing the reservoir permeability. In addition, highly alkaline drilling-fluid filtrate can shift carbonate chemical equilibria to increase the concentration of carbonate ion in the water phase. The increase in dissolved carbonate ion can result in the precipitation of calcium or magnesium carbonate scale. Younger, less mature crude oils contain significant quantities of organic acids. These organic acid components of crude can react with highly alkaline mud filtrate to form surface-active compounds that may alter reservoir wettability or enable an emulsion to form. Alteration of formation wettability from preferentially water-wet to preferentially oil-wet decreases the relative permeability to oil, and an emulsion block that forms in the pore space can reduce permeability in the near-wellbore region.

Connate formation brines are present in the reservoir along with the hydrocarbons. These connate brines contain dissolved ions that vary from one reservoir to another. Individual ionic components of connate brine can precipitate a variety of scales when mixed with drilling-fluid filtrate. Unless specifically formulated to exclude certain ions, drilling-fluid filtrates are likely to contain sodium, potassium, calcium, magnesium hydroxide, carbonate, sulfate, and chloride ions. Common components of connate formation brine that lead to precipitation of scale are calcium, magnesium, sulfate, and carbonate.

The interaction between clay mineral components of some reservoirs and drilling-fluid filtrate can reduce the permeability of the filtrate-invaded zone. Some clay-containing reservoirs are relatively unaffected by interactions with mud filtrate while others are highly sensitive to filtrate chemistry. Drilling-fluid filtrate can disperse clay that is present in sandstones into the pore space, allowing it to migrate and block off pore channels to flow. Freshwater filtrate and high-alkalinity filtrate are most likely to cause problems with clay in the reservoir section. Drilling-fluid additives for minimizing swelling or the migration of clays include cationic clay-inhibition agents, potassium salts, and sodium chloride. Special mud types that minimize the dispersion and swelling of clays include non-dispersed polymer fluids, low-alkalinity fluids, oil-based drilling fluids, and synthetic-based drilling fluids.

5-4.4 Horizontal Wells

The preceding discussion of filtration, bridging, and filtrate chemistry applies to formation damage in all types of wells. For vertical or deviated wells that are completed by cementing casing through the production interval and perforating the casing with a completion fluid in the well, these considerations are adequate for drilling-fluid design.

Horizontal wells, however, are usually completed as open holes, slotted liners, or prepacked screens, and are more impaired by damage from drilling fluids. In addition to performing its usual functions, a drilling fluid for horizontal wells should be

- Nondamaging to the formation permeability

- Compatible with the completion method that will be used for the horizontal well

- Responsive to any stimulation or clean-up techniques that will be used

The term *drill-in fluid* is applied to specialized fluids for drilling the production interval. A drill-in fluid minimizes formation damage to the reservoir formation and ensures that the completion and clean up can be performed so that the well produces at the maximum rate.

Selection of the appropriate drill-in fluid for a horizontal well is strongly constrained by the type of permeability present in the reservoir and the manner in which the well will be completed (Stephens, 1991b). The flowchart in Figure 5-2 outlines the types of completions used for horizontal wells and the basic logic for selecting different

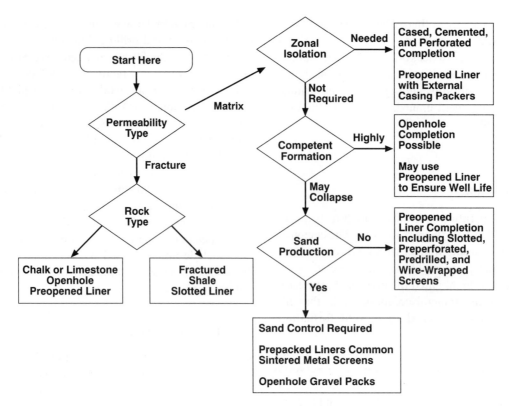

Figure 5-2   Flowchart for horizontal well completions

completion methods. The first consideration is the type of permeability in the reservoir: fracture or matrix. Fracture permeability commonly occurs in chalks, limestones, and shales. Matrix permeability is the intergranular permeability that typically occurs in sandstones, but may also occur in limestones and dolomite.

5-4.4.1 Fractured Reservoirs

Horizontal wells are commonly used to exploit fractured reservoirs because properly oriented horizontal wellbores are more likely to intersect multiple vertical and near-vertical fractures than conventional wells. Austin Chalk and many fractured limestone reservoirs are in very competent rock formations that are typically completed open-hole, although "preopened" liners with external casing packers are sometimes used for zonal isolation. A preopened liner is usually a slotted liner but can include predrilled, preperforated, and even some wire-wrapped liners. These liners are primarily used to ensure that the wellbore remains open throughout the life of the well, although some preopened liners may have a limited ability to minimize sand production.

For the drilling of fractured chalk or limestone reservoirs, a very low-solids drill-in fluid is preferred because

the fluid that invades fractures can be removed by production. High-solids loading in fluids can cause mud that enters the fracture system to "gel up." Gelled mud may be impossible to remove. Some fractured limestones or chalks also have a low matrix permeability so that fluid loss can occur from the fracture. Fluid-loss control may be important for such zones to prevent a filter cake from plugging the fractures. Austin Chalk and other fractured limestones are sometimes drilled underbalanced with solids-free brines. Underbalanced drilling permits oil to be produced as the well is being drilled, and the oil is separated from the solids-free drill-in fluid at the surface.

Fractured shales present a somewhat different problem than fractured chalks and limestones. Freshwater-based fluids can react with clay minerals in shale formations, causing them to swell and close the fractures. Oil-based, invert-emulsion drill-in fluids have been successfully used for drilling in the fractured Bakken shale and have resulted in high production rates. This shale does not swell in oil-based fluids. Inhibitive water-based systems that use potassium chloride have also been used on Bakken shale wells, but well productivity has been lower than the levels experienced with oil-based drill-in fluids. Fractured shales are usually completed with a slotted liner to protect the well from cavings.

Table 5-2 shows formation types, completion methods, suitable drill-in fluids, and clean-up or stimulation treatments for horizontal wells into vertically fractured formations.

5-4.4.2 Matrix Permeability

A wide variety of completion methods is used for horizontal wells in formations with matrix permeability, such as sandstones. Drill-in fluid selection for wells with matrix permeability depends on the type of completion (Stephens, 1994). The flowchart in Figure 5-2 shows the completion methods used for horizontal wells. If zonal isolation is needed, a cased, cemented, and perforated completion may be performed. Preopened liners with external casing packs and sliding-sleeve isolation devices have also been used for zonal isolation. If zonal isolation is not needed and the reservoir formation is highly competent mechanically, then an openhole completion could be used. (Some operators do not use openhole completions even for competent formations, but use preopened liners to ensure well life even for remote possibilities that wellbore collapse could occur.) If the formation is not highly competent and does not produce sand, then a preopened liner completion will be used. A very limited amount of sand control can be obtained with narrow slots in a slotted liner or a simple wire-wrapped screen. If sand production is likely to occur, then a sand-control completion will be required. Usually for horizontal wells, a prepacked liner or a sintered metal screen will be used. These filters can prevent sand from filling the liner or production tubing. In a few instances, openhole gravel packs have been placed in horizontal wells. However, these latter techniques, while successful in excluding sand, may also become damaged with precipitous reduction in well performance. (See Chapters 18 and 19 for sand-production control techniques.)

For cased, cemented, and perforated horizontal wells, a wide range of water-based or oil-based fluids can be used successfully. The perforations ordinarily penetrate the skin of drill-in fluid solids surrounding the wellbore, making the perforated completion relatively insensitive to the type and amount of drill-in fluid solids. For cased and cemented wells to achieve zonal isolation, it is critical to clean debris and cuttings from the well prior to running casing. Any accumulation of debris on the bottom of the wellbore will interfere with a good cement job. In addition, any hole that is larger than gauge will be difficult to clean. Careful attention to hole-cleaning properties and potential wellbore erosional properties of the drill-in fluid are essential for fluids that will be used in horizontal wells that will be completed with casing, cement, and perforations.

Horizontal wells in competent formations with matrix permeability are often completed open-hole or "barefoot," with no liner through the production zone. Completions with preopened and prepacked liners share many characteristics with openhole completions. These wells differ from perforated completions because near-wellbore skin damage from the drill-in fluid affects production. To prevent damage in openhole completions, the bridging properties of the drill-in fluid must be excellent. Leakoff or spurt losses from the drill-in fluid should be minimized.

The ability to remove the filter cake formed by the drill-in fluid can be important for obtaining an undamaged openhole completion. Low drawdown pressures associated with horizontal well production may not provide adequate "lift-off" pressure for removing filter cakes. Remedial clean-up procedures are sometimes planned to remove formation damage caused by filter-cake solids. Several methods can make the drill-in fluid more responsive to chemical clean-up or stimulation. Acid-soluble calcium carbonate or water-soluble sized-salt bridging particles can make the filter cake more responsive to clean-up. Organic polymers that can be degraded by oxidation, acid, or enzymes are frequently used in the place of clay-based viscosifiers to ensure bet-

Table 5-2 Vertically fractured reservoir drilling-fluid selection chart for horizontal wells (from Stephens, 1994).

Formation Type	Completion Method	Drill-in Fluids	Cleanup/Stimulation
Fractured Limestone and Chalk	Open Hole Slotted Liner	Clear Water Polymer/Water/Salt Very Low Solids	Acid Treatment
Fractured Shale	Slotted Liner	Oil-Based Fluid Inhibitive Water-Based Fluid	

ter clean-up. For horizontal wells with openhole completions, the ability to wash, scratch, and scrape the filter cake to assist in removal can be important factors in obtaining an undamaged well.

Preopened liner completions are common in horizontal wells. Like openhole wells, these completions require excellent bridging. The size of the openings in the preopened liners is large enough that bridging particles from the drill-in fluid can be produced through the opening. The preopened liner restricts access to the well face and limits the ability to scrape or scratch once the preopened liner is in place. The ease with which the filter cake can be removed or degraded becomes critical for preventing damage when preopened liners are used. Organic polymer-based fluids allow oxidative washes that help remove the filter cake. Acid-soluble calcium carbonate solids are frequently used in drill-in fluids for preopened liners so that acid can be used for clean-up if necessary.

Reservoir formations that require sand control make special demands on the drill-in fluid. To prevent production of sand, horizontal wells in these reservoirs are completed with prepacked screens, sintered metal screens, or openhole gravel-packs to prevent sand production. Openhole gravel-packs are used infrequently, and present more clean-up problems than prepacked liners or sintered metal screens. Prepacked liners have a consolidated sand pack that provides an in-depth filter surrounding a preopened liner. Sintered metal screens are made of porous sintered metal that is several millimeters thick; these screens provide a porous medium surrounding the liner. Both prepacked liners and sintered metal screens can be plugged by drill-in fluid particulate materials. For this reason, acid-soluble calcium carbonates or water-soluble sized-salts are often used as bridging agents for drill-in fluid in wells that will be completed with prepacked liners or sintered metal screens. During the completion, these bridging agents can be dissolved to clean up the filter cake without damaging the screen. Several alternatives to using soluble particles exist. One alternative is to control the size of bridging particles carefully so that they can be produced through the screen or pack. The disadvantage of using only fine or ultrafine particles is that the bridging properties of the fluid may be compromised. Because unconsolidated sands form an important class of reservoirs for horizontal drilling projects, the development of specialized drill-in fluids that can be used with prepacked liners and other sand control devices is an active area of drilling-fluids research. One innovation is to use acid-sensitive emulsifiers and wetting agents for oil and synthetic fluids that allow acids to convert bridging solids to a water-wet state so acids

can dissolve the solids more effectively. This innovation could result in the wider use of oil- and synthetic-based drill-in fluids. Another technique involves using a "solids-free" drill-in fluid that does not contain bridging particles but achieves filtration control using the rheology of a polymer solution that can be broken with oxidizing solutions (Svoboda, 1996).

Openhole gravel-pack completions provide perhaps the greatest challenge to the development of adequate drill-in fluids. The problems of providing good protection to the formation in a lengthy horizontal interval while simultaneously preventing damage to an openhole gravel-pack are significant. The drill-in fluid must provide filtration control during drilling. Leakoff control must also be provided while the gravel pack is being emplaced. Once the gravel pack is in place, reaching the filter cake with chemical solutions that can break down polymers or dissolve solids becomes difficult. Fluids based on ultrafine bridging solids that can be produced through the gravel pack, self-dissolving bridging agents (Hodge *et al.*, 1995), or "solids-free" polymer loss-control additives may be beneficial in the future. Formation damage issues from drill-in fluids with openhole gravel-pack completions have not yet been fully resolved.

Table 5-3 shows the completion methods and characteristics of suitable drill-in fluids and clean-up or stimulation methods available for horizontal wells with matrix permeability.

5-5 TYPES OF DRILLING FLUIDS

The drilling process relies on successively protecting and isolating troublesome intervals through the use of casing strings that are run in the hole and cemented in place using a telescopic scheme. For most wells, drilling conditions change and, consequently, drilling-fluid properties also change, sometimes significantly, for each casing interval. Drilling efficiency depends largely on matching the drilling fluid to the formations being drilled. In many cases, one drilling fluid is completely displaced by another fluid formulated for different drilling conditions expected in the next interval. No single formulation is suitable for all situations. The drilling fluid of choice depends on specific well requirements (Hutchison and Anderson, 1974).

Drilling-fluid types usually are grouped according to their primary component. Four basic types of drilling fluids are available (Figure 5-3): water-based, oil-based, synthetic-based, and pneumatic. Synthetic-based muds

Table 5-3   Sand and sandstone reservoir drilling-fluid selection chart for horizontal wells (from Stephens, 1994)

Completion Method	Drill-In Fluid	Cleanup/Stimulation Fluid
Cased Hole With Perforation	Conventional Drilling Fluid Elevated Low-Shear-Rate Rheology	Acid Frac Fluids
Open Hole	Bridging Particles Polymer Additives	Acid Polymer Breakers
Preopened Liner	Bridging Particles Polymer Additives Easily Removed Filter Cake	Acid Polymer Breakers
Prepacked Liner Or Screen	Bridging Particles Polymer-Based Fluid Soluble Filter Cake Ultrafine Particles	Acid Undersaturated Brine Polymer Breakers
Gravel Pack	Polymer-Based Fluid Soluble Filter Cake	Acid Undersaturated Brine Polymer Breakers

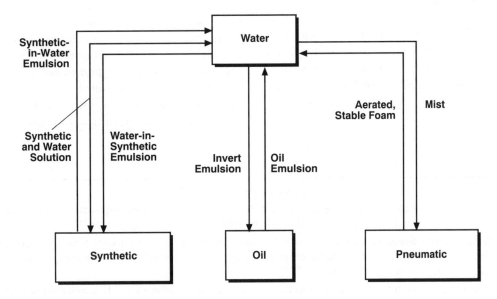

Figure 5-3   Basic types of drilling fluids

are sometimes called pseudo-oil-based muds. Pneumatic fluids include air and natural gas. It is not uncommon for different base fluids to be present in the same drilling fluid. Drilling-fluid names are traditionally formed by concatenating the terms identifying the base fluid, distinguishing additives, and important physical and chemical characteristics. Various combinations are also included in Figure 5-3. For example, water-based muds containing emulsified oil drops are called *oil-emulsion muds*. Oil-based muds containing emulsified drops of brine are called *invert-emulsion muds*.

Most drilling fluid components are associated with one of three phases. The *continuous phase* is the base fluid in which chemicals, minerals, formation solids, and other fluids are dissolved, suspended, and/or emulsified. The continuous phase is nearly always the principal component of the drilling-fluid system and filtrate. The *solids phase* consists of particles suspended in the continuous phase. Solids are typically categorized by their specific gravity (low or high) and/or by their reactivity with water (active or inert). Fluids emulsified in the continuous phase are the *discontinuous phase*. The solids and discontinuous phases are the main sources of viscosity and filter cake in a drilling fluid.

The great number of drilling fluids and specialty additives that exists today result from the wide variety of drilling conditions and problems encountered during drilling, and the fact that few drilling-fluid types or additives are ever retired permanently. Hazardous products, those containing asbestos or certain heavy metals for example, have been banned for health and/or environmental reasons. However, drilling fluids that are generally considered to be outdated continually reappear because they may offer unique advantages or provide economic benefits. In some instances, outdated fluids may be more compatible with new technology, or in other cases, they are the personal preference of the decision maker.

Trade journals regularly catalog drilling-fluid additives according to their primary use in major mud systems. Recent listings also include environmental information on products (*World Oil,* June 1996; *Offshore/Oilman,* Sept. 1995). Details on how to formulate, mix, test, and run various generic and proprietary mud systems are more difficult to find. The best sources are traditionally the technical manuals, product brochures, and material safety data sheets provided by drilling-fluid companies and vendors.

5-5.1 Water-Based Muds

The vast majority of drilling fluids are water-based. Water-based muds range from clear water to muds highly treated with chemicals. The base liquid can be fresh water, seawater, saltwater, or saturated saltwater, depending on the availability of make up water and the necessary mud properties. Common water-based mud additives are listed in Table 5-4. API specifications are available for barite, hematite, bentonite, attapulgite, sepiolite, carboxymethyl cellulose (CMC), and starch (API Spec 13A, 1993).

The large, diverse class of water-based muds is difficult to classify without overlap. Clear water and brines, the simplest water-based muds, are suitable for many competent and nonreactive formations. Lightly treated, noninhibitive muds are usually inexpensive, low-toxicity muds that provide hole-cleaning capability and some filtration control to drill routine top-hole sections. More footage has been drilled with these muds than any other type. Common lightly treated muds include muds viscosified with native formation clays or commercial bentonite (gel), clay-based muds dispersed with phosphates or low concentrations of organic thinners, and seawater and saltwater muds viscosified with attapulgite clay. Saltwater muds are most often the result of using seawater, brackish water, or field brines for make up. Salt is

Table 5-4   Common water-based mud additives

Weight Materials:	barite, hematite, calcium carbonate
Viscosifiers:	bentonite, attapulgite, sepiolite, beneficiated bentonite, biopolymers (xanthan and welan gum), guar gum, hydroxyethyl cellulose (HEC), mixed-metal hydroxide (MMH)
Dispersants and Deflocculants:	lignite (standard, causticized, potassium, chrome), lignosulfonate (chrome, chrome-free, calcium), tannin (extract blend, chrome, chrome-free), polyacrylate, sodium tetraphosphate (STP), sodium acid pyrophosphate (SAPP)
Filtration-Control Agents:	starch, cellulose polysaccharide, carboxymethyl cellulose (CMC), polyanionic cellulose (PAC), polyacrylate (SPA), resinated organic polymer, lignite
Shale Stabilizers:	polyacrylamide (PHPA), potassium acetate, asphalt (blown, sulfonated), cationic PHPA, polyamino acid, quaternary ammonium compound, polyvinyl alcohol (PVA)
Lubricants:	petroleum-distillate based, polyglycol, fatty acid blend, diesel oil, mineral oil, graphite, silicon, solid beads (plastic, glass)
Defoamers:	alcohol blend, mineral-oil/alcohol blend, aluminum stearate with diesel
Corrosion Inhibitors:	water and brine-dispersible blended amines, persistent filming amine, phosphate organic, zinc-based sulfide scavenger, organic biocide, ammonium bisulfite
Commercial Chemicals:	sodium chloride (NaCl), potassium chloride (KCl), calcium chloride ($CaCl_2$), caustic soda (NaOH), potassium hydroxide (KOH), magnesium oxide (MgO), sodium bicarbonate ($NaHCO_3$), soda ash (Na_2CO_3), lime ($Ca(OH)_2$), gyp ($CaSO_4 \cdot 2H_2O$), sodium bichromate, diesel, mineral oil, citric acid ($H_3C_6H_5O_7$)
Lost-Circulation Materials:	nut shells, mica, shredded wood fiber, cane fiber, cottonseed hulls, diatomaceous earth, shredded paper, plastic cups, cellophane flakes

sometimes added to inhibit bentonitic shales. High salt levels up to saturation are used to prevent washout of massive salt formations and prevent the formation of gas hydrates, which can interfere with rig safety equipment in deepwater offshore wells. Less common salts, such as potassium formate, are used in special applications to formulate high-density systems without the use of inert weight materials.

Inhibitive muds reduce the chemical interaction between the mud and water-sensitive formations. When added to the muds, sodium, potassium, and calcium ions minimize hydration and the swelling of clays and reactive shales. Lime muds are calcium-based muds that can maintain low viscosities and gel strengths in high-solids environments. Additionally, their high alkalinity effectively neutralizes corrosive acid gases, such as H_2S. Gypsum (gyp) muds, originally developed for the drilling of massive anhydrite and gyp formations, also rely on calcium ions to provide shale inhibition. The viscosity and gel strengths of lime and gyp muds are controlled by strong deflocculants that disperse fine solids in the mud.

When the most common deflocculant, chrome ligno-sulfonate, is used in fresh water and saltwater, the resulting muds are known as *lignosulfonate muds*. High concentrations of lignosulfonate and lignite improve inhibition, solids tolerance (more dispersion), filtration control, temperature stability, and resistance to salt, anhydrite, and cement contamination. In areas where chrome compounds cannot be used, chrome-free ligno-sulfonates are used with only a slight reduction in performance. Oil-emulsion muds contain up to 10% emulsified oil, which improves penetration rates, lubricity, filtration control, and shale inhibition. To reduce mud toxicity, however, the oil has been systematically replaced by oil-like synthetic materials and glycol/glycerol additives.

Polymers, both natural and synthetic, are routinely used in many water-based drilling fluids for viscosity, filtration control, shale inhibition, high-temperature stability, total or selective flocculation, and/or deflocculation. Most polymers are effective at very low concentrations and can be formulated for low toxicity. Polymer muds depend heavily on these materials to achieve desired properties. Polymers in low-solids, non-dispersed muds lower solids levels by enhancing the performance of bentonite and by providing filtration and rheological characteristics without using conventional dispersants. Partially hydrolyzed polyacrylamide (PHPA) muds adsorb on shale surfaces to minimize dispersion. PHPA muds are particularly effective in combination with a potassium source for additional inhibition and a biopolymer for improved suspension and viscosity. The biopolymer either enhances or completely replaces commercial clays. Cationic muds, which use cationic polymers to achieve high inhibition levels, are considered among the most inhibitive water-based muds. However, because of their high polymer consumption rates, they are still too costly for most applications.

Rheologically engineered fluids form a special class that has been optimized for use in horizontal wells and difficult hole-cleaning situations. In addition, these systems are environmentally friendly and nondamaging. Two notable examples are clarified xanthan-gum biopolymer fluids, which contain no bentonite, and mixed-metal hydroxide (MMH) systems, which use mixed-metal hydroxide salts to achieve unique rheological properties from bentonite slurries. Unlike conventional systems, these fluids are formulated on the basis of specific rheological characteristics first. Thereafter, any additives must accomplish their primary function without disturbing the original rheological behavior. Both systems exhibit excellent characteristics for horizontal and multilateral/lateral wells.

5-5.2 Oil-Based Muds

The unique performance characteristics of oil-based drilling fluids serve a wide range of applications, some of which cannot currently be performed adequately by any other mud type. Oil-based muds are highly inhibitive, resistant to contaminants, stable at high temperatures and pressures, highly lubricious, noncorrosive, and flexible. As such, they are particularly effective for the drilling of (1) highly reactive shale and evaporite formations, (2) extended-reach wells, and (3) deep, high-pressure, high-temperature (HPHT) H_2S wells. They are also used for special applications, such as freeing stuck pipe. The economics of oil-based muds can be attractive when considering the high drilling rates achieved with modern bits, the reduction and elimination of mud-related problems, and the reduced overall drilling costs. However, oil-based muds are highly toxic and often their use or disposal is restricted. Disposal costs should always be included in economic evaluations of oil-based muds.

Petroleum oils are used for the continuous phase of oil-based muds. The crude oil that was initially used for this purpose was first replaced by diesel oil and, more recently, both have been replaced by low-toxicity mineral oils. Unlike the early muds, which were intolerant of water, oil-based muds today can contain up to 60% emulsified brine. The volume percentages of oil and water are expressed as an oil/water ratio, which can range from 100:0 to 40:60. Low ratios (60:40 to 40:60) can cause very high viscosities, but they help minimize the oil retained on cuttings. Because the term "oil emulsion" already had been chosen for oil-in-water emulsion muds, the term "invert emulsion" was coined to describe water-in-oil emulsion muds. The emulsified water droplets help suspend weight material and lower fluid loss. Additionally, high salinity levels in the water phase improve wellbore stability by creating osmotic pressures that dehydrate and harden reactive shales.

The basic components of an invert-emulsion mud include oil, brine (usually calcium chloride), primary and supplementary emulsifiers, oil-wetting agents, viscosifier/gellants (oil-dispersible bentonite), filtration-control additives, and slaked lime. Sometimes, a rheology modifier is added for improving hole cleaning in directional and large-diameter wells. Fatty acid soaps are the most common emulsifiers used in oil-based muds. Polyamines, polyamides, imidazolines, and other cationic emulsifiers are also effective. Not all modern oil-based muds contain emulsified brine. Some all-oil muds have demonstrated exceptional performance and have been suitable for special disposal methods (Carter and Faul, 1992).

Toxicity is the most serious, and perhaps insurmountable, drawback of oil-based muds. Environmental regulations strictly control and even forbid their use and disposal in some areas. While great strides have been made to reduce toxicity, the very characteristics that result in the superior performance of oil-based muds can also contaminate the environment. A wide variety of refined mineral oils has been developed for use in low-toxicity oil-based muds to reduce environmental problems and improve working conditions. These oils have lower aromatic contents than diesel and reduced acute toxicity to various organisms. However, even the cleanest of these oils is considered toxic, and mud and contaminated cuttings must be disposed of in accordance with local environmental regulations.

5-5.3 Synthetic-Based Muds

Synthetic-based systems represent the latest technology for providing high performance with minimal environmental impact. The goal of synthetic-based drilling fluids is to provide the performance and inhibitive properties of oil-based fluids without the objectionable toxicity and

environmental impact. Drilling performances have been exceptional, easily equaling those of muds based on diesel and mineral oil. In many offshore areas, regulations that prohibit the discharge of cuttings drilled with oil-based muds do not currently apply to synthetic-based fluids. While unit costs are still high, synthetic-based systems have proven economical in many offshore applications.

Conceptually, synthetic-based drilling fluids are mixed and run in the field much like conventional oil-based muds. In certain geographical areas, they are called pseudo-oil-based muds, a term unacceptable in the U.S. because of environmental regulations. Ironically, synthetic-based muds are really replacements for water-based muds. Oil-based muds are typically chosen for difficult wells if regulations permit their use and hauling and disposal costs are affordable.

Synthetics are man-made, nonaqueous liquids which have molecules that are wholly dissimilar to the raw materials used for manufacture. The continuous phase of a synthetic-based drilling fluid is at least 99% synthetic material. The first compounds used as the base liquid for synthetic-based drilling fluids were esters, ethers, polyalphaolefins (PAOs), and acetals. Second-generation synthetics have been characterized by lower costs and lower kinematic viscosities, but slightly higher toxicity levels. These synthetics include isomerised olefins (IOs), linear alphaolefins (LAOs), and linear alkylbenzenes (LABs). The low-cost LABs, after a very brief appearance in the North Sea, became defunct when unfavorable results were obtained from a seafloor study. Although some experts might consider low-toxicity linear paraffins (LPs) to be synthetics, LPs are highly refined from crude oil. Therefore, LPs are more appropriately classified as low-toxicity mineral oils. Tables 5-5 and 5-6 (Friedheim, 1996) compare typical properties of first- and second-generation synthetics.

5-5.4 Pneumatic Drilling Fluids

Pneumatic drilling fluids are used in special applications, primarily to minimize damage to productive formations, prevent loss of circulation, and achieve very high penetration rates. These fluids can handle normal drilling-fluid functions adequately, except for cuttings suspension, filter cake deposition, and control of subsurface pressures. As a result, their use is limited to competent, low-permeability formations, such as limestones and dolomites.

Table 5-5 Typical properties for first-generation synthetic liquids used to formulate drilling fluids (from Friedheim, 1996)

Properties	PAO	Ester	Ether	Acetal
Density (sg)	0.80	0.85	0.83	0.84
Viscosity at 40°C (cSt)	5.0 to 6.0	5.0 to 6.0	6.0	3.5
Flash Point (°C)	>150	>150	>160	>135
Pour Point (°C)	<(−55)	<(−15)	<(−40)	<(−60)
Aromatic Content	No	No	No	No

Table 5-6 Typical properties for second-generation synthetic liquids used to formulate drilling fluids (from Friedheim, 1996)

Properties	LAB	LP	LAO	IO
Density (sg)	0.86	0.77	0.77 to 0.79	0.77 to 0.79
Viscosity at 40°C (cSt)	4.0	2.5	2.1 to 2.7	3.1
Flash Point (°C)	>120	>100	113 to 135	137
Pour Point (°C)	<(−30)	(−10)	(−14) to (−2)	(−24)
Aromatic Content	Yes	Yes	No	No

The principal component of a pneumatic drilling fluid is air or natural gas, although use of either is commonly referred to as air drilling. Air or gas must be circulated just like conventional liquid drilling muds. The circulation pressure for air is provided by large air compressors installed as part of the surface drilling equipment; gas pressure is usually obtained from a high-pressure field source. Both require a surface rotating seal around the annulus to direct the return flow and cuttings a safe distance away from the rig. Extra safety precautions are required because of the inherent dangers of explosions and fires on the surface or downhole. Risks of a downhole explosion are greater when air is used, since the combination of air and formation hydrocarbons can readily form an explosive mixture.

The basic forms of air drilling are dusting, mist, stable foam, and aerated mud. Dusting, which involves the circulation of dry air or gas at very high velocities, can be used only when formations are nearly dry. Small amounts of water can be tolerated (mist), but injection of dilute mixtures of foaming agents and polymers in water may be required to keep cuttings from sticking together. Superior foaming agents form a stable foam with the consistency of shaving cream. Stable foams can provide excellent hole-cleaning and still maintain the benefits of reduced-density drilling. Some of the advantages of air drilling can be obtained by the aeration of conventional drilling muds. Aerated muds are used for severe lost-circulation problems when other forms of air

drilling are not possible. Nitrogen generated at the well-site has been used instead of air to reduce the high corrosion rates associated with aerated muds.

5-6 ENVIRONMENTAL CONCERNS

Drilling fluids, like many industrial chemicals, can be hazardous to humans if not used properly and with appropriate protective equipment. Contamination of the natural environment could also occur if fluids and formation cuttings are not disposed of properly. Regulations vary by country and by jurisdictional area within a country. For example, in the Norwegian sector of the North Sea, cuttings generated with a low-toxicity, oil-based mud can be discharged into the sea if the oil-on-cuttings is less than 10 g/kg. However, discharge of any oil-based mud cuttings is prohibited in the Gulf of Mexico, and regulations in Mobile Bay strictly prohibit all discharges (zero discharge).

Hydrocarbons, chlorides, and heavy metals are the principal sources of toxicity in drilling fluids. Products containing these contaminants have been used in past designs of drilling mud. However, these contaminants also occur naturally and are sometimes incorporated into the mud during the drilling operation. Examples include crude oil found in productive formations, salt from massive salt formations, and trace metals contained in organically rich shales that are drilled out. Heavy metals include chrome, lead, zinc, arsenic, barium, mercury, and cadmium. Common sources include drilling-fluid additives, corrosion inhibitors, pipe dope, and subsurface formations (Candler *et al.*, 1992).

The main means of reducing environmental impact is to address the combined drilling-fluid issues of performance, economics, composition, disposal, and environmental impact. Although environmental problems and regulations at times have been a source of consternation to the drilling industry, they have stimulated and significantly advanced drilling-fluid developments (Bleier *et al.*, 1993). These developments have involved (1) toxicity testing, (2) waste minimization techniques, and (3) treatment/disposal options.

5-6.1 Toxicity Testing

Toxicity of a drilling fluid is determined by its composition and measured by the test protocol (bioassay). Bioassays measure the acute toxicity (lethality) on a test population of organisms. Results are used for determining whether mud and cuttings can be discharged in offshore waters. Mysid shrimp *(Mysidopsis bahia)* is the species specified by the US Environmental Protection Agency for use in drilling-fluid toxicity tests (Jones *et al.*, 1987). The shrimp are exposed for 96 hr to several concentrations of the suspended particulate phase of the effluent. The effluent is the test additive mixed in one of eight generic muds and diluted 1:9 with seawater. The 96-hr LC_{50} value (lethal to 50% of the population) determines the toxicity. Bioassays are required on monthly, change-of-mud, and end-of-well basis. Currently, an LC_{50} value less than 30,000 ppm prevents discharge of mud and cuttings in the Gulf of Mexico. Other countries use similar tests with different species and discharge limits.

Some countries (U.S.A, Canada, and European countries) and oil companies now require all chemicals intended to be used or discharged to be tested for toxicity and health effects. In addition to acute toxicity for organisms representing different trophic levels, chemicals may undergo testing for bioaccumulation, biodegradation, byproduct and degradation effects, duration of impact, and rate of seabed recovery.

5-6.2 Waste Minimization

The most desirable method of controlling pollution is to minimize or eliminate it at the source. The two most common ways to achieve this goal are product substitution and changes in operating practices. With regard to product substitution, new products are continually formulated and reformulated to meet performance requirements without the use of objectionable materials. Substituting chrome-free lignosulfonate for chrome lignosulfonate is an example of a product substitution that minimizes heavy-metal contamination. In some cases, low-toxicity synthetics can substitute for petroleum-based lubricants, while polyglycols are now used instead of alcohol- and petroleum-based defoamers. Environmentally acceptable synthetic-based muds now provide performance that is similar to oil-based mud performance.

Improvements in drilling and drilling-fluid technology have also achieved source reduction through operational practices. Use of inhibitive mud systems (for gauge holes) and improved solids-control equipment have led to dramatic decreases in the volume of solids and liquids that need to be discharged into the environment. Other examples of waste minimization include systematic segregation of contaminated waste from uncontaminated waste, the use of closed-loop solids-control systems, and new packaging methods for drilling-fluid products.

Recycling is an alternative to control pollution for situations where source reduction is limited or not feasible. After the well has been completed, the remaining drilling fluid may be stored in tanks until needed on a subsequent well. Recycling is an economic advantage for oil-based and synthetic-based muds. Another example of recycling involves the conversion of drilling fluids for other uses at the wellsite. Muds can be treated with blast furnace slag (Nahm *et al.*, 1993), converted to cement, and used as a replacement for conventional Portland cement. While cuttings generated from the wellbore are difficult to reuse because of detrimental effects on drilling-fluid performance, they sometimes are recycled for use as landfill cover or construction materials.

5-6.3 Treatment/Disposal Options

Drilling-mud and formation-cuttings disposal must be economical yet still limit long-term liability. Disposal options for offshore drilling are few. If allowed by local regulations, mud and cuttings can be discharged directly into the sea, the cuttings can be returned to the formation through the use of annular injection, or the mud could be converted to cement and used in place of conventional oilfield cement. Otherwise, the waste must be hauled to shore for treatment and ultimate disposal through the use of one of the previously mentioned methods.

Onshore, the concerns are oil and grease, chlorides, and heavy metals. The following are some of the currently available disposal processes:

- **Onsite separation** Waste-material dewatering, which uses water-treatment methods used at municipal and industrial plants, yields water and a dry mudcake as the two end products.

- **Solidification** Cementing compounds such as fly ash, kiln dust, and Portland cement are mixed with waste fluids in the reserve pit and allowed to dry.

- **Bacterial degradation** Hydrocarbons are removed from the drilling-fluid waste (commonly oil-based mud) by special bacteria which "eat" the hydrocarbons off the solids. The remaining solids are then solidified.

- **Dewatering/backfilling** Water in the reserve pit is allowed to evaporate. Alternatively, the water is pumped out after solids are flocculated chemically and allowed to settle. Top soil from the reserve pit dike is then backfilled to complete the reserve-pit closure.

- **Landfarming** An even distribution of mud and cuttings is mechanically mixed with the soil and tilled to accelerate biodegradation of hydrocarbons (Carter and Faul, 1992).

- **Washing** Free oil on cuttings is removed with a solvent or washing solution and mixer.

- **Landfill disposal** Mud and formation cuttings are first treated to remove free liquids and then buried in a secure landfill.

- **Annular injection** Cuttings are slurried and injected through a casing annulus into downhole formations protected by casing (Minton *et al.*, 1992).

- **Experimental techniques** Experimental waste disposal techniques include incineration, distillation, and critical-fluids extraction.

5-7 PERFORMANCE TESTING

Drilling-fluid properties are continually measured and adjusted at the wellsite to satisfy requirements efficiently and economically. For practical reasons, wellsite tests cannot be lengthy, overly complicated, or require delicate or sophisticated test equipment. Fortunately, basic physical and chemical tests in concert with field experience serve to monitor the drilling-fluid condition and guide its optimization, despite the great complexity of drilling-fluid chemistry and physical behavior. The American Petroleum Institute (API) issues recommended practices that specify equipment and procedures for testing and monitoring water-based (API RP 13B-1, 1990) and oil-based drilling fluids (API RP 13B-2, 1991). Current API field tests are listed in Table 5-7. Additional tests are sometimes run in the field for the evaluation of special properties.

Density (mud balance) and relative viscosity (Marsh funnel) are two basic measurements taken on every

Table 5-7 Standard API drilling-fluid field tests

Tests Common to Water- and Oil-Based Fluids	**Water-Based (RP 13B-1)**
Mud Weight (Density)	Sand
Viscosity and Gel Strength	Methylene Blue Capacity
Marsh Funnel	pH
Direct-Indicating Viscometer	Chemical Analysis
Filtration	Alkalinity and Lime Content
Water, Oil, and Solids	Chloride
Shear Strength	Total Hardness as Calcium
Oil-Based (RP 13B-2)	Calcium
Chemical Analysis	Magnesium
Whole Mud Alkalinity	Calcium Sulfate
Whole Mud Chlorides	Formaldehyde
Whole Mud Calcium	Sulfide
Electrical Stability	Carbonate
Oil and Water Content from Cuttings	Potassium
Aqueous-Phase Activity by Electrohygrometer	Resistivity
Aniline Point	Drill-Pipe Corrosion Ring Coupon

well. Other key parameters include filtration, pH, solids and sand content, and chlorides. Rheological parameters such as plastic viscosity and yield point are necessary as mud-conditioning indicators to guide mud treatments, calculate frictional pressure losses, and optimize drilling hydraulics.

For critical wells, drilling-fluid tests are run in field-service, research, and specialty laboratories to support field operations and provide a higher level of expertise. Many of these tests require sophisticated analytical equipment and/or carefully controlled environments. These tests include the following:

- **Toxicity** In some areas, mud systems and mud additives must be tested for toxicity and health effects before they are used. Bioassays based on protocol set by local regulations measure acute toxicity to determine if whole mud and/or cuttings can be discharged. Chemicals may also undergo testing for bioaccumulation, biodegradation, byproduct and degradation effects, duration of impact, and rate of seabed recovery.

- **Shale stability** Several tests are used for measuring (or monitoring) unfavorable physio-chemical reactions that may occur between water-based drilling fluids and representative samples of shale formations. Shale dispersion and swelling tests are conducted near ambient conditions, while triaxial tests evaluate mud-shale interactions at confining pressures above 1500 psi and temperatures above 150°F to simulate downhole conditions.

- **HPHT rheology** Rheological properties and gel strengths (thixotropy) are measured at the wellsite using API-recommended rotational viscometers that operate at low temperature (<180°) and atmospheric pressure. For HTHP wells, especially those using oil and synthetic-based drilling fluids, rheological measurements should be obtained from laboratory viscometers, some of which can be run at 450°F and 20,000 psi (API RP 13D, 1995).

- **High-temperature aging** Concerns in high-temperature wells include severe gelation when the mud is static and extreme viscosity during circulation. These two conditions are simulated by testing mud samples in pressurized aging cells, typically for 16 hours (overnight). Cells placed in conventional ovens simulate static conditions; rotating or rolling ovens simulate circulating conditions. Both tests can be run at the wellsite, if necessary

- **Dynamic filtration** Filter cakes formed on permeable formations under static and dynamic conditions do not exhibit the same characteristics. Static filtration measurements are routinely taken in the field under low and high temperatures and pressures. However, dynamic tests require more sophisticated equipment: either a stirred-cell unit or a rotating-shaft device. Unlike their static counterparts, dynamic units can run filtration tests on sandstone or ceramic disks to provide a closer approximation of downhole formations than standard filter paper.

- **Lubricity** Quantitative correlations between lubricity measurements and field results have not been very successful. The most commonly used test measures metal-to-metal friction, appropriate for extreme-pressure lubrication and somewhat suitable for drillpipe/casing friction. Different units can measure metal-to-sandstone friction under ambient test conditions. However, reasonable results are now being obtained by a new class of devices capable of lubricity measurements under pressure and temperature, using metal, sandstone, or shale for the simulated wellbore (Toups *et al.*, 1992).

- **Particle-size distribution** The standard sand-content test is really a particle-size measurement (percentage of solids >200-mesh). However, true particle-size distribution is important for filtration, formation damage, and rheological characteristics. Distribution measurements can be obtained quickly and accurately in the laboratory with electronic-sensing devices and laser-beam analyzers.

- **Return permeability** As discussed previously, permeability impairment of a reservoir-rock core is generally evaluated in the laboratory before a damage-sensitive formation is drilled. These tests can be run with a static or dynamic drilling fluid.

5-8 PROBLEMS RELATED TO DRILLING FLUID

Many wells are drilled under circumstances that do not require advanced technology. However, some wells must be drilled through troublesome intervals, including "gumbo" shales, unconsolidated sands, massive salt zones, high formation pressures, long lateral sections, and formations containing mud contaminants such as hydrogen sulfide (H_2S) and carbon dioxide (CO_2). Drilling conditions are most difficult when multiple hazards occur in the same well, and the combined efforts of numerous drilling technologies are required.

Some mud-related problems can delay, suspend, or even cancel a drilling project. Stuck pipe, lost circulation, and wellbore instability continue to top the list of drilling problems on the basis of lost time and money. Depleted sands, barite sag, poor hole cleaning, H_2S contamination, and gas hydrates are also among the most common and/or serious problems.

5-8.1 Stuck Pipe

Stuck pipe is generally considered the most expensive and greatest lost-time problem in drilling. A survey by a major drilling contractor estimated that 36% of reported drilling problems worldwide over a 15-month period were due to stuck pipe (Jardine *et al.*, 1992). Stuck-pipe problems cost the industry more than $250 million annually (Bradley *et al.*, 1991).

Traditionally, stuck-pipe causes are categorized either as differential or mechanical. Differential sticking results when the drillstring is held tightly by differential pressure against a permeable formation. Mechanical sticking can be caused by key seating, inadequate hole cleaning, wellbore instability, and/or an undergauge hole. The percentage of incidents in either category depends on well type and drilling conditions. Bradley *et al.* (1991) estimated that in their company's Gulf of Mexico wells, differential sticking accounted for 61% of stuck-pipe costs, while in their North Sea wells, mechanical sticking accounted for 70%.

The mechanics of differential sticking are well understood (Outmans, 1958). When a nonmoving drillstring becomes embedded in a thick filter cake on a permeable formation, the cake can act as a pressure seal. Differential pressure plus any side forces (pipe weight) can hold the pipe with such great force that it cannot be pulled free. Inadequate fluid-loss control, poor filter-cake characteristics, excessive solids content, and high overbalance pressures exacerbate cake thickness and the severity of the problem. In water-based muds, lubricants may help free stuck pipe. Time is critical—sometimes the pipe can be worked or jarred free if action is taken quickly. Reduction in mud hydrostatic pressure is an alternative, but this must be done carefully so that it does not compromise well control. Finally, spotting a fluid designed to penetrate the filter cake seal is recommended (full circulation is usually possible). A soak time of at least 12 hours should be allowed. After about 24 hours, the probability of success diminishes rapidly.

Mechanical sticking usually requires working or jarring to free the pipe. Thereafter, the cause of the sticking needs to be corrected, and the stuck zone needs to be reamed or otherwise reconditioned. More mud weight may be required to handle wellbore instabilities caused by formation stresses; reactive shales may require a more inhibitive mud, or a fresh water pill may need to be spotted to free the pipe grabbed by plastic salts.

5-8.2 Depleted Sands

Differential sticking is likely when drilling severely depleted sands, such as those reported by Newhouse (1991). Weakley (1990) determined that the chance for differential sticking increased dramatically when the differential pressure across permeable sands drilled with water-based muds exceeded 2000-1000 times the sine of the hole angle for vertical and directional wells. Because of the number of casing strings required to satisfy these strict threshold limits, some wells have been drilled with differential pressures >3500 psi.

Proper filtration control, the use of deformable bridging agents, and thin, compressible filter cakes can greatly minimize stuck-pipe problems in depleted sands. Abrams (1977) found that effective plugging could be achieved by 5% (by volume) bridging particles equal to or slightly larger than 1/3 the median pore size. Significant improvements can be realized by also considering particle-size distribution.

Oil-based muds are often selected for drilling depleted sands because of their inherent low filtration rates and thin filter cakes. However, water-based muds can be used when economic or environmental constraints prevent the use of oil-based muds. Newhouse (1991) recommends using special water-based mud additives such as asphaltite (up to 4 lb/bbl) and cellulosic fibers (3 to 8 lb/bbl) for providing proper bridging and plugging. Deformable and compressible fluid-loss products including starch and PAC materials are also helpful. A filtration unit designed for high differential pressures (2,500 psi) should be used for pilot testing to determine optimum mud formulations.

5-8.3 Lost Circulation

Lost circulation is a perennial drilling problem characterized by loss of whole mud into downhole formations. Other hole problems including stuck pipe, hole collapse, and loss of well control can also affect the likelihood or extent of lost circulation. Drilling costs can be excessive, especially when using synthetic-based drilling fluids. Lost circulation can occur in formations that are (1) fractured (natural or induced), (2) highly permeable and/or porous (massive sands, gravel beds, reef deposits, shell beds), or (3) cavernous/vugular (limestone, dolomite, chalk). Induced fractures are caused by excessive annular pressures.

Lost circulation in naturally occurring voids and unconsolidated formations cannot typically be avoided; however, the following preventive measures apply in other cases:

- Casing should be set so that it protects weak formations and provides a formation fracture gradient sufficient to support drilling-fluid imposed pressures in the annulus.

- Minimal mud weight should be maintained for the drilling conditions.

- Excessive downhole pressures should be avoided, including those caused by improper rheology/hydraulics, high flow rates, thick filter cakes, surge pressures during tripping in the hole, bridges, high shut-in surface pressures, or sloughing shales.

Proper control of lost circulation involves keeping the hole full to prevent a kick, avoiding stuck pipe, sealing off the loss zone, and cautiously regaining circulation. Leakoff often can be corrected by the addition of lost-circulation materials (LCMs) to the mud. These LCMs include sized calcium carbonate or cellulosic fibers. For partial losses, the bit should be pulled safely above the loss zone, the hole should be kept full with low-weight mud or base fluid and allowed to stand full for 4 to 8 hr. Then, the bit should be returned to bottom carefully. If returns are still not achieved, an LCM pill or a high-fluid-loss slurry squeeze should be mixed and circulated. For oil-based muds, a gunk squeeze (organophilic clay in water) is recommended. Complete losses usually require a high fluid-loss slurry squeeze or a hard plug such as cement, cement-bentonite, cement-gilsonite, or diesel-bentonite-cement.

LCM bridging agents that are larger than normal mud solids are used to seal off lost-circulation zones. Blends of particle sizes are often necessary for proper sealing. LCMs are classified as fibrous (paper, cottonseed hulls), granular (nut shells), or flakes (cellophane, mica). Preblended products usually include all three types of LCM in coarse, medium, and fine particle sizes.

5-8.4 Wellbore Instability

Wellbore instability can be caused by mechanical factors and/or physico-chemical interactions between the mud and formation. Chapter 6 describes several problems related to wellbore stability and rock mechanics. Most, but not all, wellbore instability problems occur in shales. Mechanical factors include formation stresses (low mud weight), erosion, surge/swab pressures, and pipe whip, as well as unconsolidated sands and plastic salt flows. Physico-chemical interactions are related to hydration phenomena that cause shale swelling and/or dispersion.

Oil- and synthetic-based drilling fluids are the most inhibitive and the fluids of choice for drilling most troublesome, hydratable shales because free water does not contact the shales. In fact, since such fluids have high-salinity brines as their internal phase, they can create "balanced-activity" osmotic pressures that prevent water from entering the shale (Mondshine and Kercheville, 1966; Chenevert, 1970).

Continual improvements have been accomplished in water-based mud technology to prevent shale hydration, but water-based muds cannot match the inhibition levels of oil- and synthetic-based fluids. Soluble salts added to water-based muds help control swelling. "Encapsulating" polymers minimize dispersion. The potassium ion (from KCl, for example) is generally regarded as the most effective because of its low hydration energy and its small size, which enables it to fit between silica layers in the clay crystal and reduce interlayer swelling (Darley and Gray, 1988). Fortunately, potassium is effective at relatively low concentrations (3 to 5% as KCl), because higher levels will fail some bioassays. PHPA, one of the more effective encapsulating polymers, is highly compatible with KCl, and together they form the nucleus for an excellent, inhibitive water-based mud system.

Cationic (water-based) drilling fluids are highly inhibitive, but not completely practical. Initial high-toxicity and incompatibility problems have been solved, but the high consumption rates of the expensive polymers have proven costly. However, recently developed cationic compounds added to conventional (anionic) muds have significantly improved inhibition levels (Stamatakis *et al.*, 1995).

Silicate-based muds have made several attempts to gain acceptance as the most inhibitive water-based muds for drilling troublesome formations like microfractured shales and chalks. van Oort *et al.* (1996) now claim that previous problems with silicate muds have been corrected. In addition to the muds themselves having better rheological control, polymer-based rheology modifiers and fluid-loss agents are now available. These muds have also benefited from advances in solids-removal equipment, and a better understanding of wellbore stability problems in general. However, early tests show that silicates may cause formation damage, and the lubricity of silicate-based muds is suspect.

Unconsolidated sands lack natural cohesion among individual grains. Control of drilling-fluid properties is essential when drilling these formations (Stephens and Bruton, 1992). Some degree of overbalance should always be maintained without the use of excessive mud weights. Properly controlled filtration, spurt loss, and filter-cake properties are critical for the prevention of hole erosion and stuck-pipe problems; bridging solids are especially important. Nondamaging fluids should be used if the unconsolidated formation is a potential producer. Finally, mud rheology should be set so that proper hole cleaning can be provided at relatively low flow rates to minimize hydraulic erosion.

5-8.5 Hole Cleaning

Practically speaking, hole cleaning in most vertical wells can be improved by increasing the mud viscosity (yield point) and flow rate (annular velocity). However, hole cleaning in directional wells (extended reach, horizontal, and multilateral) can be an order of magnitude more difficult. In inclined intervals, the combination of skewed annular velocity profiles, unusual settling patterns, and force imbalances over the annular cross section can promote formation of troublesome cuttings beds on the low side of the hole. Angles between 30° and 60° are the most difficult to clean, because beds formed over this angle range can slide downward toward the bottom of the well. Poor drilling practices, low flow rates, and inadequate mud viscosity and suspension properties exacerbate the problem.

It is generally agreed that annular velocity is the key parameter affecting hole cleaning; however, mud viscosity, pipe rotation, and pipe eccentricity can be critical and even rival velocity in importance. Hole cleaning is also affected by specific hole conditions, formation properties, and, to some extent, personal preferences. For example, many of the small-diameter horizontal wells drilled in the highly competent Austin Chalk formation in Texas have been drilled in turbulent flow with water, brine, or lightly treated muds. As a result, hole cleaning in many Austin Chalk wells has been relatively insensitive to viscosity, pipe eccentricity, and pipe rotation.

However, turbulence may not be desirable or even possible in large-diameter formations and/or less-consolidated formations, especially when fluid-loss control is required.

Elevated low shear-rate viscosities and enhanced suspension capabilities, such as those achieved in rheologically engineered fluids (Beck *et al.*, 1993; Fraser, 1990), have successfully improved hole cleaning in high-angle wells. The conventional yield point is not a good measure of hole-cleaning capability. Better indicators include (1) a field viscometer reading at 3 or 6 rpm, (2) a low shear-rate yield point determined from 3- and 6-rpm readings, or (3) an ultra-low shear-rate viscosity measured with nonstandard viscometers.

The complexity of the hole-cleaning problem has made modeling difficult at best. Many current field practices are still based on empirical data and flow-loop studies, and are distributed as practical guidelines rather than analytical solutions (Zamora and Hanson, 1991). Several models have been proposed, but the flow-rate prediction charts developed by Luo *et al.* (1992) have been particularly useful. Recent advancements in hole-cleaning models have been based on fuzzy logic concepts.

5-8.6 Barite Sag

Barite sag can lead to serious problems in high-angle wells drilled with weighted muds. Barite sag is a significant variation in mud density caused by the settling of barite or other weight material. The greatest density variation occurs during the first bottoms-up after a trip or other operation where the mud has been static for a period. Therefore, sag was originally believed to be primarily a static problem. However, Hanson *et al.* (1990) proved that three key mechanisms were involved: dynamic settling, static settling, and slumping.

Most of the barite bed forms on the low side of the inclined hole under dynamic conditions. Low annular velocities, stationary pipe (no rotation), and poor low shear-rate rheology exacerbate the settling. Additional settling occurs when circulation stops, although gels formed in the mud can significantly slow down this process. Studies by Bern *et al.* (1996) showed that the most difficult angle range is 60° to 75°. The biggest problem occurs when the barite bed slumps downward toward the bottom of the hole. Slumping, which rarely occurs at angles greater than 75°, causes the characteristic density variation and can lead to problems including lost circulation, stuck pipe, induced wellbore instability, and loss of well control.

5-8.7 H₂S Contamination

The first line of defense against hydrogen-sulfide contamination is to sustain a positive overbalance to prevent an influx from H_2S-bearing formation fluids. It is equally critical to maintain a high pH or mud alkalinity using lime to neutralize (ionize) the H_2S into less harmful hydrosulfide (HS^-) and sulfide (S^{2-}) ions. Chemical scavengers that react with the soluble sulfides must then be used to form insoluble precipitates and prevent reformation of the H_2S if the pH level drops. Zinc oxide, basic zinc carbonate, zinc chelate, and iron oxide are the most commonly used scavengers.

The two fluid systems used most often for drilling in H_2S environments are a high-pH, dispersed, water-based mud, and a very stable oil-based mud. Each should contain sufficient excess lime to provide a buffered alkalinity and neutralizing potential. Scott (1994) recommends that water-based muds should have a pH >11, a mud alkalinity >3 mL through the use of lime, and a filtrate alkalinity >1 mL through the use of caustic soda. Oil-based muds should have a mud alkalinity >10 mL, an electric emulsion stability >1500 v, and, preferably, an oil/water ratio greater than 85:15. Oil-based muds offer additional protection because the oil-wetting of metal surfaces minimizes the corrosive effects of H_2S. However, sulfide scavengers react more slowly, and gas is more soluble in oil-based muds. Synthetic-based muds act much like oil-based muds, although some synthetic-based fluids are not stable at high alkalinity levels, and they cannot tolerate excess lime.

5-8.8 Gas Hydrates

Barker and Gomez (1987) described well-control problems in deepwater operations in which gas hydrates plugged subsea risers, blowout preventers, as well as choke and kill lines. Seabed temperatures were above 40°F. Although few such incidents have been reported, the potential hazards are great.

Gas hydrates are ice-like crystalline solids formed by the physical reaction between gas and water at temperatures well above the freezing point of water. The necessary elements for gas-hydrate formation are gas, water, low temperature, high pressure, and time; gas hydrates will not form if one or more of these elements is missing. Unfortunately, the conditions that result when a deepwater well (>1500-ft water depth) is shut in on a gas kick are precisely those most conducive to gas-hydrate formation.

Although hydrates form more readily in water-based muds, the water content of a typical oil-based mud is sufficient to cause their formation, but the reaction is considerably slower. Much research has been conducted regarding the formulation of water-based muds suitable for use in deepwater environments. The current industry standard for water-based fluids is a PHPA-polymer mud containing up to 22% NaCl (Jones and Sherman, 1986). Other salts are also effective, but they are less compatible with standard mud products. Dispersants (lignosulfonate, lignite, etc.) do not inhibit hydrate formation; rather, they tend to increase the hydrate-formation rate (Lai and Dzialowski, 1989). Methanol and ethylene glycol, two inhibiting compounds frequently added to water in other applications, have limited use in drilling fluids for environmental and safety reasons.

5-9 SPECIAL APPLICATIONS

Drilling fluids should satisfactorily meet all basic requirements regardless of the relative complexity of the project. However, some drilling-fluid characteristics must be emphasized for certain types of well-construction projects, including (1) high-pressure/high-temperature wells, (2) horizontal and multilateral wells, (3) extended-reach wells, and (4) slimhole/coiled-tubing wells. Their overall success can depend on how well the drilling fluids handle specific needs of each well type.

5-9.1 High-Pressure, High-Temperature (HPHT) Wells

No universal definition exists for HPHT wells, but wells are generally placed in this category when formation pressures exceed a 15-lb/gal equivalent, and formation temperatures are greater than 300°F. Formation pressures have required fluid densities over 22-lb/gal to control, and temperatures in oil and gas wells have been recorded above 550°F at 24,000 ft. Often, HPHT wells are also inherently linked to most of the drilling problems discussed in the previous section. In some areas, such as the North Sea, classification as "HPHT" can automatically elevate the well status to "critical and difficult."

Drilling-fluid density and rheology are key concerns in HPHT wells. Water-based muds are sensitive to temperature, but relatively insensitive to pressure. Oil- and synthetic-based fluids are very sensitive to both. This was clearly illustrated by offshore well data collected by White *et al.* (1996) on a synthetic-based mud. The hydrostatic pressure during the use of an oil- or synthetic-based mud cannot be accurately determined without correcting the surface-measured density for downhole conditions. Sorelle *et al.* (1982) and others have proposed practical models for ESD (equivalent static density) that account for volumetric changes in the solids, oil (or synthetic), and water phases. The models are strongly dependent on the compressibility and thermal behavior of the oil (synthetic), and the temperature profile of the mud.

True downhole rheological properties affect, among other parameters, equivalent circulating density (ECD), hole cleaning, barite sag, surge/swab pressures during tripping, pump pressures, and bit hydraulics. Gelation and excessive viscosity are major concerns at temperatures above about 300°F. Downhole properties can be estimated from surface measurements if the general rheological behavior of the fluid is fully characterized, or if HPHT rheological measurements are available (Zamora, 1996).

Oil-based muds and some synthetic-based muds are inherently more temperature stable than water-based muds. Although water-based muds have been used in HPHT wells at temperatures above 500°F, temperatures above 350°F clearly favor oil-based muds. Even well-treated lignosulfonate muds are difficult to run economically above 350°F. Thermal stability of water-based muds is achieved by minimizing the concentration of active solids (especially bentonite) and using polymeric additives for viscosity and suspension control. These practices reduce the possibility of high-temperature flocculation of active clays and viscosity increases caused by saltwater, salt, and/or acid-gas contamination. (Similarly, organophilic clays used for structural viscosity in oil-based muds degrade above 350°F and should be eliminated or used sparingly.) Low molecular-weight polymers, such as sodium polyacrylate (SPA) and the sodium salt of sulfonated styrene maleic anhydride copolymer (SSMA), can be used as deflocculants for maintaining low rheological properties in high-temperature environments. Special long-chain polymers, including vinyl sulfonated copolymers and vinylamide/vinylsulfonated terpolymers, are required for filtration control at high temperatures.

Densities up to 20 lb/gal can be reached through the use of barite in either water- or oil-based muds. Solids loading and the availability of free water or oil are the controlling parameters. The use of hematite instead of barite can help, but sag problems may occur when a directional well is drilled. Oil-based muds are preferred at ultra-high densities, because the buildup of fine solids is less probable.

5-9.2 Horizontal and Multilateral Wells

Key mud-related concerns for the drilling of horizontal (and multilateral) wells can be broadly grouped into two categories: drilling and completion. Hole cleaning, torque and drag, wellbore stability, and stuck pipe impact the drilling effort. Formation damage, filtration characteristics, mud cleanup, and compatibility with the completion method highlight completion aspects. Some of these concerns are addressed in Chapter 3.

Horizontal wells are seldom completed with perforations through casing, which is relatively forgiving with regard to formation damage. For this reason, drilling and completion aspects must be integrated to avoid damage. Generally, the reservoir section is drilled with special drill-in fluids that improve well productivity. Drill-in fluids, in addition to performing their basic functions as drilling fluids, should be (1) nondamaging, (2) compatible with the completion method, and (3) responsive to stimulation or clean-up techniques. For horizontal wells, the drilling fluid used for drilling the overlying, nonproductive formations is often displaced with a nondamaging drill-in fluid.

Selection of the optimum drill-in fluid depends on the reservoir type and permeability, either natural fracture or matrix. Stephens (1994) developed practical guidelines for selecting drill-in fluids for horizontal wells depending on formation type, completion method, and cleanup/-stimulation technique. Table 5-2 provides selection information for vertically fractured reservoirs (limestones, chalks, brittle shales); Table 5-3 provides selection information for sand and sandstone formations.

Generally, horizontal wells are drilled with "thin" muds in turbulent flow or "thick" muds in laminar flow. Laminar flow is usually characterized by elevated low shear-rate viscosities and gel strengths; turbulent flow is more likely in competent formations such as the Austin Chalk. The build section of a horizontal well is typically the most difficult to clean. Cuttings beds formed in intervals between 30° and 60° can slide downward and cause packoffs, stuck pipe, high torque and drag, and cementing problems. Ironically, high-angle intervals (>75°) are inherently easier to clean. Mud lubricity is important, especially in long horizontal sections, but torque and drag can be exacerbated by the presence of a thick cuttings bed.

Wellbore stability problems in horizontal wells often can be attributed to formation stresses in shale intervals. Mud weights from 0.2 to 1 lb/gal higher than normal may be required to prevent compressive failure and subsequent hole collapse. Wellbore instability caused by stresses can be easily misdiagnosed as a primary hole-cleaning problem. Higher mud weights can aggravate stuck pipe, so filter cake buildup must be minimized on permeable formations.

5-9.3 Extended-Reach Wells

Extended-reach wells, particularly those with ultra-long horizontal displacements, represent some of the greatest challenges for drilling fluids. Stuck pipe, barite sag, excessive torque and drag, shale instability, poor hydraulics, and inadequate hole cleaning are among the potential limitations. Formation damage may also be critical, especially if the productive interval is drilled horizontally and completed open hole. Torque and drag problems, and hydraulics problems are most often linked with extended-reach wells.

Oil- and synthetic-based muds are the drilling fluids of choice for most extended-reach wells, primarily because of high levels of lubricity and inhibition. Well design and optimization require extensive modeling for torque and drag. Payne and Abbassian (1996) derived representative friction factors by analyzing historical data (Table 5-8). The friction factors vary with mud type and wellbore material (open hole or casing). Friction factors can also be measured in the laboratory, but correlations with field results have been difficult.

Many extended-reach wells are drilled with downhole motors. Some have horizontal departures approaching 25,000 ft, and sail angles on most vary from 65° to 75°. For these conditions, optimizing well hydraulics for penetration rates, hole cleaning, and barite sag can be difficult. Minimum high-shear-rate viscosity (plastic viscosity) is necessary, both for reasonable flow rates without exceeding pump horsepower limits and for maximizing bit hydraulic horsepower. Plastic viscosity increases with mud weight and solids content. Elevated low-shear-rate viscosities improve hole cleaning and minimize barite sag.

Table 5-8 Representative friction factors for different muds (from Payne and Abbassian, 1996)

Mud Type	Cased-Hole Friction Factor	Openhole Friction Factor
Water-based mud	0.24	0.29
Oil-based mud	0.17	0.21
Brine	0.30	0.30

5-9.4 Slimhole/Coiled-Tubing Drilling

Slimhole drilling is a re-emerging technology expected to make step-improvements in the efficiency and economy of well construction. Wells are normally classified as slimholes if the production interval is intentionally drilled with a bit diameter less than 4.75 in. (Hough, 1995). McCann *et al.* (1993) suggested a "narrow gap" criterion (a drillstring-to-hole diameter ratio greater than 0.8) to distinguish slimhole from reduced-bore and conventional wells. Either definition is acceptable.

The primary slimhole drilling techniques include (1) continuous coring, (2) conventional rotary drilling, (3) downhole-motor drilling, and (4) coiled-tubing drilling. They share several mud-related traits, including small circulating and pit volumes, small pumps, and high frictional pressure gradients (inside the drillstring and/or annulus). The key mud-related differences among the methods are the rotary speed of the drillstring and the annular gap, both of which can vary widely.

Continuous coring, adapted from the hard-rock mining industry, requires the use of solids-free drilling fluids. High-speed rotation (up to 1000 rpm) creates extraordinary centrifugal forces that can cause mud solids to cake on the inside walls of the drill rod (drillstring) and interfere with wireline recovery of the core barrel. Innovative mud formulations based on formate brines (primarily potassium formate, up to 1.6 sg) have provided the best results (Downs, 1993). High-density brines made with formate salts are nonhazardous and compatible with both conventional oilfield polymers and formation waters containing sulfates and carbonates. Continuous-coring operations also typically use very narrow annular gaps, some less than 0.25 in. Annular pressure losses, which can uncharacteristically exceed drill-rod pressure losses, become critical for lost-circulation and well-control concerns. Pipe rotation also increases annular pressure losses (McCann *et al.*, 1993). The resulting high overbalance may contribute to formation damage, but insufficient data are available at this time.

Rotary speeds for conventionally drilled slimholes (Sagot and Dapuis, 1994) are considerably lower (usually less than 350 rpm), minimal for downhole-motor drilling, and nonexistent for coiled-tubing drilling. Ultra-low solids content is not required, although solids (especially fines) should be minimized and carefully controlled for maximum performance.

Different types of drilling fluids have been used successfully for slimhole wells. However, biopolymer/brine fluids have distinct advantages. Brines provide solids-free density, inhibition, and improved temperature stability. Biopolymers, such as xanthan and welan gum, are compatible with brines and provide excellent viscosity, suspension, and drag reduction in turbulent flow (at low solids content). Starch or starch derivatives can be added to control fluid loss and acid-soluble, sized calcium carbonate can provide bridging, if required.

Goodrich *et al.* (1996) reported successful use of solids-free xanthan/brine drilling fluids for a coiled-tubing project in Prudhoe Bay, Alaska. Claims include increased horizontal reach, improved hole cleaning, better formation stability, lower pump pressures, and reduced pipe-sticking tendencies. Fluid loss is controlled by penetration of a highly viscous filtrate into the formation rather than a conventional filter cake.

5-10 FUTURE DEVELOPMENTS

Pressures on the petroleum industry to find and produce more oil and natural gas should intensify well into the next century. Unfortunately, drilling for oil and gas is not easy, inexpensive, or sanitary. Clearly, the interactions among performance, economics, and environmental concerns will continue to dominate. Health and environmental regulations will become stricter and more global. Commercial aspects will maintain high profiles. New drilling challenges will require new and improved drilling-fluids technology. Historically, drilling-fluids technology has always responded to new challenges, albeit some solutions have been more forthcoming than others. Real progress has been made in all three areas; more is needed.

The highest research priority in drilling fluids is the elusive, nontoxic, inexpensive, water-based mud that performs like an oil-based mud. Advancements in polymer technology continue to be encouraging, and new classes of synthetic-based muds are available that already match the performance levels of oil-based muds. Currently, these fluids are costly and not yet fully embraced by the worldwide environmental community. Then again, maybe the final answer will be less revolutionary and found in the collection of individual products being developed as a matter of course to replace lesser performing, more expensive, and/or environmentally objectionable counterparts.

REFERENCES

"1995-96 Environmental Drilling and Completion Fluids Directory," *Offshore/Oilman* (Sept. 1995) 33–56.

API RP 13B-1, "Standard Procedure for Field Testing Water-Based Drilling Fluids," 1st ed., Am. Pet. Inst. (June 1, 1990).

API RP 13B-2, "Standard Procedure for Field Testing Oil-Based Drilling Fluids," 2nd ed., Am. Pet. Inst. (Dec 1, 1991).

API RP 13D, "Recommended Practice on the Rheology and Hydraulics of Oil-Well Drilling Fluids," 3rd ed., Am. Pet. Inst. (June 1, 1995).

API Spec 13A, "Specification for Drilling-Fluid Materials," 15th ed., Am. Pet. Inst. (May 1, 1993).

Abrams, A.: "Mud Design To Minimize Rock Impairment due to Particle Invasion," *JPT* (May 1977) 586–92.

Bacho, J.: "Developments in Drilling Fluids Technology and Impact on Industrial Mining," paper SME 94–115, 1994.

Barker, J.W., and Gomez, R.K.: "Formation of Hydrates During Deepwater Drilling Operations," paper SPE 16130, 1987.

Beck, F.E., Powell, J.W., and Zamora, M.: "A Clarified Xanthan Drill-In Fluid for Prudhoe Bay Horizontal Wells," paper SPE 25767, 1993.

Bern, P.A., Zamora, M., and Slater, K.S.: "The Influence of Drilling Variables on Barite Sag," paper SPE 36670, 1996.

Bleier, R.D., Leuterman, A.J.J., and Stark, C.L.: "Drilling Fluids: Making Peace With the Environment," *JPT* (Jan. 1993) 6–10.

Bradley, W.B., Jarman, D., Plott, R.S., Wood, R. D., Schofield, T.R., Auflick, R.A., and Cocking, D.: "A Task Force Approach to Reducing Stuck-Pipe Costs," paper SPE 21999, 1991.

Candler, J.E., Leuterman, A.J.J., Wong, S.-Y.L., and Stephens, M.P.: "Sources of Mercury and Cadmium in Offshore Drilling Discharges," *SPEDE* (Dec. 1992) 279–283.

Carter, T.S., and Faul, G.L.: "Successful Application of the AOBM System in a Deep West Texas Well," paper SPE 24590, 1992.

Chenevert, M.E.: "Shale Control with Balanced Activity Oil-Continuous Muds," *JPT* (Oct. 1970) 1309–1316.

Clark, J.A., and Halbouty, M.T.: *Spindletop,* Gulf Pub. Co., Houston (1980).

Darley, H.C.H. and Gray, G.R.: *Composition and Properties of Drilling and Completion Fluids,* 5th ed., Gulf Pub. Co., Houston (1988).

Downs, J.D.: "Formate Brines: Novel Drilling and Completion Fluids for Demanding Environments," paper SPE 25177, 1993.

Fraser, L.: "Effective Ways to Clean and Stabilize High-Angle Holes," *Pet. Eng. Int.* (Nov. 1990) 30–35.

Friedheim, J.E., and Conn, H.L.: "Second Generation Synthetic Fluids in the North Sea: Are They Better?" paper SPE 35061, 1996.

Goodrich, G.T., Smith, B.E., and Larson, E.B.: "Coiled Tubing Drilling Practices at Prudhoe Bay," paper SPE 35128, 1996.

Hanson, P.M., Trigg, T.K., Rachal, G., and Zamora, M.: "Investigation of Barite 'Sag' in Weighted Drilling Fluids in Highly Deviated Wells," paper SPE 20423, 1990.

Hodge, R.M., MacKinley, W.M., and Landrum, W.R.: "The Selection and Application of Loss Control Materials to Minimize Formation Damage in Gravel Packed Completions for a North Sea Field," SPE 30119, 1995.

Hough, R.: "Slimhole Wells Present Tremendous Economic Opportunity," *Pet. Eng. Intl.* (July 1995) 22–27.

Hutchison, S.O., and Anderson, G.W.: "What To Consider When Selecting Drilling Fluids," *World Oil* (Oct. 1974) 83–94.

Jardine, S.I., McCann, D.P., and Barber, S.S.: "An Advanced System for the Early Detection of Sticking Pipe," paper SPE 23915, 1992.

Jones, F.V., Moffitt, C.M., and Leuterman, A.J.J.: "Drilling Fluids Disposal Regulations: A Critical Review," *Drilling* (Mar./Apr. 1987) 21–24.

Jones, R.D., and Sherman, J.: "Controlling Hydrates in Deep Water," *Offshore* (July 1986) 27–29.

Lai, D.T., and Dzialowski, A.K.: "Investigation of Natural Gas Hydrates in Various Drilling Fluids," paper SPE 18637, 1989.

Luo, Y., Bern, P.A., and Chambers, B.D.: "Simple Charts to Determine Hole Cleaning Requirements in Deviated Wells," paper SPE 27486, 1992.

McCann, R.C., Quigley, M.S., Zamora, M., and Slater, K.S.: "Effects of High-Speed Pipe Rotation on Pressures in Narrow Annuli," paper SPE 26343, 1993.

Minton, R.C., Meader, A., and Wilson, S.M.: "Downhole Cuttings Injection Allows Use of Oil-Base Muds," *World Oil* (Oct. 1992) 47–52.

Mondshine, T.C., and Kercheville, J.D.: "Successful Gumbo-Shale Drilling," *Oil & Gas J.* (Mar. 28, 1966) 194–205.

Nahm, J.J., Javanmardi, K., Cowan, K.M., and Hale, A.H.: "Slag Mix Mud Conversion Cementing Technology: Reduction of Mud Disposal Volumes and Management of Rig-Site Drilling Wastes," paper SPE 25988, 1993.

Newhouse, C.C.: "Successfully Drilling Severely Depleted Sands," paper SPE 21913, 1991.

Outmans, H.D.: "Mechanics of Differential Pressure Sticking of Drill Collars," *Trans.*, AIME **213** (1958) 265–274.

Payne, M. L., and Abbassian, F.: "Advanced Torque and Drag Considerations in Extended-Reach Wells," paper SPE 35102, 1996.

Sagot, A.M., and Dupuis, D.C.: "A Major Step in Ultra Slimhole Drilling," paper SPE 28299, 1994.

Scott, P.: "Drilling Fluids with Scavengers Help Control H2S," *Oil & Gas J.* (May 23, 1994) 72–75.

Sorelle, R.R., Jardiolin, R.A., Buckley, P., and Barios, J.R.: "Mathematical Field Model Predicts Downhole Density Changes in Static Drilling Fluids," paper SPE 11118, 1982.

Stamatakis, E., Thaemlitz, C.J., Coffin, G., and Reid, W.: "A New Generation of Shale Inhibitors for Water-Based Muds," paper SPE 29406, 1995.

Stephens, M.P.: "Drilling Fluid Design Based on Reservoir Characterization" in Lake *et al.* (eds.) *Reservoir Characterization II,* Academic Press (1991a).

Stephens, M.P.: "Drilling Fluid Key Factor in Preventing Damage," *Am. O&G Rep.* (Aug. 1991b) 47–51.

Stephens, M.P.: "Strategies For Preventing Formation Damage to Water-Sensitive Sands," *Can. Asso. of Drilling Eng. News* (Nov. 1993) 4.

Stephens, M.P.: "Selection of Non-Damaging Drill-In Fluids for Horizontal Wells," 1994 Conf. Horizontal Well Tech.

Stephens, M.P., and Bruton, J.R.: "Fluid Selection and Planning for Drilling Unconsolidated Formations," paper OTC-7021, 1992.

Svoboda, C. F.: "New 'Solids-Free' Drill-In Fluid Enhances Horizontal Wells," *Am. O&G Rep.* (Aug. 1996) 68–72.

Toups, J., Dzialowski, A., and Slater, K.: "New Device Measures Drilling Mud Lubricity Under Simulated Downhole Conditions," ASME ESTC, 1992.

van Oort, E., Ripley, D., Chapman, J. W., Williamson, R., and Aston, M.: "Silicate-Based Drilling Fluids: Competent, Cost-Effective and Benign Solutions to Wellbore Stability Problems," paper SPE 35059, 1996.

Weakley, R.R.: "Use of Stuck Pipe Statistics to Reduce the Occurrence of Stuck Pipe," paper SPE 20410, 1990.

White, W.W., Zamora, M., and Svoboda, C. F.: "Downhole Measurements of Synthetic-Based Drilling Fluid in Offshore Well Quantify Dynamic Pressure and Temperature Distributions," paper SPE 35057, 1996.

"World Oil's 1996 Drilling, Completion, and Workover Fluids," *World Oil* (June 1996) 87-126.

Zamora, M.: "On the HPHT Rheology and Hydraulics of Synthetic-Based Drilling Fluids," *Trans.,* Nordic Rheological Soc (1996) **4,** 16–19.

Zamora, M., and Hanson, P.: "Rules of Thumb to Improve High-Angle Hole Cleaning," *Pet. Eng. Intl.* Part 1 (Jan. 1991) 44–51, and Part 2 (Feb. 1991) 22–27.

6 Rock Mechanics in Wellbore Construction

Hazim Abass
Halliburton Energy Services

Justo Neda
Intevep

6-1 INTRODUCTION

Creating a circular hole and introducing drilling and completion fluids to an otherwise stable formation is the reason for a series of phenomena that result in wellbore instability, casing collapse, perforation failure, and sand production.

The circular hole causes a stress concentration that can extend to a few wellbore diameters away from the hole. This stress concentration, which differs from the far-field stresses, could exceed the formation strength, resulting in failure. The circular hole also creates a free surface that removes the natural confinement, which can, depending on the mechanical properties of the formation, reduce formation strength and trigger inelastic and time-dependent failure. Thus, a circular hole causes several important effects around a wellbore:

- Creation of a stress concentration field

- Removal of the confinement condition

- Inelastic and time-dependent displacement caused by the creation of a free surface

The severity of these effects and subsequent hole failure depend on the stress magnitudes and mechanical properties of the formation.

Similarly, introducing foreign fluids to the formation

- Disturbs the pore pressure, creating a localized, elevated pore pressure

- Reduces the cohesive strength of the formation, depending on the fluid interaction with the formation matrix

- Changes capillary forces

6-2 ROCK MECHANICAL PROPERTIES

Rather than discussing basic rock mechanics properties, this section will discuss only those specific properties relevant to near-wellbore activities. These activities include wellbore instability associated with wellbore drilling, cement failure, rock mechanics aspects of perforations, hydraulic fracture initiation, near-wellbore fracture geometry, and sand production. The mechanical properties discussed will include loading-unloading characteristics and cyclic loading, poroelasticity, viscoelasticity, pore collapse, and fracture toughness.

6-2.1 Loading-Unloading Characteristics

Wellbore construction consists of loading and unloading the formation and cement sheath. Therefore, studying the characteristics of a complete loading and unloading cycle is important. Additionally, evaluating cement and formation samples for the effect of cyclic loading can be beneficial during wellbore construction.

The stress-strain relationship describes the way the formation's framework of granular material responds to the applied load. This relationship indicates whether the material exhibits elastic or plastic, brittle or ductile, strain-softening or strain-hardening behavior during loading.

6-2.2 Poroelasticity

Within the proximity of the wellbore, poroelasticity can be examined based on the effective stress concept introduced by Terzaghi (1943) and Biot (1941). This concept suggests that pore pressure helps counteract the mechanical stress carried through grain-to-grain contact. The efficiency of the pore pressure, p_r, effect is measured by the poroelastic factor α; the relationship is

$$\sigma' = \sigma - \alpha p_r \qquad (6\text{-}1)$$

where σ' is the effective stress and σ is the total (absolute) stress.

The poroelastic constant is

$$\alpha = 1 - \frac{c_{ma}}{c_b}, \qquad 0 \le \alpha \le 1 \qquad (6\text{-}2)$$

where C_{ma} is the compressibility of the rock matrix, with the bulk compressibility c_b given by

$$c_b = \frac{3(1 - 2v)}{E} \qquad (6\text{-}3)$$

where v is Poisson's ratio, and E is Young's modulus.

If the rock has no porosity, the rock matrix compressibility, c_{ma}, is equal to c_b, and α becomes zero. Conversely, with high porosity, the matrix compressibility is small compared to the bulk compressibility, and α approaches unity. Although the poroelastic constant can be evaluated in the laboratory, the following technique explains physically how α can be evaluated from a given failure envelope obtained from dry samples.

6-2.2.1 Example 1: Determining the Poroelastic Constant

Available Data

The following data have been obtained in the laboratory with dry samples of sandstone:

Confining Pressure, psi	Ultimate Strength, psi	
0	13,500	
500	22,643	Tensile Strength
10,000	72,000	= 662 psi

One sample was saturated with oil and tested under a confining pressure of 7500 psi and a pore pressure of 7000 psi. The ultimate strength under these conditions was 28,409 psi. Calculate α.

Solution

Using the dry samples data, construct the Mohr envelope as shown in Figure 6-1. Since the sample had a failure stress of 28,409 psi, a Mohr circle was drawn so that it touches the failure envelope. An effective confining stress of 1250 psi is calculated from the left end of the Mohr circle, suggesting that the sample was exposed to an effective stress of 1250 psi. Then, α can be calculated with Equation 6-1:

$$1250 = 7500 - \alpha\,(7000)$$

and therefore $\alpha = 0.89$.

6-2.3 Viscoelasticity

"Creep" is a viscoelastic property or time-dependent phenomenon that significantly contributes to many instability problems related to wellbore construction. The following three applications in which creep is encountered will be discussed in Section 6-4: wellbore stability, resin-coated proppant used in hydraulic fracturing, and fracture closure in acid fracturing.

Creep tests study rock deformation under constant stress as a function of time. The total displacement obtained from applying a constant stress is the sum of two components:

$$\varepsilon_t = \varepsilon_e + \varepsilon(t) \qquad (6\text{-}4)$$

where ε_t is the total strain and ε_e is the elastic strain.

The creep function, $\varepsilon(t)$, characterizes the rheological properties of a rock. A general equation for the creep function can be very complex, and it is best described by experimental data for a given range of stress and temperature. Section 6-5 will demonstrate the use of this property.

6-2.4 Fracture Toughness

Fracture toughness is a material property that reflects the rock resistance for an existing fracture to propagate for a given fracturing mode. For a radial or "penny-shaped" fracture to propagate, the following condition must be satisfied at the fracture tip:

$$K_I \ge K_{IC}$$
$$K_I = (p_f - \sigma_{H,\min})\sqrt{\pi l} \qquad (6\text{-}5)$$

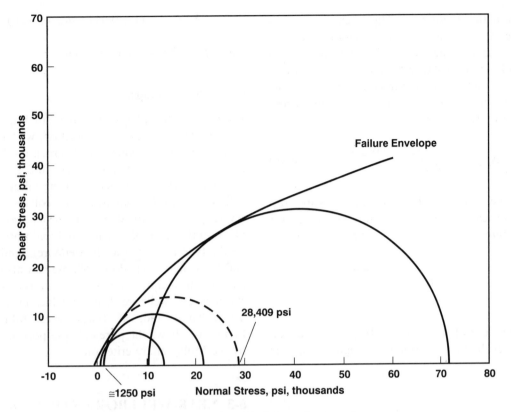

Figure 6-1 Graphical determination of α

where $\sigma_{H,\min} = \sigma'_{H,\min} + p_r$, and p_f is the pressure inside the fracture.

When linear elastic fracture mechanics is used, fracture toughness can be related to tensile strength, T (assuming static condition), as follows (Economides and Nolte, 1987):

$$T = \frac{K_{IC}}{\sqrt{\pi l}} \qquad (6-6)$$

where l is the half-length of an existing crack.

Other useful correlations between fracture toughness and tensile strength, T, Young's modulus, E, and compressive strength, C_0, are given by Whittaker *et al.* (1992):

$$K_{IC} = 0.27 + 0.107T \qquad (6-7)$$

$$K_{IC} = 0.336 + 0.026E \qquad (6-8)$$

$$K_{IC} = 0.708 + 0.006C_0 \qquad (6-9)$$

where K_{IC} is in MPa $\sqrt{m}$, E is in GPa, and C_0 is in MPa.

6-2.4.1 Example 2: Fracture Toughness vs. Cracks

Available Data

Sand and shale samples were laboratory-tested for tensile strength and fracture toughness with the following results:

Formation	Tensile Strength, psi	Fracture Toughness, psi $\sqrt{in}$
Sand	845	553
Shale	1155	784

Determine the size of the largest crack in the samples.

Solution

When the given data is applied to Equation 6-6, the following results are calculated:

- The size of the largest defect in the sand is 0.126 in.

- The size of the largest defect in the shale is 0.14 in.

6-2.4.2 Apparent Fracture Toughness

In hydraulic fracturing, the fracturing fluid does not fill the fracture completely, and a nonwetting tip region always exists at the fracture tip. This phenomenon has been repeatedly observed in the laboratory during hydraulic fracturing experiments. The nonwetting tip causes an additional resistance to fracture propagation, which leads to a higher apparent fracture toughness. The magnitude of this additional non-wet pressure drop, Δp_{nw}, is given by Knott (1973):

$$\Delta p_{nw} = \frac{1}{\sqrt{\pi l}} \left(\frac{E \sigma_{H,\min} W_{nw}}{1 - v^2} \right) \qquad (6\text{-}10)$$

where w_{nw} is the non-wet width, while the pressure drop in the fracture, caused by rock fracture resistance for Mode I fracture and a specific fracture geometry, is given by

$$\frac{K_{IC}}{\sqrt{\pi l}} \qquad (6\text{-}11)$$

A modified ring test (Thiercelin and Roegiers, 1986) can be used for evaluation of fracture toughness in the laboratory.

6-2.4.3 *Fracture Toughness vs. Effective Confining Pressure*

Roegiers and Zhao (1991) have shown that fracture toughness is a function of confining pressure, which may explain the difficulty of matching fracturing pressure on the basis of fracture toughness measured under unconfined conditions. Table 6-1 shows selected values of fracture toughness that were determined experimentally. Muller (1986) showed experimental results on sandstone specimens that followed the equation below:

$$K_{IC}(p_c) = (1 + 0.037 p_c) K_{IC} (p_c = 0) \qquad (6\text{-}12)$$

where p_c is the confining pressure.

6-2.5 Pore Collapse

Although pore collapse becomes an important factor during later-time reservoir depletion, wellbore construction design must account for this rock mechanical property as a means of preventing future near-wellbore problems such as subsidence, casing failure, perforation collapse, and other damage. Pore collapse is caused by severe pore-pressure depletion, which can create problems in the reservoir, such as reduced porosity and permeability (especially near the wellbore), sand production, and fines migration. Pore collapse can also cause compaction throughout the overburden layers, leading to subsidence or wellbore shearing, which may result in casing collapse and, at times, even tubing collapse. Figure 6-2 shows the pore collapse portion as described by the Mohr failure criterion.

6-3 NEAR-WELLBORE STRESS FIELD

A wellbore drilled through a rock formation introduces a new stress field at the wellbore proximity that may be great enough to cause failure. Additionally, when a wellbore is actively loaded (pressure in the wellbore is less than the reservoir pressure) or passively loaded (pressure in the wellbore is higher than the reservoir pressure), another stress effect could cause formation failure. If we assume a homogeneous, isotropic, linearly elastic rock mass being stressed below its yield limit, a stress field expressed in polar coordinates as vertical, tangen-

Table 6-1   Selected values of fracture toughness

Rock Type	Confining Pressure (MPa)	L_{IC} MPa $\sqrt{m}$ (Roegiers and Zhao, 1991)	K_{IC} MPa $\sqrt{m}$ (from Equation 6-12)
Chalk	0	0.73	0.73
Chalk	24.13	2.22	1.38
Chalk	48.26	2.32	2.04
Limestone	0	1.44	1.44
Limestone	24.13	2.12	2.73
Limestone	48.26	4.92	4.01
Sandstone	0	1.36	1.36
Sandstone	24.13	2.62	2.57
Sandstone	48.26	4.96	3.79

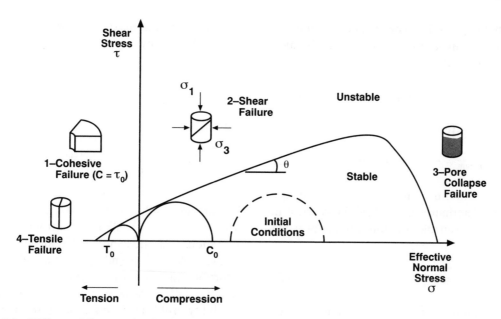

Figure 6-2 Different failure mechanisms and their stress-field locations within the Mohr-Coulomb failure envelope

tial, and radial is given by the Kirsch solution (Jaeger and Cook, 1979):

$$\sigma_v' = g \int_0^H \rho_b dH - \alpha p_r \qquad (6\text{-}13)$$

where ρ_b is the bulk density of the overburden layers and H is the depth.

$$\sigma_{rr}' = \frac{1}{2}\left(\sigma_{H,\max}' + \sigma_{H,\min}'\right)\left(1 - \frac{r_w^2}{r^2}\right)$$
$$+ \frac{1}{2}\left(\sigma_{H,\max}' - \sigma_{H,\min}'\right)\left(1 - \frac{4r_w^2}{r^2} + \frac{3r_w^4}{r^4}\right)\cos 2\theta$$
$$+ \frac{r_w^2}{r^2}(p_w - p_r) \qquad (6\text{-}14)$$

$$\sigma_{\theta\theta}' = \frac{1}{2}\left(\sigma_{H,\max}' + \sigma_{H,\min}'\right)\left(1 + \frac{r_w^2}{r^2}\right) -$$
$$\frac{1}{2}\left(\sigma_{H,\max}' - \sigma_{H,\min}'\right)\left(1 + \frac{3r_w^4}{r^4}\right)\cos 2\theta -$$
$$\frac{r_w^2}{r^2}(p_w - p_r) \qquad (6\text{-}15)$$

where p_w is the well bottomhole pressure and p_r is the reservoir pressure.

$$\tau_{r\theta} = \frac{1}{2}\left(\sigma_{H,\max}' - \sigma_{H,\min}'\right)\left(1 + \frac{2r_w^2}{r^2} - \frac{3r_w^4}{r^4}\right)\sin 2\theta \quad (6\text{-}16)$$

where the compressive stresses are positive, and θ is the angle measured from the direction of $\sigma_{H,\max}$. At the wellbore, $r = r_w$, and assuming that a mudcake differentiates the wellbore pressure, p_w from the reservoir pressure, p_r, then

$$\sigma_{rr}' = p_w - p_r \qquad (6\text{-}17)$$

$$\sigma_{\theta\theta}' = \sigma_{H,\max}' + \sigma_{H,\min}' - 2\left(\sigma_{H,\max}' - \sigma_{H,\min}'\right)$$
$$\cos 2\theta - (p_w - p_r) \qquad (6\text{-}18)$$

and

$$\tau_{r\theta} = 0 \qquad (6\text{-}19)$$

If we consider two cases where $\theta = 0\,(\sigma_{H,\max})$ and $\theta = 90\,(\sigma_{H,\min})$, we get

$$\sigma_{\theta=0} = 3\sigma_{H,\min}' - \sigma_{H,\max}' - p_w + p_r \qquad (6\text{-}20)$$

and

$$\sigma_{\theta=90} = 3\sigma_{H,\max}' - \sigma_{H,\min}' - p_w + p_r \qquad (6\text{-}21)$$

To initiate a tensile failure, as is the case in hydraulic fracturing, $\sigma_{\theta=0}$ should become a negative value of the tensile strength ($\sigma_{\theta=0} = -T$). The breakdown pressure p_{bd} required for initiating a fracture can be readily calculated from Equation 6-20:

$$p_{bd} = 3\sigma'_{H,min} - \sigma'_{H,max} + T + p_r \qquad (6\text{-}22)$$

or, in terms of total stresses,

$$p_{bd} = 3\sigma_{H,min} - \sigma_{H,max} + T - p_r \qquad (6\text{-}22a)$$

As explained in Section 6-5, the stresses $\sigma_{H,min}$, $\sigma_{H,max}$, and σ_v should be determined in the field; however, if no field information is available, the following approximation can be used:

$$\sigma'_v = 1.1H - \alpha p_r \qquad (6\text{-}23)$$

Based on the assumption that the formation is elastic, tectonically relaxed, and constrained laterally, the effec-tive minimum and maximum horizontal stresses can be estimated as follows:

$$\sigma'_{H,min} = \frac{v}{1 - v}\left(\sigma'_v\right) \qquad (6\text{-}24)$$

and

$$\sigma'_{H,max} = \frac{\sigma'_v + \sigma'_{H,min}}{2} \qquad (6\text{-}25)$$

The previous solution is valid only for elastic rocks. However, when a wellbore is introduced to an intact formation, a plastic region is developed in the proximity of the wellbore, extending a few wellbore diameters before an in-situ elastic region prevails, as shown in Figure 6-3. The plastic region can create many wellbore instability problems during drilling. In the payzone, as a result of production, the plastic region may propagate

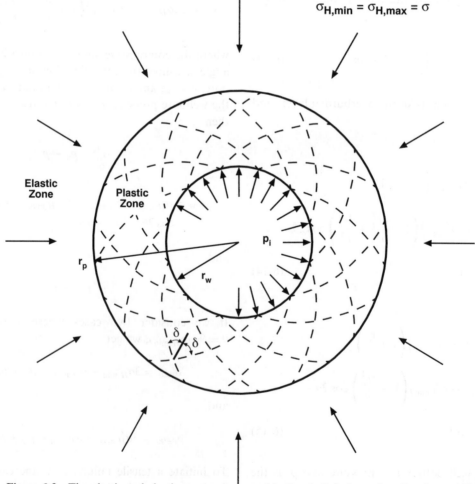

Figure 6-3   The plastic and elastic zones as assumed in Bray's Solution, after Goodman, 1980

deeper into the reservoir, causing sand production. In fractured rocks, wellbore collapse may result, unless a high drilling-fluid weight is used. The size of the plastic region must be known for such applications as wellbore stability, perforations, and sand production.

Bray (1967) assumed that fractures exist in the plastic zone with log spirals inclined at d degrees with the radial direction as shown in Figure 6-3 (Goodman, 1980). If we apply the Mohr-Coulomb theory, the radius of the plastic zone, r_{pl}, is

$$r_{pl} = r_w \left[\frac{2\sigma' - C_0 + \left(1 + \frac{1 + \sin\phi}{1 - \sin\phi}\right)C_j \cot\phi_j}{\left(1 + \frac{1 + \sin\phi}{1 - \sin\phi}\right)\left((p_w - p_r) + C_j \cot\phi_j\right)} \right]^{1/Q} \tag{6-26}$$

where

$$Q = \frac{\tan\delta}{\tan(\delta - \phi_j)} - 1 \tag{6-27}$$

and where C_j is the cohesion for jointed rocks (experimentally determined) and ϕ_j is the internal friction angle for jointed rocks.

In Equation 6-26,

$$\sigma'_{H,\min} = \sigma'_{H,\max} = \sigma'$$

Within the elastic and plastic zones, Bray's solution is given by

$$\sigma'_r = \sigma' - r_{pl}\left[\frac{\left(\frac{1 + \sin\phi}{1 - \sin\phi} - 1\right)\sigma' + C_0}{r^2\left(\frac{1 + \sin\phi}{1 - \sin\phi} + 1\right)} \right] \tag{6-28}$$

$$\sigma'_{\theta\theta} = \sigma' - r_{pl}\left[\frac{\left(\frac{1 + \sin\phi}{1 - \sin\phi} - 1\right)\sigma' + C_0}{r^2\left(\frac{1 + \sin\phi}{1 - \sin\phi} + 1\right)} \right] \tag{6-29}$$

for the elastic zone, and

$$\sigma'_{rr} = \left((p_w - p_r) + C_j \cot\phi_j\right)\left(\frac{r}{r_w}\right)^Q - C_j \cot\phi_j \tag{6-30}$$

$$\sigma'_{\theta\theta}\left((p_w - p_r) + C_j \cot\phi_j\right)\frac{\tan\delta}{\tan(\delta - \phi_j)}\left(\frac{r}{r_w}\right)^Q - C_j \cot\phi_j \tag{6-31}$$

for the plastic zone.

The following demonstration of the equations was taken from Goodman (1980).

6-3.1 Example 3: Plastic-Zone Size

6-3.1.1 Available Data

Consider a plastic zone created around a wellbore with fractures that can be described as $\phi_j = 30$, $C_j = 0$, and $\delta = 45$. The mechanical properties of the virgin rock mass are $C_o = 1300\,\text{psi}$ and $\phi = 39.9°$. The in-situ stresses are given by $\sigma_{H,\min} = \sigma_{H,\max} = 4000\,\text{psi}$ and $p_w - p_r = 40\,\text{psi}$. Determine the stresses in the elastic and plastic regions using Bray's solution. Determine the stresses using Kirsch's solution.

6-3.1.2 Solution

From the data above, the following is obtained: From Equation 6-27, $Q = 2.73$; from Equation 6-26, $r_{pl} = 3.47\,r_w$. Therefore, from Equations 6-30 and 6-31,

$$\sigma'_{rr} = 40\left(\frac{r}{r_w}\right)^{2.73} \quad \text{and} \quad \sigma'_{\theta\theta} = 149\left(\frac{r}{r_w}\right)^{2.73}$$

for the plastic zone, and from Equations 6-28 and 6-29,

$$\sigma'_{rr} = 4000 - 33,732\frac{r_w^2}{r^2} \quad \text{and} \quad \sigma'_{\theta\theta} = 4000 - 33,732\frac{r_w^2}{r^2}$$

for the elastic zone.

Figure 6-4 shows a plot of the results. When Bray's solution is used, the radius of the plastic zone is 3.47 times the wellbore radius. These data are important if the plastic zone must be consolidated in a poorly consolidated formation. With well-testing, the zone can be examined as a skin caused by the plastic deformation of the rock. Kirsch's solution for the same data is given in Figure 6-4.

6-4 FAILURE CRITERIA

To understand a failure mechanism, we must apply a specific and compatible failure criterion. Granular material, such as sand, fails in shear, while for soft material

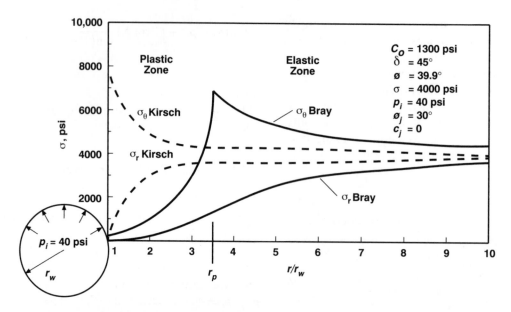

Figure 6-4   Kirsch and Bray Solutions, from Goodman, 1980

such as clays, plastic compaction dominates the failure mechanism. The following failure mechanisms can cause wellbore and near-wellbore instability problems:

- Shear failure without appreciable plastic deformation, such as a breakout

- Plastic deformation and compaction, which may cause pore collapse

- Tensile failure, causing the formation to part

- Cohesive failure, equivalent to erosion, which can cause fines migration and sand production

- Creep, which can cause a tight hole during drilling

- Pore collapse, which is comprehensive failure of the infrastructure of the matrix framework that can happen during the later time of production

Many empirical criteria have been developed that predict rock failure. It is imperative to understand the physical interpretation of those criteria before they are applied for problems associated with wellbore construction. Such criteria are empirical, and engineers should carefully select the appropriate ones for a given problem. Generally, failure criteria are used to generate failure envelopes, usually separating stable and unstable zones. Although some engineers attempt to linearize the failure envelope, such linearization is artificial. Three criteria

that are useful to wellbore and wellbore proximity failure are presented in the following sections.

6-4.1 Mohr-Coulomb Failure Criterion

The Mohr-Coulomb failure criterion relates the shearing resistance to the contact forces and friction and the physical bonds (cohesion) that exist among the grains. A linear approximation is given as follows:

$$\tau = C + \sigma'_n \tan \phi \qquad (6\text{-}32)$$

where τ is the shear stress, C is the cohesive strength, ϕ is the angle of internal friction, and σ'_n is the effective normal stress acting on the grains. The factors C and ϕ are coefficients for the linearization and should be determined experimentally. A deviation from a straight line is very common during attempts to interpret other failure mechanisms with this criterion, which is solely based on shear failure. Therefore, this criterion should be applied only to situations for which it is valid. The failure envelope is determined from many Mohr circles. Each circle represents a triaxial test where a sample is subjected to lateral confinement ($\sigma_2 = \sigma_3$), and axial stress (σ_1) is increased until failure. The envelope of Mohr circles represents the basis of this failure criterion.

Sometimes, it is more convenient to express the Mohr-Coulomb linear envelope in terms of σ_1 and σ_3. This expression becomes

$$\sigma_1' = \frac{1 + \sin\phi}{1 - \sin\phi}\sigma_3' + C_0 \qquad (6\text{-}33)$$

$$C_0 = 2C\tan\left(\frac{\pi}{4} + \frac{\phi}{2}\right) = 2C\frac{\cos\phi}{1 - \sin\phi} \qquad (6\text{-}34)$$

Equation 6-34 is also equivalent to

$$\sigma_1' = C_0 + \sigma_3'\tan^2\left(\frac{\pi}{4} + \frac{\phi}{2}\right) \qquad (6\text{-}35)$$

Once the failure envelope is determined, stability can be analyzed by calculations of the normal and shear stresses for a given situation assuming the shear angle 45°

$$\sigma_n' = \frac{\sigma_1 + \sigma_3}{2} - \alpha p_r \qquad (6\text{-}36)$$

and

$$\tau_{\max} = \frac{\sigma_1 - \sigma_3}{2} \qquad (6\text{-}37)$$

These points are then plotted on the failure envelope, and the rock is evaluated for stability. If the failure envelope exhibits nonlinear behavior, the linearization attempt should be exercised for the stress range for which a given problem is applied.

6-4.2 Hoek-Brown Criterion

The Hoek-Brown criterion (Hoek and Brown, 1980) is also empirical and applies more to naturally fractured reservoirs. The criterion states that

$$\sigma_1' = \sigma_3' + \sqrt{I_m C_0 \sigma_3' + I_s C_0^2} \qquad (6\text{-}38)$$

where I_m is the frictional index and I_s is the intact index. Both indices are material-dependent properties.

This criterion matches reasonably the brittle failure, but it gives poor results in ductile failure. Therefore, it is used for predicting failure in naturally fractured formations. The parameters I_m, I_s, and C_0 are measured in the laboratory. For weak rock, I_m is less than 0.1 and I_s is less than 0.0001; however, for hard rock, I_m ranges from 5 to 15 and I_s is equal to 1.

6-4.3 Drucker-Prager Criterion (Extended Von Mises)

The Drucker-Prager criterion is based on the assumption that the octahedral shearing stress reaches a critical value:

$$\alpha I_1 + \sqrt{J_2} - K = 0 \qquad (6\text{-}39)$$

where

$$J_2 = \frac{1}{6}\left[(\sigma_1' - \sigma_2')^2 + (\sigma_2' - \sigma_3')^2 + (\sigma_3' - \sigma_1')^2\right] \qquad (6\text{-}40)$$

and

$$I_1 = \sigma_1' + \sigma_2' + \sigma_3' \qquad (6\text{-}41)$$

which is the first invariant of stress tensors.

The material parameters, α and K, are related to the angle of internal friction, ϕ, and cohesion, C, for linear condition, as follows:

$$\alpha = \frac{2\sin\phi}{\sqrt{3}(3 - \sin\phi)} \qquad (6\text{-}42)$$

and

$$K = \frac{6C\cos\phi}{\sqrt{3}(3 - \sin\phi)} \qquad (6\text{-}43)$$

A plot of $\sqrt{J_2}$ vs. I_1 at failure conditions allows evaluation of a given problem related to rock failure. This criterion fits the high stress level.

6-5 APPLICATIONS RELATED TO WELLBORE-PROXIMITY MECHANICS

6-5.1 Wellbore Stability During Drilling

Most wellbore stability problems occur in shale formations. Unfortunately, shale properties range from very soft to very hard, and from very laminated to very intact. Several mechanisms cause wellbore instability problems (Figure 6-5); chemical and mechanical effects will be discussed in this section.

6-5.1.1 Chemical Effects

Ion-exchanging clays, such as illite, mica, smectite, chlorite, mixed-layer clays, and zeolites, can cause many wellbore instability problems. Engineers may erroneously try

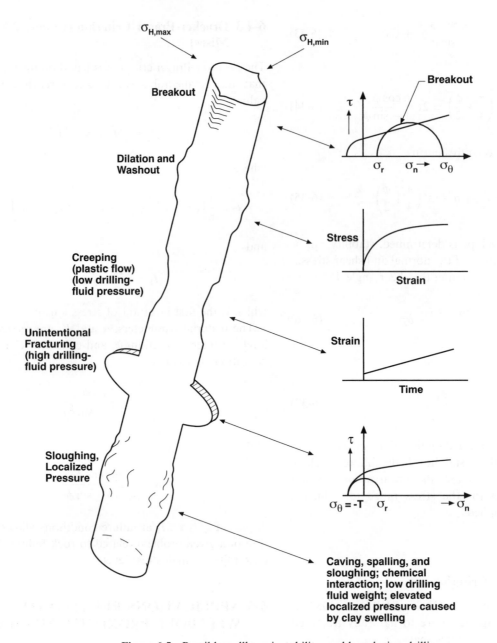

Figure 6-5   Possible wellbore instability problem during drilling

to model failure mechanisms with analytical or empirical mechanical models while the main mechanism may be failure because of chemical effects. The following failure mechanisms during wellbore construction can be related to chemical causes.

Clay Swelling (Hydration) and Migration

Most shale formations contain water-sensitive clay materials such as smectite, illite, and mixed-layer clays, which absorb water that induces an elevated localized pressure.

This pressure reduces the effective stress around the wellbore, which causes the shale matrix to swell, disintegrate, and collapse (Dusseault and Gray, 1992). Mody and Hale (1993) developed a model that incorporated mechanical and chemical effects to evaluate the drilling fluid on shale stability.

Ion Exchanging

Brines such as KCl can control clay swelling, but illite, chlorite, smectite, and mixed-layer clays can change the brine through ion-exchanging mechanisms and swell afterward.

Cementation Deterioration

When examining sand formations, engineers must study the degree and type of cementation. Mineralogical analysis, thin-section petrography, and fluid compatibility are viable testing methods for evaluating sand production.

Near-Wellbore Damage

Near-wellbore formation damage can occur because of paraffin deposits, scale deposits, fines migration caused by kaolinite and illite clays, asphaltene precipitation, sand production, emulsions induced by iron, formation of oil emulsions by acid in combination with soluble iron, iron-compound precipitation, and even emulsions formed from fracturing fluids during stimulation.

To overcome these chemical effects, engineers should select the type of drilling-fluid system based on its effect on formation strength. The effects of the chemical and physical properties of the drilling-fluid system on formation stability should be lumped and the drilling-fluid system should be evaluated on the basis of rock mechanics. The following example shows an easy technique that would lump all chemical effects and evaluate the drilling-fluid systems based on rock mechanics.

6-5.1.2 Example 4: Drilling-Fluid Selection—A Rock Mechanics Perspective

Available Data

Six drilling-fluid samples are being evaluated for use through a shale segment that exhibited many instability problems during drilling. Show how these systems can be evaluated on the basis of rock mechanics.

Solution

The core samples should be preserved after coring, and no cleaning or drying processes should be conducted. The samples should be brought to their estimated in-situ stress fields at a suggested value of the confining pressure ($p_c = p_w - p_r$). The same strain rate (for example, $10^{-4}\,\text{sec}^{-1}$) or the same loading rate (for example 5 to $10\,\text{psi/sec}$) should be used for all tests. The samples should be cut in the same manner and same direction (vertically or horizontally). Two samples for each drilling-fluid system and two samples with no drilling fluid should be tested for repeatability. Two samples should be saturated in each drilling-fluid system for the same length

of time (for example, 1 month) in addition to the two other samples that should be left in closed jars as the base samples. This procedure is applied to optimize the drilling-fluid system based on the compressive strength and Young's modulus reduction obtained for these systems.

The results are given in Figure 6-6. System 5 was selected to drill the given shale section. Notice how both the compressive strength and Young's modulus were affected by the different drilling-fluid systems.

6-5.1.3 Mechanical Effects

Tensile and shear failure mechanisms should be considered for wellbore stability evaluation during drilling.

Tensile Failure

The effective stress at the wellbore exceeds the tensile strength of the formation and causes tensile failure. Therefore, an induced fracture can result because of drilling-fluid loss if

$$p_w \geq p_r + \sigma'_{\theta\theta} + T \qquad (6\text{-}44)$$

For an elastic medium, this is given by (Haimson and Fairhurst, 1967)

$$p_w = 3\sigma'_{H,\min} - \sigma'_{H,\max} + T + p_r \qquad (6\text{-}45)$$

However, if a natural fracture exists, then the tensile strength, *T,* should be assumed to be zero.

Shear Failure

Once a wellbore is drilled and a stress concentration field is established, the rock will either withstand the stress field or yield, resulting in a near-wellbore breakout zone that causes spalling, sloughing, and hole enlargement. An appropriate failure criterion should be used for evaluation of this type of failure.

Drilling-Fluid Weight

Drilling-fluid weight should be calculated as a means of preventing the initiation of tensile and shear (plastic) failures. In some formations, the drilling-fluid weight should prevent creeping in viscoplastic formations, such as salt rock. Drilling-fluid weight is an important consideration for treating wellbore instability problems. The drilling-fluid weight is limited by two boundaries:

- The upper boundary is the pressure that causes tensile failure and drilling-fluid loss. This pressure can be determined in the field based on Equation 6-44.

- The lower boundary is the pressure required to provide confining stress, which is removed during drilling. The confining stress prevents shear failure, the creation of a plastic zone, and plastic flow (creep).

The upper boundary is estimated from the in-situ stress field, and the tensile strength is measured in the laboratory. While the lower boundary is estimated from the in-situ stress field, mechanical properties of the formation are estimated from one of the failure criteria described in Section 6-3 that best models a given formation. The Mohr-Coulomb failure criterion, which is a two-dimensional model, can be used for determining the drilling-fluid weight required to prevent shear failure (Equation 6-33).

Near the wellbore, the stress σ_3' can be represented as σ_{rr}', as given in Equation 6-17.

However, the term σ_1' can be represented by the tangential stress $\sigma_{\theta\theta}'$ which is described by Equations 6-20 and 6-21. Of the two equations, 6-21 represents the point at the wellbore that will be exposed to the most tangential stress. This stress should be considered for drilling-fluid weight calculations. Therefore, the term σ_1' is given by

$$\sigma_1' = 3\sigma_{H,\max}' - \sigma_{H,\min}' \qquad (6\text{-}46)$$

By substituting these results in Equation 6-33, we receive

$$3\sigma_{H,\max}' - \sigma_{H,\min}' = \frac{1 + \sin\phi}{1 - \sin\phi}(p_w - p_r) + C_0 \qquad (6\text{-}47)$$

or

$$p_w = \frac{3\sigma_{H,\max}' - \sigma_{H,\min}' - C_0}{\dfrac{1 + \sin\phi}{1 - \sin\phi}} + p_r \qquad (6\text{-}48)$$

If we assume that $\sigma_{H,\max}' = \sigma_{H,\min}' = \sigma'$, then

$$p_w = \frac{2\sigma' - C_0}{\dfrac{1 + \sin\phi}{1 - \sin\phi}} + p_r \qquad (6\text{-}49)$$

The drilling-fluid weight should be calculated based on p_w as given in Equations 6-48 or 6-49.

A different model can be used based on the radius of the plastic zone created as a result of drilling a wellbore, as given by Equation 6-26.

To prevent the generation of a plastic zone, r_{pl} must equal r_w. Substituting this condition in Equation 6-26, we get

$$2\sigma' - C_0 = \left(1 + \frac{1 + \sin\phi}{1 - \sin\phi}\right)(p_w - p_r) \qquad (6\text{-}50)$$

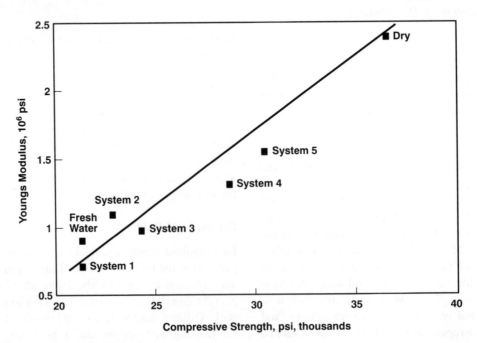

Figure 6-6   The effect of different drilling fluid systems on the mechanical properties of shale samples

Then

$$p_w = \frac{2\sigma' - C_0}{1 + \dfrac{1 + \sin\phi}{1 - \sin\phi}} + p_r \qquad (6\text{-}51)$$

The drilling-fluid weight should be determined based on p_w. A comparison of Equations 6-49 and 6-51 suggests that the two models will provide different results for the drilling-fluid weight. The following example demonstrates the use of these models to determine the drilling-fluid weight constraints that should be considered for the prevention of wellbore instability.

6-5.1.4 Example 5: Drilling-Fluid Weight Window

Available Data

For the same data discussed in Example 4, determine the drilling-fluid weight window to drill through the shale section using the two methods previously discussed.

Solution

$$\sigma_{av} = \frac{\sigma_{H,min} + \sigma_{H,max}}{2} = \frac{8000 + 10,000}{2} = 9000$$

$$\sigma'_{av} = 9000 - 0.6(6000) = 5400\,\text{psi}$$

$$\sigma'_{H,min} = 8000 - 0.6(6000) = 4400\,\text{psi}$$

$$\sigma'_{H,max} = 10,000 - 0.6(6000) = 6400\,\text{psi}$$

From Equation 6-48, $p_w = 9011\,\text{psi}$ and therefore $\rho_{\text{drilling fluid}} = 15.8\,\text{lb/gal}$. From Equation 6-51, $p_w = 7363\,\text{psi}$, and therefore $\rho_{\text{drilling fluid}} = 12.9\,\text{lb/gal}$.

Based on the assumption that tensile strength is zero and a natural fracture intersecting the well extends beyond the near-wellbore stress field, the pressure required to propagate the fracture would be equal to the reservoir pressure plus the minimum horizontal stress (neglecting fracture toughness), and thus $p = 4400 + 6000 = 10,400\,\text{psi}$, which corresponds to $\rho_{\text{drilling fluid}} = 18.2\,\text{lb/gal}$.

If we assume that the wellbore pressure calculated from the plastic radius method provides conservative results, the drilling-fluid weight window is

$$15.8 < \rho_{\text{drilling fluid}} < 18.2\ \text{lb/gal}$$

Any model used should be calibrated in the field as a means of examining its application to the formation.

Other models can be used to develop a procedure to determine the drilling-fluid weight window for a given reservoir.

6-5.1.5 Example 6: Evaluation of Wellbore Stability

Available Data

A shale section with a vertical well drilled through it provides the following data:

$$\sigma_v = 12,000\,\text{psi}$$
$$\sigma_{H,min} = 8000\,\text{psi}$$
$$\sigma_{H,max} = 10,000\,\text{psi}$$
$$p_w = 6000$$
$$p_r = 5700$$
$$\alpha = 0.6$$
$$H = 11,000\,\text{ft}$$
$$C_0 =$$
$$\phi = 32° \ \text{(friction angle)}$$

Several samples were tested, and the Mohr-Coulomb failure envelope was constructed for the evaluation of borehole stability along the entire wellbore circumference (Figure 6-7).

Solution

Select four locations along the borehole circumference so that $\theta = 0°, 30°, 60°,$ and $90°$, which covers one quarter of the wellbore; the analysis is repetitive for the other locations.

$$\sigma'_{rr} = p_w - \alpha p_r = 2580\,\text{psi}$$
$$\sigma'_{\theta\theta} = \sigma_{H,max} + \sigma_{H,min} - 2(\sigma_{H,max} + \sigma_{H,min})\cos 2\theta$$
$$\qquad - (p_w - \alpha p_r)$$

Thus,

$$\sigma'_{rr} = 18,000 - 4000\cos 2\theta - 2580$$
$$\sigma'_{\theta\theta} = 15,420 - 4000\cos 2\theta$$

and therefore

$$\sigma'_{\theta\theta} = 11,420 \ \text{at}\ \theta = 0°,$$
$$\sigma'_{\theta\theta} = 13,420 \ \text{at}\ \theta = 30°,$$
$$\sigma'_{\theta\theta} = 17,420 \ \text{at}\ \theta = 60°,$$
$$\sigma'_{\theta\theta} = 19,420 \ \text{at}\ \theta = 90°$$

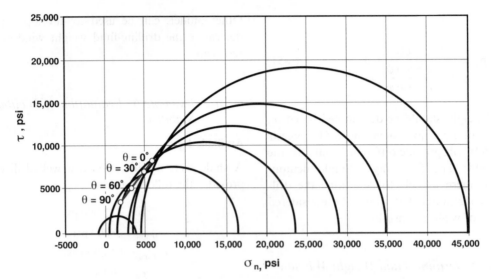

Figure 6-7 Graphical evaluation of wellbore stability using a given failure envelope. Data show stable hole but a close to stable/instable boundary

Using these results and Equations 6-36 and 6-37 where σ_1 is replaced by and $\sigma'_{\theta\theta}$ and σ_3 is replaced by σ'_{rr}, then σ'_n and τ_{max} can be calculated and tabulated below

θ	σ'_n (psi)	τ_{max}(psi)
°	5300	8420
30°	4300	7420
60°	2300	5420
90°	1300	4420

Figure 6-7 shows that all points are within the stable region of the failure envelope, but they are close to the border. If a deviated or a horizontal well is to be evaluated, then the stress field around the wellbore should be calculated based on transforming the principal stresses into the principal ones of the well. Then the same previously described procedure can be applied.

6-5.1.6 Example 7: Failure of Shales

Available Data

Assuming that the failure envelope of the shale section is given in Figure 6-8, evaluate the wellbore stability given in the previous example.

Solution

All data replotted in Figure 6-8 are in the unstable region of the failure envelope, which suggests that the wellbore

will fail in all directions. It is possible that one criterion could show a stable well while another could indicate failure. Therefore, the appropriate failure criterion must be selected and verified in the field.

If a 3D failure criterion is used, such as the Drucker-Prager criterion, then the three stresses can be used to analyze wellbore stability, and the parameters used will be as follows:

$$I_1 = \frac{1}{3}(\sigma_r + \sigma_\theta + \sigma_z) \qquad (6\text{-}52)$$

and

$$J_2 = \frac{1}{6}\left[(\sigma_z - \sigma_r)^2 + (\sigma_\theta - \sigma_z)^2(\sigma_r - \sigma_\theta)^2\right] \qquad (6\text{-}53)$$

6-5.2 Perforations

Perforations are the traditional means of allowing a cemented, cased well to communicate with the reservoir. Good, clean perforations allow sufficient, unhindered production with reasonably low drawdown, inhibiting the process of sand production. Research in this vital area of wellbore construction has resulted in the following industry practices, even though the effectiveness of some practices is still debatable:

• The use of clean, low-solids-content, compatible completion fluids

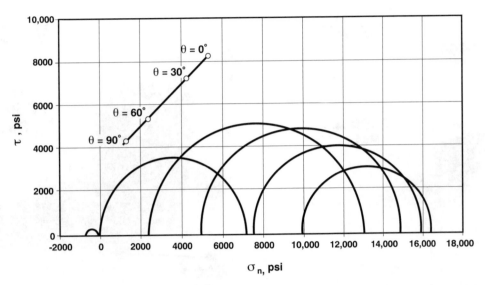

Figure 6-8   Graphical evaluation of wellbore stability with a given failure envelope. Data show instability problems in all directions around wellbore

- The use of underbalanced perforations, such as 500 psi in oil wells and 1500 psi in gas wells

- The use of small charges, minimizing the effect of the compacted area (stress cage) around a perforation tunnel, which can reduce the original permeability as much as a factor of 10 or more

McLeod (1982) showed that the skin caused by the compacted zone could be very large.

The purpose of perforating should be defined clearly and designed appropriately. The perforation design should be evaluated based on the expected well completion and stimulation activity (Morita and McLeod, 1994). The purposes of perforating are listed below:

- Perforation as a completion method only or with the intention of matrix stimulation

- Perforation with the intention to gravel-pack

- Perforation for hydraulic fracturing (proppant or acid fracturing)

The perforation process will be discussed thoroughly in Chapter 13. This chapter will only discuss oriented perforations relative to the in-situ stresses. This concept has been introduced to solve critical problems encountered during wellbore construction.

6-5.2.1 Oriented Perforations for Hydraulic Fracturing

Experimental and theoretical work indicates that perforation orientations should be designed to eliminate problems in fracturing vertical and deviated wells (Morita and McLeod, 1994; Behrmann and Elbel, 1991; Abass *et al.*, 1994; Venditto *et al.*, 1993). For a successful hydraulic fracturing treatment, perforation should be in phase with the anticipated fracture direction (the direction of maximum horizontal stress). This condition will

- Obtain maximum fracture width near the wellbore

- Create a single fracture

- Reduce the breakdown and propagation pressures

Figure 6-9 presents experimental results that show a complex fracture system in which the perforations are oriented within certain angles from the fracture direction (Abass *et al.*, 1994). The experimental study was conducted with hydrostone samples to study the effects of oriented perforations in the high and low sides of the horizontal well in the direction of the anticipated fracture. Figure 6-10 shows that for perforation angles of 0°, 15°, and 30°, the average breakdown pressure was 3200 psi; this pressure steadily increased for angles higher than 30°. Additionally, this experiment showed that fracture width is a function of perforation orientation.

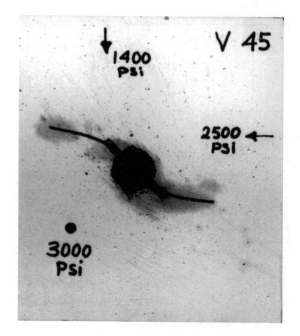

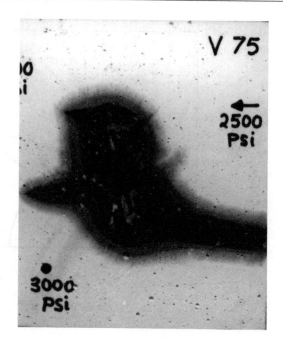

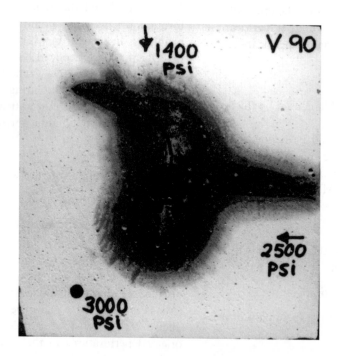

Figure 6-9   Near wellbore fracture geometry as a function of perforation orientation relative to direction of in-situ stresses

Figure 6-10 suggests that the optimal perforation phasing is 60° (equivalent to at most 30° deviation from the fracture direction) or less, at which the breakdown pressure is minimal. For an explanation of the negative widths in Figure 6-10, the reader is referred to Abass *et al.* (1994).

6-5.2.2 Clustered Perforations for Fracturing Deviated Wells

Clustered perforations can produce a transverse fracture perpendicular to the wellbore. A short perforated interval of 1 to 2 ft with 24 shots/ft can help reduce the occurrence of multiple fractures. A prefracturing stage of hydrochloric acid (HCl) in the treatment program can help establish a better communication channel between the wellbore and the main fracture.

6-5.2.3 Hydrojetting with HCl for Fracturing Horizontal Wells

Hydrojetting can ease the near-wellbore stress concentration, resulting in a successful fracturing treatment (Haigist *et al.*, 1995). Figure 6-11 presents the sequence of operations for creating a single fracture from a horizontal well.

6-5.2.4 Oriented Perforations for Sand Control

As previously explained, a circular wellbore in a rock formation creates a new stress field around the wellbore, which causes oriented failure (breakout) or total collapse (washout). Oriented perforations can be used for breakouts or unconsolidated formations.

Consolidated Formations

If breakout exists in a consolidated formation, the following steps are recommended (Figure 6-12):

- The near-wellbore area should be consolidated with a liquid resin material that is injected into the payzone

- Because the breakout is oriented in the direction of minimum horizontal stress, a 180° phasing should be performed in the direction of maximum horizontal stress

- A hydraulic fracture using a *fracpack* design should be performed

Unconsolidated Formations

Experiments showed that a wellbore should not be drilled through an unconsolidated formation because it would create a concentrated stress field around the wellbore. The hydrocarbon can be produced through

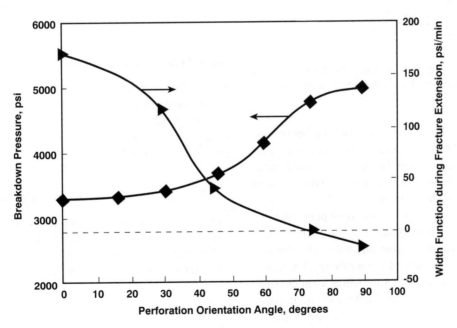

Figure 6-10 Fracture width performance and fracture initiation pressure as a function of perforation orientation relative to direction of in-situ stresses

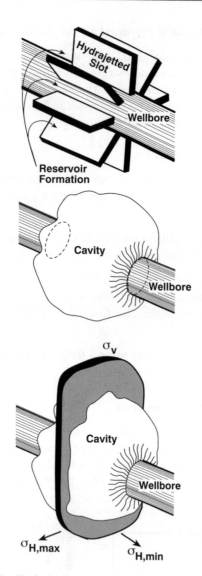

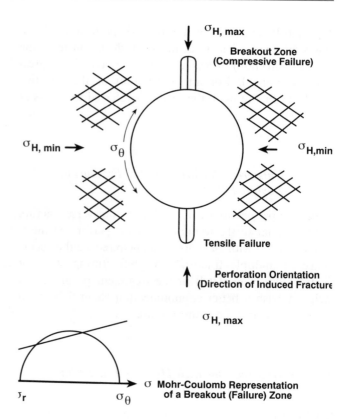

Figure 6-11 Hydrojetting and acidizing stage to initiate a plane fracture from horizontal wells in carbonate formation

Figure 6-12 Oriented perforation for sand control, where the breakout region is left undisturbed

disturbance of energy required for fracture propagation works favorably in the technique above.

6-5.3 Hydraulic Fracturing

This section will discuss only the rock mechanics aspects of hydraulic fracturing as it relates to near-wellbore proximity. The fracturing process as a stimulation technique, and its interaction with the reservoir will be discussed in Chapter 17.

6-5.3.1 Fracture Initiation and Breakdown Pressure

The fracture initiates at the wellbore as a result of tensile failure when the tangential stress at any point around the wellbore becomes tensile (negative) and equal to the tensile strength of the formation, T.

$$\sigma_{\theta\theta} = -T \qquad (6\text{-}54)$$

The breakdown pressure, p_{bd}, is given by Hubbert and Willis (1957):

hydraulic fracturing, whether in a vertical or a horizontal well. In a horizontal well, the perforation can be oriented in the lower side (Figure 6-13). Based on fracture propagation mechanics, the fracture will have two wings in homogeneous formation even if it is forced to propagate in one direction (Figure 6-14). Figure 6-14 demonstrates this concept, where a horizontal well was drilled in a homogeneous formation (hydrostone) with zero-phasing perforations in the lower side of the wellbore. A fracture was then initiated, and two fracture wings were created. However, when one wing of the fracture encountered a medium with less resistance to fracture propagation (lower fracture toughness), the upward fracture wing stopped and the lower one continued to propagate. The

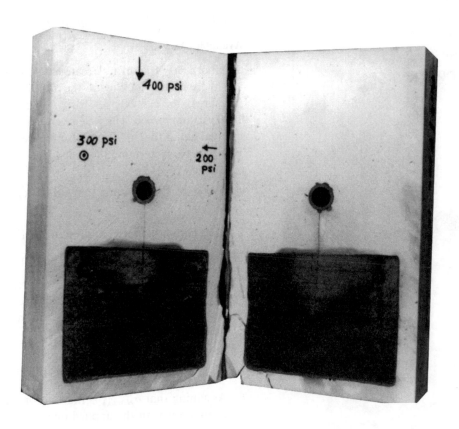

Figure 6-13 Experimental demonstration of drilling in the boundary layer and the use of oriented perforation and fracture to communicate with the poorly consolidated sandstones (sand production exclusion)

$$P_{bd} = 3\sigma_{H,\min} - \sigma_{H,\max} + T - p_r \qquad (6\text{-}55)$$

The fracture propagation pressure is given by

$$p_f = \sigma_{H,\min} + \Delta p_{\text{net}} + \Delta p_f + \Delta p_{\text{tip}} \qquad (6\text{-}56)$$

The friction pressure, Δp_f, can be divided into many terms of perforation, near-wellbore, and fracture geometry, which will determine whether the fracturing pressure will decrease or increase with time. Chapter 17 will provide detailed information.

If we assume that the fracture ceases to propagate after shut-in, then the instantaneous shut-in pressure, *ISIP*, is

$$ISIP = \sigma_{H,\min} + \Delta p_{net}; \text{ therefore,}$$
$$p_f = ISIP + \Delta p_f + \Delta p_{tip} \qquad (6\text{-}57)$$

The net pressure, Δp_{net}, corresponds to the pressure to keep the fracture open with a given fracture width. This pressure may change throughout the job, especially in tip-screenout treatments. The fracture gradient is simply the *ISIP* divided by the depth.

The term Δp_{tip} is the pressure drop required to propagate the fracture, which is a function of fracture toughness. Figure 6-15 shows the fracture initiation pressure of a deviated and horizontal well in terms of α, the deviation angle from the vertical axis and β, the azimuth angle from the direction of maximum horizontal stress. The highest pressure was observed after the fracturing of horizontal wells ($\alpha = 90°$) drilled in the direction of minimum horizontal stress ($\beta = 90°$). These results are for the stress field

$$\sigma'_v > \sigma'_{H,\max} > \sigma'_{H,\min},$$

where

$$\sigma'_v = 3000 \, \text{psi}$$

$$\sigma'_{H,\max} = 2500 \, \text{psi, and } \sigma'_{H,\min} = 1400 \, \text{psi}.$$

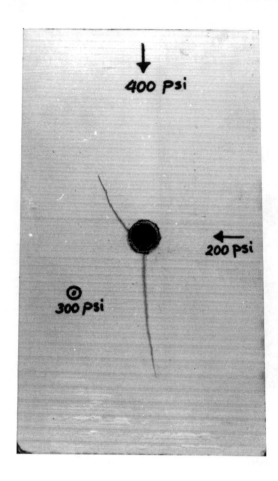

Figure 6-14   Two-wing fracture propagating from one perforation tunnel

6-5.3.2 Example 8: Fracturing Treatment

Available Data

A laboratory fracturing treatment was performed on an hydrostone block with the following stresses and pressures: $\sigma'_v = 3000$ psi, $\sigma'_{H,max} = 2500$ psi, $\sigma'_{H,min} = 1400$ psi, and $p_r = 0$. The well is vertical with an openhole interval equal to 2 in. The fracturing data is provided in Figure 6-16. Determine the following:

1. Estimate the fracture toughness, K_{IC}, for the rock tested.

2. Determine the pressure drop at the tip, Δp_{tip}.

3. Determine the friction-pressure drop, Δp_f.

4. Verify whether the breakdown pressure observed from the test matches that of the theory. The tensile strength of the sample tested was measured in the laboratory as 780 psi.

5. Determine the net pressure, which is used to develop the fracture width.

6. Assuming that $\sigma'_{H,max}$, is not known, calculate it and compare it with the applied one.

Solution

From Figure 6-16, the following data can be determined: $p_{bd} = 2540$ psi $ISIP = 1750$ psi, and $p_f = 2150$ psi at the end of the fracturing time, which corresponds to the fracture length of 4 in., which was observed later.

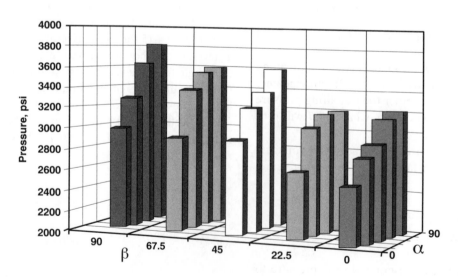

Figure 6-15   Laboratory results of fracture initiation pressures in deviated and horizontal wells for the stress field $_v < \tau_{H,min} < \tau_{H,max}$

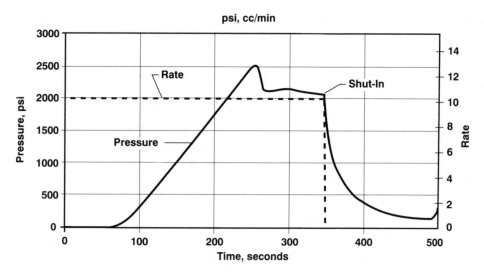

Figure 6-16   Experimental rsults of fracturing a rock block with a vertical well

Fracture toughness can be estimated from the following equation for a circular fracture, as is the case for this test (Shlyapobersky *et al.*, 1988):

$$K_{IC} = \frac{2}{\sqrt{\pi}} \Delta p \sqrt{R} \qquad (6\text{-}58)$$

where R is the fracture radius, which is equal to 2 in., and Δp can be estimated as $ISIP - \sigma_{H,min}$. Therefore,

1. $K_{IC} = \dfrac{2}{\sqrt{\pi}}(1750 - 1400)\sqrt{2} = 558\,\text{psi}\,\sqrt{\text{in.}}$

2. $\Delta p_{tip} = \dfrac{K_{IC}}{\sqrt{\pi L_f}}$, where $L_f = 4\,\text{in.}$ (fracture length) = $157\,\text{psi}$

3. $\Delta p_f = p_f - ISIP - \Delta p_{tip}$

   $\Delta p_f = 2150 - 1750 - 157 = 243\,\text{psi}$

4. $p_{bd} = 3\sigma_{H,min} - \sigma_{H,max} + T + p_r$

   $p_{bd} = 3(1400) - 2500 + 780 + 0 = 2480\,\text{psi}$ (as predicted by the elastic theory)
   The measured p_{bd} was 2540 psi.

5. $\Delta p_{net} = ISIP - \sigma_{H,min}$

   $\Delta p_{net} = 1750 - 1400 = 350\,\text{psi}$

6. $\sigma_{H,max} = 3\sigma_{H,min} - p_{bd} + T + p_r$
   $\sigma_{H,max} = 3(1400) - 2540 + 780 + 0 = 2440\,\text{psi}$,   while the applied $\sigma_{H,max}$ was 2500 psi.

Note that the preceding was a laboratory test and that the fracture is very small. For this reason, the Δp_{tip} was almost 50% of the Δp_{net}. As the fracture becomes longer, the term Δp_{tip} gets smaller and becomes negligible.

6-5.3.3 *Fracturing Deviated and Horizontal Wells*

Nonplanar Fracture Geometry

The term "nonplanar" refers to any fracture plane that does not follow the conventional singular planar fracture (Abass *et al.*, 1996). Nonplanar fracture geometry is created when a deviated wellbore is hydraulically fractured. Nonplanar fractures can be multiple parallel fractures, reoriented fractures, or T-shaped fractures (Figure 6-17). This type of fracture geometry can be responsible

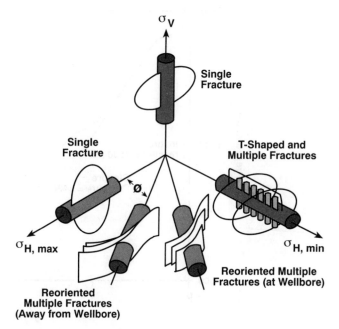

Figure 6-17   Non-planar fracture geometry for different wellbore orientations relative to the direction of in-situ stresses

for premature screenout and excessive fracturing pressure, leading to unsuccessful hydraulic fracturing treatments. Experimental results have shown that fracture initiation pressure is a function of the wellbore deviation and azimuth. Field experience suggests that fracture reorientation is the main reason for any excess net pressure observed during the hydraulic fracturing of deviated wellbores (Economides *et al.*, 1989).

6-5.3.4 Dual Closure Pressure After Shut-in

Dual closure stress has been observed in the field during the fracturing of vertical wells in which layers with different horizontal stresses were drilled (Warpinski *et al.*, 1985). The authors presented a valid explanation that during shut-in, the zone with higher stress closes first, followed by another closure of the zone with lower stress. Dual closure pressure has also been observed in the laboratory during fracturing from samples containing horizontal wells (Abass *et al.*, 1996). Figure 6-18 shows a microfrac analysis of a laboratory test where two closure stresses were observed: 1800 psi and 1430 psi. If such a phenomenon is observed in the field and the well is deviated or horizontal, it indicates near-wellbore nonplanar fracture geometry; therefore, the second closure should be used for estimating the minimum horizontal stress.

6-5.3.5 Proppant Selection, Proppant Flowback, and the Use of Resin-Coated Proppant

Proppant crushing, embedment, and flowback could result in the total failure of a stimulation treatment; proppant must be strong enough to support the closure stress. If embedment is expected, the proppant bed concentration should be increased so that it will compensate for conductivity loss.

The major problem of proppant flowback is often solved through the use of resin-coated proppant. To show the combined effect of stress and temperature as a function of time on the stability of a resin-coated proppant, a stability envelope can be constructed as shown in Figure 6-19 (Dewprashad *et al.*, 1993). The envelope separates two zones: (1) the area inside represents a stable resin under in-situ conditions of temperature and closure stress while (2) the outside area reflects a failure condition dictated by creep. This envelope was obtained from several long-term tests performed on several core samples. Figure 6-20 shows a typical creep test performed at 150°F on consolidated Ottawa sand (20/40 US mesh) coated with an epoxy resin. As shown, progressive loads were applied and maintained constant as the resulting deformation was measured. The strain rate was small until a stress level of 4250 psi was applied.

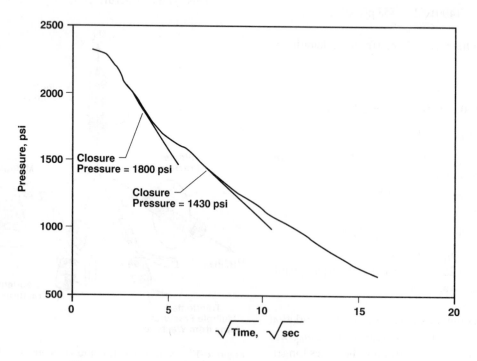

Figure 6-18   Dual closure pressure observed in fracturing horizontal wells

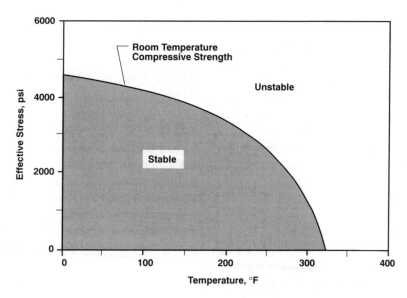

Figure 6-19 Stability envelope of a given formulation of resin-coated proppant

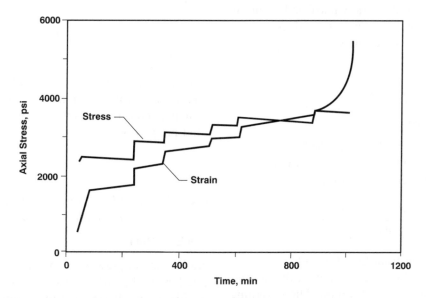

Figure 6-20 Laboratory demonstration of creep testing

6-5.3.6 Nonlinear Stress-Strain Behavior in Fracturing

The general equations to calculate fracture width at the wellbore for different models is given by

$$w_{\max} = \frac{2(1 - v^2)(p_f - \sigma_{H,\min})}{E} \begin{bmatrix} 2x_f \\ h_f \\ 4R/\pi \end{bmatrix} \begin{matrix} \text{KGD} \\ \text{PKN} \\ \text{Radial} \end{matrix} \quad (6\text{-}59)$$

For nonlinear behavior, which occurs in coal and poorly consolidated sands, the mechanical properties change with pressure. For small pressure ranges, these changes are not important. However, for large pressure ranges, which occur in tip-screenout designs, the issue of pressure change is significant. Let us only consider the nonlinear behavior of Young's modulus and assume that Poisson's ratio stays constant. A numerical solution can be adopted to include nonlinear behavior.

$$w_{\max} = 2(1 - v^2) \begin{bmatrix} 2x_f \\ h_f \\ 4R/\pi \end{bmatrix} \left[\sum \frac{\Delta p}{E} \right] \begin{matrix} \text{KGD} \\ \text{PKN} \\ \text{Radial} \end{matrix} \quad (6\text{-}60)$$

6-5.4 Sand Production

When a well is drilled in a poorly consolidated formation, an unconfined condition that creates a plastic zone that can grow with time is established around the wellbore. When the well is cased and cemented, confinement is restored, but the plastic zone is not reversible. Therefore, when perforations are introduced in the formation, the plastic zone is exposed to unconfined conditions again. When the problem of formation solids production is considered, we must differentiate between sand and fines production.

- Fines migration (mobilized clay minerals, quartz, mica, carbonate, etc.) results from drag forces and incompatible completion fluids. If fines are being produced, they should not be stopped by screening techniques such as gravel-packing. Fines should be produced to prevent permeability impairment.

- Sand migration is a result of mechanical and chemical effects and can be eliminated with several techniques that will be discussed in the following paragraphs.

6-5.4.1 Mechanical Effects

Shear and Tensile Failures

Perkins and Weingarten (1988) applied the Mohr-Coulomb failure criterion to predict the near-wellbore pressure drawdown, which triggers sand production:

$$\Delta p = \frac{C(1 + 3\sin\alpha)}{\tan\alpha(1 - 3\sin\alpha)} \tag{6-61}$$

where α can be obtained from Equation 6-42.

This equation is based on the assumption that a sand arch is developed around a perforation. The governing stress equation to keep the cavity stable is

$$\frac{d\sigma_{rr}}{dr} = \frac{2(\sigma_{rr} - \sigma_{\theta\theta})}{r} = 0 \tag{6-62}$$

Bratli and Risnes (1981) presented the following equation for spherical flow around each perforation, based on the assumption that the pressure drawdown is the same for all perforations:

$$\Delta p = 2C\tan\left(\frac{\pi}{4} + \frac{\phi}{2}\right) \tag{6-63}$$

Since

$$C_0 = 2C\tan\left(\frac{\pi}{4} + \frac{\phi}{2}\right) \tag{6-64}$$

then $\Delta p = C_0$ for radial flow and $\Delta p = 2C_0$ for spherical flow.

Cohesive Failure

Cohesive failure is really an erosional effect that removes individual particles resulting from flow across the perforation tunnels. If the drag force generated from fluid velocity exceeds the cohesive strength of the rock material, erosion occurs and sand is produced. If a perforation tunnel with a radius r_p is considered, the shear stress across the internal surface can be calculated as

$$\tau = r_p\frac{dp}{dl} \tag{6-65}$$

Therefore,

$$r_p\frac{dp}{dl} = C \tag{6-66}$$

The right-hand side of Equation 6-66 is referred to as the flux in the porous medium or in the perforation tunnel. Equation 6-66 provides a relationship for the onset of sand production.

6-4.5.2 Example 9: Sand Production from Shear or Cohesive Failure

Available Data

Near-wellbore pressure drawdown is 2000 psi, the perforated interval is 20 ft with 4 shots/ft, the perforation tunnel diameter is 0.5 in., and the sample is 10 in. long. The sand formation's tensile strength is 400 psi and its cohesive strength is 150 psi. Based on this wellbore data, will sand be produced because of cohesive failure or tensile failure?

Solution

Assuming that the pressure drawdown occurs entirely near the wellbore and within the perforation tunnel, from Equation 6-65, shear stress = $(0.5/2)(2000/10)$ = 500 psi.

Thus, both tensile and cohesive failure will occur because the drawdown exceeds the tensile strength *and* the shear stress that causes erosion.

6-4.5.3 Critical Drawdown for Sand Production

On the basis of rock mechanics, we can assume that the formation is stable around the wellbore (a static circle) before production (Figure 6-21). When the wellbore is put on production, a differential pressure (Δp) is established. Based on Equations 6-14 and 6-15, this differential pressure increases the tangential stress while equally decreasing the magnitude of the radial stress. Therefore, the center of Mohr's circle remains stationary, but the radius grows. If the pressure drawdown is high enough for a failure envelope of a given formation, the new circle may touch the failure envelope. The pressure drawdown shown in Figure 6-21 then represents the maximum safe drawdown pressure for that formation, and this pressure can be determined for a given reservoir.

This failure envelope is frequently nonlinear. Therefore, only the part of the failure envelope relevant to the in-situ stress condition should be used for failure evaluation. Figure 6-21, for example, shows that two linear approximations can be obtained for different stress levels. The initial line is affected by the tensile strength circle and the uniaxial strength circle. It is artificial to include the tensile circle into the failure criteria, and therefore it should not be considered in constructing the failure envelope.

6-4.5.4 Example 10: Critical Drawdown for Sand Production

Available Data

An oil-sand reservoir has the following properties: porosity $= 0.25$, permeability $= 100$ md, Young's modulus 5×10^5 psi, Poisson's ratio $= 0.3$, cohesive strength $= 150$ psi, internal friction angle $= 30°$, and formation thickness $= 80$ ft. Perforations are at 60° phasing with 10-in. perforation tunnels. There are 6 shots/ft with $r_p = 0.5$ in. Vertical total stress $= 880$ psi, minimum horizontal stress $= 5800$ psi, maximum horizontal stress $= 7000$ psi, reservoir pressure $= 3600$ psi, the poroelastic factor $= 1.0$, and the uniaxial compressive strength $= 520$ psi.

If the previous information is not available, see the previous examples to determine how to estimate these data. Use different models to calculate the critical drawdown above which the well starts producing sand.

Solutions

1. A 3D model (Morita *et al.*, 1989) was used to estimate the pressure drawdown that triggers sand production. The estimated pressure drawdown was 1000 psi.
2. From Equation 6-63,

$$\Delta p = 2C \tan\left(\frac{\pi}{4} + \frac{\phi}{2}\right)$$

$$= 2(150) \tan\left(\frac{\pi}{4} + \frac{30}{2}\right) = 520 \text{ psi for radial flow}$$

$$= 2C_0 = 1040 \text{ psi for spherical flow.}$$

6-4.5.5 Fracpacks

From a rock mechanics perspective, the *fracpack* helps reduce near-wellbore pressure drawdown, which in turn reduces or prevents cohesive failure (erosion) and tensile failure. A combination fracpack/gravel pack can often effectively control sand production in many areas.

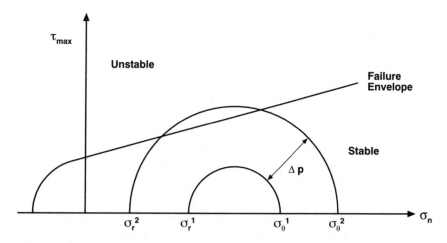

Figure 6-21 Mechanical representation of the critical pressure drawdown that triggers sand production

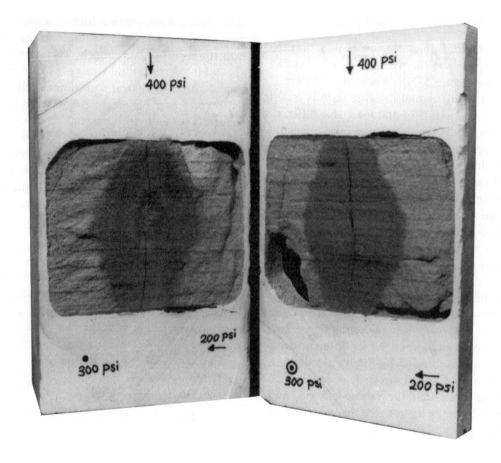

Figure 6-22   Laboratory demonstration of fracturing poorly consolidated sandstone formations

Figure 6-22 shows a hydraulic fracturing experiment in a poorly consolidated outcropping sample that has a Young's modulus of 377,000 psi and a compressive strength of 1037 psi. This figure shows that a poorly consolidated formation can be fractured just like any conventional formation.

The fracture length should be optimized to reduce the severe near-wellbore pressure drawdown (Abass *et al.*,1994; Fletcher *et al.*,1995).

6-4.5.6 Chemical Effects

A sandstone material's granular framework and type of natural cementation are inherent characteristics that help maintain stability; drilling a wellbore in the formation and introducing foreign fluids disturbs this natural stability. This section discusses the effect of drilling and completion fluids on the natural cementation material and describes a new means of restoring cementation during drilling.

Mineralogical analyses of most sand formations will reveal quartz, feldspar, carbonate (such as dolomite), and clay (such as chlorite, smectite). Cementation materials, such as quartz, dolomite, and chlorite, provide stability to a given formation and therefore should be maintained during drilling, completion, and stimulation phases.

Yale *et al.* (1995) studied the effects of cementation on the difference between the static and dynamic mechanical properties. The most interesting finding is the relationship between the degree of nonlinearity and the static/dynamic ratio of mechanical properties. In other words, the type of cementation controls whether the material exhibits linear elastic, nonlinear elastic, or elastoplastic behavior during loading and unloading. Since a formation is exposed to many loading and unloading cycles during wellbore construction phases, studying a formation's loading and unloading characteristics is important.

For exclusive sand control, the following techniques can be used individually or in combination depending on the failure mechanism:

- A horizontal wellbore in the boundary layer and a fracture to the formation

- Oriented perforations in the direction of maximum horizontal stress

- Fracpack and/or gravel pack

- Fracpack with resin-coated sand

- Consolidation during drilling

- Consolidation and fracpack

6-6 IN-SITU STRESS MEASUREMENTS

In-situ stresses and mechanical properties of the formation are crucial for evaluation of the many problems associated with wellbore construction. Several methods are available to measure the magnitude and direction of the in-situ stresses, and several should be performed and compared so that reasonable results can be determined (Daneshy *et al.*, 1986).

6-6.1 Microfrac Test

During a microfrac test, 1 to 2 bbl of fracturing fluid or drilling fluid is injected into a small isolated interval of the wellbore (Soliman, 1993). The pressure falloff data that occurs after injection is used to analyze the magnitude of $\sigma_{H,\min}$. An oriented core from the bottom of the well should be retrieved after the microfrac test. The fracture direction is observed from this oriented core, with the direction of $\sigma_{H,\min}$ being perpendicular to the orientation of the fracture. To confirm this direction, an anelastic strain recovery, ASR, can be performed on an unfractured portion of a retrieved core.

Although it is not recommended, a microfrac test can also be performed in deviated or horizontal wells. An initial analysis indicates that the magnitude of $\sigma_{H,\min}$ may be overestimated in such wells; therefore, a larger volume of fracturing fluid may have to be injected (McLeod, 1982).

6-6.1.1 Example 11: Determining Stresses in the Laboratory

Available Data

Figures 6-16 and 6-23 show real data from a rock sample that was fractured in the laboratory. The stresses applied were $\sigma'_v = 3000\,\text{psi}$, $\sigma'_{H,\max} = 2500\,\text{psi}$, and $\sigma'_{H,\min} = 1400\,\text{psi}$. The tensile strength of the rock is 800 psi. The well was vertical. Calculate $\sigma'_{H,\min}$ and $\sigma'_{H,\max}$ from the pressure data and compare them with the applied $\sigma'_{H,\min}$ and $\sigma'_{H,\max}$ Are they comparable?

Solution

The breakdown pressure from Figure 6-16 is 2540 psi. Figure 6-23 is a p vs. $\sqrt{t}$ plot from the shut-in pressure, which is 2110 psi. *ISIP* is determined from Figure 6-16 to be 1750 psi. Therefore, the data to be analyzed in Figure 6-23 will be the points after pressure falls to 1750 psi. The closure pressure is determined from Figure 6-23 and is equal to 1370 psi.

$$\sigma'_{H,\min} = 1370\,\text{psi from pressure data}$$

$$\sigma'_{H,\min} = 1400\,\text{psi applied during the test}$$

$$\sigma'_{H,\max} = 3\sigma'_{H,\min} - p_{bd} + T$$
$$= 3(1370) - 2540 + 800$$
$$= 2370\,\text{psi}$$

$$\sigma'_{H,\max} = 2370\,\text{psi from pressure data}$$

$$\sigma'_{H,\max} = 2500\,\text{psi applied during the test}$$

6-6.2 Geological Information

Geological considerations can provide valuable qualitative information on the stress field. Figure 6-24 shows how the geological structure can be used to infer the in-situ stress field.

6-6.3 Breakout Analysis

In a formation where the σ_θ that is created from drilling exceeds the in-situ compressive strength of the formation, a near-wellbore breakout zone may occur. Experimental work has shown that breakout can be a valuable technique that allows engineers to determine the direction of minimum horizontal stress. Figure 6-25 demonstrates experimentally how the breakout is created in the direction of minimum horizontal stress. An oriented six-arm caliper log can show the breakout phenomenon. Breakout should not be confused with wash-

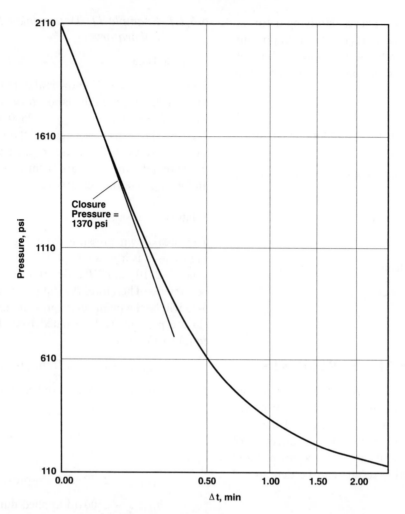

Figure 6-23 Graphical determination of closure pressure

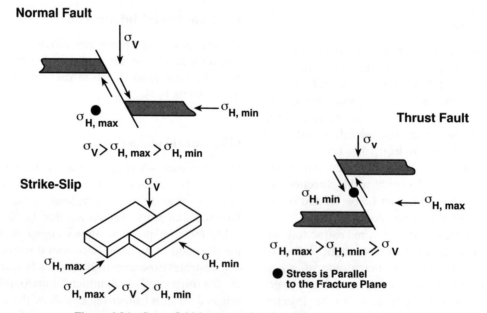

Figure 6-24 Stress field interpretation for different geological structures

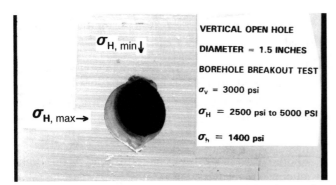

VERTICAL OPEN HOLE

DIAMETER = 1.5 INCHES

BOREHOLE BREAKOUT TEST

σ_v = 3000 psi

σ_H = 2500 psi to 5000 PSI

σ_h = 1400 psi

Figure 6-25 Experimental demonstration of breakout phenomenon

out, in which the wellbore is caving in all directions. As shown in the following example, a breakout does not have to be created during drilling.

6-6.3.1 Example 12: Breakout Analysis

Available Data

A vertical wellbore is drilled in the following stress system: $H = 9000\,\text{ft}$, $\sigma_v = 9000\,\text{psi}$, $\sigma_{H,\max} = 7000\,\text{psi}$, $\sigma_{H,\min} = 5000\,\text{psi}$, $p_r = 3800\,\text{psi}$, $p_w = 4000\,\text{psi}$, $C_0 = 7800\,\text{psi}$, and $\alpha = 0.85$.

1. Will breakout or tensile failure occur? Would either condition occur if the uniaxial compressive strength were 10,000 psi?
2. In a formation with tensile strength of 1000 psi, will tensile failure occur in the direction of maximum horizontal stress? If not, what is the bottomhole pressure required to initiate a tensile fracture (hydraulic fracture), and what is the drilling-fluid weight that can cause fracture initiation?

Solution

The tangential stress in the direction of minimum horizontal stress ($\theta = 90°$) is given by

$$\sigma_{\theta=90} = 3(7000 - 0.84 \times 3800) - (5000 - 0.85 \times 3800)$$
$$- 3800 + 4000 = 8348\,\text{psi}$$

Since the stress concentration is higher than the compressive strength, a breakout will occur (even if the effective stress is used). However, if the compressive strength of the formation were 10,000 psi, a breakout would not be expected. This result explains why the wellbore imaging

log reveals breakout in some formations while nothing appears in other locations.

The tangential strength in the direction of maximum horizontal stress is given by

$$\sigma_{\theta=0} = 3(5000 - 0.85 \times 3800) - (7000 - 0.85 \times 3800)$$
$$- 3800 + 4000 = 2280\,\text{psi}$$

Since the stress is positive, it is in compression in the direction of maximum horizontal stress; therefore, no tensile failure is expected around the wellbore.

The bottomhole pressure to initiate a fracture should bring the compressive strength to zero, then bring it to the tensile strength of the formation, which in the example would be

$$p = 2280 + 500 + 3800 = 6580\,\text{psi}$$

The drilling-fluid weight that can initiate fracture in the formation is

Drilling-Fluid Weight $= 6580 \times 8.34/\ (9000 \times 0.433)$
$= 14\,\text{lb/gal}$ and, therefore, with the drilling-fluid pressure of 4000 psi, the formation will not fail in tension.

6-6.4 Anelastic Strain Recovery (ASR)

The ASR method predicts the direction of the in-situ stresses in the formation. This method is based on the theory that a core relaxes when the in-situ stresses are removed from it (Voight, 1968; Blanton, 1983). The amount of relaxation is time-dependent and proportionally related to the direction and magnitude of the original stresses. An oriented core is usually retrieved from the well and is mounted in a device to measure strain recovery in different directions. The amount of strain collected can vary from 0 to more than 600+ microinches.

The magnitudes of in-situ stresses can be estimated from the principal strain recoveries (Blanton, 1983) as presented in the following example.

6-6.4.1 Example 13: Analysis of an ASR Test

Figure 6-26 shows a real example of an ASR test, from which the direction of the in-situ stresses was determined. The magnitude of the principle stresses is based on the following assumptions:

$$\sigma_v = 10{,}000\,\text{psi}, \quad p_r = 4000\,\text{psi}, \quad \alpha = 0.7, \text{ and } \nu = 0.23.$$

From Figure 6-26, the following strains were calculated during the 0 to 200-minute time interval:

$$\Delta\sigma_v = 57 \text{ Microstrain,}$$

$$\Delta\sigma_{H,max} = 40 \text{ Microstrain,}$$

$$\Delta\sigma_{H,min} = 20 \text{ Microstrain}$$

$$\sigma_{H,min} = (\sigma_v - \alpha p_r)\frac{(1-v)\Delta\sigma_{H,min} + v(\Delta\sigma_{H,max} + \Delta\sigma_v)}{(1-v)\Delta\sigma_v(\delta\sigma_{H,min} + \Delta\sigma_{H,max})}$$
$$+ \alpha p_r = 7506 \text{ psi}$$

$$\sigma_{H,max} = (\sigma_v - \alpha p_r)\frac{(1-v)\Delta\sigma_{H,max} + v(\Delta\sigma_{H,min} + \Delta\sigma_v)}{(1-v)\Delta\sigma_v + v(\Delta\sigma_{H,min} + \Delta\sigma_{H,max})}$$
$$+ \alpha\, p_r = 8854 \text{ psi}$$

Before the ASR data is analyzed, the following observations should be seriously considered:

- The method is based on the relaxation after the core physically detaches from the stressed rock mass. The anelastic growth of the core is a result of microcracks caused by the release of the in-situ stresses. The number of these microcracks is proportional to the state of stress before coring. Therefore, the method is highly dependent on the quality of the core retrieved and the existence of natural fractures. The sample should be carefully observed before and after the test. If the sample exhibits a natural fracture, it should not be used.

- If the natural fracture was not obvious before the test but observed after the test after the sample is cut in several planes, the data should be discarded. A natural fracture can reverse the real direction by as much as 90°.

- Many efforts have been made to analyze this small amount of recovery. If total strain is less than 50 microinches, the results could be considered unreliable. Good results are obtained when the total recovery is 200 microinches or more.

- The method for determining the stress magnitudes from the ASR method (Example 6-6.4.1) should be considered an approximation. More research is required for this method.

- Strain curves may all show continuous positive strain recovery, negative strain, or mixed mode of recovery. The most reliable data is from curves that all show positive recovery. Curves that all indicate negative recovery are dominated by the pore pressure effect; their results should be analyzed, but they should not be considered reliable. Curves that show mixed modes of recovery are caused by the directional effect of permeability (anisotropy), mechanical effects, etc. The results of these data should not be considered reliable.

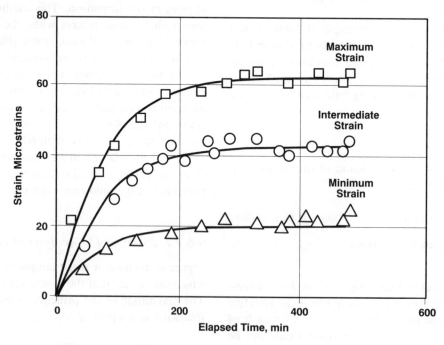

Figure 6-26   Anelastic principal sins observed from ASR testing

- The strain recovery is highly affected by the temperature variation during the test; therefore, selecting the interval that shows constant temperature is very important.

- Differential Strain Curve Analysis (DSCA) has been used when the ASR test has not been performed on site. The DSCA method is based on many assumptions, and the results should only be considered approximate.

REFERENCES

Abass, H.H. *et al.:* "Nonplanar Fracture Propagation From a Horizontal Wellbore: Experimental Study," *SPEPF* (Aug. 1996) 133-137.

Abass, H.H. *et al:* "Oriented Perforation—A Rock Mechanics View," paper SPE 28555, 1994.

Behrmann, L.A., and Elbel, J.L.: "Effect of Perforations on Fracture Initiation," *JPT* (May 1991).

Biot, M.A.: "Generalized Theory of Three-Dimensional Consolidation," *J. Applied Physics* (1941) **12**, 155–164.

Blanton, T.L.: "The Relation Between Recovery Deformation and In-Situ Stress Magnitudes From Anelastic Strain Recovery of Cores," paper SPE/DOE 11624, 1983.

Bratli, R.K., and Risnes, R.: "Stability and Failure of Sand Arches" *SPEJ* (April 1981) 236-248.

Bray, J.: "A Study of Jointed and Fractured Rock—Part II," *Felsmechanik and Ingenieurgeologic,* **V**, No. 4., 1967.

Daneshy, A.A., *et al.:* "In-Situ Stress Measurements During Drilling," *JPT* (Aug. 1986) 891–898.

Dewprashad, B., *et al.:* "A Method to Select Resin-Coated Proppant," paper SPE 26523, 1993.

Dusseault, M.B., and Gray, K.E.: "Mechanisms of Stress-Induced Wellbore Damage," paper SPE 23825, 1992.

Economides, M.J., and Nolte, G.N: *Reservoir Stimulation,* Schlumberger Educational Services (1987).

Economides, M.J. *et al.:* "Fracturing of Highly Deviated and Horizontal Wells," *CIM* (May 1989) **1**.

Fletcher, P.A. *et al.:* "Optimizing Hydraulic Fracture Length To Prevent Formation Failure in Oil And Gas Reservoirs" *Proc.,* 35th U.S. Symposium on Rock Mechanics (1995).

Goodman, R. E.: *Introduction to Rock Mechanics,* John Wiley & Sons (1980).

Haigist, P. *et al.:* "A Case History of Completing and Fracture Stimulating a Horizontal Well," paper SPE 29443, 1995.

Haimson, B.C., and Fairhurst, C.: "Initiation and Extension of Hydraulic Fracture in Rocks," *SPEJ* (1967) **7**.

Hoek, E., and Brown, E.T.: "Empirical Strength Criterion for Rock Masses," *J. Geotech. Eng. Div. ASCE 106 (GT9)* (1980) 1013-1035.

Hubbert, K.M., and Willis, D.G.: "Mechanics of Hydraulic Fracturing," *Petro. Trans.,* AIME **210** (1957) 153–166.

Jaeger, J.C., and Cook, N.W.: *Fundamentals of Rock Mechanics,* Chapman and Hall, New York (1979).

Knott, J.F.: *Fundamentals of Fracture Mechanism,* John Wiley & Sons, New York (1973).

McLeod, H.O.: "The Effect of Perforating Conditions on Well Performance," paper SPE 10649, 1982.

Mody, F.K., and Hale, A.H.: "Borehole Stability Model to Couple the Mechanics and Chemistry of Drilling-Fluid/ Shale Interactions," *JPT* (1993) 1093-1101.

Morita, N., *et al.:* "Realistic Sand Production Prediction: A Numerical Approach," *SPEPE* (Feb. 1989) 15–24.

Morita, N., and McLeod, H.O.: "Oriented Perforation to Prevent Casing Collapse for Highly Inclined Wells," paper SPE 28556, 1994.

Muller, W.: "Brittle Crack Growth in Rock," *PAGEOPH* (1986) No. 4/5.

Perkins, T.K., and Weingarten, J.S.: "Stability and Failure of Spherical Cavities in Unconsolidated Sand and Weakly Consolidated Rock," paper SPE 18244, 1988.

Roegiers, J.C., and Zhao, X.L.: "Rock Fracture Tests in Simulated Downhole Conditions," *Proc.,* 32nd U.S. Symposium on Rock Mechanics (1991) 221–230.

Shlyapobersky, J., Walhaug, W.W., Sheffield, R.E., Huckabel, P.T.: "Field Determination of Fracturing Parameters for Overpresure-Calibrated Design of Hydraulic Fracturing," paper SPE 18195, 1988.

Soliman, M.Y.: "Interpretation of Pressure Behavior of Fractured Deviated and Horizontal Wells," paper SPE 21062, 1993.

Terzaghi, K.: *Theoretical Soil Mechanics,* Wiley, New York (1943).

Thiercelin, M., and Roegiers, J.C.: "Toughness Determination with the Modified Ring Test," *Proc.,* 27th U.S. Symposium on Rock Mechanics (1986).

Venditto, J.J., *et al.:* "Study Determines Better Well Completion Practices," *O&GJ* (Jan. 1993).

Voight, B.: "Determination of the Virgin State of Stress in the Vicinity of a Borehole From Measurements of a Partial Anelastic Strain Tensor in Drill Cores," *Felsmechanik V. Ingenieureol* (1968) **6**, 201–215.

Warpinski, N.R. *et al.:* "In-Situ Stress Measurements at U.S. DOE's Multiwell Experiment Site, Mesaverde Group, Rifle, Colorado," *JPT* (March 1985) 527–536.

Whittaker, B.N. *et al.:* *Rock Fracture Mechanics Principles, Design, and Application,* Elsevier, New York (1992).

Yale, D.P., Nieto, J.A., and Austin, S.P.: "The Effect of Cementation on the Static and Dynamic Mechanical Properties of the Rotliegendes Sandstone" *Proc.,* 35th U.S. Symposium on Rock Mechanics (1995) 169–175.

7 Casing and Tubing Design

Robert F. Mitchell
Enertech

Stefan Miska
University of Tulsa

Randolf R. Wagner
Enertech

7-1 INTRODUCTION

All wells drilled for oil/gas production or for underground injection must be cased with material that has sufficient strength and adequate functionality. This chapter covers basic casing and tubing strength evaluation and design.

Casing is the major structural component of a well. Casing is needed to maintain borehole stability, prevent contamination, isolate water from producing formations, and control well pressures during drilling, production, and workover operations. Casing provides locations for the installation of blowout preventers (BOPs), wellhead equipment, production packers, and production tubing. The cost of casing is a significant part of total well costs. Therefore, the selection of casing size, grade, connectors, and setting depth is a primary engineering and economic consideration.

Tubing is the conduit through which oil and gas is brought from the producing formations to the field surface facilities for processing. Tubing must be strong enough to resist the loads and deformations associated with production and workovers. It must also be sized to support the expected rates of oil and gas production. If tubing is too small, it will restrict production; if it is too large, it will have economic impact beyond the cost of the tubing string itself, since the tubing size will influence the overall casing design of the well.

7-2 CASING TYPES

Six basic types of casing strings are available: conductor casing, surface casing, intermediate casing, production casing, liners and tieback strings.

Conductor casing is set below the drive pipe or marine conductor that is run to protect loose, near-surface formations and to enable circulation of drilling fluid. The conductor isolates unconsolidated formations and water sands and protects against shallow gas. The casing head is usually installed onto this string. A diverter or a BOP stack can be installed onto this string. This string is typically cemented to the surface or the mudline.

Surface casing provides blowout protection, isolates water sands, and prevents lost circulation. It often provides adequate shoe strength for drilling into higher-pressure transition zones. In deviated wells, the surface casing may cover the build section to prevent keyseating of the formation during deeper drilling. This string is typically cemented to the surface or the mudline.

Intermediate casing isolates unstable hole sections, lost-circulation zones, low-pressure zones, and production zones. Often, it is set in the transition zone from normal to abnormal pressure. The cement top of this casing must isolate any hydrocarbon zones. Some wells require multiple intermediate strings. If a liner is run beneath them, some intermediate strings can also be used as production strings.

Production casing isolates production zones and contains formation pressures in the event of a tubing leak. It may also be exposed to injection pressures from fracture jobs down casing, gas lift, or the injection of inhibitor oil. A good primary cement job is very critical for this string.

A *liner* is a casing string that does not extend back to the wellhead, but instead is hung from another casing string. Liners are used instead of full casing strings to reduce cost, improve hydraulic performance during deep drilling, and allow the use of larger tubing above the liner top. Liners can be either intermediate or production strings. Liners are typically cemented over their entire length.

A *tieback string* is a casing string that provides additional pressure integrity from the liner top to the wellhead. An intermediate tieback isolates a worn casing string that cannot withstand possible pressure loads if drilling is continued. Similarly, a production tieback isolates an intermediate string from production loads. Tiebacks can be uncemented or partially cemented.

Figure 7-1 shows an offshore casing program that includes each type of casing string.

7-3 PROPERTIES OF CASING AND TUBING

The American Petroleum Institute (API) has established standards for oil/gas tubing and casing. These standards are recognized by oil and service companies in most countries. In the API standards, tubing is defined as pipe with nominal diameters from 1.0 to 4.5 in., while casing sizes range from 4.5 to 20 in. Casing is classified according to five properties: the manner of manufacture, steel grade, type of joints, length range, and wall thickness (unit weight).

7-3.1 Steel Grades

Almost without exception, casing and tubing are manufactured of mild (0.3 carbon) steel, normalized with small amounts of manganese. For additional strength, manufacturers may quench or temper the steel.

API has adopted a casing grade designation to define the strength of casing steels. This designation consists of a grade letter followed by a number that indicates the minimum yield strength of the steel in ksi (1000 psi). Table 7-1 summarizes the standard API grades.

The yield strength for these purposes is defined as the tensile stress required to produce a total elongation of 0.5% of the length except in the case of P-110 casing, where yield is defined as the tensile stress required to produce a total elongation of 0.6% of the length.

Proprietary steel grades are also widely used in the industry, but these grades do not conform to API specifications. These steel grades are often used in special applications requiring high strength or resistance to hydrogen sulfide cracking. Table 7-2 lists commonly used non-API grades.

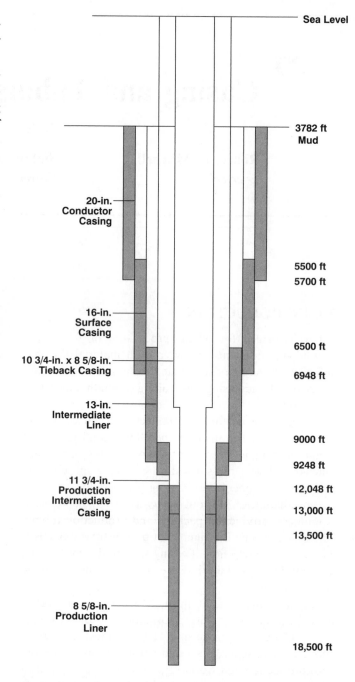

Figure 7-1   Offshore casing program showing types of casing strings

7-3.2 Pipe Strength

To design a reliable casing string, it is necessary to know the actual strength of pipe under different load conditions. Burst strength, collapse-pressure resistance, and tensile strength are the most important mechanical properties of casing and tubing.

Table 7-1 API steel grades

API Grade	Yield Stress, psi		Minimum Ult. Tensile, psi	Minimum Elongation (%)
	Minimum	*Maximum*		
H-40	40,000	80,000	60,000	29.5
J-55	55,000	80,000	75,000	24.0
K-55	55,000	80,000	95,000	19.5
N-80	80,000	110,000	100,000	18.5
L-80	80,000	95,000	95,000	19.5
C-90	90,000	105,000	100,000	18.5
C-95	95,000	110,000	105,000	18.5
T-95	95,000	110,000	105,000	18.0
P-110	110,000	140,000	125,000	15.0
Q-125	125,000	150,000	135,000	18.0

Table 7-2 Non-API steel grades

Non-API Grade		Yield Stress, psi		Minimum Ult. Tensile, psi	Minimum Elongation (%)
		Minimum	*Maximum*		
S-80	Lone Star longitudn'l	75,000 55,000	- -	75,000	20.0
modN-80	Mannes.	80,000	95,000	100,000	24.0
C-90	Mannes.	90,000	105,000	120,000	26.0
SS-95	Lone Star longitudn'l	95,000 75,000	- -	95,000	18.0
SOO-95	Mannes.	95,000	110,000	110,000	20.0
S-95	Lone Star longitudn'l	95,000 92,000	- -	110,000	16.0
SOO-125	Mannes.	125,000	150,000	135,000	18.0
SOO-140	Mannes.	140,000	165,000	150,000	18.0
V-150	U.S.Steel	150,000	180,000	160,000	14.0
SOO-155	Mannes.	155,000	180,000	165,000	20.0

7-3.2.1 Burst Strength

If the casing is subjected to internal pressure that is higher than the external pressure, it is said that casing is exposed to *burst pressure*. Burst-pressure conditions occur during well control operations, integrity tests, squeeze cementing, etc.

The burst strength of the pipe body is determined by the internal yield pressure formula (API Bulletin 5C3, 1985).

$$p = 0.875 \left[\frac{2 Y_p t}{D} \right] \qquad (7\text{-}1)$$

where p is the minimum internal yield pressure (psi), Y_p is the minimum yield strength (psi), t is the nominal wall thickness (in.), and D is the nominal outside diameter (OD) of the pipe (in.).

Equation 7-1, commonly known as the *Barlow equation*, calculates the internal pressure at which the tangential (or hoop) stress at the inner wall of the pipe reaches the yield strength of the material. The expression can be derived from the Lamé equation for tangential stress by making the thin wall assumption that $D/t \gg 1$. Most oilfield casing has a D/t between 15 and 25. The factor of 0.875 appearing in the equation represents the allowable manufacturing tolerance of -12.5% on wall thickness as specified in API Specification 5C2 (1982).

A burst failure will not occur until after the stress exceeds the ultimate tensile strength. Therefore, the use of yield strength criterion as a measure of burst strength is an inherently conservative assumption, particularly for lower-grade materials such as H-40, K-55, and N-80 which have an ultimate tensile strength to yield strength ratio that is significantly greater than that of higher-grade materials, such as P-110 and Q-125.

The effect of axial loading (tension or compression) on the burst strength is discussed in Sections 7-4.2 and 7-4.3.

7-3.2.2 Collapse Strength

If external pressure on the pipe exceeds internal pressure on the pipe, the casing is subjected to *collapse*. Such conditions may exist during cementing operations, well evacuation, etc.

Collapse strength is primarily a function of the material's yield strength and its slenderness ratio, D/t. Collapse strength as a function of D/t is shown in Figure 7-2.

The collapse strength criteria given in API Bulletin 5C3 (1985) consists of four collapse regimes that are determined on the basis of yield strength and the slenderness ratio. They are listed below in order of increasing D/t.

Yield strength collapse (Equation 7-2) is based on yield at the inner wall using the Lamé thick-wall elastic solution. This criterion does not represent a "collapse" pressure at all. For thick-wall pipes ($D/t < 15\pm$), the tangential stress will exceed the yield strength of the material before a collapse instability failure occurs

$$p_{Y_p} = 2Y_p\left[\frac{(D/t) - 1}{(D/t)^2}\right] \qquad (7-2)$$

Nominal dimensions are used in the collapse equations. The applicable D/t ratios for yield strength collapse are shown in Table 7-3.

Table 7-3 Yield collapse pressure formula range

Grade	D/t Range[*]
H-40	16.40 and less
-50	15.24 " "
J-K-55	14.81 " "
-60	14.44 " "
-70	13.85 " "
C-75 & E	13.60 " "
L-N-80	13.38 " "
C-90	13.01 " "
C-T-95 & X	12.85 " "
-100	12.70 " "
P-105 & G	12.57 " "
P-110	12.44 " "
-120	12.21 " "
Q-125	12.11 " "
-130	12.02 " "
S-135	11.92 " "
-140	11.84 " "
-150	11.67 " "
-155	11.59 " "
-160	11.52 " "
-170	11.37 " "
-180	11.23 " "

[*]Grades indicated without letter designation are not API grades but are grades in use or grades being considered for use and are shown for information purposes.

Plastic collapse is based on empirical data from 2488 tests of K-55, N-80, and P-110 seamless casing. No analytical expression has been derived that accurately models collapse behavior in this regime. Regression analysis

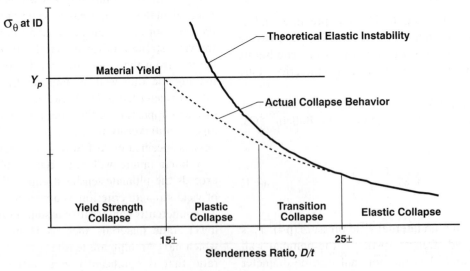

Figure 7-2 Collapse strength as a function of D/t

results in a 95% confidence level that 99.5% of all pipes manufactured to API specifications will fail at a collapse pressure higher than the plastic collapse pressure.

The minimum collapse pressure for the plastic range of collapse is calculated by the following:

$$p_p = Y_p \left[\frac{A}{D/t} - B \right] - C \qquad (7\text{-}3)$$

The factors *A, B, C,* and the applicable *D/t* range for the plastic collapse formula (Equation 7-3) are shown in Table 7-4.

Transition collapse is obtained by a numerical curve fit between the plastic and elastic regimes. The minimum collapse pressure for the plastic-to-elastic transition zone, p_T, is calculated by the following:

$$p_T = Y_p \left[\frac{F}{D/t} - G \right] \qquad (7\text{-}4)$$

The factors *F* and *G* and the applicable *D/t* range for the transition collapse pressure formula are shown in Table 7-5.

Elastic collapse is based on theoretical elastic instability failure. This criterion is independent of yield strength and applicable to thin-wall pipe ($D/t > 25\pm$).

The minimum collapse pressure for the elastic range of collapse is calculated by the following:

$$p_E = \frac{46.95 \times 10^6}{(D/t)[(D/t) - 1]^2} \qquad (7\text{-}5)$$

The applicable *D/t* range for elastic collapse is shown in Table 7-6.

Most oilfield tubulars experience collapse in the *plastic* and *transition* regimes.

Many manufacturers market "high-collapse" casing, which they claim has collapse performance properties that exceed the ratings calculated using the formulae in API Bulletin 5C3 (1985). This improved performance is the result of better manufacturing practices and stricter

Table 7-4 Formula factors and D/t ranges for plastic collapse

| Grade | Formula Factor* | | | | |
	A	*B*	*C*	*D/t* Range*
H-40	2.950	0.0465	754	16.40 to 27.01
-50	2.976	0.0515	1056	15.24 to 25.63
J-K-55	2.991	0.0541	1206	14.81 to 25.01
-60	3.005	0.0566	1356	14.44 to 24.42
-70	3.037	0.0617	1656	13.85 to 23.38
C-75 & E	3.054	0.0642	1806	13.60 to 22.91
L-N-80	3.071	0.0667	1955	13.38 to 22.47
C-90	3.106	0.0718	2254	13.01 to 21.69
C-T-95 & X	3.124	0.0743	2404	12.85 to 21.33
-100	3.143	0.0768	2553	12.70 to 21.00
P-105 & G	3.162	0.0794	2702	12.57 to 20.70
P-110	3.181	0.0819	2852	12.44 to 20.41
-120	3.219	0.0870	3151	12.21 to 19.88
Q-125	3.239	0.0895	3301	12.11 to 19.63
-130	3.258	0.0920	3451	12.02 to 19.40
S-135	3.278	0.0946	3601	11.92 to 19.18
-140	3.297	0.0971	3751	11.84 to 18.97
-150	3.336	0.1021	4053	11.67 to 18.57
-155	3.356	0.1047	4204	11.59 to 18.37
-160	3.375	0.1072	4356	11.52 to 18.19
-170	3.412	0.1123	4660	11.37 to 17.82
-180	3.449	0.1173	4966	11.23 to 17.47

*Grades indicated without letter designation are not API grades but are grades in use or grades being considered for use and are shown for information purposes.

Table 7-5   Formula factors and D/t range for transition collapse

Grade	Formula Factor* F	Formula Factor* G	D/t Range*
H-40	2.063	0.0325	27.01 to 42.64
-50	2.003	0.0347	25.63 to 38.83
J-K-55	1.989	0.0360	25.01 to 37.21
-60	1.983	0.0373	24.42 to 35.73
-70	1.984	0.0403	23.38 to 33.17
C-75 & E	1.990	0.0418	22.91 to 32.05
L-N-80	1.998	0.0434	22.47 to 31.02
C-90	2.017	0.0466	21.69 to 29.18
C-T-95 & X	2.029	0.0482	21.33 to 28.36
-100	2.040	0.0499	21.00 to 27.60
P-105 & G	2.053	0.0515	20.70 to 26.89
P-100	2.066	0.0532	20.41 to 26.22
-120	2.092	0.0565	19.88 to 25.01
Q-125	2.106	0.0582	19.63 to 24.46
-130	2.119	0.0599	19.40 to 23.94
S-135	2.133	0.0615	19.18 to 23.44
-140	2.146	0.0632	18.97 to 22.98
-150	2.174	0.0666	18.57 to 22.11
-155	2.188	0.0683	18.37 to 21.70
-160	2.202	0.0700	18.19 to 21.32
-170	2.231	0.0734	17.82 to 20.60
-180	2.261	0.0769	17.47 to 19.93

*Grades indicated without letter designation are not API grades but are grades in use or grades being considered for use and are shown for information purposes.

quality assurance programs to reduce ovality, residual stress, and eccentricity.

High-collapse casing was initially developed for use in the deeper sections of high pressure wells. The use of high-collapse casing has gained wide acceptance in the industry, but its use remains controversial among some operators (Klementich, 1995). Unfortunately, not all manufacturers' claims have been substantiated with the appropriate level of qualification testing. If high-collapse casing is deemed necessary in a design, appropriate expert advice should be obtained to evaluate the manufacturers' qualification test data, such as the length to diameter ratio, testing conditions (end constraints), and the number of tests performed.

If the pipe is subjected to both external and internal pressures as depicted in Figure 7-3, the equivalent external pressure p_e is calculated as follows:

$$p_e = p_o - \left(1 - \frac{2}{D/t}\right)p_i = \Delta p + \left(\frac{2}{D/t}\right)p_i \qquad (7\text{-}6)$$

where p_o is the external pressure, p_i is the internal pressure and Δp is equal to $p_o - p_i$.

To provide a more intuitive understanding of this relationship, Equation 7-6 can be rewritten as

$$p_e D = p_o D - p_i d \qquad (7\text{-}7)$$

where d is the nominal inside diameter (ID).

7-3.2.3 Axial Strength

The axial strength of the pipe body is determined by the pipe body yield strength formula (API Bulletin 5C3, 1985).

$$F_y = \frac{\pi}{4}\left(D^2 - d^2\right)Y_p \qquad (7\text{-}8)$$

where F_y is the pipe body axial strength (in units of force). Axial strength is the product of the cross-sectional area (based on nominal dimensions) and the yield strength.

Table 7-6 D/t range for elastic collapse

Grade	Grade*
H-40	42.64 and greater
-50	38.83 " "
J-K-55	37.21 " "
-60	35.73 " "
-70	33.17 " "
C-75 & E	32.05 " "
L-N-80	31.02 " "
C-90	29.18 " "
C-T-95 & X	28.36 " "
-100	27.60 " "
P-105 & G	26.89 " "
P-110	26.22 " "
-120	25.01 " "
Q-125	24.46 " "
-130	23.94 " "
S-135	23.44 " "
-140	22.98 " "
-150	22.11 " "
-155	21.70 " "
-160	21.32 " "
-170	20.60 " "
-180	19.93 " "

*Grades indicated without letter designation are not API grades but are grades in use or grades being considered for use and are shown for information purposes.

7-4 COMBINED STRESS EFFECTS

All previous pipe strength equations are based on a uniaxial stress state, a state in which only one of the three principal stresses is nonzero. Pipe in the wellbore, however, is always subjected to combined loading conditions.

The fundamental basis of casing design is that if stresses in the pipe wall exceed the yield strength of the material, a failure condition exists. Hence, the yield strength is a measure of the maximum allowable stress.

To evaluate the pipe strength under combined loading conditions, the uniaxial yield strength is compared with the yielding condition. Perhaps the most widely accepted yielding criterion is based on the maximum distortion energy theory, which is known as *Huber-Hencky-Mises yield condition* or simply the *von Mises, triaxial, or equivalent* stress (Crandall and Dahl, 1959).

Triaxial stress (equivalent stress) is not a true stress. It is a theoretical value that allows a generalized three-dimensional (3D) stress state to be compared with a uni-axial failure criterion (the yield strength). In other words, if the triaxial stress exceeds the yield strength, a yield failure is indicated. The triaxial safety factor is the ratio of the material's yield strength to the triaxial stress.

The yielding criterion is stated as follows:

$$\sigma_{VME} = \frac{1}{\sqrt{2}} \left[(\sigma_z - \sigma_\theta)^2 + (\sigma_\theta - \sigma_r)^2 + (\sigma_r - \sigma_z)^2 \right]^{1/2} \geq Y_p \tag{7-9}$$

where σ_{VME} is the triaxial stress, σ_z is the axial stress, σ_θ is the tangential (hoop) stress, and σ_r is the radial stress. Figure 7-4 shows the axial, tangential, and radial stress components.

The calculated axial stress, σ_z, at any point along the cross-sectional area should include the effects of self-weight, buoyancy, pressure loads, bending, shock loads, frictional drag, point loads, temperature loads, and buckling loads. Except for bending/buckling loads, axial loads are normally considered to be constant over the entire cross-sectional area.

The tangential and radial stresses are calculated using the Lamé equations (Timoshenko and Goodier, 1961) for thick-wall cylinders as follows:

$$\sigma_\theta = \frac{r_i^2 + r_i^2 r_o^2 / r^2}{r_o^2 - r_i^2} p_i - \frac{r_o^2 + r_i^2 r_o^2 / r^2}{r_o^2 - r_i^2} p_o \tag{7-10}$$

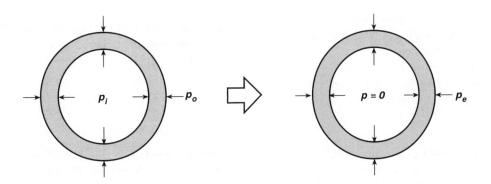

Figure 7-3 Equivalent external pressure of pipe subjected to both internal and external pressure

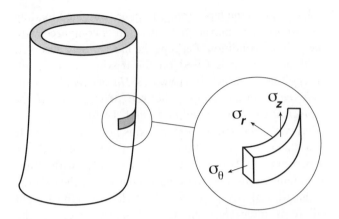

Figure 7-4   Axial, tangential, and radial stress components

$$\sigma_r = \frac{r_i^2 - r_i^2 r_o^2 / r^2}{r_o^2 - r_i^2} p_i - \frac{r_o^2 - r_i^2 r_o^2 / r^2}{r_o^2 - r_i^2} p_o \qquad (7\text{-}11)$$

where r_i is the inner-wall radius, r_o is the outer-wall radius, and r is the radius at which the stress occurs.

The absolute value of σ_θ is always greatest at the inner wall of the pipe and for burst and collapse loads where $|p_i - p_o| \gg 0$, and then $|\sigma_\theta \gg |\sigma_r|$. For any p_i and p_o combination, the sum of the tangential and radial stresses is constant at all points in the casing wall.

Substituting Equation 7-10 and Equation 7-11 in Equation 7-9 after rearrangements, yields the following:

$$\sigma_{VME} = \sqrt{(f_1 f_2)^2 + f_3^2} \qquad (7\text{-}12)$$

in which

$$f_1 = \left(\frac{r_i}{r}\right)^2 \frac{\sqrt{3}}{2}(p_o - p_i) \qquad (7\text{-}12a)$$

$$f_2 = \frac{1}{2} \frac{(D/t)^2}{(D/t) - 1} \qquad (7\text{-}12b)$$

$$f_3 = \sigma_z - \frac{r_i^2 p_i - r_o^2 p_o}{r_o^2 - r_i^2} \qquad (7\text{-}12c)$$

Equation 7-12 (a through c) permits calculation of the equivalent stress at any point of the pipe body for given pipe geometry and loading conditions. The following examples illustrate these concepts.

7-4.1 Combined Collapse and Tension

Assuming that σ_z and $\sigma_\theta \gg \sigma_r$ and setting the triaxial stress equal to the yield strength, results in the following equation of an ellipse:

$$Y_p = \left[\sigma_z^2 - \sigma_z \sigma_\theta + \sigma_\theta^2\right]^{1/2} \qquad (7\text{-}13)$$

This well-known *biaxial* criterion used in API Bulletin 5C3, accounts for the effect of tension on collapse

$$Y_{pa} = \left[\sqrt{1 - 0.75\left(\frac{S_a}{Y_p}\right)^2} - 0.5\frac{S_a}{Y_p}\right] Y_p \qquad (7\text{-}14)$$

where S_a is the axial stress based on the buoyant weight of pipe (psi). As S_a (in this case, tension) increases, the pipe's resistance to collapse pressure decreases. Plotting this ellipse (Figure 7-5) allows a direct comparison of the triaxial criterion with the API ratings. Loads within the design envelope meet the design criteria.

7-4.2 Combined Burst and Compression Loading

Combined burst and compression loading corresponds to the upper left quadrant of the design envelope. Triaxial analysis is most critical in this region because reliance on uniaxial criteria alone will not predict several possible failures.

For high burst loads (high tangential stress) and moderate compression, a burst failure can occur at a differential pressure less than the API burst pressure. For high compression and moderate burst loads, the failure mode is permanent *corkscrewing*, which is plastic deformation caused by helical buckling. This combined loading typically occurs when a high internal pressure is experienced (caused by a tubing leak or a buildup of annular pressure) after the casing temperature has been increased because of production. The temperature increase in the uncemented portion of the casing causes thermal growth, which can significantly increase compression and buckling. The increased internal pressure also results in increased buckling.

7-4.3 Combined Burst and Tension Loading

Combined burst and tension loading corresponds to the upper right quadrant of the design envelope. In this region, reliance on the uniaxial criteria alone can result in a design that is more conservative than necessary.

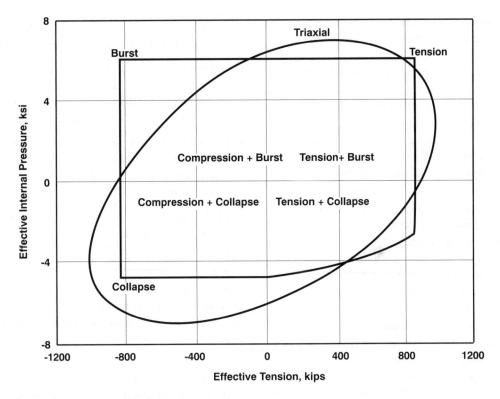

Figure 7-5 Triaxial stress analysis compared with API ratings

For high burst loads and moderate tension, a burst yield failure will not occur until after the API burst pressure has been exceeded. As the tension approaches the axial limit, a burst failure can occur at a differential pressure less than the API value. For high tension and moderate burst loads, pipe body yield will not occur until a tension greater than the uniaxial rating is reached.

By taking advantage of increased burst resistance in the presence of tension, design engineers can reduce casing costs without jeopardizing wellbore integrity. Similarly, engineers may wish to allow loads that fall between the uniaxial and triaxial tension ratings. However, great care should be taken in the latter case because of the uncertainty of what burst pressure may occur in conjunction with a high tensile load (an exception to this is the green cement pressure-test load case, Section 7-6.3.4). In addition, connection ratings could limit an engineer's ability to design in this region.

7-4.4 Use of Triaxial Criterion for Collapse Loading

For most oilfield pipe, collapse is either an inelastic stability failure or an elastic stability failure independent of

yield strength. Since the triaxial criterion is based on elastic behavior and the yield strength of the material, it should not be used with collapse loads. The one exception is for thick-wall pipes with a low D/t ratio that have an API rating in the yield strength collapse region. This collapse criterion, along with the effects of tension and internal pressure (which are triaxial effects), result in the API criterion being essentially identical to the triaxial method in the lower right quadrant of the triaxial ellipse for thick-wall pipes.

For high-compression and moderate-collapse loads experienced in the lower left quadrant of the design envelope, the failure mode may be permanent corkscrewing as a result of helical buckling. Triaxial criterion should be used in this case. Typically, this load combination occurs only in wells that experience a large temperature increase as a result of production. The combination of a collapse load that causes reverse ballooning and a temperature increase both increase compression in the uncemented portion of the string.

Most design engineers use a minimum wall for burst calculations and nominal dimensions for collapse and axial calculations. Arguments can be made for using either assumption in the case of triaxial design.

However, the choice of dimensional assumptions is not as important as the triaxial analysis results being consistent with the uniaxial ratings.

Triaxial analysis is perhaps most valuable for the evaluation of burst loads. Hence, it is sensible to calibrate the triaxial analysis to be compatible with the uniaxial burst analysis. This calibration can be done by the appropriate selection of a design factor. Since the triaxial result nominally reduces to the uniaxial burst result when no axial load is applied, the results of both of these analyses should be equivalent. Since the burst rating is based on 87.5% of the nominal wall thickness, a triaxial analysis based on nominal dimensions should use a design factor equal to the burst design factor multiplied by 8/7. This factor supports the philosophy that a less conservative assumption should be used with a higher design factor. Therefore, for a burst design factor of 1.1, a triaxial design factor of 1.25 should be used.

Figure 7-6 summarizes the triaxial, uniaxial, and biaxial limits that should be used in casing design and provides a set of consistent design factors.

Because of the potential benefits, a triaxial analysis should be performed for *all* well designs. Specific applications include the following:

- Saving money in burst design by taking advantage of the increased burst and resistance in tension

- Accounting for large temperature effects on the axial load profile in high-pressure, high-temperature (HPHT) wells. This practice is particularly important in combined burst and compression loading.

- Providing greater accuracy in determining stresses when using thick-wall pipe ($D/t < 12$). Conventional uniaxial and biaxial methods have imbedded thin-wall assumptions.

- Evaluating buckling severity, since permanent cork-screwing will occur when the triaxial stress exceeds the yield strength of the material.

Although the von Mises criterion is the most accurate method of representing elastic yield behavior, its use in tubular design should be accompanied by these precautions:

- For most pipe used in oilfield applications, collapse is frequently an instability failure that occurs before the computed maximum triaxial stress reaches the yield

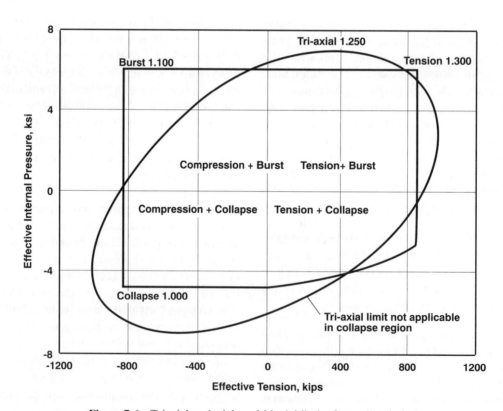

Figure 7-6   Triaxial, uniaxial, and biaxial limits for casing design

strength. As a result, triaxial stress should *not* be used as a collapse criterion. Only in thick-wall pipe does yielding occur before collapse.

- The accuracy of triaxial analysis depends upon the accurate representation of (1) the conditions that exist when the pipe is installed in the well and (2) the subsequent loads of interest. It is the *change* in load conditions that is most important in stress analysis. Therefore, an accurate knowledge of the temperatures and pressures that occur throughout the life of the well can be critical to accurate triaxial analysis.

7-5 CONNECTIONS

Threaded connections are the mechanical means of joining neighboring tubing or casing joints. These threads must hold joints together during axial tension or compression.

Selection of a threaded connection may depend on strength requirements, the ability to re-cut the thread, cost, or leak resistance. Tubing connections are designed with some level of leak resistance. For all casing sizes, the threads are not intended to be leak-resistant when made up.

While a number of joint connections are available, API recognizes three basic types: (1) coupling joints with long or short rounded threads, (2) coupling joints with assymetrical, trapezoidal buttress threads, and (3) extreme-line casing with trapezoidal threads without coupling. API 5C2 (1982) provides information on casing and tubing thread dimensions.

7-5.1 API Connection Ratings

7-5.1.1 Coupling Internal Yield Pressure

The internal yield pressure is the pressure that will initiate yield at the root of the coupling thread

$$p = Y_c \left(\frac{W - d_1}{W} \right) \qquad (7\text{-}15)$$

where Y_c is the minimum yield strength of the coupling (psi), W is the nominal OD of the coupling (in.), and d_1 is the diameter (in.) at the root of the coupling thread in the power-tight position. This dimension is based on data given in API Specification 5B and other thread geometry data.

The coupling internal yield pressure is typically greater than the pipe body internal yield pressure. The internal-pressure leak resistance is based on the interface pressure between the pipe and coupling threads as a result of makeup

$$p = \frac{ETNP_t(W^2 - E_s^2)}{2E_s W^2} \qquad (7\text{-}16)$$

where p is the internal-pressure leak resistance, E is the modulus of elasticity, T is the thread taper, N is a function of the number of thread turns from hand-tight to power-tight position, P_t is the thread pitch, and E_s is the pitch diameter at the plane of the seal as given in API Specification 5B. It is important to note that this equation accounts only for the contact pressure on the thread flanks as a sealing mechanism; it ignores the long helical leak paths filled with thread compound that exist in all API connections.

In round threads, two small leak paths exist at the crest and root of each thread. In buttress threads, a much larger leak path exists along the stabbing flank and at the root of the coupling thread. Although API connections rely on thread compound to fill these gaps and provide leak resistance, the resulting leak resistance is typically less than the API internal leak resistance value, particularly for buttress connections.

Leak resistance can be improved through the use of API connections that have smaller thread tolerances (and hence smaller gaps). However, even with smaller gaps, API connections will not typically withstand pressures greater than 5,000 psi with any long-term reliability. Applying tin or zinc plating to the coupling will also result in smaller gaps and improve leak resistance.

7-5.1.2 Round-Thread Casing-Joint Strength

The round-thread casing-joint strength is given as the lesser of the *pin* fracture strength and the *jump-out* strength. The pin fracture strength is determined from

$$F_j = 0.95 A_{jp} U_p \qquad (7\text{-}17)$$

The jump-out strength is determined from

$$F_j = 0.95 A_{jp} L \left[\frac{0.74 D^{-0.59} U_p}{0.5L + 0.14D} + \frac{Y_p}{L + 0.14D} \right] \qquad (7\text{-}18)$$

where F_j is the minimum joint strength (lbf), A_{jp} is the cross-sectional area (in.^{2}) of the pipe wall under the last perfect thread, which is equal to $\pi/4[(D - 0.1425)^2 - d^2]$, L is the engaged thread length given in API Specification 5B, and U_p is the minimum ultimate tensile strength of pipe (psi).

These equations are based on tension tests to failure on 162 round-thread test specimens. Both are theoretically derived and adjusted with statistical methods to match the test data. For standard coupling dimensions, round threads are pin-weak (i.e. the coupling is noncritical in determining joint strength).

7-5.1.3 Buttress-Thread Casing-Joint Strength

The buttress-thread casing-joint strength is given as the lesser of the fracture strength of the pipe body (the pin) and the coupling (the box).

Pipe-Thread Strength

$$F_j = 0.95A_pU_p[1.008 - 0.0396(1.083 - Y_p/U_p)D] \quad (7\text{-}19)$$

Coupling-Thread Strength

$$F_j = 0.95A_cU_c \quad (7\text{-}20)$$

where U_c is the minimum ultimate tensile strength of coupling, A_p is the cross-sectional area of plain-end pipe, and A_c is the cross-sectional area of coupling, which is equal to $\pi/4(W^2 - d_1^2)$.

These equations are based on tension tests to failure on 151 buttress-thread specimens. These equations are theoretically derived and adjusted with statistical methods to match the test data.

7-5.1.4 Extreme-Line Casing Joint Strength

Extreme-line casing joint strength is calculated as follows:

$$F_j = A_{cr}U_p \quad (7\text{-}21)$$

where A_{cr} is the critical section area of the box, pin, or pipe, whichever area is smallest.

When performing casing design, it is very important to note that the API joint strength values are a function of the *ultimate tensile strength*. This is a different criterion than that used to define the axial strength of the pipe body, which is based on *yield strength*. If care is not taken, this approach could result in a design that does not have the same level of safety for the connections as

the pipe body. This practice is not prudent, particularly since most casing failures occur at connections. This discrepancy can be countered by using a higher design factor when performing connection axial design with API connections.

7-5.1.5 Joint-Strength Equations for Tubing

The joint strength equations given in API Bulletin 5C3 (1985) for tubing are very similar to those given for round-thread casing, except they are based on yield strength. As a result, the ultimate tensile strength to yield strength discrepancy does not exist in tubing design.

If API casing connection joint strengths calculated with the above formulas are the basis of a design, the designer should use higher *axial* design factors. The logical basis for a higher axial design factor is to multiply the pipe body axial design factor by the ratio of the minimum ultimate tensile strength (U_p) to the minimum yield strength (Y_p)

$$DF_{\text{connection}} = DF_{\text{pipe}} \times \left(\frac{U_p}{Y_p}\right) \quad (7\text{-}22)$$

Tensile property requirements for standard grades are given in API Specification 5C2 (1982) and are shown in Table 7-7 for reference along with their ratio.

Table 7-7   Tensile property requirements

Grade	Y_p, psi	U_p, psi	U_p/Y_p
H-40	40,000	60,000	1.50
J-55	55,000	75,000	1.36
K-55	55,000	95,000	1.73
N-80	80,000	100,000	1.25
L-80	80,000	95,000	1.19
C-90	90,000	100,000	1.11
C-95	95,000	105,000	1.11
T-95	95,000	105,000	1.11
P-110	110,000	125,000	1.14
Q-125	125,000	135,000	1.08

7-5.2 Proprietary Connections

Special connections are used to achieve gas-tight sealing reliability and 100% connection efficiency (ratio of joint tensile strength to pipe body tensile strength) under more severe well conditions, such as high pressures (>5000 psi) and high temperatures (>250°F). Proprietary connections can also be used for sour gas wells, gas production, high-pressure gas lifts, steam wells, or large doglegs in horizontal wells.

Proprietary connections improve the connection efficiency in flush-joint (FJ), integral-joint (IJ), or other special clearance applications. These connections also improve the stab-in and makeup characteristics of large-diameter (>16-in.) pipe and reduce galling, particularly in chrome alloy applications or in tubing strings that will be reused. Proprietary connections can also prevent connection failure under high torsional loads during pipe rotation.

The improved performance of many proprietary connections results from the use of complex thread forms, resilient seals, torque shoulders, or metal-to-metal seals. Because proprietary connections are expensive, they should be used only when the application requires them. Selection of proprietary connections should be compared to optimizing the performance of API connections by using tighter dimensional tolerances, applying plating to couplings, and selecting the optimal thread compound.

When proprietary connections are to be used, they should be thoroughly tested. The following steps are recommended to verify the performance of proprietary connections:

1. Audit the manufacturer's performance test data (sealability and tensile load capacity under combined loading).
2. Audit the manufacturer's field history data.
3. Require additional performance testing for the most critical applications. When requesting tensile performance data, ensure that the manufacturer indicates whether quoted tensile capacities are based on the ultimate tensile strength or parting load (the load at which the connection will fracture) or on the yield strength (commonly called the joint elastic limit). If possible, use the joint elastic limit values in the design to maintain consistent design factors for both the pipe body and connection analyses. If only parting load capacities are available, use a higher design factor for connection axial design.

7-5.3 Connection Failures

The majority of casing failures occur at connections. These failures can be attributed to improper design, excessive torque during makeup or subsequent operations, yielding resulting from internal pressure, and

jump-outs or fractures under tensile load. Connections may fail because they do not meet the promised manufacturing tolerances. In addition to these causes, connections could be damaged during storing and handling, or they could be subjected to corrosion or wear during production operations. Regardless of the cause, connection failures such as leakage, structural failure, and galling during makeup result in costly, time-consuming repairs.

Selecting the appropriate connection is fundamental in avoiding connection failure. When selecting connections, engineers and operators should carefully consider manufacturing tolerances, how the threads will be protected during storage and handling (thread protectors, storage thread compound, handling procedures), and certain aspects of the running procedures, such as selection and application of thread compound. The overall mechanical integrity of a correctly designed casing string relies on a quality assurance program that ensures that damaged connections are not used and that operations personnel adhere to the appropriate running procedures.

7-5.4 Connection Design Limits

In addition to geometry and material properties, connection design limits are affected by manufacturing characteristics such as the surface treatment, phosphating, the metal plating (copper, tin, or zinc), and the use of bead blasting.

Connection design limits are also a function of operational variables such as (1) the type of thread compound used, (2) the makeup torque, (3) the use of a resilient seal ring (which is not recommended by many manufacturers and service companies), (4) the type of fluid exposure (drilling fluid, clear brine, or gas), (5) temperature and pressure cycling, and (6) the presence of large doglegs in medium- or short-radius horizontal wells.

7-6 LOADS ON CASING AND TUBING STRINGS

To evaluate a given casing design, a set of loads is necessary. Casing loads result from running the casing, cementing the casing, subsequent drilling operations, production, and well workover operations. Casing loads are principally pressure loads, mechanical loads, and thermal loads. *Pressure loads* are produced by (1) fluids within the casing, (2) cement and fluids outside the casing, (3) pressures imposed at the surface by drilling and workover operations, and (4) pressures imposed by the formation during drilling and production.

Mechanical loads are associated with casing hanging weight, shock loads during running, packer loads during production and workovers, and hanger loads. Temperature changes and resulting *thermal expansion loads* are induced in casing by drilling, production, and workovers, and these loads may cause buckling (bending stress) loads in uncemented intervals.

The following paragraphs comprise a fairly complete list of casing loads typically used in preliminary casing design. However, each operating company usually has its own special set of design loads for casing, based on their experience. Because so many possible loads must be evaluated, most casing design is performed with computer programs, which generate the appropriate load sets (often custom tailored for a particular operator), evaluate the results, and sometimes automatically determine a minimum cost design.

7-6.1 External-Pressure Loads

The following pressure distributions are typically used to model the external pressures in cemented intervals. The formulas are given in English units, with pressure in psig and pressure gradients in psi/ft.

7-6.1.1 Mud-Cement Mix-Water—External-Pressure Profile (Figure 7-7)

$$p(z) = \gamma_m z \qquad\qquad z < z_{\text{toc}}$$
$$p(z) = \gamma_m z_{\text{toc}} + \gamma_{\text{cem}}(z - z_{\text{toc}}) \qquad z > z_{\text{toc}}$$

where p is the external pressure, z is the true vertical depth, γ_m is the mud gradient, γ_{cem} is the cement mix-water gradient (internal pore fluid pressure gradient), and z_{toc} is the true vertical depth of the top of cement. This pressure profile is continuous with depth.

7-6.1.2 Permeable Zones—External-Pressure Profile: Good Cement (Figure 7-8)

$$p(z) = \gamma_m z \qquad\qquad z < z_{\text{toc}}$$
$$p(z) = \gamma_m z_{\text{toc}} + [p_f(z_{ft}) - \gamma_m z_{\text{toc}}](z - z_{\text{toc}}) \qquad z_{\text{toc}} < z < z_{ft}$$
$$p(z) = p_f(z) \qquad\qquad z_{ft} < z < z_{fb}$$
$$p(z) = p_f(z_{fb}) + \gamma_{\text{cem}}(z - z_{fb}) \qquad z > z_{fb}$$

where $p_f(z)$ is the formation pore-pressure profile over the permeable zone interval z_{ft} to z_{fb} true vertical depths (ft). For example, if $p_f(z)$ was a linear distribution, then

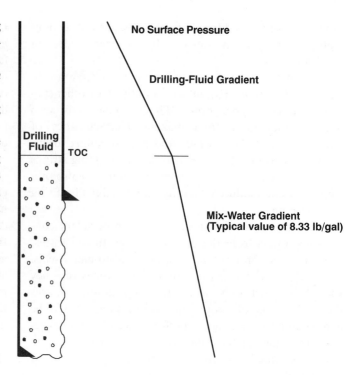

Figure 7-7   Mud cement mix-water—external pressure profile

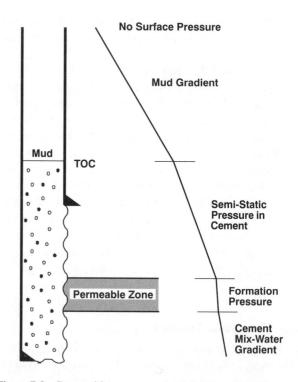

Figure 7-8   Permeable zones— external-pressure profile—good cement

$$p_f(z) = p_{ft} + (p_{fb} - p_{ft})(z - z_{ft})/(z_{fb} - z_{ft})$$

where p_{ft} is the pressure at z_{ft}, and p_{fb} is the pressure at z_{fb}. This pressure profile is continuous with depth.

7-6.1.3 Permeable Zones—External-Pressure Profile: Poor Cement—High-Pressure Zone (Figure 7-9)

$$p(z) = p(z_{tf}) + \gamma_{cem}(z_{toc} - z_{ft}) + \gamma_m(z - z_{toc}) \qquad z < z_{toc}$$
$$p(z) = p(z_{tf}) + \gamma_{cem}(z - z_{ft}) \qquad z_{toc} < z < z_{ft}$$
$$p(z) = p_f(z) \qquad z_{ft} < z < z_{fb}$$
$$p(z) = p_f(z_{fb}) + \gamma_{cem}(z - z_{fb}) \qquad z > z_{fb}$$

In this case, the formation pore pressure is felt at the surface through the poor cement. This pressure profile is continuous with depth.

7-6.1.4 Permeable Zones—External-Pressure Profile: Poor Cement—Low-Pressure Zone (Figure 7-10)

$$p(z) = 0 \qquad z < z_{md}$$
$$p(z) = p_f(z_{tf}) + \gamma_{cem}(z_{toc} - z_{ft}) + \gamma_m(z - z_{toc}) \qquad z_{md} < z < z_{toc}$$
$$p(z) = p_f(z_{tf}) + \gamma_{cem}(z - z_{ft}) \qquad z_{toc} < z < z_{ft}$$
$$p(z) = p_f(z) \qquad z_{ft} < z < z_{fb}$$
$$p(z) = p_f(z_{fb}) + \gamma_{cem}(z - z_{fb}) \qquad z > z_{fb}$$

where the mud drop z_{md} is defined as

$$z_{md} = z_{toc} - [p_f(z_{tf}) + \gamma_{cem}(z_{toc} - z_{ft})]/\gamma_m$$

This pressure profile is continuous with depth.

7-6.1.5 Openhole Pore-Pressure—External-Pressure Profile: TOC Inside Previous Shoe (Figure 7-11)

$$p(z) = \gamma_m z \qquad z < z_{toc}$$
$$p(z) = \gamma_m z_{toc} + \gamma_{cem}(z - z_{toc}) \qquad z_s > z > z_{toc}$$
$$p(z) = \gamma_{em} z \qquad z < z_s$$

where z_s is the true vertical location of the previous shoe and γ_{em} is the minimum equivalent mud weight gradient of the open hole below the shoe. This pressure profile is not continuous with depth; it is discontinuous at the previous shoe.

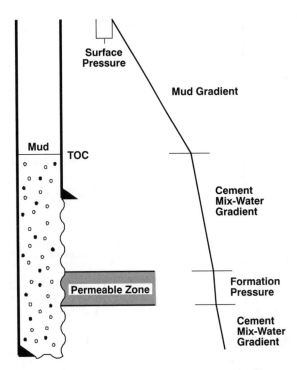

Figure 7-9 Permeable zones—external-pressure profile—poor cement (high-pressure zone)

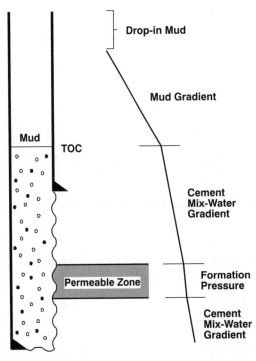

Figure 7-10 Permeable zones—external-pressure profile—poor cement (low-pressure zone)

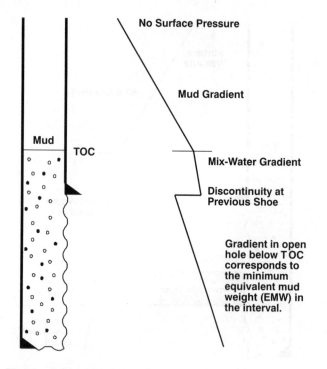

Figure 7-11 Openhole pore pressure—external-pressure profile—TOC inside previous shoe

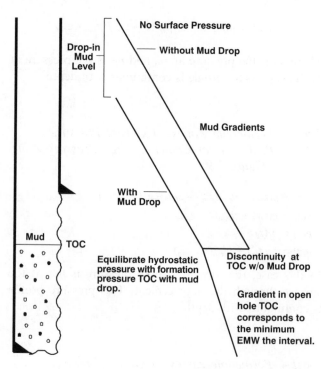

Figure 7-12 Openhole pore pressure—external-pressure profile—TOC below previous shoe (with and without mud drop)

7-6.1.6 Openhole Pore-Pressure—External-Pressure Profile (Figure 7-12)

TOC Below Previous Shoe without Mud Drop

$$p(z) = \gamma_m z \qquad\qquad z < z_{toc}$$
$$p(z) = \gamma_{em} z \qquad\qquad z > z_{toc}$$

This pressure profile is not continuous with depth; it is discontinuous at the top of cement.

TOC Below Previous Shoe with Mud Drop

$$p(z) = 0 \qquad\qquad z < z_{md}$$
$$p(z) = \gamma_{em} z_{toc} + \gamma_m(z - z_{toc}) \qquad z_{md} < z < z_{toc}$$
$$p(z) = \gamma_{em} z \qquad\qquad z_{toc} < z$$

where the mud drop z_{md} is defined as $z_{md} = z_{toc} - \gamma_{em} z_{toc}/\gamma_m$. This pressure profile is continuous with depth.

7-6.1.7 Above/Below TOC—External-Pressure Profile (Figure 7-13)

With Specified Pore Pressure In Open Hole

$$p(z) = \gamma_m z \qquad\qquad z < z_{toc}$$
$$p(z) = \gamma_m z_{toc} + \gamma_{cem}(z - z_{toc}) \qquad z_s > z > z_{toc}$$
$$p(z) = p_{spec}(z) \qquad\qquad z > z_s$$

where $p_{spec}(z)$ is the specified openhole pore pressure. This external-pressure distribution may be discontinuous at z_s, depending on $p_{spec}(z)$.

With Specified Pore-Pressure Gradient In Open Hole

$$p(z) = \gamma_m z \qquad\qquad z < z_{toc}$$
$$p(z) = \gamma_m z_{toc} + \gamma_{pp}(z - z_{toc}) \qquad z_s > z > z_{toc}$$

where γ_{pp} is the specified openhole pore-pressure gradient. This pressure distribution is continuous.

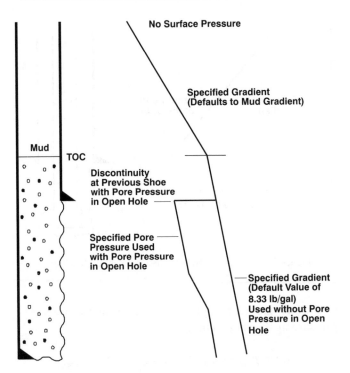

Figure 7-13 Above/below TOC —external-pressure profile— with and without pore pressure in open hole

7-6.2 Pressure Load Cases

7-6.2.1 Burst (Drilling): Gas Kick (Figure 7-14)

This load case uses an internal-pressure profile, which is the envelope of the maximum pressures that the casing experiences while circulating out a gas kick using the driller's method. It should represent the worst-case kick to which the current casing can be exposed while a deeper

interval is drilled. Typically, this means the casing will have to withstand a kick at the TD of the next hole section.

If the kick volume causes pressure to exceed the maximum fracture pressure at the casing shoe, it is reduced to the maximum volume that can be circulated out of the hole without exceeding the fracture pressure at the shoe. The maximum pressure experienced at any casing depth occurs when the top of the gas bubble reaches that depth.

7-6.2.2 Burst (Drilling): Displacement to Gas (Figure 7-15)

This load case uses an internal-pressure profile consisting of a gas gradient extending upward from a formation pressure in a deeper hole interval or from the fracture pressure at the casing shoe

$$p(z) = p_f - \gamma_g z$$

where γ_g is the gas gradient in psi/ft. This pressure physically represents a well-control situation in which gas from a kick has completely displaced the mud out of the drilling annulus from the surface to the casing shoe. This load is the worst-case drilling burst load that a casing string could experience. If the fracture pressure at the shoe is used to determine the pressure profile, it ensures that the weak point in the system is at the casing shoe and not at the surface. This precludes a burst failure of the casing near the surface during a severe well-control situation.

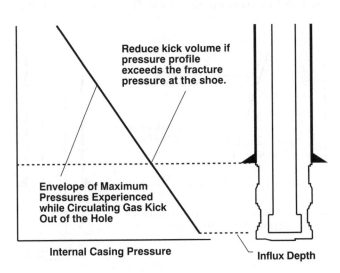

Figure 7-14 Burst (drilling): gas kick

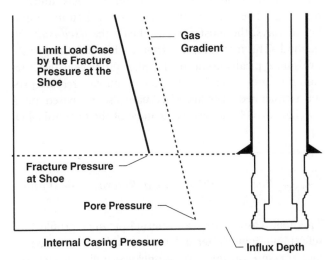

Figure 7-15 Burst (drilling): displacement to gas

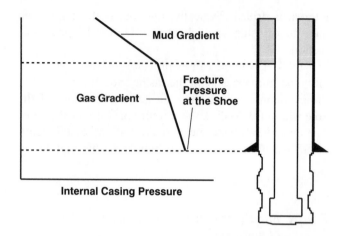

Figure 7-16   Burst (drilling): maximum load concept

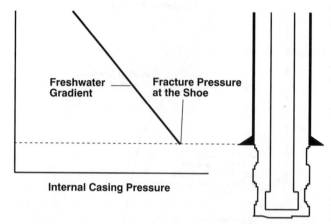

Figure 7-17   Burst (drilling): lost returns with water

7-6.2.3 Burst (Drilling): Maximum Load Concept (Figure 7-16)

This load case is a variation of the displacement-to-gas load case that is widely used in the industry and is taught in several popular casing design schools. It has been used historically because it results in an adequate but conservative design (particularly for wells deeper than 15,000 ft), and it is simple to calculate. The load case consists of a gas gradient (typically 0.1 psi/ft) extending upward from the fracture pressure at the shoe up to a mud/gas interface and then a mud gradient to the surface:

$$p(z) = p_{\text{frac}} + \gamma_m(z - z_{\text{int}}) + \gamma_g(z_{\text{int}} - z_{\text{shoe}}) \qquad z < z_{\text{int}}$$
$$p(z) = p_{\text{frac}} + \gamma_g(z - z_{\text{shoe}}) \qquad z_{\text{int}} < z < z_{\text{shoe}}$$

where p_{frac} is the frac pressure at depth of the shoe, z_{shoe} and γ_m is the mud gradient above the mud/gas interface depth z_{int}. The mud/gas interface is calculated in a number of ways, the most common being the *fixed endpoint* method. The interface is calculated based on a surface pressure typically equal to the BOP rating and the fracture pressure at the shoe, and a continuous pressure profile is assumed. The interface can also be based on a specific gas volume or a percentage of the openhole TD.

7-6.2.4 Burst (Drilling): Lost Returns with Water (Figure 7-17)

This load case models an internal-pressure profile that reflects pumping water down the annulus to reduce surface pressure during a well-control situation where lost returns are occurring. The pressure profile represents a

freshwater gradient applied upward from the fracture pressure at the shoe depth.

$$p(z) = p_{\text{frac}} + \gamma_{\text{water}}(z - z_{\text{shoe}}) \qquad z < z_{\text{shoe}}$$

A water gradient is used assuming that the rig's barite supply has been depleted during the well-control incident. This load case will typically dominate the burst design when compared to the gas-kick load case, especially for intermediate casing.

7-6.2.5 Burst (Drilling): Surface Protection (BOP) (Figure 7-18)

This load case is less severe than the displacement-to-gas criteria and represents a moderated approach to preventing a surface blowout during a well-control incident. It is

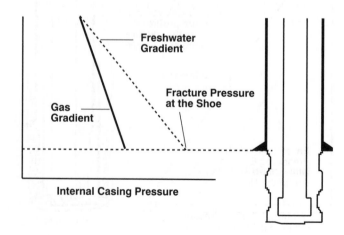

Figure 7-18   Burst (drilling): surface protection (BOP)

not applicable to liners. The same surface pressure calculated in Section 7-6.2.4 is used, but in this load case, a gas gradient from this surface pressure is used to generate the rest of the pressure profile:

$$p(z) = p_{frac} - \gamma_{water} z_{shoe} + \gamma_g z \qquad z < z_{shoe}$$

This load case represents no actual physical scenario; however, when used with the gas-kick criterion, it ensures that the casing weak point is not at the surface. Typically, the gas-kick load case will control the deeper portion of the design, and the surface protection load case will control the shallower portion of the design. As a result, the string's weak point is somewhere in the middle.

7-6.2.6 Burst (Drilling): Pressure Test (Figure 7-19)

This load case models an internal-pressure profile that reflects a surface pressure applied to a mud gradient

$$p(z) = p_{surface} + \gamma_m z$$

The test pressure is typically based on the maximum anticipated surface pressure determined from the other selected burst load cases plus a suitable safety margin (e.g. 500 psi). For production casing, the test pressure is typically based on the anticipated shut-in tubing pressure. This load case may or may not dominate the burst design, depending on the mud weight in the hole at the time the test occurs. The pressure test is normally performed before the float equipment is drilled out.

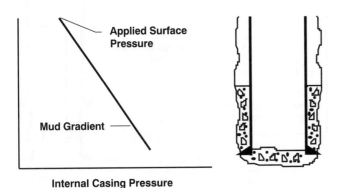

Figure 7-19 Burst (drilling): pressure test

7-6.2.7 Collapse (Drilling): Cementing (Figure 7-20)

This load case models an internal and external-pressure profile that reflects the collapse load imparted on the casing after the plug has been bumped during the cement job and the pump pressure has been bled off. The external pressure considers the mud hydrostatic column and different densities of the lead and tail cement slurries. The internal pressure is based on the gradient of the displacement fluid. If a light displacement fluid is used, the cementing collapse load can be significant.

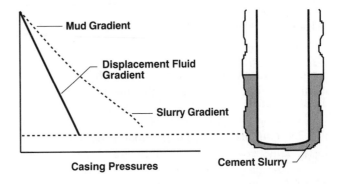

Figure 7-20 Collapse (drilling): cementing

7-6.2.8 Collapse (Drilling): Lost Returns with Mud Drop (Figure 7-21)

This load case models an internal-pressure profile that reflects a partial evacuation or a drop in the mud level caused by the mud hydrostatic column equilibrating with the pore pressure in a lost-circulation zone

$$p(z) = 0 \qquad z < z_{md}$$
$$p(z) = p_f + \gamma_m(z - z_{lc}) \qquad z_{md} < z < z_{lc}$$

where p_f is the pore pressure at the lost-circulation zone depth z_{lc}. The mud drop depth is given by

$$z_{md} = z_{lc} - p_f/\gamma_m$$

The heaviest mud weight used to drill the next hole section should be used along with a pore pressure and depth that result in the largest mud drop. Many operators conservatively assume that the lost-circulation zone is at the TD of the next hole section and that it will be normally pressured (a 0.465 psi/ft gradient). A partial evacuation of more than 5000 ft caused by lost circulation during drilling will normally not be seen. Many operators use

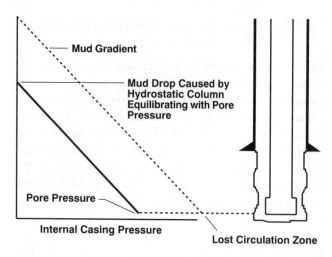

Figure 7-21   Collapse (drilling): lost returns with mud drop

a partial evacuation criterion in which the mud level is assumed to be a percentage of the openhole TD.

7-6.2.9 Collapse (Drilling): Other Load Cases

Full Evacuation

The full evacuation load case should be considered if drilling with air or foam. It may also be considered for conductor or surface casing where shallow gas is encountered. This load case would represent all of the mud being displaced out of the wellbore (through the diverter) before the formation bridges off.

Water Gradient

For wells with a sufficient water supply, an internal-pressure profile consisting of a freshwater or seawater gradient is sometimes used as a collapse criterion. This load case assumes a lost-circulation zone that can only withstand a water gradient.

7-6.2.10 Burst (Production): Gas Migration in Subsea Wells (Figure 7-22)

This load case models bottomhole pressure applied at the wellhead (subject to fracture pressure at the shoe) from a gas bubble migrating upward behind the production casing with no pressure bled off at the surface. The pressure is the minimum of the following two pressure distributions:

$$p(z) = p_{\text{frac}} + \gamma_m(z - z_{\text{shoe}})$$
$$p(z) = p_{\text{res}} + \gamma_m z$$

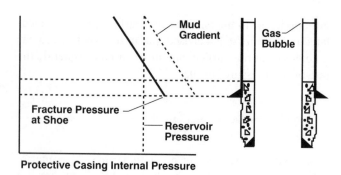

Figure 7-22   Burst (production): gas migration (subsea wells)

The load case only applies to the intermediate casing in subsea wells where the operator has no means of accessing the annulus behind the production casing.

7-6.2.11 Burst (Production): Tubing Leak (Figure 7-23)

This load case applies to both production and injection operations and represents a high surface pressure on top of the completion fluid caused by a tubing leak near the hanger. A worst-case surface pressure is usually based on a gas gradient extending upward from reservoir pressure at the perforations

$$p(z) = p_{\text{res}} - \gamma_g z_{\text{res}} + \gamma_m z$$

where p_{res} is the reservoir pressure at reservoir depth z_{res}. If the proposed packer location was determined during the casing design, the casing below the packer can be assumed to experience a pressure based on the produced fluid gradient and reservoir pressure only.

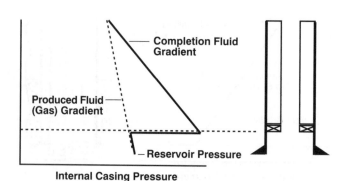

Figure 7-23   Burst (production): tubing leak

7-6.2.12 Burst (Production): Injection Down Casing (Figure 7-24)

This load case applies to wells that experience high-pressure annular injection operations, such as a casing fracture stimulation job. The load case models a surface pressure applied to a static fluid column:

$$p(z) = p_s + \gamma_m z$$

where p_s is the applied surface pressure. This load case is analogous to a screenout during a fracturing job.

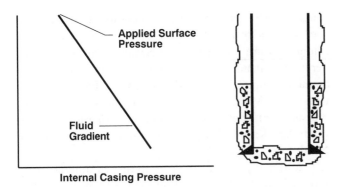

Figure 7-24 Burst (production): injection down casing

7-6.2.13 Collapse Above Packer (Production): Full Evacuation (Figure 7-25)

This severe load case has the most application in gas-lift wells. It represents a gas-filled annulus that loses injection pressure:

$$p(z) = \gamma_g z$$

Many operators use the full-evacuation criterion for all production casing strings regardless of the completion type or reservoir characteristics.

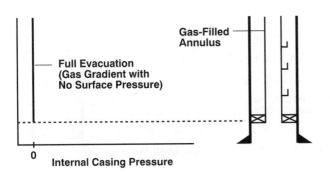

Figure 7-25 Collapse above packer (production): full evacuation

7-6.2.14 Collapse Above Packer (Production): Partial Evacuation (Figure 7-26)

This load case is based on a hydrostatic column of completion fluid equilibrating with depleted reservoir pressure during a workover operation:

$$p(z) = 0 \qquad\qquad z < z_{\text{drop}}$$
$$p(z) = p_{\text{res}} + \gamma_c(z - z_{\text{res}}) \qquad z_{\text{drop}} < z < z_{\text{res}}$$

with

$$z_{\text{drop}} = z_{\text{res}} - p_{\text{res}}/\gamma_c$$

where p_{res} is the depleted reservoir pressure, z_{res} is the depth of the reservoir, and z_{drop} is the depth of the fluid top of completion fluid, γ_c. Some operators do not consider a fluid drop, but just a fluid gradient in the annulus above the packer. This practice is acceptable if the final depleted pressure of the formation will be greater than the hydrostatic column of a lightweight packer fluid.

7-6.2.15 Collapse Below Packer (Production): Common Load Cases

Full Evacuation

This load case applies to severely depleted reservoirs, plugged perforations, or a large drawdown of a low-permeability reservoir. It is the most commonly used collapse criterion.

Fluid Gradient

This load case assumes zero surface pressure applied to a fluid gradient. A common application is the underbalanced fluid gradient in the tubing before perforating (or after perforating if the perforations are plugged). It is a less conservative criterion for formations that will never be drawn down to zero.

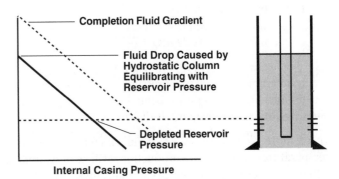

Figure 7-26 Collapse above packer (production): partial evacuation

7-6.2.16 Collapse (Production): Gas Migration in Subsea Wells (Figure 7-27)

This load case models bottomhole pressure applied at the wellhead (subject to fracture pressure at the prior shoe) from a gas bubble migrating upward behind the production casing with no pressure bled off at the surface. The pressure distribution is the minimum of the following two pressure distributions:

$$p(z) = p_{\text{frac}} + \gamma_m(z - z_{\text{shoe}})$$
$$p(z) = p_{\text{res}} + \gamma_m z$$

The load case is only applied in subsea wells where the operator has no means of accessing the annulus behind the production casing. An internal-pressure profile consisting of a completion-fluid gradient is typically used.

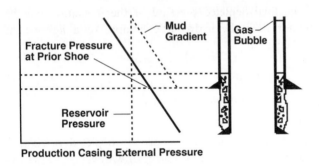

Figure 7-27 Collapse (production): gas migration (subsea wells)

7-6.2.17 Collapse: Salt Loads (Figure 7-28)

If a formation that exhibits plastic behavior, such as a salt zone, will be isolated by the current string, an equivalent external collapse load (typically assumed to be the overburden pressure) should be superimposed upon all of the collapse load cases (except cementing) from the top to the base of the salt zone:

$$p(z) = \gamma_{\text{ob}} z$$

where γ_{ob} is the overburden gradient (typically 1 psi/ft).

7-6.2.18 Annular Pressure Buildup

In offshore wells with sealed annuli, increases in fluid temperatures caused by production will cause fluid expansion, resulting in increased fluid pressures. For instance, for water at 100°F, a 1° increase in temperature will produce a pressure increase of 38,000 psi in a rigid

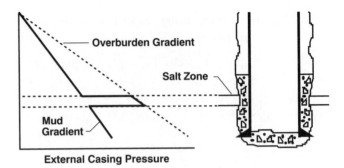

Figure 7-28 Collapse: salt loads

container. However, the casing and formation are sufficiently elastic to greatly reduce this pressure. The equilibrium pressure produced by thermal expansion must be iteratively calculated to balance fluid volume change with annular volume change (Adams and MacEachran, 1994; Halal and Mitchell, 1993). Nevertheless, the annular pressure change produced by thermal expansion has proven to be a serious design consideration, especially in the North Sea.

Traditionally, operators have mitigated annular pressure buildup by (1) perforating the casing near the shoe to provide a leak path, (2) completely cementing the annulus, (3) increasing the weight/grade of the casing, or (4) keeping the cement top below the shoe of the previous casing so that the annular pressure cannot exceed that pressure which results in leakoff at the shoe. Newer methods for relieving annular pressure include the use of wellhead pressure relief valves and the use of crushable foam wraps or crushable spheres. Crushable foam wraps or spheres provide pressure relief by increasing the annular area. Both materials can be designed to crush at a given pressure. Crushable foam wrap is installed in the top casing joints and crushable spheres are added to the mud system before the cementing job.

7-6.3 Mechanical Loads

7-6.3.1 Changes in Axial Load

In tubing and over the free length of the casing above the top of cement, changes in temperatures and pressures will have the largest effect on the ballooning and temperature load components. The incremental forces caused by these effects are given as

$$\Delta F_{\text{bal}} = 2\nu(\Delta p_i A_i - \Delta p_o A_o) + \nu L(\Delta \rho_i A_i - \Delta \rho_o A_o)$$

$$(7\text{-}23)$$

where ΔF_{bal} is the incremental force (lbf) caused by ballooning, ν is the Poisson ratio (0.30 for steel), Δp_i is the change in surface internal pressure, Δp_o is the change in surface external pressure, A_i is the cross-sectional area associated with casing ID, A_o is the cross-sectional area associated with casing OD, L is the free length of casing, $\Delta \rho_o$ is the change in internal fluid density, and $\Delta \rho_i$ is the change in external fluid density.

$$\Delta F_{temp} = -\alpha E A_s \Delta T \qquad (7\text{-}24)$$

where ΔF_{temp} is the incremental force (lbf) resulting from temperature change, α is the thermal expansion coefficient ($6.9 \times 10^{-6}\,°F^{-1}$ for steel), E is Young's modulus (3.0×10^7 psi for steel), A_s is the cross-sectional area of pipe, and ΔT is the average change in temperature over free length (°F).

7-6.3.2 Axial Load: Running in Hole

This installation load case represents the maximum axial load that any portion of the casing string experiences as it is run into the hole. This load case can include the effects of (1) self-weight, (2) buoyancy forces at the end of the pipe and at each cross-sectional area change, (3) wellbore deviation, (4) bending loads superimposed in dogleg regions, (5) frictional drag, and (6) shock loads based on an instantaneous deceleration from a maximum velocity. This velocity is often assumed to be 50% greater than the average running speed, which is typically 2 to 3 ft/sec. Typically, the maximum axial load experienced by any joint in the casing string is the load that occurs when the joint is picked up out of the slips after it is made up.

7-6.3.3 Axial Load: Overpull During Running

This installation load case models an incremental axial load applied at the surface while the pipe is run into the hole. Casing that is designed on the basis of this load case should be able to withstand an overpull force applied with the shoe at any depth if the casing becomes stuck as it is run in the hole. This load case includes the effects of (1) self-weight, (2) buoyancy forces at the end of the pipe and at each cross-sectional area change, (3) wellbore deviation, (4) bending loads superimposed in dogleg regions, (5) frictional drag, and (6) the applied overpull force.

7-6.3.4 Axial Load: Green Cement Pressure Test

This installation load case models the application of surface pressure after the plug is bumped during the primary cement job. Since the cement is still in its fluid state, the applied pressure will result in a large piston force at the float collar, and it often results in the worst-case surface axial load. This load case includes the effects of (1) self-weight, (2) buoyancy forces at the end of the pipe and at each cross-sectional area change, (3) wellbore deviation, (4) bending loads superimposed in dogleg regions, (5) frictional drag, and (6) piston force caused by differential pressure across the float collar.

7-6.3.5 Other Axial Load Cases

Two other load cases involve examining the *air weight of casing* and examining the *buoyed weight plus overpull*. Both of these criteria have calculations that are easy to perform and both normally result in adequate designs. Both load cases are still frequently used in the industry today, but since many factors are not considered, they are typically used with a high axial design factor (1.6+).

7-6.3.6 Axial Load: Shock Loads

Shock loads can occur if the pipe hits an obstruction or the slips close while the pipe is moving. The maximum additional axial force caused by a sudden deceleration to zero velocity is given by the following equation:

$$F_{shock} = V A_s \sqrt{E \rho_s} \qquad (7\text{-}25)$$

where F_{shock} is the axial force caused by shock loading, V is the instantaneous running speed, A_s is the cross-sectional area of the pipe, E is the elastic modulus, and ρ_s is the density of steel.

The shock load equation is often expressed as

$$F_{shock} = V w_{nom} \sqrt{E / \rho_s} \qquad (7\text{-}26)$$

where w_{nom} is the nominal casing weight per unit length $\approx A_s \rho_s$, and $\sqrt{E / \rho_s}$ is the speed of sound in steel (16,800 ft/sec). For practical purposes, some operators specify an average velocity in this equation and multiply the result by a factor (typically 1.5) that represents the ratio between the peak and average velocities.

7-6.3.7 Axial Load: Service Loads

For most wells, installation loads will control axial design. However, in wells with uncemented sections of casing and where large pressure or temperature changes will occur after the casing is cemented in place, changes in the axial load distribution can be significant. These changes are the result of self-weight, buoyancy forces, wellbore deviation, bending loads, changes in internal or external pressure (ballooning), temperature changes, and buckling.

7-6.3.8 Axial Load: Bending Loads

Stress at the pipe's outer diameter caused by bending can be expressed as

$$\sigma_b = \frac{ED}{2R} \qquad (7\text{-}27)$$

where σ_b is the stress at the pipe's outer surface and R is the radius of curvature.

This bending stress can be expressed as an equivalent axial force:

$$F_b = \frac{E\pi}{360} D(\alpha/L)A_s \qquad (7\text{-}28)$$

where F_b is the axial force caused by bending, and α/L is the dogleg severity (in degrees per unit length). The bending load is superimposed on the axial load distribution as a local effect.

7-6.4 Thermal Loads and Temperature Effects

In shallow, normally pressured wells, temperature will typically have a secondary effect on tubular design. In other situations, loads induced by temperature can be the governing criteria in the design.

Temperature increases that occur after the casing is landed can cause thermal expansion of fluids in sealed annuli, which results in significant pressure loads. Usually, these loads need not be included in the design because the pressures can be bled off. However, in subsea wells, the outer annuli cannot be accessed after the hanger is landed. Pressure increases will also influence the axial load profiles of the casing strings exposed to the pressures resulting from ballooning effects.

Changes in temperature will increase or decrease tension in the casing string as a result of thermal contraction and expansion, respectively. The increased axial load caused by the pumping of cool fluid into the wellbore

during a stimulation job can be the critical axial design criterion. In contrast, the reduction in tension during production caused by thermal expansion can increase buckling and possibly result in compression at the wellhead.

Changes in temperature not only affect axial loads but also influence the load resistance. Since the material's yield strength is a function of temperature, higher wellbore temperatures will reduce the burst, collapse, axial, and triaxial ratings of the casing.

In sour environments, operating temperatures can influence the types of equipment that can be used at different depths in the wellbore. Produced temperatures in gas wells will also influence the gas gradient inside the tubing since gas density is a function of temperature and pressure.

7-7 SYSTEMS APPROACH TO CASING DESIGN

To design a casing string, it is necessary to know the purpose of the well, the geological cross section, the available casing and bit sizes, the proposed cementing and drilling practices, rig performance, and safety and environmental regulations. For an optimal solution to be found, the casing must be considered as part of a whole drilling system. The following sections briefly describe the elements of the design process.

7-7.1 Design Engineer's Responsibilities

The casing design engineer must (1) establish a design that will account for all the anticipated loads that could be encountered during the life of the well (Klementich and Jellison, 1986; Prentice, 1970), (2) design strings that minimize well costs throughout the life of the well, and (3) clearly document the final design to operational personnel at the wellsite.

7-7.2 Tubular Failure

The primary goal of any casing design is to provide reliable well construction at a minimal cost, but, sometimes, failures do occur. Most documented failures occur because the pipe was exposed to loads for which it was not designed. These failures are called *off-design* failures. *On-design* failures are rather rare because casing design practices are usually conservative. Many failures occur at connections, which suggests that either field makeup practices were inadequate or the connection design basis was not consistent with the pipe body design.

7-7.3 Design Methodology

During the *preliminary design phase*, well planners and casing designers gather and interpret well data, determine the casing shoe depths and the number of strings required, select hole and casing sizes, determine the appropriate mud weight, and if applicable, consider additional aspects of directional design. The quality of the gathered data will have a significant impact on the choice of casing sizes and shoe depths and whether the casing design objective is successfully met.

During the *detailed design phase*, pipe weights and grades are selected for each string and the connections are determined. Specifically, casing designers compare pipe ratings with design loads and apply minimum acceptable safety standards (design factors). The primary goal of the detailed design phase is to meet all the design criteria while using the most economical pipe available. Figure 7-29 shows a preliminary/detailed design checklist.

7-8 PRELIMINARY DESIGN

The preliminary design establishes casing and drill bit sizes, casing setting depths and, consequently, the num-

Formation Properties

❏ Pore pressure
❏ Formation tensile strength (fracture pressure)
❏ Formation compressive strength (borehole failure)
❏ Temperature profile
❏ Location of squeezing salt and shale zones
❏ Location of permeable zones
❏ Chemical stability/sensitive shales (mud type and exposure time)
❏ Lost circulation zones
❏ Shallow gas
❏ Location of freshwater sands
❏ Presence of H_2S and/or CO_2

Directional Data

❏ Surface location
❏ Geologic target(s)
❏ Well interference data

Minimum Diameter Requirements

❏ Minimum hole size required to meet drilling objectives
❏ Logging tool OD
❏ Tubing size(s)
❏ Packer and related equipment requirements
❏ Subsurface safety valve OD (offshore well)
❏ Completion requirements

Production Data

❏ Packer fluid density
❏ Produced fluid composition
❏ Worst case loads that may occur during completion, production, and workover operations

Other

❏ Available inventory
❏ Regulatory requirements
❏ Rig equipment limitations

Figure 7-29   Casing design checklist

ber of different casing strings. The casing program (well plan) can be established based on the preliminary design. The well plan consists of three major steps. First, the mud program is prepared; next, the casing sizes and corresponding drill bit sizes are determined; and finally, the setting depths of individual casing strings are found.

7-8.1 Mud Program

Mud weight is the most critical mud program parameter. However, a complete mud program is designed on the basis of (1) pore pressure, formation strength, and lithology, (2) hole-cleaning and cuttings transport capabilities, (3) potential formation damage, stability problems, and the drilling rate, and (4) the operator's formation evaluation requirements, as well as environmental and regulatory requirements.

7-8.2 Hole and Pipe Diameters

Hole and casing diameters must be sized to provide adequate clearance for production equipment, and facilitate certain evaluation and drilling requirements, as shown in the following list:

- **Production equipment requirements** include tubing, subsurface safety valves, submersible pump diameter and gas-lift mandrel size, completion requirements (such as gravel packing), and comparing the benefits of increased tubing performance of larger tubing to the higher cost of larger casing over the life of the well.

- **Evaluation requirements** include logging interpretation requirements and tool diameters.

- **Drilling requirements** include the minimum bit diameter for adequate directional control and drilling performance, the available downhole equipment, rig specifications, and available BOP equipment.

Therefore, casing sizes should be determined from the inside outward starting from the bottom of the hole. Usually, the design sequence is as follows.

1. Based upon the reservoir and the inflow and tubing intake performance, select the proper tubing size as explained in Chapter 15.
2. Once completion requirements are considered, determine the required production casing size.
3. After carefully considering drilling and cementing stipulations, select the drill bit diameter for drilling the production section of the hole.

4. Determine the smallest casing clearance through which the drill bit will pass and repeat the process for other sections of the well.

During this portion of the preliminary design phase, large cost savings are possible through the use of smaller clearances. This practice has been one of the principal motivations in the increased popularity of slimhole drilling. Typical casing and rock bit sizes are given in Table 7-8.

Figure 7-30 depicts various selecting scenarios for drill bit (hole) and corresponding casing sizes with standard and low clearances.

7-8.3 Casing Shoe Depths and Number of Strings

Once the drill bit and the casing sizes are selected, the setting depth of individual casing strings must be determined (API, 1955). In conventional rotary drilling operations, the setting depths are primarily determined according to the mud weight and the fracture gradient, as shown in Figure 7-31. This information is sometimes referred to as the "well plan."

First, pore and fracture gradient pressure lines must be drawn on a chart that shows well depth vs. the equivalent mud weight (EMW) These gradient lines are the solid lines in Figure 7-31. Next, safety margins are introduced and broken lines are drawn that establish the design ranges. The offset from the predicted pore-pressure/fracture gradient nominally accounts for kick tolerance and the increased equivalent circulating density (ECD) during drilling.

From the highest mud weight required at the total depth (Point A in Figure 7-31), a vertical line is drawn to Point B. A protective casing string ($7\frac{5}{8}$ in.) must be set at the depth corresponding to Point B (about 10,900 ft) so that Section AB can be drilled safely. The setting depth of the next casing is determined by a horizontal line (BC) and then a vertical line (CD). The resulting Point D determines the next casing point ($9\frac{5}{8}$-in. casing at 9,300 ft.). The procedure is repeated for other casing strings.

In practice, a number of other factors can affect shoe depth design:

- **Regulatory requirements** Such requirements may dictate the protection of underground sources of drinking water or other zones, to a specific depth.

- **Hole stability** This factor can be a function of mud weight, deviation, and stress at the wellbore wall or it can be chemical in nature. Often, hole stability problems exhibit time-dependent behavior (making shoe

Table 7-8 Commonly used bit sizes that will pass through API casing

Casing Size (OD, in.)	Weight Per foot (lbm/ft)	Internal Diameter (in.)	Drift Diameter (in.)	Commonly Used Bit Sizes (in.)
$4\frac{1}{2}$	9.5	4.09	3.965	$3\frac{7}{8}$
	10.5	4.052	3.927	
	11.6	4.000	3.875	
	13.5	3.920	3.795	$3\frac{3}{4}$
5	11.5	4.560	4.435	$4\frac{1}{4}$
	13.0	4.494	4.369	
	15.0	4.408	4.283	
	18.0	4.276	4.151	$3\frac{7}{8}$
$5\frac{1}{2}$	13.0	5.044	4.919	$4\frac{3}{4}$
	14.0	5.012	4.887	
	15.5	4.950	4.825	
	17.0	4.892	4.764	
	20.0	4.778	4.653	$4\frac{5}{8}$
	23.0	4.670	4.545	$4\frac{1}{4}$
$6\frac{5}{8}$	17.0	6.135	6.010	6
	20.0	6.049	5.924	$5\frac{5}{8}$
	24.0	5.921	5.796	
	28.0	5.791	5.666	
	32.0	5.675	5.550	$4\frac{3}{4}$
7	17.00	6.538	6.413	$6\frac{1}{4}$
	20.00	6.456	6.331	
	23.00	6.366	6.241	
	26.00	6.276	6.151	$6\frac{1}{8}$
	29.00	6.184	6.059	6
	32.00	6.094	5.969	
	35.00	6.006	5.879	
	38.00	5.920	5.795	$5\frac{5}{8}$
$7\frac{5}{8}$	20.00	7.125	7.000	$6\frac{3}{4}$
	24.00	7.025	6.900	
	26.40	6.969	6.844	
	29.70	6.875	6.750	
	33.70	6.765	6.640	$6\frac{1}{2}$
	39.00	6.625	6.500	
$8\frac{5}{8}$	24.00	8.097	7.972	$7\frac{7}{8}$
	28.00	8.017	7.892	
	32.00	7.921	7.796	$6\frac{3}{4}$
	36.00	7.825	7.700	
	40.00	7.725	7.600	
	44.00	7.625	7.500	
	49.00	7.511	7.386	
$9\frac{5}{8}$	29.30	9.063	8.907	$8\frac{3}{4}$, $8\frac{1}{2}$
	32.30	9.001	8.845	
	36.00	8.921	8.765	
	40.00	8.835	8.679	$8\frac{5}{8}$, $8\frac{1}{2}$
	43.50	8.755	8.599	
	47.00	8.681	8.525	$8\frac{1}{2}$
	53.50	8.535	8.379	$7\frac{7}{8}$
$10\frac{3}{4}$	32.75	10.192	10.036	$9\frac{7}{8}$
	40.50	10.050	9.894	
	45.50	9.950	9.794	$9\frac{5}{8}$
	51.00	9.850	9.694	
	55.00	9.760	9.604	
	60.70	9.660	9.504	$8\frac{3}{4}$, $8\frac{1}{2}$
	65.37	9.560	9.404	$8\frac{3}{4}$, $8\frac{1}{2}$

Table 7-8   Commonly used Bit Sizes that will pass through API Casing *(continued)*

Casing Size (OD, in.)	Weight Per foot (lbm/ft)	Internal Diameter (in.)	Drift Diameter (in.)	Commonly Used Bit Sizes (in.)
$11\frac{3}{4}$	38.00	11.154	10.994	11
	42.00	11.084	10.928	$10\frac{5}{8}$
	47.00	11.000	10.844	
	54.00	10.880	10.724	
	60.00	10.772	10.616	
$13\frac{3}{8}$	48.00	12.715	12.559	$12\frac{1}{4}$
	54.50	12.615	12.459	
	61.00	12.515	12.359	
	68.00	12.415	12.259	
	72.00	12.347	12.191	11
16	55.00	15.375	15.188	15
	65.00	15.250	15.062	
	75.00	15.125	14.939	$14\frac{3}{4}$
	84.00	15.010	14.822	
	109.00	14.688	14.500	
$18\frac{5}{8}$	87.501	7.7551	7.567	$17\frac{1}{2}$
20	94.00	19.124	18.936	$17\frac{1}{2}$

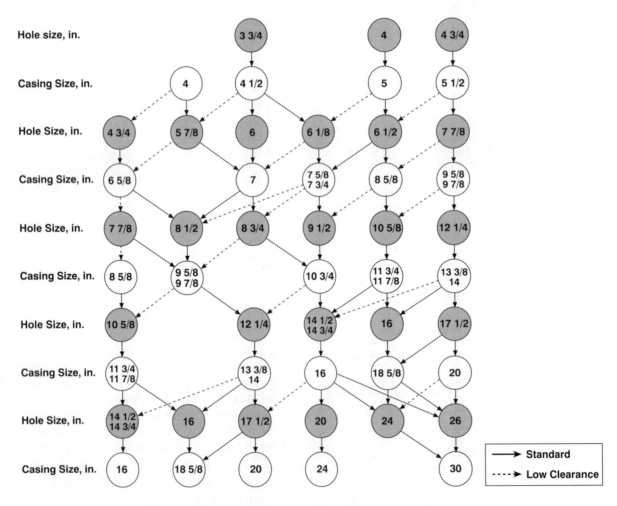

Figure 7-30   Drillbit (hole) and casing size options

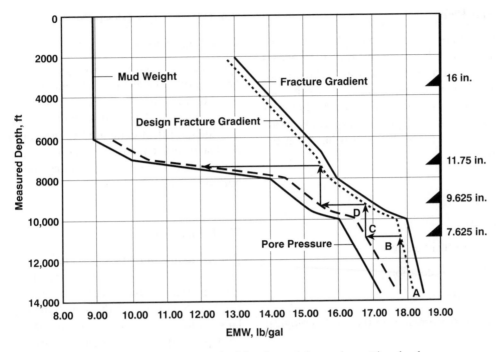

Figure 7-31 Graphical method for determining casing setting depths

selection a function of penetration rate). The plastic flowing behavior of salt zones must also be considered.

- **Differential sticking** The likelihood of differential sticking will increase with (1) increased differential pressure between the wellbore and formation, (2) increased formation permeability, and (3) increasing fluid loss or thicker mud cakes.

- **Zonal isolation** Shallow freshwater sands must be isolated from contamination, and lost-circulation zones must be isolated before a higher-pressure formation is penetrated.

- **Directional drilling concerns** A casing string is often run after an angle-building section has been drilled. This string prevents keyseating problems in the curved portion of the wellbore as a result of the increased normal force between the wall and the drillpipe.

- **Uncertainty in predicted formation properties** Exploration wells often require additional strings to compensate for the uncertainty in pore pressure and fracture gradient predictions.

Another method for determining casing setting depths is to plot formation and fracturing pressures vs. hole depth (rather than gradients as shown in Figure 7-31). This

procedure, however, is very conservative, typically yielding many strings.

Choosing casing setting depths is more complicated in exploratory wells, because of the shortage of geological information, pore pressures, and fracture gradients. In such a situation, a number of assumptions have to be made. Usually, a formation pressure gradient of 0.54 psi/ft is used if the hole depth is less than 8000 ft. If the hole depth is greater than 8000 ft, 0.65 psi/ft is used. Overburden gradients are generally taken at 0.8 psi/ft at shallow depth and 1.0 psi/ft at greater depths.

7-8.4 Top-of-Cement (TOC) Depths

Top-of-cement (TOC) depths for each casing string should be selected in the preliminary design phase because this selection will influence the axial load distributions and external-pressure profiles used during the detailed design phase. TOC depths are typically based on zonal isolation, regulatory requirements, prior shoe depths, formation strength, buckling, and in subsea wells, the annular pressure buildup. Buckling calculations are not performed until the detailed design phase, so that the TOC depth can be adjusted as a means of preventing buckling.

7-8.5 Directional Plan

For casing design purposes, a directional plan consists of determining the wellpath from the surface to the geological targets. The directional plan will influence all aspects of casing design, including (1) mud weight and mud chemistry selection for hole stability, (2) shoe seat selection, (3) casing axial load profiles, (4) casing wear, (5) bending stresses, and (6) buckling. It is based on the following factors:

- Geological targets

- Surface location

- Interference from other wellbores

- Torque and drag considerations

- Casing wear considerations

- BHA and drill bit performance in the local geological setting

To account for the variance from the planned build, drop, and turn rates that occur because of the BHAs and the operations practices used, higher doglegs are often superimposed over the wellbore. This practice increases the calculated bending stress in the detailed design phase.

7-9 DETAILED DESIGN

7-9.1 Load Cases

To select the appropriate weights, grades, and connections during the detailed design phase, design criteria must be established. These criteria normally consist of load cases and their corresponding design factors, both of which are then compared to pipe ratings. Load cases consist of burst loads, drilling loads, production loads, collapse loads, axial loads, running and cementing loads, and service loads.

7-9.2 Load Lines

Rather than compare each load case's profile to the pipe's rating on an individual basis, the approach taken for conventional burst, collapse, and axial design is to combine load cases of the same type (e.g., burst or collapse) into one load line of maximum load as a function of depth. Figure 7-32 shows a burst load line formed from two load cases used as burst criteria.

7-9.3 Design Factors

For a direct graphical comparison between the load line and the pipe's rating line, the design factor must be considered:

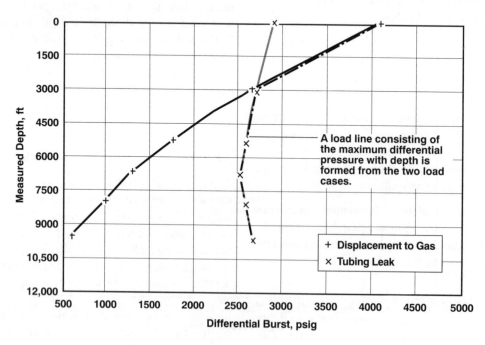

Figure 7-32   Use of load lines in casing design

$$DF = SF_{\min} \leq SF = \frac{\text{pipe rating}}{\text{applied load}}$$

where *DF* is the design factor (the minimum acceptable safety factor) and *SF* is the safety factor.

It follows that

$$DF \times (\text{applied load}) \leq \text{pipe rating}$$

Hence, by multiplying the load line by the *DF*, we can directly compare the design factor with the pipe rating. Provided that the rating is greater than or equal to the modified load line (the design load line), the design criteria have been satisfied.

7-9.4 Performing Graphical Design

Figure 7-33 shows an example of graphical design using burst. Using depth-pressure system of coordinates, the actual load, design load, and nominal casing ratings are plotted. Two other effects that impact design can be considered in graphical casing design by increasing the design load line:

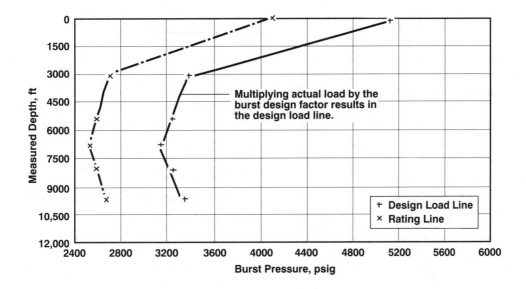

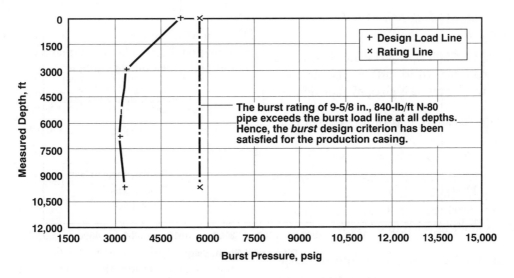

Figure 7-33 Graphical casing-design example: burst

- **The reduction of collapse strength resulting from tension (a biaxial effect)** The load line is increased as a function of depth by the ratio of the uniaxial collapse strength to the reduced strength.

- **The deration of material yield strength resulting from temperature** Like the effect of tension on collapse, the load line is increased by the ratio of the standard rating to the reduced rating.

7-9.5 Other Considerations

After burst, collapse, and axial stresses are considered, the initial design is complete. Before a final design is determined, the connections must be selected on the basis of potential wear and corrosion on both the casing and the connectors. Next, the triaxial stresses caused by combined loading (ballooning and thermal effects) must be considered. This stage of the final design is often called the *service-life analysis*. Finally, temperature effects and the likelihood of buckling must be addressed before finalizing the design.

7-9.6 Sample Design Calculations

In the following examples, burst, collapse, and uniaxial tension failure criteria are examined, and triaxial stresses are calculated for a variety of load situations. The examples illustrate the use of casing strength formulas presented in Sections 7-3 and the casing load formulas in Section 7-5.

7-9.6.1 Sample Burst Calculation with Triaxial Comparison

Assume that a $13\frac{3}{8}$-in., 72-lb/ft, N-80 intermediate casing is set at 9000 ft and cemented to surface. The burst differential pressure for this casing is given by Equation 7-1:

$$\Delta p = 0.875\,(2)(80{,}000\,\text{psi})(0.515\,\text{in.})/(13.375\,\text{in.})$$
$$= 5380\,\text{psi}$$

The load case we will test against is the *burst: displacement to gas case* given in Section 7-5.2.2, with a formation pressure of 6000 psi, the formation depth at 12,000 ft, and gas gradient equal to 0.1 psi/ft:

$$\text{Surface internal pressure} = 6000\,\text{psi} - 0.1\,\text{psi/ft}\,(12{,}000\,\text{ft})$$
$$= 4800\,\text{psi}$$

$$\text{Surface external pressure} = 0$$

$$\text{Net pressure differential} = 4800\,\text{psi}$$

According to this calculation, the casing is strong enough to resist this burst pressure. As an additional test, we will calculate the von-Mises stress associated with this case. Surface axial stress is the casing weight divided by cross-sectional area (20.77 in²) minus the pressure loads when the casing is cemented (assume a 15-lb/gal cement):

$$\sigma_{ri} = (72\,\text{lb/ft})(9000\,\text{ft})/(20.77\,\text{in}^2)$$
$$- (15\,\text{lb/gal})(0.052\,\text{psi/lb/gal})(9000\,\text{ft})$$
$$= 24{,}182\,\text{psi (tensile stresses are positive by convention)}$$

The radial stresses for the internal and external radii are the internal and external pressures:

$$\sigma_{ri} = -4800\,\text{psi (pressures are compressive stresses and negative by convention)}$$

$$\sigma_{ro} = 0\,\text{psi}$$

The hoop stresses are calculated by the Lamé formula (Equation 7-10):

$$\sigma_{\theta i} = (4800)\,\text{psi}[(6.688\,\text{in.})^2 + (6.174\,\text{in.})^2]/$$
$$[(6.688\,\text{in.})^2 - (6.174\,\text{in.})^2] = 60{,}152\,\text{psi}$$

$$\sigma_{\theta o} = (4800\,\text{psi})(2)(6.174\,\text{in.})^2/$$
$$[(6.688\,\text{in.})^2 - (6.174\,\text{in.})^2] = 55{,}352\,\text{psi}$$

The von-Mises equivalent stress or triaxial stress is given in Equation 7-9. Evaluating Equation 7-9 at the inside radius and at the outside radius, we have

$$\sigma_{vmi}\sqrt{(0 - 24{,}182\,\text{psi})^2 + (24{,}182 - 60{,}152\,\text{psi})^2}$$
$$+ (60{,}152 - 0\,\text{psi})^2 \div 2 = 52{,}426\,\text{psi}$$

$$\sigma_{vmi}\sqrt{4800\,\text{psi} - 24{,}182\,\text{psi})^2 + (24{,}182 - 53{,}352\,\text{psi}^2}$$
$$+ (53{,}352 - 4800\,\text{psi})^2 \div 2 = 47{,}905\,\text{psi}$$

The maximum von-Mises stress is at the inside of the $13\frac{3}{8}$-in. casing with a value that is 66% of the yield stress. In the burst calculation, the applied pressure was 89% of the calculated burst pressure. Thus, the burst calculation is conservative relative to the von-Mises calculation for this case.

7-9.6.2 Sample Collapse Calculation

For the collapse sample calculation, we will test the collapse resistance of a 7-in., 23-lb/ft, P-110 liner cemented from 8000 to 12,000 ft. A comparison of the 7-in. liner properties against the various collapse regimes listed in Section 7-3.2.2 reveals that transition collapse was predicted for this liner. The collapse pressure for this liner is calculated from Equation 7-4; the following values for F and G were taken from Table 7-5:

$$F = 2.066, \qquad G = 0.0532$$

The collapse pressure is then given by

$$p_c = (110,000 \, \text{psi})(2.066/(22.08) - 0.0532) = 4440 \, \text{psi}$$

Before we can evaluate the collapse of this liner, we must know the internal and external pressures. Internal pressure is determined with *the full evacuation above packer* formula given in Section 7-6.2.13:

$$p_i = 0.1 \, \text{psi/ft}(12,000 \, \text{ft}) = 1200 \, \text{psi}$$

The external pressure is based on a fully cemented section behind the 7-in. liner. The external-pressure profile is given by the *Mud Cement Mix-Water External-Pressure Profile* (Section 7-6.1.1), where the liner is assumed to be cemented in 10-lb/gal mud with an internal mix-water pressure gradient of 0.45:

$$p_o = (10 \, \text{lb/gal})(0.052 \, \text{psi/ft/lb/gal})(8000 \, \text{ft})$$
$$+ 0.45 \, \text{psi/ft} \, (12,000 - 8000 \, \text{ft}) = 5960 \, \text{psi}$$

An equivalent pressure is calculated from p_i and p_o for comparison with the collapse pressure p_c, through the use of Equation 7-6:

$$p_e = 5960 \, \text{psi} - [1 - 2/22.08](1200 \, \text{psi}) = 4869 \, \text{psi}$$

Since p_e exceeds p_c (4440 psi), the liner is predicted to collapse. We would not calculate a von-Mises stress for collapse in this case, since collapse in the transitional region is not strictly a plastic yield condition.

7-9.6.3 Sample Uniaxial Tension Calculation

For this example, a $9\frac{5}{8}$-in., 43.5-lb/ft, N-80 production casing is run in an 11,000-ft vertical well, with the top of cement at 8000 ft. The casing is run in 11-lb/gal water-based mud. The hanging weight in air for the casing is

$$F_{\text{air}} = 43.5 \, \text{lb/ft} \, (11,000 \, \text{ft}) = 478,500 \, \text{lb}$$

The casing stress at the surface is F_{air} divided by the cross-sectional area of the casing subtracted from the hydrostatic pressure at the bottom of the casing when it is cemented. If we assume 15-lb/gal cement and 11-lb/gal displaced mud, this bottomhole pressure is

$$p_{bh} = 11 \, \text{lb/gal} \, (0.052 \, \text{psi/ft/lb/gal}) \, (8000 \, \text{ft})$$
$$+ (15 \, \text{lb/gal}) \, (0.052 \, \text{psi/ft/lb/gal}) \, (11,000 - 8000 \, \text{ft})$$
$$= 6916 \, \text{psi}$$

The surface hanging stress is therefore

$$\sigma_z = 478500 \, \text{lb}/(12.56 \, \text{in}^2) - 6916 \, \text{psi} = 31,181 \, \text{psi}$$

For N-80 casing, a stress of 31,181 psi leaves a large margin of safety. Next, consider the effects of a stimulation treatment on this surface stress. Assume that the average temperature change in the 0 to 8000-ft interval is $-50°F$. The change in axial stress resulting from this temperature increase is given by Equation 7-24:

$$\Delta\sigma_z = -\alpha E \Delta T$$

where α is the coefficient of thermal expansion ($6.9 \times 10^6/°F$ for steel) and E is Young's modulus (30×10^6 psi for steel). The net surface stress in the casing is

$$\sigma_z = 31,181 \, \text{psi} - (6.9 \times 10^6/°F)(30 \times 10^6 \, \text{psi})(-50°F)$$
$$= 41,531 \, \text{psi}$$

7-10 CASING/TUBING MOVEMENT AND BUCKLING

7-10.1 Introduction

As it is installed, casing or tubing either hangs straight down in vertical wells or lays on the low side of the hole in deviated wells. Thermal or pressure loads may produce compressive loads, and if these loads are high enough, the initial configuration will become unstable.

However, since the tubing is confined within the open hole or casing, the tubing can deform into another stable configuration, usually a helical or coil shape in a vertical wellbore (Figure 7-34) or a lateral, S-shaped configuration in a deviated hole (Figure 7-35). The term *buckling* is used to refer to these new equilibrium configurations. In contrast, conventional mechanical engineering design considers buckling in terms of stability, the prediction of the critical load at which the original configuration becomes unstable (Crandall and Dahl, 1959).

Accurate analysis of buckling is important for several reasons. First, buckling generates bending stresses not present in the original configuration; if the stresses in the original configuration were near yield, this additional stress could produce failure, including a permanent, plastic deformation called *corkscrewing*.

Second, buckling causes tubing movement. It can be seen that a coiled tubing is shorter than straight tubing,

Figure 7-35   Lateral, s-shaped tubing deformation— deviated well

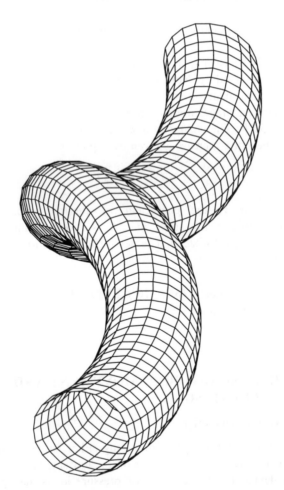

Figure 7-34   Helical or coil-shaped tubing deformation— vertical well

which is an important consideration if the tubing is not fixed in the packer. Free packers and PBRs have seals that accommodate this tubing movement; therefore, the length of these seals is an important design consideration (Chapter 14).

Third, tubing buckling relieves compressive axial loads when the packer is fixed. This effect is not as recognized as the first two buckling effects, but it is equally important. The axial compliance of buckled tubing is much less than the compliance of straight tubing. This effect can be easily demonstrated by first pushing on the end of a straight yardstick, then pushing on the end of a bowed yardstick. Tubing movement caused by thermal expansion or ballooning can be accommodated with a lower increase in axial load for a buckled tubing.

The accuracy and comprehensiveness of the buckling model is important for designing tubing. For example, the most commonly used buckling model was developed by Lubinski and Woods (Lubinski *et al.*, 1962;

Hammerlindl, 1977, 1980). This model is accurate for vertical wells, but it must be modified for deviated wells (Mitchell, 1988; Miska and Cunha, 1995). When Lubinski's solution is applied in deviated wells, it will overestimate both the tubing bending stress caused by buckling and the tubing movement. For a fixed packer, this solution will overestimate tubing compliance, which may greatly underestimate the axial loads, resulting in a nonconservative design. For a free packer or PBR, exaggerated tubing motion will require excessive seal length.

7-10.2 Buckling Models and Correlations

7-10.2.1 Analysis of Buckling

Buckling will occur if the buckling force, F_b, is greater than a threshold force, F_p, which is also known as the *Paslay buckling force* (Dawson, 1984). The buckling force is defined as follows:

$$F_b = -F_a + p_i A_i - p_o A_o \qquad (7\text{-}29)$$

where F_a is the tension-positive axial force, p_i is internal pressure, A_i is πr_i^2, r_i is the inside radius of the tubing, p_o is external pressure, and A_o is πr_o^2, where r_o is the outside radius of the tubing.

The Paslay buckling force F_p is defined as follows:

$$F_p = \sqrt{4w \sin \phi EI / r} \qquad (7\text{-}30)$$

where F_p is the Paslay buckling force, w is the distributed buoyed weight of the casing, ϕ is the wellbore angle with the vertical, EI is the pipe-bending stiffness, and r is the radial annular clearance.

Table 7-9 shows the relationship between the buckling force, the Paslay buckling force, and the type of buckling expected for the tubing (Chen *et al.*, 1990; Mitchell, 1996a).

Increased internal pressure affects the buckling force in two ways: (1) ballooning will increase F_a, which tends to decrease buckling, and (2) the $p_i A_i$ term will increase, which tends to increase buckling. Since the second effect

Table 7-9 Buckling force and type of buckling

Buckling Force Magnitude	Result
$F_b < F_p$	No buckling
$F_p < F_b < \sqrt{2}F_p$	Lateral (S-shaped) buckling
$\sqrt{2}F_p < F_b < 2\sqrt{2}F_p$	Lateral or helical buckling
$2\sqrt{2}F_p < F_b$	Helical buckling

is much greater, an increase in internal pressure will result in increased buckling.

A temperature increase will reduce the axial tension (or increase the compression). This reduction in tension will result in an increase in buckling. The onset and type of buckling is a function of hole angle. Because of the stabilizing effect of the lateral distributed force of a casing lying on the low side of the hole in an inclined wellbore, a greater force is required to induce buckling. In a vertical well, $F_p = 0$, and helical buckling will occur at any $F_b > 0$. For production tubing that is free to move in a seal assembly, the upward force resulting from pressure/area effects in the seal assembly will decrease F_a which, in turn, will increase buckling.

The following definitions are used in the correlations for tubing stresses and movement. The lateral displacements of the tubing, shown in Figure 7-36, are given by the following:

$$u_1 = r \cos \theta \qquad (7\text{-}31)$$

$$u_2 = r \sin \theta \qquad (7\text{-}32)$$

where θ is the helix angle. The quantity θ', where denotes d/dz, is important and will appear often in the following analysis. It can be related to the more familiar quantity pitch P through the following:

$$P = 2\pi/\theta' \qquad (7\text{-}33)$$

Other important quantities, such as pipe curvature, bending moment, bending stress, and tubing length change are proportional to the square of θ'. Nonzero θ' indicates that the pipe is curving, while zero θ' indicates that the pipe is straight.

7-10.2.2 Correlations for Maximum Buckling (Dogleg)

The correlation for the maximum value of θ' for lateral buckling can be expressed as follows:

$$\theta'_{max} = \frac{1.1227}{\sqrt{2EI}} F_b^{0.04} \left(F_b - F_p\right)^{0.46} \qquad (7\text{-}34)$$

for $2.8F_p > F_b > F_p$

The corresponding helical buckling correlation is

$$\theta'_{max} = \sqrt{\frac{F_b}{2EI}} \qquad (7\text{-}35)$$

for $F_b > 2.8F_p$

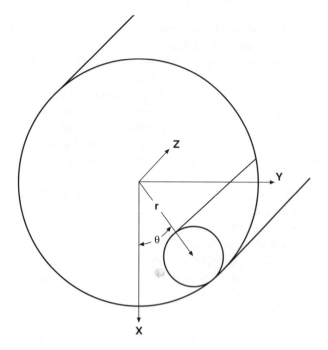

Figure 7-36 Lateral displacements of buckled tubing

The region $2.8F_p > F_b > 1.4F_p$ may be either helical or lateral; however, $2.8F_p$ is believed to be the lateral buckling limit on loading, while $1.4F_p$ is believed to be the helical buckling limit on unloading from a helical, buckled state. An important distinction between Equation 7-34 and Equation 7-35 is that Equation 7-34 is the maximum value of θ' while Equation 7-35 is the actual value of θ'.

The equation for dogleg curvature for a helix is

$$\kappa = r(\theta')^2 \qquad (7\text{-}36)$$

The dogleg unit for Equation 7-36 is radians per inch. To convert to the conventional unit of °/100 ft, multiply the result by 68,755.

7-10.2.3 Correlations for Bending Moment and Bending Stress

Given the tubing curvature, the bending moment is determined:

$$M = EI\kappa = EIr(\theta')^2 \qquad (7\text{-}37)$$

The corresponding maximum bending stress is

$$\sigma_b = \frac{MD_o}{2I} = \frac{ED_or(\theta')^2}{2} \qquad (7\text{-}38)$$

where D_o is the OD of the pipe. The following correlations can be derived with Equations 7-34 and 7-35:

$$M = 0$$

for $F_b > F_p$

$$M = 0.6302rF_b^{0.08}(F_b - F_p)^{0.92} \qquad (7\text{-}39)$$

for $2.8F_p > F_b > F_p$

$$M = 0.5rF_b \qquad (7\text{-}40)$$

for $F_b > 2.8F_p$ and

$$\sigma_b = 0$$

for $F_b < F_p$

$$\sigma_b = 0.3151\frac{D_or}{I}F_b^{0.08}(F_b - F_p)^{0.92} \qquad (7\text{-}41)$$

for $2.8F_p > F > F_p$

$$\sigma_b = 0.2500\frac{D_or}{I}F_b \qquad (7\text{-}42)$$

for $F_b > 2.8F_p$

7-10.2.4 Correlations for Buckling Strain and Length Change

The buckling "strain," in the sense of Lubinski, is the buckling length change per unit length. The buckling strain is given by the following relationship:

$$e_b = -\frac{1}{2}(r\theta')^2 \qquad (7\text{-}43)$$

For the case of lateral buckling, the actual shape of the θ' curve was integrated numerically for determining the following relationship:

$$e_{\text{bavg}} = -0.7285\frac{r^2}{4EI}F_b^{0.08}(F_b - F_p)^{0.92} \qquad (7\text{-}44)$$

for $2.8F_p > F_b > F_p$, which compares to the helical buckling strain:

$$e_b = -\frac{r^2}{4EI}F_b \qquad (7\text{-}45)$$

for $F_b > 2.8F_p$

The lateral buckling strain is roughly half the conventional helical buckling strain. To determine the buckling length change ΔL_b, we must integrate Equations 7-44 and 7-45 over the appropriate length interval:

$$\Delta L_b = \int_{z_1}^{z_2} e_b dz \qquad (7\text{-}46)$$

where z_1 and z_2 are defined by the distribution of the buckling force F. For the general case of arbitrary variation of F_b over the interval $\Delta L = z_2 - z_1$, Equation 7-46 must be numerically integrated. However, two special cases are commonly used. For constant force F_b, such as in a horizontal well, Equation 7-46 is easily integrated:

$$\int_{z_1}^{z_2} e_b \, dz = e_b \Delta L \qquad (7\text{-}47)$$

where e_b is defined by either Equation 7-44 or Equation 7-45. The second special case is for a linear variation of F_b over the interval:

$$F_b(z) = wz + c \qquad (7\text{-}48)$$

The length change is given for this case by

$$\Delta L_b = \frac{-r^2}{4EIw}\left(F_2 - F_p\right)\left[0.377IF_2 - 0.3668F_p\right] \qquad (7\text{-}49)$$

for $2.8F_p > F_2 > F_p$ and

$$\Delta L_b = -\frac{r^2}{8\,EIw}\left[F_2^2 - F_1^2\right] \qquad (7\text{-}50)$$

for $F > 2.8F_p$

7-10.2.5 Correlations for Contact Force

From equilibrium considerations only, the average contact force for lateral buckling is

$$W_n = w_e \qquad (7\text{-}51)$$

The average contact force for the helically buckled section is

$$W_n = rF^2/4EI + w_e \qquad (7\text{-}52)$$

When the buckling mode changes from lateral to helical, the contact force increases substantially.

7-10.3 Buckling in Oilfield Operations: Casing

Buckling should be avoided in drilling operations to minimize casing wear. To reduce or eliminate buckling during drilling, operators can apply a pickup force before landing the casing. In subsea wells, pressure can be held while waiting on cement (WOC) to pre-tension the string. Other methods of reducing or eliminating buckling include raising the top of cement, using centralizers, or increasing pipe stiffness.

In production operations, casing buckling is normally not a critical design issue. However, a large amount of buckling can occur as a result of increased production temperatures in some wells. During design, a check should be made to ensure that plastic deformation or corkscrewing will not occur. This check should use triaxial analysis and include the bending stress caused by buckling. Corkscrewing will only occur if the triaxial stress exceeds the yield strength of the material.

7-10.4 Buckling in Oilfield Operations: Tubing

Buckling is typically a more critical design issue for production tubing than for casing because (1) tubing is typically exposed to the hottest temperatures during production, (2) pressure/area effects in floating seal assemblies can significantly increase buckling, and (3) tubing is not as rigid as casing, and annular clearances can be quite large. Buckling in tubing is a concern because it can prevent wireline tools from successfully passing through the tubing.

To control buckling, operators can use an appropriate tubing-to-packer connection (latched or free, sealbore diameter, allowable movement in seals, etc.). Buckling control can also be achieved through the use of slackoff or pickup force at the surface or through the use of cross-sectional area changes in the tubing. Packer fluid density, pipe stiffness, centralizers, and hydraulic set pressure can further affect buckling. As in casing design, triaxial stresses analysis should be performed to ensure that plastic deformation or corkscrewing will not occur.

7-10.5 Sample Buckling Calculations

The following sample calculations are based on the buckling of $2\frac{7}{8}$-in., 6.5-lb/ft tubing inside 7-in. 32-lb/ft casing. The tubing is submerged in 10-lb/gal packer fluid with no other pressures applied. [For more information about pressure effects on buckling, see Lubinski *et al.* (1962), a good reference for application of pressure forces to practical problems, and Mitchell (1996b), for theoretical insight into pressure effects.] Through buoyancy, the packer fluid reduces the tubing weight per unit length:

$$w_e = w + A_i \gamma_i - A_o \gamma_o$$

where w_e is the effective weight per unit length of the tubing, A_i is the inside area of the tubing, γ_i is the density of the fluid inside the tubing, A_o is the outside area of the tubing, and γ_o is the density of the fluid outside the tubing. The calculation gives

$$
\begin{aligned}
w_e &= 6.5\,\text{lb/ft} + (4.68\,\text{in}^2)(0.052\,\text{psi/ft/lb/gal})(10.0\,\text{lb/gal}) \\
&\quad - (6.49\,\text{in}^2)(0.052\,\text{psi/ft/lb/gal})(10.0\,\text{lb/gal}) \\
&= 5.56\,\text{lb/ft} = 0.463\,\text{lb/in.}
\end{aligned}
$$

Other information useful for the buckling calculations are the following:

$$\text{radial clearance} = r = 1.61\,\text{in.}$$

$$\text{moment of inertia} = I = 1.611\,\text{in.}^4$$

$$\text{Young's modulus} = 30 \times 10^6\,\text{psi}$$

7-10.5.1 Sample Buckling Length Calculations

From Equation 7-30, we can calculate the Paslay force for a variety of inclinations. First, we calculate the value for a horizontal well:

$$
\begin{aligned}
F_p &= \sqrt{4(0.463\,\text{lbm/in.})(30 \times 10^6\,\text{psi})(1.611\,\text{in.}^4) \div (1.61\,\text{in.})} \\
&= 7456\,\text{lbf}
\end{aligned}
$$

This means that the axial buckling force must exceed 7500 lbf before the tubing will buckle. We can evaluate other angles by multiplying the horizontal F_p by the square root of the sine of the inclination angle. Table 7-10 provides a list of these buckling forces. Note that these buckling forces are large for relatively small deviations from vertical. For a well 10° from the vertical, the buckling forces are nearly half the horizontal-well buckling forces.

On the basis of this table, the total buckled length of the tubing can be calculated, as well as maximum and minimum lateral buckling or helical buckling. Assume a buckling force of 30,000 lbf is applied at the end of the tubing in a well with 60° deviation from vertical. The tubing will buckle for any force between 6939 lbf and 30,000 lbf. The axial force will vary as $w_e \cos \phi$, i.e.

$$w_a = w_e \cos(60) = 5.56\,\text{lbf/ft}(0.50) = 2.78\,\text{lbf/ft}$$

Therefore, the total buckled length L_{bkl} is

$$L_{bkl} = (30{,}000 - 6939)\,\text{lbf}/(2.78\,\text{lbf/ft}) = 8295\,\text{ft}$$

The maximum helically buckled length L_{helmax} is

$$L_{\text{helmax}} = (30{,}000 - 9813)\,\text{lbf}/(2.78\,\text{lbf/ft}) = 7262\,\text{ft}$$

The minimum helically buckled length L_{helmin} is

$$L_{\text{helmin}} = (30{,}000 - 19{,}626)\,\text{lbf}/(2.78\,\text{lbf/ft}) = 3732\,\text{ft}$$

Table 7-10 Sample buckling lengths

Deviation Angle Degrees	Buckling Forces (lbf) Minimum Lateral	Maximum Lateral	Minimum Helical
0	0	0	0
5	2201	6226	3113
10	3107	8788	4394
15	3793	10,729	5365
30	5272	14,913	7456
60	6939	19,626	9813
90	7456	21,090	10,545

7-10.5.2 Sample Buckling Bending Stress Calculations

The maximum bending stress caused by buckling can be evaluated with Equation 7-42:

$$\sigma_b = 0.25(2.875 \text{ in.})(1.61 \text{ in.})(30,000 \text{ lbf})/(1.611 \text{ in.}^4)$$
$$= 21,550 \text{ psi}$$

This stress is fairly high compared to tubing yield strengths of about 80,000 psi, so buckling /bending stresses can be important for casing and tubing design. At the buckling load of 19,626 lbf, both helical and lateral buckling can occur. The lateral bending stress is given by Equation 7-41:

$$\sigma_b = 0.3151(2.875 \text{ in.})(1.61 \text{ in.})/(1.611 \text{ in.}^4)(6939 \text{ lbf})$$
$$0.08(19,626 - 6939 \text{ lbf})0.92 = 10,945 \text{ psi}$$

The equivalent calculation for helical buckling gives

$$\sigma_b = 0.25(2.875 \text{ in.})(1.61 \text{ in.})(19,626 \text{ lbf})/(1.611 \text{ in.}^4)$$
$$= 14,097 \text{ psi}$$

Therefore, helical buckling produces stresses about 29% higher than the stresses produced by lateral buckling. Therefore, determining the buckling type can be important in casing designs in which casing strength is marginal.

7-10.5.3 Sample Buckling Length Change Calculations - Tubing Movement

Tubing length change calculations must consider tubing movement caused by lateral buckling and tubing movement caused by helical buckling. Equations 7-49 and 7-50 are used to calculate tubing movement, and these equations assume the minimum amount of helical buckling. A third calculation is made to show the movement due to pure helical buckling. The lateral buckling tubing movement is given by

$$\Delta L_b = -(1.61 \text{ in.})^2(19,626 - 6939 \text{ lbf})$$
$$\times [0.3771(19,626) - 0.3668(6939) \text{ lbf}]/$$
$$[(4)(30 \times 10^6 \, psi)(1.611 \text{ in.}^4)(2.78 \text{ lbf/ft})] = 0.297 \text{ ft}$$

The helical buckling tubing movement is given by

$$\Delta L_b = -(1.61 \text{ in.})^2(300,002 - 196,262 \text{ lbf}^2)/$$
$$[(8)(30 \times 10^6 \, psi)(1.611 \text{ in.}^4)(2.78 \text{ lbf/ft})] = 1.242 \text{ ft}$$

The total tubing movement is 0.297 ft plus 1.242 ft, which equals 1.539 ft. Pure helical buckling produces the following length change:

$$\Delta L_b = -(1.61 \text{ in.})^2(300,002 \text{ lbf}^2)/[(8)(30 \times 10^6 \text{ psi})$$
$$(1.611 \text{ in.}^4)(2.78 \text{ lbf/ft})] = 2.170 \text{ ft}$$

The use of pure helical buckling produces a 41% error in the calculation of tubing movement. When designing seal length in a deviated well, use of pure helical buckling can produce significant error.

REFERENCES

Adams, A.J., and MacEachran, A.: "Impact on Casing Design of Thermal Expansion of Fluids in Confined Annuli," *SPEDE* (Sept. 1994) 210.

API 5C2: "Bulletin on Performance Properties of Casing, Tubing, and Drill Pipe," 18th Edition, Am. Pet. Inst., Dallas (Mar. 1982).

API 5C3: "Bulletin on Formulas and Calculations For Casing, Tubing, Drill Pipe, and Line Pipe Properties," Fourth Edition, Am. Pet. Inst., Dallas (Feb. 1985).

API D7: "Casing Landing Recommendations," Am. Pet. Inst., Dallas (1955).

API Specification 5B:

Chen, Y.C., Lin, Y.H., and Cheatham, J.B.: "Tubing and Casing Buckling in Horizontal Wells," *JPT*, (Feb. 1990).

Crandall, S.H., and Dahl, N.C.: *An Introduction to the Mechanics of Solids*, McGraw-Hill Book Company, New York, (1959) 199.

Dawson, R., and Paslay, P.R.: "Drillpipe Buckling in Inclined Holes," *JPT*, (Oct. 1984).

Halal, A.S. and Mitchell, R.F.: "Casing Design for Trapped Annulus Pressure Buildup," paper SPE 25694, 1993.

Hammerlindl, D.J.: "Movement, Forces, and Stresses Associated With Combination Tubing Strings Sealed in Packers," *JPT* (Feb. 1977), 195-208.

Hammerlindl, D. J.: "Packer-to-Tubing Forces for Intermediate Packers," *JPT* (Mar. 1980), 195–208.

Klementich, E.F.: "A Rational Characterization of Proprietary High Collapse Casing Grades," SPE 30526, Proc. 1995 SPE Conference, Oct. 1995.

Klementich, E.F., and Jellison, M.J.: "A Service Life Model for Casing Strings," *SPEDE* (April 1986) 141.

Lubinski, A., Althouse, W. S., and Logan, J. L.: "Helical Buckling of Tubing Sealed in Packers," *JPT* (June 1962) 655–670.

Miska, S., and Cunha J. C.: "An Analysis of Helical Buckling of Tubulars Subjected to Axial and Torsional Loading in Inclined Wellbores," paper SPE 29460, (Apr. 1995).

Mitchell, R.F.: "Buckling Analysis in Deviated Wells: A Practical Method," paper SPE 36761, 1996a.

Mitchell, R.F.: "Forces on Curved Tubulars Caused by Fluid Flow," *SPEPF* (Feb. 1996b).

Mitchell, R.F.: "New Concepts for Helical Buckling," *SPEDE* (Sept. 1988), 303–310.

Prentice, C.M.: "Maximum Load Casing Design," *JPT*, July 1970, pp. 805–811.

Timoshenko, S.P., and Goodier, J.N.: *Theory of Elasticity*, third edition, McGraw-Hill Book Co., New York, (1961).

8 Primary Cementing

Kris Ravi
Halliburton Energy Services

Larry Moran
Conoco

8-1 INTRODUCTION

Primary cementing is the process of placing cement between the casing and borehole in a well. The main objectives of primary cementing are to seal the annulus and to obtain zonal isolation. The latter is accomplished if cement in the annulus prevents the flow of formation fluids. For zonal isolation to be achieved, all drill cuttings and drilling fluid must be removed from the annulus and replaced by cement slurry. Cement slurry must then undergo hydration, changing from the liquid to the solid phase and developing properties to prevent flow of formation fluids and to support the casing. The cement sheath should also be able to withstand different operations such as stimulation, perforation, production, and intervention during the life of the well.

8-1.1 Objectives of Primary Cementing

If the main objective of primary cementing (zonal isolation) is achieved, the economic, liability, safety, governmental, and other requirements imposed during the life of the well will be met. Since the zonal isolation process is not directly related to production, zonal isolation receives less attention than other well-construction activities. However, this necessary task must be performed effectively to conduct many production or stimulation operations. Thus, the success of a well depends on this primary operation.

Economics is involved in every aspect of well construction. Drilling engineers should consider the life of the well and construct a well that is cost-effective for its life. Usually, being cost-effective does not mean constructing the lowest cost well. Instead, money spent on primary cementing must be aimed at preventing more costly problems during production such as cement sheath cracking, water influx, or pressure buildup in the annulus. Otherwise, remedial treatments will have to be done throughout the life of the well or, in extreme cases, the well will have to be abandoned. Therefore, cements with fluid-loss control, gas-flow control, chemical stability, and mechanical integrity are often necessary and should be used to overcome all of these problems and provide cost-effective solutions.

Government regulations on primary cementing generally cover protection of surface waters. These regulations are designed to protect fresh water from contamination by downhole brines, oils, and gas. Normally, these regulations require that cement be pumped to surface on all casings that cover fresh water.

8-1.2 Steps to Meet the Objectives

In general, the following steps are required to obtain successful primary cement placement and to meet the objectives outlined above:

1. Analyze the well parameters; define the needs of the well, and then design placement techniques and fluids to meet the requirements for the life of the well. Fluid properties, fluid mechanics, and chemistry all influence the design used for a well. Mathematical and physical modeling of fluid flow are discussed in Section 8-2. The chemistry of cements and additives is discussed in Section 8-3.

2. Calculate fluid (slurry) composition and perform laboratory tests on the fluids designed in Step 1 to make sure they meet the job requirements. The cal-

culations and laboratory tests are discussed in Section 8-4.

3. Use necessary hardware to implement the design from Step 1; calculate the volumes of fluids (slurry) to be pumped; blend, mix, and pump fluids into the annulus. Casing attachments and equipment needed to perform these operations are discussed in Section 8-5.

4. Monitor the treatment in real time, compare it with the design from Step 1, and make changes as necessary.

5. Evaluate the results, compare them with the design from Step 1, and make changes as necessary for future jobs.

Reservoir conditions such as high temperature, high pressure, extreme depth, and deviation pose unique problems and are outlined in Table 8-1. This chapter discusses the fundamentals of these problems, the design methods and materials used to solve these problems, and issues of field implementation.

8-2 MATHEMATICAL AND PHYSICAL MODELS

It has been emphasized already that the objective of primary cementing is to obtain zonal isolation for the life of the well. This means that

- Cement slurry should be placed in the annulus so that 100% displacement efficiency is obtained.

- The slurry should then undergo hydration reaction and produce a cement sheath that will withstand all operations throughout the life of the well.

Displacement efficiency is defined as the percentage of the annulus that is filled with cement. Figure 8-1 shows a cross section of a cemented annulus. The hole diameter is D_w and the casing outer diameter is D_o. The cross-sectional area that should be filled with cement is

$$A_c = \pi(D_w^2 - D_o^2)/4 \qquad (8-1)$$

where A_c is the cross-sectional area of the annulus. In Figure 8-1, a section of the annulus still contains drilling fluid that has not been displaced by the cement. Let the cross-sectional area that is filled with cement be equal to A_{cem}. The displacement efficiency is then defined as

$$\eta = (A_{\text{cem}}/A_c)100 \qquad (8-2)$$

where η is the displacement efficiency.

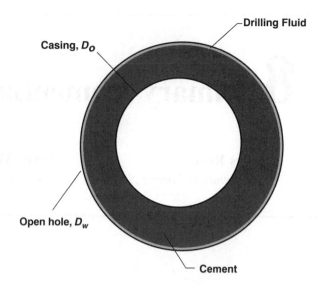

Figure 8-1 Cross section of cemented annulus

In the process of obtaining 100% displacement, the formation integrity should be maintained. This means that the pressure reached during the pumping operation must be such that the formation will not break down. Therefore, an accurate mathematical model is needed to estimate the pressure drop. The physics of the problem is illustrated with different cases in the following section. A number of simplifying assumptions are made in Case I. The rheology of fluids used in cementing operations is invariably non-Newtonian in nature and should be included in even a simplified model. Therefore, non-Newtonian models are discussed in Case I. Complexities in modeling the cementing operations are then gradually built in Cases II to VI.

8-2.1 Case I—Concentric

Figure 8-2 shows cement flowing into a wellbore filled with drilling fluid. This figure is an ideal depiction in which we assume that

- The casing is concentric with the hole.

- Cement by itself is displacing the drilling fluid out of the hole in its entirety and the fluids are stacked one over the other.

In this case, the total pressure drop is the sum of the individual pressure drops. In addition to the assumptions above, we also assume that

- The flow has attained steady state.

- The flow is only in the longitudinal or z direction.

Table 8-1 Well conditions presenting cementing problems and how to treat them

Well condition	Definition	Problems	Design Considerations
Deviated wells	Wellbore angle > 20°	Casing lying on the bottom Solids settling Free water forming Intermixing Fluids channeling	Use a complete mathematical model (Case V). Design methods and materials to prevent solids settling and prevent free water in slurry at downhole conditions
High pressure and high temperature	Temperatures in excess of 300°F	Solids settling Cement sheath cracking Micro annulus forming Drilling fluid gelling Fluids channeling	Use a complete mathematical model (Case V). Design methods and materials to prevent cement sheath cracking due to temperature and pressure cycling. Prevent solids settling, and remove drilling fluids at downhole conditions.
Deep wells	Depth > 15,000 ft	Formation damage caused by excessive friction pressures Need long casing Need long pumping time	Use a complete mathematical model (Case V). Design methods and materials for 100% displacement with no formation breakdown. Use liner to reduce casing and length; stage tools to reduce friction pressures; proper chemicals and strict quality control to ensure required pumping time.
Deep water	Water depth in excess of 1000 ft	Low temperature, low density	Use a complete mathematical model (Case V). Design methods and materials so that cement will set under these conditions and prevent fluid influx
Multilateral	Multiple laterals drilled from one parent wellbore	Integrity of the sealant in the junction	Use a complete mathematical model (Case V). Design methods and materials so that zonal isolation will not be compromised when the well is subjected to stresses from different operations.
Unconsolidated to weak formation		Loss of fluids to the formation	Optimize density and flow rate of fluids and use lost-circulation material as needed
Drilling with high fluid loss and/or in highly permeable formations		Differential sticking	Use flushes and pipe movement and minimize fluid loss of drilling fluid

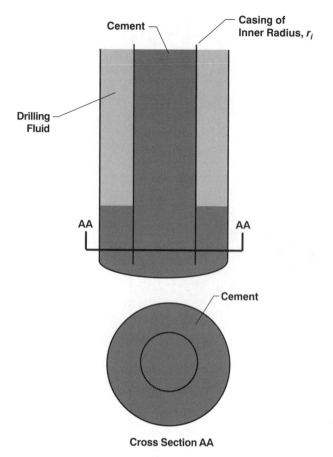

Figure 8-2   Cement flowing into wellbore filled with drilling fluid

- The fluids are incompressible.

Then, flow through the casing can be described by an equation of motion in cylindrical coordinates (Bird *et al.*, 1987) by setting v_r and v_θ equal to zero. The z component equation can then be written as

$$0 = -dp/dz - d\tau_{rz}/dr \qquad (8\text{-}3)$$

where τ_{rz} is the viscous momentum flux of z momentum in the r direction or the shear stress in z direction at a radial distance r, and dp/dz is the pressure gradient.

The boundary condition is $v_z = 0$ for $r = r_i$, where r_i is the inner radius of the casing.

8-2.1.1 Fluid Rheology

Shear stress is force per unit area and is expressed as a function of the velocity gradient of the fluid as:

$$\tau = -\mu(dv/dr) \qquad (8\text{-}4)$$

where μ is the fluid viscosity and dv/dr is the velocity gradient. The negative sign is used in Equation 8-4 because momentum flux flows in the direction of negative velocity gradient. That is, the momentum tends to go in the direction of decreasing velocity. A velocity gradient, just like the temperature gradient for heat transfer and the concentration gradient for mass transfer, is the driving force for momentum transfer. The absolute value of velocity gradient is the shear rate and is defined as

$$\dot{\gamma} = |dv/dr| \qquad (8\text{-}5)$$

Then Equation 8-4 can be written as

$$\tau = \mu\dot{\gamma} \qquad (8\text{-}6)$$

Viscosity is the resistance offered by a fluid to momentum transfer or deformation when it is subjected to a shear stress (Bird *et al.*, 1960; Eirich, 1960; Perry, 1985). If the viscosity is independent of shear stress or equivalent to the shear rate, then the fluid is called a Newtonian fluid. Water is an example of a Newtonian fluid. The shear rate is linear with shear stress for Newtonian fluids and is illustrated by Curve A in Figure 8-3. The symbol μ without any subscript is used to refer to the viscosity of a Newtonian fluid.

Most of the fluids used in cementing operations are *not* Newtonian and their behavior is discussed below. If the viscosity of a fluid is a function of shear stress or of shear

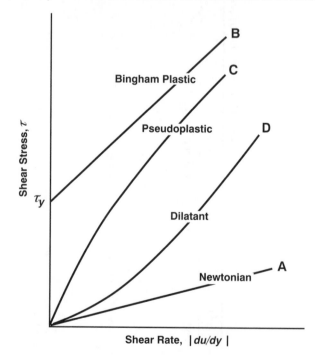

Figure 8-3   Rheology of fluids

rate, then the fluid is a non-Newtonian fluid. Non-Newtonian fluids can be classified into three general categories:

- The fluid properties are independent of the duration of shear.

- The fluid properties are dependent on the duration of shear.

- The fluid exhibits many properties that are characteristics of solids.

Time-Independent Fluids

The following three types of fluids are time-independent fluids:

- **Bingham plastic** These fluids require a finite shear stress, τ_o, below which they will not flow. Above this finite shear stress, referred to as yield point, the shear rate is linear with shear stress, just like a Newtonian fluid. The fluid is illustrated by Curve B in Figure 8-3. The shear stress can be written as

$$\tau = \tau_o + \mu_p g \qquad (8\text{-}7)$$

where μ_p refers to the plastic viscosity of the fluid. Some water-based slurries and sewage sludge are examples of Bingham plastic fluids. Most water-based cement slurries and water-based drilling fluids exhibit Bingham plastic behavior.

- **Pseudoplastic** These fluids exhibit a linear relationship between shear stress and shear rate when plotted on log-log paper, and their behavior is illustrated by Curve C in Figure 8-3. Pseudoplastic fluids are often referred to as power-law fluids. The shear stress can be written as

$$\tau = -K(dv/dr)|dv/dr|^{n-1} = K\dot{\gamma}^n n < 1 \qquad (8\text{-}8)$$

where K is the consistency index and n is the exponent, referred to as the power-law index. A term μ_a, defined as the apparent viscosity, is

$$\mu_a = K|dv/dr|^{n-1} = K\gamma^{n-1} \qquad (8\text{-}9)$$

The apparent viscosity decreases as the shear rate increases for power law fluids. Polymeric solutions and melts are examples of power-law fluid. Some drilling fluids and cement slurries, depending on their formulation, polymer content, or other qualities, exhibit power law behavior.

- **Dilatant** These fluids also exhibit a linear relationship between shear stress and shear rate when plotted on log-log paper and are illustrated as Curve D in Figure 8-3. The shear stress expression for a dilatant fluid is similar to a power-law fluid, but the exponent n is greater than 1. The apparent viscosity for these fluids increases as shear rate increases. Quicksand is an example of a dilatant fluid. In cementing operations, it is disadvantageous if fluids increase in viscosity as shear stress increases.

Time-Dependent Fluids

These fluids exhibit a change in shear stress with the duration of shear. This does not include changes resulting from chemical reaction, mechanical effects, or other events. Cement slurries and drilling fluids usually do not exhibit time-dependent behavior. However, with the introduction of new chemicals on a regular basis, the behavior of the fluids should be tested and verified.

Fluids with Solids Characteristics

These fluids exhibit elastic recovery from the deformation that occurs during flow and are called viscoelastic. Very few cement slurries and drilling fluids today exhibit this behavior. However, new polymers are being introduced on a regular basis, and tests should be conducted to verify the behavior.

The unit of viscosity, μ, is pascal-second (Pa-s) in the SI system and lb/(ft-sec) in English engineering units. One Pa-s equals 10 Poise (p), 1000 centipoise (cp), or 0.672 lb/(ft-sec). The exponent n is dimensionless and the consistency index, K, is expressed in units of Pa-s^n in the SI system and lbf-sec^n-ft^{-2} in oilfield units. One Pa-s^n equals 208.86 lbf-sec^n-ft^{-2}.

8.2.1.2 Concentric Viscometer

The rheology parameters of μ, μ_p, τ_o, K, and n are often determined from tests in a concentric viscometer. This instrument consists of concentric cylinders; one of the cylinders rotates, usually the outer one. A sample of fluid is placed between the cylinders and the torque on the inner cylinder is measured. The fluid velocity profile is illustrated in Figure 8-4.

Assuming an incompressible fluid (flow only in the θ direction) in a laminar flow regime, the equation of motion (Bird *et al.*, 1960) can be written as

$$0 = d[(1/r(d(rv_\theta)/dr))]/dr \qquad (8\text{-}10)$$

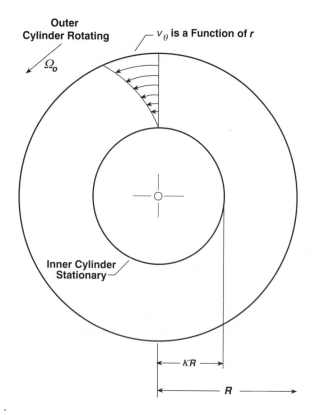

Figure 8-4 Concentric viscometer fluid viscosity profile

This equation is solved with the boundary conditions at $r = \kappa R$, $v_\theta = 0$; $r = R$, $v_\theta = \Omega_o R$. Equation 8-10 can be solved for $\tau_{r\theta}$ to give

$$\tau_{r\theta} = T/(2\pi R^2 L) \qquad (8\text{-}11)$$

where $\tau_{r\theta}$ is the θ momentum in r direction, T is the torque, R is the radius of the outer cylinder, κ is the ratio of radius of the inner cylinder to the outer cylinder, L is the length, and Ω_o is the angular velocity of the outer cylinder.

8-2.1.3 Pressure Drop Equations

In a concentric viscometer, the torque, T, is measured at different rotational speeds of the outer cylinder. Shear stress is then calculated from Equation 8-11, and shear rate is given by

$$\dot{\gamma} = 4\pi R^2 \Omega_o/(R^2 - \kappa^2 R^2) = 4\pi\Omega_o/(1 - \kappa^2) \qquad (8\text{-}12)$$

Shear stress and shear rate are then analyzed to determine the rheology model. Once the relationship between τ and $\dot{\gamma}$ is known, it can be substituted into Equation 8-3 to solve for the velocity profile. These solutions are valid

only for laminar flow because the shear stress/shear rate relationship from a concentric viscometer is valid only in laminar flow.

As an example, for a Bingham plastic fluid, Equation 8-7 should be substituted for the shear stress, τ, into Equation 8-3. This equation can then be solved for the velocity distribution in laminar flow, with boundary conditions $v_z = 0$ for $r = r_i$, as

$$v_z = (\Delta p/4\mu_p L)r_i^2[1 - (r/r_i)^2](\tau_o/\mu_p)r_i[1 - (r/r_i)]$$
$$\text{for } r \geq r_p \qquad (8\text{-}13)$$

and

$$v_z = (\Delta p/4\mu_p L)r_i^2[1 - (r/r_i)^2] \text{ for } r \leq r_p \qquad (8\text{-}14)$$

where r_p is the radius of the plug region, defined by $r_p = (\Delta p \tau_o/2L)$. The volumetric flow rate can then be estimated by integrating the velocity distribution over the cross-sectional area of the pipe.

Correlations for turbulent flow have been developed empirically from experiments conducted in a flow loop. A typical set of data is shown in Figure 8-5. Experimental data in laminar flow should be compared with estimated values from a correlation such as Equation 8-13. The non-slip boundary condition is applied in solving Equation 8-3. However, some solids-laden polymers are known to exhibit what is known as *shear-induced diffusion*, in which solids migrate away from the walls to the center of the pipe. These fluids show deviation in calculated and experimental values in laminar flow. The correlation should be modified as needed to reflect this behavior. A number of polymers are known to exhibit drag reduction in turbulent flow.

Experimental data illustrated in Figure 8-5 are usually analyzed and correlated between two dimensionless numbers f, the Fanning friction factor, and N_{Re}, the Reynolds number. An example plot of f vs. N_{Re} for flow through a circular pipe is shown in Figure 8-6. Then, the pressure drop per unit length for flow through a circular pipe of diameter D is given by

$$\Delta p/L = 2fv^2\rho/D \qquad (8\text{-}15)$$

where f is the Fanning friction factor, L is the length, v is the velocity, ρ is the density, and Δp is the pressure drop. The friction factor depends on N_{Re} and the roughness of the pipe. The Reynolds number, N_{Re}, is given by

$$N_{Re} = Dv\rho/\mu \qquad (8\text{-}16)$$

where ρ is the density of the fluid.

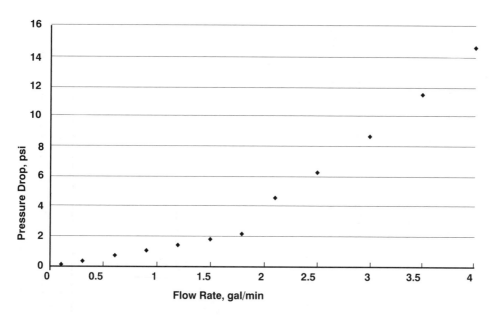

Figure 8-5   Experimental data from flow loop

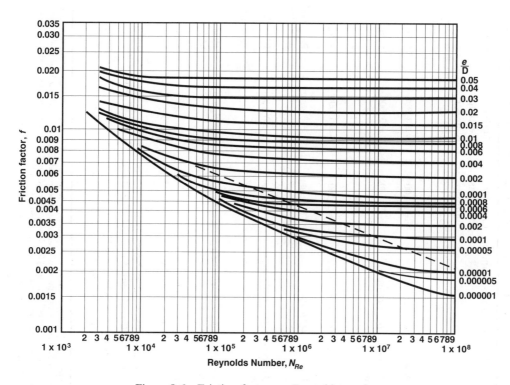

Figure 8-6   Friction factor vs. Reynolds number

Correlations for the friction factor, f, in both laminar and turbulent flow and for the critical Reynolds number are available for a number of fluids (Bird *et al.*, 1960; Govier and Aziz, 1987; Shah and Sutton, 1989). However, flow loop tests should be conducted and data compared with calculations that are based on fundamental equations for flow.

It is not uncommon to use the pipe flow equations for flow in the annulus by using equivalent diameter approximation and substituting $(r_w - r_o)$ for r_i. However, if the

Table 8-2 Summary of pressure drop equations in pipes and concentric annuli for Different fluids

Fluid Rheology	Model Used	Reynolds Number	Laminar Reynolds Number[3]	Turbulent Reynolds Number[4]	Friction Factor Laminar[1]	Friction Factor Turbulent[2]	$\Delta p/L$
Newtonian	Flow in Pipe	$D_i v\rho/\mu$	$\le 2{,}100$	$\ge 3{,}000$	$16/N_{Re}$	$1/f^{1/2} = 4.0\log(N_{Re}f^{1/2}) - 0.4$	$2f\rho v^2/D_i$
	Flow in Annulus, Pipe Model	$(D_w - D_o)v\rho/\mu$	$\le 2{,}100$	$\ge 3{,}000$	$16/N_{Re}$	$1/f^{1/2} = 4.0\log(N_{Re}f^{1/2}) = 0.4$	$2f\rho v^2/(D_w - D_i)$
	Flow in Annulus, Slit Model	$(D_w - D_o)v\rho/\mu$	$\le 2{,}100$	$\ge 3{,}000$	$24/N_{Re}$	$1/f^{1/2} = 4.0\log(N_{Re}f^{1/2}) - 0.4$	$2f\rho v^2/(D_w - D_i)$
Bingham Plastic	Flow in Pipe	$D_i v\rho/\mu_p$	$\le N_{ReBP1}$	$\ge N_{ReBP2}$	$16[(1/N_{Re}) + (N_{He}/(6N_{Re}^2)) - (N_{He}^4/(3f^3 N_{Re}^8))]$	$A(N_{Re})^{-B}$	$2f\rho v^2/D_i$
	Flow in Annulus, Pipe Model	$(D_w - D_o)v\rho/\mu_p$	$\le N_{ReBP1}$	$\ge N_{ReBP2}$	$16[(1/N_{Re}) + (N_{He}/(6N_{Re}^2)) - (N_{He}^4/(3f^3 N_{Re}^8))]$	$A(N_{Re})^{-B}$	$2f\rho v^2/(D_w - D_i)$
	Flow in Annulus, Slit Model	$(D_w - D_o)v\rho/(1.5\mu_p)$	$\le N_{ReBP1}$	$\ge N_{ReBP2}$	$16[(1/N_{Re}) + (9/8)N_{He}/(6N_{Re}^2)) - (N_{He}^4/(3f^3 N_{Re}^8))]$	$A(N_{Re})^{-B}$	$2f\rho v^2/(D_w - D_i)$
Power-law	Flow in Pipe	$D_i^n v^{2-n}\rho/(8^{n-1}[(3n+1)/4n]^n K)$	$\le (3250 - 1150n)$	$\ge (4150 - 1150n)$	$16/N_{Re}$	$1/f^{1/2} = \{[(4.0/n^{0.75})\log(N_{Re}f^{1-n/2})] - (0.4/n^{1.2})\}$	$2f\rho v^2/D_i$
	Flow in Annulus, Pipe Model	$(D_w - D_o)^n v^{2-n}\rho/(8^{n-1}[(3n+1)/4n]^n K)$	$\le (3250 - 1150n)$	$\ge (4150 - 1150n)$	$16/N_{Re}$	$1/f^{1/2} = \{[(4.0/n^{1-n/2})\log(N_{Re}f^{1-n/2})] - (0.4/n^{1.2})\}$	$2f\rho v^2/(D_w - D_i)$
	Flow in Annulus, Slit Model	$(D_w - D_o)^n v^{2-n}\rho/(12^{n-1}[(2n+1)/3n]^n K)$	$\le (3250 - 1150n)$	$\ge (4150 - 1150n)$	$24/N_{Re}$	$1/f^{1/2} = \{[(4.0/n^{n/2})\log(N_{Re}f^{1-n/2})] - (0.4/n^{1/2})\}$	$2f\rho v^2/(D_w - D_i)$

[1] $N_{He} = \tau_o\rho D_i^2/\mu_p^2$ for flow in pipe, $\tau_o\rho(D_h^2 - D_i^2)/\mu_p^2$ for flow in annulus, pipe model and $\tau_o\rho(D_h^2 - D_i^2)/1.5^2\mu_p^2$ for flow in annulus, slit model.

[2] $N_{ReBP2} = N_{ReBP2} - 866(1 - \alpha_c)$ for flow in pipe and annulus, pipe model, $N_{ReBP2} - 577(1 - \alpha_c)$ for flow in annulus, slit model.

[3] $N_{ReBP2} = N_{He}[0.968774 - 1.362439\alpha_c + 0.1600822\alpha_c^4)/(8\alpha_c)]$ for flow in pipe and for flow in annulus, pipe model, and $N_{He}[(0.968774 - 1.362439\alpha_c + 0.1600822\alpha_c^4)/(12\alpha_c)]$ for flow in annulus, slit model.
$\alpha_c = \frac{3}{2}[((2N_{He}/24{,}500) + (3/4)) - \{(2N_{He}/24{,}500)^2\}^{1/2})/(2(N_{He}/24{,}500))]$

[4] For $N_{He} \le 0.75e + 05$, $A = 0.20656$, $B = 0.3780$; For $0.75 < N_{He} \le 1.575e + 05$, $A = 0.26365$, $B = 0.38931$; For $N_{He} > 0.75e + 05$, $A = 0.20521$, $B = 0.35579$.

annulus dimensions are small ($r_o/r_w > 0.3$), a slit flow approximation should be used (Bird *et al.*, 1987; White, 1979). A summary of the equations for flow through pipes, flow through the annuli, and slit approximations are given in Table 8.2. The details of the derivations can be found in Ravi and Guillot (1997).

8-2.2 Case II—Eccentric Casing

Figure 8-1 is an idealization. The casing invariably sags to the bottom of a wellbore. Reinforcements called centralizers are used to minimize this sagging. Figure 8-7 illustrates a casing that is lying eccentrically in the wellbore. Given the number of centralizers, a correlation is available to estimate the amount of eccentricity. The eccentricity, e, is the distance between the center of the hole and the casing. The eccentricity ratio, ε, is

$$\varepsilon = e/(r_w - r_o) \qquad (8\text{-}17)$$

where r_w is the hole radius and r_o is the casing outer radius. The term *standoff* is commonly used in cementing operations to refer to the eccentricity ratio expressed as a percentage.

All the assumptions from Case I, other than the assumption of concentric casing, are still used in Case II. Then, Equation 8-3 is still applicable. However, (r_w-r_o) cannot be used uniformly as the annulus dimension. The eccentric annulus is then modeled as a slit of variable height (Tosun, 1984; Uner *et al.*, 1988). The annulus

approximated as a slit is shown in Figure 8-8. The height of the slit, h, is then calculated from

$$h = r_o[(1 - \kappa^2 \sin^2 \theta)^{1/2} + \kappa \cos \theta - r^*] \qquad (8\text{-}18)$$

where $\kappa = \varepsilon(1 - r^*)$, $r^* = r_i/r_o$, and θ is the angle.

Considering an example of Bingham plastic fluid, the boundary conditions are

$$\tau_{yz} = \tau_o - \mu_p(dv_z/dy) \text{ for } y_o \leq y \leq h/2;$$

$$dv_z/dy = 0 \text{ for } 0 \leq y \leq y_o$$

$$v_z = 0 \text{ for } y = h/2$$

where $y_o = \tau_o/(\Delta p/L)$ and $\Delta p/L$ is the pressure drop per unit length.

Equations 8-3 and 8-18 can be solved for velocity profile as

$$v_z = (\Delta p/2\mu_p L)(h/2)^2[1 - (y/(h/2))^2](\tau_o/\mu_p)(h/2)$$
$$[1 - (y/(h/2))] \text{ for } y_o \leq y \leq h/2 \qquad (8\text{-}19)$$

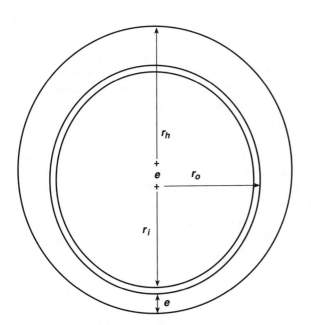

Figure 8-7   Eccentric casing

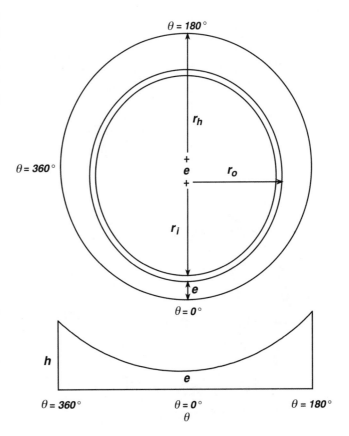

Figure 8-8   Eccentric annulus as a slit

and

$$v_z = (\Delta p/2\mu_p L)(h/2)^2[1 - (y_o/(h/2))^2](\tau_o/\mu_p)(h/2)$$
$$[1 - (y_o/(h/2))] \text{ for } 0 \leq y \leq y_o \qquad (8\text{-}20)$$

The volumetric flow rate can then be estimated by integrating the velocity distribution over the cross-sectional area of the slit.

8-2.2.1 Illustrations

A typical velocity profile of a Bingham plastic fluid in an eccentric annulus is shown in Figure 8-9. As shown in this figure, most of the fluid flows in the wide annulus. This effect will result in poor cement displacement in the

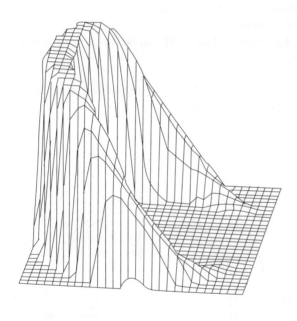

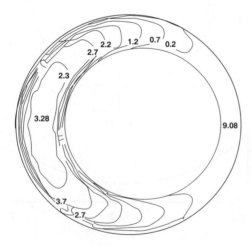

Figure 8-9   Flow profile in eccentric annulus

narrow annulus (Walton and Bittleston, 1991; Uner *et al.*, 1988; Haciislamoglu and Langlinais, 1990). Effects of an eccentric annulus on cement displacement are illustrated in Figures 8-10 to 8-12. These figures show that cement did not displace the drilling fluid in the narrow annulus. The extent of displacement will depend both on the eccentricity and the condition of drilling fluid under downhole conditions. The effects of downhole conditions on drilling fluids are discussed in Case III.

8-2.3 Case III—Filter Cake, Gels, and Solids Deposition

Figure 8-7 included eccentricity but still assumed that the cement slurry was displacing the drilling fluid in its entirety. However, in an actual well, the conditions are much different:

- Drilling fluid loses filtrate to the formation, deposits filter cake, and develops gel strength when it is left static for procedures such as logging the open hole and running the casing.

- Cement is usually incompatible with drilling fluid. This means that when the cement comes in direct contact with the drilling fluid, the properties of one or both fluids change.

- Solids in the fluid drop out in both dynamic and static conditions.

8-2.3.1 Effect of Downhole Conditions on Drilling Fluid

The deposition of filter cake and gelling of drilling fluid offers resistance to displacement by cement that was not included in the previous case. The filter cake thickness and the permeability of filter cake can be estimated by assuming that the filter cake is a packed bed, as discussed by Sutton and Ravi (1989). The resistance offered or force needed to displace the drilling fluid in the wellbore should be estimated by simulating downhole conditions.

Let τ_y be the resistance offered by the drilling fluid. Figure 8-13 shows a section of an annulus of length L and equivalent diameter D through which fluid is flowing. If buoyancy forces can be neglected, then a force balance on this section yields

$$(\pi D^2/4)(p + \Delta p) - (\pi D^2/4)(p) = \tau_w(\pi DL) \qquad (8\text{-}21)$$

where Δp is the pressure drop across length L; $\pi D^2/4$ is the cross-sectional area; τ_w is the shear stress exerted by

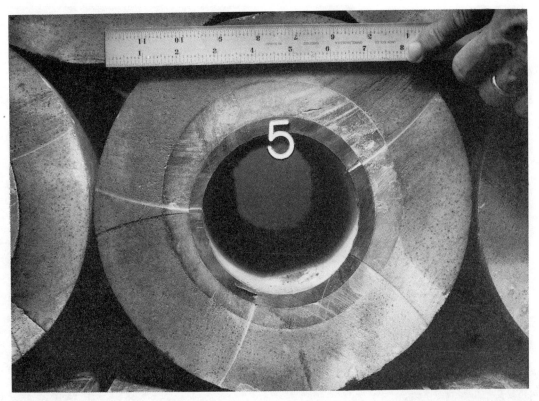

Figure 8-10   Eccentricity and displacement, I

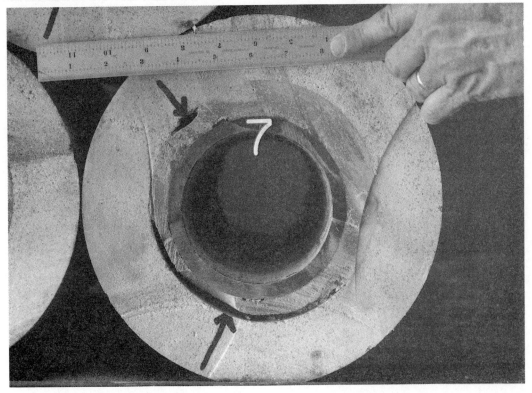

Figure 8-11   Eccentricity and displacement, II

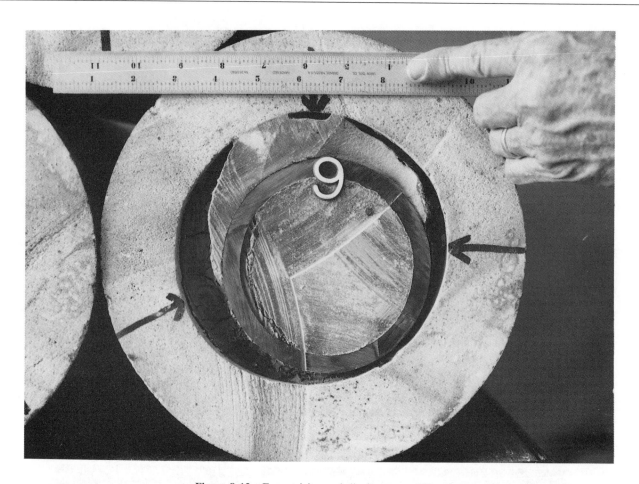

Figure 8-12   Eccentricity and displacement, III

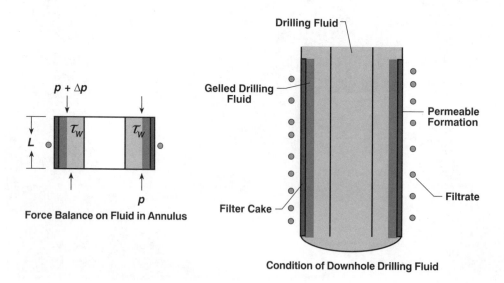

Figure 8-13   Model with forces

the fluid on the formation wall or filter cake/gelled drilling fluid; and πDL is the surface area over which the shear stress is acting. Equation 8-21 can be written as

$$\tau_w = D\Delta p/4L \qquad (8\text{-}22)$$

This equation results in the force exerted by the fluid. This force should be equal to or greater than the resistance offered by the drilling fluid in the wellbore (Ravi *et al.*, 1992a). The following condition should be satisfied:

$$\tau_w > \tau_y \qquad (8\text{-}23)$$

The resistance that is offered by the drilling fluid in the wellbore depends on its state. The exact state will depend on the following:

- *Drilling fluid type* Drilling fluids may be water-based, polymer-based, or oil-based.

- *Drilling fluid properties* Drilling fluid properties include rheology, gel strength, density, and particle concentration, among others.

- *Fluid loss to formation* The amount of fluid lost to the formation is governed by fluid loss of the drilling fluid, permeability of the formation, temperature, differential pressure, or other downhole factors.

- *Downhole pressure* The fluid pressure in the wellbore is governed by hydrostatic and friction pressures.

- *Downhole temperature* Downhole temperature is the temperature to which the drilling fluid is exposed during circulation and shutdown periods.

- *Differential pressure into the formation* The difference in pressure between the fluid pressure in the wellbore and formation pressure is the driving force for fluid loss and filter cake deposition.

- *Shutdown period* The shutdown period is the amount of time the fluid is left static, usually after the wiper trip to log the open hole and run the casing.

- *Drilling history* The drilling history reveals how well solids were controlled, the quality of the wiper trip, and other factors important to the operation.

These conditions should be simulated in the laboratory to estimate the resistance offered by the drilling in the wellbore. Figure 8-14 shows an experimental setup used to estimate these forces.

Once τ_y is known, Equations 8-22 and 8-23 can be solved to design the fluid properties needed to displace the gelled drilling fluid from the wellbore.

8-2.3.2 Incompatibility

If the cement is not compatible with the drilling fluid, then the displacement efficiency will be poor, the cement will not hydrate properly, and zonal isolation will not be obtained. A third fluid that is compatible with both the drilling fluid and cement is placed in between to separate them. This fluid is called spacer, and Equations 8-3, 8-22, and 8-23 also apply to this fluid. Spacer should overcome the resistance offered by the drilling fluid, and the cement should then overcome any resistance offered by the spacer. Equations 8-3, 8-22, and 8-23 also apply to cement.

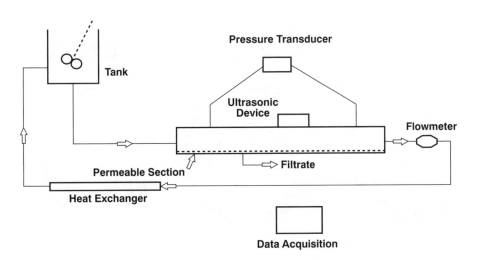

Figure 8-14 Erodibility cell

Some drilling fluids develop gel strength high enough that it is difficult to displace them by fluid circulation alone. That is, it is difficult for τ_w to exceed τ_y. In such cases, the resistance could be decreased by chemical action. Bringing the gel in contact with suitable chemical solutions will accomplish this objective (Ravi and Weber, 1996). These chemical solutions are called chemical flushes. The effectiveness of such flushes can also be simulated in an experimental setup (Figure 8-14). The casing is either reciprocated or rotated in some cases to improve cement placement. Equations discussed for concentric viscometers can be adapted to model the effect of pipe rotation. A likely effect of pipe rotation on velocity profile is shown in Figure 8-15. Pipe movement will change the pressure drop (McCann *et al.*, 1993) and should be taken into consideration.

8-2.3.3 Solids Deposition

Solids in the drilling fluid, spacer, or cement could drop out in either dynamic or static conditions. The settled solids do not provide zonal isolation and thus the objectives will not be met if the fluids are not designed to prevent solids settling (Sassen *et al.*, 1991). The deposition under dynamic condition is governed by

$$v_d = f(D, D_p, q, C_s, \rho_F, \rho_p, \mu_p, \tau_o) \qquad (8\text{-}24)$$

where v_d is the deposition velocity, D is the equivalent diameter of the geometry through which fluid is flowing, D_p is the representative diameter of the solid particle(s), q is the flow rate, C_s is the concentration of solids, ρ_F is the density of fluid, ρ_p is the density of

representative particle(s), μ_p is the plastic viscosity and τ_o is the yield point of the fluid.

A force balance should be applied on the depositing solids to determine the form of Equation 8-24. Experimental data on solids deposition from these fluids should then be used to determine the empirical correlation. The details are discussed by Shah and Lord (1990), Moran and Lindstrom (1990), and Oraskar and Turtan (1980). To prevent solids deposition, the velocity of fluids, v, should be greater than v_d:

$$v > v_d \qquad (8\text{-}25)$$

The property of the fluid that determines solids deposition under static conditions is the static gel strength of the fluid and not the yield point or plastic viscosity. Tests should be conducted to make sure that solids do not deposit under static conditions. However, solids that do not deposit under static condition may deposit under dynamic conditions because the properties of the fluids that govern the deposition of the solids under the two conditions are different.

8-2.3.4 Illustrations

An illustration of solutions to some of the mathematical models discussed is shown in Figures 8-16 to 8-18. One of the main features of these models is the calculation of flow rates at which the fluids should be pumped so that the formation will not break down. A typical wellbore schematic plot is shown in Figure 8-16. This figure shows the position of fluids after pumping a certain volume into the wellbore. It should be checked to make sure that the

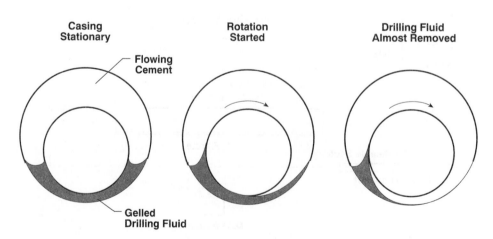

Figure 8-15   Velocity profile in pipe rotation

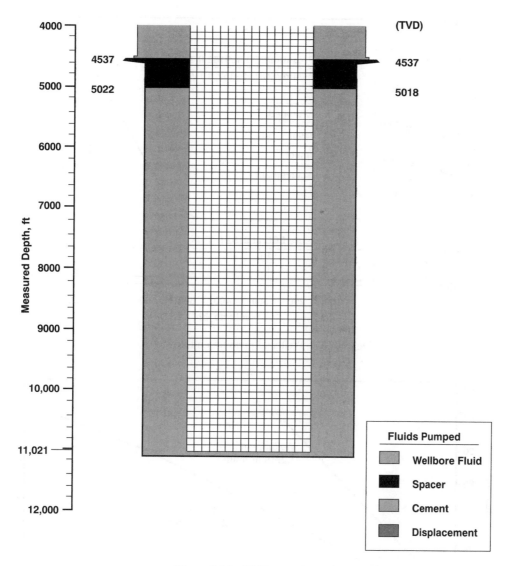

Figure 8-16   Wellbore schematic

fluids are placed where they were designed to be placed in the wellbore. A typical wellhead pressure and flow-rate plot is shown in Figure 8-17. This figure shows the rates at which the fluids should be pumped so that the formation will not break down. The extent of hole·cleaning that is expected at different centralizations is shown in Figure 8-18.

8-2.4 Case IV—Temperature of the Fluid

Case III discussed fluid properties needed to overcome the resistance of drilling fluid in the wellbore, prevent solids deposition, etc. However, rheology of fluids is a function of temperature. Rheological properties can be measured in a high-pressure/high-temperature rheometer

and can also be mathematically modeled (Ravi and Sutton, 1990; Ravi, 1991). Estimating the viscosity or other fluid properties as a function of temperature requires estimation of the temperature of the circulating fluid along the wellbore.

A heat balance should be applied to the flowing fluid to estimate the temperature. The mode of heat transfer when the fluid is circulating is mainly forced and natural convection and conduction. The following assumptions are made:

- Heat transfer in longitudinal or z direction only
- Steady-state conditions
- Negligible natural convection.

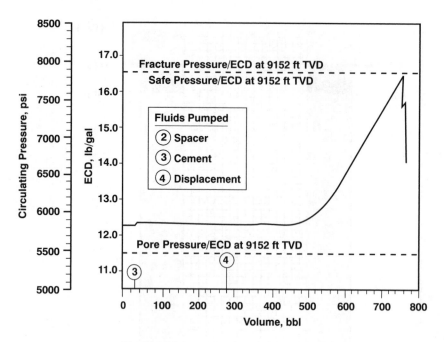

Figure 8-17   Wellhead pressure and flowrate

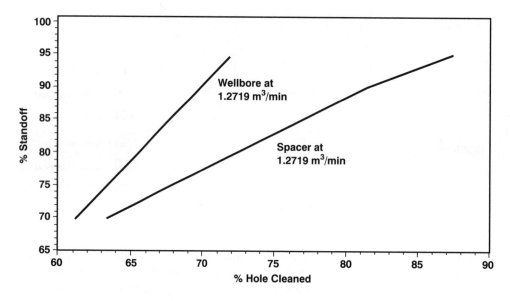

Figure 8-18   Extent of hole cleaning

Then, the heat balance around a small element of length Δz, shown in Figure 8-19, can be written for flow in the casing and annulus as

$$\dot{m}c_p(T_o - T_i) = -U_i A_i (T_c - T_a) \qquad (8\text{-}26)$$

and

$$\dot{m}c_p(T_o - T_i) = -U_o A_o (T_a - T_c) - U_h A_h (T_a - T_r) \qquad (8\text{-}27)$$

where $\dot{m}$ is the mass flow rate, c_p is the heat capacity of the fluid, T_i is the temperature of the incoming fluid, and T_o is the temperature of the outgoing fluid. T_a is the average temperature in the annulus, T_c is the average temperature in the casing, T_r is the temperature in the reservoir, U is the overall heat transfer coefficient, A is the area, and Δx is the effective thickness of formation. U_i is based on casing inner diameter, U_o is based on casing outer diameter, and U_h is based on hole diameter.

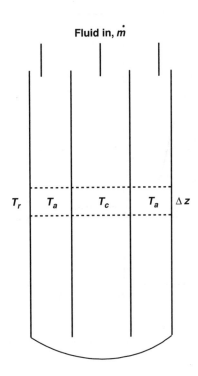

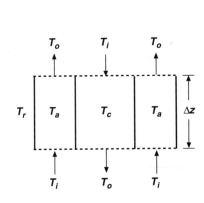

Figure 8-19 Temperature modeling

These equations are solved with boundary conditions $z = 0$, $T_i = T_{in}$; $T_r = f(z)$.

*Equations 8-26 to 8-28 are solved until temperatures converge. The overall heat transfer coefficient is a function of individual heat transfer coefficients and the thermal conductivity (Bird *et al.*, 1960; Perry, 1985). This approach is simplified, and to improve the accuracy of results, the effects of free convection, radial temperature gradient, frictional dissipation, and so forth should be considered. The heat transport equation in the longitudinal and radial directions in Bird *et al.* (1960) should be solved by a finite-differencing scheme (Wedelich and Galate, 1987; Wooley *et al.*, 1984). The calculated temperatures should be validated with field measurements (Honore and Tarr, 1993).

Once the temperatures are known, the rheology of the fluids can be estimated as a function of temperature, and the equations in Case III can be solved with the rheology values calculated as a function of temperature.

When a fluid is left static, the mode of heat transfer is mainly from conduction. Fluid is left static while logging the open hole, running the casing, and waiting on cement (WOC) to undergo hydration reaction and become a competent cement sheath.

If heat of the hydration reaction is neglected, then the heat balance for fluid under static conditions results from heat conduction only and is a function of time. The unsteady-state equation can be written as

$$0 = M c_p dT/dt - k_r A(T_r - T)/\Delta x \qquad (8\text{-}28)$$

with the boundary condition, $t = 0$, $T = T_F$, where M is the mass of fluid, c_p is the heat capacity of fluid, T is the temperature of fluid, k_r is formation thermal conductivity, T_r is formation temperature, T_F is the fluid temperature at the beginning of static time (the end of circulation), and Δx is the effective thickness of formation. Equation 8-28 can be solved by a Runge-Kutta solver for temperature as a function of time.

8-2.4.1 Illustrations

An illustration of solutions to some of the mathematical models discussed is shown in Figures 8-20 to 8-22. These solutions are based on models that include natural and forced convection in longitudinal and radial directions.

Figure 8-20 shows the temperature when drilling fluid is circulated. Among other factors, this temperature will depend on the temperature of the inlet fluid. Water available on the rig is used to mix the cement slurry and, as a result, the cement slurry inlet temperature is usually

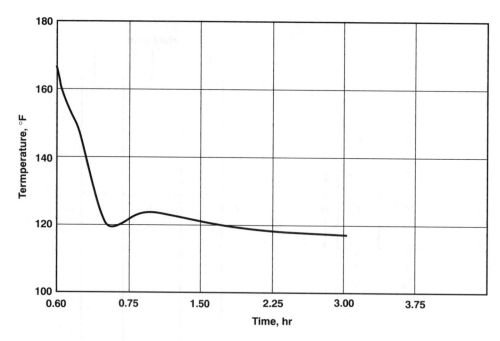

Figure 8-20   Temperature while circulating drilling fluid

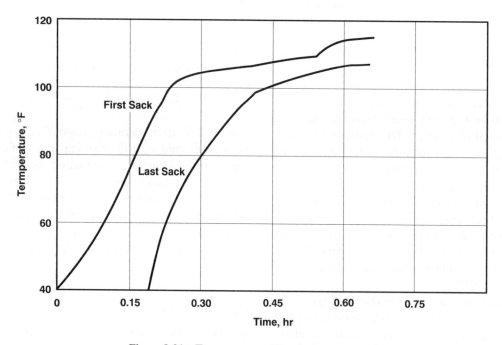

Figure 8-21   Temperature while placing cement slurry

much lower than the drilling fluid inlet temperature. This situation causes the cement bottomhole circulating temperature to be less than the drilling fluid bottomhole circulating temperature.

An illustration of bottomhole circulating temperature while circulating cement is shown in Figure 8-21. Note the difference between the circulating temperatures in Figures 8-20 and 8-21. This difference is caused by the

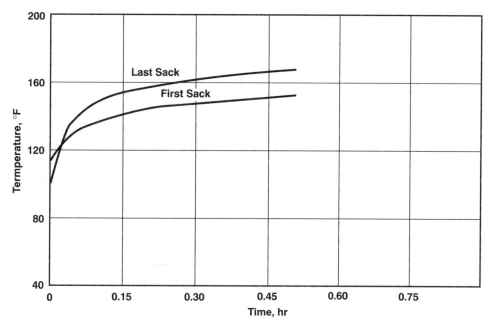

Figure 8-22 Temperature of cement slurry after placement

cement inlet temperature being about 40° cooler than the drilling fluid inlet temperature. The temperature of cement slurry as it is left static in the wellbore is illustrated in Figure 8-22.

Note the difference between the temperatures in Figure 8-21 and 8-22. This difference is caused by cement being heated by the formation when left static.

The cement circulating temperature and the static temperature are dependent on the input parameters. These temperatures influence the rheological properties of the cement slurry and also a number of other critical properties discussed in Section 8-4.

8-2.5 Case V—Intermixing and Multiple Fluids

In Case IV, the modeling of the temperature of circulating and static fluids was addressed, and methods to integrate this into cement placement were discussed. A number of assumptions were relaxed from Case I to Case IV. Each case added additional complexity to the problem. Multiple fluids of different density, different rheology, different chemical nature, and different solute concentration are involved in cementing operations. When these come into contact, there will be intermixing of fluids because of the differences in their properties. The gradients in the radial direction cannot be neglected either.

The mass, momentum, and heat balance equations must be solved simultaneously to study the effects of eccentricity, fluids intermixing, buoyancy, temperature and velocity distribution, rheology, or other fluid qualities. The differential equations can be solved by a finite differencing scheme (Barton *et al.*, 1994).

8-2.6 Case VI—Material Integrity

The objective of primary cementing is to provide zonal isolation for the life of the well. The cement sheath should be able to withstand the various operations such as tripping, milling, production, or injection during the life of the well. The effect of these operations can be modeled as stresses induced by these operations (Craig, 1996; Fjaer *et al.*, 1996; Hertzberg, 1996; Timoshenko, 1970). Properties of the cement sheath such as Young's modulus or compressive strength then determine how the cement would withstand these stresses.

Numerical modeling should be viewed as a beginning to the solution of a problem and not as an end. The results should be verified by simulating the downhole conditions in an experimental setup (Beirute *et al.*, 1991; Bittleston and Guillot, 1991; Lockyear *et al.*, 1990; Smith and Ravi, 1991). The model should be improved continuously based on laboratory and field results.

The materials that give the properties used in the mathematical model(s) discussed in the previous section are discussed next, followed by a discussion on deploying the methods and materials in the field for a successful primary cement job.

8-3 CEMENTING MATERIALS

Cement slurry used in the oil industry consists mainly of cement, additives, and water. The various types of cements and additives depend on the end application.

Portland cement is the material of choice for about 99% of all primary cement treatments. Portland cement is readily available all over the world and is inexpensive compared to most other cementing materials. Materials such as slags or pozzolans are comparatively inexpensive but not readily available in all areas. Materials such as epoxy resins are expensive compared to cement, slag, or pozzolan. Other materials such as magnesium oxychloride cement or high alumina cements are occasionally used in the oilfield but have problems that limit their use.

8-3.1 Portland Cement

Portland cement is manufactured by fusing with heat calcium carbonate (limestone) and aluminum silicates (clay). A small amount of iron may be added. The molten rock is cooled, a small amount of gypsum (calcium sulfate) is added, and the mixture ground into a powder. One of the biggest problems with Portland cement is that it is not a pure chemical (raw materials are not pure, so the final product is not pure) and the particle size is never completely reproducible. Therefore, the finished products can never be totally consistent, and much work has gone into improving the consistency of oilfield cement (see API Specification 10, 1990).

The major component of Portland cement is tricalcium silicate. Portland cement does not get hard through a drying process but through the chemical reaction when the water with which it is mixed reacts with tricalcium silicate to form calcium silicate hydrate (CSH). The calcium silicate hydrate has the physical strength that makes Portland cement hard at room temperature in several hours. Because Portland cement is hydraulic, it will set under water.

Portland cement is manufactured all over the world, but oilfield cement use compared to construction cement use is very small (< 5%). However, oilfield cement may constitute a significant part of the business of some manufacturers.

Several types of cement are manufactured for oilwell use. The American Petroleum Institute (API) has set standards for oilwell cements. The specifications are regulated by Committee 10 in the API committee for standardization. The common types of API cements are API Class A, API Class C, API Class G, and API Class H.

The chemistry of each of these grades is similar, but the quality control is most stringent for API Classes G and H. The particle size for each of these powders is different, with API Class C being the finest, and API Class A, API Class G, and API Class H in increasing order of coarseness. Increasing coarseness dictates that less water is required to wet the particles and create a pumpable slurry. Water content is a very important factor in mixing cement. The amount of water added could make the slurry either too thin, which would allow solids to settle from the slurry or too thick, which would make it difficult to pump.

Water is low in density compared to cement powder. API Class C cement requires the most water and therefore creates the lowest density neat slurry. (*Neat cement* is cement and water alone.) Conversely, API Class H requires the least water and makes the highest density neat slurry. Cement reacts with water and produces a product that is hard and provides zonal isolation. The more finely ground the material, the larger the surface area and the quicker the reaction. Therefore, API Class C is the fastest setting and the highest early strength cement of the group, while API Class H is the slowest setting of the group. Water dilutes cement; therefore, even though API Class C sets the most quickly, it will not develop as much ultimate strength as API Class H that has less water.

API Class G is the cement used in most parts of the world. In the U.S., API Class H is used in the central and the gulf coast regions and API Classes C and H are used in west Texas. These three are the common oilwell cements. API Class A oilwell cement is very close to construction cement, and in areas where the local manufacturer does not normally make an oilwell cement, construction cement can often pass the requirements of API Class A.

The transportation cost of Portland cement can be greater than the cost of the product. Therefore, local cement is normally used if it can successfully meet application objectives.

Cement is normally shipped in bulk. A sack (base unit for calculation purposes) of cement in the USA weighs 94 lb, and the bulk volume of one sack of cement is about one ft^3. Cement specific gravity can vary somewhat but it is usually about 3.14. Although cement is shipped in bulk, it is usually still specified by the number of sacks (sk). The absolute volume of cement is about 3.6 gal/sk.

However, in other parts of the world—in Europe, for example—a sack of cement is typically 100 kg.

Portland cement will set at room temperature in several hours. Hydration reaction is accelerated thermally, so at higher temperatures, cement will take less time to set.

Portland cement, when mixed with water, has a pH greater than 13, and it is not stable in low pH environments; it will degrade in flowing acid. However, Portland cement is stable or can be made stable in almost every other environment if the mechanical properties of the material are not exceeded.

8-3.2 Other Cementitious Material

There are cementitious materials other than Portland cement that are used either with or in place of Portland cement for petroleum applications. These materials are discussed in the following section.

8-3.2.1 Pozzolan

Pozzolans are siliceous and aluminous materials that possess little or no cementitious value but will, in finely divided form and in the presence of moisture, chemically react with calcium hydroxide at ordinary temperatures to form compounds possessing cementitious properties. The most common pozzolan is fly ash, which is a waste product from coal-burning power plants. These noncombustible materials are collected before going out of the smoke stacks. Naturally occurring pozzolans are normally created from volcanic activity.

When pozzolans are used in combination with Portland cement, calcium hydroxide, liberated from the hydration of Portland cement, reacts with aluminosilicates present in the pozzolans to form cementitious compounds possessing cohesive and adhesive properties. Since fly ash is a waste product, it is inherently variable. Fly ash is most commonly added to Portland cement because it is cheaper than cement, does not dilute cement, and increases the cement's compressive strength. Since it is cheaper than Portland cement, it adds volume at low cost. However, because it is inconsistent, it is normally only used at low temperatures (less than 200°F).

8-3.2.2 Ultrafine Cements

Ultrafine cements are much smaller in particle size than conventional Portland cements. The average particle size of ultrafine cements is 2μm, whereas conventional cements could be 50 to 100 μm. The main application of ultrafine cements in primary cementing is as lightweight cement with early strength development. Ultrafine cements are also used in fixing squeeze cementing, repairing casing leaks, shutting off water flows, or resolving similar problems because these cements with smaller particle size can penetrate small openings.

8-3.2.3 Epoxy Cements

Epoxy cements are specialty cements, most commonly used when the cement will be exposed to corrosive fluids. Portland cement, slag cement, and pozzolan are all soluble in acid. The solubility may be slow, but they will dissolve in flowing acid streams. Epoxy is not soluble in acid, but it is expensive. Therefore, it is normally used only for disposal or injection wells where low pH fluids are to be injected. Epoxy cements are pure products and are very consistent. The epoxy resin is a two-part, true-liquid mixture of resin and activator. The activator breaks the epoxy ring and forms a long carbon-oxygen polymer providing strength in excess of 10,000 psi. The liquid can be made into slurry by adding particulate materials that are not acid soluble to increase the volume and decrease the cost. The volume can also be increased by adding a low-cost solvent. Like the particulate materials, the solvent will dilute the epoxy and thus reduce strength. Epoxy resins are not compatible with water; a thick nonpumpable mass will form if epoxy resin and water are combined. Unlike the other cements, epoxy resins are soluble in some organic solvents.

8-3.2.4 Slag Cement

Blast-furnace slag cement is a byproduct of the steel industry. The material on top of the molten steel is removed, cooled, quenched with water, and ground. The material composition is mainly monocalcium silicate, dicalcium silicate, and dicalcium aluminosilicate. These silicates set very slowly (days or weeks required) at room temperature when mixed with water. Blast-furnace slag normally requires temperatures greater than 200°F (100°C) to react with water to form calcium silicate hydrates. By increasing the pH of the slurry of blast-furnace slag and water, the setting process can be accelerated to make the slurry set at room temperature like Portland cement. The calcium silicate hydrates formed are similar to the calcium silicate hydrate formed from Portland cement except that the product is more brittle and cracks easier, but it is less permeable when it does not crack. The use of slag in cementing depends on the purpose and availability of good quality slag. The advantage of slag is that it is tolerant to drilling fluid contamination when designed properly. However, the variation in slag properties and drilling fluid properties demand that detailed lab tests be performed to optimize the formulation when slag is used. Effects of temperature and pressure variations on the set slag sheath during hydrocarbon production must be evaluated for long-term durability.

One of the current applications for slag is to convert drilling fluid to cement. The slag can be incorporated into the active drilling fluid system. The drilling fluid system will not set unless the temperature is high or unless the pH of the drilling fluid system is increased. This is a new idea that is being thoroughly evaluated by the oil industry. Additional information is available in papers by Nahm *et al.* (1993, 1994), Benge and Webster (1994), and Cowan *et al.* (1992).

8-3.3 Portland Cement Chemistry

Cement undergoes hydration reaction in the presence of water to form CSH, calcium silicate hydrate. The major components in cement are C_3S, tricalcium silicate ($3CaO.SiO_2$); C_2S, dicalcium silicate ($3CaO.SiO_2$); C_3A, tricalcium aluminate ($3CaOAl_2O_3$); and C_4AF, a ferrite phase of average composition $4CaO.Al_2O_3.Fe_2O_3$. When cement is produced commercially, there are impurities in cement in addition to the components listed above (Papadakis-Vayenás and Fardis, 1989; Ramachandran, 1990; Taylor, 1992).

Typically, a cement clinker consists of 70 to 80% of C_3S and C_2S. The chemical reaction of C_3S can be written approximately as

$$2[3CaO.SiO_2] + 7H_2O \rightarrow 3CaO.2SiO_2.4H_2O + 3Ca(OH)_2.$$

The hydration reaction is exothermic, and the mechanism can be broadly classified into five different stages. These stages are depicted graphically in Figure 8-23. The stages are discussed below.

In the pre-induction period, there is a rapid evolution of heat as soon as C_3S comes in contact with water. This might last for about 15 to 20 minutes. In the dormant period, a very slow reaction occurs that could last for a few hours. Cement is plastic and workable during this stage. In the rapid reaction stage, active reaction accelerates with time, and the rate of heat evolved reaches the maximum; the initial set of the cement occurs during this stage. In the slow deceleration stage, the reaction rate decreases and the rate of heat evolved decreases. The

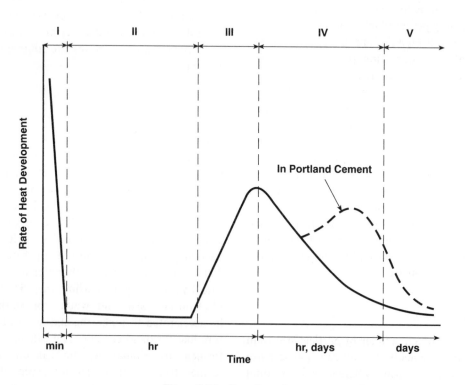

Figure 8-23   Reaction stages

slow reaction stage could last for days. As shown in Figure 8-23, Portland cement may show another peak in the fourth stage.

The reaction of C_2S is slower than C_3S and also the amount of $Ca(OH)_2$ formed is lower. The average C_3A content in Portland cement is about 4 to11%, but it significantly influences early reactions. The phenomenon of flash set is caused by the reaction of C_3A. In Portland cement, the hydration of C_3A phase is controlled by the addition of gypsum. Cement exposed to sulfate solutions should not contain more than 5% C_3A.

As a general rule, the rate of hydration of cement in the first few days proceeds in the order $C_3A > C_3S > C_4AF > C_2S$. In a fully hydrated Portland cement, the $Ca(OH)_2$ could be about 20 to 25% of the solid content. The morphology of $Ca(OH)_2$ may vary and form equidimensional crystals; large, flat, platy crystals; large, thin, elongated crystals; or a combination of them.

Chemical additives are used to change the length of these stages, especially the first two. The chemical additives are discussed in the next section.

8-3.4 Cement Additives

Cement additives have been developed to allow the use of Portland cement in many different oil and gas well applications. Cement additive development has been ongoing for decades. These additives make obtaining required performance properties relatively easy.

Typical additive classifications are listed below:

- **Accelerators**—reaction rate enhancers

    Accelerators are used to shorten the set time of cement slurries or accelerate the cement setting. They do not increase the ultimate compressive strength of cement but do increase the rate of strength development. They also shorten the thickening time. Accelerators are used at low temperatures to reduce time waiting on cement (WOC) time. The most common accelerator is calcium chloride, commonly used at 3% or less by weight of cement.

- **Retarders**—nucleation poisoners

    Retarders are used to decrease the set time of cement slurries or to retard the cement setting. They do not decrease the ultimate compressive strength of cement but do slow the rate of strength development. They also lengthen the thickening time. Retarders are used at higher temperatures to allow time for placement of the liquid slurry. The most common retarders are natural lignosulfonates and sugars. Lignosulfonates are normally used at circulating temperatures up to 200°F. Sugar compounds are normally used at circulating temperatures from 200 to 350°F. The newest retarders are made from various synthetic compounds.

- **Extenders**—water adsorbing or lightweight additives

    Extenders are a broad class of materials that reduce the density or the cost of the slurry. Water is the cheapest material that can be added to cement. Normally a sack of cement (94 lb) will give about 1 ft^3 of slurry. Adding extra water can increase the volume to 3.0 ft^3/sk. The problem is that the slurry will become too thin and the cement will settle and have free water. To prevent this problem, extenders (thickeners) such as bentonite or sodium silicate are added. The disadvantage of adding the extra water is that the strength of the set cement is lessened by the dilution.

    If low density and higher strength are required, the density of the slurry can be reduced with gas. A stable foam cement will have discrete air bubbles that lower the density of the slurry but do not dilute the strength as much as water. Hollow ceramic spheres can be used for the same purpose. The spheres will be crushed, however, if the hydrostatic pressure is too high.

    The last common extender is pozzolan (sometimes called poz). Fly ash is the most commonly used pozzolan. Fly ash alone has a density slightly less than Portland cement, so it does not reduce slurry density very much. However, because fly ash is cheaper than cement, it can reduce the cost of the composition if, for instance, the job is performed with half cement and half fly ash. Normally, some bentonite is added to a cement/pozzolan blend, which makes it possible to add more water.

- **Fluid-loss additives**—permeability plugging additives

    Fluid-loss additives reduce the rate at which water from cement is forced into permeable formations when a positive differential pressure exits into the permeable formation. Fluid-loss additives are normally polymers such as cellulose, polyvinyl alcohol, polyalkanolamines, polymers of polyacrylamides, and liquid latex such as styrene butadiene latex.

Most fluid-loss additives increase the slurry viscosity, although some retard it to some degree.

- **Lost-circulation additives**—macro plugging materials

Lost-circulation additives are used to plug zones that have a tendency to draw in fluid because they are unconsolidated or weak. Large particulates can be placed in the cement slurry to prevent fracturing or to bridge existing fractures. These particles should have a broad particle size distribution, should not accelerate or retard excessively, should have sufficient strength to keep a fracture bridged, and should be inexpensive and non-toxic. The most common materials are ground coal, ground Gilsonite, and ground walnut hull. Fibrous and flake materials are also used as lost-circulation materials. The fibrous materials are normally inert polymers such as nylon, and the flake materials are cellophane or similar materials.

- **Expansion additives**—solid state crystal growth

Expansion additives cause the exterior dimensions of set cement to grow slowly when the cement is in the presence of downhole fluids. This minor growth of the exterior dimensions of the slurry causes the cement to bond better to pipe and formation. The most common additives for this use are based on calcium sulfoaluminate and calcium oxide.

- **Dispersants**—reduce slurry consistency

These are used to reduce the viscosity of cement slurry. They are useful in designing high density slurries and also tend to improve fluid-loss control. The most common dispersant is the sodium salt of poly-naphthalene sulfonate.

- **Stability agents**—water absorbers

Preventing a slurry from settling or having free water develop is a complicated problem. Even with extenders, at elevated temperatures slurry may bleed water or settle. Therefore, special additives are required in small quantities to aid in slurry stability. Because these additives do not increase the viscosity of the slurry excessively at surface conditions, they allow easy mixing.

- **Anti-gas migration**—gel strength modifiers

Refer to Chapter 9 for information about anti-gas migration techniques.

- **Weighting agents**—high-density particulate

Weighting agents allow the formulation of high-density slurries. Slurries with densities greater than 17.5 lb/gal would be too thick to mix and pump without weighting agents. High-density slurries are required when the bottomhole pressure in a well is high. Weighting agents are heavy particulate material such as iron oxide, barite, or titanium oxide.

- **Anti-foam agents**—surfactants that alter surface tension

Many additives cause foaming problems, but if a slurry foams badly, during mixing, centrifugal pump will air lock, and mixing must be stopped. The most common anti-foam agents are polyglycols. Defoamers or foam breakers such as silicones are sometimes used because they can break and prevent foam.

- **Salt**—modifies ionic strength of the mix water

Salts are used to improve bonding to shale sections and also in cementing through immobile salt zones.

- **Special systems**—used in arctic, acid, and other environments

Portland cement can be modified for use at high temperatures or low temperatures. These are called thermal cements and arctic cements, respectively. Thermal cements have extra silica blended with the cement. Arctic cements have accelerators to hasten cement setting at these low temperatures and chemical additives to keep the water in the cement from freezing.

8-4 CALCULATIONS AND LABORATORY TESTING

Before cement slurry can be designed, the temperature it will be exposed to must be determined. Cement will set faster as temperature increases and set slower as temperature decreases. Placement time dictates the length of time cement slurry is required to be in the liquid condition. The cement should set quickly after placement to save time and money. The slurry liquid time, "pump time" or "thickening time," can be adjusted chemically, but the temperature the cement will be exposed to must be known to make the adjustment.

8-4.1 Temperature Determination

As the wellbore is extended more deeply, borehole temperature increases. The undisturbed temperature at any depth is called the static temperature. If a well is circulated, the wellbore is cooled. Many factors determine how much a wellbore will be cooled by circulating. The cooled wellbore temperature is called the bottomhole circulating temperature (BHCT). This temperature is used to determine how long cement slurry will stay liquid while pumping.

Reliable temperature gauges are available in the market to measure temperature while circulating. Validating the numerical results with gauge measurements can increase the reliability of temperature calculations based on numerical models (Tilghman *et al.*, 1990).

API Specification 10 (1990) has a set of tables to approximate circulating temperatures. However, these temperatures will work well only if the field being developed is similar to the field on which the data were collected. If a well's true vertical depth and static temperature or pseudostatic temperature are known, the tables give a circulating temperature. The API equations and tables were generated from measured static and circulating temperatures in hundreds of wells. Measured static and circulating temperatures were used by the API to derive average circulating temperatures. In established areas where the rise in temperature as a function of depth is known, temperature gradients or geothermal gradients have been mapped. This information can be used to determine a static temperature that can be used to obtain a circulating temperature. Care must be taken when using geothermal gradient maps because some areas have highly localized significant variations in earth temperatures.

The API temperature schedules not only give an average circulating temperature for a given static temperature, but they also give average heat-up and pressure-up schedules for laboratory testing of cement slurries. These heat-up and pressure-up schedules are also average values and should be modified for individual well conditions.

The API tables can be modified to give *approximate* circulating temperatures for deep water, arctic, and high-angle primary cement jobs. The API tables were not designed for this purpose and significant error can be introduced by extrapolating the correlation to situations for which they were not intended. However, since no measured data is available, the API tables may be all that is available to estimate temperatures for deep water, arctic, and high-angle primary jobs. Once measured temperatures are obtained, the measured values should be used instead of the estimated values. The measured temperatures are also affected by other factors such as fluid inlet temperature, time of circulation, and time of shutdown. Therefore, it is recommended that a fundamental approach be used and the temperatures calculated from numerical models. The results from calculations should then be verified by temperature gauge measurements.

8-4.2 Slurry Volume Calculations

Cement is mixed with water and other additives as needed to make a cement slurry that meets the required properties for the well. The water required to mix the cement depends on the type of cement and additives present. The water requirement expressed as a percentage of cement by weight is given in Table 8-3. These values are for neat slurry.

Three design parameters needed for any slurry are the density, yield, and water required. Density and other well parameters and slurry properties determine whether slurry could break down weak formations. Yield, hole size, casing size, washouts, and top of cement required, determine the quantity of cement, water, and the additives required to fill the annulus. For neat cement, the water required is known. The density and yield can then be calculated. If the cement is not neat, then the slurry density is sometimes specified and the water required and yield will need to be calculated. If water required is known, the density and yield can be calculated. Below are two typical calculations, one where density is known and one where water required is known.

8-4.2.1 Example 1 Slurry Calculations—Cement Density Known

What will be the density in lb/gal, the slurry yield in ft^3/sack, and the water requirement in gal/sk for a neat Class H cement slurry?

The amount of cement used = 100 g

Table 8-3   Water requirement for API cements

Type of cement	Water requirement
API Class C	6.3 gal/94-lb sack or 56%
API Class A	5.2 gal/94-lb sack or 46%
API Class G	5.0 gal/94-lb sack or 44%
API Class H	4.3 gal/94-lb sack or 38%

From Table 8-3, the water requirement for Class H cements is 38%.

The amount of water needed $= 100 \times 0.38 = 38$ g

The volume of 38 g of water $= 38/1 = 38\,\text{cm}^3$

The volume of 100 g of cement $= 100/3.14 = 31.85\,\text{cm}^3$

Total volume of slurry $= 38 + 31.85 = 69.85\,\text{cm}^3$

Total mass of slurry $= 100 + 38 = 138$ g

Density of cement slurry $= 138/69.85 = 1.9757\,\text{g/cm}^3$
$= 16.48\,\text{lb/gal}$

Water requirement $= 38\,\text{cm}^3/100$ g

There are 94 lb or 42,638 g of cement in a sack.

Therefore, the water requirement per sack of cement $=$ $38 \times 426.38 = 16{,}202\,\text{cm}^3$,

or water requirement $= 16{,}202/3785.4 = 4.28$ gal/sk.

Next, calculate the slurry yield. This calculation yields the amount of cement slurry obtained from one sack or 94 lb of cement.

There are $69.85\,\text{cm}^3$ of slurry obtained from 100 g of cement.

Therefore, the slurry yield $= 69.85\,\text{cm}^3/100$ g $= 0.6985\,\text{cm}^3/\text{g} = 1.05\,\text{ft}^3/\text{sack}$

8-4.2.2 Example 2 Slurry Calculations—Water Required Known

If 48% water is added to Class H cement, what will be the density of the slurry?

The amount of cement $= 100$ g

Amount of water added $= 100 \times 0.48 = 48$ g

Following the steps in Example 1,

Density of cement slurry $= 148/79.85 = 1.85\,\text{g/cm}^3$
$= 15.47\,\text{lb/gal}$

Calculating the density, water, and yield is just one part of slurry design. The slurry may not mix, may be very viscous, or may settle badly. Therefore, laboratory testing must be performed on slurry to make sure it is a viable system.

A detailed description of laboratory testing devices and testing procedures is included in API Specification 10. The importance and reasons for performing these tests are discussed below.

8-4.3 Laboratory Testing

Laboratory testing of slurry always starts with the ability of the slurry to be mixed with the specified amount and type of water. The water used on the actual job may be different from the lab water. Therefore, it is recommended that the same water that will be used on the specific job be used for at least the final testing.

Mainly, the chlorides present in the water affect the reaction of the cement. In addition, the chlorides and sulfates in water and the pH of water should be checked often. Once the slurry is mixable, the following tests need to be done on the slurry before it can be used in primary cementing.

- **Density** The density of slurry is measured in this test after removing any entrained air. Depending on the additives used to make the slurry, there could be a significant amount of air entrained into the slurry when it is mixed in the laboratory. False density readings will occur if the air is not removed before measuring the density.

- **Slurry stability** The suspending capability and uniformity of the slurry at temperature is measured in this test. Slurry stability is assessed through the combination of two tests: free water and solids settling. Cement normally has a narrow water requirement range over which it is stable. Slurry must have sufficient viscosity to prevent solids settling and free water development. If excessive water is present, the slurry will be too thin, solids will settle, and water will rise to the surface of a cement column before it sets. This settling and water separation can have adverse effects, such as bridging in the annulus, bridging of float equipment, lack of zonal isolation, and water pockets causing casing collapse if large temperature increases are encountered later in the life of the well.

- **Thickening time** The length of time cement slurry will remain pumpable at bottomhole pressure and temperature is measured in this test. This time is called the "pump time" or the "thickening time." The slurry should be designed so that it will not develop excess viscosity and become unpumpable during the time it takes to pump the slurry into the wellbore. This characteristic is an important property of the slurry and is mainly dictated by the type of cement and the amount and type of retarder used. If not properly designed, the fluid will become too viscous, and cement cannot be placed as designed. The cement will have to be drilled out, and the remedial costs are very expensive. There is no excuse for designing slurry that cannot be placed properly as a result of poorly designed laboratory thickening time tests. A safety factor of about 1 to 2 hours is commonly specified for thickening time requirements.

- **Fluid loss** The amount of filtrate lost by the fluid under bottomhole temperature and 1000-psi differential pressure is measured in this test. A differential pressure normally exits to prevent fluid flow from the formation into the wellbore, and most formations have pore throats that are too small to allow cement particles to invade the formation. However, if a differential pressure exits into the formation, the water in the cement slurry can leak into the formation. Neat cement is likely to have a fluid loss of around 1000 ml/ 30 min in an API fluid loss cell; a cement slurry with good fluid-loss control has a fluid loss of under 50 ml/ 30 min. If water is squeezed out of the cement slurry, the slurry density and viscosity will increase and the slurry volume and pump time will decrease. If the amount of water squeezed from the slurry is excessive, the loss of water can have a negative impact on the success of primary cementing.

- **Rheology** The shear stress and shear rate behavior of a slurry at different temperatures is measured in this test. The rheology of fluids has a major effect on solids settling and free water properties and also on friction pressures. This friction pressure is added to the hydrostatic pressures of the fluids, and the total pressure is divided by the height to give the equivalent circulating density (ECD). Jobs should be designed so that the ECD stays below the fracture initiation pressure. Additionally, the rheology of slurries is used to determine the flow regime of a slurry.

  Coaxial cylindrical viscometers are used for these tests. These tests can be conducted under pressure and temperature. A high-pressure, high-temperature rheometer is shown in Figure 8-24. The rheological behavior of a cement slurry vs. temperature is illustrated in Figure 8-25. This figure shows that the slurry is thinning with temperature. This behavior is the most likely of cement slurries. However, there are slurries that behave differently and show an increase in rheology as a function of temperature, at least over a certain range.

- **Compressive strength** The pressure it takes to crush the set cement is measured in this test. This test indicate whether the cement sheath will withstand the differential pressures in the well. In destructive testing, a cement slurry is poured into cubical molds and the cement cubes are then crushed to determine their compressive strength. In a non-destructive test, sonic speed is measured through the cement as it sets. This value is then converted into compressive strength.

- **Gas migration** The ability of a cement slurry to resist fluid (gas or water) flow under static conditions is measured in this test. The tests are static gel strength

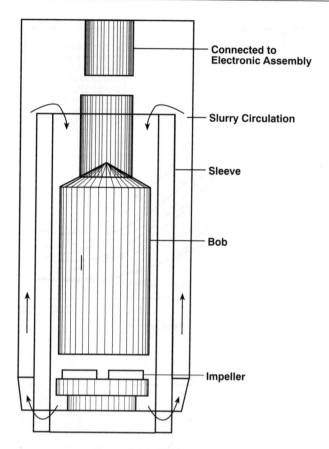

Figure 8-24 HPHT rheometer

development, zero gel time, transition time, gas flow simulations, shrinkage, expansion, and permeability. These are discussed in detail in Chapter 9.

- **Young's modulus** The ratio of stress to strain of a cement sheath is not commonly performed on a cement sheath. However, HPHT wells, multilateral wells, and deep horizontal wells have created a need for a cement sheath that is more resilient than usual (Goodwin and Crook, 1992; MacEachran and Adams, 1991). A cement sheath without any additives is brittle. Various additives are included in the cement slurry to improve its ductility. Additional information on these tests is presented by Grant *et al.*, (1990), Moran and Lindstrom (1990), Nelson (1990), Sabins and Sutton (1982), and Smith (1990).

8-5 PERFORMING THE JOB

Designing a cement job using the best mathematical model and chemistry is only the beginning of a series of steps to obtain zonal isolation. The different factors in performing the job are discussed below.

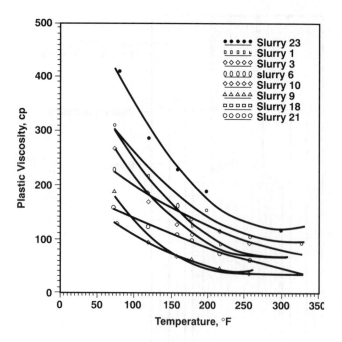

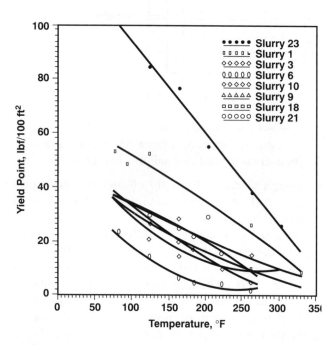

Figure 8-25   Laboratory data from HPHT rheometer

the best job of determining hole size. However, 4- to 6-arm calipers are not run as often as 1- to 2-arm calipers because of cost, availability, and ease of use with other logs. Single-arm calipers tend to find the largest diameter. Boreholes are irregular shapes; therefore, the single-arm caliper finds the widest part of the borehole. The size of the hole is then assumed to be round based on this large radius. One-arm calipers normally give too large a hole size, especially when trying to get a cement top close to the design point. The cement volume then must be estimated and this could pose a serious problem in critical conditions. If having a correct cement top is critical, run the 4- or 6-arm caliper to determine the needed volume of cement.

The drilled hole size, casing size, and cement top are used to calculate the volume of cement required in fields that are well known. An excess factor is commonly used to account for irregularities in the borehole. If a number of wells have been drilled in the area and openhole logs have been run on these wells, the excess factor can be estimated with some accuracy. The calculated cement value is then multiplied by the excess factor to determine the needed volume of cement.

If the volume of cement estimated is much in excess of what is needed to reach the planned top of cement (TOC), then this additional cement slurry will create extra pressure on the formation. This pressure can be a problem in weak formations.

Fluid calipers can also be used to determine the volume of cement required. A fluid caliper is a fluid with some type of visible marker in it. The fluid is pumped around in a well and the volume is measured until the marker shows up at the surface. The annular volume is then estimated. Fluid calipers are not very reliable. Many variables can make the marked fluid show up early at the surface and give too small an openhole size. Fluid calipers can be useful when used in the proper situations, but the validity of the caliper should be checked with calculations based on drilled hole and experience in the area.

Different types of cement treatments that depend on wellbore conditions are discussed below.

8-5.1 Volume Determination

The volume of cement must be determined before a treatment can be planned. Mechanical calipers are recommended in measuring openhole size. They physically measure the openhole size with arm(s) that make contact with the open hole. Multi-arm (4- to 6-arm) calipers do

8-5.2 Types of Primary Cement Treatments

Several different types of primary cement applications are available:

- **Conductor casing** In conductor casing, the installation of a flowline at a sufficient elevation allows drilling fluid to return to the pits. It also provides for the installation of a diverter system to divert any shallow hydrocarbons away from the rig.

- **Surface casing** The purpose of surface casing is to protect shallow, freshwater sands from contamination by drilling fluids and fluids produced during the life of the well. This string is normally cemented back to the surface so that freshwater zones have a sheath of cement and steel casing to protect them throughout the drilling and producing life of the well.

- **Intermediate casing** The basic use of intermediate casing is to provide hole integrity during subsequent drilling operations. Usually, this type of casing is set to cover high-pressure zones or exposed weak formations to the required depth. Intermediate casings cover these zones and allow continued drilling without fluid migration or lost circulation. They also protect production casing from corrosion and resist high formation pressure.

- **Production casing** The production casing string is set through the producing zone and provides a backup for the tubing string that is usually installed to produce fluids. The production casing is the casing through which the production zone will be perforated. See Chapter 13 for a discussion of perforation.

- **Liner** Liners are used in deep drilling operations to eliminate the need to run a full string of casing. Unlike regular casings that are installed back to the surface, liners terminate downhole at a point in the last casing. Special tools called liner hangers are required to complete these installations.

- **Inner string** When large diameter casing is cemented, long cement placement time and large volumes of cement are required. In such cases, the cementing may be done through the drillpipe run inside the casing to reduce contact time and volume. In addition, less cement is left to drill out. Special tools are available to perform this operation.

- **Multistage** A long casing string can be cemented in multiple stages if the casing is extremely long or if different formation conditions warrant it. This procedure reduces drilling fluid contamination and also decreases the possibility of formation breakdown. Special tools are also available to perform this treatment.

- **Reverse** Reverse cementing is the process of pumping cement down the annulus and up the casing, instead of pumping down the casing and up the annulus as is done during a conventional cementing process. The pressures will be much lower than with conventional cementing. The main advantage of this process becomes evident when the fracture gradients are low, because cement can be placed properly without breaking the formation in some weak zones.

- **Deviated/Horizontal** Wells are drilled at an angle to be able to contact more of the producing zones or to reach an entire reservoir from one offshore platform. Refer to Chapter 2 for details of drilling deviated or horizontal wells. Multilateral wells are a special subset of this well type (Chapter 3).

- **Deepwater** Exploration for production of hydrocarbons is extending offshore to water depths in excess of 600 ft and sometimes up to 6000 ft. Under these conditions, the formations through which conductor surface pipes are set are weak to unconsolidated, temperatures are low, and charged water zones could be encountered. Density and flow rate of the cement slurry should be such that the formation will not break down, yet the slurry still displaces the drilling fluid effectively and provides effective zonal isolation. Considering the low temperatures encountered, the cement slurry should develop sufficient gel strength to resist fluid migration. See Chapter 9 for a detailed discussion on fluid migration.

- **High-Pressure, High-Temperature (HPHT)** Exploration for production of hydrocarbons is extending to reservoirs that are under high pressure and high temperature. The materials used in the construction of these wells, including cement, should be able to withstand the high temperatures and pressures. The strength of cement sheath undergoes retrogression beyond about 220°F. Silica should be added to cement slurries that are expected to endure temperatures in excess of 220°F. During its life, an HPHT well will undergo much greater changes in temperature and pressure than other wells. Ordinary cement is brittle and will not be able to withstand the fatigue from cycling of pressure and temperature. Researchers are working to improve the cement slurry design to meet the needs of HPHT wells.

8-5.2.1 Example 3 Design of the Cement Slurry

A well is being constructed as shown in Figure 8-26. The surface and intermediate casing have been cemented and production casing has been set at a total depth (TD) of 9200 ft. The production casing is 23 lb/ft, 7 in., and the

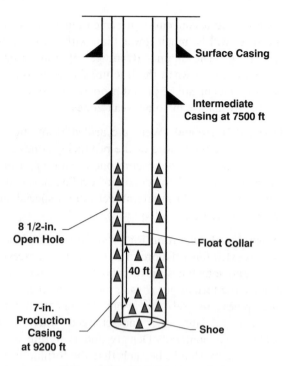

Figure 8-26   Well for Example 3

open hole is $8\frac{1}{2}$ in. Based on the openhole logs, a 20% excess is required in calculating cement volumes. The previous or intermediate casing is set at 7500 ft. Production casing needs to be cemented from 9200 to 7700 ft. The bottomhole static temperature (BHST) is 280°F. The fracture pressure, expressed in equivalent drilling fluid density, is 18 lb/gal, and the pore pressure is 13 lb/gal. The cement slurry designed for this application should have a thickening time of 4 hours at bottomhole circulating temperature (BHCT) and a fluid loss of 30 cm³/30 min at BHST and should be able to resist gas migration.

Solution—Designing the Cement Slurry

We will assume that a 16.4 lb/gal cement slurry prepared with Class H cement will be used to construct the well. In reality, the density of the slurry should be optimized so that the slurry will

- Meet the requirements discussed in Section 8-1.6

- Be placed without breaking the formation

- Displace fluid ahead of it from the entire annulus

- Resist gas migration

- Have the necessary properties for short and long term needs of the well.

The slurry discussed in Section 8-4.2.1 could be used as the starting point for design. Additives to this neat slurry need to meet the requirements discussed above.

The first step is to obtain the BHCT. The BHCT should be estimated from the temperature simulator and validated with experimental data obtained from the BHCT gauge. The validation is more critical if this is a new field. Alternatively, equations from API can be used to estimate BHCT. This value should be used with caution if the well is deviated and/or the field conditions are unique in terms of BHST, ambient temperature, or other significant factors.

The next step is to estimate the potential for gas to migrate through the unset cement. This will determine what anti-gas migration additives will have to be incorporated into the slurry design.

Appropriate retarders, fluid-loss additives, dispersants, or other additives should be added to the slurry. The slurry should be mixed and tested in the laboratory to meet all the requirements discussed above and those in Section 8-1.6.

The volume of slurry needed and the amount of additives needed should be estimated. The volume in the annulus is given by:

$$\text{ft}^3 \text{ per linear ft} = (D_w^2 - D_o^2) \times 0.005454$$

where D_w is the diameter in inches of the open hole and D_o is the casing OD in inches.

The length of open hole to be cemented is $(9200 - 7700) = 1500$ ft.

The volume of cement in the open hole $= (8.5^2 - 7^2) \times 0.005454 \times 1500 = 191 \text{ ft}^3$.

The excess cement needed is 20%; therefore, the volume of cement in the open hole $= 191 \times 1.2 = 229 \text{ ft}^3$.

There is cement left in the shoe joint. The volume of this cement is determined by the length of casing below the float collar and is given by

$$\text{volume of cement in the shoe} = 0.005454 \times D_i^2 L_{bfc}$$

where D_i is the ID of the casing and L_{bfc} is the length of casing below the float collar.

The ID of a 7-in., J-55, 23-lb/ft casing is 6.366 in.

From Figure 8-26, $L_{bfc} = 40$ ft.

Volume of cement in the shoe $= 0.005454 \times 6.366^2 \times 40 = 9 \text{ ft}^3$.

Total volume of cement needed $= 229 + 9 = 238 \text{ ft}^3$.

The slurry yield is $1.05 \, \text{ft}^3/\text{sk}$. This calculation is shown in Section 8-4.2.1.

Amount of cement needed $= 238/1.05 = 227$ sk.

The amount of additives should be calculated for 227 sk of cement. All the laboratory tests such as thickening time, and fluid loss should be repeated with the field cement and chemicals to make sure that the specifications are met. The cements vary from batch to batch, so the confirmation of the laboratory data with field sample of materials is very important.

8-5.3 Casing Attachments

Various types of hardware are attached to the casing as it is run into the wellbore to aid in cement placement. Each of these types is described briefly below. Figure 8-27 depicts these attachments:

- **Centralizers** Centralizers are attached to the casing to keep it off the borehole wall and centralize it as much as possible. Centralizing the casing helps provide a uniform fluid flow profile around the annulus and leads to better drilling fluid removal and proper cement placement. In general, there are two types of centralizers, rigid and bow-spring.

- **Bottom plug** A bottom plug is inserted ahead of the cement slurry to minimize intermixing of the slurry with the fluid ahead of it and to minimize contamination. The bottom plug also wipes any accumulated film of drilling fluid from the inner walls of the casing. When the bottom plug lands on the float collar, a small differential pressure ruptures the diaphragm in the plug and lets cement pass into the annulus.

- **Top plug** The purpose of a top plug is to minimize intermixing of cement slurry with the fluid behind it, referred to as displacement fluid, which minimizes contamination. Unlike bottom plugs, top plugs, which are inserted behind the cement slurry, are solid and seal against the bottom plug. When the top plug reaches the bottom plug, a pressure increase is observed and this pressure increase signals the end of cement job.

- **Plug container** Plug containers are used in most cementing operations to facilitate pumping fluid into the casing and also to hold and release cementing plugs at the proper time. If only one plug can be handled at a time, the container should be opened to load the second plug. This procedure could result in introduction of an air pocket in front of the second plug. To avoid this, two plug containers may be used

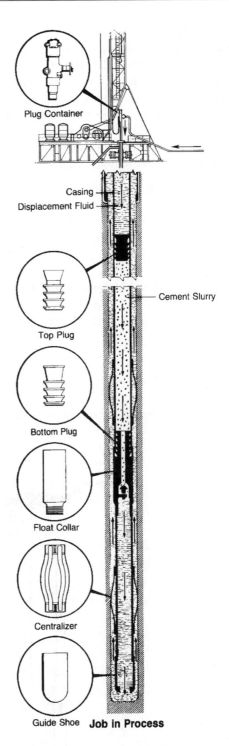

Figure 8-27 Casing attachments

so that the container need not be opened to load additional plugs.

- **Float collar** The original purpose of a float collar was to facilitate running the casing by reducing the strain on the derrick. This purpose is accomplished by preventing the drilling fluid from flowing into the casing when it is run into the hole. After the casing is run, circulation is established downward through the casing. After cement is pumped, the float collar prevents cement in the annulus from coming into the casing. The derrick design has improved over the years, and now the main purpose of the float collar is to prevent the cement from flowing back into the casing.

- **Guide/float shoe** In its simplest form, the purpose of a guide shoe, which is installed on the end of the casing, is to direct the casing away from ledges to minimize side-wall caving and to aid in safely passing hard shoulders and passing through crooked holes. These shoes are now designed with side ports to improve the jetting action and to clean the gelled fluid or other debris from the wellbore. Guide/float shoes are recommended to aid in better displacement.

8-5.4 Equipment

Equipment used to perform primary cementing is discussed below.

8-5.4.1 Mixing Equipment

Cement slurry may be mixed and pumped downhole continuously or semi-continuously, or it may be batch-mixed. The mode chosen depends on the application.

Continuous mixing refers to mixing and pumping slurry at the same time. Batch-mixing refers to mixing all of the slurry first and then pumping.

Continuous mixers were invented to eliminate cutting sacks and mixing by hand. The first of these mixers was the jet mixer. Later improvements, such as the recirculating mixer, were created. Continuous mixing sometimes allows cement that is not mixed correctly to be pumped downhole. Batch-mixing was developed to prevent this problem. However, batch-mixing requires more surface equipment, which increases the cost. The latest move is to go back to continuous mixing, but with computers automating the system to prevent improperly mixed cement from being pumped downhole. A typical continuous mixer is shown in Figure 8-28.

8-5.4.2 Pumping Equipment

Cement is normally pumped with a triplex positive-displacement pump. Triplex pumps are boosted with centri-

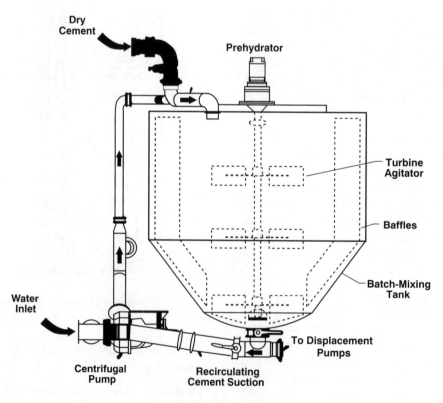

Figure 8-28 Cement mixer

fugal pumps. Centrifugal pumps impart a great deal of shear to the slurry, which is beneficial for mixing. Triplex pumps are low shear but pump at high pressures. Sometimes the displacement pressures for a cement treatment can be several thousand psi and triplex pumps do a good job in these cases. The number of pumps will depend on the required rate and anticipated pump pressure. A typical mobile pumping unit is shown in Figure 8-29.

8-5.4.3 Bulk Equipment

For land operations in areas with existing oil fields, most cement is dry blended with additives that give cement the required performance properties. The blended cement is hauled to the location, mixed with water, and pumped downhole. For remote locations and offshore operations, the additives are normally taken to the well as liquids, mixed with the water, and then mixed with the cement. When liquid additives are used, the cement mixing equipment is normally equipped with a liquid additive device that allows accurate metering of liquid additives into the mix water during cement mixing operations. A typical bulk unit is shown in Figure 8-30.

8-5.5 Fluid Circulation

Once the casing is at the bottom, the hole is normally circulated to remove the gelled drilling fluid and make it

Figure 8-29   Pump

Figure 8-30   Bulk equipment

mobile. This step is very important since the drilling fluid is more difficult to remove from the annulus. It is a common practice to thin the drilling fluid that is being pumped into the wellbore. However, thinning should be done only to the extent that the thinned drilling fluid will not channel into the drilling fluid that is already in the wellbore and also that the solids will not drop out of the fluid under downhole conditions. Normally, circulation is broken slowly to prevent fracturing the well. The drilling fluid has been static in the well for several hours and may be severely gelled.

Proper centralization of the casing will greatly improve the success of a primary cement job. As shown in Figure 8-31, the pipe is not centralized and is lying at the bottom of the hole. Figure 8-9 illustrates the velocity profile of a fluid flowing in an eccentric annulus. This figure shows that fluid circulation is ineffective in the narrow side.

Moving the pipe during hole conditioning will improve the gelled drilling fluid removal. Rotating or reciprocating the pipe tends to cause flow to go into the narrow side of the annulus. If a high degree of standoff cannot be achieved, then pipe movement becomes essential. A likely effect of pipe rotation is shown in Figure 8-15. Some rigid centralizers have flow deflectors that force flow into the narrow side of the annulus. However, these flow deflectors work only when the rigid centralizer is used in near-gauge holes. If the hole is not gauge, then the fluid will flow around the rigid centralizer and will not be deflected. Having a gauge hole greatly increases the probability of meeting cementing objectives.

Pressure changes resulting from pipe movement must be considered so that pressure changes caused by surge and swab will not change the well conditions.

8-6 DATA ACQUISITION AND SOFTWARE

In critical wells, data on the cement job must be monitored and real-time engineering calculations must be made to make real-time decisions. The pressure at the surface can be monitored using reliable and accurate pressure transducers along with the flow rate, density, rheology or other characteristics of the fluids. The measured pressure should then be compared with calculated values and real-time decisions made so that formation does not break down or experience other damage (Ravi *et al.,* 1992b). Figure 8-32 is an illustration of real-time measurements. This figure shows wellhead pressure at constant flow rate measured and plotted vs. elapsed time. The figure shows that pressure decreases as time progresses at a constant flow rate. This pressure can be converted to percent hole circulating when other parameters such as cased-hole dimensions and fluid

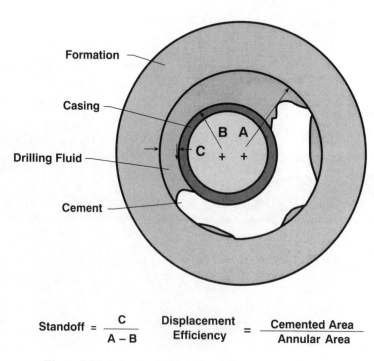

$$\text{Standoff} = \frac{C}{A - B} \qquad \frac{\text{Displacement}}{\text{Efficiency}} = \frac{\text{Cemented Area}}{\text{Annular Area}}$$

Figure 8-31 Non-centralized pipe. The concept of standoff

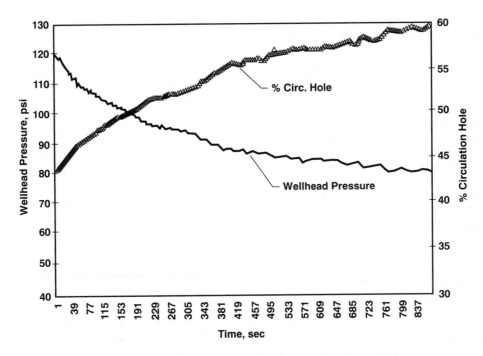

Figure 8-32   Real-time measurements and calculations

properties are known. This plot is also shown in Figure 8-32. The percent hole circulating increases with time but approaches a constant value. When a constant value is reached, one or more of the parameters should be changed to increase the percent hole circulating. The real-time performance should be analyzed and compared with the planned job, and changes should be made for future jobs.

REFERENCES

API Specification 10 (Spec 10): "Specification for Materials and Testing for Well Cements," API (1990).

Barton, N.A., Archer, G.A., and Seymour, D.A.: "Computational Fluid Dynamics Improves Liner Cementing Operation," *O&GJ* (Sept. 26, 1994).

Beirute, R.M., Sabins, F.L., and Ravi, K.: "Large Scale Experiments Show Proper Hole Conditioning: A Critical Requirement of Successful Cementing Operations," paper SPE 22774, 1991.

Benge, O.G., and Webster, W.W.: "Evaluation of Blast Furnace Slag Slurries for Oilfield Application," paper SPE 27449, 1994.

Bird, R.B., Curtis, C.F., Armstrong, R.C., and Hassager, O.: *Dynamics of Polymeric Fluids, Vol. 1, Fluid Mechanics,* John Wiley & Sons, 1987.

Bird, R.B., Stewart, W.E., and Lightfoot, E.N.: *Transport Phenomena,* John Wiley & Sons, 1960.

Bittleston, S., and Guillot, D.: "Mud Removal: Research Improves Traditional Cementing Guidelines," *Oilfield Rev.* (Apr. 1991) 44–54.

Cowan, K.M., Hale, A.H., and Nahm, J.J.: "Conversion of Drilling Fluids to Cements With Blast Furnace Slag: Performance Properties and Applications for Well Cementing," paper SPE 24575, 1992.

Craig Jr., R.R.: *Mechanics of Materials,* John Wiley & Sons, 1996.

Eirich, F.R.: *Rheology, Vol. 3,* Academic Press (1960).

Fjaer, E., Holt, R.M., Horsurd, P., Raaen, A.M., and Risnes, R.: *Petroleum Related Rock Mechanics,* Elsevier (1996).

Goodwin, K.J. and Crook, R.J.: "Cement Sheath Failure," paper SPE 20453, 1992.

Govier, G.W., and Aziz, K.: *The Flow of Complex Mixtures in Pipes,* R.E. Krieger Publishing (1987).

Grant Jr., Rutledge, J.M., and Gardner, C.A.: "The Quality of Bentonite and its Effect on Cement Slurry Performance, paper SPE 19940, 1990.

Haciislamoglu, M., and Langlinais, J.: "Non-Newtonian Flow in Eccentric Annuli," *J. of Energy Res. Tech.* (Sept. 1990) **112,** 163–169.

Hertzberg, R.W.: *Deformation and Fracture Mechanics of Engineering Materials,* John Wiley & Sons (1996).

Honore Jr., R.S., and Tarr, B.A.: "Cementing Temperature Predictions Based on Both Downhole Measurements and Computer Predictions: A Case History," paper SPE 25436, 1993.

Lockyear, C.F., Ryan, D.F., and Gunningham, M.M.: "Cement Channeling: How To Predict and Prevent," *SPEDE* (Sept. 1990) 201–208.

MacEachran, A., and Adams, A.J.: "Impact of Casing Design of Thermal Expansion Fluids in Confined Annuli," paper SPE 21911, 1991.

McCann, R.C., Quigley, M.S., Zamora, M., and Slater, K.S.: "Effects of High-Speed Pipe Rotation on Pressures in Narrow Annuli," paper SPE 26343, 1993.

Moran, L.K., and Lindstrom, K.O.: "Cement Spacer Fluid Solids Settling," paper SPE 19936, 1990.

Nahm, J.J., Kazem, J., Cowan, K.M., and Hale, A.H.: "Slag Mix Mud Conversion Cementing Technology:reduction of Mud Disposal volumes and Management of Drilling Fluid Wastes," paper SPE 25988, 1993.

Nahm, J.J., Romero, R.N., Wyant, R.E., and Hale, A.A.: "Universal Fluid: A Drilling Fluid To Reduce Lost Circulation and Improve Cementing," paper SPE 27448, 1994.

Nelson, E.B.: *Well Cementing*, Schlumberger Educational Services (1990).

Oraskar, A.R., and Turian, R.M.: "The Critical Velocity in Pipeline Flow of Slurries," *AlChEJ* (July 1980) 550–558.

Papadakis-Vayenas, C.G., and Fardis, M.N.: "A Reaction Engineering Approach to the Problem of Concrete Carbonation," *AlChEJ* (Oct. 1989) **35**, 10, 1639–1650.

Perry, J.: *Chemical Engineer's Handbook*, McGraw-Hill (1985).

Ramachandran, V.S.: *Concrete Admixtures Handbook* (1990).

Ravi, K., Beirute, R.M., and Covington, R.L.: "Erodibility of Partially Dehydrated Gelled Drilling Fluid and Filter Cake," paper SPE 24571, 1992a.

Ravi, K., Beirute, R.M., and Covington, R.L.: "Improve Primary Cementing by Continuous Monitoring of Circulatable Hole," paper SPE 26574, 1992b.

Ravi, K.: "Rheology as a Function of Temperature," 1991 AlChE Annual Conference, Los Angeles.

Ravi, K., and Guillot, D.: "Specification for Materials and Testing for Well Cements," *Appendix J, API Specification 10 (Spec 10)*, API (to be published 1997).

Ravi, K., and Moran, L.K.: "Deposition and Resuspension of Solids From Cementing Fluids," Internal Report, to be published.

Ravi, K., and Weber, L.: "Drill Cutting Removal in a Horizontal Wellbore for Cementing," paper 35081, 1996.

Sabins, F.L., and Sutton, D.L.: "The Relation of Thickening Time, Gel Strength, and Compressive Strengths of Oilwell Cements," paper SPE 11205, 1982.

Sassen, A., Marken, C., Sterri, N., and Jakobsen, J.: "Monitoring of Barite Sag Important in Deviated Drilling," *O&GJ* Aug. 26, 1991) 43–50.

Shah, S.N., and Lord, D.L.: "Hydraulic Fracturing Slurry Transport in Horizontal Pipes," SPEDE (Sept. 1990), 225–232.

Shah, S.N., and Sutton, D.L.: "New Friction Correlation for Cements From Pipe and Rotational Viscometer Data," paper SPE 19539, 1989.

Smith, D.K.: *Cementing*, SPE Monograph Series, Vol. 4 (1990).

Smith, T.R., and Ravi, K.: "Investigation of Drilling Fluid Properties To Maximize Cement Displacement Efficiency," paper SPE 22775, 1991.

Sutton, D.L., and Ravi, K.: "New Method for Determining Downhole Properties That Affect Gas Migration and Annular Sealing," paper SPE 19520, 1989.

Taylor, H.F.W.: *Cement Chemistry*, Academic Press (1992).

Tilghman, S.E., Benge, O.G., and George, C.R.: "Temperature Data for Optimizing Cementing Operations," paper SPE 19939, 1990.

Timoshenko, S.P., and Goodier, J.N.: *Theory of Elasticity*, McGraw Hill (1970).

Tosun, I.: "Axial Laminar Flow in an Eccentric Annulus: An Approximate Solution," *AlChEJ*, (1984) **30**, 5, 877–878.

Uner, D., Ozgen, C., and Tosun, I.: "An Approximate Solution for Non-Newtonian Flow in an Eccentric Annulus," *Ind. Eng. Chem. Res.* (1988) **27**, 698–701.

Walton, I.C., and Bittleston, S.H.: "The Axial Flow of a Bingham Plastic in a Narrow, Eccentric Annulus," *J Fluid Mech.* (1991) **222**, 39–60.

Wedelich, H., and Galate, J.W.: "Key Factors That Affect Cementing Temperatures," paper SPE 16133, 1987.

White, F.M.: *Fluid Mechanics*, McGraw-Hill (1979).

Wooley, G.R., Giussani, A.P., Galate, J.W., and Wedelich, H.F.: "Cementing Temperatures for Deep Well Production Liners," paper SPE 13046, 1984.

9 Formation-Fluid Migration After Cementing

Larry T. Watters
Halliburton Energy Services

Robert Beirute
Amoco Production Company

9-1 INTRODUCTION

Formation-fluid migration following primary cementing occurs when fluid moves to the surface or into another zone from a formation through pathways created in a cemented annulus. Fluid migration after primary cementing, whether water or gas, has affected the well-completion industry since the introduction of oilwell cementing. It is estimated that this migration occurs at some point in the life of approximately 25% of wells drilled. Problems caused by fluid migration include nuisance interzonal flow, pressure behind the production casing at the wellhead, blowouts, and the loss of rigs or platforms. Small or nuisance flow may require monitoring of pressure buildup and occasional bleeding off of trapped volume at the wellhead. Catastrophic flow can result in blowout and loss of the well.

The number of wells with pressure at the surface between casings is steadily increasing as well ages are increasing and regulatory bodies are enforcing tighter monitoring and control of the problem. Estimates of this problem suggest that as many as 20% of existing wells have surface pressure between casings.

Until recently, gas flow has received the most attention, but shallow water flows have also been encountered frequently during the completion of conductor pipes in deep-water wells.

Gas migration was first identified as a significant problem during completion of gas storage wells in the early 1960s (Sutton *et al.*, 1984). This awareness of gas migration was attributed to increases in well depth and the increased monitoring of gas reservoirs. Gas migration has become a significant problem with the development of deep, high-pressure reservoirs, leading to a variety of laboratory and field investigations designed to prevent or correct the problems associated with gas migration. Flow may occur almost immediately after cementing or may not begin until months after completion of the well. When flow has started, the migration problem becomes increasingly difficult to correct.

Water migration was identified by Jones and Berdine (1940) as a problem resulting from incomplete displacement of drilling-fluid filter cake during cement placement. Awareness of the severity of water migration has been heightened recently because of water flows after conductor-pipe cementing for deep-water wells (Griffith and Faul, 1997).

Although fluid migration seems to occur at various places for various causes, three fundamental factors are always common to fluid flow following cementing:

- Presence of formation fluid

- Failure to contain the fluid in the formation allows its entry into the wellbore

- Flow path in the wellbore allows access to a lower-pressure destination for the fluid.

This chapter will first present a description of the mechanism of fluid migration. The differences between immediate and long-term flow are outlined and compared. In accordance with historical nomenclature, immediate migration will be referred to as flow, and long-term migration will be referred to as leakage. Then, theories and practice of prevention are presented including example applications. Since the majority of research has focused on gas migration and the underlying theories and mechanisms for the cause and control of gas and water migration are similar, the remainder of this chapter will focus on gas migration. Recently developed mechanisms or procedures specific to shallow water flows will be discussed as warranted.

9-2 THEORY OF GAS MIGRATION

Generally, the presence and characteristics of a gas-bearing formation are known before cementing. Pressures required to keep gas from flowing into a wellbore are determined so that the well can be controlled during drilling. The density of the cement slurry and the placement process are also designed to control hydrostatic

pressure at the gas zone. Immediately after the cement slurry placement, the gas within any adjacent zones is trapped in the formation, and the well is static. Factor one of gas migration is satisfied, but no means exists for gas to enter the wellbore (Figure 9-1A).

During the next few hours or months, gas may overcome the controlling pressure, enter the wellbore, and migrate through the annulus through the following mechanisms. The mechanism of gas flow immediately after cementing is presented first, followed by mechanisms responsible for long-term gas leakage.

9-2.1 Mechanism for Gas Flow Following Cementing

The controlling forces that govern formation-gas entry into an annulus following cementing (Sabins *et al.*, 1980) are

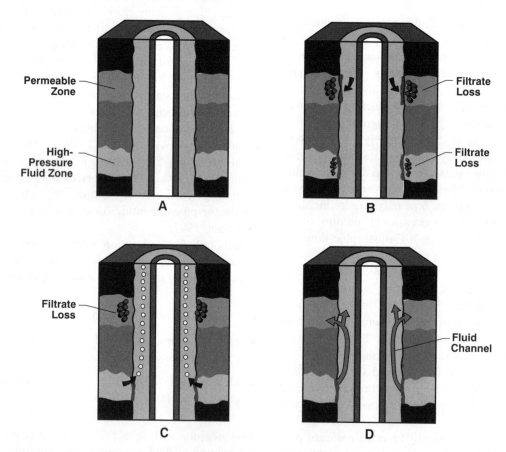

Figure 9-1 Mechanism of fluid flow following cementing: (A) Initially following cementing formation fluid is contained by hydrostatic pressure. (B) Gelation of static cement restricts hydrostatic pressure transmission. Loss of filtrate to permeable formations as well as volume reduction due to chemical hydration result in relief of controlling pressure. (C) Pressure declines sufficiently to allow fluid percolation through gelled cement. (D) Permanent flow channels form.

- Gel-strength development within the cement slurry

- Pressure differential between initial hydrostatic pressure exerted by the cement column onto the gas-bearing formation and pore pressure of the formation

- Fluid loss from the cement to adjacent formations

- Volume losses during cement hydration reactions

These controlling forces are described below and effects of their occurrence are depicted graphically in Figure 9-2. Once inside the annulus, formation gas must have a migration path and positive pressure differential to flow through the annulus and make a way for more formation gas to enter. The possible migration paths are also described below.

9-2.1.1 Gel-Strength Development

If the annular fluid column above the gas-bearing zone is designed properly, the hydrostatic force of the column is sufficient initially to overcome the pore pressure of the formation. However, the complex chemical processes of cement hydration (Chapter 8), which begin as soon as Portland cement is mixed with water, create two conditions that promote the reduction of hydrostatic pressure. The first condition is gel-strength development. Initially after placement, the cement slurry behaves as a true fluid because pressure at any point in the cement column is equal to the hydrostatic head. When the cement slurry becomes static, ionic forces can create a weak tertiary

structure that later grows stronger. This structure is manifest as gelation in the bulk slurry. With gelation, adhesion to the hole surfaces causes the cement column to lose the ability to transmit full hydrostatic pressure. This gelation continues to increase until the cement becomes a solid. The degree of gelation has been quantified by Sabins *et al.* (1980) in terms of *static gel strength* (SGS, lbf/100 ft^2). Gel strength as used here is not a complex rheological function but is rather a direct measurement of the adhesive strength of a fluid to a borehole wall or casing. This study also introduced the concept of *transition time* as the interval during the setting process from the time a cement slurry column first loses the ability to transmit full hydrostatic pressure until the time a cement develops sufficient gel strength to prevent entry or migration of formation gas. This end value was determined experimentally to be a minimum value of 500 lbf/100 ft^2.

As gelation continues to increase, sufficient internal structure develops within the cement to partially support the slurry in the annulus. Thus, gelation prevents transmission of full hydrostatic pressure from the fluid column above (Fig 9-1B and C). However, this gelation does no harm until volume losses take place.

9-2.1.2 Volume Reduction and Resulting Pressure Loss

Volume losses occur internally in the setting cement through two mechanisms: fluid loss and hydration volume reduction. These two phenomena, listed as the

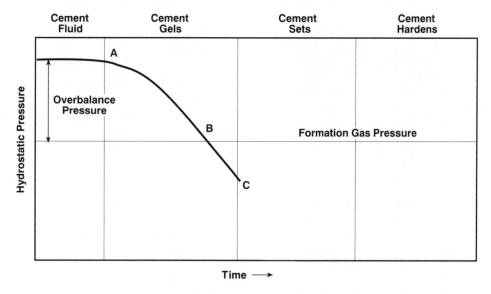

Figure 9-2 Pressure decline during cement hydration resulting in fluid flow

last two controlling forces contributing to fluid entry, are outlined in detail in Chapter 8. Fluid loss from the slurry refers to movement of the slurry's aqueous phase from the slurry into adjacent permeable formations. This movement is driven by positive differential pressure between hydrostatic pressure in the cement column, and formation pore pressure, formation permeability, and permeability of any deposited drilling fluid or cement filter cake. Hydration volume reduction describes the volume losses that occur in Portland cement during the chemical setting process. These losses result from the chemical combination of water and reactive compounds in the cement. Simply stated, this means that the absolute volume of the products is less than the volume of the reactants. Hydration volume reduction causes the formation of porosity and volumetric shrinkage in Portland cement.

The volume reductions occur during the period when cement gelation normally prevents hydrostatic pressure transmission, resulting in pressure losses in the cement column (Figures 9-1B and C). This pressure reduction reduces the well's capacity to keep pressurized formation fluids contained within formations. Sufficiently high pressure losses in the cement column may reduce trapped hydrostatic pressure to just below the pore pressure of a gas bearing reservoir, allowing this gas to invade the annulus, provided the cement is still sufficiently fluid to be displaced by the gas. Thus, the ability of a cement column to prevent formation gas invasion is governed by the rate of pressure loss as well as the rate of early three-dimensional structure development sufficient to prevent gas entry (SGS of at least $500\,\text{lbf}/100\,\text{ft}^2$).

Pressure reduction in a cemented column was demonstrated in the laboratory by several investigators, including Christian *et al.* (1975) and in laboratory and field tests by Levine *et al.* (1979). Sutton *et al.* (1984) present basic calculations for pressure drop rates based on volume reductions and pressure restrictions caused by gel-strength development. The equation for pressure drop possible from volume losses is

$$\Delta p = (\overline{V}_{FL} + \overline{V}_H)/c \tag{9-1}$$

where Δp is the maximum pressure drop possible for a specific slurry, $\overline{V}_{FL}$ is the total unit volume reduction due to fluid loss during transition time, $\overline{V}_H$ is the unit volume reduction due to chemical hydration during transition time, and c is the compressibility factor for cement slurry.

The pressure restriction possible from static gel-strength development is represented by

$$p_R = SGS \times 4L/D \tag{9-2}$$

where p_R is pressure restriction resulting from SGS, psi, SGS is the static gel strength, $\text{lbf}/100\,\text{ft}^2$, D is the diameter, in., and L is the length of interval, ft.

Setting SGS equal to the minimum value to prevent gas percolation ($500\,\text{lbf}/100\,\text{ft}^2$) reduces Equation 9-2 in standard oilfield units to:

$$p_{R,\text{max}} = 1.67 \times L/D \tag{9-3}$$

where $p_{R,\text{max}}$ is the maximum pressure reduction, psi.

Comparing the maximum possible pressure reduction to the initial overbalance pressure for the gas zone, p_{OB}, provides a means to evaluate gas flow potential:

$$GFP = p_{R,\text{max}}/p_{OB} \tag{9-4}$$

where GFP is the gas flow potential factor, dimensionless.

The magnitude of GFP for a specific well condition indicates the level of difficulty that the control of gas flow will present. Magnitudes vary from 0 to infinity, with increasing value indicating greater potential for flow. Table 9-1 lists commonly accepted levels of GFP for classification of the severity of a potential problem.

Comparing GFP with Δp calculated from cement slurry variables in Equation 9-1 but written in terms of volume reduction rates results in

Table 9-1 Evaluation of gas-flow severity: Example GFP calculation

TD	9468 ft, 5 in. casing, 6.75 in. hole
Drilling Fluid	16.5 lb/gal
Gas Zone Depth	9100 ft
Gas Pressure	7438 psi (15.7 lb/gal)
Cement TD to 7500 ft with 17.5 lb/gal cement	
Calculations:	
p_h	$= ((7500)(16.5) + (1600)(17.5)) \times 0.0519$
	$= 7875\,\text{psi}$
p_{OB}	$= 7875 - 7438$
	$= 392\,\text{psi}$
$p_{R,\text{max}}$	$= 1.67 \times 1600 \div (6.75 - 5)$
	$= 2672\,\text{psi}$
GFP	$= 2672/392$
	$= 6.8$

This indicates a moderate annular gas-flow potential.
Ranges of GFP:

0–3	Low
3–8	Moderate
8+	High

$$\Delta p = \left(\frac{d\overline{V}_{FL}}{dt} + \frac{d\overline{V}_H}{dt} \right) \frac{t_{tr}}{c} \qquad (9\text{-}5)$$

where $\dfrac{d\overline{V}_{FL}}{dt}$ and $\dfrac{d\overline{V}_H}{dt}$ are the total fluid loss and hydration volume rates, and t_{tr} is the transition time of the cement slurry.

Equation 9-5 allows tailoring of cement slurry properties to maintain pressure across a specific gas-bearing formation until the slurry is sufficiently gelled to resist gas flow. Application of these simple relationships, which will be discussed later, forms the basis of a quantitative method for analyzing the potential to encounter gas flow on a specific well and indicates the cement properties necessary to combat this gas flow.

9-2.1.3 Migration Paths

Once inside the annulus, the gas must travel along a path to the surface or into another permeable formation to complete the action of gas flow. Gas then flows through the annulus through three potential paths (Beirute and Cheung, 1989): the cement-formation interface or the cement pipe interface, the drilling-fluid channel, or the unset cement (Figures 9-1d, 9-3a, b, and c).

Microannuli can form at the cement's interface to either the pipe or formation face. As discussed in Chapters 8 and 11, microannuli (from 0.001 to 0.005 in.) may form as a result of thermal or pressure fluctuations in the wellbore through normal well operations during cementing (Goodwin and Crook, 1990). In some cases, a larger microannulus can be created at the cement-pipe interface by viscous drilling fluid bypassed during cement placement clinging to the pipe or borehole wall, or by excessive cement expansion against soft formations (Beirute *et al.*, 1988). Bol *et al.* (1997) developed a gas migration model based on physical principles, literature documentation, and an experimental program and concluded that the only practical mechanisms for gas flow were flow through the microannulus caused by shrinkage or inadequate drilling-fluid displacement. Bol's conclusions, in contrast to the conclusions of other investigators, were that behavior of the unset cement did not contribute to gas flow and that percolation through cement slurry or matrix was not a mechanism of gas flow.

Poor displacement of drilling fluid during cement placement can bypass a continuous channel of drilling fluid traversing the annulus (Chapter 8). This channel can act as a conduit for formation fluid invading the annulus

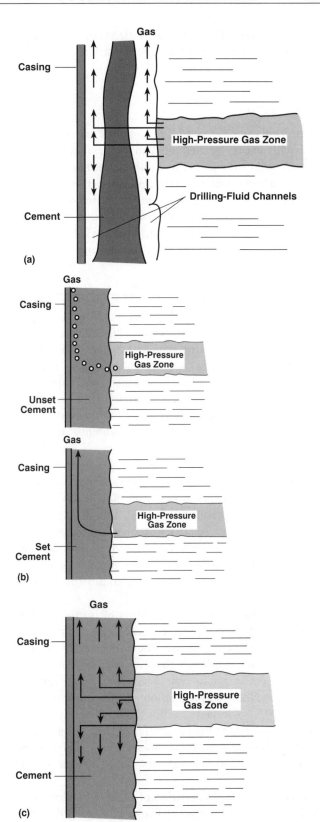

Figure 9-3 Fluid migration paths (a) through undisplaced drilling-fluid channels, (b) through a cement-casing micro annulus, and (c) through channels in the cement

since the drilling fluid develops little static gel strength and thus offers little resistance to migration.

9-2.1.4 Percolation

When formation gas invades the annulus and a potential flow path exists, percolation can occur if a sufficient pressure gradient exists. Gas bubbles will migrate along the pressure gradient to lower-pressured permeable zones or to the surface. As they migrate toward lower pressure, gas bubbles expand, pushing open a greater flow path. Eventually, this percolation through gelled fluid can result in the establishment of a permanent path as the cement sets.

Several investigators, including Rae *et al.* (1989), Cheung and Beirute (1985), and Moroni *et al.* (1997), have investigated the possibility of gas flow via a path through the unset cement permeability. As cement in the annulus gels and the cement becomes "load bearing," they theorize that gas can enter cement through the permeability of the three-dimensional structure. This theory is based on the assumption that permeability of the gelling cement is very high. However, Sutton and Ravi (1989) evaluated the possibility that this mechanism was solely responsible for gas percolation through setting cement. Using Darcy's flow equation and laboratory-measured permeability values for setting cement, they calculated potential flow rates of gas through cement permeability and found likely flow volumes to be too small to account for catastrophic gas flow. Therefore, while gas flow via a path through the unset cement permeability is one potential source of gas flow, it cannot be the sole mechanism for gas flow through the cement.

Gel strength governs the migration of formation fluid through unset cement as a homogeneous medium, as well as gas entry into the wellbore. The progressively stronger three-dimensional gelation that precedes the conversion of the cement to a solid, affects fluid migration through the cement. Once cement gelation reaches a magnitude of $500 \, \text{lbf}/100 \, \text{ft}^2$, resistance to flow through the cement is sufficient to prevent fluid migration. Below this strength, invading fluid can move the cement aside or ahead and create a flow channel through the cement (Figures 9-1D and 9-3C).

A basic numerical simulation of limits to movement of gas bubbles was developed by Prohaska *et al.* (1995). The simulation indicates limited gas-migration distance in the cemented column above the gas zone through which gas can displace gelling cement as a homogeneous mass and flow up the column. Within this distance after pressure is balanced, the driving force of the reservoir can force gas into the annulus by displacing whole slurry. Above this distance, termed the *critical distance*, pressure in the cement column is above the formation pressure so the gas will be contained in the column below.

The volume of gas that can enter the annulus at any time is equal to the volume reduction of cement caused by hydration or fluid loss within the critical distance. Expressing force balance at the critical distance in terms of pressures and gel strength reveals that the critical distance varies directly with cement gel strength. Further examination of the relationship indicates that gas influx may greatly increase when the critical distance reaches a permeable zone into which fluid loss will occur since leakoff volume changes are generally substantially greater than volume changes for hydration (Figure 9-4).

The model also analyzes the relationship between gas inflow volume and bubble migration. The model accounts for the coalescence of smaller bubbles, the detachment of a bubble from the formation wall, and the bubble velocity upward as a function of the cement's static gel strength. Laboratory results that confirm these theoretical relationships and increased velocity of subsequent bubbles caused by disruption of cement gel strength are presented in Figures 9-5 and 9-6.

Finally, the model is used to calculate gas inflow rates caused by hydration volume reduction and filtrate loss as well as to simulate the bubble motion process through the gelling cement. This model accounts for four factors:

- Pressure profile of the wellbore and filtration into permeable layers before gas migration

- Critical distance after pressure balance and cement volume changes

- Bubble detachment time and volume

- Bubble inflow, motion of the bubble front, and migration height

This model can be used to analyze individual well parameters and to assess the relative importance of various well and cement parameters to prevent gas flow.

9-2.2 Long-Term Gas Leakage

Long-term gas leakage is similar to gas flow but begins through a different mechanism. Long-term gas leakage begins weeks or months after the well is completed and is

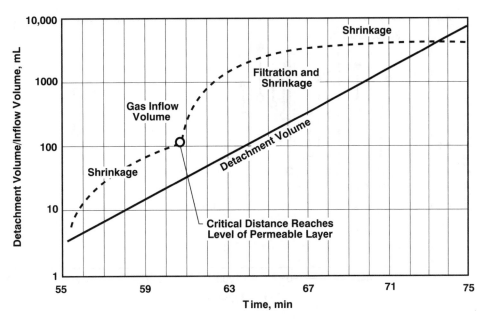

Figure 9-4   Gas in-flow and bubble detachment (after Prohaska *et al.*)

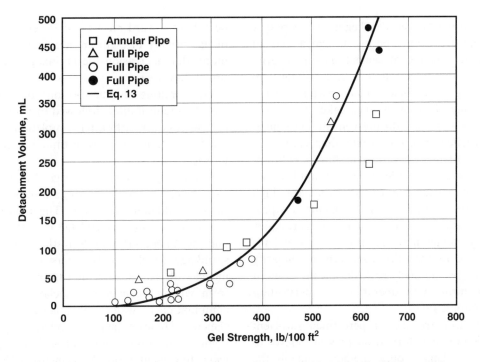

Figure 9-5   Bubble detachment volumes in cement slurries (after Prohaska *et al.)*

usually detected at the wellhead (or sea floor in the case of offshore drilling) long after the well has been completed and put on production. Although this leakage is not usually severe, it can become a nuisance. As government agencies impose stricter controls on well integrity, long-term gas leakage has become a problem that must be addressed.

The causes of long-term gas leakage are the same as those outlined earlier for gas flow. The first controlling force is the presence of gas, which would be alleviated if initial primary cementing operations had successfully contained gas in the formation. The mechanism of gas leakage depends on the state of the primary cement sheath as production of the well begins.

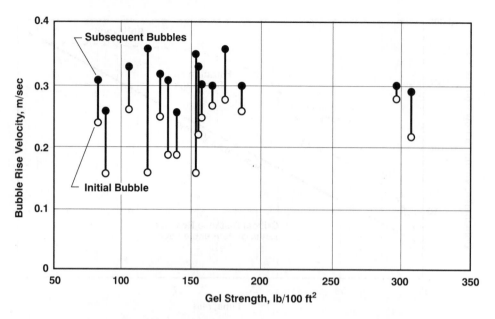

Figure 9-6   Velocities of initial and subsequent bubbles (after Prohaska *et al.*)

The gas-leakage mechanism can begin with a poorly placed cementing treatment, poorly designed cement composition, or destruction of cement sheath integrity due to well operations. In the first case, a drilling-fluid channel containing either whole fluid, partially dehydrated fluid, or loosely formed filter cake (Jones and Berdine, 1940), is bypassed during cement placement. As described in Chapter 8, the drilling-fluid channel is bypassed because of ineffective drilling-fluid conditioning, pipe decentralization, ineffective spacer application, or suboptimum placement rate. In the second case, high water-content cements or low-strength cements may possess high permeability or large volumetric shrinkage, which eventually creates flow channels for gas leakage (Rae *et al.*, 1989). In the third case, cement integrity may be destroyed by thermally or hydraulically induced pressure cycles during well operation or intervention (Goodwin and Crook 1990).

When the cementing treatment is performed, sufficient hydrostatic pressure over formation pressure exists to control in-situ gas. This trapped pressure controls the gas until the cement sets, initially locking the gas in the formation. However, in the first scenario, cement placement is incomplete, leaving a channel or microannulus of drilling fluid either as gelled fluid, partially dehydrated fluid or paste, or as thick, unconsolidated filter cake (Figure 8-11, Chapter 8). As time passes and the well is produced, the filter cake or gelled drilling fluid can dehydrate and shrink, leaving a path for trapped formation gas. As shrinkage occurs, gas enters the annulus and

percolates toward an area of lower pressure. This flow of gas usually dries the noncementitious material even more, thus increasing the shrinkage and the cross-sectional flow area.

The resulting situation is one of a seemingly well-sealed annulus initially followed by detection of gas pressure weeks or months after completion. The gas pressure build-up rate increases with time, requiring more maintenance.

Alternately, a flow path for gas leakage can occur in a well cemented annulus through intrinsic cement properties or production operations. The intrinsic cement property of dimensional shrinkage is outlined in Chapter 8 and by Chenevert and Shrestha (1991). Dimensional shrinkage, which usually occurs in cement compositions mixed with excess water and fillers or extenders, results from volume reductions associated with the chemical combination of water with reactive components in the cement. These reductions, referred to as hydration volume reductions, range up to 5%, depending on the water-to-cement ratio. All cements undergo this reduction to some extent, and hydration volume reduction is the mechanism responsible for porosity formation in set cement. In addition to porosity formation, volumetric reduction of the cement usually occurs depending on the strength of the crystalline lattice being formed. Cements with high water-to-cement ratio form a weaker crystalline lattice structure and are therefore more subject to volumetric shrinkage. It should be noted here that the shrinkage occurring from hydration volume reduction is quite unlike volume changes in a gas or liquid since the developing crystalline structure in solid cement tends to retain its original volumetric dimensions. Parcevaux and Sault (1984) performed laboratory tests indicating that a path for gas leakage may be formed in a previously completely cemented annulus by the effects of volumetric shrinkage and loss of hydraulic bond.

Operationally, sufficient stresses can be generated in the cemented annulus to destroy the integrity of the cement seal. This destruction of integrity is caused by thermal or hydrostatic forces exerted on the casing during production cycling or well intervention. These cyclic gradients in temperature or pressure can cause the casing to expand, generating tensile stresses in the cemented annulus. Since cement is a brittle material with low tensile strength, these forces create tensile cracks in the cement. This cracking can become extensive enough to provide a flow path for any gas residing in a pressurized formation.

9-2.3 Summary of Mechanisms

The ultimate result of each of the mechanisms described above is unwanted gas migration. However, the methods for controlling these mechanisms are quite different. First, the potential for each type of problem to occur must be assessed. The most difficult part of the gas flow prevention process, problem identification and quantification, is presented below.

9-3 DIAGNOSING THE POTENTIAL FOR GAS MIGRATION

As with most other problems associated with well construction, the potential for gas migration problems can be analyzed and planned for during the early stages of construction. This assessment is made using knowledge of previous activity in the local field or formation and estimates of formation and drilling conditions. Techniques to predict gas migration are reviewed below.

9-3.1 Gas Flow

Gas flow diagnosis depends on knowledge of the formation pressures to be encountered, the hydrostatic pressure exerted at the completion of the cementing process, and the rate at which the cement will develop gel strength after completion of the cementing treatment. The first two parameters are measured or calculated based on known reservoir characteristics or information obtained during drilling. These parameters also define the severity of the potential problem and indicate solution methods. Static gel-strength development and fluid loss characteristics of the specific cement composition are dependent on batch runs and additive composition, and therefore must be measured in the laboratory for specific application.

Identification of the zone bearing the highest pore pressure is the first step toward diagnosing the potential for gas flow. This zone is usually known beforehand for development wells and is measured or estimated while drilling for exploratory wells. This parameter is important while drilling so that sufficient drilling-fluid density can prevent loss of well control. The pressure at depth defines the minimum pressure to be maintained in the annular fluid column to prevent gas entry. The pressure as calculated by minimum drilling-fluid density to control is given by

$$p_f = 0.433 \rho_{min} H \qquad (9\text{-}6)$$

where p_f is the formation pressure of highest pressured gas sand, psi, ρ_{min} is the minimum density of drilling fluid to control gas, lb/ft^3, and H is the depth of the gas zone, ft.

Initial hydrostatic pressure exerted on the zones targeted for gas flow control is calculated by

$$p_h = 0.433 \sum \rho_i h_i \qquad (9\text{-}7)$$

where ρ_i and h_i are the fluid density and thickness of each layer above the target. These two parameters define the upper and lower pressure boundaries for the gas-flow problem. Figure 9-2 illustrates the maximum pressure initially controlling formation fluid which is hydrostatic pressure exerted by the fluid column after placement. Minimum pressure below which the trapped hydrostatic pressure cannot fall without influx potential is the formation pore pressure.

From this point, several techniques can be used to determine the probability of gas flow. Qualitative techniques include analyzing potential for flow based on past experience and assuming the likelihood of flow based on the level of Δp. The quantitative *GFP* calculation presented in Section 9-2.1.2 can also be used to evaluate the flow probability.

An example diagnosis of the potential of gas flow through *GFP* calculation is illustrated in Table 9-1. This example is based on a gas-well completion at a depth of 9468 ft. Pertinent well parameters required to analyze the *GFP* are listed in Table 9-1 along with *GFP* evaluation. This cementing operation has a *GFP* of 6.8, a moderate potential for gas flow, which will require a special cement design to prevent the gas-flow problem. The example will be continued in Section 9-4.1.4 with quantitative analysis of the cement composition designed for this application.

9-3.2 Long-Term Gas Leakage

Long-term gas leakage is difficult to predict because knowledge of past field experience and other subjective formation and completion practice indicators are required for analysis. The potential for long-term leakage is also governed by the formation pressure of the highest-pressured gas sand since this pressure is the driving force for delivery of gas into the system if a flow path is created. A subjective analysis of the potential for gelled, partially dehydrated drilling fluid to be deposited and bypassed during cement placement and an analysis of the potential for volumetric shrinkage of cement must be based on expert knowledge and on an analysis of the fluid compositions, testing procedures, and case histories of the formation. The analysis should also include qualitative evaluation of scenarios such as those outlined in Table 9-2.

The following example illustrates some general considerations for evaluating gas leakage. Unwanted gas flow to the surface from a shallow high-pressure sandstone reservoir has occurred in previous infield wells. The gas flow is not a problem after initial completion of the well but begins within a few weeks and gradually increases. Gas flow potential analysis indicates a low *GFP* of 2.1 which should be sufficiently controlled by the low-fluid-loss cement slurries used, but previous experience reveals a history of gas leakage.

9-4 CONTROLLING MIGRATION PHENOMENA

The basic premise of controlling unwanted gas migration is to prevent gas entry into the annulus. This requires maintaining controlling pressure on the gas-bearing formation and eliminating potential migration paths. All control or prevention measures must be complemented by optimum cementing procedures. There can be no shortcuts to the process of controlling gas migration. Proper hole conditioning and displacement practices must be employed to ensure best possible removal of drilling fluid in all forms from the hole. Proper design of slurry and set cement properties must be considered to effect a satisfactory seal in addition to controlling fluid movement. These basics are covered in detail in Chapter 8.

The following sections summarize various methods available to prevent gas entry for the two types of gas migration.

9-4.1 Gas Flow

In the short term, pressure must be maintained on the gas-bearing formation to prevent movement into the wellbore until the cement sets. Pressure can be maintained through the use of three basic mechanisms: altering cement characteristics, mechanically blocking the flow path, or increasing the magnitude of pressure applied to the gas-bearing zone. Ways to increase the magnitude of initial pressure to gas-bearing formations include applying additional pressure at the surface, or increasing density of fluids in the column. The other two mechanisms are discussed below. Theory and practice behind these techniques are summarized in Table 9-3 along with pertinent references for further review.

Table 9-2   Analysis of gas-leakage probability

Historical	Has the field exhibited a problem with long-term gas leakage in the past?
Filter Cake Analysis	Are drilling-fluid loss characteristics sufficient to prevent deposition of loosely consolidated, thick filter cakes against the formation face?
Placement Mechanics	Are cement placement practices adequate to displace drilling fluids and establish a continuous volume of cement?
Cement Composition	Are the volumetric shrinkage characteristics of the cementing composition sufficiently low to avoid shrinkage from an annular wall?

Table 9-3   Techniques to control gas flow

Technique	Property Alteration	Effect	Benefits, Drawbacks, and Application Range	Reference
Alter Cement Properties	Low fluid loss	Reduce $\overline{V}_{FL}$.	Effective only with relatively low *GFP*.	Cook and Cunningham (1976)
	Thixotropic cement	Reduce t_{tr}.	Viscous, effective only with relatively low *GFP*.	Sutton *et al.* (1984)
	Delayed gel strength	Delay and reduce t_{tr}.	Difficult to control, especially at high temperature. Improved fluid loss control. Effective at high *GFP*.	Sykes and Logan (1987)
	Increase compressibility by gas generation or entrainment.	Reduce effects of $\overline{V}_{FL}$ and $\overline{V}_H$.	Increased treatment complexity. Safety considerations. Non-shrinking. Effective at high *GFP*.	Watters and Sabins (1980)
	Shorten cement column via stage tool or lower cement volume.	Reduce effects of $\overline{V}_{FL}$ and $\overline{V}_H$.		Levine *et al.* (1979)
	Low perm during early hydration	Block gas invasion.	Improved fluid loss control. Non-settling Effective at moderate *GFP*.	Rae *et al.* (1989)
	Stabilize invading gas in fluid cement in annulus using foam-stabilizing surfactants.	Prevent upward flow if gas enters annulus.	Improved rheology and fluid loss control. Difficult to mix due to surface foaming. Effective at moderate *GFP*.	Stewart and Schouten (1988)
Increase Initial Pressure	Increase cement density	Increase Δp.	Formation breakdown. Effective at low *GFP*.	Levine *et al.* (1979)
	Apply back pressure at surface or increase drilling fluid density.	Increase Δp.	Formation breakdown. Effective at low *GFP*.	Levine *et al.* (1979)
Mechanically Block Flow Path	External casing packer	Seal annulus preventing flow path for invading gas.	Seal leaks occur. Gas invasion below packer. Effective at high *GFP*.	Voorman *et al.* (1992)
	Flush formation gas into formation away from bore hole face using high fluid-loss squeeze prior to cementing.	Displace gas with fluid less able to invade annulus.	May damage producing formation. Operationally complex.	Teichrob (1993)

9-4.1.1 Altering Cement Characteristics

Cement characteristics may be altered by interrupting or altering one of the three governing cement characteristics described earlier: gel-strength development, fluid loss, or volume reduction. Several operations have been developed to create a necessary, and often difficult, alteration. These include increasing fluid-loss control, applying pressure to the annulus after cement placement, increasing density of fluids in the hole, adjusting the thickening time of the cement to cause it to set from the bottom of the hole upward, shortening the cement column, using multi-stage cement (mechanical or through cement design), shortening the cement's transition time, increasing the cement's compressibility, or decreasing the cement's internal permeability. All these gas flow control techniques have been used with varying degrees of success.

In addition to the techniques summarized in Table 9-3, three methods of cement system alteration are discussed in more detail in this section. Employing any of these cement system alterations does not guarantee prevention of gas flow, and compositional alterations applied without ensuring a continuous cement sheath around the casing have little chance of preventing gas flow. The underlying assumption on which all compositional alteration techniques are based is that drilling fluid from the annulus must be completely displaced during cement placement.

Alteration of Gel-Strength Development   The time for gel strength to develop to a minimum of $500\,lbf/100\,ft^2$ is critical to gas flow control, and this transition time must be shortened to maintain controlling pressure on the gas bearing formation. This alteration may be performed in one of two ways:

- Accelerate the onset and duration of the gelation by creating a thixotropic cement.

- Delay the onset of gelation until the cement hydration process is sufficiently advanced to move through the transition time rapidly.

Effects of these alterations on maintenance of hydrostatic pressure are illustrated in Figure 9-7. These changes in gel-strength development are not simple or straightforward, especially at application temperatures or pressures. Various thixotropic additives such as gypsum, polymers, or chelants can be added to the cement slurry to induce thixotropy. Gel-strength suppression is more difficult and requires application of one of a series of copolymers to attain a delayed transition time (Sykes and Logan, 1987). The actual mechanism by which these copolymers function is unknown.

Permeability Modification   Several gas-flow control methods are based on the theory that some cementing compositions develop extremely high permeability during hydration, thereby creating a flow path through the cement matrix. Although Sutton and Ravi (1989) demonstrated that intrinsic cement permeability would need to be extremely high in order to initiate gas flow solely through this mechanism, several laboratory studies

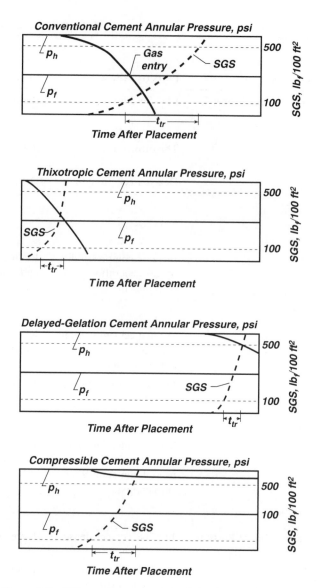

Figure 9-7   Effects of cement slurry behavior modification on pressure maintenance

indicate that intrinsic cement permeability can be a contributor. Also, permeability modifiers function effectively as gas-flow control agents in cement. These additives usually impart other desirable properties to the cement composition in addition to reducing permeability.

Use of stabilized latex polymer emulsions in the cement slurry is a widely-used method of combating gas flow. This method employs excellent cementing practices along with the latex-modified cement slurry to stop gas influx. The latex additive is one of several formulations of copolymer synthesized through emulsion copolymerization to form a highly concentrated emulsion of finely divided polymer in suspension. The addition of latex additive at a concentration of 1 to 3% by weight imparts several desirable properties to the resulting cement including excellent fluid loss control, low rheology, decreased shrinkage, decreased permeability, and resilience. These properties, along with optimum displacement practices, tend to alleviate gas flow as described above. Additionally, Drecq and Parcevaux (1988) report that the drying process of latex polymers in cement, whether by cement hydration or by gas flowing through the annulus, results in precipitation of the polymer into an impermeable film within the cement porosity.

Other materials employed to reduce permeability of cement include silica fume, carbon black (Moroni *et al.,* 1997), silica sols, and other ultra-fine materials to pack into and physically block an invading gas-flow path. An alternative to this permeability alteration method is to make invading gas less mobile by incorporating it into a more viscous medium. Stewart and Schouten (1988) describe laboratory and field evaluations of this technique using foaming surfactants mixed with cement slurry. Invading gas from the formation is stabilized into small, discrete bubbles within the cement slurry by the foaming surfactants. Thus, the gas is trapped downhole in the annulus, and upward flow is arrested.

Physical Gel-strength Disturbance This technique relies on mechanical agitation through casing movement to disturb the cement's gel-strength development. With developing gel strength disturbed, the cement column will continue to function as a fluid and will maintain hydrostatic pressure on gas bearing zones. As the cement hydration progresses to the point of solid formation, stopping the agitation results in accelerated three-dimensional structure development, which traps the formation gas in the formation before significant pressure drop occurs to allow invasion.

Mechanical agitation of the cement column during setting is achieved by casing movement. Sutton and Ravi (1991) describe laboratory and field evaluation of low-rate casing movement during the cement's transition time to destroy gel strength. Another method employs casing vibration using various techniques to disturb the cement's gel-strength development.

9-4.1.2 Mechanical Method

A more basic approach to alleviating gas-flow problems is to concede that formation gas will invade the cemented wellbore and to interrupt the potential flow path of the gas. This technique relies upon mechanical packers (Voorman *et al.,* 1992) set in the annulus to contain gas flow in the lower portions of the well. These packers usually consist of an inner casing element with an internal diameter equal to that of the casing string. The external packer element is basically the same as the elements discussed in Chapter 8. These packers are installed in the casing string to ultimately reside just above potential gas-flow zones, and the cement is placed normally. When the cement is placed and is still in the unset state, the packer element is expanded to form a mechanical seal to gas flow.

An inflatable packer employs a packer element that can be inflated with fluid injected through ports in the inner casing element, so that the elements contact the wellbore wall. The ports can then be sealed, trapping the inflation medium in place. Drilling fluid or cement is usually used as an inflation medium. Cement is employed to produce a solid seal that does not rely on the integrity of the packer element to maintain a seal throughout the life of the well. However, in many cases, the inflation medium is believed to be the cause of external casing packer seal failure. With fluid media such as drilling fluid, seepage of fluid from the packer element that relieves pressure, reduces volume, and opens gas-flow paths is one proposed mechanism for failure. With the use of setting fluids such as Portland cement, hydration volume reduction and shrinkage are blamed for the loss of seal. Mechanisms to combat this loss of seal include increasing the compressibility of the inflation medium or preventing shrinkage caused by the reduction of the hydration volume of cement.

Another mechanical method outlined by Teichrob (1993) proposes pushing formation gas back into the formation with high-fluid-loss squeeze fluid before cementing the casing. When cementing is performed, if pressure drops below formation pressure of the flushed zone, less mobile or non-migrating fluid will be forced into the annulus rather than formation gas.

9-4.1.3 Laboratory Testing Techniques

If the desired method to combat gas flow involves cement modification, the cement composition is usually designed and performance-tested in the laboratory. Many standard testing methods described in Chapter 8 are used in this process to determine a cement's ability to prevent gas flow. Additionally, all standard laboratory testing and design techniques must be used for cements designed for gas-flow application. After all, the cement must do all the things a non-gas-control cement would. However, two quantitative test techniques will be described because they can be used to measure gas control parameters under application conditions.

The first of these techniques, depicted in Figure 9-8, was described by Beirute and Cheung (1990). The device and test methods are designed to evaluate a particular cement composition's ability to contain gas in a formation under specific well conditions. This device simulates a cement column placed against a gas reservoir. Fine screens and regulated back pressures are employed to simulate fluid loss to adjacent formations. Pressure application from the top of the cell through the use of a piston simulates hydrostatic pressure from the cement column above. This pressure application is adjusted as a test progresses to simulate pressure reduction in the column above caused by gelation and hydration volume reduction. Temperatures of the cement are also simulated.

Tests with this pressure apparatus simulate cement fluid loss, gel-strength development, volume reduction, set time, and the presence of formation gas. Results of this test include the incidence or prevention of gas flow along with identification of potential path of the gas.

Pressures, temperatures, and flow rates are monitored during the test, and visual inspection of set cement in the cell indicates the path of gas flow if it occurs. This test apparatus and method provide a fairly realistic means of comparing gas-flow control characteristics of various cement compositions under realistic well conditions. An example output from a test is presented in Figure 9-9.

The laboratory measurement of a cement's gel strength under simulated downhole conditions is necessary to estimate the composition's ability to control gas flow for a specific application. The apparatus depicted in Figure 9-10 is designed to analyze a cement composition's gel-strength behavior under static conditions after simulated placement operations. The apparatus consists of a heated pressure chamber equipped with a rotating paddle driven by a magnetic coupler. The paddle can be rotated at high speeds while the temperature and pressure are elevated to bottom hole circulating temperature (BHCT) to simulate placement. After placement simulation, high-rate paddle rotation is stopped and intermittent ultra-slow rotation (2 to 5 deg/min) is initiated.

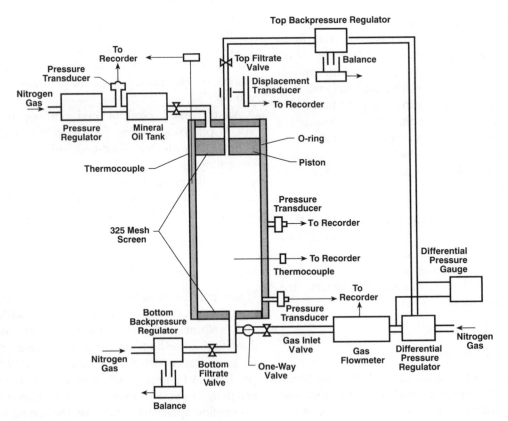

Figure 9-8 Schematic of laboratory device for testing gas-flow characteristics of cement under application conditions (after Beirute and Cheung)

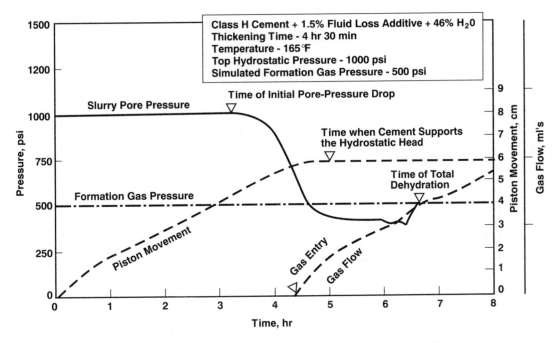

Figure 9-9 Test data from apparatus shown in Figure 9-8

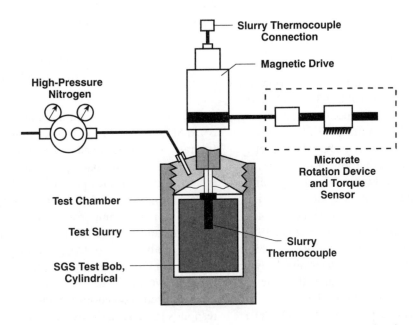

Figure 9-10 Device for measuring static gel-strength development of cements under simulated conditions

This slow rotation prevents disruption of the bulk static gel-strength development while still detecting changes in the cement's static gel strength through torque measurements. As the static cement develops gel strength, the torque required to maintain rotation speed is measured and converted into static gel strength values in $lbf/100\ ft^2$. These tests are conducted until the cement has extended the gel-strength development calculated to prevent gas

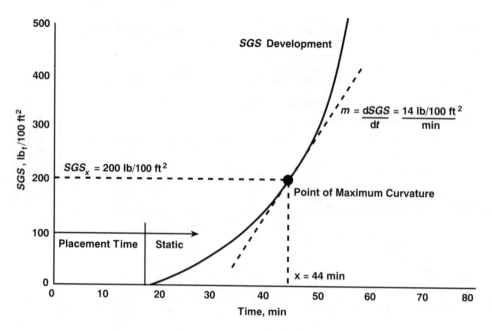

Figure 9-11   Test data from apparatus shown in Figure 9-10

flow. Figure 9-11 presents example output from this testing device.

9-4.1.4 Cement Composition Design

Methods to design cement compositions with properties to prevent gas flow, as outlined in Table 9-3, include standard generalized compositions and specific compositions tailored to a particular well's conditions. Tailoring parameters for a specific application may be based on severity of gas flow and experience, evaluation through specific physical models as depicted in Figure 9-8, or on quantitative design criteria related to the specific application. Several quantitative techniques are in use including calculation of the *slurry performance factor* (Rae *et al.*, 1989) and the *slurry response number* (Sutton and Ravi, 1989).

The slurry performance factor is determined by a cement composition's fluid-loss rate and the rate of hydration under dynamic conditions. The basis for determination of the slurry response number (*SRN*) is gel-strength development, fluid loss, and wellbore volume/area ratios. *SRN* derivation and governing parameters are discussed here along with illustration of the method's use by continuation of the example presented in Table 9-1.

SRN equals static gel-strength increase rate divided by fluid-loss rate where both terms are real-time rates measured during the composition's transition time under simulated downhole conditions. The most rigorous adherence to this method involves incorporating the effects of drilling fluid and spacers. Specifically

$$SRN = \frac{(dSGS/dt)_{max}/SGS_x}{(dl/dt)_x/(V/A)} \tag{9-8}$$

where $(dSGS/dt)_{max}$ is the maximum rate of change of *SGS*, SGS_x is the *SGS* at time of $(dSGS/dt)_{max}$, $(dl/dt)_x$ is the rate of fluid loss velocity at time of SGS_x, V is the volume of annular space per unit length, and A is the area of formation face per unit length.

Each component of this relationship has been documented to play an important role in gas-flow mechanics. The terms are arranged so that increasing *SRN* indicates improved ability to control gas flow. Specific importance of each component is outlined below.

$(dSGS/dt)_{max}$—The maximum rate of gelation increase is considered important because faster rate corresponds to better gas-flow control. As maximum rate of *SGS* development increases, *SRN* increases. Figure 9-11 illustrates the determination of this variable from laboratory gel-strength measurement testing.

SGS_x—The point in the transition time at which rate of *SGS* increase occurs indicates the duration of t_{tr}. Reaching maximum *SGS* rate early in transition time results in shorter t_{tr}. Therefore, *SRN* increases with increasing *SGS*.

dl/dt and *V/A* — Fluid-loss rate at the time of maximum *SGS* increase influences the other important factor in gas-flow control. Lower fluid-loss rates during this critical time result in lower volume reductions. Therefore, a lower value of *dl/dt* results in a higher *SRN*. The volume-to-area ratio also relates to volume reduction. A wider annulus holding a larger volume of cement per unit length will withstand volume losses with less corresponding pressure loss than will a smaller volume of the same cement in a narrower annulus. Thus, *SRN* varies directly with *V/A*.

These parameters can all be determined through a series of laboratory tests that evaluate fluid loss and gel-strength development. Sutton and Ravi (1989) outline a testing procedure for fluid loss. This procedure accounts for the effects of drilling fluid-loss control on cement fluid-loss control since the permeability of drilling-fluid filter cake deposited by the drilling fluid during drilling controls the rate of fluid leaving the cement into the formation. This testing method approximates the downhole fluid loss using standard API fluid-loss test equipment. Thus, no complicated, non-standard equipment is required for the fluid-loss portion of the evaluation. The test procedure for the *dl/dt* determination involves:

- Depositing drilling-fluid filter cake in standard cement fluid loss apparatus for 30 minutes at a differential pressure of 100 psi and BHCT

- Pouring drilling fluid from the apparatus and replacing it with cement slurry to be tested. The drilling-fluid filter cake remains in place

- Collecting fluid loss from the cement slurry for one hour at a differential pressure equal to p_{OB} and BHCT

Thus, the limiting effects of drilling-fluid filter cake and p_{OB} are taken into account in a relatively simple test. The measurement of static gel-strength development is performed using the test device described above with analysis of results illustrated in Figure 9-11.

These approximations are illustrated here and in Table 9-4. The example well conditions listed in Table 9-1 indicate a *GFP* of 6.8, which is considered a moderate problem. The operator elected to combat the problem using a cement with additives to alter the gel-strength development. The cement composition, listed in Table 9-4 also contains fluid loss additive, retarder, silica, and dispersant to create a cement well suited to best placement practices and long term stability.

Fluid loss data from Table 9-4 were used in the following equation to estimate downhole fluid loss:

$$dl/dt = \frac{\overline{V}_{DF\text{-}CEM}}{A t_{test}} \qquad (9\text{-}9)$$

where $\overline{V}_{DF\text{-}CEM}$ is the fluid loss collected for the modified test outlined above, cm^3/t_{test}, A is the API screen surface area (7.1 in.^{2} or 45.8 cm^2), and t_{test} is the test time in minutes.

Results from this analysis indicate that the *SRN* for the cement under application conditions is 185. Harris *et al.* (1990) present an empirical relationship for assessing upper and lower bounds for *SRN* based on *GFP* which indicates that the *SRN* is sufficient to prevent gas flow. The guidelines in Table 9-4 are based on this empirical relationship.

9-4.2 Gas Leakage

Just as long-term gas leakage is more difficult to diagnose than gas migration, preventing this occurrence is more difficult than for gas migration. Long-term gas leakage control relies even more extensively upon best possible cement placement practices and removal of drilling fluid in all its forms from the hole. Preventing this gas leakage requires two things: maximum possible fill of cement in the annulus and prevention of shrinkage of the cement caused by hydration or damage to the cement by well-intervention operations.

Most investigators of gas migration control phenomena emphasize the fact that all control techniques rely on the complete fill of cement in the annulus. No amount of cement modification can affect gas leakage if a channel of non-cementitious material is left during cement placement.

Prevention of shrinkage in cement through expansion was reported as a successful means to prevent gas leakage by Siedel and Greene (1985). This study of in-fill drilling in a field exhibiting long-term gas leakage indicated that expanding cement in conjunction with best cementing practices prevented leakage at the wellhead.

It should be noted here that cement expansion does not occur as it does with a gas pressurizing an elastic containment medium. Volumetric expansion in cement occurs through crystal growth between existing crystals in the cement matrix. This new crystal growth can push an existing crystalline structure apart, resulting in volumetric expansion if the existing cement matrix is not too strong. If the existing matrix is bound sufficiently to resist expansion, the result is that new crystals are crushed, reducing the porosity but not affecting the bulk volume. Thus, the crystallization reaction designed

Table 9-4 Evaluation of gas-flow control effectiveness: Calculation of *SRN* example continued from Table 9-1.

Cement Composition:	API Class G cement + 35% coarse silica + 10.3% weighting material + 3% KCl + 0.5% gel-strength modifier + 0.3% dispersant + 0.5% retarder (density = 17.5 lb/gal, yield = 1.38 ft³/sk).
API drilling-fluid loss	= 2.4 cm³/30 min
API cement fluid loss	= 96 cm³/30 min
Combination fluid loss	$\overline{V}_{DF\text{-}CEM}$ = 2 cm³/60 min
SGS (Figure 9-11)	
Zero Gel Time	= 20 min
t_{tr}	= 18 min
For 5 in casing in 6.75 in hole:	V/A = 1.93 cm
dl/dt	$= \dfrac{2 \text{ cm}^3/60 \text{ min}}{45.8 \text{ cm}^2} = 0.00073 \text{ cm/min}$
$(dSGS/dt)_{max}$ (from Fig 9-11)	= 14 lbf/100 ft²/min
SGS @ 44 min	= 200 lbf/100 ft²
SRN	$= \dfrac{1 \text{ lb/100 ft}^2/\min/200 \text{ lbf/100 ft}^2}{0.00073 \text{ cm/min}/1.93 \text{cm}}$
	= 185
SRN requirements	
For GFP = 0 − 3:	SRN range = 70–170
For GFP = 3 − 8:	SRN range = 170–230
For GFP = 8+:	SRN range = 230+

to create expansion must be timed to occur when cement is sufficiently weak to allow intercrystalline growth. If the reaction occurs too early, the cement is fluid, and volume expansion is not maintained. If the cement is too strong, new crystal growth cannot force movement in the matrix. This reaction is sensitive to specific cement batch, other additives in the formulation, and applications. Performance can only be tested with direct laboratory measurement.

The following description of well conditions continues the scenario begun in Section 9-3.2 and illustrates the focus on proper cementing practices and expansive additives to combat shrinkage when designing a cement for gas-leakage control. Because of the high permeability of shallow sandstone reservoirs and moderate fluid loss control in the drilling-fluid system, drilling-fluid filter cake build-up on the formation face is expected to be severe unless the drilling-fluid filtrate loss is controlled. When a filter cake is placed it is very difficult, if not impossible, to remove with fluid circulation alone. Once the cement slurry is deposited, the gas flow is controlled because of hydrostatic pressure and low fluid loss from the cement slurry, but after the cement has set, the drilling-fluid filter cake begins to shrink and crack because of the dry gas environment. This shrinking and cracking allows gas to migrate through the filter cake until it surfaces or reaches a formation with lower pressure. The drilling-fluid type and properties should be modified to result in a low filtrate loss value and a thin, tight filter cake. Expansive additives or additives which increase slurry compressibility may be added to the cement slurry to compensate for the plastic state shrinkage that occurs as cement slurries set. Prevention should always be considered the best solution. Remedial repair will be very difficult in most cases because of the inability to remove the dehydrated cake in the channel before attempting to pump slurry through the channel. When possible, fluid production should be allowed through the channel until the channel is clean and void of drilling fluid.

9-5 EVALUATION OF EFFECTIVENESS

Effectiveness of techniques to prevent gas flow to the surface or long-term gas leakage is simple to evaluate. If no gas appears at the surface, the treatment was successful. To evaluate the success of preventing interzonal communication, various acoustic or temperature logs, as discussed in Chapter 10, may be used. Analytical techniques may enhance the interpretation of ultrasonic logs to quantify the degree of gas invasion (Douglas and Uswak, 1991). These extended techniques can help determine the best remedial approach as discussed in Chapter 11.

9-6 WATER MIGRATION

Gas is not the only fluid known to migrate through a cemented annulus. Water flows following cementing are also encountered during construction of some wells. However, the methods used to control water flows are governed by the same mechanisms outlined for gas.

Water flows have been encountered in various locations, but these are much less prevalent than gas flows except in wells drilled in fields undergoing waterflood operations. Mitchell and Salvo (1990) describe a field evaluation of techniques to control water flows in one such field. The method chosen to control the water involved pressure application at the surface immediately after cement placement. In this application, water flow was successfully prevented.

Drilling operations in deep water are hindered by water flow occurring after cementing of conductor casings (Griffith and Faul, 1997). These conductor installations encounter high-pressured water approximately 1000 ft below the mud line in formations with inherently low fracture gradients. Solutions to this problem include the use of low-density cements designed with modified gel-strength characteristics or with increased compressibility from the addition of foam. Other operational improvements include the use of reactive sweeps ahead of the cement to aid in the removal of drilling fluid or in sealing formations, .

9-7 SUMMARY

Fluid migration after cementing is a complex problem that is manifest in several ways. However, the underlying forces governing this migration are fairly simple: formation fluid will migrate through a cemented annulus if positive pressure differential and flow paths exist. Preventing the occurrence requires maintenance of controlling pressure or closing flow paths.

Underlying all the prevention techniques is the requirement of using optimum cement placement practices. No control can occur if the annulus is not filled with cement designed to act as the fluid migration control medium.

Preventing the migration during primary cementing is highly desirable and effective because of the severity of fluid migration occurrences and the difficulty of correcting a problem once it occurs.

REFERENCES

Beirute, R.M., and Cheung, P.R.: "A Method for Selection of Cement Recipes to Control Fluid Invasion after Cementing," *SPEPE* (Nov. 1990) 443–440.

Beirute, R.M., and Cheung, P.R.: "A Scale-Down Laboratory Test Procedure for Tailoring to Specific Well Conditions: The Selection of Cement Recipes to Control Formation Fluids Migration After Cementing," paper SPE 19522, 1989.

Beirute, R.M., Wilson, M.A., and Sabins, F.L.: "Attenuation of Casing Cemented with Conventional and Expanding Cements Across Heavy Oil and Sandstone Formations," paper SPE 18027, 1988.

Bol, G.M., Bosma, M.G.R., Reijrink, P.T.M., and van Vliet, J.P.M.: "Cementing: How to Achieve Zonal Isolation," 1997 Offshore Mediterranean Conference, 1997.

Chenevert, M.E., and Shrestha, B.K.: "Chemical Shrinkage Properties of Oilfield Cements," *SPEDE* (Mar. 1991) 37–43.

Cheung, P.R., and Beirute, R.M.: "Gas Flow in Cements," *JPT* (June 1985) 1041–1048.

Christian, W.W., Chatterji, J., and Ostroot, G.W.: "Gas Leakage in Primary Cementing - A Field Study and Laboratory Investigation," paper SPE 5517, 1975.

Cook, C., and Cunningham, W.: "Filtrate Control—A Key to Successful Cementing Practices," paper SPE 5898, 1976.

Douglass, R.F., and Uswak, G.: "Detection of Gas Migration Behind Casing Using Ultrasonic Imaging Methods," paper CIM/AOSTRA 91-39, 1991.

Drecq, P., and Parcevaux, P.A.: "A Single Technique Solves Gas Migration Problems Across a Wide Range of Conditions," paper SPE 17629, 1988.

Goodwin, K.J., and Crook, R.J.: "Cement Sheath Stress Failure," paper SPE 20453, 1990.

Griffith, J.E., and Faul, R.: "Cementing the Conductor Casing Annulus in an Overpressured Water Formation," paper OTC 8304, 1997.

Harris, K.L., Ravi, K.M., King, D.S., Wilkinson, J.G., and Faul, R.R.: "Verification of Slurry Response Number Evaluation Method for Gas Migration Control," paper SPE 20450, 1990.

Jones, P.H., and Berdine, D.: "Factors Influencing Bond Between Cement and Formation," *API Drilling and Production Practice* (1940).

Levine, D.C., Thomas, E.W., Bezner, H.P., and Tolle, G.C.: "Annular Gas Flow After Cementing: A Look at Practical Solutions," paper SPE 8225, 1979.

Mitchell, R.K., and Salvo, G.S.: "A Procedure To Shut Off High-Pressure Waterflows During Primary Cementing," paper SPE 19994, 1990.

Moroni, N., Calloni, G., and Marcotullio, A.: "Gas Impermeable Carbon Black Cements: Analysis of Field Performances," 1997 Offshore Mediterranean Conference.

Parcevaux, P.A., and Sault, P.H.: "Cement Shrinkage and Elasticity: A New Approach for a Good Zonal Isolation," paper SPE 13176, 1984.

Prohaska, M., Fruhwirth, R., and Economides, M.J.: "Modeling Early-Time Gas Migration Through Cement Slurries," *SPEDC* (Sept. 1995) 178–185.

Rae, P., Wilkins, D., and Free, D.: "A New Approach to the Prediction of Gas Flow after Cementing," paper SPE 18622, 1989.

Sabins, F.L., Tinsley, J.M., and Sutton, D.L.: "Transition time of Cement Slurries Between the Fluid and Set State," paper SPE 9285, 1980.

Seidel, F.A., and Greene, T.G.: "Use of Expanding Cement Improves Bonding and Aids in Eliminating Annular Gas Migration in Hobbs Grayburg-San Andres Wells," paper SPE 14434, 1985.

Stewart, R.B., and Schouten, F.C.: "Gas Invasion and Migration in Cemented Annuli: Causes and Cures," *SPEDE* (1988) 77–83.

Sutton, D.L., and Ravi, K.M.: "Low-rate Pipe Movement During Cement Gelation to Control Gas Migration and Improve Cement Bond," paper SPE 22776, 1991.

Sutton, D.L., and Ravi, K.M.: "New Method for Determining Downhole Properties that Affect Gas Migration and Annular Sealing," paper SPE 19520, 1989.

Sutton, D.L., Sabins, F.L., and Faul, R.: "Preventing Annular Gas Flow—Two Parts," *Oil and Gas Journal* (Dec. 1984) 84-92 109–112.

Sykes, R.L., and Logan, J.L.: "New Technology in Gas Migration Control," paper SPE 16653, 1987.

Teichrob, R.R.: "Controlling Gas Migration During Primary Cementing Operations Using the Flushed Zone Theory," paper CADE/CAODC 93-603, 1993.

Voorman, D., Garrett, J., Badalamenti, A., and Duell, A.: "Packer Collar Stops Gas Migration Mechanically," *Petroleum Engineer International* (Apr. 1992) 18–22.

Watters, L.T., and Sabins, F.L.: "Field Evaluation of Method to Control Gas Flow Following Cementing," paper SPE 9287, 1980.

10 Cement-Sheath Evaluation

John W. Minear
Halliburton Energy Services

K. Joe Goodwin
Mobil Exploration and Production

10-1 INTRODUCTION

The primary functions of the cement sheath placed between the casing and the formation are to support the casing and to seal the annulus to the flow of fluid (liquid or gas). The objective of cement-sheath evaluation is to quantify the ability of the cement annulus to provide these functions. Evaluating the hydraulic sealing function of the cement annulus is the most important and the most difficult part of cement evaluation. In essence, this process involves detecting existing or potential fluid migration pathways that may exist between the casing and the formation shortly after the cement has been emplaced. These pathways are formed by material that may be movable under differential pressures between formation intervals (Chapter 9). Specific examples are a cement sheath that is completely absent, channels in the cement running axially along the borehole, and gaps between the cement and the casing or formation. Devoid of cement, these pathways are filled with water, gas, mudcake, drilling fluid gel, or formation material. It is useful to know their size, vertical extent, circumferential location, and the material with which they are filled. Ideally, it would be desirable to obtain a scan of the cement sheath analogous to the ultrasonic scans that are used to image the human body. At present, no instrument exists to perform this desired imaging. However, several technologies can provide measurements that can be interpreted in terms of the hydraulic sealing potential of the cement sheath.

This chapter presents basic measurement techniques used for cement-sheath evaluation. These include acoustic, nuclear, and temperature measurements. A brief historical synopsis of the techniques is presented in Section 10-2. Acoustic techniques are discussed in Section 10-3.

Basic physics of the measurements, calculation of the quantities used in interpretation, and strengths and limitations of each measurement are discussed. Non-acoustic techniques are discussed in Section 10-4. Logging recommendations and interpretation guidelines for cement evaluation are presented in Section 10-5. A final log example illustrating the combined use of two techniques is discussed in Section 10-6. Measurement limitations are discussed in Section 10-7. Examples are presented throughout the chapter.

10-2 HISTORICAL REVIEW

The first evaluation of the cement sheath used a temperature sensor to measure the top of the cement (Leonardon, 1936). The technique of doping cement with radioactive material, Carnotite, and then using gamma ray measurements to measure cement tops was reported by Kline *et al.* (1986). Neither of these techniques provides any information about the distribution of cement behind the casing.

A sonic method for evaluating the quality of the cement annulus was first reported by Grosmangin *et al.* (1961). The technique consisted of measuring, at a fixed distance from the source, the amplitude of an elastic wave traveling down the casing. This technique, known as the *cement bond log* or CBL, was initially used mainly for finding cement tops and for making qualitative estimates of the isolation of porous zones. Results of full-scale model experiments reported in that paper provided a relation between wave amplitude and the percentage of the circumference of the casing acoustically coupled to the cement; amplitude decreases with increasing percentage. It was also observed that the amplitude decreases with increasing cement compressive strength. Pardue *et al.* (1963) reported about a study of cement and casing variables affecting cement bond logs. An important result in this work was theoretical analysis based on symmetric plate-wave propagation (Lamb, 1917), which showed that attenuation of casing-borne waves (symmetrical plate waves) was strongly affected by the shear modulus of the cement. The concept of a bond index was also introduced. This index has been used for estimating the percentage of the casing circumference bonded to cement. Basic information obtainable from cement bond logs and calibration, operational, and interpretation problems have been considered by many authors (e.g. Winn *et al.*, 1962; Bade, 1963; McNeely, 1973; Fertl *et al.*, 1974; Fitzgerald *et al.*, 1985; Pilkington, 1988; Jutten and Corrigall, 1989).

Calibrating measured amplitudes is a major problem with single-transmitter, single-receiver cement bond tools. Computational techniques to account for some of the factors affecting amplitude (casing size, bore-fluid attenuation, and eccentering) have been developed (Nayeh *et al.*, 1984). Golwitzer and Masson (1982) reported on an improved technique for measuring attenuation effects of casing alone. This technique uses two transmitters with two receivers located symmetrically between the transmitters. As a result of this mechanical arrangement, attenuation of the casing wave can be directly measured without the effects of tool eccentering, fluid attenuation, transmitter or receiver sensitivity, or temperature drift.

The development of other techniques was motivated by the lack of circumferential resolution and the fact that a microannulus of no cement next to the casing has a detrimental effect on cement bond logs. One of these new techniques was reported by Froelich *et al.* (1981) and Havira (1982). This technique is based on the excitation of the thickness resonance of the casing. The rate of decay of the resonant energy with time can be interpreted in terms of the presence or absence of cement next to the casing. Resonant frequency can also be used for estimating casing thickness. The embodiment of this technique as reported by Froelich *et al.*, consists of eight acoustic pulse-echo transducers arranged in a helix around a logging sonde. As a result of the multiple transducers, information on the circumferential distribution of cement close to the casing can be obtained. However, information on the cement-to-formation bond provided by the waveform displays common to cement bond log presentations is not available. Because of the different types of elastic waves used in the pulse-echo technique, the measurement is less sensitive to a thin microannulus than is the CBL.

A logical extension of the pulse-echo technique using eight transducers is the use of one transducer on a rotating head—in essence, a cased-hole adaptation of the borehole televiewer (Zemanek *et al.*, 1969). This approach provides greater circumferential resolution and eliminates the necessity for calibrated transducers.

Several variations of transducers have been made in an effort to provide measurements of circumferential variation of cement-to-casing bond with basically conventional cement bond tools. These variations are based on sectoring the cylindrically omnidirectional transmitters and/or receivers used in conventional cement bond tools. As a result, casing-wave amplitude measurements can be made over azimuthal sectors of the casing and cement annular volume. Although they potentially provide an improved cement bond log, these techniques do not provide the circumferential resolution of a rotating pulse-echo transducer technique. They have the same inherent microannulus problems as omnidirectional cement bond logs.

The most recently introduced cement-sheath evaluation technique uses pad-mounted arrays of transducers that measure the attenuation of the casing waves in a manner analogous to the one used in the two-transmitter version of the cement bond technique (Lester, 1989). Multiple pads helically arranged on the tool provide compensated attenuation measurements around the circumference of the casing. Spacings between transmitters and receivers are smaller (a few inches) and the operating frequency is higher (300 kHz) than CBL tools, but in principle, the measurement physics is the same.

Techniques that measure the azimuthal variation of gamma-ray radiation around the casing have been applied to cement evaluation (Simpson and Gadeken, 1993; Wyatt *et al.* 1996). Standard neutron and acoustic logging techniques and oriented density logs have been used for evaluating the cement sheath around multiple tubing strings (Pennebaker, 1972).

Papers by Pilkington (1988) and Fitzgerald *et al.* (1985) have excellent reference lists on cement-sheath evaluation. The API Technical Report on Cement Sheath Evaluation by Goodwin *et al.* (1994) contains tool-specific information, operating procedures, and interpretation recommendations.

10-3 ACOUSTIC TECHNIQUES FOR CEMENT-SHEATH EVALUATION

Acoustic techniques are used more than other techniques for evaluating cement sheath quality because they allow a reasonably direct interpretation of the basic measurement in terms of casing-to-cement bond, and they provide the highest spatial resolution measurements. It should be noted, however, that none of the acoustic techniques currently available directly measure the presence or absence of cement. This information must be inferred from the actual measurement made by a particular acoustic technique. Acoustic techniques can be divided into two categories based on the type of elastic waves used: those that use plate waves and those that use reflected waves.

10-3.1 Plate-Wave Techniques

10-3.1.1 Large-Scale Techniques

Plate Waves

Several cement evaluation techniques are based on the use of cylindrical plate waves (generally referred to as casing waves) propagating along the casing. These waves are generated by the incidence of compressional waves on the casing as depicted in Figure 10-1. A cylindrically symmetric source produces a short pressure pulse in the bore fluid between the tool and the casing. Propagation of this pressure pulse can best be described with ray theory. Figure 10-1 shows a typical CBL tool configuration. Ray paths for refracted and plate waves are shown schematically. The path of refracted waves from the formation is indicated by the ray ABCDEFG. Plate-wave ray paths are shown on the left. They are generated by the compressional wave TH exciting the steel casing. For simplicity, propagation of two-dimensional plate waves (Lamb, 1917) will be used for discussion. Energy leaving the source travels outward in a cone subtended by the source and strikes the casing wall at incident angles ranging from 0° to 180°. Because of the radiation pattern of the source, most energy is incident from 0° to about 45°. Energy striking the wall at the proper angle excites symmetric plate waves (Lamb, 1917), which travel along the plate (Figure 10-2).

At the midpoint of the steel layer, particles vibrate longitudinally (compressional wave motion). At the boundaries of the steel, particles vibrate in elliptical paths. Therefore, the particle motion has both compressional and shear components. The formulas for compressional (V_c), shear (V_s), and plate-wave (V_{pl}) velocities are given where K is bulk modulus, μ is shear modulus, and ρ is bulk density. The proper angle allows waves propagating inside the plate to interfere constructively at each

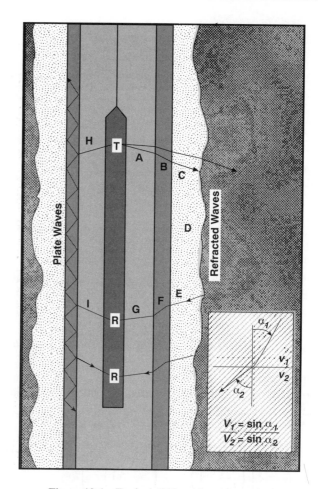

Figure 10-1   Typical CBL tool configuration

reflection from the plate boundary; this angle is represented by the zigzag ray path in Figure 10-1.

Plate waves attenuate as they propagate along a plate because energy is continually lost to the surrounding media. This attenuation depends on the wave coupling and impedance contrast between the casing and surrounding media. Particle motion at the plate surface consists of both compressional wave longitudinal motion and shear-wave transverse motion, so that the resulting motion is elliptical, as illustrated in Figure 10-2. Thus, both shear and compressional wave energy can be coupled into the surrounding media. If the surrounding media cannot propagate shear waves (fluids), then shear-wave energy will not be transferred out of the plate. Thus, attenuation of the plate wave will be decreased.

In the case of a plate surrounded by a vacuum, no energy is transferred to the vacuum because neither shear nor compressional-wave motion can be coupled into the vacuum (Ewing *et al.*, 1957). If the plate is

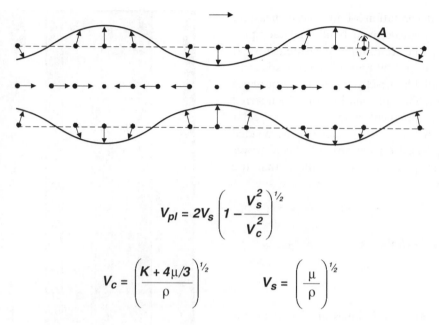

$$V_{pl} = 2V_s\left(1 - \frac{V_s^2}{V_c^2}\right)^{1/2}$$

$$V_c = \left(\frac{K + 4\mu/3}{\rho}\right)^{1/2} \qquad V_s = \left(\frac{\mu}{\rho}\right)^{1/2}$$

Figure 10-2   Symmetric plate-wave particle motion

bounded by infinite half spaces of water and solid, attenuation is approximated by

$$\alpha = \frac{52.2\rho_2/\rho_1}{\left[\left(\dfrac{V_{pl}^2}{V_{c2}^2} - 1\right)^{-1/2} + \left(\dfrac{V_{pl}^2}{V_{s2}^2} - 1\right)^{1/2}\right]h} \qquad (10\text{-}1)$$

where ρ_2 is the density of the solid, ρ_1 is the density of the plate, V_{c2} is the compressional velocity in the solid, V_{s2} is the shear-wave velocity in the solid, V_{pl} is plate velocity (for a plate bounded by water) given by

$$V_{pl} = 2V_{s1}\left(1 - \frac{V_{s1}^2}{V_{c1}^2}\right)^{1/2}$$

and h is the plate thickness. With densities in gm/cm³, velocities in ft/sec, and thickness in in., attenuation is in dB/ft (Pardue *et al.*, 1963). Using water and 16-lb/gal Class H cement parameters, the attenuation is 13 dB/ft. Water and lightweight (9-lb/gal) cement parameters give an attenuation of 10 dB/ft. For comparison, the attenuation of a plate bounded by water is less than 1 dB/ft. Note that attenuation is dependent on plate thickness. For plates bounded by finite thickness materials, attenuation also depends on the material thickness for thicknesses less than about 0.75 in. At greater thicknesses, the material behaves as an infinite half space.

The condition that produces the very large difference in energy loss between cement and water boundary layers is the shear stress boundary condition between the plate and the bounding layers. If shear stress can be transferred across the boundary, shear energy can be radiated to the boundary layer. Shear stress cannot be transferred across a fluid boundary layer, even a very thin one, so only compressional wave energy can be lost. Because of the large impedance contrast between steel and water, very little compressional wave energy can be lost. Therefore, a fluid boundary layer even a few microns thick effectively isolates the plate from the bounding layer.

Intrinsic attenuation of the plate produces some attenuation in addition to that caused by radiation into the boundary layers. This effect is caused by the elastic energy of vibration being converted into heat. This attenuation is frequency-dependent and is about 0.1 dB/ft at the frequencies used in cement-sheath evaluation with plate waves.

Refracted Waves

In addition to exciting plate waves, the energy generated by the source is also transmitted through the casing and into the formation. A ray leaving the source is reflected and refracted at each acoustic interface. Pressure waves incident on the casing are refracted according to Snell's law

$$V_1/\sin\alpha_1 = V_2/\sin\alpha_2 \qquad (10\text{-}2)$$

where V_1 and V_2 are the compressional wave speeds in Medium 1 and Medium 2, and α_1 and α_2 are the incident and refracted angles shown in Figure 10-1. If α_2 is 90°, then the wave propagates along the boundary between Layers 1 and 2. The incident angle for which the angle of refraction is 90° is termed the angle of critical refraction.

As waves propagate along the boundary between two media, energy is continually radiated away from the boundary at the critical angle. The wave propagating along the cement-to-formation interface in Figure 10-1 radiates energy along raypath E. This ray is refracted through the steel and water back to the receiver along path FG.

Waveform Data

Recorded waveforms provide the basic data for evaluating the quality of logging measurements based on plate-wave attenuation. Waveform data are the time series of pressure variations caused by the radiation of elastic waves refracted from the casing, cement, and formation into the bore fluid. These data can be displayed as variable density, wiggle, shaded wiggle, or combinations of wiggle and variable density, as shown in Figure 10-3. In this figure, the vertical axis is depth and the horizontal axis is time, increasing to the right. The leftmost panel, which is called a wiggle plot, shows wave amplitude vs. time. The time at which the transmitter fires is the start of each waveform. The time for the first wave to arrive at the receiver corresponds to the flat section at the beginning of each waveform. Two types of waves are shown. The high-frequency waves arriving at a constant time are the plate waves traveling through the casing. The lower-frequency waves showing the curved arrival time are the refracted compressional formation arrivals. Their arrival time varies because the formation velocity varies.

Note that at the top, formation arrivals precede the casing arrivals, indicating that the formation velocity is faster than the plate-wave velocity in the casing. Note the interference of casing and formation arrivals after the formation arrivals. The next two panels show variable-density presentations. Positive amplitudes are gray-scale coded and negative amplitudes are shown as white. The second panel is not scaled properly, so all positive amplitudes appear black; all amplitude information is lost. This problem is common in black-and-white, variable-density presentations. The fourth panel shows a combination of wiggle and variable-density presentations.

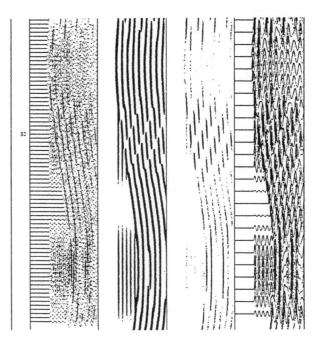

Figure 10-3 Graphical presentations of waveforms recorded by CBL tools

Wiggle or shaded-wiggle presentations generally provide the most information. Variable-density presentations tend to obscure amplitude variations. However, color-coded variable-density presentations, if coded properly, preserve amplitude information and provide the greatest density of waveform information.

Waveforms represent the raw data from which transit time and amplitude are computed and provide a valuable check and aid in the interpretation of these curves (Section 10-5). Waveform data are usually recorded by a receiver located 5 ft from the transmitter, whereas amplitude and transit-time measurements are made at a distance of 3 ft. Acoustic parameter values for common fluids and rocks are included in Table 10-1.

Measurement of Amplitude, Attenuation, and Transit Time

Amplitude and Attenuation

Two basic measurements are made with cement bond-log tools: casing-wave attenuation and transit time. Amplitude is measured at one receiver for CBL tools and at two receivers for attenuation tools. The near receiver, is generally located 3 ft from the transmitter; the far receiver is generally 5 ft from the transmitter.

Table 10-1 Acoustic parameter values

Material	V_p (km/s)	V_s (km/s)	ρ_b (gm/cc)	Z (MRayls)
Fresh water	1.52	0	1.00	1.52
Salt water (200 Kppm)	1.74	0	1.14	1.98
Diesel oil	1.25	0	0.80	1.00
Mineral oil	1.45	0	0.83	1.2
Free gas (mostly methane)	0.38	0	0.001	0.1
Water-based drilling fluid (8 lb/gal)	1.44	0	0.96	1.38
Water-based drilling fluid (16 lb/gal)	1.40	0	1.92	2.69
Oil-based drilling fluid (8 lb/gal)	1.34	0	0.96	1.29
Oil-based drilling fluid (16 lb/gal)	1.20	0	1.92	2.30
10% Porosity sandstone	4.66	2.91	2.49	11.60
30% Unconsolidated sands	3.31	1.94	2.16	6.42
10% Porosity limestone	4.91	2.73	2.54	12.47
10% Porosity dolomite	5.24	3.06	2.68	14.04
Class H cement (12 lb/gal)	3.1	1.8	1.55	4.8
Class H cement (16.6 lb/gal)	3.20	1.90	1.94	6.21
Lightweight cement (9 lb/gal)	3.10	1.80	1.55	4.81
Steel	5.90	3.23	7.70	45.43

If casing-wave amplitude is measured at a distance x_1 from the source, attenuation over distance x is

$$\alpha = \frac{20}{x_1} \log \frac{A_1}{A_0} \qquad (10\text{-}3)$$

where A_1 is the measured amplitude at x_l and A_0 is the amplitude at the source. For reference, an attenuation of 6 dB/ft means that the amplitude decreases by one-half when the wave travels a distance of 1 ft. Figure 10-4A depicts the measurement of casing-wave amplitude. The upper figure illustrates the detection of the first arrival. If the first peak is above the specified detection threshold, it will be detected. If its amplitude is decreased for some reason, the first arrival could be missed and a later peak will be detected. This effect is known as *cycle skipping*. Also note that even if the first peak is detected, the exact time at which amplitude exceeds the threshold depends on the peak amplitude. A time gate is specified over which to search for the beginning of the casing-wave arrival. When the first positive peak of the casing wave exceeds a preset threshold level, the amplitude of the peak, E_1, is measured as shown in Figure 10-4.

Eccentering of the tool produces variations in casing-wave amplitude because of the averaging of waveforms around the casing. Waveforms from opposite sides of the casing are shown for centered and eccentered tools in Figure 10-4B. When eccentering occurs, the waves from the side of the tool closest to the casing (TCDR1) will arrive earlier than those from the side of the tool farthest from the casing (TABR1), which results in an early detec-

tion point. Amplitude and transit time of the eccentered waveforms from opposite sides of the bore are different because of different path lengths in the bore fluid. Consequently, the amplitude of the casing wave averaged around the bore will vary because of eccentering.

A problem with measuring amplitude at only one distance from the transmitter is that source strength, coupling of the pressure wave energy into the casing, and bore fluid attenuation may vary with depth. Although techniques have been developed to correct for drilling-fluid attenuation (Nayeh *et al.*, 1984) accurate casing-wave attenuation measurements require measurements made between two points with two transmitters. Golwitzer and Masson (1982) describe a technique that can make such a compensated attenuation measurement (Figure 10-5). A_{ij} is the amplitude of the casing wave generated by transmitter i and measured at receiver j. A_{ij} is given by

$$A_{ij} = P_i S_j 10^{-\frac{\alpha d}{20}} \qquad (10\text{-}4)$$

where P_i is the pressure amplitude of source i, S_j is the receiver sensitivity of receiver j, and d_j is the source-receiver distance of receiver j. The use of similar expressions for both transmitters and receivers yields for the attenuation α

$$\alpha = \frac{-10 \log (A_{12} A_{21})}{(d_2 - d_1)(A_{11} A_{22})} \qquad (10\text{-}5)$$

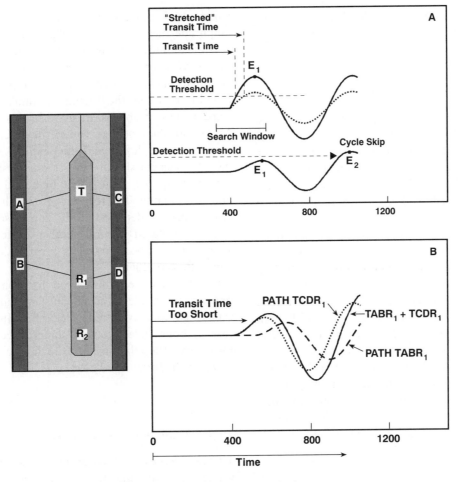

Figure 10-4 Amplitude and transit-time measurements

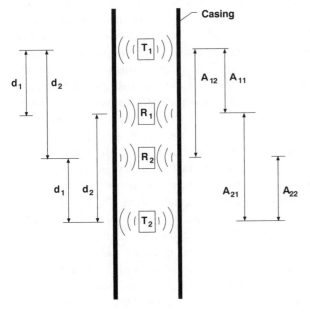

Figure 10-5 Compensated attenuation measurement technique (after Golwitzer and Masson, 1982)

Note that transmitter and receiver effects have been removed and variations in bore-fluid attenuation (included in P and S) have canceled out. Attenuation determined with this technique is not affected by drilling-fluid attenuation, temperature effects, or receiver and source variations. It is also less sensitive to eccentering because, as discussed above, eccentering causes amplitude variation of the casing wave. If this variation occurs in amplitudes at both receivers, the effect will cancel if the eccentering is the same for all transmitters and receivers (the tool is not tilted in the casing). Because attenuation is measured between receivers, vertical resolution is greater than in a CBL measurement.

Sources and receivers for plate-wave measurements have been designed with either cylindrically symmetric or azimuthally directional radiation patterns. Cylindrically symmetrical transducers provide measurements of the average characteristics of the annular volume of cement sheath between transmitter and receiver. Azimuthally directional or sectored transmitters and/or receivers (Schmidt, 1989) provide some information on the azimuthal variation of cement-sheath properties.

Transit Time

Transit time is affected by the lengths of ray paths A, I, G, and H (Figure 10-1), which are all affected by the position (centering) of the tool in the well. Transit time is also affected by the amplitude of the first casing arrival. Variations in amplitude can stretch or compress transit time by as much as $15\,\mu s$. Transit time is used as an indication of tool centralization. Generally, transit time is measured as the time at which the amplitude exceeds a specified threshold, as shown in Figure 10-4. However, other measurement points can be used, such as the time of the peak or the zero-crossing time following the peak. Eccentering always causes the value of transit time to decrease from that of a centralized tool. Therefore, considerable effort has been made to design tools that produce stable transit-time measurements.

If the amplitude of the first casing-wave peak is lower than the threshold, the second or later peak may trigger the threshold indicator. This triggering results in a jump in transit time equal to the period (or multiple periods) of the waveform. This phenomenon is referred to as cycle skipping.

High-Velocity Formations

Some formations have velocities greater than steel plate-wave velocity (17,740 ft/sec, 5.41 km/s). In this case, refracted formation compressional waves may arrive before the casing wave. Hence, measurements of transit time and amplitude will be made on the compressional wave instead of the casing wave. Tools with a receiver placed very close to the transmitter have been developed. If transmitter-to-receiver offset is small enough, the casing wave will arrive before the formation wave since the formation wave must travel a longer path than the casing wave. The waveform display is the best indicator of a fast formation arrival (Figures 10-3 and 10-7).

Bond Index and Cement Compressive Strength

The bond index, originally suggested by Grosmangin *et al.* (1961), represents an attempt to quantify the amount of annulus filled by cement. It is given by

$$BI = (A_{fp} - A)/(A_{fp} - A_{100}) \qquad (10\text{-}6)$$

where A_{fp} is amplitude measured in free pipe, A_{100} is amplitude measured in a 100% cemented section, and A is the amplitude measured at some depth. By the definition, BI is zero when $A = A_{fp}$ and one when $A = A_{100}$. Attenuation can be used instead of amplitude. BI is supposed to represent the fraction of annular space filled by cement, but problems exist with this interpretation. Many variables besides volumetric fraction of cement may affect amplitude and attenuation. These variables include cement-sheath thickness, casing thickness, cement properties that affect bulk and shear moduli, and microannuli. The relation of BI to hydraulic seal is also poorly understood. Still, zonal isolation criteria have been based on the number of feet over which some form of bond index is greater than a specified value (Pickett, 1966).

Charts have been developed for converting single-receiver amplitude measurements into attenuation and cement compressive strength (Pardue, 1963). As in the case of bond index, cement compressive strength may have very little relationship to hydraulic seal (Goodwin *et al.*, 1994; Jutten *et al.*, 1987).

10-3.1.2 Small-Scale Techniques

The most recent development in techniques that use plate waves for evaluating the cement sheath is essentially a small-scale version of the previously described compensated attenuation technique (Lester, 1989). Ultrasonic transmitters and receivers are mounted on pads that contact the casing wall in the configuration shown in Figure 10-6. Solid lines represent an actual pad arrangement on flattened casing. Pads actually form a helix around the inside of the casing so that the pad with transducers R_6 and T_6 is next to the pad with transducers T_1 and R_1. Dotted lines represent virtual pads to show the six compensated attenuation measurement paths (solid lines) around the casing. In effect, each of the attenuation measurements uses the same geometry as the compensated attenuation measurement shown in Figure 10-5. For example, the measurement using T_1-R_2-R_3-T_4 is the same geometry as that in Figure 10-5. The lower part of Figure 10-6 indicates the amplitudes measured between transducers on different pads. Transmitter-to-receiver distance is about 6 in., and the transmitter operating frequency is about 100 kHz. Attenuation is determined as in the case of the two-transmitter, two-receiver CBL tool. The equation for calculating attenuation is the same as Equation 10-5.

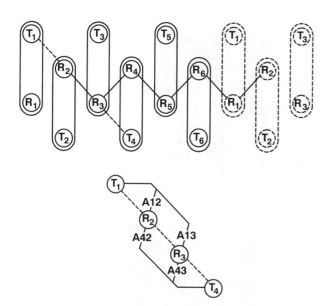

Figure 10-6 Pad arrangement for small-scale plate-wave attenuation measurement (after Lester, 1989)

Because multiple pads are used, two advantages of this technique are azimuthal information and very low sensitivity to drilling fluid effects and eccentering. Attenuation estimation is improved over that of a one-transmitter CBL for the same reasons as discussed for the two-transmitter CBL technique.

Wavelength-to-casing thickness determines the propagation effects of the casing on the elastic waves traveling in the casing. Attenuation is frequency dependent because the dissipation of energy by a propagating wave is determined by the number of cycles of vibration per length. Wavelength-to-casing thickness and wavelength-to-transmitter-to-receiver offset are given in Table 10-2. In both small- and large-scale measurement techniques, λ/casing thickness is greater than one and λ/offset is about the same. This means that the mode of wave propagation will be the same in both cases, i.e. a symmetric plate wave. Attenuation measured with both techniques should be the same because the number of cycles over the two offsets is about the same (Table 10-2). Attenuation is dependent on casing thickness as in the case of large-scale measurements.

Table 10-2 Parameter comparison for large and small scale measurements

Parameter	Large Scale (in.)	Small Scale (in.)
λ	12	2.4
λ/casing thickness	24	4.8
λ/offset	0.33	0.4

Present configurations with small-scale refraction techniques use six pads. Variations in cement-to-casing bond can thus be resolved in 60° increments, which is the location resolution. However, variations smaller than the resolution can be detected. This is the azimuthal detection resolution. Detection resolution depends on the magnitude of the effect on measured attenuation between two adjacent pads. Small-scale attenuation measurements are generally combined with CBL measurements to provide formation compressional and shear-wave information.

10-3.1.3 Log Examples

Figure 10-7 presents several log examples of large-scale plate-wave measurement techniques. Figure 10-7A is a CBL log example in free pipe. Track 1 contains gamma ray (wavy solid), transit time (dash), and casing collar (straight solid) curves; Track 2 shows casing-wave amplitude; Track 3 shows waveforms displayed in a variable-density presentation. The gamma ray curve indicates the shale content of the formation and is useful because shale often affects cement emplacement. The gamma ray curve is also often used to correlate one formation interval to another. The casing-collar curve simply indicates the location of casing collars. Collars are thicker than the casing; therefore, they change the casing wave-guide geometry. This change causes slight changes in the transit time and reductions in the casing-wave amplitude, as indicated by the casing-wave amplitude curve. These changes also reflect the casing waves producing W or chevron-shaped patterns in the waveform display. Typical free-pipe signatures are (1) a straight amplitude curve except for collar "spikes," (2) straight casing wave arrivals ("railroad tracks") on the waveform display, and (3) strong chevron- or W-shaped reflections at the collars.

Figure 10-7B shows a CBL response in thin-wall casing that is fully cemented with high-strength, uncontaminated cement. The curves are the same as in Figure 10-7A except the collar locator curve and an expanded-scale amplitude curve (dashed curve) are in Track 2. The casing wave cannot be seen on the waveform plot because it has been almost completely attenuated by the cement layer. Consequently, casing-wave amplitudes are very low. The X10 amplified amplitude curve in Track 2 reads less than one-scale division. The normal amplitude curve lies along the left edge of the track. The erratic transit-time curve is a result of the very low casing-wave amplitude. Cycle skipping has occurred because

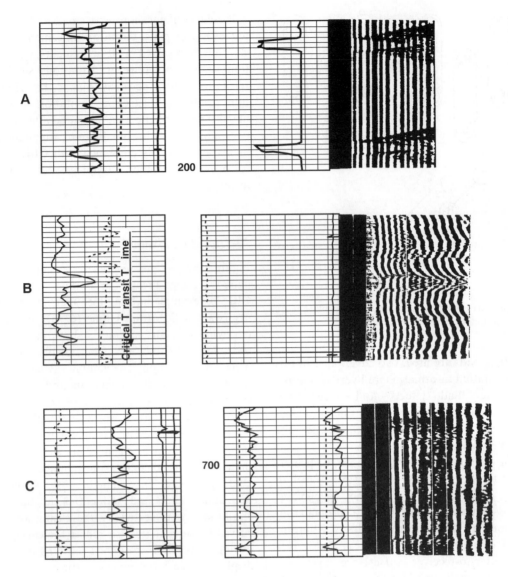

Figure 10-7 CBL log examples

the first peak is below the detection threshold. Formation arrivals are very strong, indicating good cement-to-formation cement contact. Typical signatures for well-bonded casing shown in this figure indicate (1) no casing signal on the waveform display but strong, clean formation compressional and shear waves, (2) very low amplitude and expanded-scale amplitude curves and (3) cycle stretch and cycle skip on transit time.

Figure 10-7C shows CBL response in thick-walled casing that has been fully cemented with low-strength, uncontaminated cement. The heavy, solid curve in Track 1 is the gamma curve, the thin, solid curve is the casing collar curve, and the dashed curve indicates transit time. An amplitude curve (leftmost solid) and an attenuation curve (rightmost solid) are shown in Track 2. The attenuation curve is computed from the amplitude curve and shows almost exactly the same behavior. Low-strength cement has lower velocity and density than regular Class H cement. Consequently, the impedance is lower, so less casing-wave energy is lost to the cement. Casing waves are less attenuated in the thicker-walled casing. Log curves will, therefore, appear intermediate to those of Figures 10-7A and 7B. Typical signatures are (1) casing and collar signals apparent in the waveforms but no chevrons in waveforms, (2) formation shear waves apparent but compressional waves obscured by casing waves, and (3) moderate values of amplitude and attenuation. Variable amplitude and attenuation values indicate crystalline material around the casing. If the material surrounding the casing were drilling or formation fluid, amplitude and attenuation would be almost constant.

Figure 10-8 shows a typical log presentation for small-scale plate-wave techniques. Track 3 contains attenuation curves from six attenuation measurements around the casing; the attenuation increases to the right, and each curve scales from 0 to 15 dB/ft. Track 2 shows a casing-collar curve. A pseudo-image of attenuation is shown in Track 1. This track represents a casing sliced down one side and flattened out so that the full 360° is displayed as a strip. Attenuation values from each of the six measurements are displayed with a gray scale. Gray-scale values are interpolated between the measurements to produce the pseudo-image. High attenuation values indicate bonding between casing and cement; low values indicate poor-quality bonding. Darker shading represents well-bonded casing, and light shading represents poorly bonded casing. Color representations analogous to the gray scale are also available. The gray-scale levels associated with attenuation values must be known before gray-scale presentations can be correctly interpreted. Well-bonded, low-impedance (lightweight) cement will produce less attenuation than well-bonded, higher-impedance cement.

10-3.2 Microannulus

A thin gap may exist between the casing and the cement. This gap is generally caused by pressure variations in the casing fluid after cement has set. These pressure changes cause the casing to expand or contract. Contraction causes the casing radius to be smaller than the inside diameter of the surrounding cement sheath. Casing expansion can occur as a result of heating caused by the heat of hydration generated during cement curing. A small gap will result if the cement sets before the casing cools. Gap thicknesses resulting from pressure are shown in Figure 10-9. These small gaps between casing and cement are referred to as *microannuli*. Short lines at the top of the figure represent the annular gap thickness created by pressure inside various sizes and weights of casings. The left ends of the lines correspond to the thickest casing and the right ends of the lines correspond to the thinnest casing. The horizontal line indicates that extremely thin gaps appear on CBL logs as if no cement exists behind the casing. Note that even small gaps created by pressure changes in casing will indicate no casing-to-cement bond on CBL logs.

As discussed in Section 10-3.1, breaking of the shear bond between casing and cement, even without a gap, will impede the transmission of shear-wave energy

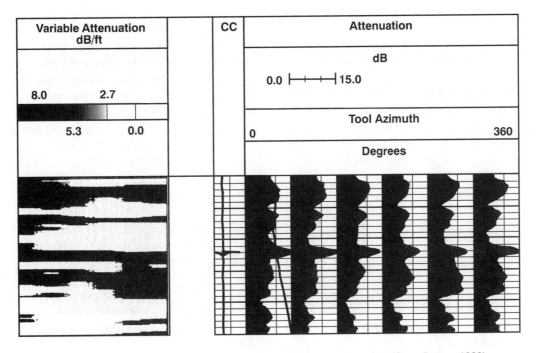

Figure 10-8   Small-scale plate-wave measurement log presentation (from Lester, 1989)

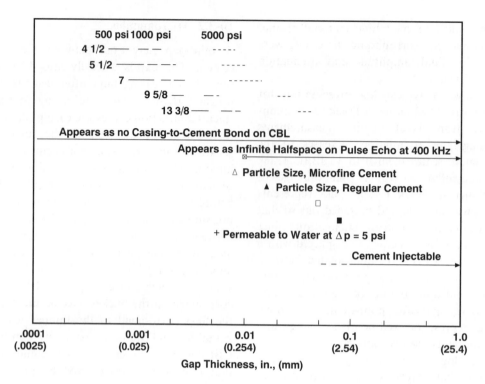

Figure 10-9 Annular gap thickness relation to cement-sheath evaluation

between casing and cement. Consequently, attenuation of the casing wave is decreased substantially. Amplitudes obtained in the presence of microannulus gaps are essentially the same as those obtained in free pipe. Figure 10-10 shows amplitude and expanded amplitude in Track 1 and waveforms in Track 2. With no casing pressurization, casing arrival is clearly observable in the waveform plot and on the amplitude curve. With pressurization, casing-wave amplitudes decrease or disappear, depending on microannulus thickness. Formation signals are easily recognizable before and after pressurization. Because the wave-length-to-microannulus thickness ratio is much greater than one for large- and small-scale measurement techniques, sensitivity to the microannulus will be approximately the same. Comparative tests (Lester, 1989) have verified this.

Pressurizing the casing to a value slightly over the maximum pressure after the cement sheath sets will close the microannulus (Figure 10-10). An overpressure of 500 psi is recommended. Casing-wave attenuation can then be measured without the microannulus effect. Because the gap thickness is very small, it is generally not an effective conduit for fluids (Figure 10-9).

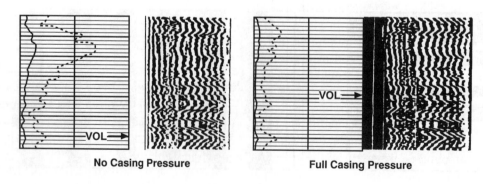

Figure 10-10 Microannulus effect with and without casing pressure

10-3.3 Strengths and Limitations of Plate-Wave Techniques

Large-scale CBL techniques provide an estimate of casing-to-cement bond averaged around the casing. CBL amplitude response is very susceptible to microannulus effects, although these effects can be eliminated in most cases by proper pressurization of the wellbore before logging. Sectored CBL techniques0 are inferior to small-scale techniques for providing azimuthal information. Small-scale techniques with pad-mounted sensors provide some azimuthal information on cement-to-casing bond. These techniques are not significantly affected by borehole fluids.

CBLs provide one unique measurement: the waveforms representing the average cement annulus and formation effects from the transmitter to the receiver. This measurement is invaluable for interpreting the presence of cement at the formation boundary and the interference of formation waves with the casing wave. However, the annular average of formation signals precludes identification of a channel at the cement-formation interface.

None of these techniques detect channels unless the channels are in contact with the casing. All techniques are adversely affected by microannuli.

10-3.4 Pulse-Echo Techniques

Pulse-echo cement-sheath evaluation techniques are based on a different wave propagation mode than that used by plate-wave techniques. As their name implies, they are based on emitting an acoustic pulse from a transducer and then receiving the reflected energy or echo with the same transducer. Pulse-echo investigations have been used for years in nondestructive testing for flaws and cracks in various types of materials (Krautkramer and Krautkramer, 1977). The first use of such techniques in cement-sheath evaluation was reported in 1981 by Froelich *et al.*

The objective of the pulse-echo technique is to create a thickness resonance mode in the casing. In Figure 10-11, the top diagram shows reflected and transmitted waves generated by an incident wave on a steel plate (casing). Wave 1 is the first reflected wave that would be received at the transducer. Wave 2 is the wave reflected from the back side of the plate. Wave 3 has undergone two reflections from the backside and one from the inside front side, etc. Ratios of reflected-to-incident and transmitted-to-incident wave amplitudes are given by the reflection and transmission coefficients respectively. For normal incidence, these coefficients are

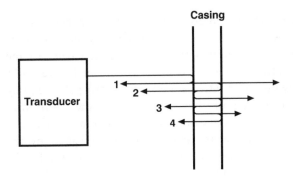

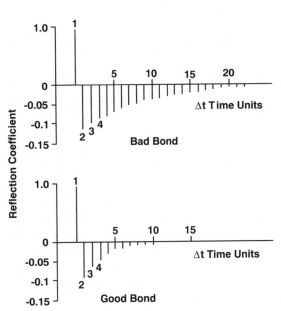

Figure 10-11 Schematic of pulse-echo technique (after Havira, 1982)

$$\text{Reflection coefficient} = R = \frac{Z_n - Z_m}{Z_n + Z_m} \qquad (10\text{-}7)$$

and

$$\text{Transmission coefficient} = T = \frac{2Z_n}{Z_n + Z_m} \qquad (10\text{-}8)$$

where Z_n is the acoustic impedance given by

$$Z_n = \rho_n V_n \qquad (10\text{-}9)$$

where ρ_n is bulk density and V_n is compressional wave velocity. The medium from which the wave is incident is denoted by m. Representative values of R and T are given in Table 10-3.

Because of the large reflection coefficient values at the casing boundaries, once energy enters the casing, it is largely trapped and bounces back and forth between the walls. At each reflection from the inner casing wall, some energy is transmitted through the casing back to the transmitter. The series of vertical lines in Figure 10-11 represents the relative amplitudes of the pulses exiting the inner wall of the casing toward the transmitter (only the pulses transmitted through the inner-casing wall back to the transmitter can be measured by the transducer). These pulses are separated in time by the two-way travel time in the casing, Δt. Note that the amplitudes decay exponentially with time. Decay rate is dependent on the reflection coefficients at the inner and outer casing walls. Because borehole fluid impedance varies little in a well, and steel impedance is known, decay rate provides a means of measuring the impedance of the material outside the casing. Amplitude of the first line (reflection from the fluid side of the inner casing wall) is about 10 times the other lines. This reflection amplitude is affected by both bore fluid and casing impedance contrasts and by the surface condition of the casing.

The larger the difference between steel impedance and the impedance of the material behind the casing, the less energy is transmitted into the material. This physics is the basis of pulse-echo cement-sheath evaluation. If cement ($Z = 6.2$) is behind the casing, more energy is lost per reflection than if water ($Z = 1.5$), oil ($Z = 1.0$), or gas ($Z = 0.2$) is behind the casing.

The series of amplitude lines in Figure 10-11 represents the pulse sequence that would be received at the transmitter for an infinitely narrow source pulse. Narrow pulses (with significant high-frequency content) approximate the reflection coefficient series. However, high-frequency pulses are attenuated in the bore fluid to such an extent that they cannot be used. Moving the transducer closer to the casing wall is not an option since reverberations between the transducer and the casing wall and transducer ringdown will interfere with the multiple

internal reflections. Drilling-fluid attenuation limits the upper frequency to about 500 to 600 kHz. Exciting the casing at its fundamental resonance will produce an exponentially decaying waveform with the resonant frequency of the casing (Figure 10-12). Casing resonance frequency is given by $f_0 = 1/\Delta t = V_c/2t_c$, where V_c is bulk velocity of the casing, and t_c is casing thickness. For a casing thickness of 5 mm, f_0 is 590 kHz; at a thickness of 10 mm, f_0 is 295 kHz. Thus, a frequency range of 200 to 600 kHz will accommodate most casing thicknesses.

Lower-frequency pulses are less affected by microannuli; layers that are thin compared to a wavelength are acoustically transparent. The lower the frequency, the less the effect of a microannulus. The lowest usable frequency is, therefore, a compromise between the desired insensitivity to the microannulus and the requirement to excite fundamental casing resonance in thin casings.

Logging tools based on the pulse-echo technique either use a series of individual transducers spaced so that they provide several measurements around the casing, or the tools use a rotating transducer that provides almost complete azimuthal coverage (Froelich *et al.*, 1981; Hayman *et al.*, 1991).

10-3.4.1 Measurement of Energy Ratios and Impedance

The basic pulse-echo measurements are energy ratios over time windows selected to enhance the resonant response and possible formation reflections. These windows are shown on the waveforms in Figure 10-12A. The top diagram shows computer-simulated waveforms for gas, water, and cement behind casing. Note the increase in decay rate of the signals from gas to water to cement. Time windows, W_1, W_2, and W_3 are used in quantitatively estimating the decay rate of the waveforms. Ratios of the energy in W_2 and W_1 plotted vs. impedance, show a smooth decrease with impedance. This ratio quantitatively indicates the decay rate of the reflected wave energy. The ratio, R_2, vs. impedance, illustrates the mapping of window energy to impedance. W_2 window energy is a function of both amplitude and decay rate. Ratios of energy in each window, normalized to free-pipe and casing reflection amplitude, are computed as

$$R_2 = (E_2/E_1)(1/R_{2fp}) \qquad (10\text{-}10)$$

and

$$R_3 = (E_3/E_1)(1/R_{3fp})$$

Table 10-3 Transmission and reflection coefficients

Interface	Reflection coefficient, R	Transmission coefficient, T
Water to steel	+0.93	1.93
Steel to water	−0.93	0.07
Steel to cement (16 lb/gal)	−0.76	0.24
Steel to light cement (13 lb/gal)	−0.81	0.19
Steel to 10% porosity limestone	−0.57	0.43
Steel to unconsolidated sand	−0.75	0.25

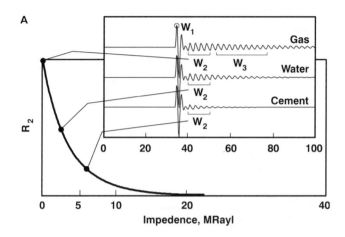

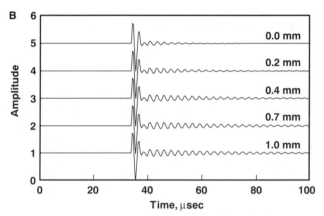

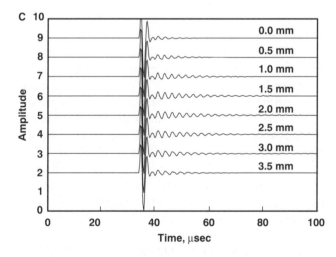

Figure 10-12 (A) Computer-simulated pulse-echo waveforms for different impedance materials outside casing; (B) Computer-simulated pulse-echo response for a water layer (between steel and cement) of increasing thickness; (C) Computer-simulated pulse-echo response for a cement layer (between casing and formation) of increasing thickness.

where E_2 and E_3 are the energy in windows W_2 and W_3, E_1 is the energy at the peak of the casing reflection, and R_{2fp} and R_{3fp} are the ratios E_2/E_1 and E_3/E_1 measured in free pipe. A variation of R_2 indicates an impedance variation of the material contacting the outer casing wall. Variation in R_3 indicates signal amplitude variation in time window W_3, which *may* be related to formation reflections. However, interference of reflected signals with the decaying resonance waveform may produce considerable variation in window W_3. Reflections can interfere constructively or destructively, as shown in Figure 10-12C. Cement layer thickness is indicated on the curves. Note the cyclical decay rate pattern caused by the constructive and destructive interference of the reflections from the cement-to-formation interface. This variation indicates the difficulty in estimating decay rate and hence impedance of the material behind casing in the presence of reflections from the formation. Gas behind casing, because of its high impedance contrast, also results in high amplitudes in window W_3 (see the gas signal in Figure 10-12A). Values of R_3 greater than a specified level are used to set formation or gas flags on log presentations.

From the discussion in the previous section, R_2 is directly related to the acoustic impedance, Z_m, of the material in contact with the outside of the casing. The inset curve of R_2 vs. Z_m in Figure 10-12A was determined from computer-generated waveforms similar to those shown in the figure. Similar tool response relationships have been published (Froelich *et al.*, 1981; Sheives *et al.*, 1986). Z_m can also be computed from the complete waveform through the use of an iterative forward modeling technique (Hayman *et al.*, 1991). Cement compressive strength vs. acoustic impedance has been measured for various types of cement. Consequently, Z_m can be converted into compressive strength. Casing thickness can also be obtained from the resonant frequency observed in time window W_2.

10-3.4.2 Microannulus Effect

The effect of a microannulus on pulse-echo measurements can be understood by considering the energy reflected from a thin layer between two half spaces (Krautkramer and Krautkramer, 1977). Reflected wave amplitudes as a function of gap thickness are shown in Figure 10-13. The curves are for a thin water or air layer between either steel or aluminum plates. The percentage of reflected energy from the layer, R, is shown for air and water layers. For air-filled gaps, almost all of the energy

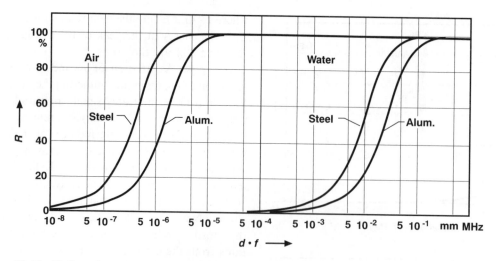

Figure 10-13   Reflected energy vs. layer thickness times frequency (from Krautkramer and Krautkramer, 1977)

is reflected for a layer thickness of only 0.25×10^{-4} mm at a frequency of 200 kHz. For a water layer, almost all the energy is reflected for a layer thickness of 0.25 mm at 200 kHz. Thus, microannuli must be very thin not to affect pulse-echo amplitude. At 200 kHz, a 0.25-mm gap filled with water produces almost total reflection. Experiments have shown that a gap this thick provides a path for liquids. Thinner gaps are conductive for gas. Therefore, at 200 kHz, microannuli that are thinner than those that permit fluid communication produce large reflections. This effect is also illustrated in Figure 10-12B with computer-generated pulse-echo waveforms with a water layer between the casing and the cement. Water-layer thickness is indicated on the curves in the figure. Note the decrease in decay rate as water-layer thickness increases. As water-layer thickness increases, the waveform approaches that for an infinitely thick water layer or free-pipe condition. Note that a gap of only 0.2 mm produces nearly the same reflected signal as an infinitely thick gap.

10-3.4.3 Log Examples

In tools with multiple transducers, impedances from each transducer are generally displayed in individual tracks similar to small-scale attenuation measurements (Figure 10-8). A similar presentation can be made for rotating, single-transducer tools by displaying impedance for azimuthal sectors around the borehole as shown in Figure 10-14. Impedance or compressive strength values may be displayed as gray-scale or color-coded pseudo-images which are referred to as

impedance maps, cement maps, or bonding index maps. However, map presentations may be misleading because of an improper choice of color ranges (Goodwin *et. al*, 1994). The conversion from impedance to compressive strength is also highly suspect because of the many unpredictable changes that may occur to the cement during and after placement. These changes may greatly affect strength but have little effect on impedance, or the opposite may occur.

A typical pulse-echo impedance log displaying maximum and minimum impedance curves is shown in Figure 10-14. Maximum and minimum impedances are simply the maximum and minimum impedance values measured in a particular sector. The data were acquired with a rotating pulse-echo transducer, and impedance curves were generated at 40° intervals around the casing, yielding nine separate tracks of impedance measurements. Each track is scaled from 0 to 5 MRayls. Casing collar spikes are easily discernible. Liquid behind casing produces a relatively constant impedance around the casing since its properties are essentially constant. Impedance should be close to that of water, 1.5 MRayls. Thus, maximum and minimum impedance curves close together, straight, and near 1.5 MRayls indicate water behind casing (600 m to 615 m in Track C). Cement signals will show impedance values of approximately five, but the signals will generally show some variation because of differences in cement impedance. Track E (655 m to 680 m) indicates good quality cement in Sector E. Impedance curves over other depth intervals and sectors in the figure are very erratic, and maximum and minimum curves are widely separated. These results

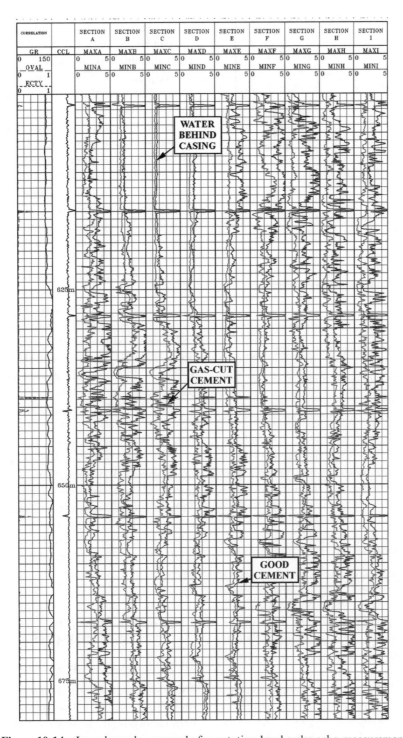

Figure 10-14 Impedance-log example for rotating-head pulse-echo measurements

indicate gas-cut cement, which has large variations in acoustic impedance because of large variations in gas concentration in the cement. The wider the separation of maximum and minimum impedance curves, the greater the gas-cut variation.

CBL and pulse-echo measurements may be effectively combined to provide improved interpretation of cement properties, as illustrated in the crossplot shown in Figure 10-15. If only CBL information were used, distinguishing gaseous cement from normal cement would be difficult.

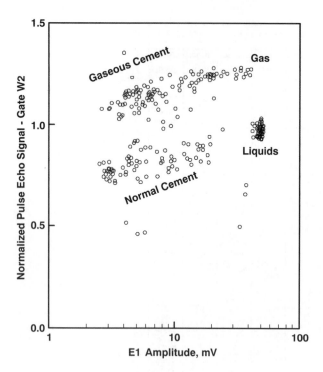

Figure 10-15   Crossplots of pulse-echo window W_2 vs. CBL amplitude (from Cantala *et al.*, 1984)

If only pulse-echo information were used, it would be difficult to distinguish liquids from either normal or gaseous cement. The combination of both information types allows a less ambiguous definition of the different materials behind casing.

10-3.5 Noise Measurements

Noise measurements use microphones or piezoelectric receivers to receive sound generated by the flow of liquid behind casing. McKinley *et al.* (1973) reported the results of laboratory experiments with the objective of characterizing the amplitude vs. frequency response of noise generated by fluid flow. Fluid was throttled across a porous plug into a cylindrical chamber filled with crushed marble. An acoustic sensor was located in a small tube running along the axis of the cylinder. Amplitude spectra of the noise generated by the fluid flow and recorded by the sensor were obtained. It was found that above about 200 Hz, single-phase flow (either air or water) spectra matched those from wind-tunnel turbulence tests. Two-phase flow spectra exhibited spectra significantly different from single-phase spectra; a major spectral component is present in the 200- to 600-Hz range. Figure 10-16 shows observed noise spectra from the McKinley experiments. The upper figure indicates water flowing into water. The lower figure indicates air flowing into air. The upper figure can be applied to single-phase water flow into a well. Note the single-peak character of the amplitude curves; the frequency of the peak increases with the pressure differential between the formation and the well. Single-phase gas flow also exhibits single-peak spectra, and the frequency of the peaks increases as the pressure differential increases. Noise amplitude for single-phase flow was found to be linearly correlated with the product of pressure differential across the orifice and flow rate (which is the energy dissipation rate in turbulent flow). Therefore, if the pressure differential is known, the flow rate can be estimated from the noise amplitude. Noise amplitude for two-phase flow in the 200- to 600-Hz range was linearly correlated with flow rate.

Several example noise logs made with an experimental tool were presented by McKinley *et al.* (1973). Early field results were also presented by Robinson (1976). The usual log presentation is to present the peak noise amplitude measured in four frequency ranges: above 200 Hz, above 400 Hz, above 600 Hz, and above 1000 Hz. However, full spectral information is the best diagnostic for noise log interpretation.

Noise log information can be used to establish the existence of flow communication behind casing if the flow rate is sufficient and if the orifice of the communication pathway is small enough to produce measurable noise levels. However, estimating the size and distribution of the pathway is generally very difficult or impossible. Therefore, it is impossible to determine if remedial cementing can be used to close the pathway.

Noise and temperature logs (Section 10-4.2) are primarily used to identify the existence of fluid flow behind casing.

10-4 NON-ACOUSTIC TECHNIQUES FOR CEMENT-SHEATH EVALUATION

10-4.1 Nuclear Measurements

Nuclear cement-sheath evaluation techniques are based on treating the cement with radioactive tracers that emit gamma rays. The most quantitative work (Kline *et al.*, 1986) on primary cement evaluation is based on the increase of the gamma ray signal from a before-placement gamma ray logging run to an after-placement run. With the assumption of a uniform composition cement sheath, the thickness of the cement sheath was computed from gamma ray measurements. Qualitative evaluations of cement quality can be obtained on the basis of gamma ray spectroscopy techniques that were first developed to

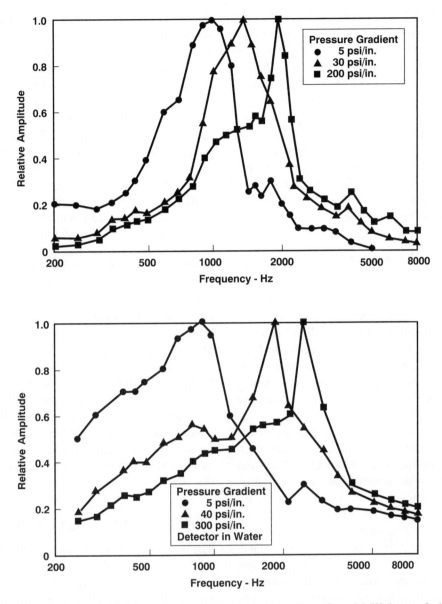

Figure 10-16   Noise amplitude vs. frequency for single-phase flow (from McKinley *et al.*, 1973)

monitor the effectiveness of fracturing operations (Gadeken *et al.*, 1991). A relative distance parameter obtained from Compton downscattering effects in the gamma-ray spectra contains cement diameter information that is independent of the gamma ray signal amplitude. New directional gamma ray logging tools (Simpson and Gadeken, 1993) can measure the azimuthal distribution of gamma radiation around the casing. When used effectively, imaging techniques can display directional gamma ray data that helps users visualize the coverage and quality of cement behind casing (Wyatt *et al.*, 1996). The most accurate evaluations are obtained when the

tracer isotope is uniformly dispersed throughout the cement volume (Bandy *et al.*, 1993).

The physics of the gamma ray transport and detection processes results in volumetric averaging. Therefore, voids in the cement that are smaller than a few inches cannot easily be distinguished from thin cement or cement mixed with other materials (drilling fluid, gas, formation). However, voids are indicated by the absence of radioactivity on the basis of imaged directional gamma ray data. Microannuli that would allow significant fluid flow (Figure 10-9) do not produce observable gamma ray signatures.

These same tracer and logging technologies can be used for evaluating remedial cementing operations. Channels or microannuli that could allow significant fluid communication are still below the detection threshold except in those instances where the remedial cement has flowed into such voids. Nuclear techniques do not provide information about the type of material that may be filling cement voids (drilling fluid, gel, water, gas) unless such fillers have been specifically tagged with an appropriate radioactive species. Future developments include the application of deconvolution techniques to gamma ray spectroscopy and directional data to resolve smaller features than is now possible.

10-4.2 Temperature Measurements

Formation temperature increases by the geothermal gradient for a particular area. Temperature of the wellbore fluid will reach the geothermal gradient if the fluid is not circulated for some period of time. Fluid entering the wellbore or the annulus between the casing and the formation will be in equilibrium with the geothermal gradient at the point of entry. However, as it moves up the wellbore, it will be relatively warmer than the temperature prescribed by the geothermal gradient. Gas may be relatively cooler because of the expansion of the gas as it exits the formation and moves to lower pressure. Measurement of the temperature in and around the wellbore can provide information on the influx, location, and movement of fluids. For cement-sheath evaluation, only flow behind the casing is of interest. Hence, the objective is to measure the temperature perturbation in the wellbore produced by a small volume of fluid behind the casing. In general, this fluid volume does not completely surround the casing; it is generally located in channels. A technique to measure the azimuthal distribution of the bore-fluid temperature was reported by Cooke (1979). This technique consists of a temperature-sensitive resistor that can be rotated in the wellbore. A logging tool is stopped at a particular depth and scans through 360° to produce a plot of temperature vs. azimuth.

10-5 INTERPRETATION GUIDELINES FOR CEMENT EVALUATION PRESENTATIONS

10-5.1 Interpretation Objectives

The primary reason for running a cement evaluation tool is to determine the cement sheath's ability to isolate the fluid/gas producing horizons exposed in the borehole as a result of drilling. Once cement is placed completely around the casing and has fully reacted to form a solid in the annulus, it is entirely academic what the compressive strength of the cement is. After the cement sets in the annulus, it cannot be removed regardless of its strength. A great deal of effort is spent trying to estimate compressive strength of the cement. A variety of contaminants (gas, oil, water influx, or drilling fluid) mix with the cement before it sets, thereby changing the strength and appearance of the cement significantly. Normal bond logs (CBL) yield erroneous results in cases such as gas cut or foamed cement, because the shear strength and density of the cement are not sufficient to attenuate the casing signal. However, the contaminated cement probably still provides a hydraulic seal and, even if it does not, it cannot be replaced by remedial cementing operations.

It is the movable materials in the annulus between casing and formation that must be identified and characterized during cement-sheath evaluation. These movable materials are liquids and gases. Drilling-fluid cake is a solid that cannot be replaced by remedial cement. Certain materials can be moved by fluids migrating from one zone to another and by remedial cementing operations. Therefore, *the primary objective of cement-sheath evaluation is to differentiate between solids and liquids.* Interpretation should concentrate on measurements and log presentations that are directly related to this objective.

10-5.2 General Logging Recommendations

The best practices for cementing are discussed in Chapter 8. These practices will, in most cases, ensure the placement of cement circumferentially around the casing over the entire cemented interval with no inclusion of liquid-filled channels. However, overpressuring the casing after the cementing operation, decreasing pressure after the cement has set, and gas migration into the cement before it is set (Chapter 9) may produce communication channels in the cement sheath. Also, the best practices are not always followed.

The following general recommendations should be used for cement-sheath evaluation logging:

- Common bond logs are not recommended in casing larger than $9\frac{5}{8}$ in. Thicker casing increases casing-wave amplitude, and bore-fluid thickness decreases signal amplitude. Transit time may also be increased to the extent that formation waveform information is not recorded.

- If any type of bond log (common, compensated, sectored, or segmented) is used for cement evaluation, a pressure pass and a nonpressure pass must be made. The casing pressure used should be equal to any changes in hydrostatic pressure (or test pressure) that have been created after the cement has set, plus and additional 500 psi.

- A bit and scraper should be run to clean the casing wall before logging occurs. Coatings or scale inside the casing decrease amplitude and, thus, increase casing-wave attenuation for CBL measurements. Casing reflection amplitude variations exacerbate pulse-echo interpretation.

- Cement should have a minimum compressive strength of 250 psi (from laboratory tests) at the top of the cement column (at the temperature at the top of the cement column) before a cement evaluation log is run.

- If a cement evaluation log is actually required, bond logs are generally not recommended because they average the reflected signal from 360° of the casing. Any of the evaluation logs that segment the annulus into definable results is recommended so that if a channel actually exists, its location can be determined.

- Common, compensated, or sectored bond logs should not be run for evaluating foamed cement or if gas contamination could exist in the cement column. This recommendation also applies when the casing is cemented with "gas-generating" additives, or additives for low-density slurries, when that additive contains any volume of trapped air. However, the segmented bond logs do a respectable job of identifying gas-impregnated cement and a liquid channel in the annulus.

10-5.3 Operational Parameter and Log Presentation Recommendations

Parameters and log output for the different types of cement evaluation tools are given in Table 10-4. Information for specific tools is given in Goodwin *et al.* (1994).

10-5.4 Common Problems in Cement Evaluation

10-5.4.1 Formation Bonding

Bonding to the formation is generally inferred on a bond log by the presence of strong, clear, bright, refracted formation *P* (compressional) and *S* (shear) waves on the log's waveform display. However, formation signals may either be missing or extremely faint. Generally, four conditions could cause the loss of formation signals:

- **Thick, soft mudcake** Signal velocity is so slow through poorly compacted mudcake that it does not have enough time to enter the formation and be refracted back to the receiver. A soft mudcake also greatly attenuates the signals. These types of drilling-fluid filter cakes may often be visible on an openhole caliper survey as decreases in hole size across permeable sections.

- **Noncemented hole enlargement** Failure to clean the hole properly before and during the cement placement will sometimes leave a "washout" full of drilling fluid instead of cement. If poorly bonded sections conform to hole enlargement on the openhole caliper survey, the enlargements are generally filled with drilling fluid. They may or may not present an annular isolation problem, depending on downhole conditions.

- **Soft and poorly consolidated to unconsolidated formations** Sonic velocity in very soft formations can be slow enough that formation refractions will not have time to reach the measurement window of the tool. Marine shales, soft salt or anhydrite beds, poorly consolidated sands, and unconsolidated sands will generally present this type of interpretation problem. These situations do not generally pose an annular isolation problem.

- **Foamed or gas-cut cement** Adding gas to a cement system, whether by design (foamed cement) or naturally (formation gas influx), decreases density and velocity and increases attenuation. The decreased velocity delays the arrival time on a bond log sufficiently that formation signals are generally lost.

Gas entrainment in the cement sheath does not necessarily pose an annular isolation problem, but loss of the formation signal creates major confusion in interpreting the quality of the cement sheath. In Figure 10-17, Track 1 has a gamma curve (thick solid), a transit-time curve (dash), and a casing collar curve (thin solid). The amplitude curve is shown in Track 2. Typical signatures indicate (1) low to moderate amplitude, (2) a weak casing signal or a stronger signal for thicker-walled casing, (3) evident collar signals but no chevrons on the waveforms, and (4) weak formation arrivals or no formation arrivals.

Table 10-4 Operation parameter and log presentation recommendations

Type of Measurement	Plate Wave Large Scale			Small Scale	Non-Rotating Pulse Echo		Rotating Pulse Echo	
Specific Commercial Tool	Attenuation CBL	Sectored CBL Sectored Bond(1)	Compensated Attenuation CBT(2)	SBT(3)	PET(4)	CET(5)	USI(6)	CAST-V(7)
Parameter								
maximum pressure (psi)	max*+500	max*+500	max*+500	max*+500	max*+500	max*+500	max*+500	max*+500
maximum temperature (C)	350	350	350	350	350	350	350	350
recommended logging speed (ft/hr)	1800	1800	2100	2400	3600	4000	note 1	note 1
maximum mud weight (lb/gal)		18		no limit	wbm, obm-16	wbm-12.5 obm-12	wbm, obm-14	wbm-i6, obm-14
maximum casing diameter (in)	$9\frac{5}{8}$	$9\frac{5}{8}$	$9\frac{5}{8}$	16	$13\frac{3}{8}$	$9\frac{5}{8}$	$9\frac{5}{8}$-$13\frac{3}{8}$	$9\frac{5}{8}$-$13\frac{3}{8}$
maximum eccentering (us)	4**		4**	NA	note 2	note 2	note 2	note 2
Recommended Log Presentations								
Transit Time	100 us span	100 us span	100 us span	N/A	N/A	N/A	N/A	N/A
Amplitude Scale***	0–100 mv	0–100 mv	0–100 mv	N/A	N/A	N/A	N/A	N/A
Amplified Amplitude Scale	0–20 mv	0–20 mv	0–20 mv	N/A	N/A	N/A	N/A	N/A
Attenuation Scale	0–20 dB/ft	N/A	0–20 dB/ft	0–9 dB/ft	N/A	N/A	N/A	N/A
Waveform Display	VD or WF	VD or WF	VD or WF	VD or WF	N/A	N/A	N/A	N/A
Acoustic Impedance (MRayles)	N/A	N/A	N/A	N/A	0–5	0–5	0–5	0–5
Other	Gamma Ray Col Locator	Gamma Ray Col Locator	Gamma Ray Col Locator	Gamma Ray Col Locator Rel Bearing Eccentering Csg ID	Gamma Ray Col Locator Rel Bearing Eccentering Csg ID	Gamma Ray Col Locator Rel Bearing Eccentering Csg ID	Gamma Ray Col Locator Rel Bearing Eccentering Csg ID	Gamma Ray Col Locator Rel Bearing Eccentering Csg ID

1 = Sectored Bond Tool—Computalog
2 = Cement Bond Tool—Schlumberger
3 = Segmented Bond Tooll—Atlas Wireline
4 = Pulse Echo Tool—Halliburton Energy Services
5 = Cement Evaluation Tool—Schlumberger

6 = Ultrasonic Scanning Imaging Tool—Schlumberger
7 = Circumferential Acoustic Scanning Tool—Halliburton Energy Services

*max indicates maximum pressure applied to casing previous to cement evaluation
**in terms of shift in transit-time measurement (1us = 0.06 in)
***enter free-pipe amplitude in log heading

Note 1: Function of azimuthal and vertical sampling; desired azimuthal resolution from 5–10; minimum vertical sampling 1.5 in

Note 2: 4% of casing OD; set transducer standoff for casing size

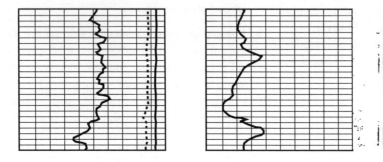

Figure 10-17 CBL response to gas-infiltrated cement

10-5.4.2 Channels and Microannuli

Weak to clear casing signals on the microseismogram of a bond log may be caused by a microannulus or a channel that is in contact with the casing surface. Current cement evaluation tools cannot identify a channel either in the body of the cement sheath or between the cement sheath and the borehole wall. To differentiate between a microannulus or a channel at the pipe-cement interface, the log is run with and without internal casing pressure. Sufficient casing pressure to expand the casing against the cement sheath will cause the casing signals to disappear if a microannulus exists. Casing signals will not change with casing pressure if a channel exists at the pipe-cement interface or if the pressure is not sufficient to close the microannulus gap.

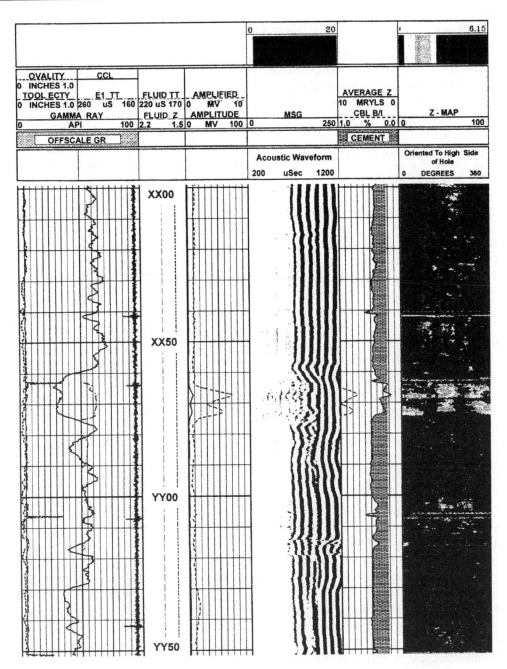

Figure 10-18 Combination of CBL and pulse-echo information in a single log

10-5.4.3 Fast Formations

Limestones and dolomite formations are commonly referred to as "fast formations" because the velocity of sound through those types of rocks can be much greater than the velocity of sound through sandstone, shales, cement, or steel. If formation velocity is greater than plate-wave velocity, the refracted formation signal may, and quite often does, arrive at the tool receiver faster than the casing signal. Early formation arrivals are quite often mistaken for casing signals, causing misinterpretation errors that result in unnecessary remedial repair of the cement sheath. Fast formations (and outer casing strings) also affect pulse-echo measurements, because the excess noise reflected from the borehole wall during logging through these formations (and outer casing strings) is often interpreted as poor-quality cement (Figure 10-12C).

10-5.4.4 Casing in Casing

Cement evaluation in casing-to-casing annuli (liner laps) is extremely difficult. When common bond logs (or such variations as compensated bond logs, sectored bond logs, or small-scale plate-wave techniques) or pulse-echo logs are used, the excess noise reflected by the outer casing string is interpreted to indicate a very weak cement bond or no cement bond.

10-6 A FINAL EXAMPLE

Figure 10-18 shows an example of the combination of information from both CBL and rotating pulse-echo logging tools. Tool eccentering (eccentricity), casing ovality, gamma ray, E_1 transit time, and casing collar curves are in Track 1. Amplified and normal amplitude curves are in Track 2. A gray-scale waveform plot is in Track 3. Track 4 contains the average impedance curve. Shading between the curve and an impedance value of about 2.25 schematically represents the amount of cement behind casing. A bond index curve computed from the CBL amplitude is also displayed in Track 4. Track 5 contains a gray-scale representation of the impedance values around the casing. The orientation of the impedance map is such that 0° is at the high side of the hole. This combination is commonly used for cement-bond interpretation. Casing ovality is simply the difference between the greatest and least diameters based on the first reflection transit time of the pulse-echo signals. Eccentricity is also determined from the diameter data; it is the difference in the average greatest and least measured diameters. If the tool is exactly centralized, eccentricity will be zero.

Cement bond quality is so good over most of the interval that the transit time cannot be computed because the casing arrival is so small. Only in the interval from XX62 to XX90 is the transit-time curve displayed. Amplitude curves in Track 2 are very small, which indicates excellent bond quality over most of the interval. Casing amplitude increases over the interval from XX62 to XX75, which indicates decreased bond quality. However, the lack of a strong casing arrival on the waveform plot and the relatively low amplitude curve values indicate that some cement is contacting the casing; free pipe is not present.

The bond index (Track 4) computed from CBL amplitude is 100% over the interval except from XX65 to XX73. Impedance computed from pulse-echo measurements and averaged around the casing is between 4 and 5

(in the range of Class H cement; see Table 10-1) over the interval except from XX62 to XX73. These results agree well with the bond index and amplitude curves. The light gray area across the impedance map (Track 5) shows the reduction of impedance around the entire casing from XX62 to XX73, which also agrees with the CBL amplitude. The cement quality in this zone is less than the quality for the remainder of the interval. However, free pipe does not exist. A clue to the cause of the lower-quality cement is given by the gamma curve, which indicates a relatively non-shaly formation, probably a relatively clean sand, behind casing over the zone of poor-quality cement. This formation is most likely a gas sand, which could be confirmed by checking openhole logs, such as the density and neutron logs. The cause of the low-impedance cement is probably gas contamination of the cement.

Another set of features on the impedance map is interesting. These are the oval patterns near the bottom and just below the gas-cut cement zone. Note that they are located on the low side of the hole, which is deviated about 23°. Because of the well deviation, casing is probably eccentered, with the casing being closer to the bottom side of the hole. The openhole caliper shows a smaller hole below the gas-cut zone. Therefore, the cement is almost certainly thinner between the casing and the low side of the hole. Consequently, reflections from the formation may interfere with the pulse-echo signal (Section 10-3.4.1) and effectively reduce the estimated cement impedance. Variations in the interference pattern result in the ovals.

The impedance above the gas-cut zone is slightly less than below it. This effect may be caused by upward gas migration from the sand contaminating the cement above it. Another possibility is that slightly thicker casing was installed above the gas-cut zone. Note the short casing joint indicated by the casing-collar curve just above the gas-cut zone.

10-7 MEASUREMENT LIMITATIONS

The basic question posed for interpretation with respect to remedial cementing is whether a fluid-filled void can be filled with cement. Cement particle size defines an absolute lower bound on the void thickness that can be filled. The smallest particle size for specially ground (microfine) cement is about 0.2 mm, and the smallest particle size for A, G, or H cement is about 0.4 mm. A more realistic estimate may be the assumption used in gravel packing—that the perforation diameter must be six times the sand-grain diameter (open and solid squares in figure). These lower bounds for void thickness are shown in Figure 10-9. Backpressures associated with viscous fluids may require an even greater minimum gap thickness for cement injection. Such viscous flow effects may be the controlling factors for nonparticulate remedial treatment fluids such as resins.

From Figure 10-9, approximately 5000 psi is required to close a 0.7-mm microannulus in lightweight 95/8-in. casing. A 0.2-mm thick gap appears as an infinitely thick layer with a 200-kHz pulse-echo measurement. Hence, if a gap is 0.9 mm thick, it will appear as infinitely thick water even with substantial pressure applied to the well. Gap thickness must be greater than approximately 1.2 mm (6×0.2 mm) to accept the finest cement (open square in figure). Even very thin gaps (microannuli) appear as infinitely thick water layers behind casing when plate-wave measurement techniques are used. Consequently, current cement evaluation techniques cannot distinguish between gaps next to the casing that are "squeezable" and those that are not.

Channels present another problem. All cement evaluation techniques will indicate where cement is completely missing in the annulus. However, current cement-evaluation techniques provide essentially no information about voids inside the cement annulus or the cement-to-formation interface. Channels and voids must (1) be very close to the casing (less than about 2 cm for small-scale plate-wave measurements or 1.2 mm for pulse-echo measurements) or (2) intersect the casing to be detected by pulse-echo or small-scale plate-wave measurements.

REFERENCES

Bade, J.F.: "Cement Bond Logging TechniquesHow They Compare and Some Variables Affecting Interpretation," *JPT* (Jan. 1963) 17–28.

Bandy, T.R., Read, D.A., and Wallace, E.S.: "Radioactive Tracing With Particles," U.S. Patent 5,182,051 (1993).

Cantala, G.N., Stowe, I.D., and Henry, D.J.: "A Combination of Acoustic Measurements to Evaluate Cementations," paper SPE 13139, 1984.

Cooke, C.E.: "Radial Differential Temperature (RDT) Logging—A New Tool for Detecting and Treating Flow Behind Casing," paper SPE 7558, *JPT* (Jun. 1979).

Ewing, W.M., Jardetsky, W.S., and Press, F.: *Elastic Waves in Layered Media*, McGraw-Hill (1957).

Fertl, W.H., Pilkington, P.E., and Scott, J.B.: "A Look at Cement Bond Logs," *JPT* (Jun. 1974) 607–617.

Fitzgerald, D.D., McGhee, B.F., and McGuire, J.A.: "Guidelines for 90% Accuracy in Zone Isolation Decisions," *JPT* (Nov. 1985) 2013–2022.

Froelich, B., Pittman, D., and Seeman, B.: "Cement Evaluation ToolA New Approach to Cement Evaluation," paper SPE 10207, 1981.

Gadeken, L.L., Gartner, M.L., Sharbak, D.E., and Wyatt, D.F.: "The Interpretation of Radioactive Tracer Logs Using Gamma Ray Spectroscopy Measurements," *Log Analyst* (Jan.-Feb. 1991) **32**, 2534.

Golwitzer, L.H., and Masson, J.P.: "The Cement Bond Tool," paper SPWLA Y, 1982, 1–15.

Goodwin, K.J., Wilson, W.N., Domanique, E., Butsch, R.J., and Graham, W.L.: "Cement-sheath evaluation," API Technical Report (1994).

Grosmangin, M., Kokesh, F.P., and Majani, P.: "A Sonic Method for Analyzing the Quality of Cementation of Borehole Casings," *JPT* (Feb. 1961).

Havira, R.M.: "Ultrasonic Cement Bond Evaluation," paper SPWLA N, 1982.

Hayman, A.J., Hutin, R., and Wright, P.V.: "High-Resolution Cementation and Corrosion Imaging by Ultrasound," paper SPWLA KK, 1991.

Jutten, J.J., and Corrigall, E.: "Studies With Narrow Cement Thicknesses Lead to Improved CBL in Concentric Casings," *JPT* (Nov. 1989) 1158.

Jutten, J.J., Parcevaux, P.A., and Guillot, D.J.: "Relationships Between Cement Slurry Composition, Mechanical Properties, and Cement Bond Log Output," paper SPE 16652, 1987.

Kline, W.E., Kocian, E.M., and Smith, W.E.: "Evaluation of Cementing Practices by Quantative Radiotracer Measurements," paper SPE 14778, 1986.

Krautkramer, J.K., and Krautkramer, H.K., *Ultrasonic Testing of Materials*, Springer-Verlag (1977).

Lamb, H.: "On Waves in an Elastic Plate," *Proc Roy Soc* (London), **A, 93** (1917).

Leonardon, E.G.: "The Economic Utility of Thermometric Measurements in Drill Holes in Connection with Drilling and Cementing Problems," *Geophysics* (1936) **1**, No. 1, 115–126.

Lester, R.A.: "The Segmented Bond Tool: A Pad-Type Cement Bond Device," *Can. Well Logging Soc.*, Paper A (1989) 1–13.

McKinley, R.M., Bower, F.M., and Rumble R.C. "The Structure and Interpretation of Noise from Flow Behind Cemented Casing," *JPT* (March 1973) 329–338.

McNeely, W.E.: "A Statistical Analysis of the Cement Bond Log," paper SPWLA BB, 1973.

Nayeh, T.H., Wheelis Jr., W.B., and Leslie, H.D.: "The Fluid-Compensated Cement Bond Log," paper SPE 13044, 1984.

Pardue, G.H., Morris, R.L., Gollwitzer, L.H., and Moran, J.H.: "Cement Bond Log—A Study of Cement and Casing Variables," *JPT* (May 1963) 545–555.

Pennebaker, E.S.: "Locating Channels in Multiple Tubingless Wells With Routine Radioactive Logs," *JPT* (Apr. 1972) 375–384.

Pickett, G.R.: "Prediction of Interzone Fluid Communication Behind Casing by Use of the Cement Bond Log," paper SPWLA J, 1966.

Pilkington, P.E: "Pressure Needed to Reduce Microannulus Effect on CBL," *Oil & Gas J.* (1988) **86**, No. 22, 68-74.

Robinson, W.S.: "Field Results From the Noise-Logging Technique," *JPT* (Oct. 1976) 1370–1376.

Schmidt, M.G.: "The Micro CBL—A Second Generation Radial Cement Evaluation Instrument," paper SPWLA Z, 1989.

Simpson. G.A., and Gadeken, L.L.: "Interpretation of Directional Gamma Ray Logging Data for Hydraulic Fracture Orientation," paper SPE 25851, 1993.

Sheives, L.N., Tello, L.N., Maki Jr., V.E., Standley, T.E., and Blankinship, T.J.: "A Comparison of New Ultrasonic Cement and Casing Evaluation Logs With Standard Cement Bond Logs," paper SPE 15436, 1986.

Winn, R.H., Anderson, T.O., and Carter, L.G.: "A Preliminary Study of Factors Influencing Cement Bond Logs," *JPT* (Apr. 1962) 369–379.

Wyatt, D.F., Gadeken, L.L., and Grossman, R.: "Cement Evaluation Using Radioactive Tracers and Directional Gamma Ray Logs," *Cementing the Future: Supplement to Pet. Eng. Int.* (1996).

Zemanek *et al.* "The Borehole Televiewer," *JPT* (June 1969) 762–774.

11 Remedial Cementing

Richard Jones
ARCO Exploration and Production Technology

Larry Watters
Halliburton Energy Services

11-1 INTRODUCTION

The most successful and economical approach to remedial cementing is to avoid it by thoroughly planning, designing, and executing all drilling, primary cementing, and completion operations. The need for costly remedial cementing to restore a well's operation indicates that primary cementing planning and execution were ineffective.

Remedial cementing requires as much technical, engineering, and operational experience as any primary cementing operation. Remedial cementing jobs can be further complicated because many applications are performed under adverse conditions when well conditions are unknown or out of control, and when lost rig time and escalating well costs force poor decisions and excessive risks to be taken. Because of continued remedial job failures, many theories and misconceptions still exist concerning these applications, and much of the industry still regards remedial cementing as more an art than a science. In reality, many scientific improvements in well diagnostic and logging tools, cementing compositions and additives, mixing and pumping equipment, downhole tools, placement techniques, and computer simulators have greatly enhanced remedial cementing. Proper use of these technologies, coupled with thorough knowledge, planning, and experience can improve success rates and reduce job costs.

Many remedial cementing applications fail because the problem is either misdiagnosed or not detected early enough for it to be corrected before excessive damage occurs. Before any remedial application, the existing downhole conditions, the cause and magnitude of the problem, and the expected results from the application must be determined. Then the necessary planning, design, and placement procedures can be engineered that will match downhole conditions and enable the operation to be completed. Whenever downhole problems, wellbore conditions, and expected results cannot be defined or controlled, time and money will be wasted and the repair will probably be unsuccessful. In addition, more well damage or total well loss may occur if the wrong decisions are made.

Remedial cementing operations consist of two broad categories: squeeze cementing and plug cementing. This chapter defines and describes the various types of operations for each of these two broad categories and provides fundamental procedures and operational practices.

11-2 SQUEEZE CEMENTING

Squeeze cementing is the process of placing a cement slurry into all necessary wellbore entry points under sufficient hydraulic pressure to dehydrate or "squeeze" water from the cement slurry, leaving a competent cement that will harden and seal all voids. The slurry's rate of dehydration is controlled by the pressure exerted on the cement slurry, the rate of filtrate loss from the slurry, and the size of the squeeze interval or permeability of the formation. Filtrate loss is usually controlled by fluid-loss additives added to the cement slurry design, discussed in this chapter as well as in Chapter 8. Squeeze cementing can be applied during drilling, cementing, or completion operations in a well, or later when workover operations are performed. Squeeze cementing is often applied to change wellbore conditions or correct downhole problems for the reasons outlined in Table 11-1 (Bradford and Cowan, 1991).

In each of these cases, proper diagnosis of the problem, understanding and controlling the wellbore conditions, and thorough planning, design, and engineering of the application are required for a successful squeeze treatment. The general success of squeeze cementing is fairly low; estimates of successful jobs range from 50% for perforation squeezes (Krause and Reem, 1992) to much lower percentages of success for other types of squeeze operations. At times, misdiagnosis of a well problem may result in unnecessary squeeze treatments. However, numerous case studies, such as those documented by Krause and Reem (1992) and Bour *et al.* (1990), report much greater success when certain guidelines and practices are used in specific squeeze applications or locations. These guidelines and practices are usually derived from the application of squeeze theory and the design procedures presented in this chapter.

11-2.1 Squeeze-Cementing Theory and Mechanics

During squeeze cementing, hydraulic pressure is used to force cement slurry into formation voids or to squeeze

Table 11-1 Reasons to perform a cement squeeze treatment

Repair a faulty primary cement job
- Repair a weak casing shoe
- Stop gas cutting or influx into cemented annulus
- Complete annular fill in casing or liner tops
- Seal high-side or low-side channels caused by improper drilling-fluid displacement or poor casing centralization
- Seal microannuli between the casing/cement or formation/cement interfaces

Isolate formation intervals
- Temporarily abandon production zone to test other intervals
- Permanently abandon nonproductive zones
- Stop lost circulation in open hole during drilling
- Seal off lost-circulation zones opened during cementing or completion
- Seal off depleted zones from production intervals in multizone completions

Alter formation characteristics
- Reduce water-oil ratio
- Change oil-gas ratio

Repair casing problems
- Repair parted or split casing, or leaking joints
- Patch holes worn in casing from running operations
- Seal eroded or corroded casing encountered during workovers

water from a cement slurry into formation permeability, forming a densely packed, impermeable filter cake. The cement or filter cake then hardens into a permanent seal against fluid movement into or from the wellbore. All conditions outlined in Table 11-1 relate to unwanted fluid movement, and squeeze-cementing operations form a barrier to this flow.

Squeeze cementing was first recognized as a remedial cementing treatment for sealing downhole water flows (Millican, 1961). In initial usage, high pressure was used to place cement in the formation to seal the flow of unwanted water. Howard and Fast (1950) further described the theory and application of various squeeze techniques to isolate producing formations. These early applications were performed at pressures higher than the fracture initiation pressure of the formation, thus creating cement-filled hydraulic fractures rather than leaks sealed with filter cake. Later studies (Huber and Tausch, 1953) determined that pressures lower than formation fracture initiation pressures were required to create an actual seal.

The theory of low-pressure squeeze cementing involves the dehydration of an aqueous slurry to form a dense filter cake. This filter cake, resulting in cement grain-to-grain contact, is described by Harris and Carter (1964). Filter cake forms when a pumpable cement slurry is placed adjacent to the permeable formation that will be sealed, and pressure is applied to the cement slurry from the surface. As long as pressure inside the wellbore is greater than formation pore pressure, fluid is lost from the slurry, and filter cake is formed. Filter-cake properties vary with formation permeability and cement slurry fluid loss. Figure 11-1 shows the differences in filter-cake appearance resulting from variations in fluid-loss control, while Figure 11-2 shows the different filter-cake nodes created by cement squeezes under varying degrees of fluid-loss control.

Thus, the mechanics of squeeze cementing involve slurry fluid-loss control (Chapter 8) and controlled pressure application. The magnitude of the hydraulic pressure applied depends on formation permeability and fracture initiation pressure.

11-2.2 Problem Diagnosis

Squeeze operations are frequently performed to repair faulty primary cement jobs. However, many of these cement jobs are totally unnecessary and often unsuccessful because the controlling factors are unknown, or the

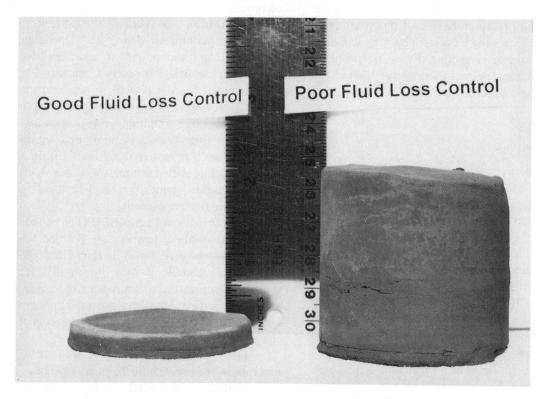

Figure 11-1   Effect of fluid-loss control on volume of filter-cake formation

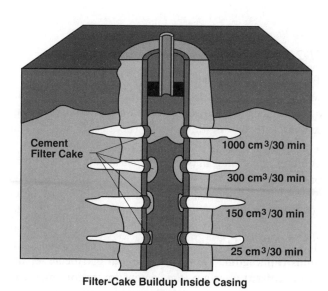

Figure 11-2   Filter-cake buildup inside casing perforations

bond logs used for primary job analysis are misinterpreted (Smith, 1991).

Squeezing a casing shoe is a common process that is often performed when a leakoff test (LOT) or pressure-integrity test (PIT) produces a pressure gradient that is too low for drilling to continue safely. Poor hole-volume estimates, partial or complete loss of returns during primary cementing, high filtrate loss from the cement column, or cement fallback (U-tubing) caused by float equipment failure produces incomplete annular fills at the top of the casing and weak or empty liner laps. Improper slurry designs tend either to gel or to have prolonged transition-to-set times or high filtrate losses that promote fluid or gas influx as well as gas migration in the cement column. These effects create channels in the annulus. Channels can also be created as a result of (1) incomplete drilling-fluid displacement and drilling-fluid/cement contamination caused by poor placement practices, (2) improper spacer design or insufficient spacer volumes, (3) poor casing centralization, or (4) hole swabbing and pressure surging caused by excessive pipe reciprocation or rotation. As discussed in Chapter 8, each of these problems can be avoided, but they can also be corrected by squeeze cementing if they are properly diagnosed. In addition, the controlling factors that affect primary cementing also affect squeeze applications.

11-2.2.1 Temperature

Wellbore temperatures affect cement slurry properties and performance more than any other factor. Before a primary cement application can be properly evaluated or a successful squeeze operation can be designed, temperature behavior in the wellbore must be thoroughly understood. An accurate bottomhole circulating temperature (BHCT) is necessary for the proper design and testing of such slurry properties as rheology, thickening time, fluid loss, etc. If the BHCT prediction is too high, cement slurries may be over-retarded or overdispersed, extending the thickening time and promoting slurry stratification, fluid commingling, and gas influx into the annulus (Tilghman, 1991). If the BHCT prediction is too low, pumping times may be shortened and slurry viscosity may increase, which could result in the application ending prematurely.

As little as a 20°F variance in BHCT can either reduce thickening time by as much as 50% or increase it by 100% (Jones, 1986). Fluid loss also varies with temperature. If the wrong BHCT is used, the slurry may not have sufficient fluid-loss control, which will affect the annular volume as well as the casing and formation seals.

An accurate bottomhole static temperature (BHST) provides information regarding (1) the rate of compressive-strength development, (2) the ultimate cement strength, and (3) the waiting-on-cement (WOC) time for pressure-testing or bond-logging. Other important temperature parameters that are often overlooked are listed below:

- Reduction in temperature from initial BHCT at the start of the treatment to the circulating temperature at the end of placement

- Rate of temperature increase from the lowest circulating temperature in the wellbore, obtained while pumping the cement slurry, to BHST where compressive strength develops

- Difference between cooldown and heatup rates at the bottom of the cement column (casing shoe) and the top of the cement column (casing or liner lap)

The time required to reach BHST may range from 24 to 36 hours. During this period, the cement may not develop much compressive strength, depending on the type of retarders and polymeric additives used in the design. After the BHST is reached, another 1 to 2 days may be needed for the cement to set hard enough to withstand pressure testing or to attenuate a bond-log signal with enough amplitude for proper analysis or interpretation (further discussion of bond-log interpretation techniques is presented in Chapter 10). In addition, any commingling with incompatible fluids, such as drilling fluid with the cement slurry, may further retard compressive-strength development.

American Petroleum Institute (API) test conditions do not duplicate wellbore temperature profiles; thus, compressive strengths determined at forced heatup rates to BHST in a laboratory may be totally unrealistic. Downhole-temperature measurement tools and computer simulators (Chapter 8) can provide temperature profiles that result in a much better understanding of cement behavior in the wellbore.

If a primary cement application is tested or logged prematurely, it can easily be diagnosed as faulty because the evaluation parameters (transit time, attenuation rate, bond index, etc.), often exceed cement quality, as discussed in Chapter 10. As a result, a decision to squeeze-cement may be made when nothing requires repair and no voids exist in the annulus to accept the squeeze slurry. Understanding wellbore temperature behavior is necessary for determining the need for squeeze cementing and for successfully designing the application.

11-2.2.2 Pressure

Typically, as pressure increases, the thickening time for a cement slurry decreases and compressive-strength development increases. However, pressure exerted on a setting cement at the wrong time can be very detrimental. When a pressure test is performed on a weak cement in the annulus, the casing may expand elastically, which compresses the cement in the annulus. Once the pressure is relieved, the casing will return to its original size; however, the low-strength cement will deform plastically and thus not return to original size with the casing, a condition which could break the bond and create a microannulus (Pilkington, 1988). Microannuli can occur at the casing/cement interface, the cement/formation interface, or within the cement itself. Excessive pressure may also force additional filtrate from the setting cement into the formation, which (1) reduces the cement volume and annular fill and seal, (2) promotes gas influx, and (3) may damage producing formations. In addition, if excessive pressure is applied on a weak primary cement or on a squeeze cement that has been dehydrated in the annular voids or perforation tunnels, the cement may be fractured, destroying the seal.

Premature pressure testing or the disturbance of primary cement before it has thoroughly cured may create a need for squeeze cementing. When an LOT or PIT is performed at a casing shoe and the cement has not completely hardened because the WOC time was not properly planned, a weak casing shoe is often diagnosed. Squeezing a high filtrate-loss or neat cement slurry to cure this problem merely reinforces the existing cement with packed cement solids as total dehydration of the slurry occurs. This compaction increases the density of the cement around the shoe and provides additional compressive strength. This condition may or may not increase the LOT or PIT enough on the first squeeze attempt for drilling to continue, and the technique may have to be repeated several times. If the casing is set near a production interval, formation damage from cement filtrate or complete formation breakdown could occur during the casing-shoe squeeze. In some cases, additional WOC time could be more cost-effective and could eliminate the need for an unnecessary squeeze application.

11-2.2.3 Bond-Log Interpretation

A cement bond-log (CBL) or evaluation tool (Chapter 10) can measure annular cement fill, compressive strength, and the bond integrity of the cement. A signal is transmitted through the cement and recorded. The amplitudes of the signal and transit time are used for determining the compressive strength; however, the attenuation rate is often set to measure a hard cement (>3,000 psi compressive strength). When the log is run prematurely and the WOC time has not been long enough for the cement to harden, or the cement has become contaminated with drilling fluid, the cement bond-log may show absolutely no cement in the annulus (Goodwin, 1989). In this case, the attenuation rate may be set too high, and a simple reduction in the attenuation rate may show that a competent cement is in place. In many cases, a compressive strength of less than 500 psi will support and protect the casing and permit drillout.

Some logs may show competent cement above or below a permeable or fractured zone but a very poor bond across the zone. In this case, job designers must compare the wellbore conditions across the interval, the spacer program, the cement slurry design, and the placement practices to determine the integrity of the cement and properly interpret the log. Bond logs can be easily misinterpreted if they are analyzed independently from the cementing operation (Jones *et al.*, 1994). An inaccurate bond-log analysis frequently results in the costly decision to squeeze-cement the well; often, an unsuccessful squeeze operation is the outcome.

When these diagnostics are used repeatedly, confidence and experience are added to the decision process. When the outcome clearly indicates that a squeeze-cementing application should be performed, it can usually be done successfully when it is properly planned, designed, and executed.

11-2.3 Job Planning

Once the need for a squeeze cement job has been confirmed, planning, design, and execution procedures can be combined into a successful program. The fourth and final step in any squeeze-cementing operation is proper evaluation of the application's success. The guidelines presented in Table 11-2 can be used to complete this process.

Table 11-2 Process of designing a cement squeeze treatment

Planning	• Determine the problem and reason to squeeze • Analyze the wellbore conditions • Select the appropriate squeeze technique
Designing	• Select the proper squeeze tools • Select all other fluids used in the process, i.e., perforating fluids, acids, cement spacers, completion fluids, etc. • Design the cement slurry to correct the problem • Prepare a detailed job procedure
Executing	• Prepare the wellbore and clean all voids to be squeezed • Set the squeeze tools at the desired locations • Set equipment and mix the slurry properly • Place the cement and apply the squeeze pressure • Hold the cement in place until it hardens • Allow enough WOC time to test or log the squeeze job
Evaluating	• Perform positive and negative pressure tests • Perform bond log

Different problems require different squeeze application techniques for cement slurry to fill all voids and obtain a proper seal.

11-2.3.1 Wellbore Analysis

Conditions of the surrounding formations, wellbore fluids, and tubulars must be analyzed immediately before a squeeze is performed. If excessive time has lapsed between the time the problem was discovered and the time of the actual squeeze operation, wellbore conditions may have changed dramatically. For example, a problem diagnosed several days or weeks earlier in the operation may have worsened, or the problem may no longer exist. The only way to determine wellbore conditions is to re-test the well immediately before the squeeze application.

Formation Characteristics

If squeeze cement will be placed between the casing and the formation, job designers must know the formation type, formation sensitivity, permeability, pore pressure, and fracture pressure to design the right type of squeeze application. In addition to being more permeable than carbonates, limestones, or dolomites, sandstones are subject to higher filtrate losses. Carbonates and limestones will seldom accept cement filtrate, but usually contain naturally occurring fractures that can be sealed only when they are bridged by cement or other materials during the squeeze operation. Shales and salts are almost always impermeable; however, exposure to fresh waters, light salt brines, or filtrates may cause these formations to swell, soften, or even deteriorate. Most formations contain some clays or silts that can swell or migrate in the presence of water or cement filtrate. These clays or silts reduce permeability and may possibly damage production intervals. Induced clays from drilling-fluid losses or naturally occurring clays in the formations can usually be dissolved with hydrochloric/hydrofluoric (HCl/HF) acid blends, which can clean and stimulate the formation and enhance the squeeze-cement operation. Unconsolidated formations could flow into the wellbore if hydrostatic pressures become too low, or if the work-string is pulled too quickly after the squeeze, resulting in excessive swabbing pressure.

Knowing the pore pressure and fracture pressure of surrounding formations is vital. When a squeeze is placed against a producing formation, pressure exerted at the squeeze point will stop the flow from the formation as long as the squeeze pressure exceeds the pore pressure. However, if the squeeze pressure or hydrostatic pressure

is reduced before the dehydrated cement has attained enough strength, the formation may flow. Pressure reduction can also result (1) from swabbing action, (2) from solids bridging above the formation, reducing hydrostatic pressure, or (3) through communication with a lower-pressured or depleted zone above or below the cement. This condition can create more channels in and around the cement or push/pull it out of the squeeze tunnel. Conversely, when squeeze pressure exceeds the fracture pressure, whole cement slurry is forced into the formation. This condition can either occur when squeeze pressures are miscalculated, fluid loss is high, and excessive filter-cake buildup causes more pressure applied at the surface, or when excessive pressure is exerted during the reversing-out process after the squeeze.

Wellbore Fluid Behavior

Understanding wellbore fluid properties and behavior while taking time to diagnose and plan the squeeze application is very important. For example, some drilling fluids become viscous and extremely difficult to condition when exposed to static temperatures and pressures for extended periods. This condition may make wellbore circulation and cleaning very difficult before the squeeze application. Likewise, if drilling fluid is trapped in the annulus or has commingled with cement as a result of the cement bypassing the fluid during primary cementing, it may be possible to circulate the drilling fluid or contaminated cement out the annulus. A low-pressure squeeze can then be conducted. However, after the drilling fluid or contaminated cement has dehydrated and hardened further with time and temperature, the channel may be difficult to open and may require fracturing. Dehydration in the annulus may also widen and extend the channel, creating a bigger problem. Any fluid or gas movement behind the casing between zones may change the annular fill or seal, or fluid migration between perforated intervals may cause dilution and subsequent settling and possible compaction of solids in the wellbore. Replacing drilling fluids with solids-free, compatible fluids may be beneficial. Maintaining wellbore circulation during the waiting period can also stabilize the wellbore fluids and the wellbore temperature. Time, temperature, and pressure may heal certain problems but magnify others.

Wellbore Tubulars

Rust or scale on the casing may dissolve when acid is used to clean perforations or formations for a squeeze application. As the acid spends, ferrous hydroxide or other reaction products precipitate and contaminate the formation or cement slurry. Where casing has rusted or corroded during prolonged storage or has been exposed to corrosive brines or gases such as H_2S or CO_2 in the wellbore, the casing may rupture if excessive pressure is applied to it. Irregular dimensions, pinch points, burrs, or etched surfaces may inhibit the proper placement and operation of squeeze tools. Excessive wear on collars or worn threads may also cause leaks or completely separate the collars from the tubulars when the casing is moved up or down in the wellbore. Regardless of the squeeze technique applied, wellbore conditions must be compatible with the procedure, or the results could be unsuccessful.

11-2.3.2 Squeeze Technique Selection

When selecting a squeeze technique, job designers must consider

- The pressure required to place the cement and cause a squeeze

- The method used to pressure-isolate the interval

- The placement procedure needed to fill the voids and seal off the problem area

Pressure

Squeeze cementing can be performed at pressures below or above fracturing pressure. Equations for calculating squeeze pressure are presented later in this section. Applications below fracture pressure (*low-pressure squeezes*) are the most common. Normally, low-pressure squeezes can be performed whenever clean wellbore fluids can be injected at a reasonable rate into a formation (permeable sandstone, lost-circulation zone, fractured limestone, etc.) or void (channel, casing split, perforation tunnel, etc.). Under such conditions, filtrate from the cement slurry can easily be displaced into the formation at low pressures, and the dehydrated cement will remain to seal the leak. Typically, only small slurry volumes (50 to 100 ft^3) are needed for a low-pressure squeeze. Unless a fracture is created or opened in the process, whole cement slurry will not invade most formations.

When circulation is possible in the annulus between two openings or sets of perforations, a low-pressure or "circulating" squeeze can be applied. Slurry volume requirements may be much greater in these cases, depending on the height of the zone or distance between perforations.

Squeeze applications performed at pressures above formation fracturing pressure (*high-pressure squeezes*) are performed whenever fracturing is required to displace the cement and effectively seal off low-permeability formations or to establish communication between channels and perforations. Slurry volumes and filtration rates will vary depending on the size of the fractures, the depth of the cement fill, and the leakoff to surrounding formations.

"Block" squeezing is the process of squeezing off permeable sections above and below a potentially productive zone to isolate it for production or stimulation. High pressure is also normally required to widen natural fractures to enable cement placement. This process is commonly used in limestone, shales, or salts to seal off sensitive or nonproductive zones, or to prevent gas or fluid migration through natural fractures. Block squeezes may also be used to change oil/water or oil/gas contacts, and to isolate and seal off unwanted or depleted production intervals. A vertical fracture is usually created during a high-pressure squeeze just as in other hydraulic fracturing operations. In some cases, when the overburden pressure above the zone is less than the fracturing pressure, a horizontal fracture may be created, but this condition is only possible in very shallow applications.

Isolation

The squeeze location can be isolated with the "Bradenhead" technique or through the use of squeeze tools such as packers, retainers, and bridge plugs. In the Bradenhead technique, a workstring is run into the wellbore, and cement slurry is pumped down the workstring and spotted across the squeeze interval. Once cement is in place, the string is pulled above the top of the cement, and the backside of the string is sealed off at the surface with the blowout preventer (BOP) or wellhead control valves. Fluid is then displaced down the workstring to squeeze the cement. Displacement continues until the necessary amount of cement slurry is squeezed, or a predetermined squeeze pressure is obtained. This method is used most frequently in coiled tubing applications, shallow wells, and lost-circulation zones, or to set cement plugs. A disadvantage of the Bradenhead technique is that the pressure from the surface is applied not only to the cement slurry but to the entire wellbore. If the pressure exceeds the burst pressure of the casing up the hole, casing rupture or formation fracture could occur.

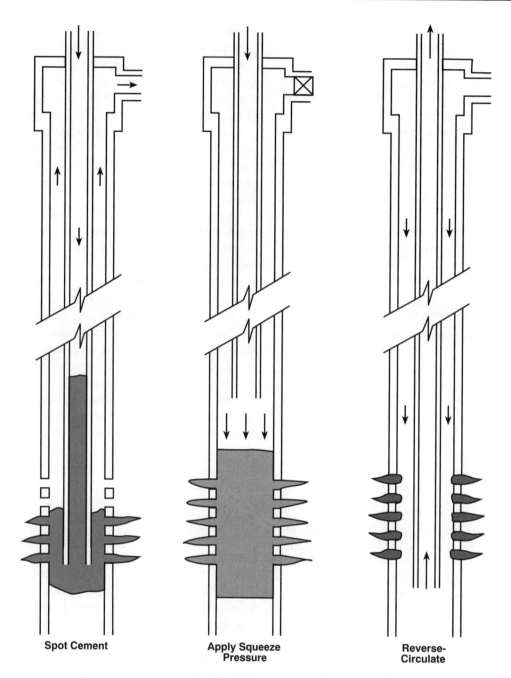

Spot Cement   **Apply Squeeze Pressure**   **Reverse-Circulate**

Figure 11-3   Typical wellbore setup for Bradenhead squeeze application

Figure 11-3 shows the typical wellbore setup for a Bradenhead squeeze application.

Squeeze tools isolate pressure much closer to the squeeze interval for better pressure control. Isolation is achieved by a bridge plug set below the squeeze zone and a packer or retainer set above the zone. A workstring can be either attached to the packer when it is run in the hole or later inserted into the packer or retainer after it has been set. Cement slurry is then pumped down the workstring and squeezed into the required interval between the packer/retainer and bridge plug. Squeeze pressure is isolated between the upper and lower tools, which minimizes the risk of breakdowns in other formations or casing collapse, and it allows the process to be more easily controlled. Figures 11-4 and 11-5 show the placement of isolation tools and typical surface-pressure response for this type of squeeze application.

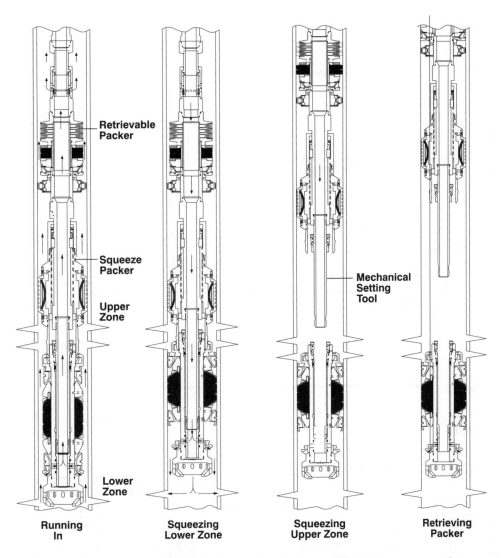

Figure 11-4 Squeeze-tool placement and operational sequence for isolation squeeze of two zones

Placement Method

A packer/retainer can be set just above the zone, and cement slurry can be continuously pumped into the squeeze interval until the necessary squeeze pressure is obtained. Continuous pumping operations are referred to as *running* or *walking squeezes*. Although the running squeeze is easier to design and apply, controlling the rate of pressure increase or obtaining the necessary squeeze pressure during the job is sometimes difficult. Many running squeeze applications designed for low-pressure squeezes have become high-pressure applications because of poor slurry designs and lack of control.

Squeeze tools can be set high enough above the squeeze interval to permit the entire cement slurry volume to clear the squeeze tool before pressure is applied from the surface. Setting height is easily calculated from casing dimensions and slurry volume. This method is preferred when retrievable tools are used, since calculating the setting height will reduce the risk of the tools becoming cemented in place in the wellbore. A running squeeze can still be performed, but this method is most suited for a "hesitation" squeeze job, which consists of alternating pumping and static sequences until a squeeze is obtained. Figure 11-6 illustrates typical surface pressure response for a running or hesitation squeeze.

When squeeze pressures are slow to build during the pumping sequence, the pump can be stopped to allow the slurry to build gel strength, which further reduces leakoff

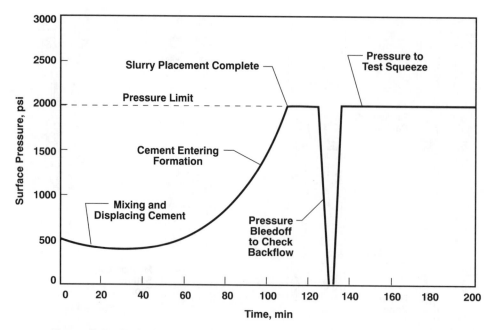

Figure 11-5   Typical surface-pressure response for walking squeeze application

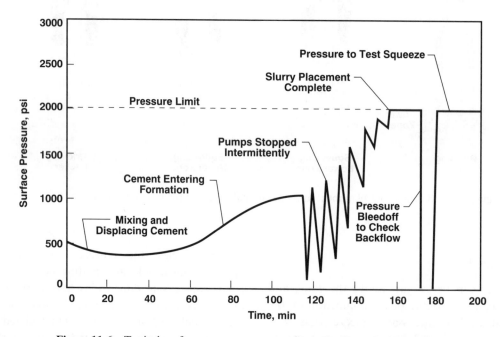

Figure 11-6   Typical surface-pressure response for hesitation squeeze application

rates and controls filter-cake deposition. This technique results in a better foundation to squeeze against, especially across lost-circulation zones, long perforated intervals, or extended casing splits. Pumping sequences may vary between 2 and 5 bbl of cement slurry pumped at rates ranging from $\frac{1}{2}$ to 1 bbl/min, followed by a 5 to 15-minute shutdown period. This sequence may be repeated several times before the required squeeze pressure is obtained. Thixotropic slurries or low fluid-loss slurries are often used in this process, as well as lost-circulation materials.

When downhole problems prevent the hole from being circulated and cleaned, or the problem cannot be addressed by conventional perforating and squeezing, the cement may have to be *bullheaded* down the annulus. Although bullheading is not a true squeeze technique, it may be the only way to seal off a troublesome zone. Bullheading pushes everything ahead of the cement into the formation; therefore, care must be taken to ensure that incompatible fluids are not forced into potential producing formations. Often, low-density, high-yield, inexpensive slurries are used, because large slurry volumes and multiple attempts may be necessary to complete this process.

11-2.4 Job Design

Many factors affect squeeze applications; however, when the problem is diagnosed properly and a good plan is prepared, a squeeze cement design can be a fairly simple process.

11-2.4.1 Tool Selection

Tool selection and placement are critical since use of the wrong tool may cause more problems to occur after the squeeze. Two basic types of tools are used for squeeze cementing: retrievable and nonretrievable (drillable).

A wireline or mechanical bridge plug seals off the zone below the squeeze interval to keep the slurry from being pumped, or from freefalling down the casing. Most plugs are drilled after the job; however, some can be retrieved if they are not cemented in place. Sand is often placed on top of the plug to prevent it from being cemented and to allow retrieval after the job. This technique is especially useful for squeezing multiple zones.

When they plan to reverse-circulate excess cement after placement, operators use a retrievable packer above the zone. Retrievable packers are equipped with bypass valves that permit complete circulation of the wellbore (1) while the toolstring is being run in the hole, (2) after the packer has been set, or (3) during reverse circulating after the job. This circulation capability prevents excessive pressure from being applied to the formation while the toolstring is being run into in the hole, and it prevents the well from being swabbed when the packer is released and retrieved. Retrievable packers can be either tension- or compression-set and can be run on wireline or a workstring. A fiberglass or aluminum tailpipe can be attached to the bottom of the packer to allow the cement slurry to be spotted close to the zone while the packer is set well above the zone (Hempill and Crook, 1981).

Drillable packers or retainers can also be used; however, double check-valves are built into these tools to prevent unwanted flow from either direction. Once the packer is set, a workstring is stung into the retainer, then cement is pumped and squeezed below the retainer. The workstring is then pulled out, and the retainer is closed. Retainers are often used to help prevent pressure reductions in the wellbore from causing cement flowback either after the squeeze job or when the cement is separating zones that have a high pressure differential. Setting a retainer between the zones isolates the intervals and permits operators to squeeze the lower zone and the upper zones in one application.

11-2.4.2 Fluid Selection

As discussed in Chapter 8, many drilling fluids are incompatible with cement slurries and can cause excessive gelation and delayed compressive-strength development. As shown in Chapter 12, drilling fluid is also damaging to many producing formations. Whenever possible, drilling fluids should be circulated out the wellbore and replaced with clean, solids-free fluids such as water, KCl, and NaCl brine before the squeeze is performed. Completion fluids can also be used, but some completion fluids such as zinc bromide, calcium chloride, or calcium bromide brines are also incompatible with cement slurries, and they will cause a very rapid set.

When problem areas behind the casing will be squeezed, perforations must be shot in the casing to establish a communication path to the problem. Whenever possible, the casing should be filled with clean fluid during the perforating process (Chapter 13). If drilling fluid is left in the hole during perforation, the fluid and any solids remaining in perforation tunnels should be removed before the squeeze. Otherwise, they might block the flowpath or be pushed into the formation ahead of the cement slurry.

Often, perforations become clogged and are very difficult to open before the squeeze. Perforations cannot always be opened by the simple application of pressure; other methods may be required. Diverting agents or perforation ball sealers are frequently used to divert fluid as pressure builds and to open new perforations. Perforation wash tools are also recommended. Surge tools may also help open stubborn perforations. HCl or HCl/HF acids are frequently used to clean perforations and dissolve any rust, scale, or clays behind the casing or in the formation. Surfactants and chelating agents increase acid penetration, water-wet the surfaces, and trap metallic salts as the acid spends and the pH rises. Diesel, xylene, or aromatic solvents are frequently used to remove oil-based drilling fluids, asphaltenes, and paraffins around the perforations. Unless perforations are open and cleaned, they will seldom accept the cement slurry, and an unsuccessful squeeze may result.

Spacer fluids or preflushes should be used ahead of the cement slurry. These fluids separate the slurry from the drilling fluid when the squeeze is performed with drilling fluid in the wellbore. Spacers and preflushes should (1) be compatible with the drilling fluid and cement slurry, (2) effectively sweep drilling fluid from the wellbore, and (3) water-wet the formation surfaces to enhance cement bonding.

11-2.4.3 Slurry Design

Key properties of any squeeze slurry design are density, viscosity, thickening or pumping time, and fluid loss (Smith, 1990). Compressive strength is not that critical since the slurry is dehydrated in place. As dehydration occurs, the cement density climbs quickly and may easily exceed 20 lb/gal, a density that usually produces tremendous compressive strength. API test conditions are generally the basis for cement slurry designs; however, the best designs simulate actual well conditions, and again, temperature and pressure are the main factors to control. The same additives used in primary cementing are required under the same conditions in squeeze cementing.

The following basic design guidelines are commonly used:

Basic Slurries	light, thixotropic Class A, G or H cement	11 to 13 lb/gal
Extenders	bentonite, sodium silicate, calcined gypsum, normal thixotropic- cement, calcined gypsum	13.8 to 15 lb/gal
	Class A, G or H cement	15.6 to 17.5 lb/gal
Additives	retarders, dispersants, fluid-loss additives 35% silica at BHST >230°F, barite or hematite at densities >17.5 lb/gal	

Fluid loss is probably the most important parameter to consider (Beach *et al.*, 1961; Binkley *et al.*, 1958). Based

on API high-pressure, high-temperature (HPHT) tests (1000 psi, 325-mesh screen), regular cement mixed with water has a fluid loss greater than 1000 cm³/30 min (API Spec 10, 1990). Almost total dehydration of the cement slurry occurs in seconds, creating a thick filter cake. When cement slurry is used without fluid-loss control, bridging can occur in the workstring, perforation tunnels, or annulus. Once cement particles pack off, the excessive pressure needed to fracture the cement may also fracture the formation. Many squeezes are performed with moderate fluid-loss control (200 to 500 cm³/30 min API). This level of fluid-loss control provides rapid dehydration but reduces the rate of filter-cake development and will seldom bridge a casing or workstring. For maximum slurry penetration, a 50-cm³/30 min API fluid-loss control level is preferred. Dehydration is very slow, and only a moderate filter cake develops, but this filter cake is adequate to seal perforation tunnels. Although extremely low fluid loss (>20 cm³/30 min) is possible with modern additives, the resultant filter cakes may be too thin to seal off perforation tunnels or too weak to withstand pressure; therefore, they may be easily pumped away or swabbed off after the squeeze. A variety of solid and liquid fluid-loss additives is available for every type of cement design condition. Recommended squeeze cement slurry design guidelines are presented in Table 11-3.

Cement composition and execution procedures can be calculated by commercial squeeze-cementing design simulators that are similar to those used for designing primary treatments. Figure 11-7 provides an example of simulator input and output. As described by Bour *et al.* (1990), simulators compare theoretical surface pressure predictions to actual pressure responses and relate these responses to the downhole progress of the squeeze.

Experience and repeated successes furnish the best design guidelines for squeeze cementing; however, a primary cementing design and procedure that eliminates the need for squeeze cementing is more desirable.

11-2.5 Squeeze-Job Procedures

Thus far, this chapter has presented a technical review of problem diagnostics, types of applications, slurry specifications, and job design considerations for avoiding or applying squeeze cementing. Careful review of this information will provide a better understanding of squeeze cementing; however, training and experience are necessary to master this complex process. Numerous publications on this topic have been written in the past 40 years,

Table 11-3 Cement composition properties for various types of squeeze application

Problem	Recommended Slurry Properties
Casing-shoe squeeze	API density, high fluid-loss, rapidly dehydrating slurry or moderate fluid-loss slurry followed by high fluid-loss slurry
Liner top squeeze	Low fluid-loss slurry to prevent bridging in tight annulus
Water zone or lost-circulation interval	Thixotropic cement, no fluid-loss control, LCM additives; or may need barrier to squeeze against, i.e., CaCl$_2$ brine and sodium silicate spearhead
High water flow in zone	Slurry Oil Squeeze or Gunk Squeeze, diesel oil:bentonite $\pm$ cement combination, or larger water seal-off cement design with silicate spearhead and LCM
High-permeability zone	Low fluid-loss slurry
Vugs or fractured formation	Thixotropic slurry followed by moderate fluid-loss slurry
Low-permeability zone	Low-to-moderate fluid-loss slurry for greater penetration in high-pressure squeeze
Squeeze lower perforations	Low fluid-loss slurry to prevent dehydration in casing and penetration into the zone; do not use LCMs
Split casing	No fluid-loss control for short sections, and thixotropic slurries for longer sections
Corroded casing	Low fluid-loss slurry, or thixotropic slurry followed by low fluid-loss slurry
Long perforation interval	Low fluid-loss slurry, or low fluid-loss slurry followed by a high fluid-loss design
Failed connection	Extremely low fluid-loss if split will take whole cement epoxy resin, or very fine cement (8000 to 9000 cm^3/g)

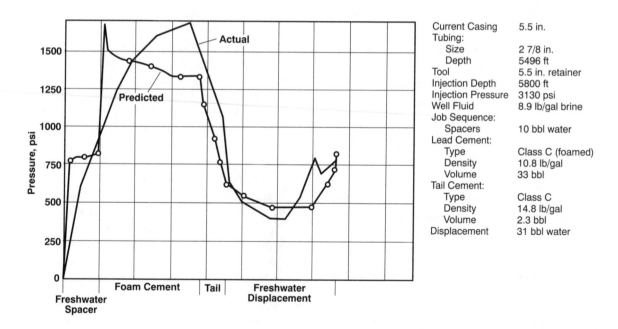

Figure 11-7 Comparison of actual surface pressures during a foam cement squeeze with predicted surface pressure

and the science of squeeze cementing has been greatly improved in the last decade (Rike *et al.*, 1982; *Cementing*, 1992). The following example illustrates the various aspects of preparing a squeeze-cementing procedure. Although every type of application cannot be presented in this book, the following example clearly shows the steps involved and the complexity of this application.

11-2.5.1 Diagnostics

A primary cement job has been performed on a 7-in. OD, 26-lb/ft production casing. After 48 hours WOC time, the bond log and the interpretations of other logging tool results, when compared to the primary cementing program, confirm that a channel exists in the annulus across a principal production interval. The casing depth is 10,000 ft total depth (TD), and the zone of interest ranges from 9750 ft to 9850 ft. The BHCT is 140°F. The formation is a low-permeability sandstone that has an 0.8-psi/ft fracture gradient. The zone at TD is weaker than the production interval and has a fracture gradient of only 0.7 psi/ft. A 10-lb/gal polymer drilling fluid was used to drill to TD and displace the primary cement. Isolation was necessary to test and produce this zone, and later to test secondary zones located up the hole. A "block" squeeze will be necessary to seal off the channel. The following squeeze plan and cementing procedure are recommended to solve this problem.

11-2.5.2 Squeeze-Cementing Plan

Because a low fracture gradient exists at the bottom of the well, a bottom squeeze tool or bridge plug should be used. This tool will prevent possible overpressuring and fracturing out the bottom of the well as the attempted squeeze occurs at the production interval.

1. The interval is perforated with a tubing-conveyed gun run on $2\frac{7}{8}$-in. OD EUE tubing. Perforations are made across the top 50 ft of the interval (9750 to 9800 ft), 4 shots/ft at 360° phasing.

2. Since the 10-lb/gal polymer drilling fluid may be damaging to the formation, the wellbore is circulated, and 10-lb/gal NaCl completion brine is loaded into the hole for perforating.

3. After perforating is completed, the hole is circulated to clean the perforations. If necessary, a perforation wash tool can be run in the hole to jet and clean all perforations.

4. The tubing is then tripped out of the hole, and a retrievable bridge plug is installed, run in the hole, and set at a depth of 9850 ft. The tubing is then removed from the bridge plug and pulled up. Then, a column of approximately 30 ft of sand is slowly pumped with the brine and allowed to settle on top of the plug. This sand prevents the bridge plug from being cemented in the hole and brings the false bottom of the well up to 9820 ft (or 20 ft below the bottom perforation).

5. A circulating, retrievable packer equipped with a 200-ft tailpipe is run in the hole on the $2\frac{7}{8}$-in. tubing to a packer depth of 9575 ft. This packer extends the tailpipe into the middle of the perforated interval.

With circulation ports open on the packer, the hole is circulated.

6. Next, 20 bbl of 7.5% HCl: 1.5% HF acid + 1 gal/bbl nonionic surfactant is pumped down the tubing to "break down" the formation. Once acid is displaced to the end of the tailpipe with the completion brine, the packer is set and the circulation ports are closed. Pump pressure is built up slowly until formation breakdown occurs and acid injection begins. All pressures and flow rates are recorded.

7. Acid is displaced into the perforations and the injection rate and pressure are monitored. An injection rate of 2 bbl/min is usually sufficient to clean the perforations, clear the channel, and treat the formation near the wellbore. Alternating rates or pressure surging may help open and clean any partially clogged perforations. Once complete, the pumps are shut down and the instantaneous shut-in pressure (ISIP) is recorded.

8. The packer is then released (unseated), pulled up the hole, and reset at a level high enough above the bottom perforation to allow complete cement slurry placement below the packer and across all perforations before the squeeze is started. If the packer were set lower (for example, at the perforations), the cement would have to be reverse-circulated out of the tubing if the total volume was not displaced during the squeeze. Setting the packer high and using the tailpipe allows all cement to be placed below the packer before the squeeze begins and eliminates problems that may be caused by reverse-circulating. The cement that remains in the casing after the squeeze is drilled out later. Setting depth has been calculated at 9285 ft. When the packer is set, the circulation ports are opened and completion brine is circulated.

9. Because the cement slurry is greatly accelerated if it is commingled with the completion brine, 30 bbl of 11-lb/gal compatible spacer is used to separate cement slurry from brine. A total of 25 bbl of spacer is pumped ahead of the slurry and 5 bbl are pumped behind the slurry.

10. Because of the great length of this interval and the possibility for cement to be displaced above and below the perforated interval behind the casing, 100 sacks of cement + 0.5% fluid-loss additive + 0.2% retarder is batch-mixed on the surface. The cement slurry density is 15.8 lb/gal; the slurry has a

thickening time of 3 hours and an API fluid loss of 100 to 200 cm³/30 min.

11. The following pumping sequence will occur: 25 bbl spacer, 20 bbl cement slurry, 5 bbl spacer, 10 lb/gal completion brine. Pump rates are controlled, and displacement occurs until the cement slurry has cleared the packer. Once the slurry is properly placed, the circulation ports are closed, and the squeeze process begins.

12. The cement is displaced at $\frac{1}{4}$ to $\frac{3}{4}$ bbl/min and the surface gauge pressure is monitored. Once pressure builds to the calculated squeeze pressure, pumping is stopped for 5 to 10 minutes; then, the pumps are slowly engaged and the pressure is observed. If all voids are full, the pressure should immediately increase to squeeze pressure. If squeeze pressure is not achieved, displacement is continued until the pressure builds. The process is repeated until the squeeze is complete.

13. The packer is slowly unseated and pulled up the well-bore until it is clear of all cement (approximately 10 casing joints). The wellbore is then reverse-circulated and brine is pumped down the backside of the tubing to clean the packer. Pump rates should be carefully controlled; otherwise, excessive downhole pressure could damage the squeeze.

14. Once reverse circulation is completed, the well is shut down for 6 to 8 hours of WOC time. At this time, the bottomhole assembly (BHA) and bit are made up, and then the bit is tripped into the hole to drill out any excess cement across the perforations. If the cement tagged is soft or drills too easily, the bit is pulled up, the hole is circulated, and the WOC time is extended until the cement can be safely drilled out.

15. Once drillout is completed, pressure is applied to the squeezed interval as a means of testing the squeeze. At this stage of the job, the squeeze must not be overpressured; excessive pressure could fracture the curing cement in the perforations or in the channel. Generally, 1000 to 1500 psi should be sufficient for testing, provided this range does not exceed the previously established injection pressure.

If the job has been successful, operations are continued as planned, and drilling resumes to the sand-fill on top of the bridge plug. If the bridge plug will be retrieved or moved and reset for future work, operators must trip out, make up the assembly, run into the hole, and unseat the bridge plug. If not, the bridge plug can be drilled out.

If the job has not been successful, the interval must be retested, the job must be replanned, and the interval must be resqueezed.

11-2.5.3 Calculations

The following data will be used for the squeeze-cementing calculations in this section.

Capacity of $2\frac{7}{8}$-in. EUE tubing:
(0.005794 bbl/ft or 172.6 linear ft/bbl)
Capacity of 7-in. OD (26-lb/ft casing):
(0.3826 bbl/ft or 26.135 ft/bbl)

Capacities are available from any pumping service company engineering handbook.

Perforated interval:
9750 to 9780 ft
Perforation density:
4 shots per foot, 360° phasing
Formation fracture gradient:
0.8 psi/ft
Completion-fluid pressure gradient:
10 lb/gal x 0.052 psi/ft = 0.52 psi/ft
Bridge-plug depth:
9850 ft (+ 30 ft sand-fill for false bottom of 9820 ft)

1. **Slurry volume and packer setting depth** Based on the assumption that 100 sk of cement will be used, the following volume and fill equations are used:

Volume Equation

$$V_{\text{slurry}} = 0.1781SY \qquad (11\text{-}1)$$

where V_{slurry} is the volume of the slurry, bbl; S is the number of sacks of cement, and Y is the yield, ft³/sk.

Fill Equation

$$H = V_{\text{slurry}}C \qquad (11\text{-}2)$$

where H is the height of cement column, ft, and C is the hole capacity, ft/bbl.

Given that 100 sk of slurry at 15.8-lb/gal density has a yield of 1.15 ft³/sk, from Equation 11-1, $V_{\text{slurry}} = 20.5$ bbl. From Equation 11-2, $H = 535$ ft of cement fill in the 7-in. casing.

Therefore, the packer setting depth that would allow complete cement placement below the packer and across all perforations before the squeeze sequence is begun would then be 9820 ft (false bottom) − 535 ft = 9285 ft.

A 200-ft tailpipe will help displace the cement across the entire perforated interval.

2. Displacement volume Equation 11-3 is used for calculating job displacement.

$$\text{Disp.Vol.} = \text{tubing volume} + \text{casing volume}$$
$$\text{(between top and packer)} \quad (11\text{-}3)$$

where

$$\text{tubing volume} = (0.005794\,\text{bbl/ft}\ 9285\,\text{ft})$$
$$= 53.8\,\text{bbl}$$
$$\text{casing volume} = [0.03826\,\text{bbl/ft} \times (9750\,\text{ft} - 9285)]$$
$$= 17.8\,\text{bbl}$$

and

$$\text{displacement volume} = 53.8\,\text{bbl} + 17.8\,\text{bbl}$$
$$= 71.6\,\text{bbl}.$$

Approximately 71 bbl is required to prevent overdisplacement.

3. Surface breakdown pressure Equation 11-4 is used for calculating the surface "breakdown" pressure, p_{bd}, to pump into the formation.

$$p_{bd} = H(FG) - \Delta p_h \quad (11\text{-}4)$$

where FG is the fracture gradient and Δp_h is the hydrostatic pressure, which is equal to the completion fluid gradient × depth. Therefore, $p_{bd} = (0.8\,\text{psi/ft} \times 9750\,\text{ft}) - (0.52\,\text{psi/ft} \times 9750\,\text{ft}) = 2730\,\text{psi}$. This pressure is the optimum breakdown pressure. Breakdown can occur at lower pressures depending on the number of opened perforations, increased efficiency with higher viscosity, etc.

4. Estimating squeeze pressure Several recommended guidelines are available for estimating squeeze pressure. One of the simplest is to add 1000 psi to the breakdown pressure; in this case, 2730 psi + 1000 psi = 3730 psi. This pressure is the pressure at

the perforations—not the surface gauge pressure. Another equation commonly used is

$$(p_{\text{squeeze}} + \Delta p_h)/H = 1\,\text{psi/ft} \quad (11\text{-}5)$$

$$\Delta p_h = \Delta p_{\text{cement}} + \Delta p_{\text{spacer}} + \Delta p_{\text{completion brine}} \quad (11\text{-}6)$$

Since

$$\Delta p_{\text{cement}} = (0.052\,\text{psi/ft} \times 20\,\text{bbl} \times 26.135\,\text{ft/bbl}$$
$$\times 15.8\,\text{lb/gal})$$
$$= 429\,\text{psi},$$

$$\Delta p_{\text{spacer}} = (0.052\,\text{psi/ft} \times 5\,\text{bbl} \times 172.6\,\text{ft/bbl}$$
$$\times 11.0\,\text{lb/gal})$$
$$= 494\,\text{psi}$$

and completion brine column = 9750 ft − 523 ft cement fill in casing − 863 ft spacer in tubing = 8364 ft brine fill

$$Dp_{\text{completion brine}} = (0.052\,\text{psi/ft} \times 8364\,\text{ft}$$
$$\times 10.0\,lb/gal)$$
$$= 4349\,\text{psi}$$

Thus, $\Delta p_h = 4329 + 494 + 429 = 5272\,\text{psi}$, and from Equation 11-5, $p_{\text{squeeze}} = 4478\,\text{psi}$.

The difference between these two calculation methods is 750 psi. In this case, the use of 1000 psi above breakdown pressure provides an additional safety margin for successful completion.

5. Pressure at the injection point The following equation is used for calculating the pressure at the injection point:

$$p_z = \Delta p_h + p_p - \Delta p_f$$

or

$$p = \Delta p_h$$
$$+ ISIP \text{ determined during breakdown} \quad (11\text{-}7)$$

where p_z is the pressure at injection point (top perforation), Δp_h is the hydrostatic pressure from fluid column (5272 psi), Δp_f is the friction pressure (usually negated at low pump rates), and p_p is the pump pressure applied from surface equipment.

Fracture pressure $= 0.8 \, \text{psi/ft} \times 9750 \, \text{ft} = 7800 \, \text{psi}$

The p_z should be kept below this value by at least 1000 psi to prevent this interval from being fractured. Therefore,

7800 psi (fracture pressure) $- 5272 \Delta p_h = 2528 \, \text{psi}$

If pump pressure or gauge pressure at the surface exceeds 2500 psi, the formation may fracture, so pump pressure should be kept at or below 1500 psi. In this case, attempting to monitor squeeze pressure on the surface would result in fracturing the formation interval. Because of the low pump rates, friction pressures are negligible.

11-2.5.4 Procedure

Once the plan has been established and the calculations have been double-checked, the last phase of the squeeze program is the squeeze procedure itself.

Note: In the following example, the 10-lb/gal drilling fluid has already been changed to 10-lb/gal NaCl completion fluid. The casing has already been perforated from 9750 to 9800 ft, and the bridge plug has been set in the casing at 9850 ft. A 30-ft column of sand has been dumped on top of the bridge plug. The well has been circulated at 3 to 5 bbl/min to stabilize conditions.

1. All $2\frac{7}{8}$-in. tubing is inspected, and a drift (a device with OD similar to tubing ID) is run inside the tubing to ensure that the tubing is clean and unobstructed.

2. All equipment and valve assemblies are positioned where personnel can observe them. Verbal communication will be used; however, hand signals may also be necessary. All communication equipment is checked.

3. All emergency safety equipment is positioned in easy-access areas close to the operation.

4. The working condition and pressure ratings of all surface valves, lines, and connections are examined.

5. The pumping equipment is isolated and all lines are pressure-tested to exceed squeeze pressure.

6. If field test equipment is available, a test series is run on a sample of the cement blend to verify that data are repeatable. The service company's cement mixing instructions are reviewed and the volumes of cement, additives, and mix water are verified. Representative samples of each material are collected and stored for after-action testing in case of job failure.

7. The BHA, consisting of the retrievable packer and tubing tester, is made up on the $2\frac{7}{8}$-in. tubing. The workstring should be pressure-tested when it is halfway in the wellbore.

8. A length of 200 ft of tailpipe is run below the packer assembly, and the $2\frac{7}{8}$-in. tubing is run into the 7-in. casing at a run-in rate of approximately 90 ft/min.

9. When the tubing is approximately halfway down the casing (4900 ft), the packer is set, and the annulus and tubing are pressure-tested to the maximum reverse-out pressure for the entire cement column. In this case, the cement column consists of 20 bbl × 172.6 ft/bbl or 3452 ft of cement. The column produces a differential pressure of (15.8 lb/gal × 0.052 psi/ft) $= 0.822 - (10 \, \text{lb/gal} \times 0.052 \, \text{psi/ft}) = 0.52$, so (0.822–0.52) × 3452 ft $= 1042 \, \text{psi}$. (This pressure is the maximum anticipated casing pressure for circulating all the cement out of the tubing.)

10. The packer is then run to depth at 9575 ft, and circulation is established. The packer is set and the tubing and annulus are again pressure-tested. The packer's circulation bypass ports are then opened and a 20-bbl breakdown of HCl/HF acid is displaced down the tubing. Once the fluid is displaced just above the packer, the tubing tester is released and acid is injected, which breaks down the formation and cleans the channel. At this time, the breakdown pressure is recorded. Displacement into the formation continues until an injection rate of approximately 2 bbl/min is established. The circulation bypass ports are then opened and 10-lb/gal brine is circulated in the hole.

11. The packer is unseated and slowly pulled up to the required setting depth (9285 ft) for the squeeze-cement job. The packer is set and the tubing and annulus are pressure-tested again. The test pressure must be kept below the recorded breakdown pressure.

12. Next, 20 bbl of cement slurry is batch-mixed to 15.8 lb/gal. Density is checked with a pressurized mud balance. Then, 30 bbl of 11-lb/gal spacer is made up and its density is checked.

13. With the packer set and the bypass open, the following pumping sequence is performed: (a) 25 bbl of spacer, (b) 20 bbl of cement slurry, (c) 5 bbl of

spacer, and (d) 10 lb/gal of completion brine for displacement. The cement is displaced down the $2\frac{7}{8}$-in. tubing until the cement volume has cleared the packer (the total fluid pumped to this point equals 9285 ft ÷ 172.6 ft/bbl or 54 bbl). At this point, 5 bbl of spacer should be located just above the packer, and the entire openhole section below the packer should be full of cement since the heavier 15.8-lb/gal cement will easily displace the brine and spacer (up the wellbore) ahead of it.

14. The packer bypass is then closed, and the cement is squeezed into the perforations at $\frac{1}{4}$ to $\frac{1}{2}$ bbl/min. During this operation, the pressure must not exceed 1500 psi on the surface.

15. Once squeeze pressure is obtained, the pumps are stopped for 5 to 10 minutes. At this point, the packer should be clear of all cement, and the 5 bbl of spacer and displacing completion fluid should be in or through the tailpipe. Cement will still be outside the tailpipe.

16. Pumping is slowly resumed and pressure is observed. If the squeeze has been successful, the pressure should rapidly return to its previous reading. Pressure is then held on the tubing for 15 minutes for testing. If additional breakdown occurs at any perforations, the pressure may bleed off slightly. If bleedoff occurs, pumping must be resumed.

17. Once the squeeze has been tested successfully, the packer is unseated, the bypass valves are opened, and 10 joints of tubing (approximately 400 ft) are slowly pulled up to clear the tailpipe. If the tailpipe becomes stuck, it can be broken off below the packer and drilled out later. Reverse-circulation occurs until the packer is clear.

   Note: In this procedure, the cement in the casing is not removed by reverse circulation through the packer. Instead, a safer approach is used, in which the packer is pulled above the cement and the cement is drilled out of the casing.

18. After 12 to 24 hours, the drillstring is run in hole, and the cement is drilled out. Then, either the squeeze can be tested or a bond log can be run on the treated section. If the treatment was successful, the bridge plug isolating the lower wellbore can be drilled out, retrieved, or moved to a new location for other treatments.

19. All records, laboratory reports, pressure charts, and other pertinent data are kept for job evaluation and improvements for future applications.

11-2.6 Job Evaluation

The primary evaluation method for a squeeze treatment is the application of hydraulic pressure to the treated interval. Increased pressure response indicates that a seal was obtained. This pressure can be applied either by pumping into the well (positive differential into the well) or by reducing hydrostatic pressure in the wellbore through the use of swabbing or introducing a lighter fluid into the wellbore (negative differential into the well). Additionally, squeezes to recement or to fill voids behind casing can be evaluated through the use of bond logs.

11-3 PLUG CEMENTING

Plug cementing refers to the placement of cement slurry in a wellbore to create a solid seal or "plug." This form of remedial cementing can be conducted at any time during the life of the well for a variety of reasons, including those outlined in Table 11-4. As with squeeze cementing, the wellbore problem must be correctly diagnosed, the wellbore condition must be well understood, and the application must be properly planned, designed, and controlled. However, the problem diagnosis for plug cementing is more straightforward than for squeezing.

11-3.1 Plug-Cementing Theory and Mechanics

The theory of plug cementing appears to be simple and straightforward: a plug location is selected, and the cement is mixed, pumped, and allowed to set. In reality,

Table 11-4   Reasons to set a cement plug

Abandonment
- Sealing a dry hole
- Sealing a depleted well

Zonal isolation
- Protecting low-pressure zone during workover
- Sealing a depleted zone

Directional drilling
- Sidetracking around a fish
- Performing offshore well deviation

Creating a base for an openhole test tool
Sealing lost circulation during drilling

however, plug-cementing applications are significantly more complicated because of the unknown nature of the environment in which the plug will be placed, the unstable nature of static slurries, and the inability to sufficiently displace resident well fluid.

Since plugs can be set throughout the life of the well, the wellbore temperature at the point of plug placement may range from a static temperature for zones that have long been undisturbed to temperatures that are well below the normal circulating temperature for lost-circulation zones. As discussed earlier for squeeze applications, accurate temperature estimates are essential for the proper set control of the cementing composition. Additionally, the condition of resident fluids varies widely, and the displacement and conditioning methods discussed in Chapter 8 are required to ensure that the hole is completely filled with uncontaminated slurry. These two criteria require no new theory or mechanics beyond those previously described. Additional considerations are covered in Section 11-3.3.

The unstable nature of static slurries is second only to fluid contamination as the leading cause of cement-plug failure. Since cements and most other fluids in the wellbore are slurries, these fluids must suspend the solid particles they contain even when fluids are static. Variations in density or slurry consistency can create instability whereby a denser fluid on top flows through a lighter fluid below.

Beirute (1978) conducted a theoretical study of the movement of unset cement plugs after placement. This study combines analytical model development from the fundamental equations of motion for the internal phase (slug of plug slurry) and external phase (wellbore fluid below the plug) with confirming laboratory modeling. The resulting mathematical model analyzed the potential for plug stability or movement for field applications when the geometry, rheology, and density of the fluids were known.

Results from the modeling were summarized into the following qualitative practices for unwanted plug movement:

- Matching plug cement slurry density to resident fluid density as closely as possible while still accomplishing plug performance demands

- Using thixotropic cements with high yield points

- Using the minimum possible retardation (accurate placement temperature must be known)

- Maximizing plug length

- Using small-diameter tubing to place a balanced plug and retrieve the tubing as slowly as possible

Calvert *et al.* (1995) conducted additional laboratory investigations of plug-cement stability in various geometries with various fluid properties. This experimental work confirmed previous work and resulted in empirical guidelines for density differentials and minimum drilling-fluid rheology. The authors cited the boycott effect (Davis and Acrivos, 1986) as the instability mechanism for cement migrating through low-rheology drilling fluid. This effect describes the migration of particles from each slurry, which creates a downward movement of the heavier cement through the less dense, thinner drilling fluid. However, as supporting fluid (drilling fluid) becomes thicker, downward migration occurs as the "roping" of whole cement slurry through the lighter drilling fluid. A rough correlation between the density differential and the drilling-fluid yield point is proposed to indicate plug stability.

This study revealed that horizontal cement plugs would form easily if the yield point of drilling fluid is sufficiently high (30 to 40 lb/100 ft^2). Under these circumstances, the cement would require a high yield point and a low volume of free water to create a plug seal.

11-3.2 Problem Diagnosis

Cement plugs can correct a wide array of problems. Each problem requires a different plug type or procedure. These plug variations are described below.

11-3.2.1 Lost Circulation

According to a 1991 API survey, lost circulation occurs during drilling on about 20 to 25% of wells drilled worldwide. Lost circulation is caused by all or part of the drilling fluid flowing up the annulus and entering a weak formation through natural or induced fractures or vugs. In severe cases, lost circulation can result in the total loss of fluid introduced into the wellbore. Minor lost-circulation problems can usually be alleviated with the addition of lost-circulation material (LCM) to the drilling fluid or the reduction of drilling-fluid density where possible. In severe cases, the leak path must be sealed by a cement plug. The severity of the fluid loss is based on its economic impact, its effect on cuttings removal, and its ability to maintain well control.

Lost-circulation plugs usually consist of low-density, quick-setting cement. Although non-Portland cement materials can be used to combat lost circulation, Portland cement is the most durable material, and it offers the best chance for success for permanent lost-circulation control. LCM is usually added to keep slurry from migrating into lost-circulation zones. LCMs range from fast-setting hydraulic materials such as blast-furnace slag or gypsum cement, to organic or inorganic materials that can be triggered to gel or precipitate bridging material. These materials are useful for many applications, and some offer advantages of compatibility, speed, or cost.

11-3.2.2 Kickoff Plugs

If debris or objects lost in the hole must be sidetracked, or if directional drilling is required, a kickoff plug must be used. In this operation, which can occur either in open or cased hole, the plug is set just below the point at which the directional hole will be started. Next, a window is milled in the casing if applicable. Then, a bit with a bent (deviated) sub is run in the hole to drill against the plug, which provides resistance to drilling. This resistance forces the bit to drill into the side of the wellbore, creating a new path for the well. Kickoff plug cements must be dense, high-strength cements that are resistant to drilling. In addition to strength and density, other factors affecting drillability include reduced water requirements because of dispersant addition and increased ductility through polymer modification. The addition of silica or inert weighting materials has no positive effect on the cement's drillability.

11-3.2.3 Well Abandonment

When production from a zone falls below commercial levels, and the operator wants to produce from other zones up the well, a cement plug can be placed across the zone to seal it from the wellbore. This operation requires cement to be placed across the depleted zone, creating a fluid-tight seal as a result of dehydration and node creation, just as in a squeeze application.

Similarly, cement plugs are used to seal a well being abandoned. Procedures and plug locations are often specified by regulatory authorities, but the function of the cement plug is the same. Plugs set for abandonment must consist of durable, nonshrinking, low-permeability cement. Cement compositions for this application are usually of normal density. Depending on the well configuration and local regulations, ten or more plugs may be required to abandon the well.

11-3.2.4 Temporary Plugs

If the borehole wall is too weak to support a sidewall anchor, cement plugs can provide a temporary solid base for test-tool applications. These plugs must be strong enough to support the tool's weight and to seal the lower portion of the well. Since these plugs are set in open hole, the hydrostatic pressure generated by the plug and the plug cement's density must be considered in the design.

Other types of less permanent plugging materials seal producing formations from the wellbore for workover operations. These materials include silicate gels, barite, sand, and polymer gels. A special type of acid-soluble cement, described by Vinson *et al.* (1992), forms a solid seal across a producing zone for workover isolation that can later be removed with an HCl treatment. This acid-soluble cement allows positive sealing characteristics without near-wellbore damage associated with regular Portland cement.

11-3.2.5 Plug Effectiveness

Failure of a plug attempt typically results in another attempt. Sometimes, the procedure or mechanics of the treatment are altered, but often, the same technique is attempted until success is achieved. The average estimated success rates for different types of plugs range from 35% for kickoff plugs (Heathman *et al.*, 1994) to 5 to 20% for lost-circulation plugs (Bugbee, 1953). These low success rates can be increased significantly if the plugs are properly designed and installed.

11-3.3 Controlling Factors

Since the materials used in plug cementing are usually Portland cement-based, the design criteria described in Chapter 8 affect plug cement behavior. However, these criteria may have a slightly different impact on the application, just as they did for squeeze cementing.

The application temperature for plug placement relative to static temperature varies widely with the well condition and the time since last circulation. Because the success of a plug placement ultimately depends on the cement curing in a relatively narrow time window, accurate knowledge of placement temperature is mandatory for successful design. Low estimates of placement temperature result in under-retarded slurries that may hydrate too quickly; high estimates of placement temperature can result in over-retarded slurries that may not fully set within the planned WOC time. Because cementing the drillstring in the hole during plug placement is a catastrophic event, most plug cement designers are cautious concerning placement temperature and thickening-time requirements. As a result, many plug cement slurries are over-retarded, resulting in plugs that have strengths lower than originally planned for the remedial operation.

The length of the plug is dictated by the severity of the problem requiring attention, the depth of the wellbore to be treated, and government regulations. Minimum lengths are approximately 10 ft, but plug lengths of 500 ft are not uncommon. As discussed in Section 11-3.1, the length of the plug influences its stability.

When a cement plug is placed into a wellbore, the fluids occupying the hole must be displaced. Poor displacement can result in the primary problem encountered in plugging operations: cement contamination. Standard design practices dictate the use of a minimal volume of cement to create the plug. Additionally, minimal spacer volumes or hole conditioning operations are used. However, these small volumes of cement must remain uncontaminated if they are to achieve the designed performance characteristics. Therefore, excellent displacement and compatibility measures are required to sweep the wellbore fluids from the hole and segregate them from the cement. To minimize contamination, plug designers must often deviate from minimum cement volumes, apply best practices to condition the fluid in the wellbore, and use realistic fluid volumes.

As stated earlier, cement-plug placement is difficult because the cement slurry is usually more dense than the wellbore fluid being displaced. Thus, the plug must balance on a column of lower-density fluid, making fluid migration an unwanted possibility. The downward migration of unset cement plugs because of density differences is second only to cement contamination as a cause for cement-plug failure. In addition to density differential, the problem is aggravated because the cement and drilling fluid are both slurries. Movement of particles from the slurries at the fluid interface initiates instability, which can result in the entire plug flowing to the bottom of the hole while the lighter drilling fluid comes up the hole from below. Solving this problem requires both careful placement techniques and the design of a cement slurry that will maximize integrity. These criteria are outlined in Section 11-3.1, and specific guidelines are covered in Section 11-3.6.

Other mechanical problems are created by plug placement in an eccentric annulus. Usually, the annular radius between the wellbore and workstring is greater than that described for primary cementing in Chapter 8. This size difference is especially true when coiled tubing is used as the workstring (a practice that is steadily increasing in workover operations). A large, eccentric annulus aggravates displacement mechanics because resident fluid is more easily bypassed by cement moving at a low velocity. This low-velocity condition is further aggravated when coiled tubing is used, because the flow rate through the tubing is limited by the reduced diameter and pressure limits of the tubing.

Wellbore deviation also produces complex geometry for the placement of a stable plug (Calvert *et al.*, 1995). As deviation increases, gravity and density effects make the formation of a complete wellbore seal very difficult. Thixotropic slurry designs that minimize static movement, extended plug lengths, and mechanical packers that "stack" the slurry will improve the chances of successful plug placement.

11-3.4 Placement Techniques

One of the following placement techniques can be used to place cement slurry for plug formation:

- Balanced-plug method
- Dump bailer method
- Two-plug method
- Mechanically supported method (with jet-hole cleaning)

Additionally, several variations of each method can be applied to improve chances of achieving plug placement. These methods are described thoroughly by Smith (1991). The method chosen is based on the volume of depth of the plug's location, the difficulty associated with setting a stable plug, and the economic or operational importance of placing a competent plug. Because of the low success rate of cement plugging operations, every possible procedural enhancement should be used to ensure successful placement.

11-3.4.1 Balanced-Plug Method

For this most prevalent method of cement-plug placement, cement slurry is pumped down open-ended drillpipe into the wellbore with the annulus open at the surface. When the height of cement inside the drillpipe is calculated to be equal to the height of the cement that has exited the drillpipe and traveled up the annular space, displacement is stopped. Thus, when spacer-fluid density is taken into consideration, the fluid inside and outside the drillpipe is calculated to be at a balance point. Once the cement is placed, the drillpipe is slowly pulled from the plug (90 to 140 ft/min) to minimize disturbance of the cement. The drillpipe is retrieved to a point 10 to 15 stands above the calculated top of the cement, and reverse circulation is started to displace any cement clinging to the pipe. This method (Figure 11-8) places cement in the wellbore in a balanced state with minimized disturbance, which allows the cement to form a homogeneous plug and to hydrate.

Gelling chemical formulations or mechanical tools can be used to provide a base in the wellbore for a balanced plug, thus eliminating downward migration caused by density variation. As discussed in Section 11-3.1, cement and most other fluids in the wellbore during well construction are actually solids-laden slurries. Gravity, wellbore deviation angle, and fluid suspension characteristics govern a slurry's ability to remain stable in static condition. If unstable solids in the slurry migrate downward, or heavier slurry migrates through lighter slurry below it, plug placement becomes more difficult since a higher-density cement is usually placed on lighter drilling fluid. Mechanical tools create a solid support for the plug, prevent inversion and dilution of fluids, and ensure the accurate location of the plug in the wellbore.

Chemical plug systems used as a base for a cement plug can either consist of highly gelled thixotropic fluids, such as bentonite or organic polymers, or they can be composed of materials that exist as fluids until, when commingled, they react chemically to form a gelatinous mass. This second type, described by Bour *et al.* (1986) can provide a relatively solid base for plug stability, supporting density differentials of over 7 lb/gal.

11-3.4.2 Dump Bailer Method

A dump bailer is a cylindrical container run into the well on wireline. A mechanically or electrically triggered mechanism on the bottom of the device can open the bailer, releasing its contents into the wellbore at any specified depth. This method (Figure 11-9) usually requires a mechanical packer, which is placed at the point of the plug bottom to minimize migration of the small volumes of cement. Multiple bailer runs may be necessary to achieve the desired fill.

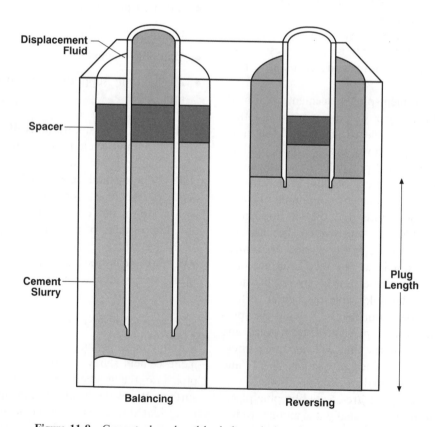

Figure 11-8   Cement plug placed by balanced-plug placement method

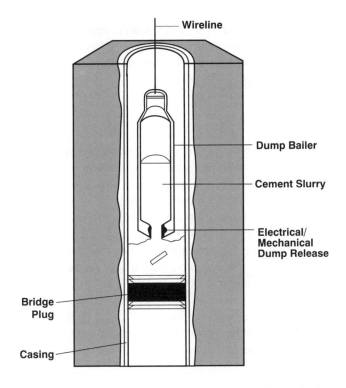

Figure 11-9   Cement plug placed by dump bailer method

For optimum dump bailer performance, especially at high temperatures, operators must use specially designed cement slurries that remain fluid after long static times at elevated temperature. Not only must setting reactions be retarded, but the slurry's static gelation properties must be modified, as discussed in Chapters 8 and 10.

11-3.4.3 Two-Plug Method

For this method, workstring wiper plugs isolate the cement from well fluids in the drillpipe and provide a positive indication of cement placement. As shown in Figure 11-10, the bottom plug precedes the cement and wipes the inside of the workstring, preventing contamination. When the bottom plug reaches the end of the workstring, it exits into the wellbore, and the cement moves up the annulus. As the last of the cement exits the plug catcher, the top plug seats in the plug catcher, resulting in a pump pressure increase, which indicates that cement placement is complete. The plug latches into the catcher to prevent cement flowback into the workstring.

Next, the workstring is raised to the desired depth of the plug's top. Pressure-actuated ports are then opened, and reverse circulation is initiated to clean the inside of the workstring and to dress off the plug top at the correct height.

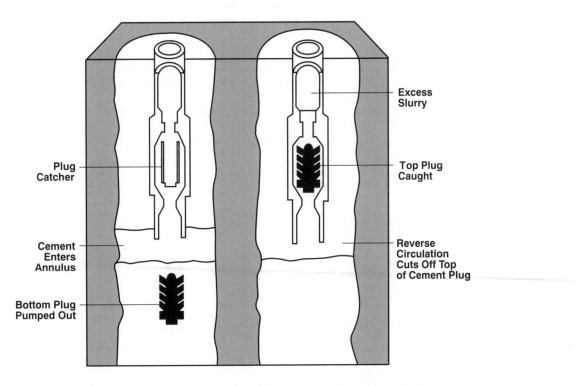

Figure 11-10   Cement plug placed by two-plug placement method

11-3.4.4 Mechanically Supported Method with Jet-Hole Cleaning

This adaptation of the balanced-plug method offers the best guarantee of plug location available. For this procedure, an inflatable-set packer provides a solid bottom for the plug. The packer is equipped with a diverter tool that thoroughly washes the wellbore by means of circulation as the tool is run into the hole. Above the packer, drillable aluminum tailpipe is installed to the required height of the plug.

Once the tool is placed at the necessary location, the inflatable packer element is filled with cement and expanded in the hole to form a permanent mechanical bottom for the cement slurry, which will form the plug. Next, ports in the tailpipe just above the packer are opened, and cement slurry is circulated into place just as it was in the balanced-plug method. The tailpipe is then disengaged and left in the plug.

This method is costly and is only used in critical situations involving well control or on offshore rigs where costs associated with failed cement plugs are great. However, Heathman et al. (1994) report complete success for kickoff plugs when this method is used.

11-3.5 Planning

Once the need for a plug is established, a program detailing the design, execution, and evaluation procedures must be developed. The same general process outlined in Table 11-2 for conducting squeeze treatments also applies for plug treatments. Several key design parameters for plugs are discussed below.

11-3.5.1 Plug Type

The type of plug to be placed dictates plug length, cement composition performance parameters, and placement technique. Plug length mainly varies with placement techniques. The most common type of placement technique, the balanced-plug method, requires the most volume since the method is the most prone to cement contamination. Lost-circulation plugs and whipstock plugs should be designed with lengths of up to 500 ft of fill. The dump bailer method, usually used for temporary zonal isolation, requires short lengths since cement is accurately placed over a solid base created by the bridge plug. Cement composition for various plugging applications is outlined in Table 11-5.

11.3.5.2 Temperature

The importance of temperature, discussed earlier in this chapter as well as in Chapter 8, is magnified for plug cementing because of the unstable condition of the plug, which can be caused by lost-circulation flows or density differences. In every case, the cement must be

Table 11-5 Cement composition properties for various plug applications

General	Adequate retardation for placement but rapid strength development
	Nonshrinking
	Rheology adequately low for placement without excessive pump pressure but sufficient consistency to prevent settling
	Minimum density to produce desired ultimate strength
Abandonment	
open hole	Fluid-loss control
	Low permeability
cased hole	Low permeability
Directional drilling	Resistance to drilling
	High strength, ductility
Zonal isolation	Thixotropic
temporary	Filtrate non-damaging to formation
	Removable
permanent	Low permeability
Lost circulation	Thixotropic
	Low density
	Particulate LCM
Formation testing	High intrinsic strength
	High bond strength

designed to set as quickly and safely as possible. This criterion requires accurate knowledge of the application temperature at the time of placement. Although this information is difficult to obtain, it is the most critical variable in the design of any plug treatment.

11.3.5.3 Wellbore Fluids

The condition of wellbore fluids before cementing and the role of spacers were both emphasized earlier in this chapter and in Chapter 8. The same general guidelines apply for plug cementing. Additionally, low placement flow rates and widely variable resident fluid conditions may be encountered in a plugging operation. As a result, larger amounts of spacer volume should be considered, as well as the use of such mechanical tools as centralizers and flow diverters.

11.3.5.4 Deviation

As deviation increases, the length of the plug should increase proportionally, with the length of a plug designed for a horizontal well being at least twice as long as a plug designed for a comparable straight-hole application. Mechanical or chemical aids to prevent migration improve the chance of success as deviation increases.

11.3.5.5 Evaluation Procedures

Evaluation techniques for most cement plugs are quick and straightforward. Plugs to isolate zones or wellbores are pressure-tested to ensure seal integrity. The success of a lost-circulation plug is based on the degree of circulation gained after the plug is drilled out. The success of sidetrack kickoff plugs is based on the plug's resistance to drilling and the successful initiation of the sidetrack hole. Plug tops are tagged to confirm that the plug is correctly placed.

11-3.6 Designing the Cement-Plug Procedure

The following section briefly describes the process of designing a cement plug.

11-3.6.1 Tools

Mechanical aids provide a positive base for the plug and improve the displacement of resident fluids. Bridge plugs, discussed in Chapter 8, are commonly used to create a positive base and prevent static cement slurry from possibly moving. Placement in the well is calculated to the desired location of the plug bottom. Centralizers and scratchers, also described in Chapter 8, are used on the placement tubing for balanced-plug methods. They enhance displacement of resident fluids by providing a concentric annulus and lowering fluid rheology by pipe movement with the scratchers. Flow diverters that jet cement out the borehole wall rather than straight down also improve displacement. Plugs and baffles used in the two-plug method isolate cement, thus preventing inter-mixing and dilution.

11-3.6.2 Cement Composition

Cement composition design is discussed in Chapter 8. Table 11-5 presents specific considerations for various types of plug applications.

11-3.6.3 Procedures

1. The location and length of the cement plug are determined.

2. The tools and the placement method are selected. Whenever possible, mechanical aids are used. Although these devices add cost to the procedure, they greatly improve the chances for success.

3. Equation 11-2 or 11-8 is used for calculating cement volume. A plug length of 300 to 500 ft is generally used. For openhole plugs, as much as 100% more cement is used. For cased-hole plugs, as much as 50% more cement is used.

4. Equation 11-13 is used for calculating displacement.

5. The cement composition is then designed according to the guidelines in Table 11-5 and Chapter 8.

6. The resident fluid is conditioned according to procedures outlined in Chapter 8.

7. Spacer is pumped ahead and behind the cement. The spacer must be compatible with resident fluids, as discussed in Chapter 8.

8. If possible, pipe movement should be used during cement placement.

9. After the cement is allowed time to set, the plug is tagged and pressure-tested.

11-3.6.4 Plug Calculations

1. **Sacks of cement needed for required fill**   Equation 11-8 calculates the quantity of cement needed to fill the treatment zone

$$s = \frac{H \times C}{Y} \qquad (11\text{-}8)$$

The feet of fill for a given number of sacks is then calculated from Equation 11-2.

2. **Height of balanced plug**   Equation 11-9 calculates the height of the balanced plug before the pipe is withdrawn.

$$H = \frac{S \times Y}{C_a + C_p} \qquad (11\text{-}9)$$

where C_p is the capacity of drillpipe or tubing (sk/ft), and C_a is the capacity of the annulus (sk/ft). The capacity of the annulus can be measured between the drillpipe or tubing and the hole or casing. Some frequently used tables do not show the annular capacity for large hole sizes. In such cases, C_a is the capacity of the hole (sk/ft) minus the capacity of the hole (sk/ft) that is the same size as the drillpipe OD.

3. **Volume of spacers pumped ahead and behind cement**   Equation 11-10 is used for calculating the spacer volume pumped ahead and behind cement for a balanced-plug design.

$$V_{s2} = \frac{C_a}{C_p} V_{s1} \qquad (11\text{-}10)$$

where V_{s2} is the volume of spacer ahead of cement (bbl), and V_{s1} is the volume of spacer behind cement (bbl). To use Equation 11-10, select V_{s1}, the spacer needed in the drillpipe or tubing (usually 1 or 2 bbl) and solve for V_{s1}, the volume of spacer to be used in the annulus. This calculation will provide the same height of spacer both inside and outside the pipe.

4. **Volume of displacing fluid to pump balanced plug**   Equation 11-11 is used for calculating the amount of displacing fluid to pump the balanced plug into place.

$$V = (L_p - H)C_p \qquad (11\text{-}11)$$

where V is the total volume of mud and spacer required (bbl), and L_p is the length of drillpipe or tubing (ft).

11-3.6.5 Examples

The following two examples are of typical cement-plug applications. The first example concerns a severe lost-circulation occurrence, and the second covers setting a kickoff plug through the use of the balanced-plug method. All pertinent design parameters are discussed.

Lost-Circulation Zone

During drilling, a lost-circulation zone was encountered at 6500 ft. The decision was made to spot a balanced plug of cement 300 ft long in an $8\frac{3}{4}$-in. open hole. The drilling-fluid density at the time of the partial loss of returns was 12.5 lb/gal. A cement slurry will be designed near the density of the drilling fluid to minimize losses. This thixotropic slurry will contain particulate material to minimize losses into the formation. A volume of 20 bbl of water will be pumped ahead of the cement slurry as spacer, and a sufficient volume of water spacer will be pumped behind the cement to provide a balanced plug. The plug will be spotted across the lost zone, and the drillpipe will be slowly pulled out of the plug. The plug will be allowed to set before drillout.

Capacity of 5-in. (19.5-lb/ft) drillpipe:
    (0.01776 bbl/ft or 56.3 ft/bbl)
Capacity of $8\frac{3}{4}$-in. open hole:
    (0.0744 bbl/ft or 13.4454 ft/bbl)
Capacity of drillpipe/openhole annulus:
    (0.0501 bbl/ft or 19.9644 ft/bbl)
Bottom of plug:
    6500 ft

1. **Calculated slurry volume**

    ft plug × 0.0744 bbl/ft in open hole = 22.32 bbl

2. **Calculated height of plug with drillpipe in place**

    bbl/(0.01776 bbl/ft + 0.0501 bbl/ft) = 328.9 ft

3. **Calculated water spacer volume to pump after the cement slurry**

    Lead water spacer annular height
        = 20 bbl × 19.9644 ft/bbl = 399.3 ft

    Water spacer volume after cement
        = 399.3 ft × 0.01776 bbl/ft = 7.1 bbl

4. Calculated drilling-fluid displacement to spot and balance plug

Total displacement = (6500 ft − 328.9 ft)
  × 0.01776 bbl/ft = 109.6 bbl
Drilling-fluid displacement = 109.6 bbl − 7.1 bbl
            = 102.5 bbl

Whipstock Plug Example

A portion of the BHA is lost in the hole and cannot be retrieved by fishing operations. A whipstock plug will be set and the portion of the wellbore containing the fish will be abandoned and sidetracked. Offset well caliper log data indicate that the $7\frac{7}{8}$-in. hole size will be washed out to $9\frac{1}{2}$ in. in the section of the hole where the plug will be set. A 500-ft balanced plug will be placed with 16.4-lb/gal cement slurry (1.06-ft^3/sk yield). Next, 50 bbl of viscosified spacer with a slightly greater density than the drilling fluid will be pumped ahead of the cement slurry, and a sufficient volume of the same spacer will be pumped behind the cement to provide a balanced plug.

Plug setting depth:
  9000 ft
Bottomhole temperature at plug:
  200°F
Capacity of $4\frac{1}{2}$-in. (16.60-lb/ft) drillpipe:
  (0.01422 bbl/ft or 70.32 ft/bbl)
Capacity of $9\frac{1}{2}$-in. open hole:
  (0.0877 bbl/ft or 11.4063 ft/bbl)
Capacity of drillpipe/openhole annulus:
  (0.0680 bbl/ft or 14.7059 ft/bbl)

1. Calculated cement volume

ft × 0.0877 bbl/ft × 25% excess = 54.8 bbl (minimum) bbl × 5.6146 ft^3/bbl / 1.06 ft^3/sk = 290.26 sk of cement (round to 300 sk)
sk × 1.06 ft^3/sk × 0.1781 bbl/ft^3 = 56.6 bbl total slurry

2. Calculated height of the plug with the drillpipe in place

Plug height = 56.6 bbl / (0.01422 bbl/ft
  + 0.0680 bbl/ft) = 688.4 ft

3. Calculated spacer volume to pump behind the cement slurry

Lead spacer annular height = 50 bbl × 14.7059 ft/bbl
            = 735.3 ft

Spacer volume after cement = 735.3 ft × 0.01422 bbl/ft
            = 10.5 bbl

4. Calculated drilling-fluid displacement to spot and balance the plug

Total displacement = (9000 ft − 688.4 ft) ×
0.01422 bbl/ft   = 118.2 bbl

Drilling-fluid displacement = 118.2 bbl − 10.5 bbl
(spacer)   = 107.7 bbl

The previous examples are very basic job designs; they do not include recommended placement procedures and design improvements that would increase the likelihood of a successful plug after the first application. Heathman *et al.* (1994) discusses job design and execution recommendations that have greatly improved the success rate for whipstock plug applications. In addition to the job parameters discussed here and in Chapter 8, these parameters include proper slurry design and volume, proper hole conditioning, plug stability, and plug placement through the use of a diverter wash tool and tailpipe assembly.

REFERENCES

API Specification 10, Fifth Edition, No. 130.114503 (July 1990).

Beach, H.J., O'Brien, T.B., and Goins, W.C.: "Formation Cement Squeezed by Using Low-Water-Loss Cements," *O&GJ* (May 29 and June 12, 1961).

Beirute, R.M.: "Flow Behavior of an Unset Cement Plug in Place," paper SPE 7589, 1978.

Binkley, G.W., Dunbauld, G.K., and Collins, R.E.: "Factors Affecting the Rate of Deposition of Cement in Unfractured Perforations During Squeeze Cementing Operations," *Trans.*, AIME (1958) **213**, 51–58.

Bour, D.L. *et al.*: "Development of Effective Methods for Placing Competent Cement Plugs," SPE 15008, 1986.

Bour, D.L., Creel, P., and Kulakofsky, D.S.: "Computer Simulation Improves Cement Squeeze Jobs," paper CIM/SPE 90–113, 1990.

Bradford, B., and Cowan, M.: "Remedial Cementing," *API Worldwide Cementing Practices*, First Edition (Jan. 1991) 83–102.

Bugbee, J.M.: "Lost Circulation a Major Problem," *API Drill. and Prod. Prac.* (1953).

Calvert, D.G. *et al.*: "Plug Cementing: Horizontal to Vertical Conditions," paper SPE 30514, 1995.

Cementing, SPE Reprint Series, No. 34, (1992).

Davis, R.H., and Acrivos, A.: "Fluid Dynamics," *Physics Today* (Jan 1986).

Goodwin, K.J.: "Guidelines for Ultrasonic Cement Sheath Evaluation," paper SPE 19538, 1989.

Harris, F. and Carter, G.: "To Squeeze Those Perforations: Use a Chemical Wash and a Low Fluid Loss Cement," *Drilling* (Jan. 1964) 25, No.3.

Heathman *et al.*: "Quality Management Alliance Eliminates Plug Failures," paper SPE 28321, 1994.

Hemphill, R.P., and Crook, R.J.: "Thixotropic Cement Improves Squeeze Jobs," *World Oil* (May 1981).

Howard, G. C., and Fast, C.R.: "Squeeze Cementing Operations," *Trans.*, *AIME* (1950) **189**, 53–64.

Huber, T.A., Tausch, G.H., and Dublin, J.R.: "A Simplified Cementing Technique for Recompletion Operations," paper SPE 313-G, 1953.

Jones, R.R., and Wydrinski, R.: "Integrated Approach to Cement Bond Log Interpretation Saves Time and Money," paper SPE 28441, 1994.

Jones, R.R.: "A Novel Economical Approach for the Determination of Wellbore Temperatures," paper SPE 15577, 1986.

Krause, R.E., and Reem, D.C.: "New Coiled-Tubing Unit Cementing Techniques at Prudhoe Developed To Withstand Higher Differential Pressure," paper SPE 24052, 1992.

Millican, C.V.: "Cementing," Chapter 7, *History of Petroleum Engineering*, API Div of Production, Dallas (1961).

Pilkington, P.E.: "Pressure Needed to Reduce Microannulus Effect on CBL," *O&GJ* (May 30, 1988) 68–76.

Rike, J.L. *et al.*: "Squeeze Cementing: State of the Art," *JPT* (Jan. 1982) 37–45.

Smith, D. *et al.*: *Squeeze Cementing*, SPE Monograph 4 (1990) 123–138.

Smith, R.C.: "Cement Job Planning," *API Worldwide Cementing Practices*, First Edition (Jan. 1991) 19–32.

Tilghman, S.E. *et al.*: "Temperature Data for Optimizing Cementing Operations," paper SPE 19939, 1991.

Vinson, E.F. *et al.*: "Acid Removable Cement System Helps Lost Circulation in Production Zones," paper SPE 23929, 1992.

12 Completion Fluids

Mike Stephens
M-I L.L.C.

Hon Chung Lau
Shell E&P Technology Co.

12-1 INTRODUCTION

The completion process prepares the well for production, and the fluids used at this stage of well construction are called completion fluids. In the early days of drilling, the drilling fluid also served as the completion fluid, but after perforating became a common practice, it was determined that using clean, solids-free fluids for perforating would increase well productivity (Klotz *et al.*, 1973). These clear fluids, frequently composed of brine, became known as "completion" brines. During the 1950s, 1960s, and 1970s, a set of well completion practices that consisted of cementing casing in place, displacing the drilling fluid with a completion brine, and then perforating became widespread. This completion design was supplanted in the 1980s and 1990s as directional drilling technologies led to horizontal and multilateral well designs, and sand control considerations became more critical for offshore wells in poorly consolidated formations. As a result, completion technology became more complex with a number of new methods for completing wells. As completion technology has evolved, specialized completion fluid systems have been developed to optimize these completion practices. In the future, well completion designs will become even more diverse as will the fluid systems used in completions.

12-1.1 Definition and Functions of Completion Fluid

The concept of a completion fluid is a recent development in the history of the petroleum industry. In the 1940s, little or no distinction was drawn between drilling and completion fluids (Radford, 1947). The definition of a completion fluid is a bridge between the traditional classifications of drilling fluids and stimulation fluids. Both narrow and broad definitions of a completion fluid are worth considering.

Sometimes it is helpful to use a narrow definition of completion fluid as the clear brine that service companies sell into the completion market. For the traditionally designed well that will be cased and perforated, this brine frequently serves as the completion fluid. In other applications, the clear brine may serve as a base fluid to which other components are added to produce a completion fluid with the required properties (although specially formulated fluids are also used). The narrow definition of completion fluid as clear brine permits discussion of issues pertaining to the composition and economics of clear brine completion fluids. The narrow focus is traditional in the sense that completion brines are frequently used as perforating or gravel-pack fluids, and guidelines for their selection are given in the section pertaining to basic selection criteria.

A broader definition of completion fluid is needed to understand the role that completion fluids play in the success or failure of a well. The completion fluid contacts the reservoir rock and interacts with reservoir components. If the result of this interaction is a decrease in permeability or blockage of flow from the reservoir, then production from the well can be impaired.

The phases in well construction can be subdivided into a drilling phase, a completion phase, and an optional stimulation phase as shown in Table 12-1. In the broadest sense, the completion phase includes every step in the operation from the time the bit cuts the reservoir rock until the well is producing. Even after the well has been

Table 12-1 Phases in well construction

Operations Using Fluids					
	Drilling		Completion		Stimulation
Overburden drilliing	×				
Reservoir drilling		×			
Perforation			×		
Gravel-pack			×		
Cleanup			×		
Emplacing Equipment			×		
Acid wash				×	
Matrix acid					×
Fracpack				×	
Hydraulic frac					×

producing for a period of time, any workover operations that occur may be considered part of the well completion process. The fluids used in any of these operations could be considered completion fluids.

Typical operations that occur during the completion of a well include perforation, gravel-packing, well cleanup, and emplacing tubulars, packers and pumps. The fluids used during these operations are all completion fluids, and these operations do not ordinarily overlap with drilling or stimulation operations.

There are a number of cases where completion fluids might also be considered to be either drilling fluids or stimulation fluids. For example, drilling the well involves both drilling the non-productive rock above the reservoir and drilling the reservoir horizons. The completion of the well properly begins as soon as the reservoir is penetrated. The fluid that is used to drill the reservoir section can be classified both as a drilling fluid and as a completion fluid. The term "drill-in" fluid has been coined to describe fluids that are specifically formulated for drilling the reservoir section of the hole. Ideally, these drill-in fluids function as efficient drilling fluids, while still protecting the reservoir as though they were completion fluids.

Some of the operations conducted on a well after it has been drilled have traditionally been considered stimulation operations. A good example of this is matrix acidization, in which acid is injected into the reservoir to stimulate the well. When applied in matrix acidization, the acid is a stimulation fluid. However, in many cases, the well is simply washed with acid to clean up calcium carbonate-based drill-in fluids that have deposited a filter cake. The acid solution in this case could be considered a completion fluid rather than a stimulation fluid.

A similar analogy can be made between hydraulic fracturing fluids and fracpack fluids. The fluid used to do the massive hydraulic fracture is a stimulation fluid in that the purpose of the fracture is to increase the production from a low-permeability reservoir. The fracpack is a method used to increase effective wellbore diameter and control sand production, and the fluid used to carry out a fracpack could be considered to be a completion fluid.

A broad definition of completion fluid is any fluid that contacts the reservoir. This broad definition is useful because it allows all the fluids that will contact the reservoir to be evaluated as part of the well completion process. In the case of complex completion procedures that involve exposing the reservoir to a series of fluids (as may occur, for example, in some horizontal wells with sand control problems), this evaluation prevents incompatibilities from arising that could ruin the completion and prevent formation damage during drilling and stimulation.

12-2 BASIC SELECTION CRITERIA FOR CLEAR BRINE COMPLETION FLUIDS

Clear brine fluids are widely used in completion operations. While there are a number of fluids that may serve as completion fluids, clear brine is certainly the most important. A number of types of clear brine are commercially available, including ammonium chloride, sodium chloride, sodium bromide, potassium chloride, calcium chloride, calcium bromide, zinc bromide-calcium bromide, sodium formate, and potassium formate. Of these, the halide brines are most commonly used because they usually represent the lowest cost for a given density. There are three basic selection criteria for a completion brine: density requirement for well control, crystallization temperature requirement for storage, and chemical compatibility between the completion brines and the formation. Each of these will be discussed in the following sections.

12-2.1 Density Requirements

The primary performance requirement for a completion brine is pressure control. The density must be sufficient to produce a hydrostatic pressure in the wellbore high enough to control formation pressures. Typically, an overbalance of 200 to 300 psi above bottomhole reservoir pressure is used for well control. The procedure to calculate the required brine density at surface temperature depends on whether the effects of temperature and pressure on brine density are important for the operation. Brine densities are commonly reported at a reference temperature of 70°F (21.1°C).

12-2.1.1 No Temperature or Pressure Correction on Brine Density

In shallow, low-temperature formations, the effects of temperature and pressure on brine density may be neglected. In this case, the bottomhole hydrostatic pressure p_h (psi) exerted by a completion brine of average density ρ_{avg} (lb/gal) is given by

$$p_h = 0.052\rho_{avg}H \qquad (12\text{-}1)$$

where H (ft) is the true vertical depth. However

$$p_h = p + \Delta p_{ob} \qquad (12\text{-}2)$$

where p (psi) is the bottomhole reservoir pressure and Δp_{ob} (psi) is the overbalance pressure. Substituting Equation 12-2 into Equation 12-1 results in

$$\rho_{avg} = (p + \Delta p_{ob})/0.052H \qquad (12\text{-}3)$$

For example, consider a 200-psi overbalanced perforating operation. A well with a bottomhole reservoir pressure of 4800 psi and a 10,000 ft TVD will require a brine with bottomhole density of $(4800 + 200)/(0.052 \times 10,000) = 9.62$ lb/gal. Since we assume no effect of pressure and temperature on brine density, ρ_{avg} is also equal to the density at the surface.

12-2.1.2 Temperature Correction Only

Of course, brine density decreases with increasing temperature because of thermal expansion, and increases with increasing pressure because of compressibility. However, in most cases, the temperature effects dominate the pressure effects as the depth of the well increases. If compressibility effects are neglected, the brine density at 70°F, ρ_{70}, is related to the density at bottomhole temperature ρ_T by the following equation (Schmidt *et al.*, 1983):

$$\rho_{70} = \rho_T + (T_{avg} - 70) \times E_f \qquad (12\text{-}4)$$

where $T_{avg} =$ average well temperature = (surface temperature + bottomhole temperature)/2 and E_f (lb/gal/°F) is the brine expansibility factor (Figure 12-1 and Table 12-2).

For example, if in the previous example, the bottomhole temperature is 230°F, the required density of the brine at 70°F can be calculated from Equation 12-4:

$$\rho_{70} = 9.62 + (150 - 70) \times 0.0024 = 9.81 \, \text{lb/gal}$$

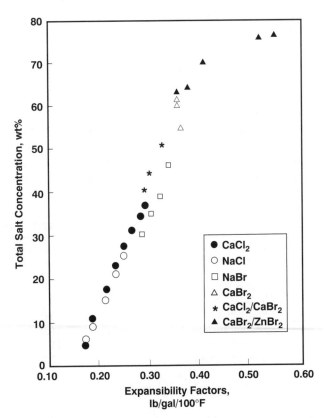

Figure 12-1 Expansibility factors for various brines (after Schmidt *et al.*, 1983)

Table 12-2 Expansibility and compressibility factors for weighted brine (Hudson and Andrews, 1986)

Brine density, lb/gal	Expansibility factor, E_f lb/gal/°F	Compressibility factor, C_f lb/gal/psi
9.0–11.0	0.24×10^{-2}	0.19×10^{-4}
11.1–14.5	0.33×10^{-2}	0.17×10^{-4}
14.6–17.0	0.36×10^{-2}	0.22×10^{-4}
17.1–19.2	0.48×10^{-2}	0.31×10^{-4}

because $T_{avg} = (70 + 230)/2 = 150°F$; $E_f = 0.0024$ lb/gal/°F for 9.62 lb/gal NaCl brine with a total salt concentration of 20% (Figure 12-1). An alternate way to calculate the expansibility factor can be found in API Recommended Practice 13J (Mar. 1996).

12-2.1.3 *Temperature and Pressure Correction*

When both brine expansibility and compressibility are taken into account, the following equation can be used to calculate the required brine density at 70°F:

$$\rho_{70} = \rho_T + (T_{avg} - 70) \times E_f - (0.5 \times p_h \times c_f) \quad (12\text{-}5)$$

where c_f (lb/gal/psi) is the liquid compressibility factor (Table 12-2) and p_h (psi) is the hydrostatic pressure.

If compressibility is taken into account in the previous example, then

$$\rho_{70} = 9.81 - (0.5 \times 5000 \times 0.000019) = 9.76 \text{ lb/gal}$$
$$(12\text{-}6)$$

Temperature and pressure corrections are generally needed for deep wells to control pressure and avoid excessive overbalance pressures.

Table 12-3 gives the maximum densities of clear brines and densities of stock solutions commonly used in the oilfield. As the brine density increases, so does the cost (Figure 12-2).

12-2.2 Crystallization Temperature Requirements

After density, the crystallization temperature is the second most important selection criterion for a completion brine. The crystallization temperature is the temperature at which the brine is saturated with respect to one of the salts that it contains. At the crystallization temperature, this least-soluble salt becomes insoluble and precipitates. The crystals can be either salt solids or freshwater ice. Cooling the brine below the crystallization temperature results in even more precipitation of salt solids.

Precipitation of salt solids in the brines at or below the crystallization temperature can lead to a number of rig problems. If the salt crystals settle in the tank, the density of the brine pumped downhole may be too low to control formation pressures. As more and more salt crystals form, brine viscosity increases. Eventually, the viscosity can become so high that the brine appears to be frozen solid. It cannot be pumped, and the lines are plugged. Therefore, crystallization of brines on location can result in considerable inconvenience, lost rig time, and expense. In deep offshore waters, the low temperatures near the sea bottom must be taken into account to prevent crystallization of brine when it is pumped downhole.

All experimental methods for measuring the crystallization temperature of brine involve alternately cooling and heating a sample of the brine. Figure 12-3 is a representative cooling curve for a high-density brine. Measured temperature of the brine is plotted against time while the brine is alternately cooled and heated.

Table 12-3 Density of Clear Brines at 70°F

Brine	Maximum density, lb/gal	Density of stock solution, lb/gal
Ammonium chloride	9.4	sacked ammonium chloride
Potassium chloride	9.7	sacked potassium chloride
Sodium chloride	10.0	sacked sodium chloride
Potassium bromide	10.9	sacked potassium bromide
Sodium formate	11.0	sacked sodium formate
Calcium chloride	11.8	11.6
Sodium bromide	12.8	12.5
Potassium formate	13.3	13.3
Calcium bromide/calcium chloride	15.1	none
Calcium bromide	15.5	14.2
Zinc bromide/calcium bromide	20.5	19.2

Note: The maximum density may vary slightly, depending on how the brine is prepared

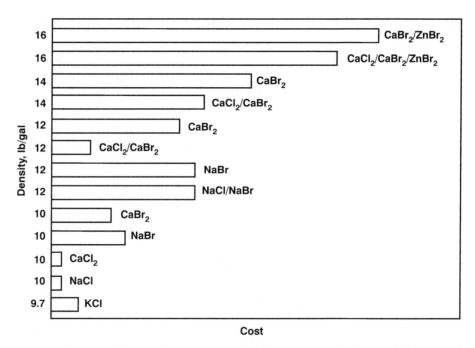

Figure 12-2   Relative cost for standard production brines (Foxenberg and Smith, 1996)

Three experimental measures of crystallization temperature are delineated in Figure 12-3. They are defined as follows:

- **First crystal to appear (FCTA)**   FCTA is the temperature at which visible crystals start to form. FCTA will generally include some supercooling effect. It appears at the minimum in the cooling curve.

- **True crystallization temperature (TCT)**   TCT is the maximum temperature reached following the supercooling minimum or the inflection point if no supercooling occurs.

- **Last crystal to dissolve (LCTD)**   LCTD is the temperature at which crystals disappear or the inflection point on the heating curve.

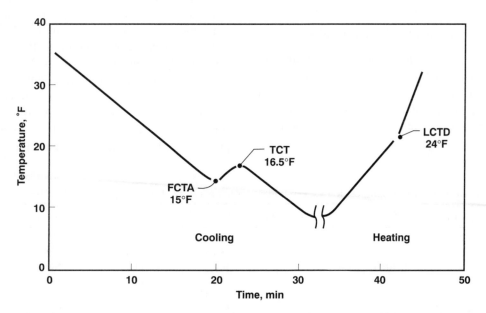

Figure 12-3   Crystallization curve for 19.2-lb/gal ZnBr$_2$/CaBr$_2$ brine (Schmidt *et al.*, 1983)

The TCT is the measured crystallization temperature nearest the temperature at which a brine will crystallize in tanks, pumps, or transfer lines in the field. In the field, supercooling effects are minimized by the slow rate of cooling that occurs in large-volume tanks from cool weather. In addition, the large brine volumes and high surface area in contact with the brine will provide abundant nucleation sites to prevent supercooling. Therefore, TCT is the best measure of crystallization temperature of a brine. The TCT of a brine can and should be measured on location and can be done using the method described in the API Recommended Practice 13J (Mar. 1996).

The crystallization temperature of a heavy brine at a given density can be varied by adjusting the concentration of the different salts in the brine. Table 12-4 gives an example of the relationship between brine composition and crystallization temperature. Brines of a given density may be formulated with various crystallization temperatures. As a rule, the lower the TCT, the more costly the brine. Therefore, choosing a brine with an excessively low crystallization temperature can be very costly. A cheaper, high-density brine with too high a crystallization temperature, on the other hand, can increase the cost of operations because of the rig problems discussed earlier.

Typically, the crystallization temperature is specified a few degrees lower than the lowest anticipated cold-weather temperature to prevent crystallization of salt solids in the brine.

12-2.3 Formation Compatibility Requirement

The third selection criterion is whether the completion brine is chemically compatible with the formation. The term "formation" means formation rock, water, and hydrocarbons. Incompatibilities can cause formation damage that means lost productivity or the need for remedial treatment.

12-2.3.1 Compatibility with Formation Clays

The main concern with formation clay compatibility is whether the completion brine, if in contact with the reservoir rock, will cause swelling and/or deflocculating of formation clays (Scheuerman and Bergersen, 1990). Clay swelling can directly block pore openings by increasing the size of clay aggregates. Both mechanisms cause clay particles to detach from each other and from pore walls. Migration of—and pore-throat plugging by—deflocculated clays is the most common impairment mechanism related to clay. To prevent clay swelling, the completion brine must meet a minimum salinity requirement. This requirement is typically 3% NH_4Cl or 2% KCl. Completion brines that do not meet this requirement can cause clay swelling and permanent formation damage (Azari and Leimkuhler, 1990). In a few cases, where extremely dirty sands are present, nearly any aqueous-based fluid can cause formation damage; oil-based fluids have been used as completion fluids in these zones.

The compatibility of heavy completion brines with formations has been studied recently. Morgenthaler (1986) has conducted laboratory core flow tests to determine the formation damage potential of brines with densities between 13.2 and 19.2 lb/gal in various core materials. Unfavorable fluid/rock interactions are not evident except in the case of calcium brines with densities higher than 14 lb/gal, where precipitation of acid-soluble calcium salts was observed. He proposed that brines with densities higher than 14 lb/gal should be formulated with a minimum of 8% $ZnBr_2$ to lower pH and to prevent precipitation. Baijal *et al.* (1991) have shown that incorporating a small amount of surfactant in the brine can mitigate this problem by inhibiting the growth of calcium salt crystals at temperatures up to 280°F. Houchin *et al.* (1991) reported that formation damage by high-density brines increases at temperatures above 350°F in low permeability cores (<100 md). At these high temperatures, the soluble silica formed by dissolution of silica grains may react with the calcium ions in the calcium-

Table 12-4 Brine composition vs. crystallization temperature (Foxenberg and Smith, Jan. 1996)

Density, lb/gal	$CaCl_2$, wt %	$CaCl_2$, lb/bbl	$CaBr_2$, wt %	$CaBr_2$, lb/bbl	H_2O, wt %	H_2O, lb/bbl	TCT, °F
13.2	28.1	155.8	22.1	122.5	49.8	276.1	48
13.2	0	0	46.1	255.6	53.9	298.8	−36
14.2	22.9	136.6	33.4	199.2	43.7	260.6	58
14.2	11.6	69.2	45.3	270.2	43.1	257.0	49
14.2	0	0	51.9	309.5	48.1	286.9	0
15.1	18.8	119.2	42.3	268.3	38.9	246.7	63

containing brines to form calcium silicates. Formation of calcium silicate crystals can block pore throats, leading to formation damage.

When high-density completion brines are used, it is recommended that actual core tests be conducted with formation sand to ensure that the heavy brine is compatible with the formation sand. There are a number of ways these core tests are done. Unfortunately, there is currently no standardized way to conduct these experiments. In one way of conducting core tests, actual formation sand is saturated with simulated formation water at reservoir temperature, and the base permeability of the sand is determined. Then, the completion brine is flowed through the sand for up to 24 hours. The core is then flooded with many pore volumes of simulated formation water to determine the final steady-state permeability. The ratio of final-to-initial permeability, often called the regained permeability or return permeability, is a measure of compatibility between the brine and the rock.

12-2.3.2 Compatibility with Formation Water

The main concern with formation water compatibility is the formation of scales caused by incompatible completion brines and formation water. Scales are deposits of inorganic minerals. The scales most commonly found in oilfield waters are calcium and iron carbonate, calcium, barium and strontium sulfates, sodium chloride, iron sulfide, and silicates. Scales can be formed through the mixing of incompatible waters, solubility change with temperature, solubility change with pressure, and water evaporation. For example, unfavorable mixing ratios of calcium-based completion brine with formation water can cause scale formation and subsequent formation damage (Ali *et al.*, 1994a; Lejon *et al.*, 1995; Azari and Leimkuhler, 1990; Ezzat, 1990).

Table 12-5 shows the compatibility data for 13.5-lb/gal $CaBr_2/CaCl_2$ completion brines mixed with saturated (10-lb/gal) NaCl at 70°F and 190°F. The data indicate that using the 60°F TCT brine with formation waters

saturated with NaCl would result in precipitation of NaCl, while the −10°F and −37°F TCT brines would remain free from precipitation of NaCl at downhole temperatures. Formation waters saturated with NaCl can be found near salt domes.

Because of either changes in temperature and pressure or changes in chemical composition (loss of gas and mixing of waters) oilfield waters will sparingly deposit soluble salts (such as $CaCO_3$). To prevent scale formation, compatibility of completion brines and formation water should be checked in the laboratory at various mixing ratios both at reservoir and surface conditions since fluids may be compatible under higher temperatures downhole, but precipitation may occur up tubing or in surface equipment (Ali *et al.*, 1994b). In addition, when more than one completion brine is used in a well operation, it is important that their compatibility be checked in the laboratory before use.

Carbonate scales can be removed by acid treatments. Calcium sulfate may be removed with moderate difficulty with chelating agents such as EDTA (Cikes *et al.*, 1988; Lejon *et al.*, 1995). There are commercial computer programs available to predict scale formation (Morgenthaler *et al.*, 1991).

12-2.3.3 Compatibility with Formation Crude and Natural Gas

The main concern with formation crude and natural gas compatibility is the formation of oil-water emulsion and/or sludge which may block pores and cause formation damage. The emulsion, if allowed to be produced, may also cause upsets in service facilities. Brine-crude incompatibility is especially important when heavy brine is used, the brine is acidic, or during acid stimulation. To prevent an emulsion from forming, the compatibility between the crude and the brine and/or treatment fluid should be checked in the laboratory at reservoir temperatures (Ali *et al.*, 1994a; Foxenberg *et al.*, 1996a; Foxenberg *et al.*, 1996b). Incompatibility may be

Table 12-5 Compatibility of calcium-based brine systems saturated with sodium chloride formation water (Ali *et al.*, 1994b)

Calcium-based brine system, 13.5 lb/gal	TCT, °F	Saturated 10 lb/gal NaCl at 70°F	Saturated 10 lb/gal NaCl at 190°F
$CaCl_2/CaBr_2$	60	PPT	PPT
$CaCl_2/CaBr_2$	−10	Tr	N
$CaCl_2/CaBr_2$	−37	N	N

PPT = Significant precipitation of NaCl.
Tr = Trace precipitation of NaCl.
N = No precipitation of NaCl.

resolved by reformulation of the brine and/or treatment fluid or the incorporation of surfactants and mutual solvents in the brine and/or treatment fluid. Natural gas may contain significant quantities of CO_2, which cause calcium carbonate to precipitate if mixed with a high-pH, brine containing calcium.

12-3 ENGINEERING AND TESTING ISSUES

Completion-fluids engineering is the measurement and control of the properties of completion fluids. In most cases, fluid properties are controlled by adding small quantities of chemical additives that impart corrosion protection, rheology, fluid-loss control, surfactancy, or other properties that may be needed. Engineering these properties often means that these properties need to be tested in the field. Some of the engineering and testing issues that arise in completion fluids are discussed in this section.

12-3.1 Corrosion

The electrically conductive nature of the brines that make up most completion fluids provides an environment favorable for corrosion of metals to occur. From a practical standpoint, the rate of corrosion that occurs will determine whether corrosion is a problem. Completion fluids with corrosion rates in excess of 10 kg/m^2-y according to corrosion coupon tests are generally considered corrosive enough that corrosion issues should be addressed. For packer fluids, considerably slower corrosion rates would be required. Corrosion rates may be high if the fluid has low pH or contains dissolved oxygen. High temperatures will also accelerate corrosion rates.

For sodium-, potassium-, calcium chloride-, or bromide brine-based completion fluids, dissolved oxygen is the primary agent causing corrosion. The solubility of oxygen in these brines decreases as saturation with the salt is approached. Even though the brine may initially contain dissolved oxygen, if the brine is not circulated during the completion in a manner that will replenish the dissolved oxygen, the oxygen present in the brine will quickly deplete and the corrosion rate will consequently decrease as well. For alkali and alkaline earth-halide brine fluids that will not be circulated, oxygen scavengers are not normally needed. For brine that is circulated, injection of an oxygen scavenger such as ammonium bisulfite into the flowstream with a proportional pump is suggested, along with increasing the pH to about 8.5.

The organophosphate-type corrosion inhibitors are not ordinarily used in completion fluids because a lack of circulation may cause a discontinuous film to develop and result in formation of corrosion cells and the development of pitting. Brine-soluble, filming, amine-type corrosion inhibitors may be used in completion brines, but some of these (particularly the quaternary ammonium-type that are strongly adsorbed to silicate surfaces) may also lead to alteration of surface wettability of mineral grains in the reservoir and consequent formation damage.

Very dense zinc-bromide mixed brine (consisting of calcium chloride, calcium bromide, and zinc bromide) is sometimes used when very high density is needed for well control. These brines have a low pH, meaning that they are somewhat acidic. Raising the pH can cause precipitation of solid materials from these brines. The acidity of zinc-based brine can cause severe corrosion if a corrosion inhibitor is not present. Most commercially available zinc-based completion brine contains a thiocynate (or other thio-family chemical) corrosion inhibitor that forms a protective film on the surface of steel or iron in the zinc-based brine environment.

Bacterial growth in completion fluids can generate acid conditions and hydrogen sulfide. The presence of sulfate ion in a completion fluid can allow sulfate-reducing bacteria to produce hydrogen sulfide, which can cause health and safety problems as well as corrosion problems. Water-soluble polymer additives that are used to viscosify and control fluid losses from completion fluids are subject to bacterial degradation. When these products are metabolized by bacteria, they lose their effectiveness, and acid is generated. Addition of biocide is needed to prevent bacterial growth in many completion fluid formulations. When using surface waters from lakes, streams, bays, or seas as a makeup or dilution fluid for completion fluids, it is important to consider possible bacterial contamination and to use biocides.

There are several situations that may arise in completion fluid technology that require special attention to corrosion issues. Completion fluids that are foamed with air can be highly corrosive and it may be impossible to control the corrosion rates from such fluids with ordinary steel tubulars at elevated temperature. Possible solutions include using nitrogen instead of air to produce the foam or using special corrosion inhibitors (Scott *et al.*, 1995). High-temperature completion environments, especially those above 150°C, may require special attention to corrosion issues.

12-3.2 Filtration

Completion with clear brine should ensure that the fluid that actually contacts the formation is as particle-free as possible. Sources of particles in brines used for completion include the following:

- Suspended solids in makeup waters
- Insoluble impurities in salts
- Rust, scale, or dirt in containers used in transporting brines
- Rust, scale, or dirt in tanks used for storage
- Solids from mud and cement incorporated into the brine during displacement

To obtain solids-free brine, it is often necessary to filter the brine either at a brine storage plant or in the field as it is being used.

Filtration can be performed either in a mixing plant or in the field. Brines prepared by dissolving salt in a make-up water do not contain oil contamination. Mixing-plant filtration of these brines can be accomplished using a two-stage filtration process: first, the fluid is passed through a 10-micron, string-wound filter cartridge, and then it is filtered with a 2-micron, pleated-paper cartridge to achieve the final cut (Wedel *et al.*, 1992). Brine that is reprocessed in the field during, or in the brine plant after, field use typically contains oil as well as solids contamination. This oil and solids contamination can reduce cartridge life of string-wound and pleated-paper cartridges. Wedel *et al.* (1992) suggest preprocessing the field brine with a diatomaceous earth filter, then using an acrylic fiber cartridge for the coarser filtration, and finally a 2-micron polypropylene cartridge.

Quality control of filtration is an important issue. Brine clarity cannot be determined simply by looking at it. Even a sample that appears clear to the unaided eye can contain significant amounts of fine particles that could cause formation damage. Turbidity measurement of brine clarity and gravimetric measurement of total suspended solids are used in the field (Solee *et al.*, 1985). In addition, automated particle-size analyzers are used in the laboratory to determine particle-size distributions. The degree of clarity needed may vary with different reservoir zones or types of completion operations that will be conducted. As measured in the field, typical standards range from 20 to 50 NTU units for turbidity (Foxenburg and Smith, 1996b).

Among the issues sometimes overlooked is the quality control needed for clean pits, tanks, lines, and other equipment used to transport or store completion fluids. A comprehensive program to ensure that pits, tanks, and lines are clean is an essential part of any completion operation that uses clear brine.

12-3.3 Weighting Up and Weighting Down

It is quite common for the density of a completion brine to need adjustment during a well operation. Weighting up can usually be done with spike fluids, which are usually stock fluids consisting of 11.6 lb/gal $CaCl_2$, 14.2 lb/gal $CaBr_2$, or 19.2 lb/gal $ZnBr_2$ /$CaBr_2$ solutions. As a general rule, only brines containing the same cation should be mixed, as mixing of brines containing different cations may cause precipitation. If in doubt, the brines to be mixed should be checked for compatibility before mixing. Weighting up with solid salts can also be done. However, some grades of sacked $CaCl_2$ may contain a high degree of impurities, and adding them to a brine may turn the brine cloudy because of these impurities. It will take time for the solids to settle. If brine is weighted up with solid salts with impurities, it may be necessary to filter the brine before use. This filtering is not necessary for $CaBr_2$ where the impurity is usually water. Adding $CaCl_2$ or $CaBr_2$ to a solution will generate heat. Generally, weighting up with liquid is preferable because it can be done rapidly with no need to filter. However, logistics on location, such as the availability of tanks and the requirements for large quantities of dry solids, may dictate which method is used.

Weighting down can be done by diluting with fresh water or a lower-density brine. However, dilution with seawater is not encouraged because seawater introduces undesirable ions and bacteria. A strategy to weight up and down should be planned before the completion brine is brought to location.

12-3.4 Fluid-Loss Control

Large overbalance pressure, high formation permeability and long perforated or openhole intervals can cause substantial loss of completion brine in the formation. Loss of large quantities of completion brines, especially heavy brines, can be expensive and may increase formation damage potential. In addition, if fluid is lost to the formation at a faster rate than can be replenished from the wellhead, a well-control problem may occur. To prevent or minimize this, fluid-loss control materials are sometimes added to the completion brine. Fluid-loss control materials can generally be classified into two types: solids and viscous materials. Solid fluid-loss control materials control fluid loss by forming a low permeability filter cake at the formation sandface. Some commonly used solid fluid-loss control materials include sized NaCl (Mondshine, 1977, 1979), sized CaCO$_3$ (Johnson, 1994), oil-soluble resins, and crosslinked polymers (Blauch *et al.*, 1989; Himes *et al.*, 1994).

The latter are usually crosslinkable polymers to which a crosslinking agent is added just before pumping downhole. Crosslinking occurs as the polymer is pumped downhole. These fluids are designed so that the fluid is fully crosslinked by the time it reaches the perforated zone. A fully crosslinked polymer solution behaves like a solid. If a differential pressure is applied between it and the formation, a leathery skin of very low permeability may form at the sandface. This leathery skin acts like a filter cake in solid fluid-loss control materials. Solid fluid-loss control materials are usually very effective in reducing and even stopping fluid loss altogether. However, they may be difficult to remove once formed. Depending on the fluid-loss control material used, removal may depend on contact by acid, a breaker solution, or crude oil. Inadequate contact, especially over long perforated intervals, can result in incomplete removal and impaired well productivity. Furthermore, when oil-soluble resins are used, the resin solubility may vary significantly among crude oils. Laboratory tests are recommended to determine that the resin used is soluble in the produced oil.

Measurement of fluid loss is carried out using API fluid-loss cells, HPHT fluid-loss cells designed for drilling fluids, or PPT cells that were originally designed to measure pore plugging for prevention of seepage losses. The API fluid-loss cell and the HPHT fluid-loss cell employ filter paper as the medium on which the filter cake is deposited. The PPT cell uses a porous ceramic or aloxite disk as the medium on which the filter cake is deposited. Other types of measurements may use core samples of some standard rock such as Berea sandstone or core samples from the reservoir.

Solid fluid-loss control materials need to be properly sized if they are to bridge properly and form a filter cake at the surface of the formation. One approximate rule of thumb for sands and sandstones (with many exceptions) is that

$$\text{pore size} \approx \sqrt{\text{permeability}} \qquad (12\text{-}7)$$

where pore size is in micrometers and permeability is in millidarcies. The pore size in a 100-md sand would therefore be about 10 microns. According to this rule of thumb, particles capable of plugging a pore 10 microns in diameter would be needed to bridge and form a filter cake on a 100-md sand. Pore sizes can be measured from thin sections of reservoir core materials using microscopic methods, or mercury-injection curves can be used to calculate pore-size distribution. In addition, tests such as the PPT test are used to evaluate fluids to determine if the particle size distribution and other additives will control fluid losses.

Viscous fluid-loss control materials are typically water-soluble polymers, such as hydroxyethylcellulose (HEC) or biopolymers (e.g. xanthan gum or succinoglycan). These polymer solutions are usually shear thinning; their viscosity increases with decreasing shear rate. Typically, a viscous fluid-loss pill is made on location by adding either a dry or liquid polymer to the completion brine. After the brine has fully viscosified, it is pumped downhole and spotted across the perforated or openhole interval. The viscosified brine will enter the formation because of overbalance. As the viscosified brine invades the formation near the wellbore, its viscosity increases since the shear rate drops inversely to the distance from the wellbore. As viscosity goes up, the ability of the pill to control fluid-loss increases. If the fluid-loss control polymer is a power-law fluid, the volume of polymer required to reduce the fluid-loss rate to an acceptable level given the overbalance, formation permeability, interval length, and power-law indices can be calculated using Darcy's law for power-law fluid (Lau, 1994).

The most commonly used viscous fluid-loss control material is HEC. Viscosifying of low-density brines by powder HEC requires proper pH adjustment. Powder HEC is often coated with a film of hydration retarder which dissolves in alkaline pH. Before the addition of powder HEC, the pH of the brine is usually lowered by addition of an acid, such as citric acid. The acid prevents the HEC from immediate hydration before it is thoroughly dispersed in the brine. Premature hydration can lead to the formation of fisheyes or microgels, which are partially hydrated HEC particles that can cause formation damage. After the powder HEC is well-dispersed in the brine, the pH of the brine is raised by addition of an alkali, a dilute sodium hydroxide solution. The alkaline pH dissolves the coating of the HEC and allows hydration of the HEC powder. In low-density brines, hydration of powder HEC is very quick at alkali pH. Continued mixing is usually needed to aid hydration. After the pill has reached its final viscosity, it is often sheared and filtered to remove any solids and partially hydrated HEC before pumping downhole. Viscosifying NH$_4$Cl brines with HEC requires special precautions. Since NH$_4$Cl brines resist pH change because of their buffering capacity, the water is first viscosfied with HEC before NH$_4$Cl is added. Adjustment of pH in calcium- and zinc-containing brines may result in formation of precipitates and is generally not practiced.

Powder HEC is usually not effective in viscosifying heavy brines. Continued shearing and/or heating may be required to viscosify heavy brines (Scheuerman, 1983). To aid the hydration of HEC in heavy brines, service companies have recently introduced the use of liquid HECs. They are typically powder HEC without an anticaking coating suspended in an organic medium, such as mineral oil, to which an organic gellant is added to increase the viscosity needed for suspension. Since the HEC powder is well dispersed in the organic medium, no pH adjustment is needed for hydration.

Occasionally, "activated" HECs have been used to viscosify heavy calcium and zinc bromide brines. Typically, they consist of HEC which is at least partially dissolved in a mutual solvent and suspended in an organic medium. However, these activated HECs may have a limited shelf life.

Viscosifying heavy calcium and zinc brines is not simple. The final viscosity depends not only on the HEC concentration but also the brine composition, specifically, whether the brine is a single- or two-salt brine. The ability of a liquid HEC to viscosify a particular heavy brine should be tested in the laboratory before being recommended for field use. For example, heavy brines with densities between 15.3 and 16.7 lb/gal and containing between 1 and 7% zinc are known to be either very difficult to viscosify with HEC, or if they do viscosify, they may not be thermally stable. Operators should ask the service company to reformulate the brine, and the ability of HEC to viscosify the brine should be checked before field use.

One limitation of HEC as a fluid-loss control material is that its viscosity decreases with increasing temperature and the polymer degrades at high temperatures, limiting its usage at very high temperatures. The advantage of HEC is that it is compatible with many brines since it is nonionic and relatively inexpensive.

Besides HEC, biopolymers such as xanthan gum have also been used as viscous fluid-loss control materials. Biopolymer solutions are usually much more shear thinning and stable at a higher temperature than HEC solutions and are therefore more effective fluid-loss control agents. Like HEC, xanthan gum is available both in powder and liquid forms. Xanthan gum is stable to above 220°F. However, dissolution of xanthan gum in heavy CaCl$_2$ (above 10.5 lb/gal) is difficult. Furthermore, xanthan gum is incompatible with CaBr$_2$ and ZnBr$_2$ brines. Xanthan gum is available in both clarified and unclarified forms. The unclarified xanthan gum contains insoluble cellulosic material that may cause formation damage, while the clarified xanthan has been processed to remove the insoluble materials.

Recently the use of another biopolymer, succinoglycan, has been introduced for fluid-loss control (Lau, 1994). Succinoglycan has a sharp transition temperature above which it loses viscosity rapidly. This transition temperature varies with the type and density of the brine. Properly designed succinoglycan solutions can be more temperature stable and more effective than fluid-loss pills prepared from HECs. However, the transition temperature of the pill should be above the reservoir temperature for the pill to be effective. An internal breaker, such as hydrochloric acid, can be incorporated in the succinoglycan pill to allow it to self-degrade after a predetermined amount of time (Bouts *et al.*, 1996).

12-3.5 Rheology

The viscosity of the completion brine is needed to calculate the friction pressure drop in the workstring during pumping. Figures 12-4 to 12-6 give the viscosity of various completion brines as a function of temperature. The figures show that the viscosity of a heavy brine can be many times that of a low-viscosity brine. When brine viscosity data are not available, experiments should be conducted to measure the friction pressure drop under simulated pumping operations. When the friction pressure drop exceeds horsepower or safety requirements of the tubulars, the completion brine can be reformulated to give a lower viscosity. Usually for a given brine density, the formulation having more water gives a lower downhole viscosity. However, this formulation is usually more expensive. Another remedy is to use drag-reducing agents with the brine. These are typically polymeric solutions.

Rheological properties of the fluid may need to be adjusted to suspend solids if they are used to help control fluid losses or if the completion fluid is used to clean sand and debris from the wellbore.

Rheology is tested with the Fann V-G meter that is commonly used in the drilling-fluid industry. This rotating-cylinder viscometer measures shear rate vs. shear stress, and data can be fit to either a Bingham-plastic or power-law rheological model. The Bingham-plastic

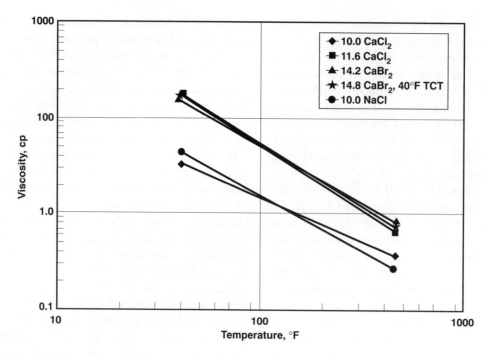

Figure 12-4   Viscosity of NaCl, CaCl₂, and CaBr₂ brines as a function of temperature (Foxenberg and Smith, 1996a)

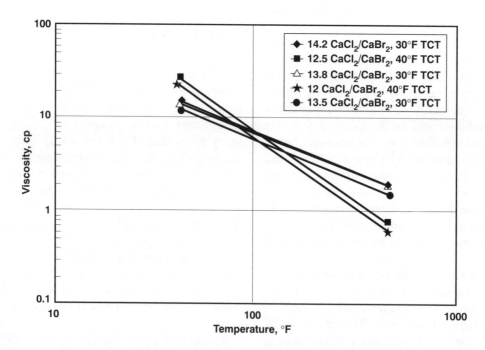

Figure 12-5   Viscosity of CaCl₂/CaBr₂ brines as a function of temperature (Foxenberg and Smith, 1996a)

model is commonly used in field operations because it can be determined with a two-speed V-G meter that measures rheology at 1022 and 511 sec⁻¹ shear rates. The Bingham model represents the rheology as two numbers: the plastic viscosity and the yield point. If shear stress is plotted on the *y*-axis and shear rate is plotted on the *x*-axis, the slope of the line (corrected to centipoise) is the plastic viscosity and the intercept at zero shear rate is the yield point. The Fann V-G meter is constructed so that subtracting the dial reading at 300 rpm (511 sec⁻¹) from

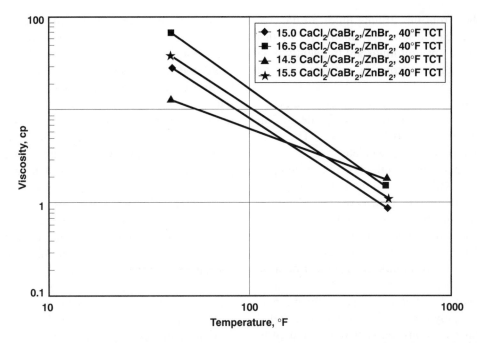

Figure 12-6 Viscosity of $CaCl_2/CaBr_2/ZnBr_2$ brines as a function of temperature (Foxenberg and Smith, 1996a)

the reading at 600 rpm ($1022 \, sec^{-1}$) gives the plastic viscosity, and subtracting the plastic viscosity from the 300-rpm dial reading gives the yield point. In many cases, the minimum yield point of a fluid is specified and sufficient polymer is added to maintain at least this yield point. Yield point values in the range of 10 to 25 lb/100 ft^2 are commonly specified.

Recent advances in design of horizontal drill-in fluids use measurement of rheology at very low shear rates to achieve enhanced suspension properties of the fluid (Beck *et al.*, 1993). A Brookfield viscometer is used to measure the low-shear-rate viscosity at shear rates as low as $0.06 \, sec^{-1}$. Xanthan gum is extremely effective in raising the low-shear-rate viscosity. When this system of measuring rheology is used, typical specifications call for values around 40,000 to 100,000 cp at $0.06 \, sec^{-1}$.

12-3.6 Emulsion Prevention

Mixing crude oil with certain compositions of brine can result in the formation of a stable emulsion. These emulsions can block production and, once formed, can be difficult to remove from the formation. Bottle tests in which crude oil and completion fluid are mixed, shaken, and observed to determine separation are commonly used to evaluate emulsion problems. Mixtures that do not easily separate are likely to form stable emulsions in the formation. Demulsifiers are sometimes added to completion fluids to prevent the formation of stable emulsions.

12-4 APPLICATIONS

In many situations, completion fluids can be used to help prepare the well for production. In the introduction to this chapter, the development of completion fluid technology as a bridge between traditional drilling fluid and stimulation fluid technologies was described. A few common applications and issues related to these applications are discussed in this section. As completion technology further evolves to solve problems related to sand control, water production, and horizontal well performance, new applications for completion fluids will certainly arise.

12-4.1 Perforated Completions

Perforated completions are very common because they offer excellent zonal isolation for stimulation treatment and water production control. The typical procedures leading up to a perforated completion involve drilling the well with conventional drilling fluid, running casing past the production interval, and cementing the casing into the well. At this point, drilling fluid (and cement) fill the volume inside the casing. Conventional drilling fluid contains clay, barite, and drill solids as well as other drilling-fluid additives. If the well is perforated with drilling fluid, the solids from this drilling fluid can damage the perforation and lead to lower productivity. For this reason, clear brine completion fluids are frequently used to displace the drilling fluid from the reservoir interval before the well is perforated.

Choosing the displacement method to ensure that clear brine is opposite the interval to be perforated is an important decision in the completion process. There are two basic methods: partial and complete.

In the partial-displacement method, a pill of completion fluid can be spotted across the interval to be perforated. This method is less expensive and may be appropriate in remote areas where transporting large volumes to the rig is impractical. The disadvantage of this method is that it is not possible to ensure that the brine is solids-free in the downhole environment because there may be some mixing of drilling fluid and completion fluid in the partial displacement. To preform a partial displacement, a polymer viscosifier is typically added to the brine to make a viscous spacer that will be used to push the drilling fluid up the hole several hundred feet above the interval to be perforated. Several spacers, including ones that contain surfactants, may be used in this process. After the spacers are pumped, the pill of clear brine is pumped so that it covers the interval to be perforated, and the perforation is made.

The full-displacement method circulates completion fluid through the entire well. This method is more costly than the partial-displacement method, but it also allows the returns to be monitored for clarity and quality. This method can ensure that the perforation is carried out with clear brine across the reservoir interval. As in the partial-displacement method, spacers containing viscosifiers and surfactants are commonly used to help clean up any drilling fluid that may be gelled up or solids that may cling to tubulars. To obtain a good displacement, it is important to pump the entire hole volume continuously without shutting down the pump. Accomplishing this often requires preplanned storage and having the entire hole volume of completion brine on hand so that the displacement will not be interrupted. One variation that is commonly used to obtain a better displacement is to reverse circulate so that the completion fluid is pumped down the annulus of the well and the returns are through the tubing or drill-pipe. This reverse circulation process prevents solids from settling in the annulus of the well where flow velocities are slower than they are inside the tubing or drillpipe. After the drilling fluid has been displaced, it is a common practice to circulate and filter the completion brine until it meets a specification for clarity. After the completion fluid is clear, the well is perforated.

In some instances, fluids other than clear brine are used as perforating fluids. Some production intervals are sensitive to water and brine. Field crude oil has been used with the philosophy that the crude oil cannot harm the formation. Crude oils can precipitate asphaltenes, paraffins, and other components as they are produced. They may not go back into solution and can make the use of crude as a completion fluid risky. Diesel oil or mineral oil completion fluids are sometimes used and probably impose fewer risks of damage than crude oil. For depleted zones where pressures are low, foam-based completion fluids are sometimes used.

Fluid-loss control from brine-based completion fluids used for perforating is another issue that sometimes needs attention. In many cases, fluid-loss control is not needed because the rate of fluid loss through the perforations is slow enough to be manageable. Perforations that intersect fractures or that are located in coarsely granular high-permeability zones may need fluid-loss control. A section of this chapter deals with methods of achieving fluid-loss control in brine-based completion fluids.

12-4.2 Openhole and Slotted Liner Completions

Openhole (where the formation is left open to the wellbore) and slotted-liner completions do not offer the zonal isolation possible with perforated completions but may be cheaper and may offer higher productivity rates. Openhole and slotted-liner completions are common in horizontal wells. Wells completed with slotted liners, where the slotted-liner is not intended to be a sand control device, are similar to other types of preopened liners including preperforated, predrilled, and even some wire-wrapped screens because the holes in these devices are large enough for drilling-fluid particles to pass through (Stephens, 1991).

Before completing the well, casing has typically been set somewhere above the reservoir interval, and the reservoir has been drilled. The drilling fluid for drilling the reservoir can be either a conventional drilling mud or a special drill-in fluid. The conventional drilling mud may contain clay, drill solids, and barite and may be the same drilling fluid that was used for drilling the unproductive rock above the reservoir. The special drill-in fluid, if it is used, will be a fluid that is used only to drill the reservoir interval and it will have many of the characteristics of a completion fluid. These special drill-in fluids typically contain brine for density instead of barite, polymeric additives for rheology instead of clay, and sized-salt or calcium carbonate for filter-cake solids instead of clay and barite.

In some cases, the drilling fluid that is used to drill the well also serves as the completion fluid; in other words, no special completion fluid is used. The slotted liner, wire-wrapped screen, or preopened liner can be placed into the well with the drilling fluid in the hole. The production flow of the well is used to clean up the mud, the filter cake, and any associated formation damage.

In other cases where damage from the mud or the filter cake will damage production, there are numerous ways that completion fluids are used to limit damage. Some examples are listed bellow:

- The drilling fluid is displaced from the hole before running the liner or wire-wrapped screen. If fluid losses to the reservoir do not need to be controlled, clear brine can be used, and spacers that will help remove the filter cake from a conventional drilling mud are sometimes used. If fluid-loss control is important, polymeric fluid-loss control additives are typically added to the brine before running the liner or screen.

- The hole has been drilled with a special drill-in fluid. The drill-in fluid is displaced with a clear brine containing polymeric fluid-loss control additive. The screen or liner is run into the hole. Brine containing a polymer breaker is spotted across the reservoir to degrade the polymer fluid-loss control additives and assist in clean up.

- The well has been drilled with an oil-based drilling fluid to prevent water-sensitive clays in the reservoir formation from being exposed to water or brine. Gelled oil is used as a completion fluid to displace the solids-bearing oil-based drilling fluid before running the liner or screen. Production is used to clean up the filter cake.

Creative use of completion fluid systems is a driving force in obtaining less costly and more productive open-hole and slotted-liner completions.

12-4.3 Cased-Hole Gravel Pack

Gravel-packing as a sand control technique is described in Chapter 18. There are two key requirements for a completion brine used as a carrier fluid for cased-hole gravel-packing: density and viscosity. The density requirement depends on whether a squeeze or circulating gravel pack is performed. In the former case, the gravel-pack slurry is pumped into the wellbore with the gravel-pack tool in the squeeze position (no fluid return is taken). Since no return is taken on the back side of the well, the completion brine in the workstring/casing annulus will not be diluted by the carrier fluid. Therefore, the carrier fluid need not be weighted for well-control purposes. In this case, a low-density brine can be chosen as long as it meets the minimum salinity requirement to prevent clay swelling. Typically, a squeeze gravel pack can be performed with 2% KCl or 3% NH_4Cl as the base fluid of the carrier fluid.

In the case of a circulating gravel pack, the gravel-pack slurry is circulated into the perforated interval with the gravel-pack tool in the circulating position. Fluid return is allowed to flow up the well in the workstring/casing annulus. In this case, the density of the carrier fluid must be high enough to maintain well control and therefore should be chosen in the same way as outlined in the basic selection criteria section.

The viscosity requirement of a gravel-pack fluid depends on whether a viscous pack vs. a water or low-viscosity pack is being performed. In a typical viscous pack, the carrier fluid is viscosified with a polymer (60 to 80 lb HEC/Mgal) so that it can suspend 15-lb/gal sand of a given size. If a heavy brine is used as the carrier fluid, the HEC concentration can be reduced to 50 to 60 lb/Mgal, since HEC solution in these brines is more viscous than in light brines (Scheuerman, 1984). It should be noted that viscosifying heavy brine with powder HEC could be tedious and may require heating and continuous mixing. In these situations, the use of liquid HEC is more appropriate. Some operators incorporate an internal breaker to reduce the viscosity of the HEC solution to enhance well cleanup. A variety of acid, oxidative, and enzyme polymer-degrading viscosity breakers is used for this purpose. In these cases, the compatibility of these breakers with the formation should be checked. Recently, biopolymer succinoglycan (Sanz *et al.*, 1994) and viscous surfactant solutions (Lehmer, 1988) have been used as alternatives to HEC solutions for viscous packs. These solutions possess unique rheologies and are commonly used without a chemical breaker. To minimize formation-damage potential, viscous gravel-pack fluids should be sheared and/or filtered before use.

Viscous slurry packs have several disadvantages. Inadequate leakoff of the viscous carrier fluid may lead to formation of voids in the gravel pack. Partially hydrated polymer solutions can also cause formation damage. Viscosifying heavy brines may be difficult on location. To avoid these problems, a number of operators have been using water or low-viscosity packs. In these packs, either no or very low concentrations of viscosifier are added to viscosify the carrier fluid. Gravel at a loading of 1 to 3 lb/gal is added to the carrier fluid on-the-fly using a gravel infuser or a specialized blender. The low viscosity of the fluid allows quick sand settling and a compact gravel pack. Recently, these water or low-viscosity packs have been pumped at a relatively high pump rate under squeeze mode, sometimes immediately following a high-rate acid job, to aid in keeping the perforations open and in placement of gravel into them (Bruner et al., 1996).

12-4.4 Horizontal Wells

Horizontal wells have gained wide acceptance worldwide in recent years. Many operators have reported significant gains in productivity by completing horizontal wells instead of vertical wells. The unique challenge for a drill-in fluid for a horizontal well is that it must perform satisfactorily both as a drilling and a completion fluid simultaneously (Dobson et al., 1996).

The drill-in fluid is used to drill the lateral hole and must possess the typical requirements of a drilling fluid including low plastic viscosities, proper rheology for cuttings removal, tight fluid-loss control, thin filter cake, lubricity, and ability to inhibit shale. Among the properties that horizontal well drill-in fluids need to modify from conventional drill-in fluids are rheology and lubricity. In horizontal well drilling, cuttings beds will form if the fluid is unable to completely suspend the cuttings. These cuttings beds can be difficult to remove from the wellbore. Experiments by Zamora et al. (1993) have shown that elevated low-shear-rate rheology of 40,000 cp at $0.06 \sec^{-1}$ (measured on a viscometer) is able to provide adequate hole cleaning for high-angle and horizontal wells. Lubricity is important because it is often a limiting factor in how far the lateral hole can be drilled before friction prevents further sliding of the drillstring. Maintaining the coefficient of friction at about 0.2 or less is generally considered critical for drilling long horizontal sections. Various water-based and oil-based horizontal drill-in fluids have been used. Commonly used water-based drill-in fluids include shear-thinning polymeric solutions using sized salt or sized carbonates as bridging materials. Oil-based drill-in fluids are typically based on a mineral oil or synthetic liquid with sized calcium carbonate as weighting and bridging particles.

After the lateral hole is drilled, various completion methods can be employed, depending on particular reservoir considerations. These methods include use of pre-packed screen, slotted liner, wire-wrapped screen, damage-tolerant screen, perforated pipe with or without a screen, a screen with gravel-packing around it, and openhole.

For openhole, slotted-liner, perforated pipe, and wire-wrapped screen completions, the particles that make up the filter cake from the drill-in fluid can be produced. One consideration in selection of the drill-in fluid is whether the formation consists of rock with matrix permeability such as sandstone, or rock with fracture permeability such as the Austin Chalk (Stephens, 1991).

For rock with fracture permeability, it is important to use solids-free or very-low-solids, non-gelling drilling fluid to prevent gelled mud or filter cake from entering the fractures and blocking production. Thermally stable weighted muds with minimal clay content have been used to drill fractured Austin Chalk wells. Underbalanced drilling techniques have also been used in Austin Chalk horizontal wells, and production is allowed to occur as the well is drilled.

For rock with matrix permeability and little risk of sand production, either water-based or oil-based drill-in fluids that form a filter cake are typically used. The filter cake prevents fine particles present in the drilling fluid from entering the pore network and damaging production. Properly sized salt or calcium carbonate particles bridge pore openings at the formation surface and prevent mud and solids from entering the pore spaces. When oil-based drill-in fluid is used, it is common to allow natural cleanup to occur when oil is produced. In some cases, the oil-based fluid is displaced from the well with specialized surfactant solutions during the completion. Water-based polymer mud has been used more frequently than oil-based fluids because cleanup is easier and can be assisted with polymer breakers or acid washes that help clean up the filter cake. For highly competent rock in Canada, underbalanced drilling techniques that use fluid aerated with nitrogen to achieve underbalance have minimized formation damage (Churcher et al., 1996).

For sandstones where sand production is expected to occur, prepacked screens are typically used to complete the wells. It is crucial that the drill-in fluid and its filter cake be non damaging to the formation and the screen. The problem is that drill-in fluid particles used to bridge the formation and prevent losses may not pass through the screen. Biopolymer viscosified water-based fluids with sized salt or sized calcium carbonate have been used as drill-in fluids. Once the lateral hole is drilled, the drill-in fluid in the openhole section is typically displaced either with a clean drill-in fluid or with a clear brine viscosified with biopolymer. The drill-in fluid in the vertical wellbore is usually displaced with a clear brine. This is done before running the screen to prevent drill-in fluid solids from plugging the screen. After the screen is placed in the horizontal well, the clean drill-in fluid or viscosified brine is displaced out with clear brine and an acid-breaker solution is spotted in the openhole section to dissolve the filter cake at the formation sandface. The base brine used in the drill-in fluid, the viscosified brine, and the clear brine must meet the density, TCT, and formation compatibility requirements mentioned earlier in this chapter. In some instances, synthetic fluids and oil-based fluids have been used to drill unconsolidated sands where prepacked screens will be used. Special surfactant and acid solutions are used to dissolve the filter cakes. The surfactant solutions help make the filter cake water-wet and accessible to acid cleanup fluids.

To prevent formation damage, permeability experiments are often performed to ascertain that the drill-in fluid, its filter cake, filtrate, and the acid breaker solutions are non-formation damaging.

12-4.5 Openhole Gravel Pack

In openhole or external-casing gravel packs, a drill-in fluid is used to underream a larger hole underneath the casing. Sometimes a section of an existing casing needs to be milled out. This step is followed by placing a screen in the open hole and gravel-packing around it. Both the drill-in fluid and the gravel-pack fluid must be tested for formation compatibility and compatibility with the completion brines used before or after the openhole gravel-packing operations.

12-4.6 Hydraulic Fracturing

Hydraulic fracturing, perhaps the most common well-stimulation technique, is described in Chapter 17. Fracturing fluids are essential to the success of the operation. These fluids are typically polymeric solutions used primarily to carry the proppant down the wellbore and into the fracture. Usually an internal breaker is incorporated into the fluid to ensure substantial viscosity reduction after proppant placement. Since fracturing is conducted in squeeze mode without taking returns, the density of the brine used to build the fluid is not a key concern. However, the fluid must still satisfy the minimum salinity requirement. Quite often, 2% KCl is used as the base brine to build the polymeric fracturing fluid. Typically, in hydraulic fracturing of a low-permeability formation, the filter cake is formed at the fracture/formation interface, which minimizes the leakoff of the fluid into the formation. However, return permeability experiments should be conducted to ascertain that the fluid, its filter cake, filtrate and the breaker should be nondamaging to the proppant pack, and to a far lesser degree, the formation. The same idea applies to fluids used in high-permeability hydraulic fracturing or *fracpack*. However, fracture face damage (leakoff damage) is far more critical in high-permeability fracturing. Because a filter cake may not be formed at the fracture formation interface, invasion of fracturing fluid into the formation is likely, and usually a more aggressive breaker schedule is used to ensure degradation of the fluid.

12-4.7 Sweep Pills and Cleanup Fluids

Sweep pills help displace drilling fluids or solids in completions. To keep the drilling fluid from stringing out as it is displaced, sweep pills with high viscosity are used as spacers in displacements. The elevated low-shear-rate rheology available from biopolymer additives makes them excellent components of sweep pills. One widely used formulation contains a minimum of 3 lbm/bbl of xanthan gum in brine, giving an effective low shear rate (0.06 sec^{-1}) viscosity of more than 80,000 cp as measured with a Brookfield viscometer.

Cleanup fluids fall into four basic chemical categories:

- Surfactant-based fluids that are used to clean and suspend dirt and solids. These fluids are often used as spacers to aid in the displacement of drilling fluid by completion fluid.

- Oxidizing agent or enzyme-based cleanup fluids are used to break polymer additives that have been used in drill-in or completion fluids. Lithium hypochlorite in 1% to 2% concentrations is often used as an oxidizing agent to break polymer viscosity and degrade polymers.

- Cosolvent fluids are sometimes used to help clean up oil-based drilling fluids. Mixtures of chemicals such as xylene and isopropanol along with proprietary additives help clean up oil-based fluids and break the water-in-oil emulsions of these fluids. These cosolvent fluids can be used as spacers in brine displacements or spotted as pills to assist in oil-based mud cleanup.

- Acid solutions are commonly used during the completion. One technique to prevent rust and scale from inside the tubing from damaging the formation is called "pickling" the tubing. In this process, an acid solution that contains a corrosion inhibitor is pumped down the tubing and reverse-flowed out before leaving the tubing. Another common use of acid during completion is to dissolve calcium carbonate and break polymer solutions that were used for fluid-loss control. In many cases, an acid somewhat more dilute than the standard 15% HCl is used for these applications to provide a slower reaction rate and reduce the need for acid-diverting agents.

The use of spacers and cleanup fluids varies widely from company to company, and standard practices in their use have not been established.

12-4.8 Non-Brine Completion Fluids

In addition to the commercially available brine solutions, a number of other fluids have been used in completions. These include seawater, bay water, produced brine, crude oil, diesel oil, mineral oil, and foam.

Seawater and bay water may have undesirable characteristics when applied as completion fluids. Sea water and bay water may have salinity too low to prevent clay hydration and migration. In addition, these fluids may contain significant amounts of suspended solids and need to be filtered. Most brine additives for fluid-loss control and rheology will work in seawater and bay water. Bay water and seawater contain sulfate ions and provide a medium that allows bacterial growth to occur. Seawater and bay water should be treated with a biocide.

Generally, produced brine from the zone of interest has enough salinity to prevent clay migration, but the brine that is produced and stored can change composition and precipitate solids. These solids need to be filtered out of the produced brine. In many cases, additives to control fluid loss and rheology of commercial brine systems will not function properly in produced brines.

Crude oil is sometimes used as a completion fluid when water-based completion fluids will swell or disperse clay from production zones with very high clay content. It is important to test crude oil to determine if it makes a suitable completion fluid because many crude oils will precipitate solid asphaltene or paraffinic compounds when they are produced and stored. These asphaltenes and paraffins can cause formation damage that is extremely difficult to remove. Diesel oil or mineral oil is sometimes used as a completion fluid for clay-rich zones. These fluids are generally cleaner and present fewer problems than crude oil.

Zones with extremely low pressure are sometimes completed using foam as the completion fluid. Stable foams can be made either with air or nitrogen as the gas phase. Nitrogen is expensive, but air can cause excessive corrosion.

12-5 CONCLUSION

Proper selection and use of completion fluids can make the difference between a well that is damaged and one that is commercial. Proper engineering of the completion fluid is essential to obtain an optimum production rate, the maximum ultimate production, and therefore the economic success of the well.

The completion engineer should visualize completion fluids as tools to assist in preparing the well for production. Commercially available clear brines are the most commonly used completion fluids because they offer a range of densities, they do not typically cause excessive formation damage, and they can be easily monitored for quality and compared to a set of specifications. Attention needs to be paid to issues of density, crystallization point, and compatibility with the reservoir and other fluids used in well construction.

The completion process should be designed to minimize damage to well productivity. This means that every step from the time the drill bit contacts the reservoir rock until the well is producing needs to be analyzed to ensure that the fluid contacting the reservoir will not cause a reduction in well productivity. In the typical well that is completed by perforating a liner, there are several fluids that contact the reservoir: the drilling fluid, cement, clean up fluids, and the clear brine completion fluid. The interactions among these fluids and with the reservoir should be assessed as part of the engineering process.

A number of non-perforated completion options have become more widely used in recent years, especially for high-angle and horizontal wells. These completions often require the use of specialized drill-in fluids that must function both as drilling fluids and as non-damaging completion fluids. Creative solutions to problems such as sand control in these wells may require new applications of completion-fluid systems. At the present time, completion fluid systems are evolving as new drilling and completion techniques that aim to extract a greater percentage of the oil or gas in place are developed.

REFERENCES

API RP 13J, "Testing of Heavy Brines," 2nd ed, Am. Pet Inst. (Mar. 1996).

Ali, S.A., Durham, D.K., and Elphingstone, E.A.: "Test Identifies Acidizing-Fluid/Crude Compatibility Problems," *O&GJ* (March 1994) 47–51.

Ali, S.A., Javora, P.H., and Guenard, J.H.: "Test High-Density Brines for Formation Water Interaction," *Pet. Eng. Int.* (July 1994b) 31–37

Azari M., and Leimkuhler, J.: "Completion Fluid Invasion Simulation and Permeability Restoration by Sodium- and Potassium-Based Brines," paper SPE 19431, 1990.

Baijal, S.K., Houchin, L.R., and Bridges, K.L.: "A Practical Approach to Prevent Formation Damage by High-Density Brines During the Completion Process," paper SPE 21674, 1991.

Beck, F.E., Powell, J.W., and Zamora, M.: "A Clarified Xanthan Drill-in Fluid for Prudhoe Bay Horizontal Wells," paper SPE 25767, 1993.

Blauch, M.E., Broussard, G.J., Sanclemente, L.W., Weaver, J.D., and Pace, J.R.: "Fluid-Loss Control Using Crosslinkable HEC in High-Permeability Offshore Flexure Trend Completions," paper SPE 19752, 1989.

Bouts, M.N., Trompert, R.A., and Samuel, A.J.: "Time Delayed and Low-Impairment Fluid-Loss Control Using a Succinoglycan Biopolymer with an Internal Acid Breaker," paper SPE 31085, 1996.

Bruner, S., Lau, H.C., Morgenthaler, L.N., Bernardi, L.A., and Kielty, J.M.: "Long-Zone, High-Angle, Squeeze Gravel Packs in Geopressured Reservoirs in the Gulf of Mexico," paper SPE 31092, 1996.

Churcher, P.L., Yurkiw, F.J., and Bietz, R.F.: "Properly Designed Underbalanced Drilling Fluids Can Limit Formation Damage," *O&GJ* (April 29, 1996), 50.

Cikes, M., Vranjesevic, B., Tomic, M., and Jamnicky, O.: "A Successful Treatment of Formation Damage Caused by High Density Brine," paper SPE 18383, 1988.

Dobson, J. Harrison, J.C., Hale, A.H., Lau, H.C., Bruner, S.D., Bernardi, L.A., Kielty, J.M., and Albrecht, M.E. S.: "Laboratory Development and Field Application of a Novel Water-based Drill-In Fluid for Geopressured Horizontal Wells," paper SPE 36428, 1996.

Ezzat, A.M.: "Completion Fluids Design Criteria and Current Technology Weaknesses," paper SPE 19434, 1990.

Foxenberg, W. F., and Smith, B. E.: "Solids-Free Completion Fluids Optimize Rig Operations," *Pet. Eng. Int.* (Jan. 1996a) 27–30.

Foxenberg, W. F., and Smith, B. E.: "Solids-Free Completion Fluids Optimize Rig Operations," *Pet. Eng. Int.* (Feb. 1996b) 63–69.

Foxenberg, W.F., Syed, A.A., and Ke, M.: "Effects of Completion Fluid Loss on Well Productivity," paper SPE 31137, 1996.

Himes, R.E., Ali, S.A., Hardy, M.A., Holtmyer, M.D., and Weaver, J.D.: "Reversible, Crosslinkable Polymer for Fluid-Loss Control," paper SPE 27373, 1994.

Houchin L.R., Baijal, S.K., and Foxenberg, W.E.: "An Analysis of Formation Damage by Completion Fluids at High Temperatures," paper SPE 23143, 1991.

Hudson, T.E., and Andrews, P.W.: "Users' Guide To Weighted Brines," *Pet. Eng. Int.* (Oct. 1986) 33–35.

Johnson, M.H.: "Completion Fluid-Loss Control Using Particulates," paper SPE 27271, 1994.

Klotz, J.A., Krueger, R.F., and Pye, D.S.: "Effect of Perforation Damage on Well Productivity," paper SPE 04654, 1973.

Lau, H.C.: "Laboratory Development and Field Testing of Succinoglycan as a Fluid-Loss-Control Fluid," *SPEDC* (Dec. 1994) 221–226.

Lejon, K., Thingvoll, J.T, Vollen, E.A., and Hammonds, P.: "Formation Damage due to Losses of Ca-Based Brine and How It Was Revealed Through Post Evaluation of Scale Dissolver and Scale Inhibitor Squeeze Treatments," paper SPE 30086, 1995.

Mondshine, T.C.: "Completion Fluid Uses Salt for Bridging, Weighting," *O&GJ* (Aug. 1977) 124–128.

Mondshine, T.C.: "Well Completion and Work Over Fluid and Method of Use," US Patent 4,175,042 (Nov. 20, 1979).

Morgenthaler, L.N.: "Formation Damage Tests of High-Density Brine Completion Fluids," paper SPE 14831, 1986.

Morgenthaler, L.N., Khatib, Z.I., French, R.N., and Cox, K.R.: "Chemical Simulator for Scale Problems in Oil and Gas Production," *Materials Performance* (Apr. 1991) 37–42.

Nehmer, W.L.: "Viscoelastic Gravel-Pack Carrier Fluid," paper SPE 17168, 1988.

Radford, H.E.: "Factors Influencing the Selection of Mud Fluid for Completion of Wells," *API Drill. and Prod. Prac.* (1947) 23–28.

Sanz, G.P., Gunningham, M.C., Lau, H.C., and Samuel, A.J.: "Use of Succinoglycan Biopolymer for Gravel Packing," *SPEDC* (June 1994) 139–143.

Scheuerman, R.F.: "Guidelines for Using HEC Polymers for Viscosifying Solids-Free Completion and Workover Brines," *JPT* (Feb. 1983) 306–314.

Scheuerman, R.F.: "A New Look at Gravel Pack Carrier Fluid Properties," paper SPE 12476, 1984.

Scheuerman, R.F., and Bergersen, B.M.: "Injection-Water Salinity, Formation Pretreatment, and Well-Operations Fluid-Selection Guidelines," *JPT* (July 1990) 836–845.

Schmidt, D.D., Hudson, T.E., and Harris, T.M.: "Introduction to Brine Completion and Workover Fluids, Part 1— Chemical and Physical Properties of Clear Completion Brines," *Pet. Eng. Int.* (Aug. 1983) 80–96.

Scott, S.L., Wu, Y., and Bridges, T.J.: "Air Foam Improves Efficiency of Completion and Workover Operations in Low-Pressure Gas Wells," *SPEDC* (Dec. 1995) 219–225.

Sollee, S.S., Elson T.D., and Lerma, M. K.: "Field Application of Clean Completion Fluids," paper SPE 14318, 1985.

Stephens, M.: "Drilling Fluid Key Factor in Preventing Damage," *Am. O&G Rep.* (Aug. 1991) 47–51.

Wedel, F.L., Haagensen, R.V., and Poch, H.F.: "Oilfield Brine Filtration: Seven Years of Procedures, Problems, Solutions and Results," *Fluid Particle Sep. J.* (1992) **5**, 37–43.

Zamora, M., Jefferson, D.T., and Powell, J.W.: "Hole-Cleaning Study of Polymer-Based Drilling Fluids," paper SPE 26329, 1993.

13 Perforating

James Barker
Halliburton Energy Services

Phil Snider
Marathon Oil Company

13-1 INTRODUCTION

After an oil or gas well is completed, the wellbore is isolated from the surrounding formation by casing and cement. Establishing fluid communication between the wellbore and formation, for either production or injection, requires some perforating operation. Perforating is the process of creating holes in the casing that pass through the cement sheath and extend some depth into the formation. The formation penetration can range from essentially zero to several inches, depending on the perforator used and the mechanical and physical properties of the materials being penetrated. The holes may be dispersed in an angular pattern around the interior of the wellbore; this dispersion is called phasing. The number of shots per linear foot can vary, typically ranging from 1 to 24 (or more if a zone is perforated several times); this number is referred to as shot density. Figure 13-1 shows the geometry of a typical perforated wellbore.

Different methods can be used to create perforations in wellbores; bullet perforating is among the oldest. Hilchie (1990) reports that the original bullet perforator was conceived and patented in 1926, but it was not until the 1930s that the method started to gain widespread use. The bullet gun performed most perforated completion operations into the early 1950s, but little new development work has been done on bullet perforators for the past two decades (Bell *et al.*, 1995). In the bullet-perforating method, propellant-driven bullets are shot through casing and cement into the formation. A steel carrier, called a gun, is used to convey the bullet penetrators downhole. Firing is accomplished by sending an electrical signal down a wireline to ignite the propellant. The burning propellant accelerates a bullet through a short barrel (2-in. or shorter) to velocities up to 3300 ft/sec. This velocity is sufficient for the hard, armor-piercing bullet to penetrate casing, cement sheath, and formation. However, bullet-penetration performance decreases substantially in high-strength formations and when very high-strength casing is used (Bell *et al.*, 1995; Thompson, 1962). Currently, bullet perforators are used infrequently, but they still have applications in soft formations, brittle formations, or where consistently round holes in casing (for ball sealers) are needed (King, 1995).

Another perforating method involves the use of high-pressure water jets or sand-laden slurries to abrade a hole into the casing, cement, and formation. The slurry is pumped down the tubing and turned at the bottom by a deflector and nozzle arrangement that allows the fluid stream to impinge directly on the casing (King, 1995). Holes and slots can be made, and the casing can even be cut completely by manipulation of the tubing. In another method, a tool that uses a pump to force high-pressure fluid through a flexible, extending lance is conveyed downhole. In this process, the lance jets its way into the formation. A major advantage of this method is that very clean tunnels can be created with little or no formation damage. The major drawback, however, is that the process is slow and expensive, and the holes must be created one at a time. Thus the process is impractical for long intervals.

A third method to perforate wellbores, called jet perforating, involves the use of high explosives and metal-lined shaped charges. Jet perforating is by far the most widely used technique to create perforations in wellbores. Bell *et al.* (1995) report that over 95% of all perforating

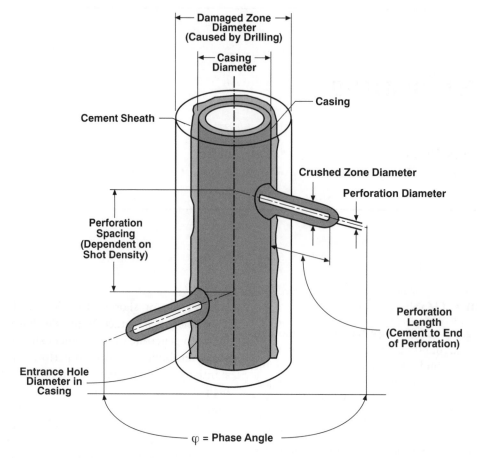

Figure 13-1   Perforated wellbore geometry (after Bell *et al.*, 1995)

operations are performed with shaped-charge jet perforators. Jet perforators can be conveyed downhole by a variety of means: slickline, electric line, coiled tubing, and production tubing. Since jet perforating is the most popular method used today, the remainder of this chapter will be devoted to describing the technique and the hardware associated with it.

13-2 THE EXPLOSIVE TRAIN

Jet-perforating systems comprise different explosive components that are linked to form an "explosive train." Figure 13-2 shows schematically a typical perforating gun and the explosive train inside it. The pertinent explosive elements of the train are (1) an initiator, or detonator, that is used to start the explosive process, (2) a detonating cord that is used to transmit detonation along the longitudinal axis of the gun, and (3) shaped charges that perforate the casing and penetrate the cement sheath and formation. The explosives used in

these components are called "high explosives." When initiated, high explosives react supersonically in a process called detonation. By comparison, "low explosives" react subsonically in a process called deflagration. Propellants and gunpowders are examples of low explosives, while TNT is an example of a high explosive. Low explosives are generally not used in jet-perforating applications, but are used instead in other oilfield situations, such as setting plugs and packers and taking core samples from the formation sidewall.

13-2.1 Primary and Secondary High Explosives

The family of high explosives can be subdivided into two categories: primary and secondary. Primary high explosives are used in initiators only; their sole purpose is to start the detonation reaction with a small energy input (usually by electrical heating of a filament wire or by impact). Examples of primary explosives are lead azide and lead styphnate. They are sensitive to energy inputs from heat, flame, friction, impact, and static discharge.

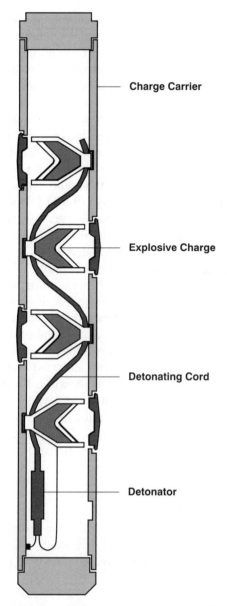

Figure 13-2   Perforating gun and explosive train

Tucker (1972) reports that some primary explosives are sensitive to energy inputs as low as a few ergs; this presupposes that they must be used with great care. The oil industry still uses devices that contain primary explosives, but it is rapidly moving away from them because of their sensitive nature.

Secondary high explosives are used in all three components (detonators, detonating cord, and shaped charges) of the explosive train. The secondary explosives are much less sensitive to external stimuli than are primary explosives, and therefore, they are much safer to handle. Because of their insensitivity, they are somewhat difficult to initiate, but once initiated, they release tremendous amounts of chemical energy in microseconds. Although TNT is a common secondary explosive, its relatively low thermal stability prevents it from being a viable oilfield explosive. For oilfield use, the most widely used secondary explosives are RDX, HMX, HNS, and PYX. These four explosives differ in thermal stability and are necessary because of the wide range of downhole temperatures encountered worldwide. Table 13-1 gives additional information about these explosives.

13-2.2   Thermal Decomposition

The thermal stability of explosives is important because explosives are energetic materials with decomposition rates that are exponential functions of temperature. At room temperature, where the decomposition rate is extremely small, the effective shelf life of an explosive can be a million years—but the same material will react within microseconds at 1500°F (Army Material Command, 1972). Other decomposition rates and corresponding lifetimes exist between these two extremes.

The decomposition of explosives is a process that generates heat and releases gaseous byproducts (Cooper,

Table 13-1   Secondary explosives used in oilfield applications

Explosive	Chemical Formula	Density (g/cm^3)	Detonation Velocity (ft/sec)	Detonation Pressure (psi)
RDX Cyclotrimethylene trinitramine	$C_3H_6N_6O_6$	1.80	28,700	5,000,000
HMX Cyclotetramethylene tetranitramine	$C_4H_8N_8O_8$	1.90	30,000	5,700,000
HNS Hexanitrostilbene	$C_{14}H_6N_6O_{12}$	1.74	24,300	3,500,000
PYX Bis(picrylamino)-3,5-dinitropyridine	$C_{17}H_7N_{11}O_{16}$	1.77	24,900	3,700,000

1996). This decomposition is commonly called "thermal outgassing," and if the heat generated by decomposition can be balanced by heat dissipation to the surroundings, then the explosive quietly decomposes until none remains. If, however, the heat generated by decomposition is not removed quickly enough, then it is possible for the process to become unstable, and the reaction can accelerate uncontrollably until an explosion occurs (sometimes called "thermal runaway"). The process can be stated in simple terms as follows:

$$\begin{array}{l}\text{Rate of} \\ \text{temperature} \\ \text{rise in the} \\ \text{explosive}\end{array} = f\left[\begin{array}{l}\text{Rate of heat} \\ \text{generation} \\ \text{due to} \\ \text{decomposition}\end{array} - \begin{array}{l}\text{Rate of heat} \\ \text{loss to the} \\ \text{surroundings due} \\ \text{to conduction}\end{array}\right]$$

The rate of heat generation caused by decomposition is an exponential function of temperature as mentioned before, while the conductive heat loss is not. As temperature increases, the heat generated by decomposition quickly begins to dominate and can result in a variety of outcomes, including catastrophic thermal explosion. To aggravate the process further, it is possible for gaseous byproducts generated by decomposition to serve as catalysts to the reaction, thereby increasing the rate even more.

Figure 13-3 depicts a set of time-temperature curves that have been experimentally generated for various explosives. They provide guidelines about the probability of quiet decomposition versus violent events. As long as a particular explosive stays below its time-temperature curve, it will function properly. If the time-temperature relationship is exceeded, quiet decomposition may or may not take place, which means it is entirely possible that a violent event can occur. Thus, the time-temperature limits should never be exceeded; further, it is important to recognize that no safety factor has been built into the curves, and that this safety factor must be accounted for when planning downhole jobs requiring the use of energetic materials. One must always consider the accuracy of the downhole temperature measurement or estimation and the amount of time the explosives will remain at that temperature under worst-case conditions, and adjust plans accordingly. Past experience with various exposure times has shown that a minimum safety factor of 50% should be applied when choosing the explosive type. For example, if the estimated time on bottom is 60 hours, then 30 more hours (90 hours total) should be added when selecting an explosive from the time-temperature chart.

The curves are applicable for conditions where the explosive is exposed solely to the effect of temperature. In the case of gun systems where the explosive components are exposed to both temperature and pressure, the time-temperature relationship is different. As an example, HMX detonating cord is normally rated to 400°F for one hour at ambient pressure as shown in Figure 13-3, but laboratory tests show it can undergo violent reaction after only 8 minutes when subjected to the simultaneous conditions of 400°F and 15,000 psi. Thus, pressure serves to accelerate the decomposition reaction.

13-2.3 Initiators

Initiators currently used with perforating systems are of two general types: electric and percussion. For wireline conveyed systems, the most common mode of initiation

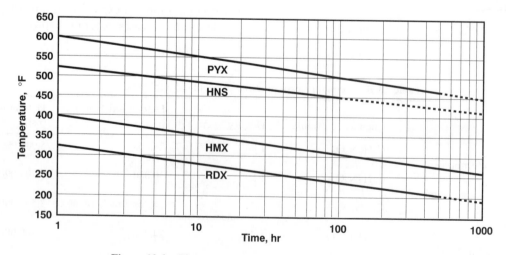

Figure 13-3 Time-temperature curves for example explosives

is an electrical detonator, more generally referred to as an electroexplosive device, or EED. Hot-wire detonators are one common type of EED and Figure 13-4A shows such a device in its simplest form. Safety improvements have been made by detonator manufacturers such as the incorporation of safety resistors in the legwires (Figure 13-4B). These so-called "resistorized" detonators can impede stray current flow (up to a small level) that might be picked up in the detonator circuit from unin-

tentional sources. The stray current is converted to heat by I^2R heating in the resistors rather than the bridgewire, resulting in a safer device. Other recent improvements in EEDs, such as the elimination of sensitive primary explosives, have also added to the safety of the devices. Exploding foil initiators (EFIs), exploding bridgewire detonators (EBWs), and deflagration-to-detonation transition (DDT) detonators are examples of EEDs that do not contain primary explosives.

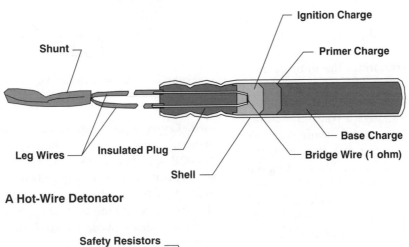

A Hot-Wire Detonator

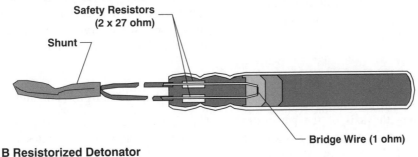

B Resistorized Detonator

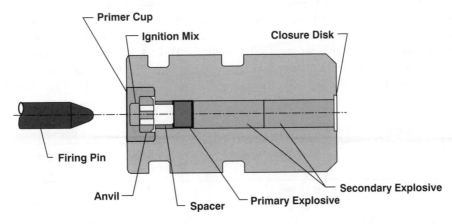

C Percussion Initiator

Figure 13-4 Initiators for perforating systems

The relative safety among the different types of EEDs can be evaluated, to a first approximation, by comparing the threshold instantaneous powers to fire the devices. Table 13-2 shows these powers; those for EBWs and EFIs are extremely high and their safety is unquestionable; however, they require downhole firing modules and special firing panels to function. The hot-wire and resistorized detonators are much simpler to fire, but are also the most susceptible to hazards; thus, their use is dependent on operational restrictions such as radio silence. The RF-protected DDT devices are between these two extremes and attempt to offer greater margins of safety without the need for special ancillary equipment (Motley and Barker, 1996).

For tubing-conveyed perforating, the initiator most commonly used is a percussion-initiated device (Figure 13-4C). Percussion devices function by having an appropriately contoured firing pin strike a relatively sensitive part of the initiator. The pinching and shearing of explosive inside the initiator generates a flash that reacts with primary and secondary explosives to achieve detonation. Because percussion detonators do not have legwires or bridgewires, they are not susceptible to electrical hazards. Nonetheless, the percussion devices must still be handled with appropriate care because they are designed to function by impact. Typical impact energies for functioning are approximately 5 to 7 ft-lb.

13-2.4 Detonating Cord

As mentioned previously, detonating cord is used to transmit detonation along the axis of the perforating gun, sequentially initiating each charge as the detonation wave passes by. Figure 13-5 shows a cross-sectional view of a detonating cord. It is simply a core of secondary explosive that is enclosed by a protective sheath. The sheath can be a single-component material such as lead or aluminum, or it can be composed of layered materials such as an extruded plastic jacket over a woven fabric braid. Detonating velocities can differ among cords primarily because of the properties of the explosive being

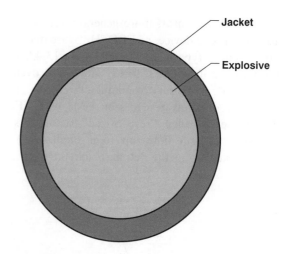

Figure 13-5   Cross-sectional view of a detonating cord

used. Cords made from HNS and PYX are typically slowest, with velocities ranging from 20,000 to 21,000 ft/sec. The RDX and HMX cords are faster, with velocities approaching 26,000 ft/sec. Velocities just over 29,500 ft/sec have been achieved by preconditioning HMX cords with pressure (the applied pressure increases explosive density inside the cord, thus boosting detonation pressure and velocity). This characteristic is sometimes used when trying to avoid shot interference (described later) in ultra-high-shot-density systems.

13-2.5 Shaped Charges (DP and BH)

The shaped charge, or jet perforator, is the explosive component that actually creates the perforation. It is based on the same technology used by armor-piercing weapons developed during World War II. The history of the shaped charge and its principle of functioning is well documented and is described in detail by several recent authors (Walters and Zukas, 1989; Carleone, 1993; Halleck, 1995). The shaped charge is a simple device, containing as few as three components, as shown in Figure 13-6. The complication, however,

Table 13-2   Threshold powers to function various electroexplosive devices

Device Type	Threshold Power	Relative Safety	Uses Primary Explosives
Hot wire	0.25 W	1	Yes
Resistorized	13.8 W	55	Yes
RF-protected DDT	187 W	748	No
Exploding bridgewire	125 KW	5×10^5	No
Exploding foil	1.5 MW	6×10^6	No

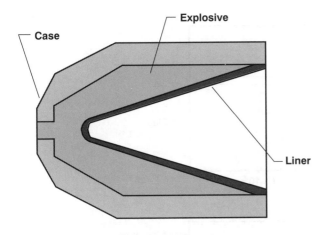

Figure 13-6   Shaped charge perforator

resides in the physics of liner collapse and target penetration; the extreme dynamic conditions that exist during collapse and penetration involve the disciplines of elasticity, plasticity, hydrodynamics, fracture mechanics, and material characterization (Walters and Scott, 1985). It is only since the introduction of powerful computer codes, called hydrocodes, that the mechanics of the collapse and penetration processes can be studied in detail (Walker, 1995; Regalbuto and Gill, 1995). Much insight into the mechanics of the collapse and penetration processes has recently been gained with the development of hydrocodes because they allow the viewing of explosive events on a microsecond-by-microsecond basis. Figure 13-7 shows the predicted penetration of a jet in a steel target compared to experiment. The agreement is virtually exact,

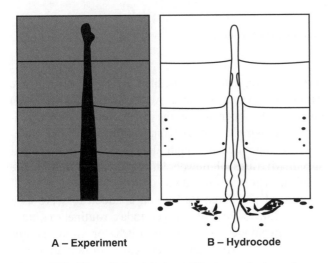

A – Experiment   **B – Hydrocode**

Figure 13-7   Experimental hole profile versus hydrocode prediction (after Regalbuto and Gill, 1995)

and it serves to illustrate that the hydrocodes correctly model the physics of the penetration process as long as the dynamic material properties for both the shaped charge and the target are well known.

The process of liner collapse and jet formation begins with the initiation of the explosive at the base of the charge. A detonation wave sweeps through the explosive, chemically releasing the energy of the solid explosive. The high-pressure gases at the detonation front (approximately 3 to 5 million psi) impart momentum, which forces the liner to collapse on itself along an axis of symmetry. Depending on the shape and material of the liner, different collapse and penetration characteristics will result. If the liner geometry is conical, a long, thin, stretching jet will be formed as shown in the hydrocode results of Figure 13-8a to 8e. The penetration of the jet into the target is relatively deep and the hole diameter is small. Walters (1992) reports that strain rates of 10^4 to 10^7 (in./in.)/s occur during collapse and that peak pressures along the centerline reach approximately 29 million psi, decaying eventually to about 2.9 million psi. The jet-tip velocity can reach values up to 26,000 ft/sec for a copper liner. The fraction of the liner that constitutes the jet is around 20% (Walters and Zukas, 1989); the remaining 80% is in the form of a slower-moving slug that generally does not contribute to penetration.

If the liner is parabolic or hemispherical as shown in Figure 13-9a to 9e, a much more massive, slower-moving jet will be formed that has shallow penetration. However, the hole diameter created during penetration will be relatively large. For hemispherical liners, the jet-tip velocity is around 13,000 to 20,000 ft/sec and the jet represents about 60 to 80% of the total liner mass (Carleone, 1993). The remaining 20 to 40% constitutes the slug. Therefore, the liner design has a tremendous influence on the penetration characteristics of a shaped charge, which is why the shape of the liner is used to categorize a jet perforator as either deep-penetrating (DP) or big-hole (BH). Typical DP charges create hole diameters between 0.2 and 0.5 in., with penetrations up to several dozen inches in concrete. They are used primarily for perforating hard formations. By comparision, the BH charges are generally used for perforating unconsolidated formations. They typically produce hole diameters between 0.6 and 1.5 in., and their penetrations seldom exceed 8 in.

The hydrocode results of Figure 13-9a to 9e also explain an often-confusing phenomenon that occurs with big-hole charges; namely, that the hole in the gun is smaller than the hole in the casing. The hydrocode graphics illustrate how this can happen: the jet exits the gun at a relatively small diameter and continues its devel-

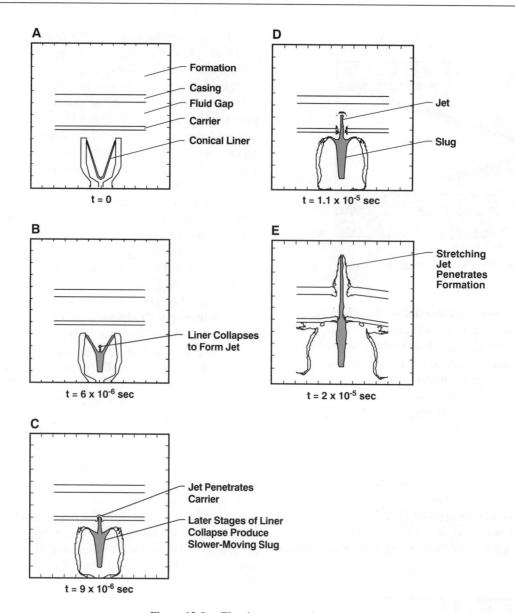

Figure 13-8   The deep penetrating (DP) charge

opment and expansion in the annular gap between gun and casing. The presence of fluid in the annulus encourages the jet to mushroom because of the fluid's impeding nature. If designed properly, the jet will have attained a proper balance between its diameter and velocity just at the moment of casing impact. Thus, for a given charge design, there is an optimal "focal length" for big-hole charges that will produce the most uniform holes with the largest diameter in casing. Many service companies therefore recommend that big-hole gun systems be centralized whenever possible.

Many of the recent advances in shaped-charge technology are related to manufacturing techniques and metal-lurgy. The importance of the liner in charge performance cannot be overemphasized; Walters and Zukas (1989) consider the liner to be the most important shaped-charge design element. Attributes of a good liner material are high density, high ductility, and high sound speed. Mixtures of powdered copper and tungsten are excellent choices for oilwell DP perforators; a small amount of powdered lead or tin is added to serve as a binder. Liners for DP charges today routinely contain high percentages of tungsten (55% or more is not uncommon); experimental evidence shows that the higher density of tungsten generally results in deeper penetration in hard targets. Liners for big-hole charges

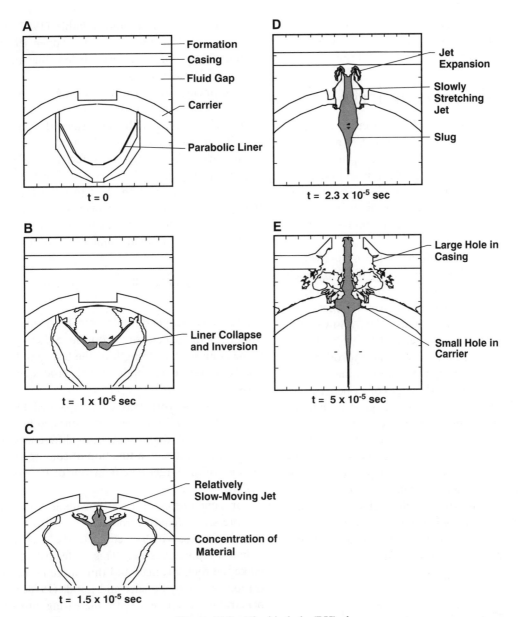

Figure 13-9 The big-hole (BH) charge

are usually fabricated from wrought materials. Copper was the most popular material until recently, when the industry began to shift from copper to brass alloys. Experimental results indicate that the presence of alloying elements (primarily zinc) in the brass cause the jet to break up and leave less liner debris in the perforation tunnel.

13-2.6 Safety and Accidents

With the use of explosives, perforating operations have the potential to be dangerous. However, when the rules of explosive safety are followed, the industry has a very good safety record. Nonetheless, accidents have occurred and their outcome can be very unforgiving. Human error frequently causes industrial accidents, and much design effort is expended to reduce the frequency of these errors. Accidents have occurred in all parts of the life cycle of oilfield explosives, ranging from the manufacture of bulk powder to the final disposal of unusable product.

Within the oil and gas industry, document API RP-67 (1994) provides recommended guidelines for the safe handling and use of explosives at the well site. These guidelines cover operational practices in the field and design recommendations for surface and downhole explosive-related equipment.

13-3 THE PENETRATION PROCESS

Jet penetration from a shaped charge occurs by the jet pushing material aside radially. The process is analagous to a high-pressure water stream penetrating a block of gelatin—the shear strength of the gelatin is so low that it offers essentially no resistance to the water stream. The water stream pushes the gelatin aside radially as it penetrates, leaving a hole. Material is not removed, but is only displaced.

Essentially the same penetration process occurs with a shaped charge jet penetrating a target, only at much higher pressures (Figure 13-10). The earliest penetration models (circa 1940s) assumed the jet and target were incompressible, or Bernoulli, fluids. This allowed the strengths and viscosities of the jet and target materials to be neglected, an assumption called the hydrodynamic approximation. This hydrodynamic approximation is a rational assumption for the early stages of jet penetration since the impact pressure far exceeds the yield strength of most materials. For example, the impact pressure generated by a copper jet against casing is approximately 15 to 30 million psi, whereas the yield strength of typical high-strength casing is only around 100,000 psi. Thus, even a material as strong as steel is weak when compared to the stresses resulting from high-velocity impact. Temperature plays a negligible part in the penetration process. Not only is the time frame too fast for any significant heat transfer to occur, but the temperature

is too low. Von Holle and Trimble (1977) experimentally measured temperatures of copper jet tips to be around 750° to 1100°F, well below the melting point of materials such as steel.

Experimental evidence soon confirmed, however, that the hydrodynamic approximation was not valid for all stages of penetration. In fact, since the jet possesses a velocity gradient (which is why it stretches), there is a point where the velocity is insufficient to allow material strength properties to be neglected. For the oil industry, Thompson (1962) was the first to publish data showing that perforator penetration decreased as a function of increased formation compressive strength. His data correlated to the semi-logarithmic expression:

$$\ln l_{pf} = \ln l_{ps} + 0.086(C_1 - C_2)(10^{-3}) \qquad (13\text{-}1)$$

where l_{pf} is the penetration into the producing formation, l_{ps} is the penetration into the test sample in in., C_1 is the compressive strength of the test sample, and C_2 is the compressive strength of the producing formation in psi.

Thompson's work was later expanded by Berhmann and Halleck (1988) where they confirmed his conclusions and also noted that penetration differences in various targets were not only a function of compressive strength, but also of the target type (concrete or Berea) and charge design.

For penetration of rock downhole, not only is the compressive strength important but also the downhole effective stress, which is equal to the overburden stress minus the pore pressure. This "effective stress" acts to make the rock stronger and thus more resistant to penetration. Work by Saucier and Lands (1978) showed that penetration decreased with increasing in-situ effective stress, up to a plateau of 5000 to 6000 psi. Penetration models were refined further when it was realized that compressibility effects should be considered when predicting penetrations in porous targets such as formation rock. Furthermore, if the jet possesses a higher sound speed than the target, it is possible for localized shocks to occur during the penetration process, reducing penetration. Regalbuto *et al.* (1988) proposed using the bulk sound speed of the target as a simple means to account for compressibility and shock effects. Later work by Halleck *et al.* (1991) used the bulk modulus to account for compressibility and found that penetration decreased with increasing bulk modulus.

In summary, several factors are known to reduce jet penetrations in formation rock under downhole condi-

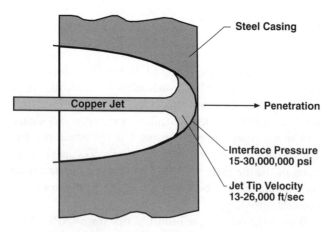

Figure 13-10   Schematic of penetration process (after Walters and Zukas, 1989)

tions when compared to penetrations obtained in concrete surface shots. Table 13-3 gives laboratory data that shows how these penetrations decrease for a typical charge.

Finally, Ott *et al.* (1994) proposed a model for predicting penetration in downhole formations based on concrete penetration in surface test shots. The simple method is used in several popular nodal analysis programs. An underlying assumption in the model is that penetration of stressed Berea targets is about 70% that of concrete. From Table 13-3, if the concrete penetration of 15.49 in. is multiplied by 0.7, a predicted Berea penetration (in stressed rock) of 10.84 in. is obtained. This is 18% greater than the experimental result of 9.21 in. Thus, some researchers believe that the model is too simple for reasonably accurate results, while other researchers maintain that it is more than adequate for practical engineering purposes.

13-4 THE DAMAGED ZONE

During the jet-penetration process, some damage occurs to the rock matrix surrounding the perforation tunnel as shown in Figure 13-11. The altered zone is called the damaged (or crushed or compacted) zone and results from the high-impact pressures that occur during perforating. The damaged zone consists of crushed and compacted grains that form a layer approximately $\frac{1}{4}$- to $\frac{1}{2}$-in. in thickness, which surrounds the perforation tunnel (Asadi and Preston, 1994; Pucknell and Behrmann, 1991). Later work by Halleck *et al.* (1992) shows that the damaged zone is nonuniform in thickness, and it actually decreases down the length of the perforation tunnel. The greatest damage occurs near the entrance hole where the impact pressure from jet impingement is greatest, while the least damage occurs at the tip of the tunnel where the impact stress is least. There is some evidence that big-hole charges (especially those with explosive loads exceeding

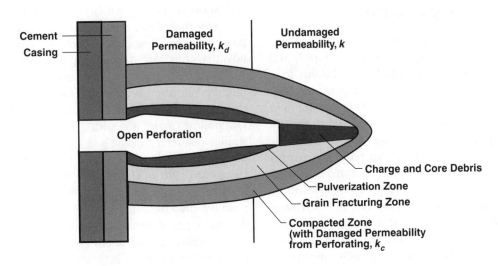

Figure 13-11 Perforation-damaged zone (from Bell *et al.*, 1995)

Table 13-3 Penetration reductions for a $2\frac{1}{8}$-in. capsule charge

Target	Compressive Strength (psi)	Effective Stress (psi)	Penetration (in.)	Comments
Concrete	6600	0	15.49	Benchmark surface shot in concrete
Berea sandstone	7000	100	10.25	Reduction caused by change in target material
Berea sandstone	7000	1500	9.21	Reduction caused by increased effective stress
Nugget sandstone	13000	100	6.68	Reduction caused by higher strength target

23 g) may cause damaged zone layers that approach 1 in. Laboratory studies indicate that the permeability of the damaged zone can be 10 to 20% of the surrounding virgin formation (Bell *et al.*, 1972). To restore effective communication between the reservoir and wellbore, several techniques are used in an attempt to remedy the effects of the damaged zone. One technique is underbalanced perforating, in which the operator tries to remove damage by backflushing or surging the formation. Another technique is extreme overbalanced perforating in which an attempt is made to circumvent perforation-induced damage by creating microfractures that pass through the crushed zone. The overbalance pressure can come from either a pressurized wellbore or from downhole propellant charges. Additionally, shaped-charge research today is concerned with developing ways to make the penetration process less damaging through new liner geometry and metallurgy.

13-5 TYPES OF GUN SYSTEMS

Figure 13-12 shows the three general types of gun systems used in the industry today: ported, scalloped, and capsule. Both ported and scalloped guns belong to a family called hollow carrier guns. These guns are relatively heavy-walled, sealed tubular conduits that are used to convey the explosive train inside them, and to protect the explosives from the surrounding wellbore environment. The port plug gun possesses an economic advantage over the scalloped gun because it can be used as many as 100 times. However, the port plug gun must also make an individual seal at each charge, increasing the likelihood of fluid leaks when it is run downhole. If a hollow carrier gun leaks and the explosives are surrounded by fluid during detonation, the gun will rupture upon firing and stick inside the casing. Therefore, port plug guns are generally used where wellbore conditions are not as demanding, such as in wells with moderate hydrostatic pressures and temperatures. Recommended limits for reliable operation of these guns is around 20,000 psi and 325°F; however, with special attention they can be run to 400°F. Port plug guns are generally run on wireline and are fired from the bottom using a "fluid-disabled" electric detonator. This type of detonator is designed to disable itself if submerged in fluid, preventing the problem of a ruptured gun in the event of an inadvertent leak.

The scalloped hollow carrier guns are used in the most demanding well conditions, and are the mainstay of tubing-conveyed perforating (TCP) operations. Scalloped

guns are available in a wide range of sizes, from small-diameter guns ($1\frac{9}{16}$ in. OD) that will pass through $2\frac{3}{8}$-in. tubing to large-diameter (7-in. OD) guns that are made to run in $9\frac{5}{8}$-in. casing. Ratings up to 30,000 psi and 500°F are available, and the guns can be run on either wireline or tubing. Because these guns are much less susceptible to leaks than are port plug guns, they are often top-fired. Gun intervals of more than 2700 feet have been shot with these types of guns, and a wide variety of mechanical, hydraulic, and electrical firing mechanisms are available to mate with them. Hollow carrier guns also absorb explosive shock when the charges are detonated, protecting the casing from potential damage. They also retain a significant amount of charge and gun debris after detonation.

Capsule perforating guns consist of charges that are enclosed, or encapsulated, by a protective cap as shown in Figure 13-13. The charges are affixed to a carrier, which is generally a set of wires or a flexible strip, and are usually conveyed by wireline. Since there is no thick-walled carrier present with a capsule gun, the charges can be larger than those used in hollow carrier guns and still pass through downhole tubulars. This is especially advantageous in through-tubing operations where small-diameter guns are necessary for passage, yet maximum penetration is desired. As an example, for running through $2\frac{7}{8}$-in. tubing, an operator can run either a 2-in. OD hollow carrier gun with 7.5-g explosive charges or a similar diameter capsule gun with 15-g charges. Because of its larger size, the capsule charge penetrates nearly twice the depth of the smaller charge, and the hole size in casing is approximately 20% bigger. However, the primary drawback to capsule gun systems is that the detonator and detonating cord are exposed to the effects of the wellbore-fluid environment. Fluid intrusion can occur at the interface between the detonator and detonating cord, which necessitates the use of special sealing techniques. Also, as mentioned earlier, the time-temperature relationships that are valid for hollow carrier guns are not applicable for capsule gun systems because the explosive train is exposed to the simultaneous effects of temperature and pressure.

One unique variation of the capsule gun uses "elongated" charges that articulate, or pivot, on a carrier through a 90° rotation. This pivoting allows the charges to remain folded and to pass through downhole tubulars, and to open up when it is time to fire. The charge performance for this type of system is impressive, but the mechanical complexities of the equipment and the large amount of debris left downhole after firing may be undesirable.

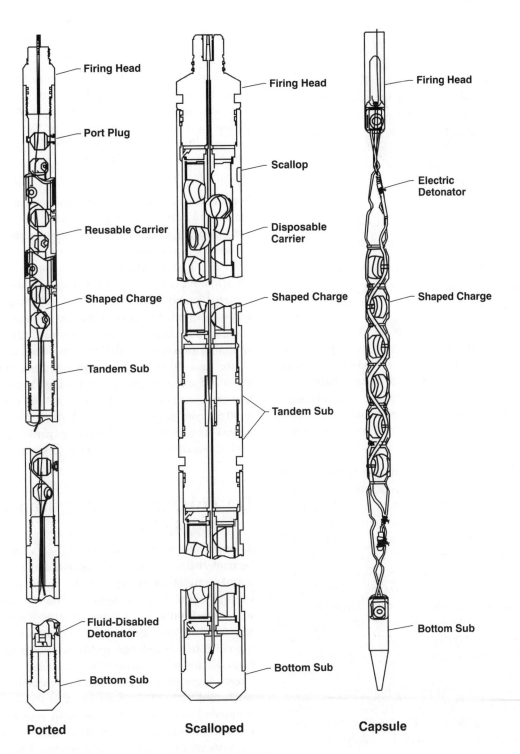

Figure 13-12   Gun system types

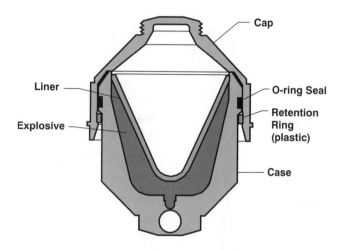

Figure 13-13   Capsule charge

13-6 PERFORMANCE MEASURES OF CHARGES AND GUN SYSTEMS (API RP-43)

To evaluate the performance of jet perforators and gun systems, a series of standard tests was developed by the industry and is described in API RP-43 (1991). These tests allow a fair comparison of different charges and gun systems so that operators can choose the appropriate explosives and hardware for conducting perforating operations. The four sections of tests are described in the following paragraphs.

Section 1:   This test evaluates the performance of multiple-shot perforating systems in concrete targets. The minimum concrete compressive strength is 5000 psi and the casing is grade L-80. The test is conducted at ambient temperature and atmospheric pressure; recorded data include both hole diameter and target penetration. Because multiple shots are fired, Section 1 testing can account for the effects of charge-to-charge interference. Interference occurs whenever the case fragments of a detonated charge impact a neighboring charge and "interfere" with jet development. Interference is especially likely to occur in high-density gun systems where the charges are very close to each other. Since the Section 1 test is the most economical of the four to conduct, it is the most widely used. However, it is the least simulative of downhole conditions since the target is not a fluid-filled formation rock under stressed conditions.

Section 2:   This test evaluates perforator penetration in a hydraulically stressed (3000 psi) Berea sandstone target. It is generally used only for deep penetrating charges. Two target diameters are specified to account for exter-

nal boundary effects on penetration: 4-in. diameter for charges 15 g or less and 7-in. for charges exceeding 15 g. This is a single-shot test and is conducted at ambient temperature.

Section 3:   This test evaluates how perforator performance may be changed after exposure to elevated temperature and time. It is a multiple-shot test conducted at atmospheric pressure using steel targets. After firing, the penetration in steel is compared to the penetration of charges that have not been thermally conditioned, which gives a measure of charge degradation after exposure to elevated temperature.

Section 4:   This test provides a measure of flow performance of a perforation tunnel under simulated downhole conditions. With respect to downhole perforator performance, it is the most representative of the four tests; however, it is also the most costly and is therefore used on a limited basis. The test is a single shot test and the target is quarry rock or an actual well core. Three pressures are applied as part of the test: (1) pore pressure to simulate reservoir formation pressure, (2) an external, confining pressure on the core to simulate overburden stress, and (3) a wellbore pressure that simulates the effects of underbalanced perforating. Typical values of the three pressures might be 1500 psi pore pressure, 4500 psi confining pressure, and 1000 psi wellbore pressure. This combination of pressures provides an effective rock stress of 3000 psi and 500 psi underbalance.

13-7 COMPLETION TECHNIQUES FOR HARD ROCK

Selection of a perforating system for a completion must be made with a reasonable understanding of what additional stimulation, if any, will be required for the well. Certain stimulation methods may not be possible if the well is perforated in a manner that did not take future activities into consideration. For example, some formations are stimulated during hydraulic fracturing with a limited number of perforations to divert fluids into different layers of a formation. This technique usually is conducted with 30 or fewer holes in the intervals to be stimulated—not in wells where entire intervals were previously perforated with high shot-density perforator systems. Perforating planning is therefore difficult; a system to maximize initial well productivity is not necessarily the best system for subsequent well stimulation. It is critical for the completion engineer to have a good understanding of the formation characteristics and discuss future well activities with other individuals (such as the geologists) to determine a mutually agreeable approach.

13-7.1 Natural Completion Perforating

"Natural" completions can be defined as those wells with sufficient reservoir permeability and formation competence to produce the desired volumes of reservoir fluids without the need for additional stimulation. On these wells, effectively communicating the undamaged formation to the wellbore is critical. The more important perforating parameters include perforator depth of penetration, charge phasing, effective shot density, percentage of the productive interval that is perforated, and underbalance levels. Perforator hole size in the casing string is less important, except in wells with extremely high deliverability (high permeability) potential.

Well geometry does not allow all perforating parameters to be optimized economically, and numerous technical articles have been written about how to predict inflow performance from a wellbore based upon the perforating parameters (Locke, 1981; McLeod, 1983; Hong, 1975; Karakas and Tariq, 1988). Many operators and service companies have developed computer programs based upon these technical articles to assist the engineer in gun systems selection based upon specific well criteria. One of the standard comparison methods of perforator systems is the use of productivity index ratios, which are defined as J_p/J, where J_p is the productivity index of the perforated well and J is the productivity index of an open-hole well.

13-7.2 Calculation of the Perforation Skin Effect

A mechanism to account for the effects of perforations on well performance is through the introduction of the perforation skin effect, s_p, in the well production equation. For example, under steady-state conditions

$$q = \frac{kh(p_e - p_{wf})}{1412B\mu\left(\ln\frac{r_e}{r_w} + s_p\right)} \quad (13\text{-}2)$$

Karakas and Tariq (1988) have presented a semianalytical solution for the calculation of the perforation skin effect, which they divide into components: the plane-flow effect, s_H; the vertical converging effect, s_V; and the wellbore effect, s_{wb}. The total perforation skin effect is then

$$s_p = s_H + s_V + s_{wb} \quad (13\text{-}3)$$

Below, the method of estimating the individual components of the perforation skin is outlined.

Calculation of s_H

$$s_H = \ln\frac{r_w}{r_w'(\theta)} \quad (13\text{-}4)$$

where $r_w'(\theta)$ is the effective wellbore radius and is a function of the phasing angle θ:

$$r_w'(\theta) = \begin{cases} \dfrac{I_{\text{perf}}}{4} & \text{for } \theta = 0 \\ a_\theta(r_w + I_{\text{perf}}) & \text{for } \theta \neq 0 \end{cases} \quad (13\text{-}5)$$

The constant a_θ depends on the perforation phasing and can be obtained from Table 13-4. This skin effect is negative (except for $\theta = 0$), but its total contribution is usually small.

Calculation of s_V To obtain s_V, two dimensionless variables must be calculated:

$$h_D = \frac{h_{\text{perf}}}{l_{\text{perf}}}\sqrt{\frac{k_H}{k_V}} \quad (13\text{-}6)$$

where k_H and k_V are the horizontal and vertical permeabilities, respectively, and

$$r_D = \frac{r_{\text{perf}}}{2h_{\text{perf}}}\left(1 + \sqrt{\frac{k_V}{k_H}}\right) \quad (13\text{-}7)$$

The vertical pseudo-skin is then

$$s_V = 10^a h_D^{b-1} r_D^b \quad (13\text{-}8)$$

with

$$a = a_1 \log r_D + a_2 \quad (13\text{-}9)$$

and

$$b = b_1 r_D + b_2 \quad (13\text{-}10)$$

The constants a_1, a_2, b_1, and b_2 are also functions of the perforation phasing and can be obtained from Table 13-4. The vertical skin effect, s_V, is potentially the largest contributor to s_p; for small perforation densities, that is, large h_{perf}, s_V can be very large.

Table 13-4 Constants for perforation skin effect calculation[a]

Perforation Phasing	a_θ	a_1	a_2	b_1	b_2	c_1	c_2
0° (360°)	0.250	−2.091	0.0453	5.1313	1.8672	1.6E-1	2.675
180°	0.500	−2.025	0.0943	3.0373	1.8115	2.6E-2	4.532
120°	0.648	−2.018	0.0634	1.6136	1.7770	6.6E-3	5.320
90°	0.726	−1.905	0.1038	1.5674	1.6935	1.9E-3	6.155
60°	0.813	−1.898	0.1023	1.3654	1.6490	3.0E-4	7.509
45°	0.860	−1.788	0.2398	1.1915	1.6392	4.6E-5	8.791

[a]From Karakas and Tariq, 1988.

Calculation of s_{wb} For the calculation of s_{wb}, a dimensionless quantity is calculated first:

$$r_{wD} = \frac{r_w}{l_{perf} + r_w} \quad (13\text{-}11)$$

Then

$$s_{wb} = c_1 e^{c_2 r_{wD}} \quad (13\text{-}12)$$

The constants c_1 and c_2 also can be obtained from Table 13-4.

13-7.2.1 Example: Perforation Skin Effect

Assume that a well with $r_w = 0.328$ ft is perforated with 2 shots/ft, $r_{perf} = 0.25$ in., (0.0208 ft), $l_{perf} = 8$ in. (0.667 ft), and $\theta = 180°$. Calculate the perforation skin effect if $k_H/k_V = 10$. Repeat the calculation for $\theta = 0°$ and $\theta = 60°$.

If $\theta = 180°$, show the effect of the horizontal-to-vertical permeability anisotropy with $k_H/k_V = 1$.

Solution: From Equation 13-5 and Table 13-4 ($\theta = 180°$)

$$r_w'(\theta) = (0.5)(0.328 + 0.667) = 0.5 \quad (13\text{-}13)$$

Then, from Equation 13-4

$$s_H = \ln \frac{0.328}{0.5} = -0.4 \quad (13\text{-}14)$$

From Equation 13-6 and remembering that $h_{perf} = 1/$ shots/ft

$$h_D = \frac{0.5}{0.667} \sqrt{10} = 2.37 \quad (13\text{-}15)$$

and

$$r_D = \frac{0.0208}{(2)(0.5)} \left(1 + \sqrt{0.1}\right) = 0.027 \quad (13\text{-}16)$$

From Equations 13-9 and 13-10 and the constants in Table 13-4,

$$a = -2.025 \log(0.027) + 0.0943 = 3.271 \quad (13\text{-}17)$$

and

$$b = (3.0373)(0.027) + 1.8115 = 1.894 \quad (13\text{-}18)$$

From Equation 13-8,

$$s_V = 10^{3.271} 2.37^{0.894} 0.027^{1.894} = 4.3 \quad (13\text{-}19)$$

Finally, from Equation 13-11

$$r_{wD} = \frac{0.328}{0.667 + 0.328} = 0.33 \quad (13\text{-}20)$$

and with the constants in Table 13-4 and Equation 13-12

$$s_{wb} = \left(2.6 \times 10^{-2}\right) e^{(4.532)(0.33)} = 0.1 \quad (13\text{-}21)$$

The total perforation skin effect is then

$$s_p = -0.4 + 4.3 + 0.1 = 4 \quad (13\text{-}22)$$

If $\theta = 0°$, then $s_H = 0.3$, $s_V = 3.6$, $s_{wb} = 0.4$, and therefore $s_p = 4.3$.

If $\theta = 60°$, then $s_H = -0.9$, $s_V = 4.9$, $s_{wb} = 0.004$, and therefore $s_p = 4$.

For $\theta = 180°$, and $k_H/k_V = 1$, s_H and s_{wb} do not change; s_V, though, is only 1.2 leading to $s_p = 0.9$, reflecting the beneficial effects of good vertical permeability even with relatively unfavorable perforation density (2 shots/ft).

13-7.2.2 Example: Perforation Density

Using typical perforation characteristics such as $r_{perf} = 0.25$ in. (0.0208 ft), $l_{perf} = 8$ in. (0.667 ft), $\theta = 120°$, in a well with $r_w = 0.328$ ft, develop a table of s_V versus perforation density for permeability anisotropies $k_H/k_V = 10$, 5, and 1.

Solution: Table 13-5 presents the skin effect s_V for perforation densities from 0.5 shots/ft to 4 shots/ft using Equations 13-6 to 13-10. For high perforation densities (3 to 4 shots/ft), this skin contribution becomes small. For low shot densities, this skin effect in normally anisotropic formations can be substantial. For the well in this problem, $s_H = -0.7$ and $s_{wb} = 0.04$.

Table 13-5 Vertical contribution to perforation skin effect

shots/ft	s_V		
	$k_H/k_V = 10$	$k_H/k_V = 5$	$k_H/k_V = 1$
0.5	21.3	15.9	7.7
1	10.3	7.6	3.6
2	4.8	3.5	1.3
3	3.0	2.1	0.9
4	2.1	1.5	0.6

13-7.2.3 Near-Well Damage and Perforations

Karakas and Tariq (1988) have also shown that damage and perforations can be characterized by a composite skin effect

$$(s_d)_p = (s_d)_0 + \frac{k}{k_s} s_p \qquad (13-23)$$

if the perforations terminate within the damage zone $l_{perf} < r_s$). In Equation 13-23, $(s_d)_0$ is the open-hole equivalent skin effect by Hawkins' formula. If the perforations terminate outside the damage zone, then

$$(s_d)_p = s_p' \qquad (13-24)$$

where is s_p' evaluated at a modified perforation length l_{perf}' and modified radius r_w'. These are

$$l_{perf}' = I_{perf} - \left(1 - \frac{k_s}{k}\right)r_s \qquad (13-25)$$

and

$$r_w' = r_w + \left(1 - \frac{k_s}{k}\right)r_s \qquad (13-26)$$

These variables are used in Equations 13-4 to 13-12 for the calculation of the skin effects contributing to the composite skin effect in Equations 13-23 and 13-24.

Perforating past the near-wellbore damage can greatly increase wellbore productivity. Drilling fluid solids and filtrate invasion, cement solids and filtrate invasion, created emulsions, and destabilized formations caused by drilling and vibrations can all reduce near-wellbore permeability permanently or temporarily. Figure 13-14 is an example showing that productivity index ratios can be greatly improved if the perforating charge can penetrate past the depth of damage. In most cases, however, the depth of damage surrounding the wellbore is uncertain, but openhole resistivity data can be used as a first approximation. The engineer should compare the shallow, medium, and deep resistivity curves (after determining the depth into the formation that each measure is made) and this should provide some sense of depth of at least fluid invasion. For example, if the medium and deep resistivity curves measure essentially the same value and are significantly different from the shallow resistivity curve (which only measures approximately 1 in. from the wellbore), the depth of invasion is probably somewhere between the shallow and medium measurement points.

Perforator penetration is also believed to be a critical parameter in wells where natural fractures are prevalent. In this instance, charges with the deepest penetration are more likely to encounter the natural fracture system.

Charge phasing is an often overlooked variable that has significant importance in many formations. Many through-tubing wireline perforator systems are designed as "zero-degree" phased guns to maximize formation penetration and entry hole size by decentralizing the gun and placing all perforations in the same plane from the wellbore. Productivity ratios can be reduced (typically 5 to 10%) because of the more tortuous path that the formation fluids have to travel to enter the wellbore. Where natural fractures are prevalent in the formation, perforating gun systems phased to penetrate at different directions are also desirable; there is a higher probability of contacting the fractures, and gun systems with phasings of 45°, 60°, 90°, 120°, and 180° are commonly used. Zero-degree-phased guns also have the additional drawback that, by placing all of the perforations in the same plane, the casing is more susceptible to failure due to reduced yield strength (King, 1989).

Effective shot density is an important perforating parameter not only to increase the number of flow paths to the wellbore, but also to contact thinly laminated hydrocarbon layers. Many completion engineers believe that four shots per foot will always provide com-

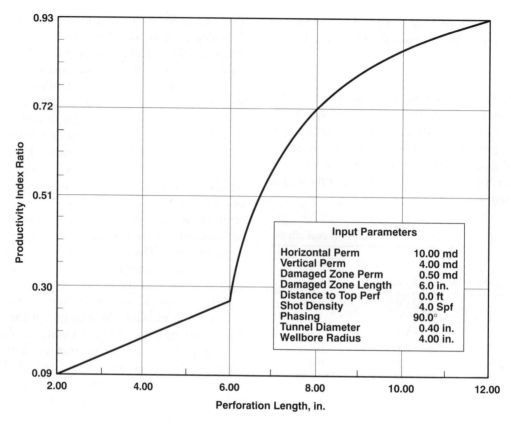

Figure 13-14 Improved productivity index ratios when perforating charge penetrates past the depth of damage

parable inflow performance to an openhole completion, but this "rule of thumb" can be quite misleading. Higher shot densities become more advantageous in natural completions with high formation permeabilities and, as discussed later in this chapter, high shot densities are critical in cased-hole, gravel-packed wells. It is important for the completion engineer to realize that perforating systems are mechanically complex and that not all charges that are run into a well automatically result in *effective* perforations being created or contacting productive formations. Studies within industry (Keese, and Oden, 1976; Hushbeck, 1986) suggest that 50 to 70% of the perforations are actually effective. This factor should be considered when designing the completion.

Some wells are perforated in only a portion of the productive pay because of geologic and drilling constraints or as an effort to reduce water and gas influx; this technique is referred to as a partial completion. Well productivity from a partial completion is impaired, and lower productivity will result because the formation fluids must converge to enter the wellbore (Cinco-Ley *et al.*, 1975).

13-7.3 Underbalanced Perforating

Underbalanced perforating is defined as perforating with the pressure in the wellbore lower than the pressure in the formation. When used properly, this technique has effectively provided higher productivity completions. Numerous technical articles have been written to compare this technique with overbalanced perforating both in the laboratory and in field studies (Allan *et al.*, 1985; Bell, 1984; Bonomo *et al.*, 1985; Behrmann, 1995; Halleck and Deo, 1989; and King *et al.*, 1985, among others). Underbalanced perforating creates an environment where formation fluid flow can immediately begin to enter the wellbore, rather than have the well in an overbalance condition where completion fluids and other particles continue to be lost into the formation. At the instant of perforating, the pressure differential to the wellbore is believed to help clear the perforations and remove crushed rock, debris, and explosive gases from the formation. Formation fluid type and reservoir permeability are the two primary factors influencing the amount of underbalance level to remove a portion of the crushed rock and other damage mechanisms from the near-wellbore area.

King *et al.* (1985) and others published the results of a large number of underbalanced perforating jobs in which initial well productivity was compared to subsequent well productivity after acidizing. Figure 13-15 shows the results of this study and plots the formation permeability versus total *minimum* underbalance necessary to achieve clean perforations in oil zones in sandstones. Figure 13-16 shows similar data for gas zones. Higher drawdowns are necessary in gas wells, probably because gas has a lower drag coefficient and cannot move particles as readily. As seen in these two figures, a few hundred psi of drawdown may be sufficient in a high-permeability formation such as Berea Sandstone, but low-permeability formations of a millidarcy or less may require a total underbalance of several thousand psi to effectively clean the perforations. Subsequent laboratory studies (Behrmann, 1995) suggest that even higher underbalance levels than those in Figures 13-15 and 13-16 are necessary. In unconsolidated formations, sand influx can potentially stick the perforating guns, and articles (Crawford, 1989) suggest the maximum underbalance

that should be used based upon log data from the adjoining shales.

The studies cited above and others suggest that the instantaneous underbalance must be followed with continued sustained flow of several gallons per perforation to further clean the perforation and to remove the crushed rock and other materials that have been loosened. This point is critical and well documented, yet is often overlooked on many jobs. A large influx of hydrocarbons into the well is undesirable because of the increased complexity it will cause if other well activities are planned. Underbalanced perforating is operationally much easier with tubing-conveyed perforating systems or if a single wireline perforating run is possible. Proper underbalance levels and continued flow are often not effectively used on wireline operations where multiple gun runs are required. Achieving appropriate underbalance levels with other intervals contributing flow and backpressure is operationally difficult. Continued wellflow after perforating concerns some operators because of the possibility that debris can be produced above the perforating guns and wireline being retrieved from the wellbore.

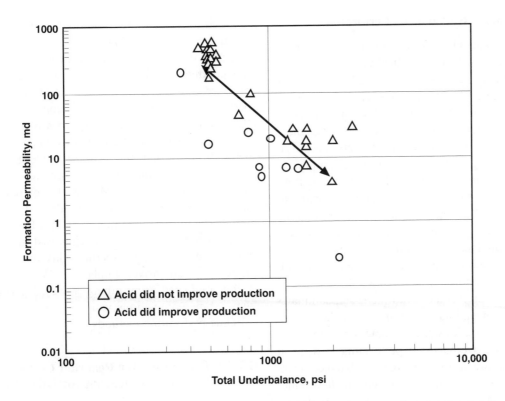

Figure 13-15 Formation permeability vs. minimum underbalance to achieve an undamaged completion in oil zones (after King, 1989)

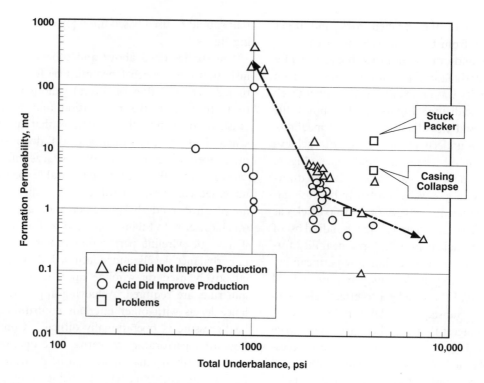

Figure 13-16 Formation permeability vs. minimum underbalance to achieve an undamaged completion in gas zones (after King, 1989)

13-7.4 Extremely Overbalanced Perforating

In many low-permeability formations, remaining reservoir pressures are insufficient to effectively clean the perforations, as suggested by King *et al.* (1985) and others. In other instances, formation competence is questionable, and the risk of sticking the perforating guns with high underbalance levels makes the use of underbalanced perforating methods an operational risk. Extremely overbalanced perforating is a near-wellbore stimulation technique (Handren *et al.*, 1993; Pettijohn and Couet, 1994; Snider and Oriold, 1996) used in conjunction with the perforating event. The method has gained popularity within the past few years because of the large number of wells that could not be effectively perforated using underbalance techniques. Extreme overbalance perforating also provides perforation breakdown in preparation for other stimulation methods, and therefore, eliminates the need for conventional breakdown methods, e.g. breakdown of perforations in carbonate formations.

During extreme overbalanced perforating jobs, most of the tubing is pressurized to high overbalance levels with compressible gases above relatively small volumes of liquids. The gases have high levels of stored energy; upon expansion at the instant of detonation, the gases

are used to fracture the formation and to divert fluids to all intervals. The high flow rate through relatively narrow fractures in the formation is believed to enhance near-well conductivity. Field data also suggest that high initial pressures are more likely to create fractures within the perforated interval and to limit height growth. Perforating systems have also recently been developed to release proppant material downhole with the gun detonation so that the extremely overbalanced fluids and nitrogen rushing to the formation carry erosive and propping materials. Most of the extremely overbalanced perforating jobs are currently designed with pressure levels set at a minimum of 1.4 psi/ft of true vertical depth. The technique is also being used to obtain a production test in very low-permeability formations before more large, expensive stimulations. While most jobs are being conducted using tubing-conveyed perforating systems, some completions with short intervals have used wireline perforating methods.

13-7.5 Perforating for Remedial Cementing, Hydraulic Fracturing, or Other Stimulations

In many instances, perforating operations are conducted primarily to create a conductive path to allow remedial cement squeezing or for stimulation of the formation with treating fluids such as massive hydraulic fractures and acidizing (discussed in Chapters 11 and 17 respectively). In each case, it is important to fully understand the objective of the subsequent activities before selecting a perforating system because the selection criteria are quite different.

For remedial cementing, it may be assumed that the perforating objective is to allow cement to subsequently be placed in a non-cement-filled void behind the casing. If it is believed that no cement or solids bed exists behind the casing, a few perforations placed in a single plane will be sufficient, but in most instances the downhole wellbore configuration is much more complex. It is important to visualize the well and also to realize that most wells have deviation that must be considered. In many instances, the casing will be uncentralized and lying on the low side of the hole either against the formation or against drilled solids and other mud filtrates. If a liquid-filled channel exists, it is more likely to be on the high side of the hole. Expensive methods are available to orient the perforator in the direction of the channel (if known), generally, the best choice of a perforator system is a high-shot-density perforator system that creates holes at least every 30° around the wellbore. Specialized gun systems with specific-application shaped charges are available if it is desired to create holes in an inner string of casing without damaging an outer string of casing. These gun systems generally require the perforating charge to be placed directly against the inner casing wall.

For hydraulic fracturing with proppants, the first important parameter to consider is perforator entry hole size in the casing to eliminate proppant bridging. Entry hole size in the casing should be at least six times the average grain diameter of the proppant (Haynes and Gray, 1974). It is important to realize that technical data suggests that the hydraulic fracture rarely exits the wellbore region from the perforations (Behrmann and Elbel, 1992). The perforation event creates a damaged area that is difficult to break down; in most instances, the fracture will initiate between the cement and the casing and travel a certain distance around the casing before the fracture extends. This tortuous path around the wellbore not only creates additional frictional forces, but it can lead to premature screen-outs of the hydraulic fracture stimulation. These types of issues are addressed in additonal detail in Chapter 6. In fields where fracture orientation is well-known, operators who began perforating with 180-degree phased perforating systems and oriented the perforations to be in the same plane as the fracture plane report good success. Perforating with 60° or 120° phasing is recommended to improve the chances of the perforations being in line with the fracture plane if fracture orientation is unknown or it is economically unjustifiable to obtain the information. The effect of perforation orientation on fracture initiation and propagation relative to the in-situ stress direction is discussed in Chapter 6.

Sufficient perforations should be placed to reduce shearing of the fracturing fluids transporting the proppant. Although it is not recommended, some wells continue to be stimulated with a limited number of perforations used to divert flow to different intervals. Several problems that exist with this limited-entry method can be reduced but not eliminated. In addition to the problems of shearing the fracturing fluids through the perforations, one of the biggest problems noted in the field is effectively communicating each perforation of known size with the desired fracture plane. It is recommended to (1) verify the perforation diameter with the shaped charge manufacturer for the specific weight and grade of casing and clearance; (2) centralize or decentralize the perforating gun such that each hole has the same clearance and subsequently should be of the same diameter; and (3) always acidize or break down each perforation and attempt to remove near-wellbore tortuousity prior to the limited-entry fracture stimulation. Limited-entry perforating is a good application for which to consider the use of bullet guns; the created hole is of a known size and less damage to the sand face outside the wellbore may occur.

For acid stimulations in carbonate formations, perforation hole size is less important because proppants are not normally used. As stated previously, perforator system selection should be conducted with subsequent operations in mind. If a "ball-out" acid job is planned, specially designed shaped charges or bullets that create no burr on the casing wall can be used; this technique will improve the ability of the ball to seal on the casing wall.

13-8 COMPLETION TECHNIQUES FOR UNCONSOLIDATED SANDSTONES

13-8.1 Perforating for Cased Hole Gravel-Packing Operations

Unlike natural completions, perforating for gravel-packing requires a completely different philosophy to optimize well productivity. Chapter 18 of this textbook addresses gravel-packing operations in greater detail, but a brief overview is appropriate here. A wellbore cross section for a natural completion and a cased-hole gravel-packed completion are shown in Figure 13-17. In a natural completion, formation fluids entering the perforation tunnel can flow unimpeded into the wellbore. In a gravel-packed completion, a primary objective is to prevent unconsolidated formation sands from entering the wellbore, and a series of "filters" is created to hold back the formation sand while still allowing fluid flow. As shown in Figure 13-17, fluid flow entering the perforation tunnel of a gravel-packed well still must flow linearly through the sand and gravel in the perforation tunnel and inside the annulus of the well before entering the gravel-pack screen. Although this linear flow path is only a few inches long, the materials inside it have a tremendous impact on well productivity. Equation 13-27 is the inflow performance equation for a cased hole gravel pack:

$$p_{wfs} - p_{wf} = \frac{qB\mu l}{1.1271 \times 10^{-3} k_g A} + \frac{9.107 \times 10^{-13} \beta (qB)^2 \rho l}{A^2}$$

$$(13\text{-}27)$$

which, for a specific well, simplifies to

$$p_{wfs} - p_{wf} = \frac{C_1 q}{k_g A} + \frac{C_2 q^2}{A^2} \qquad (13\text{-}28)$$

Ultimately, the two key parameters to well productivity (q) for a gravel-packed completion are area open to flow (A) and permeability of the gravel in the perforation tunnel (k_g).

The area open to flow (A) is essentially the number of perforations multiplied by their respective cross-sectional area. Most completion engineers perforate with at least 12 holes per foot in casing sizes of 7 in. and larger. Recognizing that area is a function of diameter to the second power, it is important for the completion engineer to realize that small increases in perforation diameter can increase the total area substantially. Big-hole perforating charges are generally designed to be shot in a centralized condition inside the wellbore, and substantial increases in cross-sectional area can be achieved by simply shooting the perforating gun in that manner. As can be seen in Figure 13-18, a 26% increase in cross-sectional area can be achieved by centralizing the gun, and uniform hole

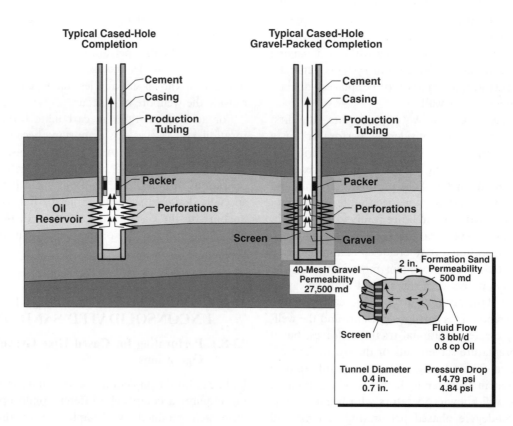

Figure 13-17   Wellbore cross section for a natural completion and a cased-hole gravel-packed completion

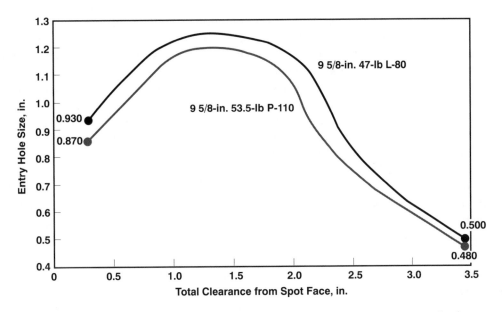

Figure 13-18   Loss in big-hole charge performance if the guns are not centralized

sizes are created. If a perforating gun were to become stuck because of formation-sand influx upon detonation, it would also be easier to fish from the wellbore if it were centralized. Uniform hole size is also operationally desirable for completion and production operations, so there are few negative aspects to using centralized guns for big-hole perforating operations if the centralization system is properly engineered.

Again referring to Equation 13-28 in its simplified form, the other key parameter to cased-hole, gravel-pack well productivity is k_g, permeability of the gravel in the perforation tunnel. Gravel-pack sand permeability will be discussed in Chapter 18, and permeability values in excess of 40,000 millidarcies are common. This number is extremely high and is as much as two orders of magnitude greater than many formation sand permeabilities. The key to perforating strategies for gravel packing is therefore to help ensure that the high-permeability gravel-pack sand can be placed in the perforation tunnel, which means that removal of perforating debris and other activities that impair injectivity is essential. Most completion engineers attempt to perforate under-balanced and flow the well to remove crushed rock and perforating debris before gravel-packing.

Removal of all of the perforating debris and crushed rock with subsequent flow is operationally difficult and is ineffective in some cases. Many current perforating-industry studies are focused on injectivity subsequent to perforating, rather than flow testing. Blok and Behrmann (Blok *et al.*, 1996) and other researchers

have published technical articles describing how conventional big-hole perforating charges crush a significant amount of formation material that essentially creates an excellent filter cake, thereby reducing injectivity. It is very likely that the types of perforating systems used before injection for stimulation, including gravel-packing operations, will change in the next few years as a result of these studies. Reduction in explosive gram weights of the charges, liner material types, penetration requirements, and other charge design parameters are likely, and efforts now focus on injectivity subsequent to perforating.

13-8.2 Perforating Unconsolidated Sands without Gravel Packs

In some instances, it is desirable to complete wells without gravel-packing the well, yet sand production could occur. Some engineers have assumed the use of big-hole charges as an approach because the larger holes have less flow velocity per perforation, but most studies and field tests indicate the opposite. Perforating these wells with high-shot-density guns (12 or more shots per foot) with deep penetrating charges is recommended as the best method to reduce sand production potential. The deep penetrating charge destroys a smaller radius around the perforating tunnel in the near-wellbore area. Big-hole charges can destroy the integrity of a much larger area around each tunnel and even reach the point where the damage areas between perforations become interconnected. When the perforating holes and the surrounding damaged rock are interconnected, continuous sand production is more likely to occur. Deep-penetrating charges also provide greater depth of penetration into undamaged formation material away from the wellbore. Studies within industry suggest this approach is less likely to produce sand than using big-hole charges.

In field operations in unconsolidated sandstones, stable arch bridges occur at the set producing rate. If the producing rate is adjusted, sand production may occur for a short period of time until a different-shaped stable arch occurs. It is important for field personnel to realize that, when increasing a well's producing rate, sand production may be only temporary and will cease after a few hours.

13-9 DEPTH CORRELATIONS

One of the most important and often overlooked perforating considerations is depth correlation. Desired perforation depths are always related to openhole logs, and it is important to realize that drilling depths and the depths of cased-hole logs will all vary. Drilling depths are based upon pipe tallies, and wireline depths are influenced by line stretch, line size, wear of the counter wheels, and other measurement devices. Use of a gamma-ray log on the perforating run is one of the best methods to ensure proper depth control, but this approach is not always available or economically feasible, so correlations using cased-hole casing collar logs, (CCLs), is common. The depth correlation is to be made to openhole logs, and different corrections may be necessary if perforating more than one interval. It is also important to measure the distance from the measurement device (CCL or gamma ray) to the top perforation and to make that adjustment to correlations. New engineers should develop a standard technique for these depth corrections, but it is advisable to write the corrections down and to ensure that they are made properly. A common mistake is to correct the wrong way, for example adding 7 ft to the cased-hole logging depth instead of subtracting the 7 ft to get on depth with the open-hole logs. Correlating cased-hole logs is greatly enhanced by having one or more short casing joints and/or a radioactive marker joint in the casing string. Completion engineers should ensure that well programs are written with the specific requirement to put some type of correlation markers near potential pay intervals.

The common depth correlation strategy is to first run the open hole logs and to always use the depths recorded on one of these logging runs (such as the resistivity logging suite) as the basis for future reference. This same logging suite also contains a gamma-ray log that measures formation radioactivity. After casing is set, a subsequent gamma-ray log and casing-collar log are run in combination. When the gamma-ray logs are cross-referenced, the casing-collar log can ultimately be compared to the resistivity log depths.

13-10 PERFORATING SUMMARY

In summary, the importance of perforating strategies is often underestimated during completion operations, leading to potential problems. The perforation provides the entire conduit between the wellbore and the formation and can affect inflow performance issues and well-stimulation and completion-design issues. The perforating strategy needs to be planned as early as possible in the well design, since it may ultimately even affect the size of casing and tubulars used in the well. When perforations are placed in the well, they are difficult, if not impossible, to plug effectively without damaging the adjacent formation. Placing perforations in a well is similar to the simple analogy of cutting a board; it is difficult to add unperforated casing back into the well, just as it is impossible to add wood to a board that has been sawn too short.. More seriously, however, it is important to respect the explosives used in perforating operations. They are hazardous, and accidents can occur if they are not handled carefully or if proper procedures are not followed. It is not easy to determine a perforating strategy for the well, and all of the individual topics discussed in this chapter deserve consideration. The ultimate objective is to maximize well performance, but poor perforating strategies can easily reduce productivity and revenue streams 10 to 50%.

REFERENCES

API RP-43: "Recommended Practices for Evaluation of Well Perforators," Fifth Edition (Jan. 1991).

API RP-67: "Recommended Practices for Oilfield Explosives Safety," First Edition (March 1, 1994).

Army Material Command: *Engineering Design Handbook-Principles of Explosive Behavior*, AMCP 706–180 (Apr. 1972).

Allan, J.C., Moore, P.C., and Weighill, G.T.: "Experience of Perforation Under Drawdown using Tubing-Conveyed Guns on the Beatrice Field," paper SPE 14012, 1985.

Asadi, M., and Preston, F.W.: "Characterization of the Jet Perforation Crushed Zone by SEM and Image Analysis," *SPEFE* (June 1994) 135–139.

Behrmann, L.A., and Halleck, P.M.: "Effect of Concrete and Berea Strengths on Perforator Performance and Resulting Impact on the New API RP-43," paper SPE 18242, 1988.

Behrmann, L.A., and Elbel, J.L.: "Effect of Perforations on Fracture Initiation," paper SPE 20661, 1992.

Behrmann, L.A.: "Underbalance Criteria for Minimum Perforation Damage," paper SPE 30081, 1995.

Bell, W.T., Brieger, E.F., and Harrigan Jr., J.W.: "Laboratory Flow Characteristics of Gun Perforations," *JPT* (Sept. 1972) 1095–1103.

Bell, W.T.: "Perforating Underbalanced Evolving Techniques," paper SPE 13413, 1984.

Bell, W.T., Sukup, R.A., and Tariq, S.M.: *Perforating,* SPE Monograph 16, Richardson, TX (1995).

Blok, R.H.J., Welling, R.W.F., Behrmann, L.A., and Venkitaraman, A.: "Experimental Investigation of the Influence of Perforating on Gravel Pack Impairment," paper SPE 36341, 1996.

Bonomo, J.M., and Young, W.S.: "Analysis and Evaluation of Perforating and Perforation Cleanup Methods," paper SPE 12106, 1985.

Carleone, J.: "Mechanics of Shaped Charges," *Tactical Missile Warheads,* J. Carleone, ed. (1993) **155**, 315–366.

Cinco-Ley, H., Ramey Jr., H.J., and Miller, F.G.: "Pseudoskin Factors for Partially Penetrating Directionally Drilled Wells," paper SPE 5589, 1975.

Cooper, P.W.: *Explosives Engineering,* VCH Publishers, Inc. (1996).

Crawford, H.R.: "Underbalanced Perforating Design," paper SPE 19749, 1989.

Halleck, P.M., Atwood, D.C., and Black, A.D.: "X-Ray CT Observations of Flow Distribution in a Shaped-Charge Perforation," paper SPE 24771, 1992.

Halleck, P.M., and Deo, M.: "Effects of Underbalance on Perforation Flow," paper SPE 16895, 1989.

Halleck, P.M., Wesson, D.S., Snider, P.M., and Navarette, M.: "Prediction of In-Situ Shaped-Charge Penetration Using Acoustic and Density Logs," paper SPE 22808, 1991.

Halleck, P.M.: "Lab Tests Expand Knowledge of Fractured Reservoir Perforating Damage," *GasTIPS* (Winter 1995/1996).

Haynes, C.D., and Gray, K.E.: "Sand Particle Transport in Perforated Casing," paper SPE 4031, 1974.

Handren, P.J., Jupp, T.B., and Dees, J.M.: "Overbalance Perforation and Stimulation Method for Wells," paper SPE 26515, 1993.

Hilchie, D.W.: *Wireline-A History of the Well Logging and Perforating Business in the Oilfields,* Douglas W. Hilchie, Inc., Boulder, CO (1990).

Hong, K.C.: "Productivity of Perforated Completions in Formations With or Without Damage," paper SPE 4653, 1975.

Hushbeck, D.F.: "Precision Perforation Breakdown for More Efficient Stimulation Jobs," paper SPE 14096, 1986.

Karakas, M., and Tariq, S.M.: "Semianalytical Productivity Models for Perforated Completions," paper SPE 18247, 1988.

Keese, J.A., and Oden, A.L.: "A Comparison of Jet Perforating Services: Kern River Field," paper SPE 5690, 1976.

King, G.E., Anderson, A., and Bingham, M.: "A Field Study of Underbalance Pressures Necessary to Obtain Clean Perforations Using Tubing Conveyed Perforating," paper SPE 14321, 1985.

King, G. E.: "The Effect of High Density Perforating on the Mechanical Crush Resistance of Casing," paper SPE 18843, 1989.

King, G.E.: *An Introduction to the Basics of Well Completions, Stimulations, and Workovers,* George E. King, Tulsa, OK (1995).

Locke, S.: "An Advance Method for Predicting the Productivity Ratio of a Perforated Well," paper SPE 8802, 1981.

McLeod Jr., H.O.: "The Effect of Perforating Conditions on Well Performance," paper SPE 10649, 1983.

Motley, J.D., and Barker, J.M.: "Unique Electrical Detonator Enhances Safety in Explosive Operations: Case Histories," paper SPE 36637, 1996.

Ott, R.E., Bell, W.T., Harrigan Jr, J.W., and Golian, T.G.: "Simple Method Predicts Downhole Shaped-Charge Gun Performance," *SPEPF* (Aug. 1994) 171-178.

Pettijohn, L., and Couet, B.: "Modeling of Fracture Propagation During Overbalanced Perforating," paper SPE 28560, 1994.

Pucknell, J.K., and Behrmann, L.A.: "An Investigation of the Damaged Zone Created Perforating," paper SPE 22811, 1991.

Regalbuto, J.A., and Gill, B.C.: "Computer Codes for Oilwell Perforator Design," paper SPE 30182, 1995.

Regalbuto, J.A., Leidel, D.J., and Sumner, C.R.: "Perforator Performance in High Strength Casing and Multiple Strings of Casing," paper presented at API Pacific Coast Joint Chapter Meeting (Nov. 1988).

Saucier, R.J., and Lands, J.F., Jr.: "A Laboratory Study of Perforations in Stressed Formation Rocks," *JPT* (Sept. 1978) 1347–1353.

Snider, P.M., and Oriold, F.D.: "Extreme Overbalance Stimulations using TCP Proppant Carriers," *World Oil* (Nov. 1996) 41–48.

Thompson, G.D.: "Effects of Formation Compressive Strength on Perforator Performance," *API Drill. and Prod. Prac.,* Dallas (1962).

Tucker, T.J.: "Explosive Initiators—Behavior and Utilization of Explosives in Engineering Design," *Proc.,* 12th Annual Symposium, New Mexico Section, ASME (Mar. 1972).

Von Holle, W.G., and Trimble, J.J.: "Temperature Measurement of Copper and Eutectic Metal Shaped Charge Jets," BRL Report No. 2004, Ballistic Research Laboratories, (Aug. 1977).

Walker, J.D.: "Ballistics Research and Computational Physics," *Tech. Today* (Sept. 1995).

Walters, W.P., and Scott, B.R.: "The Crater Radial Growth Rate Under Ballistic Impact Conditions," *Computers & Structures* (1985) **20**, No. 1-3, 641-648.

Walters, W.P., and Zukas, J.A.: *Fundamentals of Shaped Charges,* John Wiley & Sons (1989).

Walters, W.P.: "Shaped Charges and Shock Waves," *Shock Compression of Condensed Matter*, S.C. Schmidt, R.D. Dick, J.W.Forbes, and D.G. Tasker, ed., Elsevier Science Publishers, BV (1992).

14 Completion Hardware

Shari Dunn-Norman
University of Missouri-Rolla

Charlie Williams
Halliburton Energy Services

Morris Baldridge
Halliburton Energy Services

Wade Meaders
Halliburton Energy Services

14-1 INTRODUCTION

In the early 1900s, oil and gas wells were commonly completed with only a single string of cemented casing. As deeper, multiple, and higher pressure reservoirs were encountered, it was recognized that such completions imposed limitations on well servicing and control and that downhole designs would need to be changed to meet increasing needs for zonal isolation, selectivity, wellbore re-entry, and control. This objective was achieved through the development of downhole equipment.

Today, conventional oil and gas wells are completed with a variety of downhole devices. The devices specified depend on the well's ability to produce fluids and other functional requirements of the completion.

This chapter provides an introduction to the principal types of downhole equipment, their functions, and applications. This chapter also presents typical completions that use these devices.

14-2 TUBULAR GOODS

Most wells include at least one string of tubing in the completion. Other items, such as flow couplings, circulation devices, blast joints, and packers are threaded into and run as an integral part of the completion string or tailpipe. These devices become part of the well completion. Together, tubing and other completion devices provide control of the fluids from the reservoir to the wellhead and facilitate servicing the well.

14-2.1 Tubing

Tubing refers to the pipe used to create a flow conduit inside the wellbore between the reservoir and the wellhead. This flow conduit provides control of the produced fluid and facilitates wellbore service operations such as wirelining or circulating. Typically, tubing is run inside a string of casing or a liner, but tubing can also be cemented in slimhole wells as the casing and production tubing. In slimhole wells, service tools and completion equipment will be limited.

One or more strings of tubing may be used in a completion. This decision is a function of the number of reservoirs to be produced, whether the reservoirs will be commingled or produced separately, and whether the reservoirs will be produced concurrently or sequentially.

To specify tubing for a well completion, it is necessary to determine the grade and weight required. In addition, a type of threaded connection must be selected.

14-2.1.1 Tubing Size

API has developed tubing specifications that meet the major needs of the oil industry (API Spec. 5CT, 1990). Table 14-1 shows example tubing dimensions, grades, and strength of tubing.

Tubing size, or diameter, selection is based on the flow rates or pump rates anticipated from the well. Flow rates, or pump rates are determined with a systems analysis curve. A systems analysis is based on the inflow performance of the reservoir and the tubing performance

Table 14-1 API Tubing Table

Tubing Size Nom. in.	OD in.	T&C Non-Upset lb/ft	T&C Upset lb/ft	Grade	Wall Thickness in.	Inside Dia. in.	Drift Dia. in.	Coupling Non-Upset in.	Coupling Upset Reg. in.	Coupling Upset Spec. in.	Collapse Resistance psi	Internal Yield Pressure psi	T&C Non-Upset lb	T&C Upset lb	Barrels per lin. ft	Lin. ft per Barrel
3/4	1.050	1.14	1.20	H-40	.113	.824	.730	1.313	1.660		7,200	7,530	6,360	13,300	.0007	1516.13
	1.050	1.14	1.20	J-55	.113	.824	.730	1.313	1.660		9,370	10,360	8,740	18,290	.0007	1516.13
	1.050	1.14	1.20	C-75	.113	.824	.730	1.313	1.660		12,250	14,120	11,920	24,940	.0007	1516.13
	1.050	1.14	1.20	N-80	.113	.824	.730	1.313	1.660		12,970	15,070	12,710	26,610	.0007	1516.13
1	1.315	1.70	1.80	H-40	.133	1.049	.955	1.660	1.900		6,820	7,080	10,960	19,760	.0011	935.49
	1.315	1.70	1.80	J-55	.133	1.049	.955	1.660	1.900		8,860	9,730	15,060	27,160	.0011	935.49
	1.315	1.70	1.80	C-75	.133	1.049	.955	1.660	1.900		11,590	13,270	20,540	37,040	.0011	935.49
	1.315	1.70	1.80	N-80	.133	1.049	.955	1.660	1.900		12,270	14,160	21,910	39,510	.0011	935.49
1 1/4	1.660			H-40	.125	1.410					5,220	5,270			.0019	517.79
	1.660	2.30	2.40	H-40	.140	1.380	1.286	2.054	2.200		5,790	5,900	15,530	26,740	.0018	540.55
	1.660			J-55	.125	1.410					6,790	7,250			.0019	517.79
	1.660	2.30	2.40	J-55	.140	1.380	1.286	2.054	2.200		7,530	8,120	21,360	36,770	.0018	540.55
	1.660	2.30	2.40	C-75	.140	1.380	1.286	2.054	2.200		9,840	11,070	29,120	50,140	.0018	540.55
	1.660	2.30	2.40	N-80	.140	1.380	1.286	2.054	2.200		10,420	11,810	31,060	53,480	.0018	540.55
1 1/2	1.900			H-40	.125	1.650					4,450	4,610			.0026	378.11
	1.900	2.75	2.90	H-40	.145	1.610	1.516	2.200	2.500		5,290	5,340	19,090	31,980	.0025	397.14
	1.900			J-55	.125	1.650					5,790	6,330			.0026	378.11
	1.900	2.75	2.90	J-55	.145	1.610	1.516	2.200	2.500		6,870	7,350	26,250	43,970	.0025	397.14
	1.900	2.75	2.90	C-75	.145	1.610	1.516	2.200	2.500		8,990	10,020	35,800	59,960	.0025	397.14
	1.900	2.75	2.90	N-80	.145	1.610	1.516	2.200	2.500		9,520	10,680	38,180	63,960	.0025	397.14
2 1/16	2.063			H-40	.156	1.751					5,240	5,290			.0030	335.75
	2.063			J-55	.156	1.751					6,820	7,280			.0030	335.75
	2.063			C-75	.156	1.751					8,910	9,920			.0030	335.75
	2.063			N-80	.156	1.751					9,440	10,590			.0030	335.75
2 1/2	2.375	4.00		H-40	.167	2.041	1.947	2.875			4,880	4,920	30,130		.0040	247.12
	2.375	4.60	4.70	H-40	.190	1.995	1.901	2.875	3.063	2.910	5,520	5,600	35,960	52,170	.0039	258.65
	2.375	4.00		J-55	.167	2.041	1.947	2.873			6,340	6,770	41,430		.0040	247.12
	2.375	4.60	4.70	J-55	.190	1.995	1.901	2.875	3.063	2.910	7,180	7,700	49,450	71,730	.0039	258.65
	2.375	4.00		C-75	.167	2.041	1.947	2.875			8,150	9,230	56,500		.0040	247.12
	2.375	4.60	4.70	C-75	.190	1.995	1.901	2.875	3.063	2.910	9,380	10,500	67,430	97,820	.0039	258.65
	2.375	5.80	5.95	C-75	.254	1.867	1.773	2.875	3.063	2.910	12,180	14,040	96,580	126,940	.0040	295.33
	2.375	4.00		N-80	.167	2.041	1.947	2.875			8,660	9,840	60,260		.0039	247.12
	2.375	4.60	4.70	N-80	.190	1.995	1.901	2.875	3.063	2.910	9,940	11,200	71,930	104,340	.0040	258.85
	2.375	5.80	5.95	N-80	.254	1.887	1.773	2.875	3.063	2.910	12,890	14,970	102,990	135,400	.0034	295.33
2 3/8	2.375	4.00		H-40	.167	2.041	1.947	2.875			4,880	4,920	30,130		.0040	247.12
	2.375	4.60	4.70	H-40	.190	1.995	1.901	2.875	3.063	2.910	5,520	5,600	35,960	52,170	.0039	258.65
	2.375	4.00		J-55	.167	2.041	1.947	2.873			6,340	6,770	41,430		.0040	247.12
	2.375	4.60	4.70	J-55	.190	1.995	1.901	2.875	3.063	2.910	7,180	7,700	49,450	71,730	.0039	258.65
	2.375	4.00		C-75	.167	2.041	1.947	2.875			8,150	9,230	56,500		.0040	247.12
	2.375	4.60	4.70	C-75	.190	1.995	1.901	2.875	3.063	2.910	9,380	10,500	67,430	97,820	.0039	258.65
	2.375	5.80	5.95	C-75	.254	1.867	1.773	2.875	3.063	2.910	12,180	14,040	96,560	126,940	.0034	295.33
	2.375	4.00		N-80	.167	2.041	1.947	2.875			8,660	9,840	60,260		.0040	247.12
	2.375	4.60	4.70	N-80	.190	1.995	1.901	2.875	3.063	2.910	9,940	11,200	71,930	104,340	.0039	258.65
	2.375	5.80	5.95	N-80	.254	1.867	1.773	2.875	3.063	2.910	12,890	14,970	102,990	135,400	.0034	295.33
	2.375	4.60	4.70	P-105	.190	1.995	1.901	2.875	3.063	2.910	13,250	14,700	984,410	136,940	.0039	258.65
	2.375	5.80	5.95	P-105	.254	1.867	1.773	2.875	3.063	2.910	17,190	19,650	135,180	177,710	.0034	295.33
2 1/4	2.875	6.40	6.50	H-40	.217	2.441	2.347	3.500	3.668	3.460	5,230	5,280	52,780	72,480	.0058	172.76
	2.875	6.40	6.50	J-55	.217	2.441	2.347	3.500	3.668	3.460	6,800	7,260	72,580	99,660	.0058	172.76
	2.875	6.40	6.50	C-75	.217	2.441	2.347	3.500	3.668	3.460	8,900	9,910	98,970	135,900	.0058	172.76
	2.875	8.60	8.70	C-75	.308	2.259	2.165	3.500	3.668	3.460	12,200	14,080	149,360	185,290	.0050	201.72
	2.875	6.40	6.50	N-80	.217	2.441	2.347	3.500	3.668	3.460	9,420	10,570	105,570	144,960	.0058	172.76
	2.875	8.60	8.70	N-80	.308	2.259	2.165	3.500	3.668	3.460	12,920	15,000	159,310	198,710	.0050	201.72
	2.875	6.40	6.50	P-105	.217	2.441	2.347	3.500	3.668	3.460	12,560	13,870	138,560	190,260	.0058	172.76
	2.875	8.60	8.70	P-105	.308	2.259	2.165	3.500	3.668	3.460	17,220	19,690	209,100	260,810	.0050	201.72
3 1/2	3.500	7.70		H-40	.216	3.068	2.943	4.250			4,070	4,320	65,070		.0091	109.37
	3.500	9.20	9.30	H-40	.254	2.992	2.867	4.250	4.500	4.180	5,050	5,080	79,540	103,610	.0087	114.99
	3.500	10.20		H-40	.289	2.992	2.797	4.250			5,680	5,780	92,550		.0083	120.57
	3.500	7.70		J-55	.216	3.068	2.943	4.250			5,290	5,940	89,470		.0091	109.37
	3.500	9.20	9.30	J-55	.254	2.992	2.867	4.250	4.500	4.180	6,560	6,980	109,370	142,460	.0087	114.99
	3.500	10.20		J-55	.289	2.992	2.797	4.250			7,390	7,950	127,250		.0083	120.57
	3.500	7.70		C-75	.216	3.068	2.943	4.250			6,690	8,100	122,010		.0091	109.37
	3.500	9.20	9.30	C-75	.254	2.992	2.867	4.250	4.500	4.180	8,530	9,520	149,140	194,260	.0087	114.99
	3.500	10.20		C-75	.289	2.992	2.797	4.250			9,660	10,840	173,530		.0083	120.57
	3.500	12.70	12.95	C-75	.375	2.750	2.625	4.250	4.500	4.180	12,200	14,060	230,990	276,120	.0073	136.12
	3.500	7.70		N-80	.216	3.068	2.943	4.250			7,080	8,640	130,140		.0091	109.37
	3.500	9.20	9.30	N-80	.254	2.992	2.867	4.250	4.500		9,080	10,160	159,090	207,220	.0087	114.99
	3.500	10.20		N-80	.289	2.992	2.797	4.250			10,230	11,560	185,100		.0083	120.57
	3.500	12.70	12.95	N-80	.375	2.750	2.625	4.250	4.500	4.180	12,920	15,000	246,390	294,530	.0073	136.12
	3.500	9.20	9.30	P-105	.254	2.992	2.86	4.250	4.500	4.180	12,110	13,330	208,800	271,970	.0087	114.99
	3.500	12.70	12.95	P-105	.375	2.750	2.625	4.250	4.500	4.180	17,200	19,690	323,390	386,570	.0073	136.12
4	4.000	9.50		H-40	.226	3.548	3.423	4.750			3,580	3,960	72,000		.0122	81.78
	4.000		11.00	H-40	.262	3.476	3.351		5.000		4,420	4,580		123,070	.0117	85.20
	4.000	9.50		J-55	.226	3.548	3.423	4.750			4,650	5,440	99,010		.0122	81.78
	4.000		11.00	J-55	.262	3.476	3.351		5.000		5,750	6,300		169,220	.0117	85.20
	4.000	9.50		C-75	.226	3.548	3.423	4.750			5,800	7,420	135,010		.0122	81.78
	4.000		11.00	C-75	.262	3.476	3.351		5.000		7,330	8,600		230,760	.0117	85.20
	4.000	9.50		N-80	.228	3.548	3.423	4.750			6,120	7,910	144,010		.0122	81.78
	4.000		11.00	N-80	.262	3.476	3.351		5.000		7,780	9,170		246,140	.0117	85.20
4 1/2	4.500	12.60	12.75	H-40	.271	3.958	3.833	5.200	5.563		3,930	4,220	104,360	144,020	.0152	65.71
	4.500	12.60	12.75	J-55	.271	3.958	3.833	5.200	5.563		5,100	5,600	143,500	198,030	.0152	65.71
	4.500	12.60	12.75	C-75	.271	3.958	3.833	5.200	5.563		6,430	7,900	195,680	270,040	.0152	65.71
	4.500	12.60	12.75	N-80	.271	3.958	3.833	5.200	5.563		6,810	8,430	208,730	288,040	.0152	65.71

(Chapter 15). Several different tubing sizes are investigated in a systems analysis, which aids in determining a tubing size that will optimize the production rate for some period of time.

14-2.1.2 Tubing Grade

API has designated tubing grades based on chemical composition and physical and mechanical properties of the pipe. Each grade has a designation such as J55, K55, N80, L80, C75, and P110. The alphabetical designation in the tubing grade is arbitrary, but the numerical designation reflects the minimum material yield strength. The minimum yield strength must be sufficient to withstand forces in the tubing caused by changes in pressure and temperature at depth (Chapter 7).

The tubing grade selected for a particular completion is the grade that satisfies the minimum performance requirements of the application. Some grades (for example L80) have controlled hardness, which provides resistance to sulfide stress cracking. These grades are generally specified when the partial pressures of H_2S or CO_2 exceed NACE MR0175 recommendations.

14-2.1.3 Tubing Weight

Tubing weight is normally expressed as lb/ft and is a function of the thickness of the pipe wall. This measurement is an average and includes the weight of the coupling. Generally, there are two or more weights available for each tubing size and grade (Table 14-1).

14-2.1.4 Threaded Connections

Tubing comprises multiple sections of pipe threaded together and suspended from the wellhead by a tubing hanger. The standard and commonly used API connection for tubing is EUE-8R (external upset ends, eight-round thread tubing and coupling). This connection can provide reliable service for the majority of wells and is used extensively in the oilfield.

Proprietary connections, often referred to as premium threads, offer additional sealing mechanisms, such as O-rings or self-energized metal-to-metal seals, which maximize the effectiveness of the sealing area and minimize the instance of leakage failure. Common uses include corrosion-resistant alloy tubing strings, high-pressure applications, and wells with bends and doglegs. These connections are broadly classified according to their design; for example, metal-to metal seal integral-joint

(MIJ), integral-flush-joint (IFJ), and slim-line high-performance (SLH). Klementich (1995) gives an overview of proprietary connections, and *World Oil* (1990) annually publishes a summary of threaded connections and their properties. API connection ratings are given in Chapter 7.

In selecting a threaded connection, the loads expected in the tubing string should be reliably established, and the complete performance properties of the threaded connection should be identified (Klementich, 1995). When two or more types of threads meet the performance specifications required, the lower cost thread is normally preferred, unless other considerations, such as availability and the ability to re-cut the thread, are significant.

14-2.2 Flow Couplings

A flow coupling is a short piece of pipe which has a wall thickness greater than the tubing string. Flow couplings are used to delay erosional failure at points inside a completion string, where turbulent flow is expected to occur. Figure 14-1 depicts the use of a flow coupling around a landing nipple.

Flow couplings offer wall thickness nearly twice the tubing wall thickness. Their internal diameters are the same as the tubing, but they have larger external diameters. Flow couplings are available in 3- to 6-ft and 10-ft lengths. The length selected depends on fluid flow rates (how quickly turbulent flow is expected to dissipate) and the abrasiveness of the particular fluid.

API recommended practice suggests the use of flow couplings around a subsurface safety valve (API RP14B, 1994). In addition, flow couplings are typically used downstream of landing nipples or circulation devices. A suggested rule of thumb is to include flow couplings above and below any downhole device which restricts the flow area by more than 10% of the nominal tubing ID. Various applications of flow couplings can be seen in the completions presented in Section 14-7.

14-2.3 Blast Joints

Fluids entering perforations may display a jetting behavior. This fluid-jetting phenomenon may abrade the tubing string at the point of fluid entry, ultimately causing tubing failure.

Blast joints are joints of pipe with a wall thickness greater than the tubing. These joints are run in the completion opposite the casing perforations, as shown in Figure 14-2. The blast joint delays the erosional failure

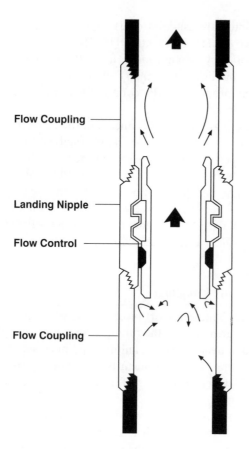

Figure 14-1   Landing nipple with flow coupling

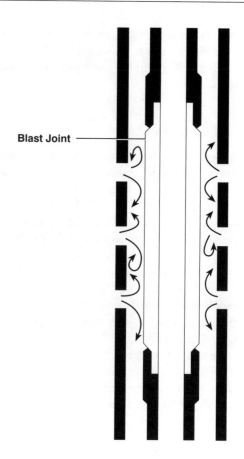

Figure 14-2   Blast joint

at the point where fluids enter the wellbore and impinge on the tubing string.

Blast joints are similar in design to flow couplings. They have the same internal diameter as the tubing, but a larger external diameter. Blast joints are normally available in 20-ft or 30-ft lengths.

Section 14-7 presents a dual completion that includes a blast joint in the design.

14-2.4 Landing Nipples

Landing nipples are short sections of thick-walled tubulars that are machined internally to provide a locking profile and at least one packing bore. The purpose of a landing nipple is to provide a profile at a specific point in the completion string to locate, lock, and seal subsurface flow controls, either through wireline or pumpdown methods.

Every subsurface control device set inside a landing nipple is locked and sealed in the profile with a locking mandrel. The profile and bore area in the nipple offer an engineered and controlled environment for the locking mandrel to form a seal. For this reason, the lock mandrel

must conform to the profile of the nipple; for example, an "X" nipple requires an "X" lock profile and an "R" profile requires an "R" lock mandrel.

Landing nipples can be used at virtually any point in the completion string. Typically, they are used in conjunction with a wireline subsurface safety valve at an intermediate point in the tubing above a packer to pressure test the tubing or set a flow-control device, immediately below a packer for packoff above perforations in multi-zone completions, and at the bottom of the tubing string for setting a bottomhole pressure gauge. Applications of landing nipples are shown in completions presented in Section 14-7.

There are three principal types of landing nipples. These are no-go nipples, selective nipples, and subsurface safety valve nipples. The characteristics of these nipples are discussed below.

14-2.4.1 No-Go Landing Nipples

A no-go landing nipple includes a no-go restriction in addition to the profile and packing bore. The no-go restriction is a point of reduced diameter, i.e. a shoulder.

This no-go shoulder is used to prevent the passage of larger diameter wireline tools and offers the ability to positively locate subsurface control devices in the nipple. Figure 14-3 shows a no-go landing nipple and associated lock mandrel.

No-go profiles may be included either above or below the packing bore of the landing nipple. When the no-go

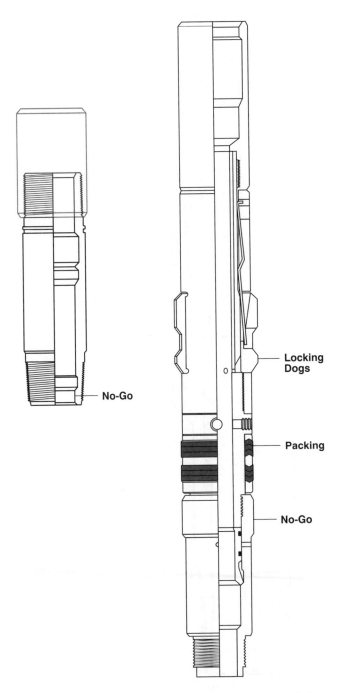

Figure 14-3   No-go landing nipple and lock mandrel

No-Go

Locking Dogs

Packing

No-Go

occurs below the packing bore, it is referred to as a bottom no-go. In this case, the minimum internal diameter is the diameter of the no-go restriction. When the no-go occurs above the packing bore, it is referred to as a top no-go. In a top no-go, the no-go diameter is the same as the packing-bore diameter.

The no-go restriction determines the largest size of wireline equipment or other devices that can be run through the device. Therefore, in completions that include many no-go profile devices, each successively deeper profile device must have a smaller internal diameter than the one above, so that the wireline equipment for the deeper profile can pass through the no-go of the profile device immediately above it. This design is often referred to as "step-down" sizing.

No-go profiles aid in positive setting for the wireline equipment because the wireline tool physically bumps against the no-go shoulder when the tool is landed in place. However, if many profiles are required in the design, the step-down sizing phenomenon may mean that the bottom no-go will be too small, either for the desired production rate or for well servicing. For this reason, many profile systems include only one no-go at the bottom, or deepest set, profile device.

14-2.4.2 Selective Landing Nipples

Landing nipples that do not include a no-go, or diameter restriction, are referred to as selective. In a completion equipped with selective nipples, it is possible to "select" any one of the nipples to install a flow-control device in it. Figure 14-4 shows a selective landing nipple and associated lock mandrel.

For a given tubing size, all selective nipples run in the tubing string will have the same internal diameter. Unlike the no-go nipples, there is no progressive restriction of the minimum diameter.

Selective nipples are divided into two categories. One type is a profile-selective landing nipple. The second type is a running-tool-selective landing nipple.

Profile-selective landing nipples have the same minimum internal diameter, the same locking profile, and the same packing bore. However, these nipples include a different locating profile. When setting a control device in one of these nipples, the lock mandrel used with the control device must be equipped with a locator assembly that matches the locating profile of the nipple in which it will be installed.

Running-tool-selective landing nipples are selected according to the running tool that is used to install the

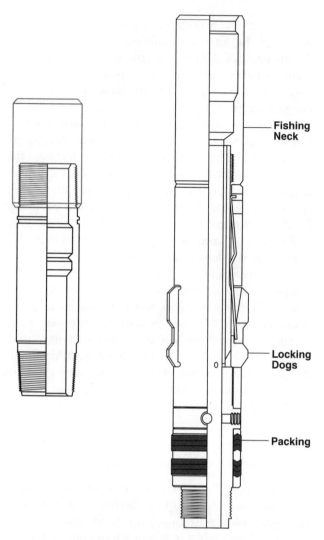

Figure 14-4 Selective nipple with lock mandrel

landing nipple because the nipple body is adapted to accept a $\frac{1}{4}$-in. hydraulic control line. The body is bored so that the control line pressure feeds into the safety valve, which is sealed between an upper and a lower packing bore (Figure 14-6).

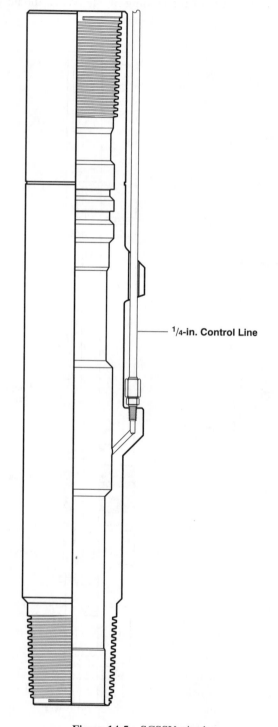

Figure 14-5 SCSSV nipple

lock mandrel and the control device in that nipple. Each running-tool-selective landing nipple of a given size and type in the tubing string is identical, allowing a virtually unlimited number of running-tool-selective landing nipples to be used in a single tubing string.

Selective nipples offer the advantage of providing a profile without restricting the minimum diameter of the completion string. However, it may be more difficult to positively set equipment in selective profiles, particularly in highly deviated wells.

14-2.4.3 Subsurface Safety Valve Landing Nipples

A subsurface safety valve landing nipple is a special type of nipple designed to hold a wireline-retrievable safety valve (Figure 14-5). The nipple differs from a standard

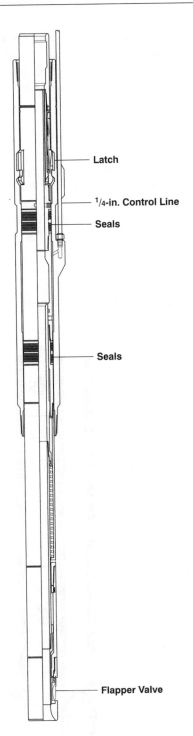

- Latch
- 1/4-in. Control Line
- Seals
- Seals
- Flapper Valve

Figure 14-6 Valve latched in safety valve landing nipple

Subsurface safety valve nipples are available with a sliding sleeve feature incorporated. The sleeve seals off the control line when running and retrieving the valve body by wireline. This feature keeps the well fluids and pressure off the control line at all times.

14-2.5 Tubing Annulus Communication: Circulating Devices

Tubing annulus communication refers to an opening or access between the inside of the tubing string and the tubing casing annulus. Such access is required to circulate fluids in a well, to treat a well with chemicals, to inject fluids from the annulus into the tubing string, or to produce a zone that is isolated between two packers.

Two devices provide communication between the inside of the tubing string and the tubing casing annulus, either above a packer or between two packers. These devices are the sliding sleeve and the side-pocket mandrel.

The sliding sleeve, discussed below, is the principal tubing annulus communication device. Sliding sleeves provide the ability to circulate a well and selectively produce multiple reservoirs.

Some operators set a dummy valve inside a side-pocket mandrel (SPM) with the intent of using this device for circulating the well. It is possible to use an SPM as the principal circulation device. However, the SPM has a restricted flow area that limits the rate of circulation. For this reason, most operators use a sleeve as the principal circulation device and include an SPM as an emergency, or backup, method for killing a well. Side-pocket mandrels are discussed in Section 14-2.6 because the principal use of the SPM is for gas lift.

In addition to these two devices, some operators choose to punch a hole in the tubing when circulation is required or pull an overshot or extra-long tubing seal receptacle (ELTSR) assembly off the packer to circulate. These methods have significant drawbacks. If a hole is punched in the tubing, the operator must replace this joint after circulation is no longer required. Furthermore, it is only possible to back off an ELTSR to circulate if such a device exists in the completion string.

14-2.5.1 Sliding Sleeves

A sliding sleeve is a cylindrical device with an internal sleeve mechanism. Both the inner sleeve and outer body are bored to provide matching openings. The inner sleeve is designed to move upward and downward through the use of a wireline shifting tool. When the sleeve is shifted to the open position, the sleeve openings mate with the openings in the outer body, thereby establishing tubing/annulus communication. When the sleeve is shifted to the closed position, the sleeve openings are displaced from the outer body openings, which are then isolated by the inner sleeve wall. An example of this device is shown in Figure 14-7.

Sliding sleeves are versatile and flexible components. They can be used at virtually any point in the completion string where circulation, injection, or selective production is required. The device can be opened and shut by wireline and therefore does not require the completion to be retrieved after circulation is established. Depending on the design of the communication ports, the device can offer a circulation area even greater than the internal diameter of the tubing.

Sliding sleeves may open in either direction, that is, by jarring upward or jarring downward. The main consideration in selecting a jar-up or jar-down sleeve is the amount of force required to shift the sleeve. A jar-down situation limits the force created to the weight of wireline tool string; it is not possible to compress the wireline to add more force. In the jar-up to open, it is possible to pull tension on the wireline up to the strength of the wire.

Sliding sleeves include a nipple profile above the inner sleeve. This profile is often used to set a blanking sleeve inside the device, to provide a means of shutting off flow if the sleeve is stuck in the open position.

Sliding sleeves are typically set above the shallowest packer to provide for well unloading and circulation (Figure 14-8) or between two packers to provide selective well production from multiple zones (Figure 14-9).

14-2.6 Side-Pocket Mandrels

A side-pocket mandrel, or SPM, is a special receptacle with a receiving chamber parallel to the flow chamber (Figure 14-10). The tubular side of the device mates with the tubing string and leaves the bore of the mandrel fully open for wireline tools to pass without interruption. The parallel receiving chamber is offset from the string. This chamber is used to house a number of flow control or sealing devices.

The side-pocket mandrel was originally designed to house retrievable gas-lift equipment, and this function

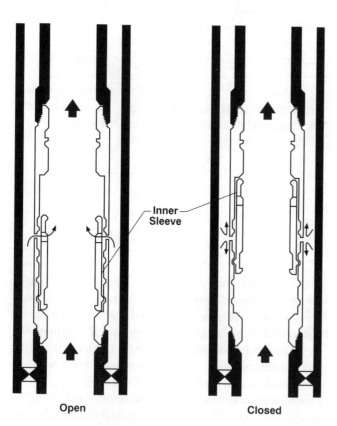

Figure 14-7   Sliding sleeve

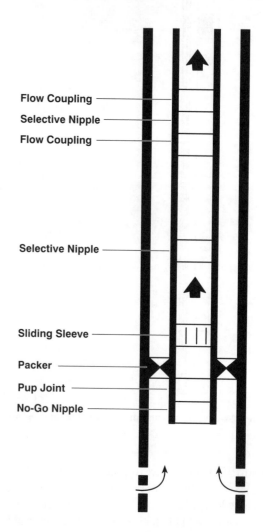

Figure 14-8   Single sliding sleeve for circulation

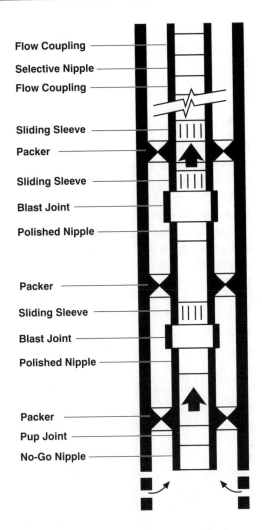

Figure 14-9   Sliding sleeve for multi-zone selective completion

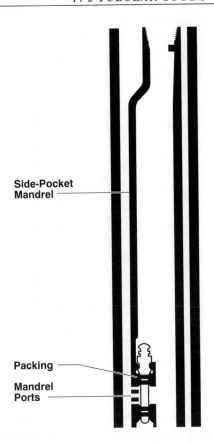

Figure 14-10   SPM with dummy valve in place

is still the primary use of the component. However, at least two other uses of the SPM are now recognized. These include providing an emergency kill device and providing a means of circulating fluids.

To provide an emergency-kill capability, the SPM is fitted with a dummy valve. This valve is normally closed and prevents communication between the tubing and the annulus. However, the valve is set to shear at a predetermined pressure and can, therefore, be opened by applied pressure to the annulus or the tubing. This configuration is considered to be an emergency-kill capability because kill fluid can be circulated without running wireline beforehand.

Since the dummy valve can be sheared open by applied pressure to the annulus, SPMs can also be used as a means of circulation. The SPM is one of the least preferable means of circulation, since the flow area through the

SPM is small and restricts the pumping rate. In addition, the valve must be replaced once circulation is completed.

14-2.7 Travel Joints

A travel joint is used to allow tubing movement or travel while maintaining pressure integrity. A travel joint consists of two concentric tubes that telescope relative to one another (Figure 14-11). Seal elements on the inner tube isolate annulus pressure and fluids from the tubing string as the travel joint strokes open and close.

A travel joint allows the completion string to expand or contract freely with changes in downhole pressure and temperature, thereby solving the problems associated with tubing contraction and elongation in producing wells, injection wells, or disposal wells. Typically, a travel joint is placed in the completion string above the shallowest packer to accommodate tubing movement. A travel joint may also be placed between packers or near the surface to facilitate well space out.

Standard stroke lengths for travel joints are 2, 4, 8, 10, 15, 20, and 25 ft. The travel joint can be shear-pinned fully closed or partially open. The position of the travel

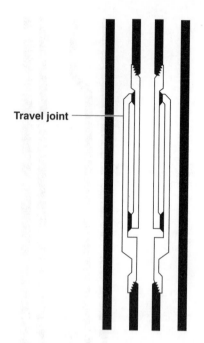

Figure 14-11   Travel joint

joint when it is run in the hole is a function of the expected tubing movement. For example, if future well operations will introduce both tubing expansion and contraction over time, then the joint can be run partially opened to allow for both types of movement downhole.

Travel joints can rotate freely, or can provide the capability to translate tubing rotation to the packer. A swivel travel joint offers continuous 360° rotation, but cannot translate this motion downhole. Swivel joints with a clutch are similar to the regular swivel travel joints except that in the collapsed position (or fully extended position) torque can be transmitted to packers or other equipment located below. Splined travel joints are used when transmission of torque through the travel joint is required.

14-2.8 Adjustable Unions

Adjustable unions are similar to travel joints in that they consist of concentric tubes designed to open and close. However, adjustable unions are threaded and open or close by rotation rather than by travel or stroking of the device.

Adjustable unions are used to facilitate spacing out at the surface and between packers and other subsurface components where spacing is critical. They offer 12 to 24 in. of extension.

Adjustable unions with keys are designed to allow tubing torque to packers and other equipment below. Adjustable unions without keys are used when tubing torque below is not required.

14-2.9 Tubing Hanger

The surface tubing hanger is a device run in the completion string, which lands inside and becomes an integral part of the wellhead (Figure 14-12). It is used to provide a seal between the tubing and the tubing head (wellhead) or to support the tubing and to seal between the tubing and the tubing head (Bradley, 1987)

The tubing hanger is a solid piece of metal that has one or more bores. The tubing string mates with one bore to continue the flowpath to the surface. If an electrical submersible pump is used in the completion, the tubing hanger must also have a bore to accommodate the motor cable.

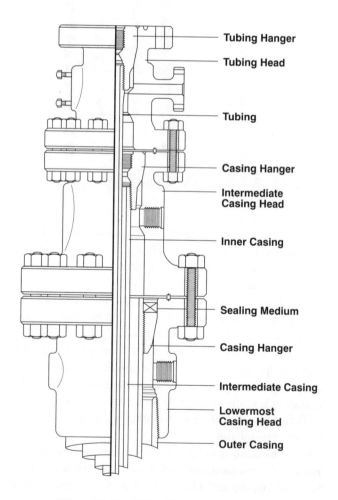

Figure 14-12   Tubing hanger and wellhead

Several different types of tubing hangers are available, including wrap-around hangers, polished-joint hangers, boll-weevil, and stripper-rubber hangers (Bradley, 1987). Both solid and split hangers are available for dual completions. The advantage of the split hanger is that each string can be manipulated separately.

Subsurface hanger assemblies that sit inside the casing are also available. These assemblies include a shear joint in the tubing string above the hanger. Such assemblies provide the ability to pull tubing-retrievable safety valves without removing the tubing string, disturbing the packer, or killing the well.

14-3 PRODUCTION PACKERS

A packer is a downhole device used to provide a seal between the outside of the tubing and the inside of the production casing or liner. The packer seal is created by resilient elements that expand from the tubing to the casing wall under an applied force. When set, this seal prevents annular pressure and fluid communication across the packer.

Production packers are those packers that remain in the well during normal well production. Service packers, such as those used in well testing, cement squeezing, acidizing, and fracturing are used temporarily and then retrieved from or milled out of the well. This discussion focuses on production packers.

Production packers are specified for many reasons (Greene, 1966). For example, they are used to protect the casing from pressure and produced fluids, isolate casing leaks or squeezed perforations, isolate multiple producing horizons, eliminate or reduce pressure surging or heading, hold kill fluids in the annulus, and permit the use of certain artificial-lift methods.

On rare occasions, well completions may not incorporate a packer. For example, many high-volume wells are produced up both the tubing and the annulus and therefore will not include a packer. Packers are not normally run in rod-pumped wells. However, in offshore wells and in many other applications, it is considered a safer practice to produce with at least one packer downhole.

Once an engineer determines that a packer is required for the completion, certain physical attributes of the packer must be specified. These attributes include the type and size of packer, the number of bores required, how the packer will be conveyed and set in place, and how it will be removed from the wellbore.

14-3.1 Packer Basics

Although packers are available for a wide variety of applications, all share similar design and operational characteristics. Each packer includes a flow mandrel, resilient elements, a cone or wedge, and slips (Figure 14-13).

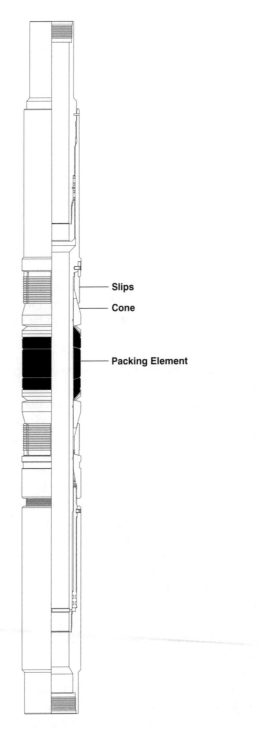

Figure 14-13 Production packer

The mandrel provides the flow conduit for production. Resilient elements form the tubing-to-annulus pressure seal. The cone assists in positioning the slips, which grip the casing wall and prevent the packer from moving upward or downward.

To set a packer, a compressive force is applied to the mandrel between the slips and the resilient elements. The force moves the slips outward to grip the casing and then transfers the compressive load to expand the packing element.

14-3.2 Packer Type: Permanent vs. Retrievable

Packers may be classified according to a number of criteria, such as their retrievability, setting mechanism, or application. Most commonly, they are classified by their retrievability.

Permanent packers are those packers that cannot be entirely retrieved and reinstalled in the well. This type of packer is normally run and set separately on electric cable or slickline, a workstring, or tubing, and the production tubing is then either stabbed into or over the packer. A permanent packer may also be run integrally with the tubing string, provided that there is a means of disconnecting the tubing above the packer. Permanent packers must be milled out to remove them from the wellbore.

Retrievable packers are those packers that are designed to be retrieved and reinstalled in the wellbore. Retrievable packers are normally run integrally with the tubing string and are set with either mechanical manipulation or hydraulic pressure. Certain retrievable packers may also be set on electric cable. They are unset either by a straight pull or by a combination of rotation and a straight pull or with a special retrieving tool on drillpipe. Once unset, the compressible packing elements and slips or hold-down buttons relax and retract, allowing the packer to be removed from the wellbore.

Permanent packers have historically provided the advantage of deeper and more accurate setting depths, larger internal diameters, and the ability to withstand larger differential pressures, whereas retrievable packers have offered the ability to remove the packer easily and nondestructively from the wellbore. However, recent developments in packers have obscured this historical distinction. There are now semi-permanent packers that are retrievable packers with performance characteristics similar to permanent packers.

14-3.2.1 Number of Packer Bores

The packer mandrel, or bore, refers to a cylindrical, machined opening in the packer. This opening is required to allow produced hydrocarbons or injection fluids to pass through the packer. Packer bores are also used to provide annular access for any electrical or instrument cables run below the packer. Normally, packers possess 1, 2, or 3 bores and are referred to respectively as single, dual, or triple packers.

The number of packer bores required is a function of the number of tubing strings in the completion and whether electrical cables will be passed through the packer. Single-string completions with no cable requirements use a single-bore packer. Dual-bore packers are run with either dual completion strings or single-string completions that require cables to be passed through the packer.

14-3.2.2 Packer Setting Method

All packers are set by an applied a compressive force to the slips and rubber packing elements. This force may be created in a number of ways, including tubing rotation, slacking off weight onto the packer, pulling tension through the tubing, pressuring the tubing against a plug, or sending an electric impulse to an explosive setting tool. The techniques for creating the setting force are referred to as packer setting methods.

Packer setting methods are classified as mechanical, hydraulic, or electric. A mechanical setting method refers to those techniques that require some physical manipulation of the completion string, such as rotation, picking up tubing, or slacking off weight. A hydraulic setting method refers to applying fluid pressure to the tubing, which is then translated to a piston force within the packer. The third method, electric line, involves sending an electric impulse through an electric cable to a wireline pressure setting assembly. The electric charge ignites a powder charge in the setting assembly, gradually building up gas pressure. This pressure provides the controlled force necessary to set the packer.

14-3.2.3 Packer Setting and Unseating Forces

Certain packers rely on applying a force to the packer to maintain the packer in a set position. These packers are referred to as tension- or compression-set packers. They are set mechanically, since the tubing string must be manipulated to provide the required compressive or tensile forces.

Compression packers require that a compressive load be continuously applied to the top of the packer. Normally, this load is supplied by slacking off tubing weight; therefore, compression packers are commonly referred to as weight-set packers. However, the compressive load may also be provided by a pressure differential across the packer; in this case, the pressure above must be greater than the pressure below the packer. As a result, these types of packers are suitable for injection wells.

A compression packer is not designed to withstand a pressure differential from below the packer, unless an anchoring device or hydraulic holddown is included. The packer may unseat if the compressive force is reduced or counteracted by a tensile tubing load. Expected wellbore operating conditions that cause the tubing to contract may cause a weight-set packer to unseat.

The tension-set packer is the opposite of the compression packer. A tensile force is pulled on the tubing to initiate packer setting and to maintain the packer in a set position. Because the packer is set with the tubing in tension, large pressure differentials from below tend to support the setting action. However, this type of packer will attempt to unseat if the tubing is in neutral or in compression. Hence, wellbore operating conditions that cause the tubing to expand may cause a tension-set packer to unseat, unless a holddown system compensates for these forces or an additional, initial tensile force is placed on the tubing.

All other packers are considered to be neutral set because their setting mechanisms allow the tubing to be in tension, neutral, or in compression as required. Generally, all packers that are set hydraulically and electrically will fall into this category. Some packers that are set mechanically, such as rotation-set packers, can also be set with the tubing in tension, neutral, or in compression.

In specifying a packer for a well completion, it is important to know the forces required for setting and unseating the packer and to ensure that these characteristics are not violated by tubing forces induced from expected operating and/or treating conditions within the well (Chapter 7).

14-3.2.4 Conveying the Packer

Packer conveying is the manner in which the packer is run to its setting depth. Conveying methods include running the packer (and any tailpipe) on wireline, running the packer on a workstring or drillpipe, or using the production tubing to convey the packer, either separately or integrally.

The packer conveying method selected must be mechanically compatible with the type of packer and the means of connecting the packer and tubing. For example, a permanent type packer is normally conveyed and set separately, and such packers are used with connections that allow the tubing string above the packer to be easily retrieved.

The manner in which the packer is conveyed to its setting depth and its manner of connecting with the tubing string are also important in analyzing and comparing completion running procedures. For example, if a permanent packer is conveyed and set on electric cable, then at least two trips will be required to run the completion, rather than a single trip for a packer run integrally on tubing. It is important to recognize these differences if design alternatives are to be assessed and compared on the basis of total cost.

14-3.2.5 Connecting the Tubing to Packer

There are four methods of connecting tubing to a packer. These include an integral threaded connection, an anchor or ratch-latch assembly, a J-latch, and a locator seal assembly. These connection alternatives are shown in Figure 14-14.

With a threaded connection, the tubing is threaded into the top of the packer and the packer becomes an integral part of the tubing string. A ratch latch, or anchor assembly, consists of a short seal stack with a ratch latch above, which engages (stabs into) an Acme-type box thread at the top of the packer. A J-slot connection is used above a seal stack. The J-slot engages or locates with either internal or external pins at the top of the packer.

The threaded, anchor, and J-latch connections fix the tubing at the packer and do not allow the tubing to move at that point.

The fourth connection is the locator seal assembly. A locator seal assembly consists of a seal stack, with either a no-go shoulder or straight slot locator positioned above the seal stack. The no-go shoulder or straight slot locator is used for positive landing and space out, but does not anchor the locator assembly to the packer. The length of the seal stack depends on the type of seal and expected service environment.

A locator seal assembly differs from the three other connections in that it allows tubing movement, such as

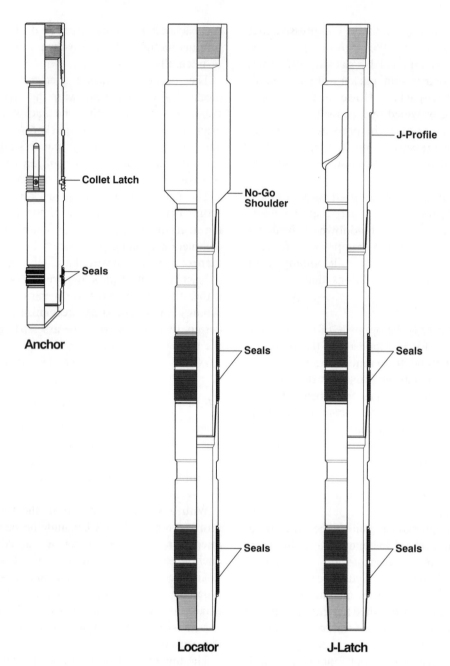

Figure 14-14   Tubing-to-packer connections

expansion or contraction, at the packer. The degree of movement allowed by a locator depends on the amount of tubing expansion or contraction expected, whether a no-go shoulder or straight-slot locating head is included in the assembly, and where the no-go shoulder is landed relative to the top of the packer when the completion is spaced out.

The tubing-to-packer connection must be consistent with the type of packer being used. Typically, permanent packers have locator seal assemblies or some anchor sys-

tem and an overshot seal assembly. Retrievable packers may be designed for a threaded connection, J-latch, or anchor. The type of connection specified depends on the requirements of the particular application.

14-3.2.6 Packer Retrieving Method

When a workover is required, part or all of the completion string is removed from the wellbore. Frequently, this requires removal of the production packer. The manner in which this is accomplished is referred to as the packer retrieving method.

The packer retrieving method depends on the type of packer. Retrievable packers are released by rotating, pulling on the tubing string, setting weight on the packer, or using a retrieving tool run on a workstring. Permanent packers normally cannot be retrieved by these means and must be milled to release the setting mechanism.

If the type of packer being used in a well completion is retrievable, it is important to consider how it will actually be removed. In cases where rotation is required, it is important to verify that it is indeed possible to rotate the completion string, since this will not be feasible for all completions.

14-3.3 Common Types of Packers and Their Applications

A number of companies manufacture packers, and each manufacturer frequently includes its own mechanical enhancements to the designs discussed above. A complete discussion of manufacturers' equipment is beyond the scope and purpose of this book. However, it is useful to recognize the main types of packers in widespread use and their range of application.

14-3.3.1 Mechanical Retrievable Packers

Mechanical-set retrievable packers are designed to be run and set on tubing, released, and moved and set again without tripping the tubing. In general, these packers are capable of BHT up to 275°F and a pressure differential of 6500 to 7500 psi.

Mechanical-set retrievable packers may have slips above and below the seal element. Depending on their internal locking mechanism, they can be set with tension, compression, or rotation and, once set, the tubing can be left in tension, compression, or neutral mode.

Certain types of mechanical-set retrievable packers can be used for production, steam injection, disposal, injection, testing and reservoir stimulation. Typically, they cannot be used in deep, deviated wells because it is difficult to transfer sufficient tubing movement to set and maintain the packer.

14-3.3.2 Hydraulic Retrievable Packers

Hydraulic-set retrievable packers are designed to be set by pressuring up the tubing string against a plugging device below the packer. Once the packer is set, the tub-ing may be put into tension or compression or left in a neutral mode. A hydraulic packer has bidirectional slips or a set of slips to resist downward movement and a hydraulic hold-down system to prevent upward movement. The packer is released by a straight pull on tension-actuated shear pins.

In general, hydraulic-set retrievable packers can normally be used in applications with a BHT up to 275°F and a pressure differential of 6500 to 7500 psi. They have a nearly universal application and are used for highly deviated wells or wells with small control lines (where rotation may be a problem). Hydraulic-set packers are also used in completions requiring multiple packers and one-trip operations.

Dual and triple hydraulic packers are installed similarly to the single hydraulic packers. Most dual hydraulic packers are run on the long string and set by hydraulic pressure applied to either the long string or to the short string. To retrieve a dual packer, the tubing is released from the short string, and the well is circulated above the packer. A straight pull on the long string shears releasing pins and places the packer in retrieving position.

14-3.3.3 Seal Bore Retrievable Packers

Seal bore retrievable packers are production packers designed for intermediate-pressure wells. They are used in applications with bottomhole pressures of up to 10,000 psi and BHTs of 275°F. These packers have the production features of permanent packers with the added feature of retrievability. External components are mill-able for cases when conventional release is not possible.

These packers are designed to be retrieved with a retrieving tool. They can be set with electric line or set hydraulically on the tubing string.

Seal bore retrievable packers may be used in many of the same applications where permanent packers are used. Their application depends on downhole pressure requirements, temperature requirements, and cost considerations.

14-3.3.4 Permanent Packers and Their Accessories

Permanent packers are production packers designed for high-pressure and/or high-temperature wells (Hopmann and Walker, 1995). Designs are available for pressure differentials and bottomhole temperatures in excess of 15,000 psi and 450°F, respectively. Once these packers are set, milling operations are required for retrieval.

The two types of permanent packers are wireline and hydraulic-set.

Wireline-set permanent packers are generally run and installed using electric wireline. When the required setting depth is reached, setting tools are activated from the surface by electric current to set the packer. These packers may also be run on tubing using hydraulic setting tools, which are activated by surface pump pressure to set the packers. These packers feature a polished bore receptacle in which a production seal unit attached to the bottom of the production tubing may be installed.

Hydraulic-set permanent packers are generally run on the bottom of the production tubing. When the required depth is reached, surface pump pressure is applied to set the packer, which features a setting piston as an integral part. Using surface pump pressure to set either type of packer requires a means of plugging the tubing at the hydraulic setting tool or below the hydraulic set permanent packer before applying surface pump pressure.

Depending on well completion requirements, a variety of latching options is available to attach the production seal unit or tubing to the top of each packer. A variety of connections is also available to attach either tubing, casing, or accessory items to the bottom of each packer.

Seal Units

Seal units consist of a stack of alternating seals and rings located at the end of the tubing string (Figure 14-14). The seal unit stabs into the top of a packer or polished bore receptacle (PBR) and forms a pressure seal between the tubing string and packer. The seal is created by compressing elastomers, plastics (nonelastomers), or rubber configurations between the smooth metal surfaces of the seal unit and the mating receptacle.

A seal stack is designed to move within with the receptacle but may only move if the tubing-to-packer connection is not a J-latch, anchor, or ratch latch.

There are several types of seal units designed to accommodate the various pressures, temperatures, and fluids that may be encountered in an oil well. Molded seal units are designed for lower temperatures and pressures and are best suited for applications where the seal unit may be moving in and out of the receptacle. Premium seal units are Vee packing seal systems designed for high temperature and pressure environments and harsh or hostile well fluids. Various seal material combinations can be used to tailor the seal unit for compatibility to the fluid

environment to be encountered. Seal units are available up to 20,000 psi and temperature rating up to 450°F.

Seal units are sized and designed to fit inside a particular size of mating receptacle. As a result, the flow area through the seal unit will be less than the tubing ID. For applications requiring a larger flow area, an overshot or ELTSR assembly is used. In this case, the seal stack is located inside the tubing extension, and the seals create an external pressure seal between the tubing and a slick joint above the packer.

Seal stacks are susceptible to damage if cycled repeatedly in and out of the packer or PBR. To maintain pressure integrity, seal stacks should be removed only when necessary.

Seal Bore Extension

A seal bore extension is a separate tubular member which is placed on the bottom of a permanent or retrievable seal bore packer. Its purpose is to extend the bore of the packer to provide a longer receptacle that can accommodate longer seal units. Packers equipped with seal bore extensions can accommodate seal stacks of 30 ft or more.

Seal bore extensions allow flexibility for extending the seal bore length for tubing movement requirements. Generally, the seal stack selected must be as long as the tubing movement expected, and a safety factor (1.25 to 1.5) is often applied to the calculated length. Depending on the magnitude and direction the anticipated movement, the seal stack may be lengthened or shortened to coordinate with landing and space-out requirements. Section 14-4 and Chapter 7 provide information on tubing movement.

Millout Extension

When a permanent packer must be milled up, either a flat or overshot mill is used to mill up the packer. The choice of mill depends on the type of packer. For example, if the slips, cones, and packer elements are designed to be milled, an overshot mill is used, and the inner body of the packer is retrieved. Alternatively, the flat mill is used when the entire packer must be milled.

After milling, the slips retract and the packer can fall to the bottom of the well when the setting mechanism is released. A fishing job is therefore required to retrieve the remaining body of the packer. If a fishing job is not desirable, a millout extension is run below the packer at the time the packer is installed. The millout extension provides the ability to run a packer-retrieving spear with the mill. This spear engages in the millout extension and prevents the packer from falling when the packer slips collapse.

14-4 TUBING MOVEMENT AND PACKER FORCES

Changes in downhole pressure and temperature can cause the tubing to elongate or contract (Chapter 7). Such phenomena must be predicted and accounted for in the selection of downhole equipment.

If the predicted tubing movement is small (a few feet), then the movement may be offset by slacking off or picking up tubing weight, or the movement may simply not be provided for. In the latter case, changes in tubing length that cannot actually occur are translated into forces on the packer. Any packer selected should be able to withstand such forces. If the packer cannot withstand the induced force, it will move or unseat.

If it is preferable to allow tubing expansion or contraction to occur downhole, then the downhole equipment must provide that capability. The downhole equipment selected to allow tubing expansion or contraction mainly depends on the type of packer and the tubing-to-packer connection.

Retrievable packers (either hydraulic or mechanical) typically have a fixed tubing connection. In this case, tubing movement must be provided at a travel joint located above the packer.

Permanent packers and retrievable seal bore packers have a locator or short stab-in seal stack as the tubing-to-packer connection. Usually, this connection is not fixed and the tubing is free to contract or expand at the packer, with the amount of permissible movement determined by landing conditions. If the expected movement is small, then the length of the short seal stack may be sufficient to accommodate the expected tubing length change. If not, a seal bore extension and a longer seal assembly can be added to the design to accommodate a longer tubing length change.

An overshot or extra-long tubing seal divider may also be used to accommodate tubing movement with seal bore packers. This device includes a seal stack which stabs over the top of the packer and strokes open and shut. These stab-over seal configurations may be used in conjunction with an anchor sub, which provides a fixed connection at the packer.

For any downhole equipment used to accommodate tubing movement, the engineer must design a system which provides the maximum expansion or contraction expected with an applied safety factor.

14-5 SUBSURFACE SAFETY VALVES

A subsurface safety valve (SSV) is a control device used to shut off production from the well in an emergency situation, for example, if the surface control system is damaged or destroyed (Figure 14-15). The subsurface safety valve's opening and closing may be controlled at

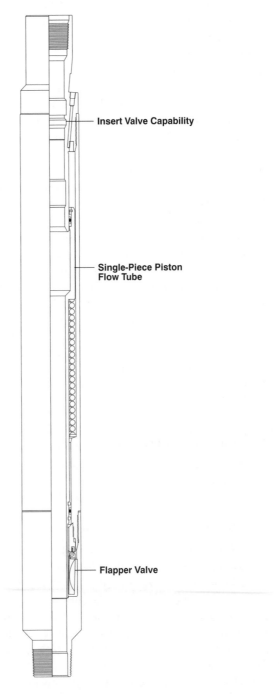

Figure 14-15 SCSSV

the surface by pressure supplied in a hydraulic line or directly by subsurface well conditions. Dines and Calhoun (1979) and Bleakley (1986) provide an overview for the selection of subsurface safety valves.

Surface-controlled subsurface safety valves (SCSSV) incorporate a piston on which hydraulic pressure acts to open the closure mechanism. A spring acts in the opposite direction on the piston to close the closure mechanism as hydraulic pressure is lost. In most SCSSV designs, well pressure acts in conjunction with the spring to oppose the hydraulic pressure and close the valve.

Subsurface-controlled subsurface safety valves (SSCSV) are operated directly by well pressures and require no hydraulic line for their operation. They are normally open while installed in the well and typically require flowing the well outside its normal production regime to close. The inability to control SSCSVs from the surface limits their use to applications that are outside the capabilities of SCSSVs.

Subsurface valves first came into prevalent use in the 1930s. However, early valves were not reliable, and there was a reluctance to voluntarily use this component. Operators were required to use these devices only if there was an obvious and imminent chance of disaster. At present, the use of subsurface safety valves is still optional in many places worldwide. Certain geographical areas, such as the US Continental Shelf, do now require a subsurface safety valve of some type.

At this time, most operators are electing to use surface-controlled valves because this type of valve affords direct control of well operation. Therefore, this discussion focuses on surface-controlled subsurface safety valves.

14-5.1 Tubing Safety Valves

The following sections provide information about tubing safety valves.

14-5.1.1 Subsurface Safety Valve Basics

Subsurface safety valves are classified according to their method of retrieving and their internal closure mechanisms. Tubing retrievable safety valves (TRSV) are valves which are an integral part of the tubing string, requiring the tubing to be removed for retrieval of the safety valve. Wireline retrievable safety valves (WLRSV) are installed inside the tubing with a locking device to secure them inside a safety valve landing nipple. WLRSVs can be installed and retrieved without removal of the tubing.

The most common types of closure mechanisms for subsurface safety valves are the ball and the flapper. Either mechanism can be used with a tubing retrievable or wireline retrievable safety valve.

To evaluate subsurface safety valves for an application, it is necessary to understand the basics of safety valve operation and application. These basics include understanding the advantages and disadvantages of the different retrieval methods and closure mechanisms. Determining the valve setting depth is another consideration. In addition, the requirement to equalize pressure across a closed valve must also be considered. These factors are discussed here.

14-5.1.2 Retrieval Method

Many factors are considered by operators when selecting the means of retrievability of their SSVs. Access for maintenance and past experience normally are strong influences.

Tubing retrievable subsurface safety valves (TRSV) are threaded into the tubing string, maintaining an inside diameter that is equivalent to that of the tubing. To provide this inside diameter and still encase a closure mechanism and valve operator, the outside diameter of the TRSV will be larger than that of the tubing. This larger outside diameter must be allowed for in the completion design and can sometimes be a limiting factor in the selection of a tubing size. Because a TRSV maintains an inside diameter equivalent to that of the tubing, it introduces no restriction to flow or tool access below the safety valve. This advantage must be weighed against the requirement to pull the tubing to retrieve and service the valve. Pulling a TRSV normally requires that a drilling or workover rig move onto the well. However, most TRSVs are designed so that a WLRSV can be run inside. Should the TRSV experience problems, a workover can be delayed and wireline can be used to set the WLRSV inside the TRSV body. This option greatly extends the use of TRSVs.

Wireline retrievable subsurface safety valves (WLRSV) are installed inside the tubing string, and, therefore, must have an outside diameter small enough to fit inside the tubing. Allowing for room to encase a closure mechanism and valve operator results in the inside diameter of a WLRSV being substantially smaller than that of the tubing. This smaller diameter acts like a downhole choke and introduces a pressure drop that may significantly reduce system flow capability. Tool access below the safety valve is not normally a major consideration because the WLRSV can be retrieved before running tools in the well and re-installed after the tools are removed. Retrieval of a WLRSV is typically done using standard wireline service methods, which are substantially lower in cost than mobilizing a rig.

Reliability and longevity have become areas of significant interest with respect to subsurface safety valves. Statistical data has been gathered in an attempt to measure reliability. These data have been reported by a number of surveys and papers, the most highly publicized articles being the "SINTEF" reports, by the Foundation for Scientific and Industrial Research at the Norwegian Institute of Technology in Trondheim, Norway (Molnes, 1990; Molnes *et al.*, 1987; Molnes and Rausand, 1986a; Molnes *et al.*, 1986b). These reports generally show a statistical reliability advantage toward tubing-retrievable safety valves. However, one should consider that tubing-retrievable safety valves are not normally operated as often as wireline-retrievable safety valves. Another factor that may have skewed the data is the fact that most TRSVs in the studies did not contain an internal equalizing mechanism, whereas most of the WLRSVs did contain such a mechanism.

14-5.1.3 Closure Mechanisms

The two primary closure mechanisms used for subsurface safety valves are the ball and the flapper. Typical flapper closure mechanisms consist of a plate (flapper) which covers a seat to shut off well flow. The flapper rotates 90° as the flow tube pushes it open and moves past it, opening the safety valve. For the safety valve to close, the flow tube must move back out of the way of the flapper, allowing a torsion spring to rotate the flapper back onto the seat.

The typical ball-type closure mechanism uses a rotating ball which covers a seat to shut off well flow. To open the safety valve, the ball is rotated 90°, opening a flow path through the ball. Closing the ball requires it to reverse its rotation to recover the seat.

The flapper-type closure is the primary choice of most operators because of its simplicity and its superior record of reliability. However, both mechanisms have specific advantages and disadvantages that should be considered in selecting a subsurface safety valve closure mechanism. These advantages and disadvantages are compared in Table 14-2.

In addition, Sizer and Krause (1968) and Sizer and Robbins (1963) provide an analytical study of the locking mandrel and an evaluation of safety valves. Calhoun

Table 14-2   Comparison of safety valve closure mechanisms

Flapper Closure Mechanism		Ball Closure Mechanism	
Advantage	Disadvantage	Advantage	Disadvantage
Lower mechanism friction results in fewer problems in operating mechanism (especially in opening).	Flapper swings onto seat, sometimes trapping debris, resulting in sealing problems.	Ball is wiped by seat as it operates reducing debris accumulation.	Greater contact surfaces result in higher mechanism friction (primarily in opening).
Flapper is easily pumped through from above should kill operations through a closed safety valve be necessary.	Flapper relies on a relatively low force torsion spring to return it to the seat during closure.	Balls are normally pulled closed with a compression spring, which generates a substantial force.	Most ball mechanisms can be pumped through from above, but offer greater pressure drops to do so.
Tool strings can normally be run past a partially open flapper valve should it experience operational problems.	The flow tube must move completely past the flapper during closure before the flapper can start rotating closed, resulting in longer closing strokes and reaction time.	Ball mechanisms start rotating closed as soon as the flow tube moves, resulting in a short closing stroke and quick response time.	Ball mechanisms must be able to open fully to pass tool string through them.
		Ball mechanisms are inherently resistant to damage during closure against flow.	

(1977) gives design considerations for SCSSV systems. Rubli (1980) cites new developments in valve technology.

14-5.1.4 Setting Depth Determination of Surface Controlled Subsurface Safety Valves

Surface-controlled subsurface safety valves are operated by a hydraulic piston that opens the closure mechanism with the application of hydraulic pressure. This piston is typically a single-acting piston that is opposed by a compression spring. When hydraulic pressure is released from the piston, the compression spring returns the safety valve to the closed position. The strength of this compression spring also plays an important role in determining the setting depth of the surface-controlled subsurface safety valve.

The hydraulic fluid is supplied to the piston of the safety valve by a line connected to a control system at the surface wellhead area. This line is typically $\frac{1}{4}$-in. stainless steel tubing. The distance from the surface wellhead to the subsurface safety valve is referred to as "setting depth." The weight of the hydraulic fluid over this "setting depth" distance produces a hydrostatic pressure on the piston of the safety valve. The compression spring must be able to overcome this hydrostatic pressure to close the safety valve. The relationship between the safety valve's compression spring force and its piston area results in a "closing pressure" of the subsurface safety valve. This relationship is Equation 14-1:

$$p_{clos} = \frac{SF_{comp}}{A_{pist}} \qquad (14\text{-}1)$$

where p_{clos} = closing pressure of the subsurface safety valve, psi; SF_{comp} = compression spring factor, lbf; and A_{pist} = piston area, in.^{2}

Equation 14-1 presents the fundamental design principle. Closing pressure of a subsurface safety valve is also affected by seal friction, part weight, and tolerances.

The maximum setting depth is the closure pressure divided by the hydraulic fluid gradient, calculated as

$$H_{max} = \frac{p_{clos}}{(\gamma_f)(0.433\,\text{psi/ft})(sf)} \qquad (14\text{-}2)$$

where 0.433 psi/ft is the gradient of water; H_{max} = maximum valve setting depth, ft; p_{clos} = closure pressure of the valve, psi; γ_f = specific gravity of the fluid in control line; and sf = safety factor (1.1 to 1.25).

Example

The following example illustrates the calculation of valve-setting depth.

Suppose the hydraulic fluid being used on an offshore platform weighs 52 lb/ft^3. The fluid's specific gravity is $52/62.4 = 0.833$. If the surface-controlled subsurface safety valve to be used has a closing pressure of 440 psi, and the setting depth safety factor is 1.20, the maximum valve setting depth is calculated with Equation 14-2 as follows:

$$H_{max} = \frac{440}{(0.833)(0.433)(1.2)} = 1016\,\text{ft} \qquad (14\text{-}3)$$

The safety valve in this example could be set shallower than 1016 ft, but should not be set any deeper unless some means for balancing the hydrostatic pressure on the piston is considered. Conventional safety valves in use today range in maximum setting depths from 500 ft to 5000 ft. Dines (1980) provides a discussion of setting safety valves in deepwater fields where setting depths are large.

14-5.1.5 Equalizing vs. Non-Equalizing Subsurface Safety Valves

When a subsurface safety valve is closed, a pressure differential often forms across the closure mechanism. This pressure differential can be from formation pressure building below the closed valve or from bleeding down all or part of the pressure above the closed valve. The pressure differential must be eliminated before opening the closure mechanism by equalizing the pressure across the safety valve.

Some valve designs incorporate an equalizing mechanism as an integral part of the safety valve; these are referred to as equalizing valves. This equalizing mechanism can be opened to allow pressure to equalize across the safety valve before opening. Safety valves that do not contain an equalizing mechanism are referred to as non-equalizing valves and must be equalized by applying pressure to the tubing above the closed valve. This pressure can be supplied from a pump/compressor or another well with equivalent or greater pressure. Actual opening procedures vary across designs, and manufacturers' actual opening procedures should be followed.

Equalizing devices can offer convenience in opening closed safety valves, but because of the nature of their task are often subject to high stresses and velocities that can lead to problems. Many operators prefer the nonequalizing valve because they typically have fewer leak paths, which boosts their reliability. This statement is especially true for tubing retrievable safety valves where retrieval of a faulty safety valve is particularly costly.

14-5.2 Annular Safety Valves

The purpose of an annular safety valve (ASV) system is to seal the annulus between the tubing and the casing immediately below the wellhead. This procedure protects surface facilities and personnel from any gas in the tubing/casing annulus in the event that wellhead integrity is compromised. The ASV system is usually set shallow and near the wellhead to limit the volume of annular gas that would escape in the event of wellhead failure.

The primary application of annular safety valves is offshore platforms where there is a concentration of personnel and surface facilities in the immediate vicinity of production wellheads.

Numerous completion configurations can be used to achieve annulus control. One configuration uses a dual completion packer with dual tubing strings to surface and conventional safety valves in each string. One string serves as a gas injector below the dual packer and the other as the production string. While this system can use standard completion tools, it is not a desirable system for most applications.

An alternative is to use an ASV packer with a single tubing string connecting the wellhead to the packer (Figure 14-16). This arrangement maximizes the production tubing flow area and the injection capabilities of the annulus. The remainder of this text will address these applications.

Gas-lift completions and completions where an ESP is used and vent gas is a byproduct are the two primary completions in which an ASV is included. In gas-lift completions, gas is injected into the annulus and the large volume of gas compressed in the annulus poses a danger. In these cases, an ASV is included to shut in the annulus in the event of wellhead failure. In an ESP completion, gas which breaks out of the ESP is vented through a bottomhole packer and enters the annulus. An ASV located high in the well provides protection from and control of the gas in the annulus.

The ASV system usually consists of a production packer, an annular safety valve, a tubing safety valve, an optional travel joint, and an optional, engineered tension member (Robison and Parker 1991; Figure 14-16). The safety-valve components are controlled by

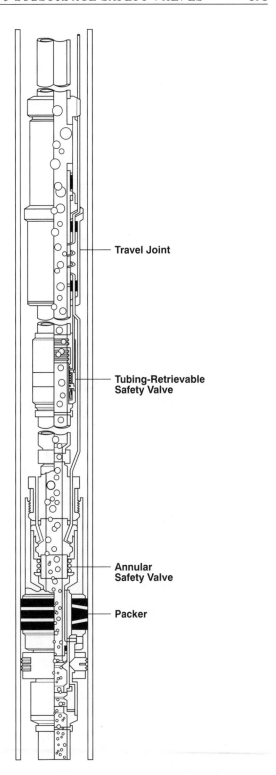

Figure 14-16   Annular safety valve

surface-supplied hydraulic pressure and are designed to fail into a closed position when hydraulic pressure is removed.

The ASV packer has multiple functions as part of the system. The packer must provide the annular flow path for the lift gas or vent gas. The packer is usually located at a shallow depth to limit gas volume. As a result, the entire tubing hang weight must be supported by the packer. The casing in which the packer is set may not be supported. Specially designed slip systems must be used to load the unsupported casing evenly so as not to permanently deform and compromise casing strength (Robison and Parker, 1991). Because of the size of the annulus area created by the casing and the tubing (which can be greater than normal completions), the axial loads resulting from the pressure differential on the packer can be great. The elastomers used on the ASV packer should be rated for temperature cycles as high as 120°F.

An alternative to a packer in the ASV system would be a specially designed tubing hanger with a dedicated casing nipple profile with a polished bore receptacle. This design offers the attributes mentioned above but cannot be run into a wellbore equipped with standard casing.

The annular safety valve functions as a surface-controlled, fail-closed device for the annular flow path created by the ASV packer. Depending on the specific application, the ASV and tubing safety valve can have a common hydraulic system or independent hydraulic systems. The ASV should feature a dependable metal-to-metal closure mechanism. The ASV should also have a pump-through feature which will allow kill fluid to be pumped into the annulus volume below the ASV packer. In addition, the annular safety valve needs to be connected to the ASV packer to interface directly with the annular flow area feature of the ASV packer. Annular safety valves can be either tubing retrievable or wireline retrievable.

A tubing-retrievable ASV includes a concentric annular poppet closure mechanism. The closure mechanism should have a conical geometry with surface treatment to enhance performance and longevity of the metal-to-metal seal. The tubing-retrievable ASV should also have a large flow area and be debris tolerant. With these features, the tubing-retrievable ASV is suited for gas-lift installations where large injection rates are required.

The wireline-retrievable ASV has a smaller flow area and is less debris-tolerant than the tubing retrieval version. The wireline retrievable ASV could be a side-pocket device positioned on top of an ASV packer with twin flow construction. Standard wireline kickover tool methods would be used to run and retrieve the ASV. The wireline retrievable ASV is normally used where either moderate gas injection is required or for ESP vent-gas applications with low flow rates.

The tubing safety valve should be compatible with the complete ASV safety system. The tubing safety valve is usually positioned just above the ASV in the completion.

Tubular members between the ASV and wellhead are an important consideration. Because of the thermal cycles that occur to the upper part of a subsea well completion, it is common to use a travel joint in the tubing string above the tubing safety valve (Robison and Parker, 1991). If it is undesirable to introduce a potential leak path associated with the seals in the travel joint, the tubing members must be rated to withstand the stresses from thermally induced tubing movement. Another possible addition in the tubular members between the ASV and the wellhead is an engineered tension member (Robison and Parker, 1991). If it is desirable to limit the upward shock loads on the ASV packer and system components associated with shearing off the wellhead, then a tension member can be designed to fail before the ASV packer is put into critical loading.

The future applications for ASV systems will increase as operators continue to explore the application of downhole separation and re injection of fluids within a single completion.

14-6 EQUIPMENT SELECTION

The following sections provide information about selecting equipment.

14-6.1 Design Requirements

Each downhole equipment item included in a completion is selected on the basis of some functional requirement of the completion (Chapter 21). For example, it will generally be necessary to establish flow from wells that have been completed and are standing full of heavy completion fluids. This situation creates some requirement to circulate out the completion fluid. This requirement would be satisfied by including a sliding sleeve in the completion above the shallowest packer.

Another functional requirement might be to provide the capability for initiating gas lift at some point in the future. In this case, the requirement for future gas lift is met by including one or more SPMs in the completion, equipped with dummy valves.

These two examples illustrate how requirements of the completion dictate the use of various downhole devices. Once a decision is made to include a device in the completion design, then all of the detailed attributes for that component must be specified. Packer selection is discussed here as an example. These details regarding packer selection exemplify the level of detail required in selecting and specifying all other downhole equipment.

Choosing the correct packer first requires careful examination of the well conditions and required operational capabilities. Many packer manufacturers publish packer selection charts that provide detailed information regarding the correct application of their packers. By identifying well conditions and the operational requirement of the packer factors before reviewing different packer designs, selection is accurately based on the needed capabilities instead of packer features. Long-term planning should also be considered when making a packer selection.

Once well conditions (pressure, temperature, fluids, etc.) and operational requirements are identified, packer capabilities should be ranked according to priority because there may be trade-off features to consider for different packers. Operational considerations for making packer selections can be divided into three general categories:

- running, setting, and tubing space-out

- production and treating

- packer retrieval

Methods for running and setting the packer will mostly depend on whether the packer is to be run in on tubing or electric line. Permanent packers can be run and set on tubing, electric line, or downhole power units after the appropriate setting adapter kit is selected. Mechanical and hydraulic set retrievable packers are typically set on tubing since they require tubing manipulation or tubing pressure to set.

The packer features must also be compatible with the required space-out requirements. Setting the production tubing in tension, neutral, or in compression will have different results on each packer type. Packer features to consider when the tubing is in tension are:

- Does the packer have a tension-shear-release feature?

- Does the packer have an emergency tension-shear-release feature?

- Will the tension reduce the setting force in the rubber packer elements?

- What effect will there be on seal bore assemblies and other accessories?

Setting the production tubing in compression may be desirable to compensate for tubing movement caused by thermal and pressure variations. This step will be very important if the production packer has a straight-tension, shear-release feature. It may also be important when setting multiple packers over short intervals where some tubing movement may be required to set the packer.

Thermal and pressure variations over the well's life can occur as the reservoir depletes, but they are most severe during well treatment or during pumping into an injection well. Thermal cycles will result in changes at the packer from tubular expansion or contraction. Thermal cycling can also have an effect on the performance of the packer sealing elements. A hydraulic retrievable packer with re-settable features may be required to compensate for the thermal changes that may affect the packer element performance.

Retrieval considerations differ among straight-tension release, minimal tubing rotation and straight-pull release, mill out, and—for some packer types—chemical or mechanical cut to release. Permanent packers are typically milled out, and retrievable packers are generally either straight- tension or shear-release or require a one-third turn and a straight pull to release. Some retrievable packer types constructed with permanent element packages can be retrieved by using chemical or mechanical cutters to release the packer.

14-6.2 Well Conditions

The following sections provide information about well conditions that influence equipment selection.

14-6.2.1 Deviation

In the course of drilling a well, the drill bit will drift from the true vertical position and may also take a helical path. This divergence from the vertical position is called deviation and is measured in degrees. The term *straight hole* or *vertical hole* defines a well with very low deviation and that generally will present no problems in landing the completion equipment. In wells with high deviation, sharp changes in the wellbore referred to as doglegs can occur, thus creating a tight area for long, stiff tool assemblies. The same situation can occur in directionally drilled or horizontally drilled wells where the kick-off point from the vertical progresses through a short radius. A short radius similar to a dogleg may create problems in long, stiff tool assemblies.

14-6.2.2 Hole Conditions/Completion fluids

To facilitate setting a well completion on-depth, casing ID restrictions should be minimized or eliminated and the well fluids should be circulated and conditioned before the completion equipment is run.

14-6.2.3 Casing ID Restrictions

Casing ID restrictions may result from mudcake build-up, cement scale, over-torque of casing connections, and pipe scale build-up.

Mudcake build-up results from solids in the drilling fluid adhering to the casing ID. Mudcake can develop into a semi-solid, thus preventing the successful installation of completion equipment. Cement scale build-up can cause the same problem. After drilling the well, the casing is cemented to the wellbore by pumping cement down the casing and back up the annulus. Even though wiper plugs should be used, some cement residue can remain in the casing ID. This problem may be more severe with heavy or thixotropic slurries.

Excessive over-torque of casing threads can cause the pin connection to crimp or flare inward, thus reducing the casing ID. A gauge ring run before running completion equipment should detect this problem. If detected, a string mill may be required to correct the pin ID.

In the case of existing wells, re-completion efforts can encounter problems if pipe scale build-up is present. Similar to mudcake and cement scale, pipe scale can be corrected by making a bit-and-scraper run.

14-6.2.4 Well Fluids

Non-conditioned well fluids may develop gel strength or conversely may not be able to keep solids in suspension. In either case, a viscous, coagulated, semi-solid may develop downhole, making it difficult if not impossible to run completion equipment. A bit-and-scraper run may be necessary to circulate out and recondition the wellbore fluids.

14-6.3 Material Considerations

The following sections provide information about the factors that influence materials selection.

14-6.3.1 Pressure and Temperature Considerations for Standard Service

When selecting completion equipment for downhole service, it is necessary to specify equipment that is suitable for use with the production tubulars in burst, collapse, and tension (Chapter 7). Deviating from this requirement may be costly if equipment failure downhole occurs.

Standard-service equipment typically meets the mechanical requirements of API L-80, which is also suitable for sour or corrosive service, regardless of temperature. Service temperatures greater than 275°F may cause degradation of elastomers over a period of time. While the upper limits of the service temperature are sometimes debated by different suppliers, it is certain that typical nitrile compounds begin to lose their original resiliency, allowing a possible leak path.

Higher performance elastomeric compounds such as propylene-tetrafluorethylene copolymer (Aflas), fluorocarbon (Viton), perflourors (Chemrez and Kalrez), and ethylene/propylene diene (EPDM) are capable of higher working temperatures but, in high pressure applications, may require containment or backups to prevent extrusion of the elastomer. Table 14-3 gives service temperatures and pressures applicable to elastomers in specific fluid environments.

14-6.3.2 Material Considerations for Harsh Environments

The following sections discuss the factors that influence the choice of materials for harsh environments.

Metal Corrosion

The presence of waters, either reservoir water or condensed, is closely related to corrosion. Corrosion cannot occur unless an electrolyte, an electrically-conductive fluid, is present. Common downhole electrolytes include mineralized (connate) waters and/or "dew-point" waters that condense inside the production tubing. These waters become corrosive when they are acidified by CO_2, H_2S, stimulation fluids, or combinations thereof. The corrosivity of the electrolyte at downhole temperatures in combination with acid gases, CO_2 and H_2S, must be assessed before the selection of downhole metallic goods. Corrosion rate increases with water production and as water pH decreases. Corrosion is mitigated by oil-wetting of the tubing or water-in-oil emulsions whenever these conditions are found.

Carbon dioxide-water (CO_2–H_2O) corrosion increases as the gas phase increases (CO_2 partial pressure), which is

Table 14-3 Guidelines for seals and packer elements

Compound (1)	Nitrile (6)	Fluoro-Carbon (6)	Aflas (4,6)	Chemraz (3)	EPDM (7)	Kalrez (3)
Service Temperature °F(°C)	32° to 275°F (0 to 135°C)	32 to 400°F (0 to 204°C)	100 to 400°F (38 to 204°C)	40 to 400°F (4 to 204°C)	40 to 550°F (40 to 288°C)	100 to 400°F (38 to 204°C)
Pressure (2) psi (MPa)	10,000 (69)	9000 (62.1)	8000 (55.2)	6000 (41.4)	3000 (20.7)	6000 (41.4)
Environments						
H_2S	NR	A	A	A	NR	A
CO_2	A	B	B	A	NR	B
CH_4 (Methane)	B	A	A	A	NR	B
Hydrocarbons (Sweet Crude)	A	A	A	A	NR	A
Xylene	NR	A	B	A	NR	A
Alcohols	A	C	B	A	B	A
Zinc Bromide	NR	A	A	A	NT	A
Inhibitors	B(5)	NR	A	A	NT	B
Salt Water	A	A	A	A	A	A
Steam	NR	NT	B	B	A	B
Diesel	B	A	B	A	NR	A
Hydrochloric Acid (HCl)	NR	A	A	A	NR	A

A: Satisfactory; B: Little or No Effect; C: Swells; NR: Not Recommended; NT: Not Tested

Note: (1) These materials are mainly used as O-rings.

(2) All pressure tests were done using 6 mil (0.006-in.) gaps; larger radial gaps will reduce pressure rating.

(3) Back-up rings must be used above 250°F (121.1°C) and 4000 psi (27.6 MPa).

(4) Back-up rings must be used above 350°F (176.7°C) and 5000 psi (34.5 MPa).

(5) Water-soluble inhibitors only.

(6) Good for O-rings, packer elements and molded seals

(7) For packer element application - EPDM compound - y267

the main factor influencing CO_2–H_2O corrosion at temperatures below 140°F. Conversely, for temperatures up to 140°F, corrosion decreases with increased salinity. Flow rate has little influence on corrosion. With temperatures above 140°F, corrosion in CO_2–H_2O mixtures decreases because of deposition of $FeCO_3$ and Fe oxide films on carbon and low-alloy steels. Increased salinity increases corrosion in CO_2–H_2O mixtures (Rice *et al.*, 1991). Operators should not rely on the formation of $FeCO_3$ films to protect steel tubulars; each case must be evaluated based on factors such as CO_2 partial pressure, corrosion history, in-situ pH, pressure of other ionic species, and presence of organic acids (Crolet, 1994).

Other important corrosion considerations when designing and selecting downhole equipment are galvanic corrosion from dissimilar metals, well stimulation services where acids are used, and corrosion induced from injecting insufficiently deaerated waters in enhanced oil recovery (EOR) operations (Bradburn *et al.*, 1983).

Metal Embrittlement

Embrittlement is a term used to characterize materials that have interacted with their environments and have become brittle or glass-like. This situation is more common when nonbrittle metals become brittle after being affected by chemicals such as H_2S and water. This phenomenon is commonly called sulfide stress cracking. The National Association of Corrosion Engineers (NACE) has issued guidelines for preventing embrittlement in sour-gas systems, most notably the document NACE standard MR0175 (NACE, 1994). Waters containing dissolved H_2S can catalyze the entry of hydrogen atoms into susceptible metallic materials, and in the presence of tensile stresses, cause embrittlement. The greater the gas pressure in the well and the concentration of H_2S in the gas phase, the greater the tendency of hydrogen to enter the metal. When a sufficient quantity of hydrogen has entered over a period of time, cracking may occur.

Factors that promote sulfide stress cracking of low-alloy steels are tensile stresses, temperatures less than 150°F, and the presence of acidic waters. Other forms of embrittlement are possible, most notably chloride stress cracking, which primarily affects corrosion-resisting alloys in hot, saline waters (Rice *et al.*, 1991).

Controlled-hardness API tubulars, such as L-80, C-90, and T-95, are typically specified for sour service. In cases where high corrosion rates are anticipated, alloys other than carbon and low-alloy steel (such as chrome) are commonly used to combat stress cracking and embrittlement. These alloys typically contain nickel, chromium, molybdenum, and major alloying constituents (Rice, *et al.* 1991).

Elastomers

Nitrile compounds have proven to be acceptable elastomers for seals and packer elements in the oil and gas industry. However, nitrile compounds have limited capabilities as the service environment becomes more harsh. Typically, nitrile will begin to degrade when service temperatures exceed 275°F over time. The presence of H_2S, aromatic solvents such as xylene or toluene, heavy completion fluids such as zinc bromide, and acids such as hydrochloric and hydrofluoric will also cause degradation of the compound.

Fluorocarbon elastomers such as Viton and Fluorel have outstanding resistance to the hostile environments that attack nitrile compounds. However, organic amine corrosion inhibitors, methanol injection systems, glutaraldehyde, and steam will have detrimental effects on viton and fluorel compounds.

Aflas has resistance to organic amine corrosion inhibitors, H_2S, zinc, and calcium bromide, with slight swelling in CO_2, methane, and steam. Aflas is not desirable for temperatures below 100°F; and for temperatures above 325°F, back-up seals must be used to help contain the compound.

Perfluoro elastomers such as Chemraz and Kalrez have outstanding resistance to heat, solvents, and harsh environments. Both elastomers can be used with organic amine-corrosion inhibitors, H_2S, CO_2, methane, aromatic solvents, and acid.

Non-Elastomers

Non-elastomer materials used in seal configurations are most commonly thermoplastic materials. Some thermoset plastics are used, but they may have limitations to high temperatures, which reduces their physical properties. Both thermoplastic and thermoset plastics have excellent chemical resistance. Thermoplastics, however, are the materials of choice because of their broad temperature capabilities, chemical resistance, ability to blend with glass fibers to enhance performance, and flexibility to be the sealing device or to be used in conjunction with elastomeric seals to enhance performance. Thermoplastics are considered high-performance materials, so they are typically used in harsh or demanding environments. Common thermoplastics are polyphenylene sulfide (Ryton), polyetherether ketone (Peek), and Teflon (Ray, 1996).

In addition to these materials, metal-to-metal seals are being used increasingly in circumstances where elastomeric and thermoplastic materials cannot be used (Morris, 1984).

14-6.4 Equipment Compatibility

The following sections provide information about equipment compatibility.

14-6.4.1 Dimensional Compatibility

In the designing phase of a completion system, it is an absolute requirement to record the tubing size, weight, grade, thread, depth to set packers, and the maximum ODs and minimum IDs of the equipment. Likewise, the minimum casing ID(s) must also be known. If a tapered casing string or liner exists in the well, the completion equipment must be compatible with the zone where the equipment is to perform. Before equipment is run into a well, each piece of equipment must be measured and recorded for maximum OD, minimum ID, and length. Balls, darts, or locating nipples used to activate or hydraulically set packers must also be measured, recorded, and compared to the minimum IDs of the completion string. Part numbers, serial numbers, and pertinent descriptions should also be recorded and related to a completion schematic.

14-6.4.2 Installation Considerations

When installing completion packers and equipment, it is best to run and set as quickly and accurately as possible. Safe run-in speeds should be determined since most completion packers have small clearances with respect to the casing ID. Running-in too fast may cause the packer elements to swab the well, which may damage the elements.

The method of conveying the packers to setting depth is dependent on the type of completion packer, depth, well deviation, and sometimes completion fluid viscosity. Permanent packers can be conveyed and set on electric cable or on tubing. Most retrievable packers require manipulation of the tubing or a pressure conduit to set the packer. Wireline-set retrievable packers require both electric cable for setting and tubing for retrieving.

Deep setting depths, 12,000 to 15,000 ft, are more difficult for properly positioning and manipulating the packer to the set and release positions because of pipe stretch. Therefore, hydraulic- or electric-set options are preferred. Wells with high deviation present a difficult situation for electric-cable-set packers since there is no tubing weight to push the packer to setting depth. Mechanical-set packers also may experience problems in transferring surface tubing manipulations downhole. Therefore, hydraulic-set packers run on hydraulic setting tools are most likely to be successful, since they require no tubing manipulation and can take advantage of pipe weight.

In a similar manner, heavy fluid viscosities like highly deviated wells, can present a challenge for electric-cable-set packers since there may not be adequate hang weight to get the packer to depth quickly. This environment would require the packer to be run and set on tubing. Mechanical-set or hydraulic-set options would be dependent on the well conditions previously discussed in this section. Other factors that influence installation techniques are heavy tailpipe loads below the packer, completions that require the packer to be set and released multiple times, and completions requiring installation of multiple packers in a single trip.

14-7 COMMON DOWNHOLE DESIGNS

Figures 14-17 through 14-20 are examples of common completion designs. Each design shown satisfies a unique set of design requirements, such as a need for artificial lift. These basic designs have been widely used and adapted to fit the specialized needs of particular reservoirs or geographical producing areas.

Figure 14-17 shows a gravel-pack completion with gas lift, which is typically used in offshore wells. In this completion, the well is first gravel-packed. Then, the production seal assembly, gas lift mandrels (SPMs), and safety valve are run at the appropriate depth in the tubing. The side pocket mandrels may be equipped with dummy valves if gas lift is not required immediately. Since there is no sliding sleeve in the design, the well would be circulated through the bottom SPM. The design

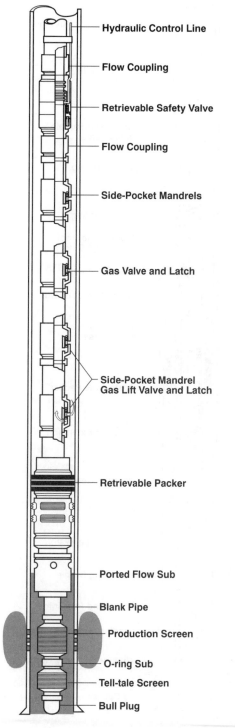

Figure 14-17   Gravel-pack completion with gas lift

includes a tubing-retrievable SCSSV, protected with flow couplings both upstream and downstream of the SCSSV.

Figure 14-18 depicts a single-string, permanent-packer installation. This completion is most typically used to

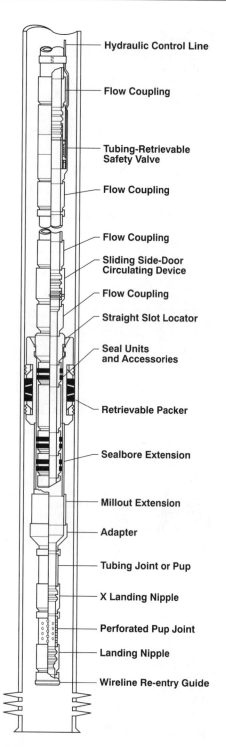

Hydraulic Control Line

Flow Coupling

Tubing-Retrievable
Safety Valve

Flow Coupling

Flow Coupling

Sliding Side-Door
Circulating Device

Flow Coupling

Straight Slot Locator

Seal Units
and Accessories

Retrievable Packer

Sealbore Extension

Millout Extension

Adapter

Tubing Joint or Pup

X Landing Nipple

Perforated Pup Joint

Landing Nipple

Wireline Re-entry Guide

Figure 14-18 Single-string permanent packer completion

produce a single zone, but multiple zones located beneath the packer could be commingled and produced with this completion. The permanent packer includes a seal bore extension to accommodate a long seal unit and a millout extension to enable the packer to be recovered,

if milled. The stab-in seal assembly would normally use a straight-slot locator to allow tubing movement, but a J-latch or ratch-latch could be used if necessary. The design includes a tailpipe with a selective landing nipple, a perforated pup joint, and a no-go nipple located above the wireline entry guide. The bottom no-go nipple allows flowing bottomhole pressure surveys with flow through the perforated pup joint.

The tailpipe assembly shown in Figure 14-18 is run into the well with the permanent packer. The tubing is then run with the other completion equipment, and the seal stack is stabbed into the permanent packer.

A sliding sleeve is included in the design for circulating the well and to provide a landing nipple profile for flow control devices. Flow couplings and a tubing-retrievable SCSSV are used as shown in Figure 14-17. Either a tubing-retrievable or wireline-retrievable valve could be used in the design.

Figure 14-19 depicts a dual-string, retrievable packer installation. This design allows production from two reservoirs through separate production strings. In this design, a permanent packer and tailpipe are run first, normally on tubing or electric line. The dual tubing strings are then run with a seal assembly on the bottom of the long string and the dual hydraulic set packer. When the dual tubing strings have been landed and spaced out at the surface, the tree can be installed. Fluid can be circulated down the tubing string and returns taken up the annulus to help unload and later initiate production. A ball is then dropped down the short string, locating in the collet catcher sub below the dual packer. Applied surface pressure sets the dual hydraulic packer, and increased surface pressure expels the ball from the collet catcher.

The uppermost zone in Figure 14-19 can be produced up the short string or through the sliding sleeve into the long string. A blast joint protects the long string opposite the zone. With this design (a polished nipple placed below the blast joints and a circulating device placed above), the circulating device can be opened and closed to produce the zone selectively. The polished nipple and circulating device provide profiles that can be used to place a straddle packoff across the blast joint in the event of failure.

Landing nipples are located in the long string and in the short string. The lowest no-go nipples enable flow testing. The landing nipple below the blast joint in the long string could be used to shut off production from the lower formation if required.

The dual completion shown in Figure 14-19 includes a safety valve landing nipple equipped with a wireline-retrievable SCSSV. Flow couplings are used both

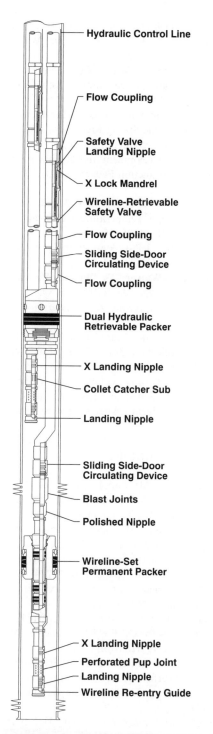

Figure 14-19 Dual-string retrievable packer completion

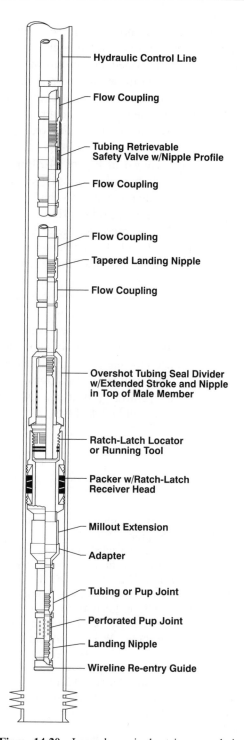

Figure 14-20 Large-bore single-string completion

upstream and downstream of the sliding sleeve and safety valve.

Figure 14-20 is similar to the completion shown in Figure 14-18, except that the tubing is stabbed over the packer with an overshot or ELTSR. As shown, this arrangement allows a larger sealbore diameter and larger ID seals. This arrangement provides for full tubing flow.

In Figure 14-20, the tubing is anchored to the short sealbore with a ratch-type locator. If tubing movement is expected, then the overshot tubing seal divider is installed immediately above the ratch latch.

A nipple profile is provided in the top of the male member to allow the well to be plugged and the tubing to be retrieved without disturbing the packer or locator seal. Location of the nipple profile at the top of the male member prevents trash from accumulating on top of the plug.

All landing nipples shown in Figure 14-20 are no-go nipples. Because of the larger tubing sizes typically run in this type of completion, all no-go nipples can be provided without significantly restricting tubing ID.

The completion designs presented here demonstrate basic arrangements of downhole completion equipment discussed in this chapter. Many modifications and enhancements of these basic designs exist and some specialized designs are reported in technical literature (Morris, 1986; McNair, 1992; Hopmann, 1993; Vinzant and Smith, 1995). Completion equipment will continue to evolve as operators devise new ways of applying basic completion equipment to meet the requirements of unique reservoir and production situations, such as high pressure and temperature (Nystrom, 1983; Boyle, 1992; Henderson, 1996; Hilts *et al.*, 1996).

REFERENCES

API Specification 5CT: "Specification for Casing and Tubing," *Am. Pet. Inst.*, Washington, DC (1990).

API RP14B: "API Recommended Practice for Design, Installation and Operation of Subsurface Safety Valve Systems," *Am. Pet. Inst.*, Washington, DC (1994).

Bleakley, W.B. (ed.): "The How and Why of Downhole Safety Valves," *Pet. Eng. Int.* (Jan. 1986) 48–50.

Boyle, C.: "Downhole Completion Equipment for High Pressure, High Temperature Wells," *Soc. for Underwater Tech.* (1992) 1-11.

Bradley, H.B. (ed.): *Petroleum Engineering Handbook*, Third printing, SPE, Richardson, TX (1987) 2-12-74

Bradburn, J.B., and Kalra S.K.: "Corrosion Mitigation—A Critical Facet of Well Completion Design," *JPT* (Sept. 1983) 1617–1623.

Calhoun, M.B., and Taylor, J.G.: "Design Considerations for Surface-Controlled Subsurface Safety Valve Systems," paper ASME 77-PET-45, 1977.

Crolet, J.L.: "Which CO_2 Corrosion, Hence Which Prediction?" working party report on Predicting CO_2 Corrosion in the Oil and Gas Industry, European Federation of Corrosion Publications, Number 13, European Federation of Corrosion, The Institute of Materials (1994) 1–29.

Dines, C.A.: "A Definitive Approach to the Selection of Downhole Safety Valves for Deep-Water Fields," *Oceanology Int.* (1980) **80**, 3–7.

Dines, C.A., and Calhoun, M.B.: "Considerations Relative to the Selection of Sub-Surface Safety Valves—A Guide to the Options," paper SPE 8162, 1979.

Greene, W.R.: "A Summary Look at Production Packers," paper presented at the Spring Meeting of the Southwestern District Division of Production (1966).

Henderson, W.D.: "Development and Qualification of an HP/HT Retrievable Production Packer," *ASME Int.* (1996) 66–70.

Hilts, R.L., Kilgore, M.D., and Turner, W.: "Development of a High Pressure, High Temperature Retrievable Production Packer," paper SPE 36128, 1996.

Hopmann, M.E.: "Nippleless Completion System for Slimhole/Monobore," paper OTC 7330, 1993.

Hopmann, M., and Walker, T.: "Predict Permanent Packer Performance," *Pet. Eng. Int.* (1995) 35–39.

Klementich, E.F.: "Unraveling the Mysteries of Proprietary Connections," *JPT* (1995) 1055–1059.

McNair, R.J.: "A Unique Completion Method for Isolating Production or Injection Zones in Slimhole Wells," paper OSEA 92230, 1992.

Molnes, E.: "Reliability of Downhole Safety Valves in a Subsea Safety," paper presented at SEN seminar Subsea Safety: Risk Assessment, Hazard Analysis and Systems Protection (1990).

Molnes, E., and Rausand, M.: "Reliability of Surface Controlled Subsurface Safety Valves—Phase II," SINTEF report (1986a).

Molnes, E., Rausand, M., and Lindqvist, B.: "Reliability of Surface Controlled Subsurface Safety Valves," SINTEF Phase II Main Report, Trondheim (1986b) 5.

Molnes, E., Rausand, M., and Lindqvist, B.: "SCSSV Reliability Tested in North Sea," *Pet. Eng. Int.* (1987) 38–42.

Morris, A.J.: "Elastomers Are Being Eliminated in Subsurface Completion Equipment," paper SPE 13244, 1984.

Morris, A.J.: "Experiences and New Developments in Well Completion," paper NPF 6983, 1986.

NACE Standard MR0175: "Sulfide Stress Cracking Resistant Metallic Materials for Oil Field Equipment," Nat. Asso. of Corrosion Engineers, Houston (1994).

Nystrom, K.O., and Morris, D.W.: "Selecting a Surface-Controlled Subsurface Safety Valve for Deep, Hot, High-Pressure, Sour Gas Offshore Completions," paper SPE 11997, 1983.

Ray, T.: *Seals and Sealing Technology*, Halliburton Energy Services, 1996.

Rice, P., Coyle, W.R., and Chitwood, G.B.: *Selecting Metallic Materials for Downhole Service*, Halliburton Energy Services (1991).

Robison, C.E., and Parker, C.: "Annular Safety Valve System Design," paper presented at the Subsea 1991 International Conference (1991).

Rubli, J.: "New Developments In Subsurface Safety Valve Technology," *Pet. Eng. Int.* (1980) 99–109.

Sizer, P.S., and Krause Jr., W.E.: "Evaluation of Surface-Controlled Subsurface Safety Valves," paper ASME 68-PET-23, 1968.

Sizer, P.S., and Robbins, K.W.: "An Analytical Study of Subsurface Locking Mandrels," paper ASME 63-PET-35, 1963.

Vinzant, M., and Smith, R.: "New Subsurface Safety Valve Designs for Slimhole/Monobore Completions," paper OTC 7885, 1995.

"Tubing Tables," *World Oil* (Jan. 1990).

TRADEMARKS

Aflas – a trademark of Asahi Glass Co. Ltd – Propylene-Tetrafluoroethylene Copolymer

Chemraz – a trademark of Greene, Tweed & Co. Inc. – Perfluoro Elastomer

Kalrez – a trademark of Dupont Co. – Perfluoro Elastomer

PEEK – a trademark of ICI Americas Inc. – Poly-Ether-Ether-Ketone

Ryton – a trademark of Phillips Petroleum Co. – Polyphenylene Sulfide

Teflon – a trademark of Dupont Co. – PTFE Polytetrafluoroethylene

Viton – a trademark of Dupont Co. – Fluorocarbon

Fluorel – a trademark of Minnesota Mining and Manufacturing Co.

15 Inflow Performance/Tubing Performance

Mohamed Soliman
Halliburton Energy Services

Michael J. Economides
Texas A&M University

15-1 INTRODUCTION

Understanding the fundamentals of fluid flow in porous media and in pipes is necessary for optimizing well and reservoir productivity. This chapter discusses not only the basic flow equations and concepts, but also fluid flow in hydraulically fractured reservoirs and in reservoirs with horizontal wells.

15-2 BASIC CONCEPTS IN FLUID FLOW THROUGH POROUS MEDIA

The following sections provide information about the basic concepts in fluid flow through porous media.

15-2.1 Mathematical Basis for Fluid Flow through Porous Media

Except near a high-gas-flow-rate well, fluid flow in a reservoir is usually laminar. Extensive analysis of this type of flow in porous media has been presented in several classic works in the petroleum literature. Coupled with knowledge of physical properties of fluids and rock, fluid flow theory forms the principal basis for the study of fluid flow in porous media. For in-depth information, please refer to Matthews and Russell (1967) and Earlougher (1977).

15-2.2 Flow of Oil (Constant Compressibility Liquid)

If the rock and fluid properties are assumed to be constant, the following partial differential equation describes the flow of an incompressible fluid:

$$\frac{\partial^2 p}{\partial r^2} + \frac{1}{r}\frac{\partial p}{\partial r} = \frac{\phi \mu c_t}{k}\frac{\partial p}{\partial t} \tag{15-1}$$

Dimensionless expressions for pressure, time and radius (in oilfield units) are used to generalize Equation 15-1:

$$p_D = \frac{kh\Delta p}{14.12qB\mu} \tag{15-2}$$

$$t_D = \frac{0.000264kt}{\phi \mu c_t r_w^2} \tag{15-3}$$

$$r_D = \frac{r}{r_w} \tag{15-4}$$

In Equations 15-1 through 15-4, the subscript D refers to dimensionless quantities. Other variables are p, pressure, t, time, r, distance, ϕ, porosity, μ, viscosity, c_t, total compressibility, k, permeability, and r_w, the well radius.

Substituting Equations 15-2 to 15-4 into Equation 15-1 produces the dimensionless flow equation:

$$\frac{\partial^2 p_D}{\partial r_D^2} + \frac{1}{r_D}\frac{\partial p_D}{\partial r_D} = \frac{\partial p_D}{\partial t_D} \tag{15-5}$$

Equation 15-5 is used to construct the general solutions of fluid flow through porous media, which are often referred to as type curves.

The governing partial differential equation (Equation 15-5) may be solved for either constant rate (an imposed constant rate at the surface) or constant bottomhole pressure. The constant rate condition is useful in analyzing pressure transient tests; the constant pressure solution has been used to predict the production rate.

15-2.3 Steady-State Solutions

The steady-state solutions assume that the reservoir has constant pressure at both the inner and outer boundaries. Thus, reservoir pressure at each point remains constant and fluid is flowing at a constant rate across the reservoir. Although a steady-state solution may not represent well performance accurately, in many cases it gives an acceptable measure of the well potential. The steady-state solution also gives a fairly reliable measure for comparing production under various well-completion options. Under steady-state conditions, the production rate for a vertical well located in a homogeneous and isotropic reservoir is given by the following equation:

$$q = \frac{kh(p_e - p_{wf})}{141.2B\mu\left(\ln\frac{r_e}{r_w} + s\right)} \qquad (15\text{-}6)$$

15-2.4 Flow of Gas

Although many equations of state are available for gases, the one that is based on the law of corresponding states has achieved wide acceptance in petroleum engineering because it can be readily applied to multi-component gases:

$$pV = ZnRT \qquad (15\text{-}7)$$

Because the compressibility and viscosity of gas are strong functions of pressure, Equation 15-1 has to be linearized before the dimensionless solution may be used to describe gas flow in a porous medium. If the reservoir pressure is relatively low, p^2 may be used to linearize the governing partial equation. If the pressure is high, the use of p in the governing equation coupled with the average viscosity-compressibility factor product would be an acceptable approach to linearize Equation

15-1. In the middle range, neither approximation is accurate.

A general approach to linearize Equation 15-1 for gas flow is to use the transformation suggested by Al-Hussainy *et al.* (1966):

$$m(p) = \int\limits_0^p \frac{2p}{\mu Z}\, dp \qquad (15\text{-}8)$$

Thus, the governing partial differential equation of gas flow in a porous medium becomes

$$\frac{\partial m^2(p)}{\partial r^2} + \frac{1}{\rho}\frac{\partial m(p)}{\partial r} = \frac{\phi m_i c_i}{k}\frac{\partial m(p)}{\partial t} \qquad (15\text{-}9)$$

The function $m(p)$ is usually called the *gas pseudopressure*.

The similarity between Equations 15-1 (for liquid flow) and 15-9 (for gas flow) indicates that the solution for Equation 15-5 (dimensionless form) can be used as a solution for the pseudopressure distribution in a gas reservoir. Furthermore, this implies that all techniques developed for analyzing and predicting the transient behavior of an oil well can be applied to a gas well if $m(p)$ is substituted for pressure.

Equation 15-9 may be transformed into a dimensionless form using dimensionless pressure and time expressions similar to the ones defined in the liquid case. Because of the pressure dependency of gas properties, the dimensionless pressure and time are defined as follows:

$$p_D = \frac{kh\Delta m(p)}{1424qT} \qquad (15\text{-}10)$$

$$t_D = \frac{0.000264kt}{\phi\mu_i c_i r_w^2} \qquad (15\text{-}11)$$

Substituting Equations 15-10 and 15-11 into Equation 15-9 yields the dimensionless form of the fluid flow equation given earlier as Equation 15-5. Thus, regardless of the type of fluid, the same general equation may be used to produce the pressure behavior with time. The effect of the type of fluid would appear through the use of the dimensionless definition of pressure and time.

15-2.5 Skin Effect

As a result of drilling and completion practices, the formation permeability near the wellbore is usually reduced. Drilling fluid invasion of the formation, dispersion of clay, presence of a mudcake, and cement tend to reduce the formation permeability around the wellbore. A similar effect can be produced by a decrease in the area of flow exposed to the wellbore. Therefore, partial well penetration, limited perforation, or plugging of perforations would also give the impression of a damaged formation. Conversely, an inclined well or inclined formation increases the area of flow near the wellbore, giving the impression of a stimulated well.

The zone of reduced (or higher) formation permeability has been called a "skin," and the resulting effect on well performance is called "skin factor." Skin factor can be used as a relative index to determine the efficiency of drilling and completion practices. It is positive for a damaged well, negative for a stimulated well, and zero for an unchanged well (Figure 15-1). Acidized wells usually show a negative skin. Hydraulically fractured wells show negative values of skin factor that may be as low as −7.

Hawkins (1956) derived the following expression relating the skin factor to wellbore radius, radius of altered permeability, and both reservoir and altered area permeabilities:

$$s = \left(\frac{k}{k_s} - 1\right) \ln\left(\frac{r_s}{r_w}\right) \qquad (15\text{-}12)$$

This expression indicates that if the area around the wellbore has a lower permeability than the original reservoir permeability—that is, a damaged well—the skin factor is greater than zero (positive value). If the permeability around the wellbore is higher than the reservoir permeability, meaning that the well is stimulated, the skin factor is lower than zero (negative value). A skin factor of zero indicates no change in permeability around the wellbore.

The effective wellbore radius is related to the wellbore radius by:

$$r_w' = r_w e^{-s} \qquad (15\text{-}13)$$

The effective wellbore radius is a term that was developed to describe the radius of an equivalent well with a skin factor of zero. Equation 15-13 suggests that if s is positive, the effective wellbore radius, r_w', is smaller than r_w. Thus, the damaged well under consideration is equivalent to a well with zero skin but with a smaller radius. Therefore, both real and equivalent wells would have the same productivity under the same pressure drop.

An equation similar to Equation 15-13 has been developed for fractured wells. In this case, an equivalent skin factor and wellbore radius may be related to the length of a vertical hydraulic fracture with infinite conductivity through

$$L_f = 2r_w e^{-2} \qquad (15\text{-}14)$$

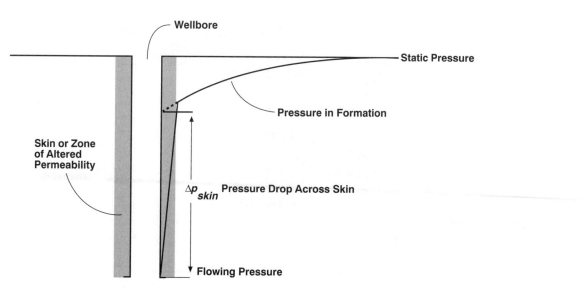

Figure 15-1   Pressure distribution in a reservoir with a skin

If r_w is equal to 3 in. and the half-length of an infinitely conductive fracture, L_f, is 75 ft, the equivalent well flowing under radial conditions would have a negative skin factor equal to 5.

As mentioned earlier, the skin factor acts as a pressure sink/source. This extra pressure drop is

$$\Delta p_{skin} = \left(\frac{qB\mu}{2\pi k h}\right)s \qquad (15\text{-}15)$$

In field units, Equation 15-15 is simply

$$\Delta p_{skin} = \left(\frac{141.2qB\mu}{k h}\right)s \qquad (15\text{-}16)$$

If the skin is positive, the flowing pressure p_{wf} will be lower than that of an undamaged well by the amount of Δp_{skin}.

15-2.6 Perforation and Skin Factor

Because most wells are cemented, cased, and perforated, the effect of perforation on the skin factor is extremely important. Perforation is arguably the process contributing most to the value of the skin factor. Figure 15-2 gives an idealized schematic of the perforation system geometry. This schematic readily indicates that the convergence of stream lines towards the perforation could increase the value of the skin factor. On the other hand, the penetration in the formation tends to give the opposite effect (stimulation effect). Which of these two effects dominates the final outcome depends on several factors that will be discussed briefly below.

Perforations, usually created using shaped charges, may pass through the steps illustrated in Figure 15-3 (Bell *et al.*, 1996). The perforation process may be underbalance, overbalance, or extreme overbalance. Each process has its own merits and applications. Either underbalance or overbalance perforations should eventually yield a perforation that may be schematically represented by Figure 15-4. The degree of perforation tunnel cleanliness may depend on optimization of both the degree of underbalance and shaped charges.

The productivity of a perforated wellbore depends on the length and diameter of the perforations, degree of damage around the perforation, perforation density (per

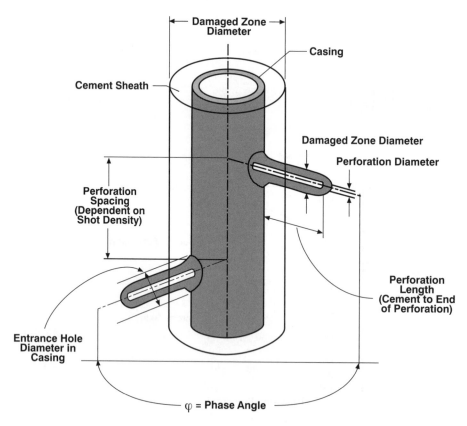

Figure 15-2 Perforation system geometry (after Bell *et al.*, 1995)

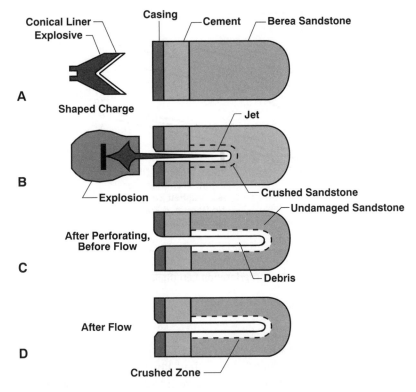

Figure 15-3   Events in perforation cleanup (after Bell *et al.*, 1996)

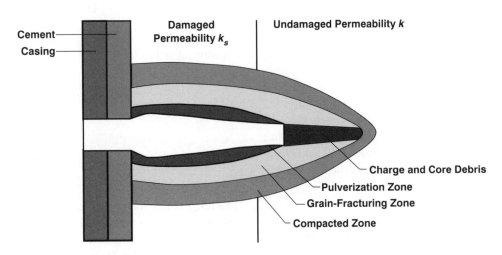

Figure 15-4   Schematic of sources of damage around wellbore and perforation

foot), and phase angle. Figure 15-5 (Tariq, 1987) summarizes the effects of phase angle, number of perforations, and perforation length on well productivity. The figure indicates that for a 90° phasing and perforations longer than 10 in., even 4 shots/ft will be superior to open hole (yielding a negative skin). Tariq (1987) also indicated that unless the perforation tunnel is very narrow, the effect of the perforation diameter is fairly small. Klotz *et al.* (1974) suggested that it is highly desirable to have the perforation extending beyond the damaged zone.

Extreme overbalance perforation, attempted recently, adds to the perforation process the benefit of creating a short fracture. Although the basic idea behind extreme

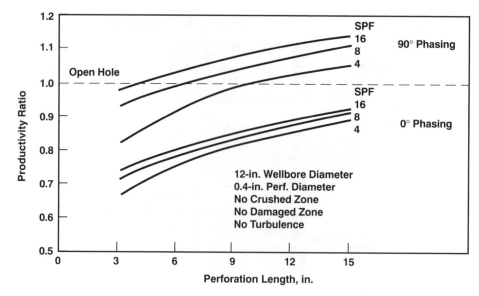

Figure 15-5   Effect of shot density and perforation length on productivity ratio (after Tariq, 1987)

overbalance perforation has been around for a few years (Branagan *et al.*, 1988), the technique in its present form is rather new. The perforation process is immediately followed by a surge of hydraulic pressure sufficient to create a fracture. This pressure is usually provided through the use of highly pressurized nitrogen.

In extreme overbalance perforating, the tubing-conveyed setup is used and liquid, gas, or a mixture of the two is used as the fracturing fluid. An additional source of gas provides the pressure gradient necessary to fracture the formation. Handren *et al.* (1993) have provided a detailed description of the technique and the setup used in this technique. Mason *et al.* (1995) described new equipment especially designed for extreme overbalance perforation, and Couet *et al.*, (1996) introduced a fracturing model for matching the pressure history during extreme overbalance perforation.

15-3 HYDRAULICALLY FRACTURED WELLS

Many reservoirs must be hydraulically fractured to become economically productive. Hydraulic fracturing involves injecting a large volume of proppant-laden fluid at a pressure sufficiently high to fracture the formation. After the fracturing fluid leaks off into the formation, the remaining proppant keeps the fracture open. Although a hydraulic fracture is narrow (a fraction of an inch in most cases), the presence of this high-permeability channel significantly enhances the productivity of the well. The presence of a fracture

alters the flow regime inside the formation as the fluid flows into the fracture and then through the fracture into the wellbore, with very little or no fluid flowing directly from the formation into the wellbore. The presence of a hydraulic fracture adds another dimension to the fluid flow in porous media and to well test design and analysis.

The presence of a fracture dictates a change and an addition of definitions:

$$t_D = \frac{0.000264 kt}{\phi \mu c_t L_f^2} \qquad (15\text{-}17)$$

$$C_{fD} = \frac{k_f w}{k L_f} \qquad (15\text{-}18)$$

The term C_{fD} is the dimensionless fracture conductivity. The dimensionless pressure is identical to the definition in Equation 15-2.

Three types of fractures have been presented in the literature: infinite conductivity, uniform flux, and finite conductivity. These fractures are discussed below.

15-3.1 Infinite-Conductivity Fracture

In this case, the fracture permeability is significantly higher than the formation permeability, causing the pressure drop inside the fracture to be negligible compared to the pressure drop in the formation. Practically, this situation is achieved when dimensionless fracture conductivity is higher than 100.

The solution for a well intersecting an infinitely conductive fracture was presented first by Gringarten and Ramey (1975). In their work, Gringarten and Ramey (1975) and Gringarten *et al.* (1975) discussed the behavior of a hydraulically fractured reservoir and its effect on well test analysis.

Because of the presence of the infinite-conductivity fracture, fluid in the porous medium immediately surrounding the fracture will start to flow into the fracture as soon as the well is put on production. Since the stream lines are perpendicular to the fracture face, the flow pattern during this early time period is linear. The dimensionless pressure is given by the following equation:

$$p_D = \sqrt{\pi t_D} \qquad (15\text{-}19)$$

As more volume of the reservoir starts contributing to production, the flow pattern becomes elliptical. After a long time period, this elliptical flow may be approximated by a radial flow pattern. This period is usually referred to as pseudoradial flow:

$$p_D = \frac{1}{2}\ln(t_D) + 1.10 \qquad (15\text{-}20)$$

Figure 15-6 is the log-log type curve of a well intersecting an infinite-conductivity fracture. The early-time period is a line with a slope of one-half, indicating the presence of linear flow in the formation. If the distance to the reservoir boundary is significantly larger than the fracture half-length, the pseudoradial flow will appear at approximately the dimensionless time of three, which means that after that time, a semi-log plot of pressure vs. logarithm of time (or logarithm of Horner time for a buildup test) would yield a straight line. The slope of the straight line may be used to calculate the formation permeability and an equivalent skin factor. However, remember that a dimensionless time of three could correspond to a very long time, depending on the fracture length, the formation, and the fluid properties. This point will be illustrated by an example when finite-conductivity fractures are discussed.

The fracture length may be calculated from the skin factor using Equation 15-14.

15-3.2 Uniform-Flux Fracture

In this case, it is assumed that the flow rate from the formation into the fracture is uniformly distributed across the fracture face. This solution was initially used to approximate fluid flow in naturally fractured formations. The use of this model for that purpose has significantly declined in recent years, in favor of the dual-porosity models. However, we have found that many of the acidized, naturally fractured reservoirs tend to follow this model, especially reservoirs located in the Middle East.

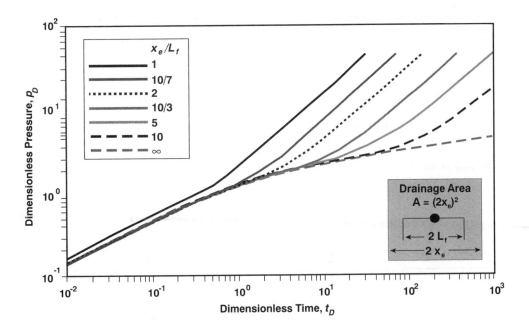

Figure 15-6   Type curve for a well intercepting an infinite-conductivity fracture (after Gringarten and Ramey, 1975)

As in the case of the infinite-conductivity fracture, the fracture length may be calculated from the skin factor. However, the equation is slightly different:

$$L_f = r_w e^{-s+1} \qquad (15\text{-}21)$$

15-3.3 Finite-Conductivity Fracture

It is more realistic to assume that a hydraulic fracture will have a finite conductivity. In this case, the pressure drop inside the fracture is not negligible when compared to the total pressure drop in the system. The flow regime in a reservoir that has a finite-conductivity fracture is significantly more complex than in the case of infinite-conductivity fractures.

The flow regimes experienced during producing or testing a finite-conductivity fracture are represented by Figure 15-7 (Cinco and Samaniego, 1982). At very early time, fluid inside the fracture expands and starts flowing towards the wellbore, forming an early linear flow inside the fracture. This period is controlled by both the conductivity and diffusivity of the fracture. This flow period will appear as a one-half slope, straight line on the type curve for a fractured well. This flow period is at very early time, and it would probably not be recorded or it would be masked by the presence of wellbore storage. Therefore, although this flow period may be of technical interest, it does not have much significance from a practical point of view.

When the pressure drop inside the fracture becomes significant, fluid starts flowing from the reservoir into the fracture. Initially, only a limited area surrounding the part of the fracture near the wellbore will contribute to production. The flow is linear in both the formation and fracture; therefore the flow regime is termed "bilinear flow." This period has several interesting characteristics:

- It appears as a one-quarter slope, straight line on the type curve.

- The tip of the fracture has not affected the pressure behavior yet. It is not possible to determine the fracture length from tests that have not been run long enough to detect the end of quarter slope. The equation describing the well behavior during this period is

$$p_D = \frac{\pi t_D^{0.25}}{\sqrt{2C_{fD}}\,\Gamma(5/4)} \qquad (15\text{-}22)$$

- In dimensional form, Equation 15-22 is independent of the fracture half-length.

- Cinco *et al.* (1978) and Cinco and Samaniego (1982) have shown that the start and the duration of the

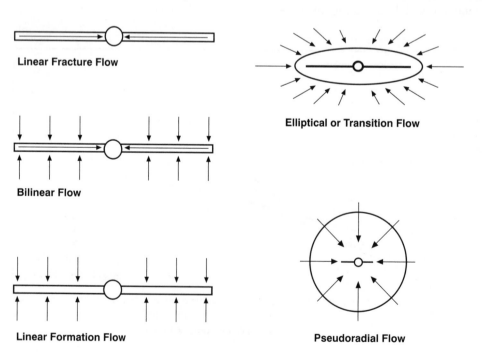

Linear Fracture Flow

Bilinear Flow

Linear Formation Flow

Elliptical or Transition Flow

Pseudoradial Flow

Figure 15-7   Fracture flow regimes (after Cinco *et al.*, 1978)

bilinear flow period depend on the value of the dimensionless fracture conductivity. Figure 15-8 provides their findings, which indicate that bilinear flow will not take place if the conductivity is excessively high or excessively low.

- Because of the essentially straight-line nature of the bilinear flow period, type-curve matching of data that are still in this flow period is very difficult, and analysis of such data will not yield a unique answer.

As the pressure drop in the formation becomes dominant, linear flow in the formation dominates well response. This flow period may or may not take place, depending on the duration of the bilinear flow. In many cases, flow regimes pass from bilinear flow directly to an elliptical or transitional period.

The last flow period that may appear is the pseudoradial flow period. It may be assumed that this flow period appears at a dimensionless time that varies from two to three, depending on the dimensionless fracture conductivity.

Log-log type curves for a well intercepting a finite conductivity fracture are provided in Figure 15-9. This figure shows the great similarity between the bilinear flow curves for different dimensionless conductivities. It is important to recognize that the pseudoradial flow may

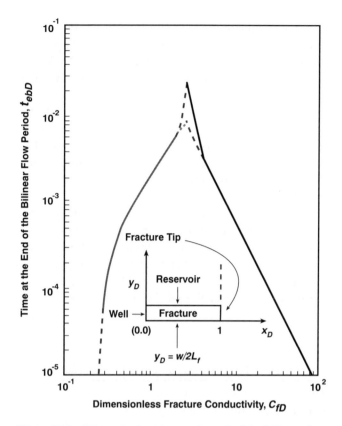

Figure 15-8   Dimensionless time at the end of the bilinear flow period; constant pressure production (after Cinco *et al.*, 1978)

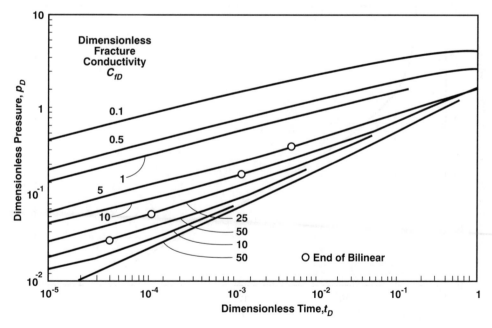

Figure 15-9   Constant rate log-log type curves for finite-conductivity fractures (after Cinco *et al.*, 1978)

take a long time to start. To illustrate this point, we will calculate the time necessary for a well to reach pseudo-radial flow (at a dimensionless time of three). Table 15-1 lists the hypothetical but typical properties of a fractured gas reservoir.

Substituting the parameters from Table 15-1 into the definition of dimensionless time, the time necessary to reach pseudoradial flow is 3712 hours (155 days). This value indicates that, up to that time, the well performance may not be approximated with a radial flow model and a negative skin factor.

From a testing point of view, the calculations above indicate that a test should be run for 1550 days to produce one cycle on the semi-log straight line. Since such a test duration is obviously impractical, semi-log analyses of tests run on fractured wells should not be used unless the permeability is high and the fracture length is very short.

Table 15-1   Typical gas reservoir properties

Formation permeability, md	0.2
Formation porosity, fraction	0.1
Gas viscosity under reservoir condition, cp	0.02
Total system compressibility, psi^{-1}	10^{-4}
Fracture half-length, ft	700

To calculate the effective skin factor, the following relationship between skin and fracture half-length may be used:

$$L_f = n r_w e^{-s} \qquad (15\text{-}23)$$

The parameter n is a function of the dimensionless fracture conductivity and may be obtained from Figure 15-10.

Many analysts use the skin factor for a hydraulically fractured well to judge the degree of success or failure of the fracturing treatment. Although such an approach is not necessarily wrong, it has some limitations and needs to be correctly understood to avoid erroneous application of the concept. The use of skin factor to approximate a fractured well behavior is only valid after the start of the pseudoradial flow regime. In the case of tight gas formations, it may take months or even years for the reservoir to reach this pseudoradial flow. Therefore early-time prediction of fractured-well behavior using the equivalent skin factor could be in error.

15-3.4 Fractures with Changing Conductivity

During the creation of a hydraulic fracture, the conductivity may vary from one place in the fracture to another. Several works have studied this effect (Bennet *et al.,*

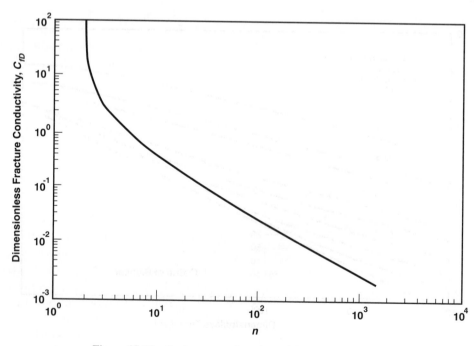

Figure 15-10   Factor *n* as a function of fracture conductivity

1983; Soliman, 1986a; Soliman *et al.*, 1987). Using both analytical and numerical simulators, Soliman (1986b) showed the choking effect of having low fracture conductivity near the wellbore as well as the effect of using a tail-in of better proppant near the end of the fracturing treatment. It was concluded that the effect of fracture conductivity distribution is more complex than was thought at the time. Soliman (1986b) has also calculated the optimum conductivity distribution inside the fracture, Figure 15-11. This optimum conductivity distribution indicates that if conductivity within the fracture follows a certain distribution pattern, a fracture may behave as if it had uniform conductivity equal to the one at the wellbore. Poulsen and Soliman (1987) presented a procedure to convert the curves in Figure 15-11 into an optimum proppant distribution, thereby producing an optimum fracture design. This procedure could lead to the use of less proppant and to a lower chance of sand-out. At the same time, it does not compromise the performance of the designed fracture.

15-3.5 Effect of Low Fracture Conductivity

As the fracture conductivity declines, the fracture becomes less efficient. Prats has studied this problem analytically, producing Figure 15-12. Before we discuss this figure, we need to define the conductivity term used by Prats:

$$C'_{fD} = \frac{\pi k(2L_f)}{4k_f w} \tag{15-24}$$

and thus,

$$C'_{fD} = \frac{\pi}{2}/C_{fD} \tag{15-25}$$

Prats found that a plot of dimensionless effective wellbore radius vs. C'_{fD} will become linear at a C'_{fD} value of 9.8. This means that increasing the fracture length would not improve fracture performance. However, increasing the fracture conductivity would improve the fracture performance. If a fracture has a conductivity lower than the value indicated by Prats, the part of the fracture that reduces the dimensionless conductivity below the Prats threshold would be ineffective. This statement explains why a long fracture would not be advised in high-permeability formations. On the other hand, it is advisable to create a long fracture in tight formations. Prats also investigated the creation of an optimum fracture assuming a fixed volume. He concluded that for a specific fracture volume, the optimum fracture would have a dimensionless conductivity of about 1.6. We must remember that Prats' study assumed steady-state conditions, and although the conclusions reached by Prats are generally valid, the exact figures may be somewhat different under unsteady-state conditions. Cinco *et al.* (1987) and Azari *et al.* (1992) have expanded on this area of research. This point is also discussed in Chapter 17.

15-3.6 Steady-State Production Increase Resulting from Fracturing

The first step in the optimization of the design of a hydraulic fracture is to predict well performance with and without the proposed fracture. This step may be done either under steady- or unsteady-state conditions. As we mentioned earlier in this chapter, steady-state solutions provide a quick and fairly reliable indication of how a well will perform. Several studies have been presented in the literature to predict the steady-state production increase as a result of fracturing. The following paragraphs are a brief discussion of these methods.

McGuire and Sikora (1960) presented the first study of production increase resulting from hydraulic fracturing. Because of the nature of their experimental work, their model represented production during pseudosteady state; therefore, their values were higher than other strictly steady-state methods. Figure 15-13 represents their results.

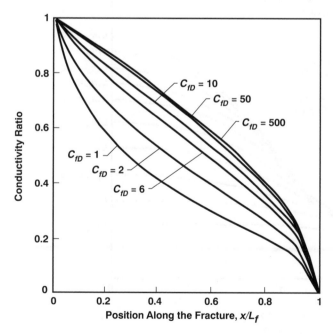

Figure 15-11 Optimum conductivity distribution inside a hydraulic fracture (from Soliman, 1987)

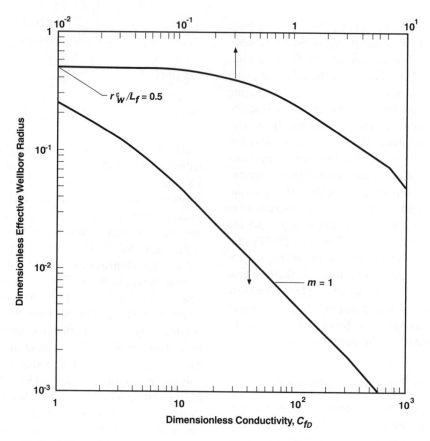

Figure 15-12   Effect of fracture conductivity on effective wellbore radius (after Prats, 1965)

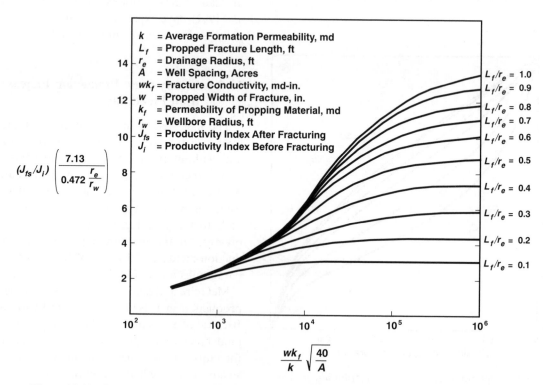

Figure 15-13   Increase in productivity from fracturing (from McGuire and Sikora, 1960)

Prats (1961) presented production increase curves based on an analytical solution. His solution considered a fully penetrating hydraulic fracture.

Tinsley *et al.* (1969) introduced the first production increase model that considered a partially penetrating model. The study was an experimental one.

Mao (1977) presented production increase curves based on an experimental study. Similar to Prats' solution, Mao only considered a fully penetrating hydraulic fracture.

Soliman (1983) discovered that the analysis performed by Tinsley *et al.* to convert the raw experimental data into a usable form was in error. Soliman did the analysis again and converted the raw data into a new dimensionless representation of the production increase curves. The new solution was verified by comparing it to Prats' and Mao's results for full-fracture penetration. Figure 15-14 gives an example of the production increase curve for fracture height to formation height equal to unity.

The steady-state production-increase curve may be used for the initial design of a particular fracture or to gain insight into the process of planning fracture length and conductivity. Figures 15-13 and 15-14 show that if the formation is tight, the fracture length should be maximized within operational, rock mechanics, and economic constraints. If the formation has a higher permeability, increasing conductivity may be an option.

For a certain fracture volume, it may be possible to design an optimum fracture length and conductivity as discussed in the previous section.

15-4 FRACTURING HIGH-PERMEABILITY FORMATIONS

Hydraulic fracturing has been recently applied in high-permeability (generally >10 md) formations for both well productivity and sand control. The reservoir engineering and fracturing aspects and justification for fracturing high-permeability formations have been extensively studied (Roodhart *et al.*, 1994; Meese *et al.*, 1994; Hunt *et al.*, 1994; Mathur *et al.*, 1995). High-permeability fracturing is treated extensively in Chapter 19.

Hydraulic fracturing is usually thought of as a technique to increase productivity or to establish production in low-permeability reservoirs. However, benefits can be realized by fracturing highly permeable formations that have formation damage and/or sand production tendencies. A well that has a reduced permeability several feet away from the wellbore can be made more productive by fracturing through this damaged zone to establish a better link with the undamaged region of the reservoir. Reservoir fluids are therefore provided an unrestricted pathway from the undamaged reservoir to the wellbore.

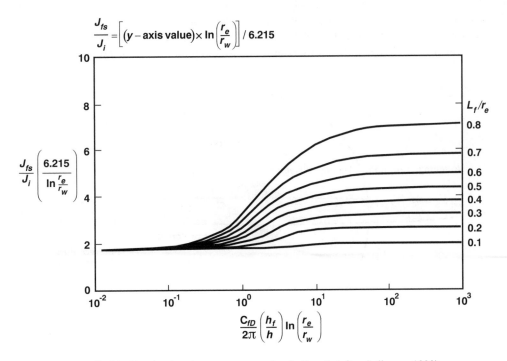

Figure 15-14 Production increase curves for $h_f/h = 1$ (after Soliman, 1983)

Encountering a very deep damage has been documented in the industry and has been discussed at length by Peterson and Holditch (1985). The conductivity within the fracture can be maximized so that the pressure drop along the fracture itself can be held to a minimum. In the case of a well with sand-producing tendencies, a hydraulic fracture decreases the pressure drop necessary to produce the well at a given rate. In addition, the hydraulic fracture changes the flow regime around the well such that sand production is minimized or even eliminated, allowing the well to be produced at a rate higher than the critical sand-producing rate for an unfractured well.

Operationally, fracturing high-permeability formations is different from fracturing low-permeability formations because of the expected high leakoff rate, which influences the fracturing pressure. In addition, because of the desired high fracture conductivity, the concept of tip screenout is applied. In tip screenout (Smith *et al.*, 1987), the fracture is designed in such a way that by the time the fracture reaches the required length, the leading pad volume has leaked off into the formation. After the pad volume has leaked off, the presence of the proppant-laden fluid at the leading edge of the fracture initiates the screenout process (Figure 15-15). Continued injection of the proppant-laden fluid causes the fracture to widen or balloon, reaching a considerably enlarged width and high proppant concentration (a packed fracture). The tip-screenout fracturing technique has been successfully applied in the field. Martins and Stewart (1992) reported that tip-screenout succeeded in more than doubling productivity of treated wells in the Ravenspurn field. Other authors (Meese *et*

al., 1994; Abass *et al.*, 1993; Dusterhoft and Chapman, 1994; Liu and Civan, 1994) have reported equally impressive results.

Many wells that are good candidates for this treatment are wells that are drilled in poorly consolidated formations. The failure behavior of such formations is different from that of competent formations. Under radial flow conditions with the anticipated high-pressure drop near the wellbore, the drag force on the sand could exceed the cohesive strength of the rock, causing formation failure and resulting in sand production (Chapter 6). This well instability causes severe operational problems and a reduction of formation permeability around the wellbore. A fracturing treatment reduces the sanding tendency of a poorly consolidated formation by reducing the pressure gradient around the wellbore and by acting as a gravel pack that supports and holds the formation sand in place.

Migration of fines toward the wellbore causes a reduction in permeability in the region around the wellbore. Drilling, completion, and production processes may also cause permeability reduction in the near-wellbore region. The resulting depth of damage may extend from as little as a few inches to as much as several feet into the formation (Martins and Stewart, 1992). During the production processes, fines migration and paraffin deposition are expected to yield a greater depth and degree of damage than the drilling and completion processes. Such deep damage cannot be removed by acidizing, leaving fracturing as the only alternative.

Fracturing high-permeability formations requires the understanding of the impact of certain parameters that

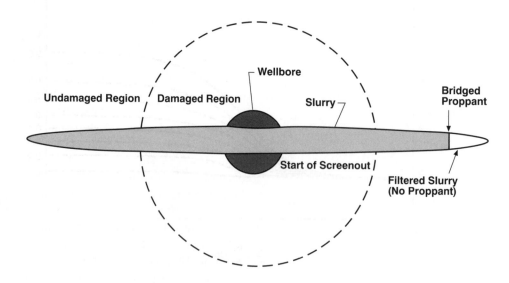

Figure 15-15   Schematic view of a tip screenout

are not usually considered when fracturing low-permeability formations. The first such parameter is the effect of the damaged region on fractured-system productivity. The damaged region can be visualized as an inner, circular region of reduced permeability extending from the wellbore to some radial distance; outside the inner region is undamaged reservoir permeability. This system is generally referred to as a radial composite system. Another parameter is the presence of a damaged fracture face with reduced permeability caused by the expected high leakoff of fracturing fluid. A third important parameter is the effect of the fracture on the pressure profile around the wellbore.

Because of the high formation permeability, it is expected that a relatively long fracture (hundreds of feet) is neither feasible nor effective. As discussed earlier, Prats (1961) has suggested that below a certain threshold dimensionless conductivity, a fracture may be ineffective. In other words, only the fraction of the fracture that yields at least that minimum dimensionless conductivity may be effective. Thus, fracturing high-permeability formations usually involves the creation of short fractures. To achieve this threshold dimensionless conductivity, high fracture conductivity is sought. Hunt *et al.* (1994) presented a detailed study of the effect of the various parameters on the performance of high-permeability fracturing.

15-4.1 Candidate Selection for Fracturing High-Permeability Formations

The discussion above indicates that the candidates for high-permeability fracturing may be loosely divided into two groups based on whether the treatment's primary goals are production enhancement or sand control.

For production enhancement, the treatment candidates are

- Moderate- or high-permeability formations with deep damage. In such formations, matrix stimulation (acidizing) would be ineffective.

- Formations that do not respond favorably to acid treatments

- Cases where a large treatment is not advised because of height or length restrictions, as would be the case in the presence of an underlying aquifer or nearby injection wells.

- Wells that have lost productivity because of pore collapse around the wellbore. A fracturing treatment would bypass the damaged zone and subsequently reduce the pressure drop necessary to produce the formation. This treatment would reduce the chance for further pore collapse.

For sand control purposes, the treatment candidates are

- Poorly consolidated formations for which a fracturing treatment may act as a form of gravel-pack without the associated likely positive skin

- Formations with low cohesive force for which it is necessary to reduce the pressure gradient associated with the required flow rate

15-4.2 Effect of Length and Conductivity on High-Permeability Fracturing

A large improvement in production occurs when a fracture propagates beyond the damaged region. This conclusion was first reached by Hunt *et al.* (1994) and was later confirmed by Chen and Raghavan (1994).

Fractures effectively extend the wellbore into the formation, and this extension is critical in low-permeability formations. In high-permeability formations, the fracture is a conduit between the well and the undamaged portion of the formation. It is important to generate enough length to penetrate into the undamaged portion of the formation. Because of the expected high leakoff, this goal may be difficult to achieve. Higher conductivity is, of course, the obvious target in high-permeability fracturing.

15-4.3 Fracture Damage

Numerically modeling the effect of fracture face damage caused by fracturing fluid leakoff, Holditch (1979) concluded that unless the damage is exceedingly high, the effect of damage on fracture performance is fairly small. Damage to the formation around the fracture may be expected when a high-permeability formation is fractured because fluid leakoff during treatment is significant. However, Hunt *et al.* (1994) suggested that the conclusion reached by Holditch (1979) for low-permeability formations is still valid for fracturing high-permeability formations. In addition, deep damage extending from the fracture may be minimized by properly designing the fracture treatment to minimize excessive fluid loss. For all practical purposes, a properly designed and conducted treatment should result in no significant productivity impairment from fracture face damage.

Mathur *et al.* (1995) developed an approximate equation to describe the combined damage resulting from both the damage around fracture face and the radial damage around the wellbore:

$$s_d = \frac{\pi}{2}\left[\frac{b_2 k_r}{b_1 k_3 + (L_\phi - b_1)k_2} + \frac{(b_1 - b_2)k_r}{b_1 k_1 + (L_f - b_1)k_r} + \frac{b_1}{L_f}\right]$$
(15-26)

The definitions of the parameters in Equation 15-26 are illustrated in Figure 15-16. Similar to the classical development of skin equations, Equation 15-26 assumes a steady-state flow condition in the damaged area, which has no storage capacity. Using this equation, Mathur *et*

al. (1995) concluded that unless the damage is very large, a fractured well will always yield a negative skin factor. This conclusion is illustrated in Figure 15-17 relating skin factor to damaged-zone thickness and permeability.

15-4.4 Post-Fracture Well Test Analysis

As discussed in the previous subsection, an effective hydraulic fracture should extend beyond the damaged zone. The behavior of a fractured well in a composite system under well-testing mode is examined in this section. Figure 15-18 presents the pressure profile of a fracture extending 50 ft into the formation characterized in Table 15-2. Because of the relatively short fracture length combined with the high formation permeability, the flow

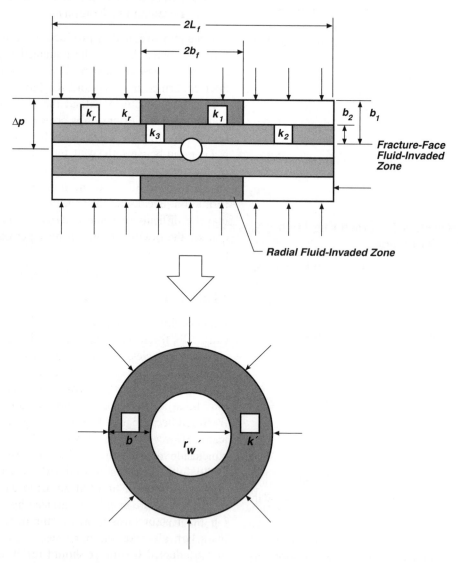

Figure 15-16   Equivalent radial skin of damaged wellbore and fracture face (after Mathur *et al.*,1995)

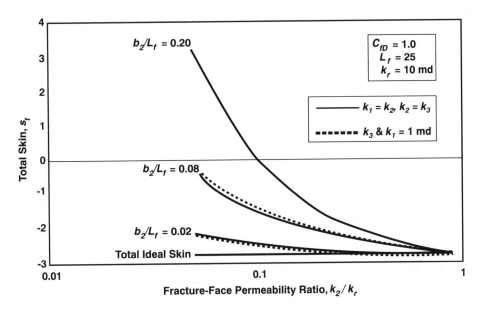

Figure 15-17 Variation of total skin with fracture-face permeability impairment ratio, k_2/k_r (from Mathur *et al.*, 1995)

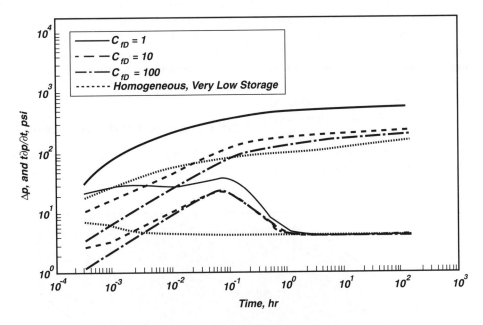

Figure 15-18 Simulated well test data: post-frac (after Hunt *et al.*, 1994)

regime reaches pseudoradial flow in a few hours. The pseudoradial flow appears as a horizontal line on the derivative plot. Thus, for the larger part of the test, the fractured system behavior closely resembles that of radial flow. At a low dimensionless-fracture conductivity ($C_{fD} = 1$), the pressure profile looks very much like the radially composite system with no fracture. However, the inner region appears to have higher permeability than the original value before fracturing. In other words, the effect of fracturing is reflected in an apparently less-damaged inner region. At higher dimensionless-fracture conductivities, the pressure profile approaches that of a homogeneous formation with the fractured inner region appearing as a skin factor.

Because of the complexity of the system and the similarity of its pressure behavior to that of a radial system, it

Table 15-2   Values used in simulator

p_i	5000 psi
ϕ	20 %
r_s	20 ft
r_w	0.4 ft
h	100 ft
q	500 STB/d
k	100 md
k_s	5 md
L_f	50 ft
C_{fD}	1, 10, and 100

Table 15-3   Analysis results from simulated data

Case	k, md	k_s, md	r_s, ft	s
$C_{fD} = 1$	105.8	22.6	19	−0.72
$C_{fD} = 10$	97.5	16.9	42	−3.26
$C_{fD} = 100$	99.6	15.8	41	−3.61

is possible to misdiagnose this fractured, composite, reservoir system. This difficulty explains why many analysts have chosen to judge the effectiveness of a high-permeability treatment based on a reduction in the skin factor, rather than in terms of fracture length and conductivity. Achieving an accurate description of the system in terms of fracture length and conductivity requires some knowledge of the inner-region permeability and depth, making a prefracture test necessary.

To illustrate the problem, the simulated post-fracture well-test data were matched against an unstimulated, radially composite system. The match was achieved using a commercially available well-test analysis software. The analyses for C_{fD} of 1, 10, and 100 are summarized in Table 15-3. Because pseudoradial flow was reached in all three tests, it is not surprising that the

calculation of the outer-region permeability is fairly accurate. Table 15-3 clearly shows that prefracture testing is a highly recommended procedure for reaching a good description of fracture length and conductivity from a post-fracturing test.

15-5 HORIZONTAL WELLS

Reservoir engineering applications for horizontal wells have centered around productivity improvements in comparison to vertical wells. The steady-state analytical solution has been used extensively to develop the current understanding of well productivity. The conclusion is that horizontal well length is the primary factor in productivity enhancement.

15-5.1 Horizontal Well Performance

Figure 15-19 is a schematic of a horizontal well in a reservoir illustrating the important variables. The main

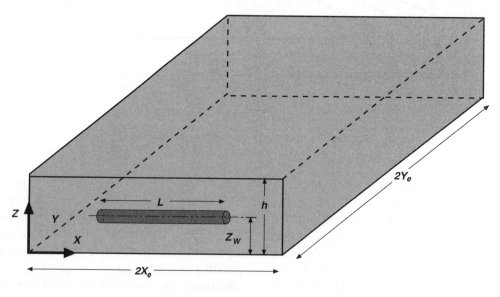

Figure 15-19   Horizontal well model

factors influencing well productivity are : the length of the well relative to the formation height, the vertical permeability, and eccentricity.

The productivity of a horizontal well having an infinitely conductive wellbore resembles that of a partially penetrating vertical fracture of infinite conductivity. The dimensionless well length, L_D, should be greater than four for an accurate comparison. In some cases, because of limited conductivity associated with fractured vertical wells, the performance of a horizontal well may be greater than that of a hydraulically fractured vertical well for equivalent wellbore and fracture lengths.

In homogeneous and isotropic reservoirs, low permeability and reduced reservoir thickness contribute to low productivity for conventional vertical wells. Drilling a horizontal well increases the length of the wellbore exposed to the formation, thereby improving the performance. This improvement becomes even more noticeable in a heterogeneous reservoir where a horizontal wellbore increases the chances of penetrating more favorable sections of the reservoir.

The steady-state productivity index of a horizontal well has been examined in the past by several investigators. Borisov (1964) studied the steady-state flow behavior of a horizontal well. That work was further developed by Giger *et al.* (1984). Until Giger's study, the investigations dealt with the qualitative prospects surrounding the performance of a horizontal wellbore. The equations were expressed in three-dimensional flow of the combined behavior of two-dimensional flow. The flow patterns were defined as radial flow along the vertical plane of the wellbore and linear flow in the horizontal plane through the thickness of the reservoir. These studies assumed a centered well in an infinite slab reservoir.

Joshi (1988) presented a form of the productivity equation that is considered now as the standard equation for horizontal well performance:

$$q_h = \frac{k_h h \Delta p / 141.2 \mu B}{\ln\left(\dfrac{a + \sqrt{a^2 - (L/2)^2}}{L/2}\right) + \dfrac{\beta h}{L} \ln\left(\dfrac{(\beta h/2)^2 - \beta^2 \delta^2}{\beta h r_w/2}\right)} \quad (15\text{-}27)$$

for $L > \beta h$, $\delta < h/2$, $L < 1.8\, r_{eh}$, where $\beta = \sqrt{k_h/k_v}$; $\delta = $ horizontal well eccentricity, ft; and

$$a = (L/2)\left[0.25 + \sqrt{0.25 + (2r_{eh}/L)^4}\right]^{1/2}$$

Equation 15-27 needs a slight modification as shown by Economides *et al.* (1991). Assuming that $\delta = 0$, the second logarithmic term in the denominator should be $\beta h/(\beta + 1)r_w$, instead of $\beta h/2r_w$, as Joshi's equation suggests. As shown by Economides *et al.* (1991), this is necessary in an anisotropic formation where the circular shape of a well would behave as if it were an ellipse.

Equation 15-27 has proven to be very useful in comparing the productivity of horizontal wells to that of vertical wells. The various parameters such as well length, permeability anisotropy, and formation thickness can be used to conduct sensitivity analyses on potential reservoirs for how much production improvement a horizontal wellbore will provide over that of a vertical well.

The influence of reservoir height on the horizontal-to-vertical well productivity ratio can be seen in Figure 15-20. This figure shows the change in productivity ratio as a function of horizontal well length and formation thickness. For a given horizontal well length, the productivity ratio increases as formation thickness decreases. The influence of k_v/k_h on productivity ratio and horizontal well length is illustrated in Figure 15-21. For a given horizontal well length, the productivity ratio increases as k_v increases. For a narrow formation, the effect of vertical-to-horizontal permeability anisotropy becomes less important. Under unsteady-state conditions, the effect of well length on productivity is illustrated in Figure 15-22.

The dimensionless parameters are defined below:

$$t_{DA} = \frac{0.000264 k_h t}{\phi \mu c_t \pi r_e^2} \quad (15\text{-}28)$$

$$q_D = \frac{141.2\, q B \mu}{k\, h \Delta p} \quad (15\text{-}29)$$

$$L_D = \frac{L}{2h}\sqrt{\frac{k_v}{k_h}} \quad (15\text{-}30)$$

Joshi (1991) has shown that under steady-state conditions, the effect of well eccentricity on horizontal well productivity is fairly small as long as the horizontal well is located $\pm h/4$ from the reservoir center. Under unsteady-state conditions, the same behavior should be expected.

Based on a numerical simulation study for modeling mud filtrate into the formation, Frick and Economides (1991) suggested that the damage distribution around a

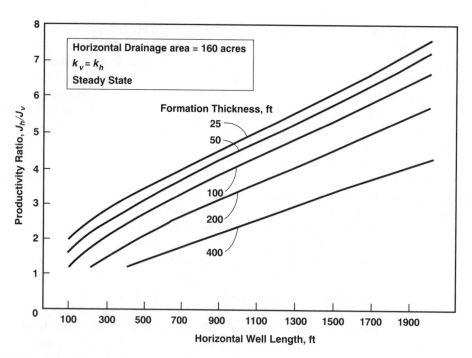

Figure 15-20   Productivity ratio of horizontal and vertical wells vs. well length for different reservoir thicknesses

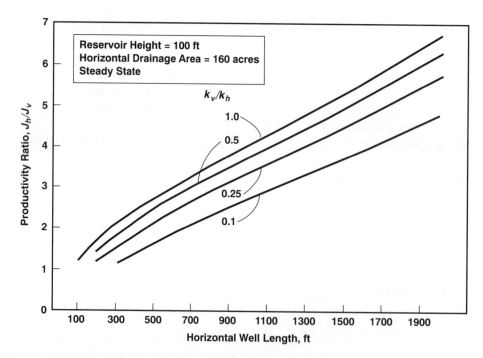

Figure 15-21   Effect of vertical permeability on productivity ratio of horizontal and vertical wells

horizontal wellbore would resemble the shapes shown in Figures 15-23 and 15-24. These two figures indicate that the shape of damage around a horizontal well is that of a truncated cone with the base of the cone at the heel of the well and the top of the truncated cone at the toe of the horizontal well. The cross section is elliptical, and its degree of ellipticity depends on the formation's degree of anisotropy. Equations for calculating skin factor for this damage configuration have been developed by Frick and Economides (1991).

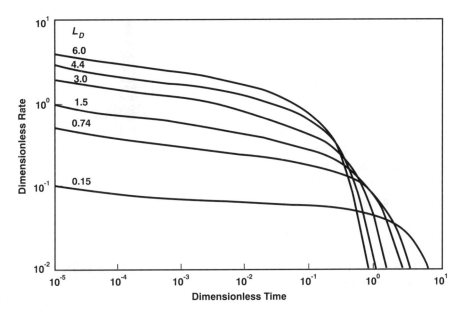

Figure 15-22 Effect of horizontal well length on transient rate (after Soliman *et al.*, 1996)

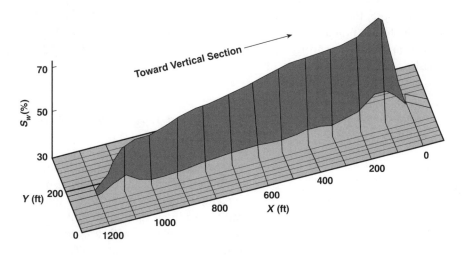

Figure 15-23 Simulated horizontal well damage profile (after Frick and Economides, 1991)

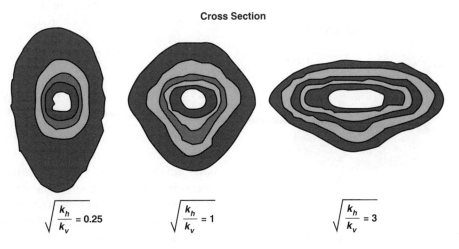

Figure 15-24 Simulated horizontal well damage profile cross section (after Frick and Economides, 1991)

15-5.2 Fractured Horizontal Wells

Depending on the well orientation with respect to the minimum horizontal stress and length of the perforated interval, either a transverse or longitudinal fracture may be created. If the horizontal well is drilled parallel to the minimum horizontal stress, the created fractures are expected to be perpendicular to the horizontal well; transverse fractures will be created (Figure 15-25). If the horizontal well is drilled perpendicular to the minimum horizontal stress, the created fracture will be longitudinal (Figure 15-26). These two cases illustrate the two limiting conditions; they are also the desired configurations.

However, if the well is at a different angle from the longitudinal (including the fully transverse), a complicated and undesirable situation may occur. In this situation, the perforated interval plays a crucial role. If the perforated interval is longer than four times the diameter of the well, an axial fracture is expected to be created at the wellbore. Away from the wellbore, the fracture reorients and becomes perpendicular to the minimum stress. The reorientation occurs as shear fractures tend to take the form of "steps" (Daneshy, 1973). The presence of

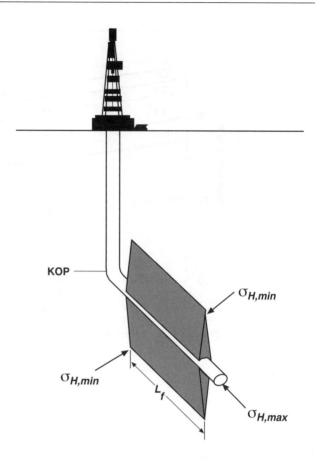

Figure 15-26   Longitudinal vertical fracture

"steps" not only contributes to abnormally high fracture propagation pressure but may also lead to a premature screenout. Thus, the perforated interval should be minimized. An interesting possibility is to "notch" the casing and cement sheath with an abrasive cutter, forcing the fracture to initiate transversely to the wellbore, limiting the near-well tortuosity.

The preferred geometry in higher-permeability reservoirs is longitudinally fractured horizontal wells; transverse fractures in a horizontal well are for a low-permeability formation.

15-5.2.1 Performance of Transverse Fractures

The main advantage of transverse fracturing is the capability to create multiple fractures intersecting the horizontal, thereby covering a large reservoir volume with fairly small fractures. Their main disadvantage from a reservoir engineering point of view is the convergence of fluid inside the fracture towards the wellbore, which creates an increased pressure drop that would not be encountered during fracturing vertical wells. Figure 15-27 is a

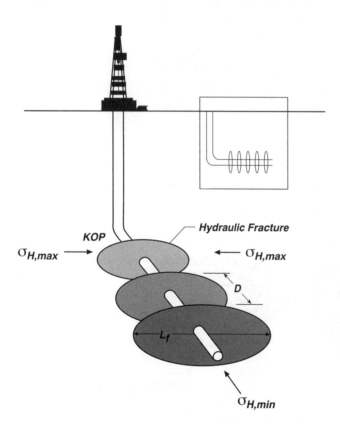

Figure 15-25   Transverse vertical fractures

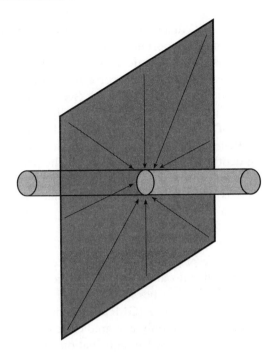

Figure 15-27 Fluids radially converge in a transverse Fracture approaching a horizontal borehole

schematic of the convergence of the stream lines into the horizontal well. Soliman *et al.* (1990) and Larsen and Hegre (1991) presented solutions for modeling this new flow regime. Mukherjee and Economides (1988), on the other hand, presented the following equations to account for the increased resistance to flow inside the fracture:

$$p_{D\,\text{total}} = p_D + (s_{\text{ch}})_c = \frac{kh\Delta p}{141.2qB\mu} \text{ for oil} \quad (15\text{-}31)$$

$$p_{D\,\text{total}} = p_D + (s_{\text{ch}})_c = \frac{kh\Delta m(p)}{1424qT} \text{ for gas} \quad (15\text{-}32)$$

$$(s_{\text{ch}})_c = \frac{kh}{k_f w}\left[\ln\left(h/r_w\right) - \frac{\pi}{2}\right] \quad (15\text{-}33)$$

Using the simplified equations above, it is possible to approximate the performance of transverse fractures using those of a fractured vertical well.

Although productivity of a system of transverse fractures increases as the number of fractures increases, the rate of this increase declines with a larger number of fractures because of interference. Simulations and field experience have shown that the optimum number of fractures is usually between 6 and 10.

15-7 INFLOW-PERFORMANCE RELATIONSHIP

Expressions such as Equation 15-2 (and the implicit solution for transient behavior), Equation 15-6 for steady state, and

$$q = \frac{kh(\bar{p} - p_{wf})}{141.2B\mu\left(\ln\dfrac{0.472r_e}{r_w} + s\right)} \quad (15\text{-}34)$$

for pseudosteady state can provide the well-production rate given the flowing bottomhole pressure.

While the latter would dictate what the reservoir can deliver, it is not independent from the hydraulics of the well. Clearly, the well production rate depends on the well depth, the tubing diameter, and the backpressure provided by the wellhead pressure.

Because well hydraulics differ depending on well configuration (plumbing), a traditional depiction of reservoir inflow is to allow the flowing bottomhole pressure to vary. A plot of the production rate (on the horizontal axis) vs. the flowing bottomhole pressure, p_{wf} (on the vertical axis), has been termed the inflow performance relationship (IPR) and has been in wide use in petroleum production engineering.

Figure 15-28 contains an IPR curve. The intercept on the y-axis is the initial reservoir pressure (for transient conditions); the constant pressure at the outer boundary, p_e (for steady state); or the average reservoir pressure, $\bar{p}$ (for pseudosteady state). The intersection with the x-axis

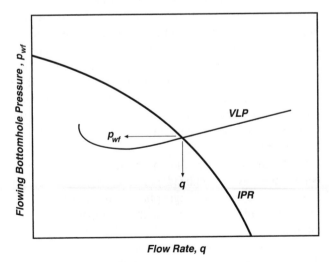

Figure 15-28 Combination of inflow performance relationship and vertical lift performance

at zero flowing bottomhole pressure is a characteristic value of the reservoir inflow and is termed the absolute open-flow potential.

The IPR curve is a straight line for steady-state liquid. It becomes curved for two-phase or gas flow, as shown in a number of petroleum production engineering text-books. The steady-state IPR remains constant throughout the life of the well. For pseudosteady-state flow, while the shape of the IPR is similar to the one for steady state, its position will shift with time parallel to the original IPR, reflecting reservoir pressure depletion. For a transient IPR, while the vertical intercept will remain constant (at p_i), the curve will swing inward for later times. (For a more detailed analysis, refer to Economides *et al.*, 1994).

15-8 VERTICAL LIFT PERFORMANCE

The fluid entering the sandface, whether into a vertical, deviated, or horizontal well, must travel to the surface through a rather complicated path. This path contains some or all of the following: slots, perforations, down-hole packers, a string of tubing(s), wellhead valves, and chokes. The fluid then continues through surface pipes, fittings, and valves into a separator.

Given either the flowing bottomhole pressure or the wellhead pressure, the other can be calculated by accounting for the pressure drop in the pipe. The total pressure drop is simply

$$\Delta p = \Delta p_{PE} + \Delta p_F + \Delta p_{KE} \qquad (15\text{-}35)$$

Δp_{PE} is the difference in the potential energy (often referred to as hydrostatic pressure drop), Δp_F is the friction pressure difference, and Δp_{KE} is the pressure difference caused by kinetic energy. The latter has an appreciable value if changes in the conduit diameter are involved and is generally negligible when compared to the other two pressure drops. In pipes carrying two-phase hydrocarbons, the problem of calculating the total pressure drop is compounded because the phases are miscible.

In the unlikely case that the bubblepoint pressure is below the flowing wellhead pressure, or in the case of neat natural gas, a single-phase calculation can be made. For all other cases, which encompass almost all oil and gas-condensate wells, two-phase correlations are needed. These correlations, prominent among which are those by Hagedorn and Brown (1965) and Beggs and Brill (1973), incorporate the mechanical energy balance and the phase behavior and phase-dependent properties

which are implicit functions of the pressures being sought. Even more important is the identification of flow regimes that cause variations in the correlations.

While the presentation of these correlations and calculations is outside the scope of this book (see Economides *et al.*, 1994 for details), the results are usually illustrated in the style of Figure 15-29, presenting the gradient curves. Plotted on the vertical axis going downward is the tubing depth; on the horizontal axis is the pressure. The intersection of the gradient curves with the upper horizontal axis is the flowing wellhead pressure. At the reservoir depth (bottom horizontal axis) are the flowing bottomhole pressures.

A calculation such as the one in Figure 15-29 is specific. It involves the production string diameter, the flow rate, the water/oil ratio and the temperature. (The last two affect the fluid properties.) A parameter allowed to vary is the gas/liquid ratio (GLR). In addition to its natural value (for which only one gradient curve would exist), other values are used in anticipation of artificial lift through the introduction of gas at the well bottom (gas lift; see Chapter 16).

For most crudes, increase in the wellhead pressure (for example, 100 psi) would require a disproportionate increase in the flowing bottomhole pressure (for example, 300 psi or higher, depending on the depth). Because a higher backpressure at the top would make the fluid column more liquid-like, an increase in the fluid density and a marked increase in the hydrostatic (or potential energy) pressure difference would result. Conversely, artificially increasing the GLR (for fixed wellhead pressure) would result in a decrease in the hydrostatic pressure and a decrease in the required flowing bottomhole pressure to lift the fluid.

However, a decrease in the hydrostatic pressure is associated with an increase in the friction pressure until eventually, for larger values of GLR, the decrease in Δp_{PE} is offset by a larger increase in Δp_F. The maximum value of the GLR at which this happens is called the limit GLR. The economic maximum GLR for gas lift may be much lower than this physical limit. (This issue is discussed in detail in Chapter 16.)

Brown *et al.* (1977) published a five-volume work containing the correlations and especially multi-thousands of pages of gradient curves for every conceivable flow conduit, flow rate, water/oil ratio, and GLR. Of course, with the proliferation of computers such graphical depictions have become largely obsolete. Now, calculations as depicted in Figure 15-29 are done to combine the IPR with the hydraulic performance of the well system. For each production rate, all pressure drops along the flow

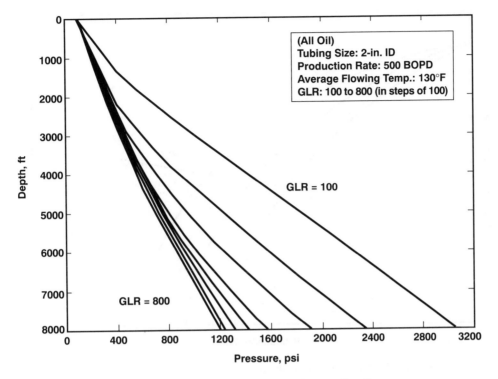

Figure 15-29   Gradient curves for an oil well

path are summed. The pressure at any point can be specified. A common and logical approach is to set the separator pressure. Then, the wellhead choke outlet pressure is calculated. This calculation would be the result of the friction pressure drop along a horizontal line. The inlet pressure of the choke is then obtained, and it becomes the flowing wellhead pressure, which is the starting point of the calculation of the pressure drop along the vertical (or deviated) well. The flowing bottomhole pressure is then obtained. In all cases, the natural or artificially induced GLR can be used.

For each production rate, the required flowing bottomhole pressure is calculated. This calculation is repeated several times and is plotted as shown in Figure 15-28, labeled VLP. On the right part of the VLP, a gentle positive slope (the slope is larger for larger GLR) reflects the increase in the friction pressure by increasing the production rate. On the left of the VLP, the upwards bend at lower production rates reflects holdup, a phenomenon that refers to liquid accumulation in the pipe. At these lower rates, the volumetric throughput of liquid is reduced, leaving a more liquid-like fluid in the well, thereby increasing the hydrostatic pressure component. The intersection of the IPR with the VLP curve is exactly the well deliverability at the actual flowing bottomhole pressure. This simultaneous solution of the two main systems, the reservoir and the well, is one of the main tasks of the petroleum production engineer.

15-8.1 Systems Analysis

A petroleum production engineer deals with modifications to the IPR (such as stimulation, reservoir pressure maintenance) or modifications to the VLP (such as artificial lift). The impact of these modifications on well performance can be assessed through variations of Figure 15-28. In addition, constructions similar to Figure 15-28 that allow an unknown variable to fluctuate (such as the number of open perforations) can be a valuable well-diagnostic tool.

Figure 15-30 is a depiction of IPR modifications for a given well configuration. For example, IPR_1 may mean the current state of the well, IPR_2 may mean the expected well performance following matrix stimulation, and IPR_3 may denote a hydraulically fractured well. The relative departure among the IPR curves is a direct indicator of attractiveness of any reservoir intervention, balanced against the costs.

Sustaining an IPR curve by maintaining the reservoir pressure (gas injection, waterflooding, etc.) instead of allowing the natural decline is another means of IPR modification (or retardation of its natural decline).

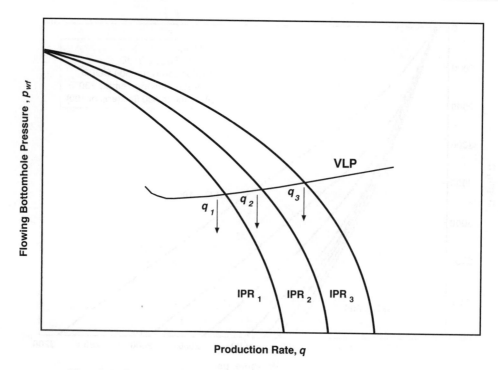

Figure 15-30   Modifications to the IPR curves: Effect on well performance

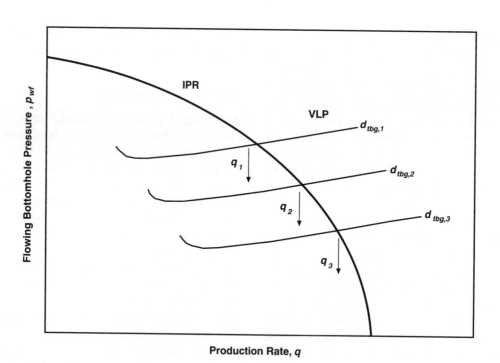

Figure 15-31   Modifications to the VLP curves: Effect on well performance

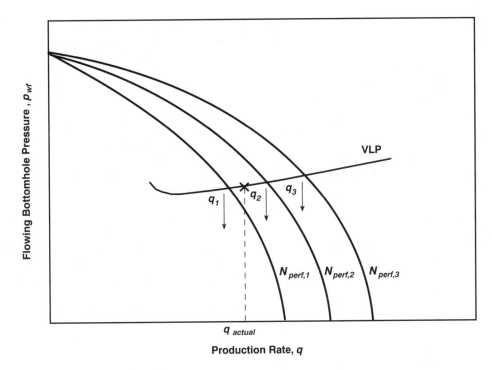

Figure 15-32 Well performance diagnosis with production systems analysis

If all possible IPR modifications have been accomplished, well performance can be affected by VLP adjustment, an example of which is shown in Figure 15-31. Means of accomplishing the VLP modification include all artificial lift methods, changing the tubing diameter, changing surface chokes, etc.

Finally, Figure 15-32 shows one of the most important uses of analyzing the system: the well performance diagnosis. Suppose that the number of open perforations is unknown. IPR curves reflecting different perforation skins can be drawn and the actual well performance can be compared with the expected performances. The number of open perforations can then be estimated.

Similar uses of systems analysis diagnosis can be envisioned for any reservoir and well configuration. Coupled with production economics, this approach is the essence of well management.

REFERENCES

Abass, H.H., Wilson, J.M., Venditto, J.J., and Voss, R.E.: "Stimulating Weak Formations Using New Hydraulic Fracturing and Sand Control Approaches," paper SPE 25494, 1993.

Al-Hussainy, R., Ramey Jr., H.J., and Crawford, P.B.: "The Flow of Real Gases through Porous Media," *JPT* (May 1966) 624-635; *Trans., AIME*, 237.

Azari, M., Knight, J.W., Coble, L.E., and Soliman, M.Y.: "Low-Conductivity and Short Fracture Half-Length Type Curves and Analysis for Hydraulically Fractured Wells Exhibiting Near Radial Flow Profile," paper SPE 23630, 1992.

Beggs, H.D., and Brill, J.P.: "A Study of Two-Phase Flow in Inclined Pipes," *JPT* (May 1973) 607–617.

Boyle, D., Bell, G., and McGinn, P.: "Coiled Tubing Stimulation Treatment in an Offshore Injection Well: A Case History," paper SPE 30427, 1995.

Bennet, C.O., Rostao, N.D., Reynold, A.C., and Raghavan, R.: "Influence of Fracture Heterogeneity and Wing Length on the Response of Vertically Fractured Wells," *SPEJ* (Apr. 1983).

Borisov, J.P.: *Oil Production Using Horizontal and Multiple Deviation Wells*, Moscow (1964). Translated by J. Strauss, The R&D Library Translation, Bartlesville, OK (1989).

Branagan, P.T., Lee, S.J., Cippola, C.L., and Wilmer, R.H.: "Pre-Frac Interference Testing of a Naturally Fractured, Tight Reservoir," paper SPE 17724, 1988.

Brown, K.E. *et al.*: *The Technology of Artificial Lift*, Pennwell, Tulsa, OK (1977).

Chen, C.C., and Raghavan, R.: "Modeling a Fractured Well in a Composite Reservoir," paper SPE 28393, 1994.

Cinco, L.H., Sameniego, V.F., and Dominiguez, A.N.: "Transient Pressure Behavior for a Well with a Finite Conductivity Vertical Fracture," *SPEJ* (Aug. 1978) 253-264.

Cinco, L.H., and Sameniego, V.F.: "Transient Pressure Analysis for Fractured Wells," *JPT* (Nov. 1982) 2655-2666.

Cinco, L.H., Ramey Jr., H.J., and Rodriguez, F.: "Behavior of Wells with Low-Conductivity Vertical Fractures," paper SPE 16776, 1987.

Couet, B., Pettijean, L.M., Abel, J.C., Schmidt, J.H., and Ferguson, K.R.: "Well Productivity Improvement by Use of Rapid Overpressure Perforation Extension: Case History," *JPT* (Feb. 1996).

Daneshy, A.A.: "A Study of Inclined Hydraulic Fractures," *SPEJ* (Apr. 1973) 61–68.

Dusterhoft, R.G., and Chapman, B.J.: "Fracturing High-Permeability Reservoirs Increases Productivity," *O&GJ* (June 20, 1994) 40–44.

Earlougher Jr., R.C.: Advances in Well Test Analysis, SPE Monograph Vol. 5, Richardson, TX (1977).

Economides, M.J., Deimbacher, F.X., Brand, C.W., and Heinemann, Z.E.: "Comprehensive Simulation of Horizontal Well Performance," *SPEFE* (Dec. 1991) 418–426.

Economides, M.J., Hill, A.D., and Ehlig-Economides, C.A.: *Petroleum Production Systems*, Prentice Hall, Englewood Cliffs, NJ (1994).

Frick, T.P., and Economides, M.J.: "Horizontal Well Characterization and Removal," paper SPE 21795, 1991.

Giger, F.M., Reiss, L.H., and Jourdan, A.P.: "The Reservoir Engineering Aspects of Horizontal Drilling," paper SPE 13024, 1984.

Gringarten, A.C., and Ramey Jr., H.J.: "An Approximate Infinite Conductivity Solution for a Partially Penetrating Line Source Well," *SPEJ* (Apr. 1975) 140–148.

Gringarten, A.C., Ramey Jr., H.J., and Raghavan, R.: "Applied Pressure Analysis for Fractured Wells," *JPT* (July 1975) 887–892.

Hagedorn, A.R., and Brown, K.E.: "Experimental Study of Pressure Gradients Occurring During Continuous Two-Phase Flow in Small-Diameter Vertical Conduits," *JPT* (Apr. 1965) 475-484.

Handren, P.J., Jupp, T.B., and Dees, J.M.: "Overbalance Perforating and Stimulation Method for Wells," paper SPE 26515, 1993.

Hawkins Jr., M.F.: "A Note on the Skin Effect," *Trans. AIME* (1956) 207, 356–357.

Holditch, S.A.: "Factors Affecting Water Blocking and Gas Flow from Hydraulically Fractured Gas Wells," *JPT* (Dec. 1979) 1514–24.

Hunt, J.L., and Soliman, M.Y.: "Reservoir Engineering Aspects of Fracturing High-Permeability Formations," paper SPE 28803, 1994.

Hunt, J.L., Chen, C., and Soliman, M.Y.: "Performance of Hydraulic Fractures in High-Permeability Formations," paper SPE 28530, 1994.

Joshi, S.D.: "Production Forecasting Methods for Horizontal Wells," paper SPE 17580, 1988.

Joshi, S.D: *Horizontal Well Technology*, Pennwell Books, Tulsa, OK (1991) 114.

Klotz, J.A., Kreuger, R.F., and Pye, D.S.: "Effect of Perforating Damage on Well Productivity," *JPT* (Nov. 1974) 1303–1314.

Larsen, L., and Hegre, T.M.: "Pressure-Transient Behavior of Horizontal Wells with Finite-Conductivity Vertical Fractures," paper SPE 22076, 1991.

Liu, X., and Civan, F.: "Formation Damage and Skin Factor due to Filter Cake Formation and Fines Migration in the Near-Wellbore Region," paper SPE 27364, 1994.

Mao, M.L.: "Performance of Vertically Fractured Wells with Finite Conductivity Fractures," PhD dissertation, Stanford U., Stanford, CA (1977).

Martins, J.P., and Stewart, D.R.: "Tip Screenout Fracturing Applied to the Ravenspurn South Gas Field Development," *SPEPE* (Aug. 1992) 252–258.

Mason, J., George, F., and Pacheco, E.: "A New Tubing-Conveyed Perforating Method," paper SPE 29820, 1995.

Mathur, A.K., Marcinew, R.B., Ehlig-Economides, C.A., and Economides, M.J.: "Hydraulic Fracture Stimulation of Highly Permeable Formations: The Effect of Critical Fracture Parameters on Oil Well Production and Pressure," paper SPE 30652, 1995.

Matthews, C.S., and Russell, D.G.: *Pressure Buildup and Flow Tests in Wells*, SPE Monograph Vol. 1, Richardson, TX (1967).

McGuire, W.J., and Sikora, V.J.: "The Effect of Vertical Fractures on Well Productivity," *Trans. AIME* (1960) 219, 401–405.

Meese, C.A, Mullen, M.E., and Barree, R.D.: "Offshore Hydraulic Fracturing Technique," *JPT* (March 1994) 226–229.

Mix, K., Bell, G., and Evans, S.J.: "Coiled Tubing Drilling Case History," paper SPE 36350, 1996.

Mukherjee, H. and Economides, M.J.: "A Parametric Comparison of Horizontal and Vertical Well Performance," paper SPE 18303, 1988.

Peterson, S.K., and Holditch, S.A.: "Analysis of Factors Affecting Drillstem Tests in Low-Permeability Reservoirs," paper SPE 14501, 1985.

Poulsen, D.K., and Soliman, M.Y.: "A Procedure for Optimal Hydraulic Fracturing Treatment Design," paper SPE 15940, 1987.

Prats, M.: "Effect of Vertical Fractures on Reservoir BehaviorIncompressible Fluid Case," *SPEJ* (June 1961) 105–118.

Roodhart, L.P., Fokker, P.A., Davies, D.R., Shlyapobersky, J., and Wong, G.K.: "Frac-and-Pack Stimulation: Application, Design, and Field Experience," *JPT* (March 1994) 230-38.

Smith, M.B., Miller II, W.K., and Haga, J.: "Tip Screenout Fracturing: A Technique for Soft, Unstable Formations," *SPEPE* (May 1987) 95–103.

Soliman, M.Y.: "Modifications to Production Increase Calculations for a Hydraulically Fractured Well," *JPT* (Jan. 1983).

Soliman, M.Y. (1986a): "Design and Analysis of a Fracture with Changing Conductivity," *JCPT* (Sept–Oct. 1986) 62–67.

Soliman, M.Y. (1986b): "Fracture Conductivity Distribution Studied," *O&GJ* (Feb. 10, 1986) 89–93.

Soliman, M.Y., Hunt, J. L., and Lee, W. S.: "Fracture ParametersCalculated Versus Designed," paper CIM 87-38-68, 1987.

Soliman, M.Y., Hunt, J.L., and El Rabaa, W.: "Fracturing Aspects of Horizontal Wells," *JPT* (Aug. 1990) 966–973.

Soliman, M.Y., Hunt, J.L., and Azari, M.: "Fracturing Horizontal Wells in Gas Reservoirs," paper SPE 35260, 1996.

Tariq, S.M.: "Evaluation of Flow Characteristics of Perforations Including Non-Linear Effects with the Finite Element Method," *SPEPE* (May 1987) 73–82.

Tinsley, J.M., Tiner, R., Williams, J.R., and Malone, W.T.: "Vertical Fracture HeightIts Effect on Steady State Production Increase," *JPT* (May 1969) 633–38.

16 Artificial-Lift Completions

James F. Lea
Amoco Production Company

Clark E. Robison
Halliburton Energy Services

16-1 INTRODUCTION

All petroleum reservoirs experiecvnce pressure declines, and most wells require artificial lift at some point, most commonly where reservoir pressure is insufficient for natural flow. Artificial lift systems may also be used to enhance production from flowing wells with a reservoir pressure that is insufficient to produce a required amount of fluid.

Extensive research and field studies have been conducted on a range of artificial-lift systems that have been developed and applied extensively to meet industry needs. These systems include beam pumping, gas lift, electrical submersible pumps (ESPs), and hydraulic lift. This chapter provides an overview of these systems and their engineering fundamentals.

16-1.1 Overview of Artificial-Lift Methods

A number of artificial-lift alternatives are used, each of which has advantages and disadvantages, and each of which has a "window" within which it operates most efficiently. The major forms of artificial lift are shown in Figure 16-1. Here beam, hydraulic, electrical submersible, gas lift, and progressive lift (progressive cavity pumps, PCPs) are illustrated. Clegg et al. (1993) found that in the U.S., a beam or rod pump is used about 85% of the time, continuous gas lift 10%, hydraulic lift 2%, ESP 4%, and PCP 1% of the time.

All of the pumping methods add energy to the fluid stream calculated by

(volumetric flow through the pump)

× (the pressure difference across the pump).

Gas lift lightens the fluid gradient to increase the flow and correspondingly lower the sand-face pressure.

One definition of lift system efficiency is found by dividing the power added to the fluid stream by the amount of power needed from external sources such as motors or compressors.

Clegg et al. (1993) noted the efficiency of artificial-lift methods ranging from 70% to 5%, with PCP being the best, followed by beam, ESP, hydraulic reciprocating and jet, continuous gas lift, and intermittent lift.

An artificial-lift methods chart by Blais et al. (1986) is summarized in Figure 16-2. This figure gives an idea of what one author indicates as best usage ranges, but there are proven applications at higher rates and deeper depths for almost all lift methods.

Selection and maintenance of the proper lift method is a major factor in the long-term profitability of most oil and gas wells. Lea and Kol (1995), Clegg et al. (1993), Brown (1982), Bennett (1980), Neely (1980), and Johnson (1968) provide details on the possible performance ranges of various methods of artificial lift and some of the factors involved in the most economic selection process. Clegg et al. (1993) contains an excellent list of references on artificial lift, organized by categories of lift. Although two methods of lift may both produce the desired rate, energy consumption, reliability, initial cost, maintenance, manpower, and other costs must be considered in the life-cycle cost of any artificial-lift method.

Beam pumping systems actuate a downhole plunger pump with traveling and standing valves. PCP pumps rotate from rods or a downhole motor that spins a metal rotor inside a rubber stator and moves a cavity of fluid upward as the pump spins. ESP systems include

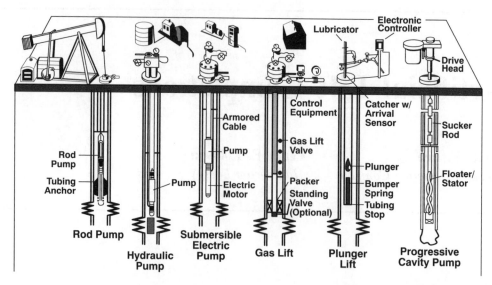

Figure 16-1   Major forms of artificial lift

a motor powered through a three-phase cable from the surface to rotate a stack of impellers inside a stack of diffusers. The hydraulic reciprocating pump is a rapidly reciprocating plunger pump with ball-and-seat check valves. An overhead hydraulic piston motor that is actuated by a continuous supply of high pressure fluid from the surface powers the system. Hydraulic jet pumps use the venturi effect from a power fluid pushed through a nozzle that creates a low-pressure area seen by well formation fluids. Gas lift relies on high-pressure gas injected at depth into the flow stream to lighten the flowing gradient so the formation will flow at a higher rate.

Table 16-1 lists advantages and Table 16-2 lists disadvantages of various artificial-lift systems. These tables have been extracted from Brown (1982). Clegg *et al.* (1993) contains more extensive tables, but these are too lengthy to include here.

16-2 BEAM PUMP COMPLETIONS

Figure 16-3 shows a schematic of a typical beam pump system. The surface pumping unit is designed to provide a reciprocating up-and-down motion at certain strokes per minute (SPM) and a given stroke length. Some design considerations for the surface unit are given in Table 16-3.

The rods are connected to the pumping unit by a polished rod (PR) hanging through the carrier bar with a clamp on the PR above the carrier bar. The PR passes through the stuffing box on the wellhead to prevent leaks.

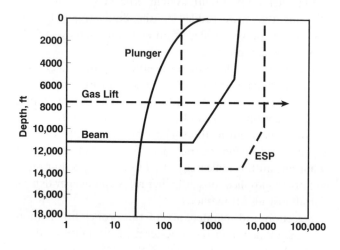

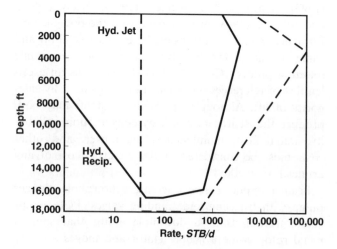

Figure 16-2   Artificial-lift methods chart

Table 16-1 Relative advantages of artificial-lift systems

Rod Pumping	Hydraulic Piston Pumping	Electric Submersible Pumping	Gas Lift	Hydraulic Jet Pump	Plunger Lift	Progressive Cavity Pumps
Relatively simple system design.	Not so depth limited– can lift large volumes from great depths.	Can lift extremely high volumes, 20,000 bbl/d (19,078 m^3/d), in shallow wells with large casing.	Can withstand large volume of solids with minor problems.	Retrievable without pulling tubing.	Retrievable without pulling tubing.	Some types are retrievable with rods.
Units easily changed to other wells with minimum cost.	500 bbl/d (79.49 m^3/d) from 15,000 ft. (4572 m) have been installed to 18,000 ft. (5,486.4 m)	Currently lifting ± 120,000 bbl/day (19,068m^3/d) from water supply wells in Middle East with 600-hp (448-kW) units; 720-hp (537-kW) available, 1000-hp (746-kW) under development.	Handles large volume in high-PI wells (continuous lift). 50,000 bbl/d (7949.37 m^3/d).	No moving parts.	Very inexpensive installation.	Moderate cost.
Efficient, simple and easy for field people to operate.	Crooked holes present minimal problems.			No problems in deviated or crooked holes.	Automatically keeps tubing clean of paraffin, scale.	Low profile.
Applicable to slim holes and multiple completions.	Unobtrusive in urban locations.	Unobtrusive in urban locations.	Fairly flexible- convertible from continuous to intermittent to chamber or plunger lift as well declines.	Unobtrusive in urban locations.		Can use downhole electric motors that can withstand sand and viscous fluid.
Can pump a well down to very low pressure (depth and rate dependent).	Power source can be remotely located.	Simple to operate.	Unobtrusive in urban locations.	Applicable offshore.	Applicable for high gas-oil ratio wells.	
System usually is naturally vented for gas separation and fluid level soundings.	Analyzable.	Easy to install downhole pressure sensor for telemetering pressure to surface by cable.	Power source can be remotely located.	Can use water as a power source.	Can be used in conjunction with intermittent gas lift.	High electrical efficiency.
Flexible-can match displacement rate to well capability as well declines.	Flexible-can usually match displacement to well's capability as well declines.		Easy to obtain downhole pressures and gradients.	Power fluid does not have to be as clean as for hydraulic piston pumping.	Can be used to unload liquid from gas wells.	
Analyzable.	Can use gas or electricity as power source.	Crooked holes present no problem.	Lifting gassy wells is no problem.	Corrosion scale emulsion treatment easy to perform.		
Can lift high- temperature and viscous oils.	Downhole pumps can be circulated out in free systems.	Applicable offshore.	Sometimes serviceable with wireline unit.	Power source can be remotely located and can pump high volumes to 30,000 bbl/d (4769.623 m^3/d).		
Can use gas or electricity as power source.	Can pump a well down to fairly low pressure.	Corrosion and scale treatment easy to perform.	Crooked holes present no problem			
Corrosion and scale treatments easy to perform.	Applicable to multiple completions.	Availability in different sizes .	Corrosion is not usually as adverse.			
Applicable to pump-off control if electrified.	Applicable offshore.	Lifting cost for high volumes generally very low.	Applicable offshore.			
Availability of different sizes.	Closed system will combat corrosion.					
Hollow sucker rods are available for slimhole completions and ease of inhibitor treatment.	Easy to pump in cycles by time clock.					
Have pumps with double valving that pump on both upstroke and downstroke.	Adjustable gear box for triplex offers more flexibility.					
	Mixing power fluid with waxy or viscous crudes can reduce viscosity.					

The rod string, composed of sucker rods, extends from the surface to the pump plunger at depth. Design considerations for the rods are summarized in Table 16-4.

Figures 16-4A and 16-4B show two types of downhole pumps. A tubing pump has the barrel connected to the tubing, and the tubing must be pulled to replace a worn pump. An entire rod pump can be pulled by pulling only the rods. Downhole pump design concerns are shown in Table 16-5.

Figures 16-4A to 16-4D show some different types of commonly used pumps. The RHA and the RWA are the heavy-walled barrel and thin-walled barrel pumps with a stationary barrel and top anchor. The RHB and the RWB are the heavy-walled barrel and thin-walled barrel pumps with stationary barrel and bottom anchor. The designation TH is for a heavy-walled tubing pump that has to be pulled with the tubing. Pump selection depends on the type and volume of fluid produced—whether there are sand, gas, scale or other associated production problems—and the mechanical strength of the pump required. API has standardized pump nomenclature as shown in Table 16-6.

One of the main problems with beam pump systems is that free gas ingested into the pump severely reduces the

Table 16-2 Relative disadvantages of artificial-lift systems

Rod Pumping	Hydraulic Piston Pumping	Electric Submersible Pumping	Gas Lift	Hydraulic Jet Pump	Plunger Lift	Progressive Cavity Pumps
Crooked holes present a friction problem.	Power oil systems are a fire hazard.	Not applicable to multiple completions.	Lift gas is not always available.	Relatively inefficient lift method.	May not take well to depletion; hence, eventually requiring another lift method.	Elastomers in stator swell in some well fluids.
High solids production is troublesome.	Large oil inventory required in power oil system, which detracts from profitability.	Only applicable with electric power.	Not efficient in lifting small fields or one-well leases.	Requires at least 20% submergence to approach best lift efficiency.	Good for low-rate wells only normally less than 299 bbl/d (31.8 m/d).	POC is difficult.
Gassy wells usually lower volumetric efficiency.	High solids production is troublesome.	High voltages (1000 V) are necessary.	Difficult to lift emulsions and viscous crudes.	Design of system is more complex.	Requires more engineering supervision to adjust properly.	Lose efficiency with depth.
Depth is limited, primarily based on rod capability.	Operating costs are sometimes higher.	Impractical in shallow, low-volume wells.	Not efficient for small fields or one-well leases if compression equipment is required.	Pump may cavitate under certain conditions.	Plunger may reach too high a velocity and cause surface damage.	Rotating rods wear tubing; windup and afterspin of rods increase with depth.
Obtrusive in urban locations.	Usually susceptible to gas interference—usually not vented.	Expensive to change equipment to match declining well capability.	Gas freezing and hydration problems.	Very sensitive to any change in backpressure.	Communication between tubing and casing required for good operation unless used in conjunction with gas lift.	
Heavy and bulky in offshore operations.	Vented installations are more expensive because of extra tubing required.	Cable causes problems with tubular handling.	Problems with dirty surface lines.	The producing of free gas through the pump causes reduction in ability to pump liquids.		
Susceptible to paraffin problems.	Treating for scale below packer is difficult.	Cables deteriorate in high temperatures.	Some difficulty in analyzing properly without engineering supervision.	Power oil systems are fire hazard.		
Tubing cannot be internally coated for corrosion.	Not easy for field personnel to troubleshoot.	System is depth limited, 10,000 ft. (3048.0 m), because of cable cost and inability to install enough power downhole (depends on casing size).	Cannot effectively produce deep wells to abandonment.	High surface power fluid pressures are required.		
H_2S limits depth at which a large-volume pump can be set.	Difficult to obtain valid well tests in low-volume wells.	Gas and solids production are troublesome.	Requires makeup gas in rotative systems.			
Limitation of downhole pump design in small-diameter casing.	Requires two strings of tubing for some installations.	Not easily analyzable unless good engineering experience.	Casing must withstand lift pressure.			
	Problems in treating power water where used.	Lack of production rate flexibility.	Safety is a problem with high-pressure gas.			
	Safety problem for high surface pressure power oil.	Casing size limitation.				
	Loss of power oil in surface equipment failure.	Cannot be set below fluid entry without a shroud to route fluid by the motor. Shroud also allows corrosion inhibitor to protect outside of motor.				
		More down-time when problems are encountered because entire unit is downhole.				

pumping capacity and can even completely gas-lock the pump. One of the best methods to combat the presence of free gas downhole is simply to land the pump below the perforations. If the pump is spaced with the traveling valve coming very close to the standing valve on the downstroke, gas that gets into the pump can be compressed out.

If the pump must be set above the perforations, then gas separators can send the gas up the casing and prevent it from coming into the pump. Figure 16-5A, after Clegg (1989), shows first the placement of the pump below the perforations. Figure 16-5B shows a packer-type separator, which works very well by letting the fluid fall back to the pump intake but requires a packer. The next four schematics (Figures 16C through 16-5F) are variations of the "poor-boy" design concept, which creates a down-

flow space where the fluid must flow downward while the bubbles (ideally) travel upward at a faster speed. Finally, the new Echometer concept (Podio *et al.*, 1995) lets the fluid come in and vents gas out of the top of the separators (Figure 16-5G). The separator is decentralized so that the intake is near the casing, and large bubbles are excluded from the intake section to flow up the large area.

For slim-hole applications, two types of beam completions may be considered. In the casing pump completion (Figure 16-6), the pump is anchored directly to the casing with no gas passage up the casing. All of the fluid (gas and liquids) must pass up the pump. The hollow-rod completion (possibly small diameter tubing or coiled tubing) directs the fluid production up the hollow rods, and the pump anchor allows the gas up the casing. Hollow rods are used for fairly shallow depths because the hollow

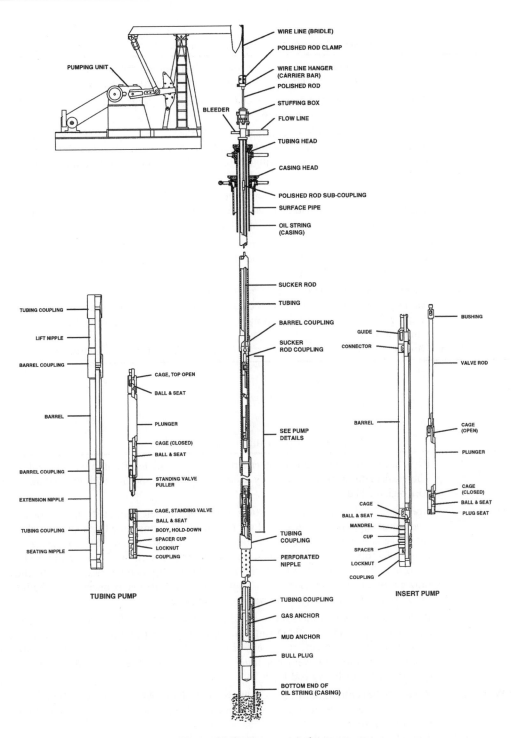

Figure 16-3 Beam pumping system

tubing cannot support heavy loads. For horizontal wells, the pump may be landed above the kick-off or placed in the curved hole with rod guides or roller rods sections.

Takacs (1993) provides a summary of most of the other considerations for beam pumping operations.

Specialty pumps can pump downward below a packer to take water out of the flow of gas wells. Specialty rods include high-strength, fiberglass, composite, and continuous rods wound on large reels to eliminate rod couplings. Viscous oil can be pumped with a traveling

Table 16-3   Beam pump surface unit considerations

Proper motor sizing	The motor should be able to carry average loads. It should not be oversized more than $1\frac{1}{2}$ to 2 times needed, as this lowers efficiency. A moderate-slip motor is preferred.
Sheave sizing to achieve Proper SPM	The motor and unit sheave can be changed to adjust speed. Motor sheaves less than about 10 in. long can cause belt wear and slip.
Gear box sizing	The gear box must be large enough to handle the required torque. The gear box is the most expensive component. For long life, it should not be overloaded, although the manufacturer should have safety factors built in. A 640-305-144 unit has a 640,000 in.-lb output torque capability at 100% loading.
Peak load	The unit must handle the peak load required. The peak polished rod loading must not exceed the middle unit designation (e.g. 640-305-144) times 100 lbs. For this designation, the peak load should not exceed 305×100 lbs.
Stroke length	Several stroke lengths are usually available. Up to a point, the longest stroke helps efficiency. It also can reduce wear to other components if the target rate is being met. A 640-305-144 unit has a maximum stroke of 144 inches.
Unit geometry	There are many geometries and types of units available. An air balance unit weighs less. Because of this, it is sometimes used offshore. A unit with a faster downstroke is more efficient with less peak torque, but can "stack" the rods on a high SPM unit. Other units include those that have an offset gear-box or various hydraulic and long-stroke units.

Table 16-4   Rod string considerations

Fatigue Loading	Rods must satisfy fatigue loading criteria usually calculated from maximum and minimum loads. These calculations may be made by using the so-called Goodman Modified diagram, described in API bulletin 11B.
Rod Size, Taper Design	Rods may be tapered with equal stress at the top of each smaller string of rods going downhole. Rods must support themselves and the fluid load. Therefore, the top rods carry the most load and should be larger than rods near the bottom of the rod string. An "86" rod string would have 1-in. rods at the top, then a section of $\frac{7}{8}$-in. rods, and then a section of $\frac{6}{8}$-in. or $\frac{3}{4}$-in. rods at the bottom. Rods are 30 ft long (25 ft in California) with threaded-pin connections on both ends, typically ranging from $\frac{1}{2}$- to $1\frac{1}{4}$-in. diameter. Couplings have at least two connections, sized for sizes $\frac{1}{2}$ to 1 in. Dimensions must be checked for various tubing sizes.
Metallurgy	Rods must be made of metallurgy suited to perform in the well fluids. Grade D rods are the most commonly used rods. Grade D rods are used for heaviest pumping loads in deep wells; Grade C rods are used for non-corrosive wells of average load conditions; Grade K rods are used for corrosive conditions in average loading conditions. Special service grades are used for very corrosive wells or heavy rod loads.
Rod and Tubing Wear	Rods may be centralized or protected to reduce tubing and rod wear and reduce buckling. Rods in wells that have severe doglegs or in deviated wells may need centralizers in certain locations to prevent rod wear. The tubing may be anchored to prevent cork-screwing. Most operators anchor the tubing to prevent tubing movement and reduce wear on the rods or tubing above the pump.
Buckling	Sinker bars may be used above the pump to reduce buckling and help rod fall. It is not uncommon to have a few large rods located on the bottom of the rod string to carry compression loads on the downstroke and reduce wear above the pump.

Table 16-5 Pump design considerations

Pump metallurgy	Pump metallurgy (barrel, plunger, balls, and seats) is selected for corrosion and wear. Materials selected to resist corrosion, perhaps in conjunction with a corrosion treatment plan, must be considered. Barrels could be carbon steel, stainless steel, brass, monel, or chrome plated, for example. Plungers could be plain steel, chrome- or nickel-plated, popular spray metal, or other options. Balls and seats can be stainless steel (440C, specially treated, 329, other) balls and seats, K monel balls and seats, bronze balls, tungsten carbide balls and seats, silicon nitride balls, or other. Cages for the balls can be of various materials.
Pump sizing	The pump size determines the pumping rate. The largest pump (lowest SPM) will provide the most energy efficiency, but also the biggest loads to the equipment. The optimum choice is the biggest pump with no overloads on the rods, unit, or motor.
Spacing	Pumps should be spaced to maximize the pump compression ratio for gas handling. The pump should be built and spaced so that the traveling valve comes close to the standing valve on the downstroke to help prevent gas lock. This may not be practical for fiberglass rods, since they are widely spaced to prevent possible compression and destruction on the downstroke.
Pump type, holddown	For deeper applications, the outside of the barrel should withstand exposure to produced fluid pressure (bottom hold down), and thicker wall (heavy wall) pumps are used. For shallower applications, the outside of the barrel should withstand exposure to suction fluid pressure (top hold down). This eliminates sediment problems around the barrel.
Production problems	Specialty pumps are designed for sand, gas, viscous oil, and other problem situations.

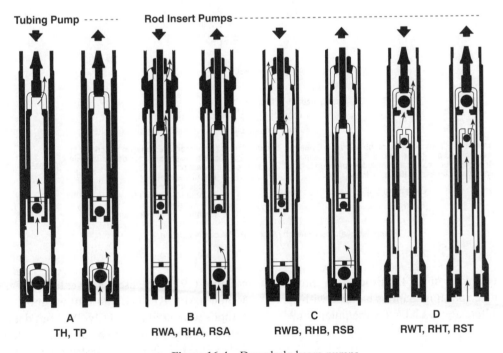

Figure 16-4 Downhole beam pumps

Table 16-6 API subsurface pump classifications

Pump Designation

The basic types of pumps and letter designations covered by this specification are as follows:

Type of Pump	Letter Designation			
	Metal Plunger Pumps		Soft-Packed Plunger Pumps	
	Heavy-Wall Barrel	Thin-Wall Barrel	Heavy-Wall Barrel	Thin-Wall Barrel
Rod Pumps				
Stationary Barrel, Top Anchor	RHA	RWA	......	RSA
Stationary Barrel, Bottom Anchor	RHB	RWB	......	RSB
Traveling Barrel, Bottom Anchor	RHT	RWT	......	RST
Tubing Pumps	TH	......	TH	......

Complete pump designations include (1) nominal tubing size, (2) basic bore diameter, (3) type of pump, including type of barrel and location and type of seating assembly, (4) barrel length, (5) plunger length, and (6) total length of extensions when used, as follows:

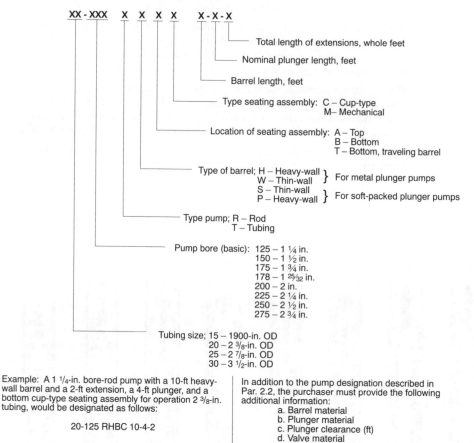

Example: A 1 ¼-in. bore-rod pump with a 10-ft heavy-wall barrel and a 2-ft extension, a 4-ft plunger, and a bottom cup-type seating assembly for operation 2 ³/₈-in. tubing, would be designated as follows:

20-125 RHBC 10-4-2

In addition to the pump designation described in Par. 2.2, the purchaser must provide the following additional information:
     a. Barrel material
     b. Plunger material
     c. Plunger clearance (ft)
     d. Valve material
     e. Length of each extension

Pump designator furnished by American Petroleum Institute

---

downhole stuffing box to divert the production into the casing and eliminate rod drag. Sand is handled with special pumps, and filters are available for the pump intake.

16-2.1 Beam Pump Design

Beam pump design is an interesting subject. Although wave equation models, such as those initially developed by Gibbs (1977), are used quite extensively now, the API method (API RP 11L) and other more approximate methods were used previously. Without detailing the theoretical background and calculations, a design example (Table 16-7) is shown to illustrate what must be determined when designing a beam pump system.

The design shown in Table 16-7 was generated with a programmed version of the API RP 11L method. The system is designed with the fluid level at the pump for

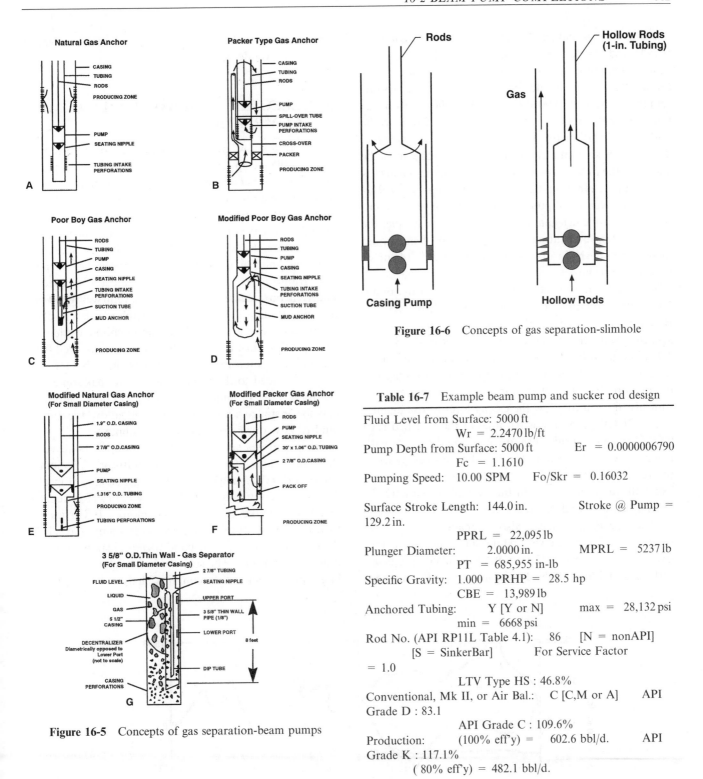

Figure 16-6   Concepts of gas separation-slimhole

Figure 16-5   Concepts of gas separation-beam pumps

Table 16-7   Example beam pump and sucker rod design

Fluid Level from Surface: 5000 ft
Wr = 2.2470 lb/ft
Pump Depth from Surface: 5000 ft   Er = 0.0000006790
Fc = 1.1610
Pumping Speed: 10.00 SPM   Fo/Skr = 0.16032

Surface Stroke Length: 144.0 in.   Stroke @ Pump = 129.2 in.
PPRL = 22,095 lb
Plunger Diameter: 2.0000 in.   MPRL = 5237 lb
PT = 685,955 in-lb
Specific Gravity: 1.000   PRHP = 28.5 hp
CBE = 13,989 lb
Anchored Tubing: Y [Y or N]   max = 28,132 psi
min = 6668 psi
Rod No. (API RP11L Table 4.1):   86   [N = nonAPI]
[S = SinkerBar]   For Service Factor
= 1.0
LTV Type HS : 46.8%
Conventional, Mk II, or Air Bal.:   C [C,M or A]   API Grade D : 83.1
API Grade C : 109.6%
Production: (100% eff'y) =   602.6 bbl/d.   API Grade K : 117.1%
(80% eff'y) = 482.1 bbl/d.

API RP11L(MODIFIED) SUCKER ROD PUMP PREDICTION
*//////////*Rev. 1.1 from LTV Energy Products, Garland, TX (214) 487-3000
Oct. 88 (C) Copyright 1987 LTVEP

maximum load and more conservatism in the design. The results are given for anchored tubing with a 2-in. pump set at 5000 ft operating at 10 SPM. An 86-rod string (1 in., $\frac{7}{8}$ in., and $\frac{3}{4}$ in.) tapered according to the API method is shown.

Using the so-called Modified Goodman Diagram (Frick and Taylor, 1962), the Grade "D" rods (the most commonly used) would be 83% loaded, but Grade "C" rods would be overloaded. The torque required is over 640,000 in.-lb, so a 912,000 in.-lb-torque pumping unit is required. The peak load is 22,095, so the middle designation for an API unit must be over 220; and the stroke input is 144 in., so the stroke must be 144 in. in the maximum or lower stroke length. An acceptable unit designation would be something like a 912-305-144 conventional pumping unit, where the first number is the torque designation of 912,000 in.-lb torque, the second is the load rating of 305 × 100 lb of maximum load at the polished rod, and the last number is 144 in. of maximum stroke.

The production in the example above, with the pump 100% filled and no leakage, is approximately 600 STB/d. Because the fill is nearly always lower and there is leakage, the production is more likely to be near 480 STB/d. To determine if the well is producing as much as possible, the fluid level is monitored or pump-off control is used to pump more than the well will produce and to periodically shut off the well to allow the casing level to rise again.

The PRHP (horsepower required at the polished rod) is 20.5, and a good rule of thumb is to size the motor with twice this value to account for downhole energy losses in the rod string, surface losses in the pumping unit, and losses through the motor. In the example given here, 50 hp is acceptable and 40 hp could be marginal.

This type of design can also be made using a wave equation model, which was first designed and used extensively by Gibbs (1977). The resultant calculated predicted surface and bottomhole dynamometer cards and gear box torque are shown in Figure 16-7 using a typical wave-equation design program.

16-3 ELECTRICAL SUBMERSIBLE PUMP COMPLETIONS

Figure 16-8 shows an overall schematic of an electrical submersible pump (ESP) system installation. At the surface, transformers bring the voltage down to a value equal to the motor nameplate voltage plus the cable voltage loss. If a variable-speed drive is used, the voltage is brought to 480 volts before the drive and brought up again before the wellhead. The switchboard is used to turn the unit on or off; to monitor voltage, current, starts, and stops; and to protect against emergencies. The vent box breaks the cable to prevent well gas from traveling though the cable to the switchboard. The cable is connected at the wellhead to the downhole cable or is packed off against rubbers. The cable seals the well from the ambient conditions and supports the tubulars. The cable is banded to the tubing and usually has some protectors to prevent damage on run-in. A check and drain may be installed above the pump to allow the tubing fluid to flow out before pulling the tubing.

The pump is a stack of centrifugal impellers keyed to the shaft running inside a stack of diffusers. The intake can be a standard intake or a rotary gas separator intake. The seal section (or protector or equalizer) is located below the intake to protect the motor. It carries the thrust of the shaft extending from the pump and allows the fluid in the motor to expand without leaking wellbore fluids to the motor. The seal can be a tortuous-path or a bag-type protector that contains shaft seals to prevent

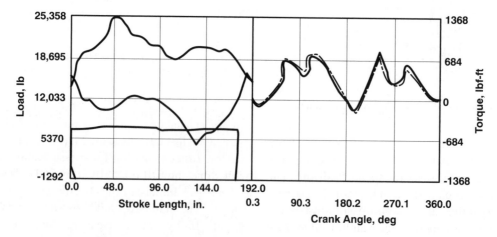

Figure 16-7 Beam-pump design example (from wave equation)

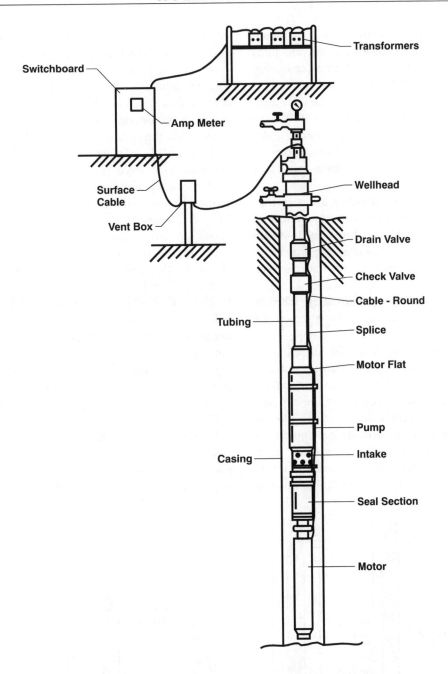

Figure 16-8   Schematic of electrical submersible pump installation

fluids from traveling along the shaft and carries the motor torque to the pump above. The motor is on the bottom of the unit so that it can be cooled by fluids traveling to the intake. A three-phase current is supplied to a two-pole, squirrel-cage motor through a cable connected to the top of the motor. The motor can be a single, a tandem, or even a triple motor, although single motors are preferred if they have enough horsepower for the job. The motor is filled with a dielectric oil, which

cools and insulates the motor. The magnet wire in the motor is insulated; some new insulations have been suggested to insulate the wire even if water gets into the motor. Instrumentation for pressure and temperature is sometimes housed on the bottom end of the motor.

Free gas diminishes the effectiveness of the centrifugal stages. The standard intake, as shown schematically in Figure 16-9A, tries to have the fluid break over sharp edges for some separation effect before entering the pump. However, the rotary separator shown in Figure 16-9B centrifuges the fluid and sends the heavier fluids into the pump and the lighter gas back into the casing. The rotary separator may not be the best choice if abrasives are present in the well.

In the type of completion shown in Figure 16-9C, the pump assembly may be set below the perforations, and a special pump and fluid path send fluid down below the motor to keep the motor cool. The annulus may then provide gas separation.

Another choice when gas is present is a device that mixes the gas and liquids and provides a component of positive displacement to the fluids before they enter the pump. This device helps the centrifugal stages provide energy to the fluids in the presence of free gas (see Figure 16-9B).

Figure 16-9 also presents some shrouding schemes which are intended to prevent gas from entering the pump.

The "Y" tool (Figure 16-10) is an old but effective completion used where space permits. This tool offsets the pump assembly and allows wireline work to extend below the pump when a plug is removed in the bypass.

If sand is present, the options are to try to filter it before it enters the pump or to provide a pump which is durable enough not to wear as it pumps sandy fluid. The completion looks the same if a hardened pump is used. If filters are used, they hang below the pump, and some are quite long to allow enough surface area to prevent high velocities. Filters are available using wire-wrapped steel elements, soft flexible-rubber elements, annulus elements of pre-packed sand or "gravel pack," and intakes which swirl the fluids before they enter the pump so the heavier sand can be collected and dropped back down the rathole (if present). With any filters, the rathole must be cleaned out periodically if sand is being pumped and filtered out.

Hardened pumps are used to pump the sand with pump elements harder than the sand, and they provide long service before wear destroys the pump. Figure 16-11 shows a pump design in which several stages in the pump are supported by a hardened thrust bearing. Figure 16-12 shows another concept where hardened inserts are used to combat sand wear by carrying radial and axial loads. In spite of all of these innovations, ESPs and sand are still not a good combination. Gas-lift and PCP pumps are probably the best options if sand is present in large amounts.

Horizontal wells are becoming more common, and several types of completions use ESPs to pump horizontal wells. The following figures are presented by Wilson (1995).

Figure 16-13 shows the ESP in the vertical. This simple arrangement can be used in vertical wells, or the vertical portion of a horizontal well, but the latter can suffer from free gas from the horizontal section.

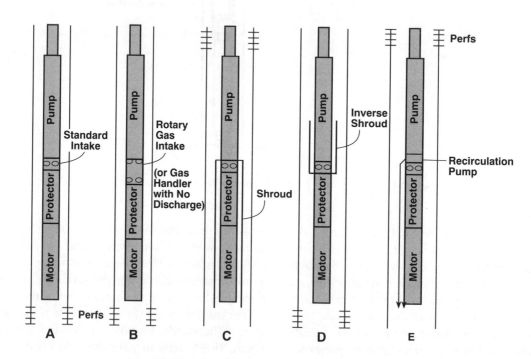

Figure 16-9   Schemes for handling gas with ESP

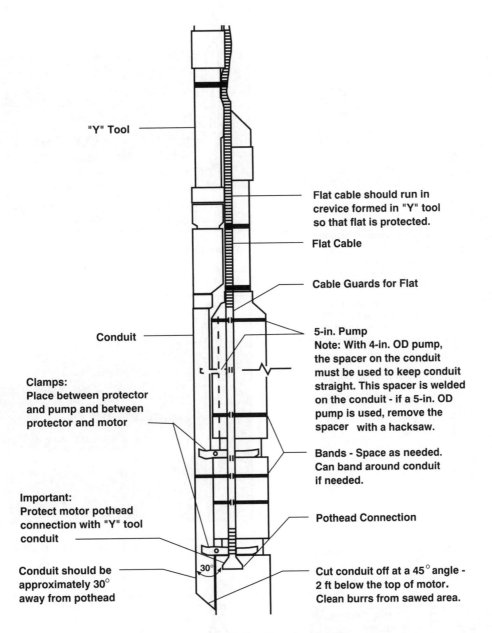

Figure 16-10   Y-tool with ESP

Figure 16-14 is the ESP in a rathole before the horizontal lateral. This is a new concept, and no record on its actual use is currently available. The motor would require special downward flow of some fluid to keep it cool, but this is possible now. This configuration would allow good gas separation.

Figure 16-15 shows the ESP in the well tangent. This allows some of the gas to pass over the pump and avoids some of the risk of entering the horizontal lateral.

However, in inclined wells, the liquid hold-up (percent liquid by volume) is greater in a near-45° section than in vertical or horizontal flow. If the dogleg is greater than 10°/100 ft, then special high strength ESPs may have to be specified.

Figure 16-16 shows an ESP in the horizontal section of the well. This method has been successfully used to help avoid gas problems. The same problem about dogleg severity still applies.

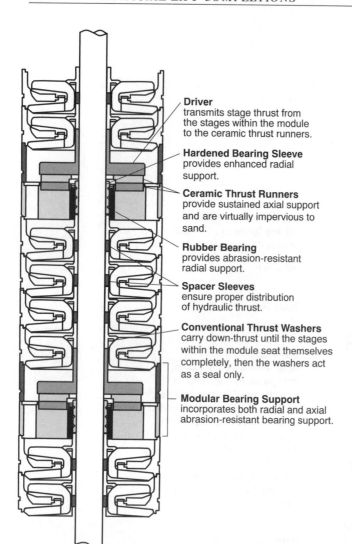

Driver
transmits stage thrust from
the stages within the module
to the ceramic thrust runners.

Hardened Bearing Sleeve
provides enhanced radial
support.

Ceramic Thrust Runners
provide sustained axial support
and are virtually impervious to
sand.

Rubber Bearing
provides abrasion-resistant
radial support.

Spacer Sleeves
ensure proper distribution
of hydraulic thrust.

Conventional Thrust Washers
carry down-thrust until the stages
within the module seat themselves
completely, then the washers act
as a seal only.

Modular Bearing Support
incorporates both radial and axial
abrasion-resistant bearing support.

Figure 16-11   ESP with hardened thrust bearings

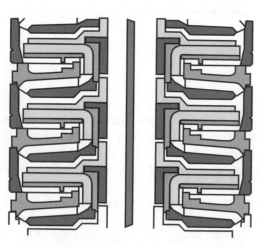

Figure 16-12   ESP with hardened inserts

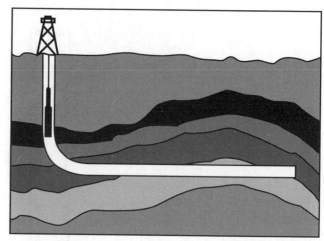

Figure 16-13   ESP in vertical section of a horizontal well

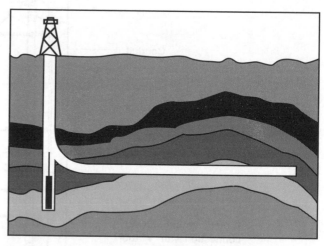

Figure 16-14   ESP in rathole of a horizontal well

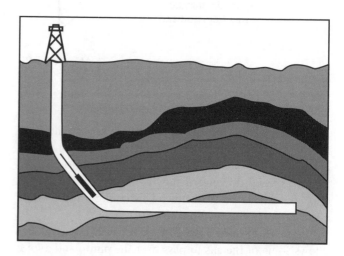

Figure 16-15   ESP in tangent section of a horizontal well

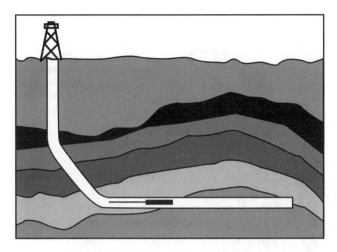

Figure 16-16 ESP in horizontal section

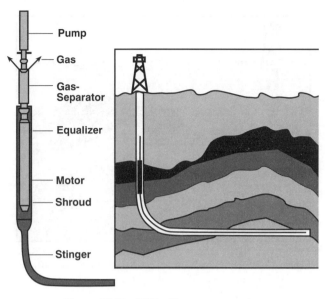

Figure 16-18 ESP with a gas separator

Figure 16-17 shows an ESP in the vertical section of the well but with a stinger inserted into the horizontal section as an attempt to avoid gas problems. The pump intake is still at low pressure, and there could be an additional amount of gas to handle as the pressure drops along the stinger to the pump intake, which would eliminate some of the effects of free gas present in the horizontal.

Figure 16-18 shows a stinger completion with a gas separator that is designed to separate gas to the casing before entering the pump. A problem with this arrangement is that gas separators might not function with a pressure differential of more than 15 psi because of the friction loss in the tubing. An extension of this idea is to

run a separate vent string for the gas separator, as shown in Figure 16-19.

Figure 16-20 shows an ESP with an inverted shroud over the pump intake. This arrangement allows gas to travel up the annulus while the fluids take a downward path to the pump intake, but would not function well in the horizontal lateral.

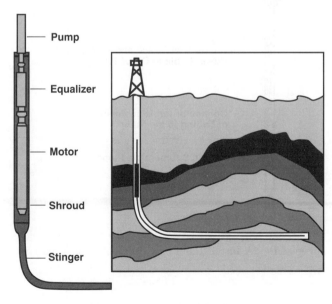

Figure 16-17 ESP with stinger

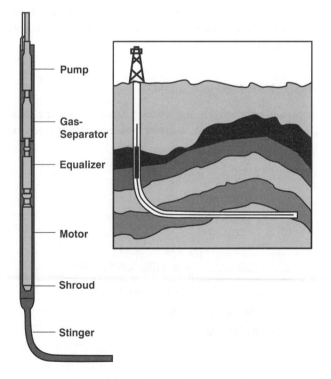

Figure 16-19 Stinger with vent string

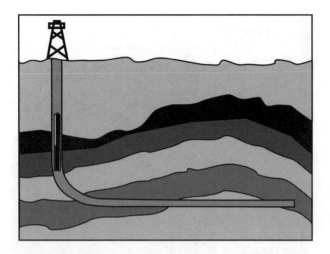

Figure 16-20   ESP with inverted shroud "portable rathole"

Other types of ESP completions are also used. Figure 16-21 shows an ESP run on coiled tubing (CT) with the cable banded to the CT as presented by Robison *et al.* (1992). A paper from the ESP workshop by Dublanko *et al.* (1995) describes the installation of ESPs using a power cable and a strength cable in the Gulf of Suez. Tovar *et al.* (1994) describes the installation of an ESP with the power cable installed inside CT. These are rare installations compared to the conventional installations mentioned above in more detail. Another special installation is to have the ESP pump downward to inject liquids in a gas well below a packer and into a water-taking zone.

16-3.1 ESP Design Example

A case is generated using water properties for the production fluid. To begin, a set of well conditions must be specified:

> Well Conditions:
> Depth: 5000 ft
> PI: 1 STB/d/psi,
> $p_r = 1200$ psi
> Tubing: $2\frac{3}{8}$ in. OD, 1.995 in. ID
> Surface Tubing Pressure: 50 psi
> Desired production: $= 1000$ STB/d
> BHT: 190°F

Select the number of stages required, motor size needed, cable size, and voltage loss.

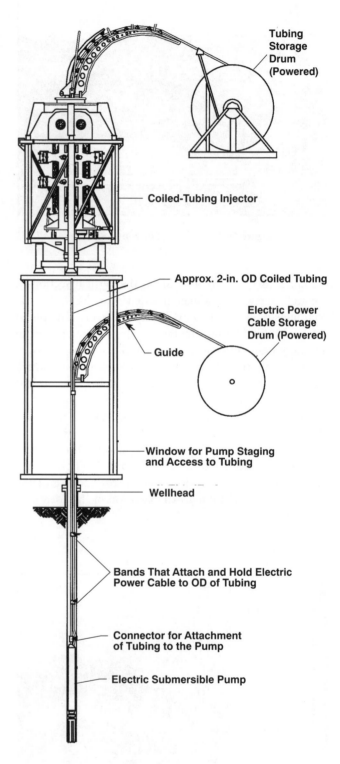

Tubing Storage Drum (Powered)

Coiled-Tubing Injector

Approx. 2-in. OD Coiled Tubing

Electric Power Cable Storage Drum (Powered)

Guide

Window for Pump Staging and Access to Tubing

Wellhead

Bands That Attach and Hold Electric Power Cable to OD of Tubing

Connector for Attachment of Tubing to the Pump

Electric Submersible Pump

Figure 16-21   Installation of a coiled tubing-deployed ESP

1. Find the pipe friction loss. Friction loss may be calculated by

$$\Delta p_f = (f)(\rho_f)\left(\frac{L}{d}\right)\left(\frac{v^2}{2g}\right) \qquad (16\text{-}1)$$

where Δp_f = friction loss, psi; f = friction factor, dimensionless; L = depth, or length of tubing, ft; d = internal diameter of tubing, in.; ρ_f = fluid density, psi/ft (0.433 psi/ft for water); v = fluid velocity; and g = gravitational constant, 32.2 lb_m-ft/(lb_f-s^2).

Most vendors provide charts for determining the friction loss for fluids flowing in a pipe, and most real designs use computer programs with multiphase flow pressure drop routines. In this example, it is assumed that the pipe roughness and viscosity and other fluid properties result in a Reynolds number in the pipe such that the friction factor f is 0.03.

The fluid velocity is calculated as the volumetric rate (1000 bbl/d) divided by the tubing area. The cross sectional area for 1.995-in. ID tubing is 0.0217 ft^2. Hence, the fluid velocity is

$$v = \frac{(1000\,\text{STB/d})(5.615\,\text{ft}^3/\text{bbl})}{(86{,}400\,\text{s/day})(0.0217\,\text{ft}^2)} = 3\,\text{ft/sec} \qquad (16\text{-}2)$$

Substituting into Equation 16-1,

$$\Delta p_f = (0.03)(0.433)\left(\frac{5000(12)}{1.995}\right)\left(\frac{3^2}{2(32.2)}\right) = 54.2\,\text{psi} \qquad (16\text{-}3)$$

2. Now calculate the pump discharge pressure, $p_\text{discharge}$

$$p_\text{discharge} = p_\text{surf} + \Delta p_{PE} + \Delta p_f \qquad (16\text{-}4)$$

where $p_\text{discharge}$ = pressure at pump discharge into tubing, psi; Δp_{PE} = potential energy or gravity component, psi; and p_surf = surface pressure, psi. Thus,

$$p_\text{discharge} = (50\,\text{psi}) + (0.433\,\text{psi/ft})(5000\,\text{ft})$$
$$+ 54.2\,\text{psi} = 2269\,\text{psi} \qquad (16\text{-}5)$$

3. Next calculate the intake pressure, p_intake

$$p_\text{intake} = q/(p_r - p_{wf}) \qquad (16\text{-}6)$$

Solving for sandface flowing pressure, p_{wf},

$$p_{wf} = p_r - \frac{q}{p_\text{intake}} = 1200 - \frac{1000}{1} = 200\,\text{psi} \qquad (16\text{-}7)$$

Note: In this example, it is assumed that the pump intake is set at the perforations. Since the pump is at the perforations, $p_{wf} = 200$ psi is also the p_intake at 1000 STB/d. If the pump is set above the perforations, p_intake will be less than p_{wf}.

4. Next find H, the head per stage, and BHP, the brake horsepower per stage at 1000 STB/d, from the example pump performance curve, Figure 16-22.

$$H = 23\,\text{ft/stage}$$

$$BHP \approx 0.3\,\text{hp/stage}$$

5. Find the number of stages, N_s required:

$$N_s = \frac{(p_\text{discharge} - p_\text{intake})}{H(0.433\,\text{psi/ft})} = \frac{(2269 - 200)}{(23)(0.433)} = 208\,\text{stages} \qquad (16\text{-}8)$$

This calculation assumes that the stages are turning at 3500 rpm, as indicated on the pump curve.

6. Find the minimum motor HHP of the motor required:

$$HHP = (N_s)(BHP/\text{stage})(\gamma_f) \qquad (16\text{-}9)$$

where γ_f is the specific gravity of the fluid

$$HHP = (208\,\text{stages})(0.3\,\text{BHP/stage})(1) = 62.4\,\text{hp} \qquad (16\text{-}10)$$

On the basis of motor specifications from a particular manufacturer, motors of 62 hp and 75 hp are available. If all of the data are certain, and if the well will not experience increase in production because of reservoir injection, then a 62-hp motor should be chosen. If there is some possibility that the data could be in error, then the 75-hp motor should be chosen to ensure that the motor will not be overloaded. Normally, a larger motor may be chosen because well data may not be accurate.

For this case, suppose that the 75-hp motor (218V, 22A at 60 Hz) is chosen. Often, the same hp motor may have different volt/amp combinations. The highest voltage rating gives the best electrical efficiency but puts more stress on the cable insulation. However, most operators choose the higher voltage.

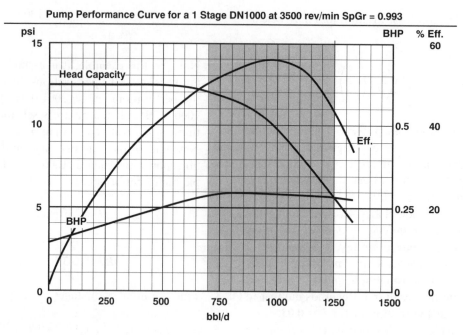

Pump Performance Curve for a 1 Stage DN1000 at 3500 rev/min SpGr = 0.993

Figure 16-22   Example pump performance curve

7. Next, the motor load should be checked by entering an appropriate motor curve showing rate vs. load and seeing what actual motor rate would be delivered. If the rate is not 3500 rpm as on the pump curve, then there is a need for iteration until the motor performance speed agrees with the pump performance.

The motor load is

$$\text{Load\%} = \left(\frac{HHP_{\text{required}}}{NPHP}\right)(100)$$

$$= \left(\frac{62.4}{75}\right)(100) = 83\% \qquad (16\text{-}11)$$

where (NPHP) is the nameplate horsepower. For the purposes of this example, it is assumed that the motor load gives a motor speed of 3500 rpm. Hence, the previous pump performance parameters from a 3500 rpm curve will not have to be adjusted although the pump curve shows that the speed could be a little lower.

8. The tubing intake curve for this design is plotted on Figure 16-23 with IPR curves for PIs of 0.75 and 1.25 in addition to the PI of 1 STB/d/psi used for the design selection. For simplicity, the friction factor of 0.03 is assumed to remain the same for the calculation of the $p_{\text{discharge}}$. For this example, the speed corrections plotted from the motor to the

pump are not included; the values are read at a particular rate and the load on the motor curve is calculated as the sum of the BHPs divided by the NPHP of 75 hp. However, a more sophisticated computer program should be able to make these corrections. The tabulated results are in Table 16-8.

Examination of Figure 16-23 shows that the design for 208 stages produces 1000 STB/d at 200 psi intake pressure. However, if the PI were 1.25 STB/d/psi, then the production would be about 1040 STB/d at about 370 psi intake pressure. If the PI were 0.75, the production would be 900 STB/d, but the intake pressure would be zero. This situation would be unacceptable, and the well would have to be choked back to a lower rate (moving the pump-tubing curve upwards and to the left) to avoid pumping off, gas interference, and inadequate pump charging pressure.

At this point, the depth, the well temperature, and the motor-required nameplate voltage and amps should be used to determine an appropriate cable size and the needed surface-voltage transformer taps. Details regarding this equipment selection can be found in manufacturer literature. Additional design details can be found in API 11S4, "Recommended Practice for Sizing and Selection of Electric Submersible Pump Installations."

The previous example is for one speed (3500 rpm) and for water. The following discussion details pump perfor-

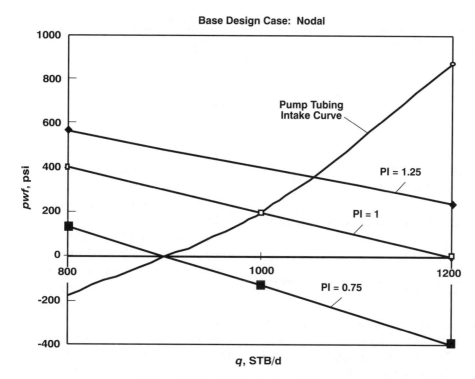

Figure 16-23   Nodal plot for ESP system

mance at different rotational speeds (rpm) and with variations in fluid properties.

An electrical submersible pump is a multi-stage stack of diffusers and centrifugal impellers. The performance of the pump is determined by

- The design of the stage

- The properties of the fluids being pumped

- The number of stages stacked in the pump housing or housings

- The rate of the pump-stage impellers

Fluid properties including specific gravity and viscosity are accounted for by correction factors to pump performance.

As the pressure is increased through the pump stages, the volume of any free gas is decreased and is finally eliminated entirely if the gas all goes back into solution as the fluid proceeds through the pump. The formation volume factor of the oil can also slightly decrease as the fluid passes through the pump. However, for a high-water-cut fluid with little or no free gas, the fluid through the pump has about the same volume into the pump as it

does leaving, so the pump performance can be considered as being at one point on a pump head curve.

Therefore, the total pump performance is determined as

$$H_{\mathrm{pump}} = (H)(N_s) \qquad (16\text{-}12)$$

where H_{pump} is the total differential head generated by the pump. Similarly, the total horsepower required by the pump is

$$HHP = (N_s)(BHP/\text{stage}) \qquad (16\text{-}13)$$

The relation of performance to speed is closely approximated by three relationships, derived from the "affinity" laws. The first of these states that flow through the pump, q, is directly related to the pump rate as follows:

$$\frac{q_2}{q_1} = \frac{RPM_2}{RPM_1} \qquad (16\text{-}14)$$

where the subscript '1' indicates the initial rate or speed, and the subscript '2' denotes the new conditions. The head produced is also proportional to the pump rate and is described by

Table 16-8   ESP Design Example

q bbl/d	p_d psi	H ft/stg	Δp psi	$p_{mp} : (p_d - \Delta p)$ psi	$p_{wf} : P_i = 1$	$p_{wf} : 1.25$	$p_{wf} : 0.75$
800	2249	27	2431	−182	400	560	133
1000	2269	23	2069	200	200	400	−133
1200	2293	15.8	1423	870	0	240	−400

$$\frac{H_2}{H_1} = \left(\frac{RPM_2}{RPM_1}\right)^2 \qquad (16\text{-}15)$$

The power required to drive the pump is also related to the rate by

$$\frac{BHP_2}{BHP_1} = \left(\frac{RPM_2}{RPM_1}\right)^3 \qquad (16\text{-}16)$$

Because the motor speed is proportional to the frequency (Hz) of the power supply, the pump performance taken from a pump curve at 3500 rpm (60 Hz) can be converted to other rates or to the proportional frequency of the power being supplied using the relationships in Equations 16-14, 16-15 and 16-16.

The relationships in Equations 16-14 and 16-15 are fairly accurate over a wide range of flows for the pump. The stages may actually show an increase in efficiency at higher frequencies, but this effect is small. A family of pump-stage operating characteristics can be obtained by using these affinity laws.

Variable-speed motors provide the ability to adjust the rotational speed of an ESP and can extend the operational range of the pump. In addition to speed control, the ESP variable drive serves to provide a "soft start," filters voltage transients from the motor, and provides a record of various diagnostics for the user. These devices are very effective, but they can be expensive, so some users have only a few for testing new wells and move them from well to well. These variable-speed motors are especially useful offshore where great expense is involved in changing out an ESP system to more closely match the system to the well.

16-4 GAS-LIFT COMPLETIONS

In a gas-lift completion, gas is injected into the tubing string at selected depths, either continuously or intermittently. The injected gas reduces the fluid gradient and, therefore, the hydrostatic head components of the pressure difference from the bottom to the top of the well.

When the hydrostatic head is lightened, fluids are able to flow to the wellhead.

A complete gas-lift system includes a source of gas; a surface injection system, including all related piping, compressors, control valves, etc.; a producing well completed with downhole gas-lift equipment; and the surface processing system, including all related piping, separators and control valves (Figure 16-24). The gas source may be gas produced from adjoining wells that is compressed and reinjected after separation. In most cases, a secondary source of gas is required to supply any shortfall in the gas from the separator.

There are two principal methods of gas lift: continuous and intermittent. For continuous gas lift, a self-regulating operating valve injects the appropriate amount of gas at the desirable tubing pressure at the injection depth. Other valves may be placed below the injection point and may be put into service during the life of the well as the reservoir pressure declines or if the water-oil ratio increases.

For intermittent gas lift, either a single injection point or multiple injection points can be used. In a single-point injection, a liquid slug is allowed to build up in the tubing above the bottom valve. The valve is opened, and gas displaces the liquid slug upward. The valve remains open until the liquid slug reaches the surface, and then the valve closes. The valve reopens when another liquid slug has formed, and the process is repeated. In multi-point intermittent injection, one or more unloading valves open as the liquid slug moves upward, and these valves also inject gas below the slug to assist in lifting the liquids to surface.

The gas-lift technique has a number of advantages over mechanical-lift techniques. It is the only lift method that can lift an already gassy well; it is not hindered by sand production; it can be used in deviated wells; and it does not require pump systems that suffer reduced effectiveness from heat, chemicals, light hydrocarbon ends, and gas. Gas lift is also often the simplest method of lift for high-volume lift, especially in remote areas, but it is not able to reduce the wellbore pressure to a pressure

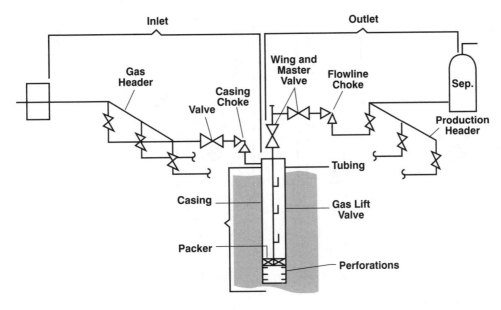

Figure 16-24

as low as many pumping systems can when they are operating to peak effectiveness.

16-4.1 Downhole Equipment

Downhole equipment for gas lift includes gas-lift valves and mandrels. Recent work such as that by Nieberding *et al.* (1993) has led to an understanding of the valve as a complex flow model instead of only a fully open orifice providing a reference point for determining flow. Brown *et al.* (1980) provides details on gas-lift valves. A brief discussion of this equipment is included here.

Gas-lift valves fall into one of three major categories: injection-pressure-operated valves, production-pressure-operated valves (Figure 16-25), and valves with an orifice rather than a closing mechanism. Orifice valves are used only as operating valves because they are always open to flow from the casing to the tubing. They are typically used in continuous-flow applications at the point of expected injection of gas and continuous operation.

Casing (tubing flow) or injection-pressure-operated valves are the most common valves used in unloading situations. Casing pressure valves are designed such that changes in the casing pressure have more effect in opening, closing, and moving the stem ball in and out of the seat in the valve than upon the tubing pressure. This allows control and diagnosis of the well by monitoring and controlling the casing pressure. However, it also requires some reduction of the casing pressure as the well unloads to lower and lower valves.

Tubing pressure-sensitive valves are used in annulus flow situations and when a particular tubing pressure is required, such as in an intermittent lift application where the valve is designed to remain closed until a sufficient fluid load is present in the tubing. The pressure-sensitive valves use either a dome charge (typically nitrogen) or a spring to provide the closing force. Normally, a bellows is used to contain the dome charge or to otherwise provide the operating area. All of the valves will include a reverse-flow check valve which prevents backflow. Tubing pressure valves may be used in places where casing pressure is limited. However, fluctuations in tubing pressure can cause periodic movement of the injection point up and down between lower valves, leading to inefficient lift or multi-point injection in some cases.

For a well where minimal intervention is a design objective, such as in a subsea completion, the spring-loaded valve may be the most reliable. With a spring-loaded valve, the closing force of the valve is either supplemented or provided entirely by a spring instead of a gas-charged bellows. If a bellows ruptures, the spring will keep the stem on the seat and the valve will remain closed. Spring-loaded valves are also desirable in wells where the temperature fluctuates because the pressure of the gas in a bellows-charged valve is not dependable.

Gas-lift mandrels are placed in the tubing string to position the gas-lift valve at the desired depth. Conventional mandrels accept threaded gas-lift valves mounted externally, and the valves are only retrievable by pulling the tubing, so they are not run where work-

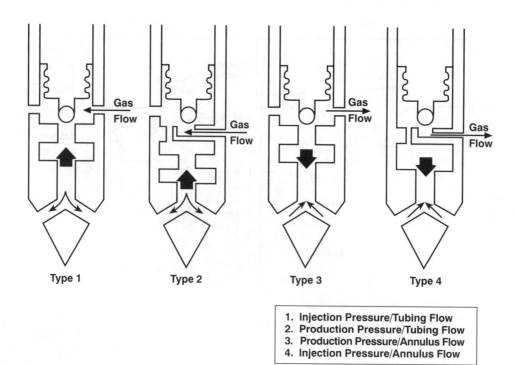

1. **Injection Pressure/Tubing Flow**
2. **Production Pressure/Tubing Flow**
3. **Production Pressure/Annulus Flow**
4. **Injection Pressure/Annulus Flow**

Figure 16-25   Basic types of gas-lift valves

over costs are prohibitive. Side-pocket mandrels (SMPs) have an internal, offset pocket that accepts a slickline-retrievable gas-lift valve. The pocket is accessible from within the tubing using a positioning or kickover tool (Figure 16-26), and the gas-lift valves use locking devices that lock into mating recesses in the SPM. Both types of mandrels are installed in the well in the same manner, but only the SPM system is serviceable with slickline operations for post-completion repair or well maintenance. Conventional mandrels are generally preferred in low-cost installations.

16-4.2 Unloading

When a well is completed, the completion fluids are present in the tubing and the annulus created between the tubing and casing. The fluid in the annulus must be displaced with the lift gas before the gas can reach the optimum point of injection, which is as deep in the well as possible. The surface pressure required to displace the fluid from the annulus into the tubing with the lift gas may be prohibitive, depending on the depth of the well and on the density of the completion fluid. The casing must be able to withstand the high pressure of the lift gas, and many operators are unwilling to subject aging casing to these pressures.

The most common alternative solution to this dilemma is to insert unloading valves at prescribed depths in the tubing. These valves are held open by the pressure from the completion fluid and provide a flowpath for the completion fluids to flow from the annulus into the tubing. When a valve starts to circulate the lift gas, the pressure in the tubing at and above the valve is reduced at the valve until it eventually reaches a preset closing pressure. The unloading valves close from the top down as the point of injection moves deeper into the well, finally reaching the depth of the operating valve. The depth that makes the most of the injection gas to achieve the lowest pressure at the sand face is an optimized rate of gas injection to a depth as near the perforations as possible.

The spacing of the unloading valves is critical to ensure that the system is operating in the most efficient mode, which occurs when the injection gas enters the tubing at the lowest point possible. Incorrect valve settings may cause an unloading valve to reopen and the injection point to move up the well, reducing well production. In addition, if the spacing of the valves is too wide, the pressure of the injection gas will not allow the gas to reach from valve to valve, and the unloading of the well will not proceed to the lowest valve. If valves are spaced too close together at the bottom, multiple-point

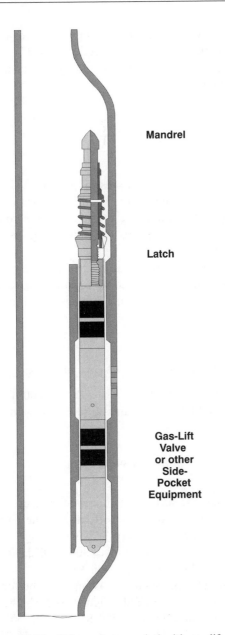

Figure 16-26 Side-pocket mandrel with gas-lift valve

16-4.3 Continuous Gas Lift

In continuous gas lift, gas is injected continuously through a single, self-regulating operating valve, and a series of unloading valves is generally included for initiating production. The depth of the operating valve, the injection gas pressure, gas-liquid ratio (GLR), and volume of gas to be injected can be calculated from well conditions such as desired rate, natural GLR, reservoir pressure, temperature gradient, and fluid composition.

A number of multiphase flow correlations (Chapter 15) can be used to generate pressure traverses, which are plots of pressures versus depth for a range of GLRs (Figure 16-27). Such curves can be represented by

$$p_{tf} + \Delta p_{\mathrm{trav}} = p_{wf} \qquad (16\text{-}17)$$

where Δp_{trav} is the pressure traverse and is a function of flow rate, the GLR, reservoir depth, and fluid properties.

Equation 16-17 can be expressed in terms of the flowing pressure gradient in the well

$$p_{tf} + \frac{dp}{dz} H = p_{wf} \qquad (16\text{-}18)$$

The pressure gradient, dp/dz is not constant, as indicated by the nature of the gradient curves. However, for very small or large values of GLR (nearly constant slope), this gradient can be read directly from the traverse curves.

Pressure traverse curves provide the relationship between p_{wf}, depth, and GLR for a specific producing rate and fluid properties. An understanding of this relationship is fundamental in designing gas-lift systems.

Equation 16-18 can be expanded as follows to express the addition of gas injection at some depth, H_{inj}:

$$p_{wf} = p_{tf} + H_{\mathrm{inj}} \left(\frac{dp}{dz}\right)_a + (H - H_{\mathrm{inj}})\left(\frac{dp}{dz}\right)_b \qquad (16\text{-}19)$$

where $(dp/dz)_a$ is the flowing pressure gradient above the point of gas injection and $(dp/dz)_b$ is the gradient below the injection point.

Figure 16-28 depicts the concept of continuous gas lift. With the available flowing bottomhole pressure and the flowing gradient $(dp/dz)_b$, fluids could not flow to surface, as indicated by the dashed line projecting $(dp/dz)_b$. For fluids to be produced to the surface, gas must be injected to reduce the gradient, shown as $(dp/dz)_a$. The alteration of the gradient allows fluids to be produced to the surface with a flowing tubing pressure, p_{tf}.

injection will occur, causing inefficient use of injection gas. Section 16-4.4 includes an example gas-lift design with five valves spaced in the completion.

Expected future well conditions must also be considered in spacing the mandrels and valves so they will last as long as possible without major alterations to the well.

Unloading valves are also used in the production sequence in wells producing with multipoint intermittent lift. In this case, the unloading valve opens as the liquid slug passes, allowing gas to enter into the tubing and assist in lifting the liquid slug to the surface.

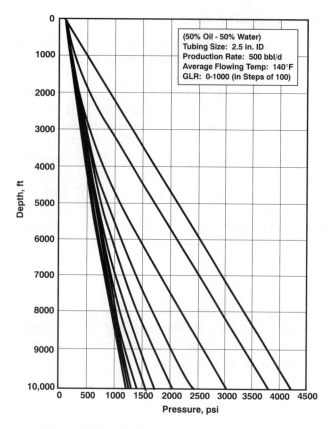

(50% Oil - 50% Water)
Tubing Size: 2.5 in. ID
Production Rate: 500 bbl/d
Average Flowing Temp: 140°F
GLR: 0-1000 (in Steps of 100)

Figure 16-27 Gradient curves for $q_L = 500$ STB/d

where p_{surf} is the is the surface injection pressure in psi and H_{inj} is the depth of gas injection in feet.

Finally, the horsepower required for gas compression can be calculated by

$$HHP = 2.23 \times 10^{-4} q_g \left[\left(\frac{p_{surf}}{p_{in}} \right)^{0.2} - 1 \right] \quad (16\text{-}22)$$

where p_{in} is the inlet compressor pressure. The energy required by the compressors is an important consideration in gas lift and must be considered when comparing alternative gas-lift designs or comparing gas lift to mechanical-lift systems.

The following example illustrates the use of pressure traverse curves and the preceding equations to determine the depth of gas injection and gas-injection rate.

16-4.3.1 Example: Point of Gas Injection

Suppose that a well at a depth of 8000 ft and a GLR = 300 SCF/STB drains a reservoir with an IPR given by

$$q_1 = 0.39(\bar{p} - p_{wf}) \quad (16\text{-}23)$$

What should the surface gas injection pressure be if the gas-lift valve is at the bottom of the well and $p_{inj} - \Delta p_{valve} = p_{wf} = 1000$ psi? [The average reservoir pressure, $\bar{p}$, is 3050 psi for $q_l = 800$ STB/d, as can be calculated readily from Equation 16-23.]

What should the point of gas injection be for a production rate of 500 STB/d? Figure 16-27 is the tubing performance curve for $q_l = 500$ STB/d with 50% water and 50% oil. Note from Figure 16-27 that $(dp/dz)_b = 0.33$ psi/ft for GLR = 300 SCF/STB between $H = 5000$ and $H = 8000$ ft. Use $\Delta p_{valve} = 100$ psi.

Solution. From Equation 16-21 and rearrangement and noting that $p_{inj} = 1100$ psi,

$$p_{surf} = 1100 / \left(1 + \frac{8000}{40,000} \right) = 915 \text{ psi} \quad (16\text{-}24)$$

For $q_l = 500$ STB/d, then from the IPR relationship (Equation 16-23)

$$p_{wf} = 3050 - \frac{500}{0.39} = 1770 \text{ psi} \quad (16\text{-}25)$$

The balance point shown in Figure 16-28 is the point in the tubing where the downhole pressure of the injected gas p_{inj} is equal to the pressure in the tubing. Normally, p_{inj} is reduced by 100 psi to allow a pressure drop across the gas-lift valve. Hence, to allow for the pressure drop across the valve, the actual point of injection is higher than the balance point as shown.

The volume of gas required in a gas-lift design can be estimated by

$$q_g = q_1(GLR_2 - GLR_1) \quad (16\text{-}20)$$

provided that the natural GLR, GLR_1, the working GLR, GLR_2, and the producing liquid rate, q_1, are known.

Gilbert (1954) provided the following relationship for estimating the injection pressure at the point of injection, p_{inj}:

$$p_{inj} \cong p_{surf} \left(1 + \frac{H_{inj}}{40,000} \right) \quad (16\text{-}21)$$

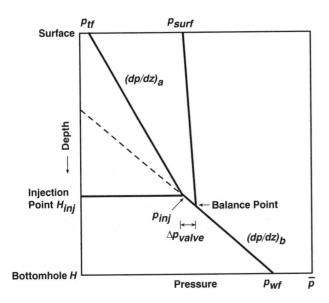

Figure 16-28 Concept of continuous gas lift (depth vs. pressure)

The injection point must be located where the pressure between the injected gas and the pressure in the production string must be balanced. Thus,

$$915\left(1 + \frac{H_{inj}}{40,000}\right) - 100 = 1700 - 0.33(8000 - H_{inj}) \tag{16-26}$$

and therefore $H_{inj} = 5490\,\text{ft}$.

Finally, the p_{inj} at H_{inj} is

$$p_{inj} = 915\left(1 + \frac{5490}{40,000}\right) \approx 1040\,\text{psi} \tag{16-27}$$

Inside the tubing, the pressure is 940 psi, since $\Delta p_{valve} = 100\,\text{psi}$.

The intersection between $p = 94\,\text{psi}$ and $H_{inj} = 5,490\,\text{ft}$ is at GLR $= 340\,\text{SCF/STB}$ (see Figure 16-27). Therefore, the gas-injection-rate from Equation 16-20 would be $q_g = 2 \times 10^4\,\text{SCF/d}$.

16-4.3.2 Limit GLR

The amount of gas injected in a gas-lift system can be increased if the gas is available. Increasing the amount of gas injected will increase the GLR, reduce the hydrostatic pressure component, and thus reduce the p_{wf}. If the reservoir pressure is constant (i.e. considering a "snapshot in time" of the reservoir), then the total liquid

rate, q_l, should increase. If the reservoir pressure is declining, then the reduction in p_{wf} associated with the increased GLR may sustain the production rate at a particular level.

There is, however, a limit value of GLR where the flowing pressure gradient is minimum. If the GLR is larger than the limit GLR, the flowing gradient will increase, mainly because of friction. If the GLR is smaller than the limit GLR, the gradient will increase because of the larger hydrostatic pressure. There is, therefore, a fixed quantity of injection gas which will lead to a physically optimum gas-lift performance for the producing system.

An optimum gas-lift performance curve can be drawn (Figure 16-29) using traverse curves and the limit GLRs. These curves depict the minimum flowing bottomhole pressures (minimum pressure intake curve) determined from the limit GLR for each producing rate. The intersection of the optimum gas-lift performance curve with the IPR defines the maximum producing rate with gas lift installed.

Although it is possible to determine the optimum gas-lift performance, there may not be sufficient gas supply for the optimum injection rate, or economics may dictate that a lower rate is more cost effective. The following example illustrates the economics of gas-lift design.

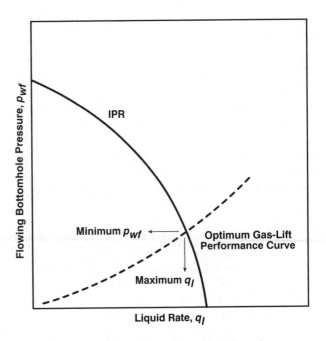

Figure 16-29 IPR with optimum gas-lift performance

16-4.3.3 Example: Economics of Gas Lift

Calculate the well deliverability for 0.5, 1, and 5×10^5 SCF/d gas injection rates, respectively, and compare them with the injection rate at the limit GLR, which is equal to 2.89×10^6 SCF/d. If the cost of gas injection is \$0.5/HHP-hr, the cost of separation is \$1/(MSCF/d), and the price of oil is \$30/STB, develop a simple economic analysis (benefits minus costs) versus available gas rate. Use the IPR curve for the example in Section 16-4.3.1, $\bar{p} = 3050$ psi, and a natural GLR = 300 SCF/STB.

Solution. If a gas injection rate of only 5×10^4 SCF/d is available, and if, for example, $q_l = 500$ STB/d, the GLR would be

$$\text{GLR} = \frac{5 \times 10^4}{500} + 300 = 400 \, \text{SCF/STB} \qquad (16\text{-}28)$$

From Figure 16-27 at GLR = 400 SCF/STB, the flowing bottomhole pressure (with $p_{tf} = 100$ psi) is 1500 psi. Similar calculations can be made with the gradient curves for other flow rates. The resulting gas-lift performance curve is shown in Figure 16-30. In addition to $q_g = 5 \times 10^4$ SCF/d, the two other curves, corresponding to 1 and 5×10^5 SCF/d, are also drawn. Intersections with the IPR curve would lead to the expected production rates and flowing bottomhole pressures. With the help of Equation 16-21, the surface pressure can be calculated (assuming 100 psi pressure drop across the valve). Finally, from Equation 16-22, the horsepower requirements are calculated (using $p_{in} = 100$ psi). The results are shown in Table 16-9.

An economic comparison among the various options is made below, considering the 5×10^4 SCF/d injection rate as a base case. Incremental benefits between this and the 1×10^5 SCF/d rate are $\Delta q_l \times (\$/\text{STB}) = (660 - 615) \times 30 = \$1350/\text{d}$. Incremental costs are $\Delta HHP \times 24 \times (\$HHP\text{-hr}) = (14.5 - 7.5)(24)(0.5) = \$84/\text{d}$ and $\Delta q_g \times \$1/MSCF = [(1 \times 10^5) - (5 \times 10^4)](1/1000) = \$50/\text{d}$.

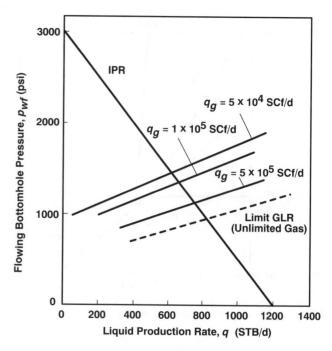

Figure 16-30 IPR with well performance for various gas injection rates

Therefore, the incremental benefits are $1350 - 134 = \$1220/\text{d}$.

Similarly, for $q_g = 5 \times 10^5$ SCF/d, incremental benefits are \$2900/d. Finally, for the limit GLR ($q_g = 2.8910^6$), the incremental benefits are −\$720/d; that is, the incremental production rate cannot compensate for the incremental compression and separation costs.

This type of calculation, of course, is always indicated. It depends heavily on local costs and, like everything else in production engineering, especially on the price of oil.

Properly designed gas-lift completions can provide flexibility for future lift as the reservoir pressure declines. Chapter 21 includes an example which illustrates planning a completion to accommodate gas lift in the future. This example shows how to determine the reservoir pressure at which gas lift is needed, and the injection depth to sustain a specific producing rate.

Table 16-9 Well performance with unlimited and limited gas injection

q_g (SCF/d)	q_l (STB/d)	p_{wf} (psi)	p_{surf} (psi)	HHP
5×10^4	615	1475	1310	7.5
1×10^5	660	1360	1220	14.5
5×10^5	750	1130	1025	66
2.89×10^6	825	935	865	348

16-4.4 Computer Generated Gas-Lift Design Example with Valve Spacing

Several software packages are available to assist in the spacing of the unloading valves, selecting the best depth for the operating valve, sizing the orifice in the operating valve, choosing the desired injection rate, and analyzing the performance of the well to optimize the recovery. A design sample of the spacing process is included below. Winkler and Smith (1983) is recommended reading for a more in-depth understanding of the design and optimization process, and other gas-lift references are listed in Clegg *et al.*, (1993). The *API Gas Lift Book 6* of the Vocational Training Series also gives a good overview of most gas-lift concerns.

A computer-generated gas-lift design example is shown here, based on the conditions given in Table 16-10.

The results are shown in Table 16-11, which summarizes the valve locations, and Figure 16-31, which shows the tubing-spacing gradient and casing pressure with depth.

Figure 16-31 (top) is a "feasibility" plot. The plot shows that the static fluid level extends to a depth a little below 1200 ft. It also shows that an "objective" gradient with, in this case, about 400 scf/STB added to the tubing will give about 950 psi flowing bottomhole pressure. It can also be seen that with the PI of 1 for the well, the objective gradient flowing bottomhole pressure is about equal to the pressure from the well perforations with the objective rate of 1000 STB/d total liquid. This means that the injected gas must reach near the bottom of the well to achieve the objective rate. Many wells have the design produced without the benefit of having the PI data as plotted in Figure 16-31 (bottom).

The design valve spacing is shown in Figure 16-31 (bottom). The first valve is typically located on either the fluid level or the maximum depth that the kickoff

Table 16-10 Gas-lift design example

Depth to perforations: 6000 ft	Separator pressure: 100 psi
Casing size: 7 in.	Water Cut: 50%
Tubing Size: currently $2\frac{7}{8}$ in	API: 30
GLR: 200 scf/STB	Water Gravity: 1.03
BHT: 200°F	Gas Gravity: 0.7
Surface flowing temp:	10°F
Injection pressure:	1100 psi kickoff
	1000 psi operating
IPR: Shut-in p_{wf}:	2000 psi
PI: 1 STB/d/psi	
Objective rate is 500 STB/d of oil or 1000 STB/d of total liquid	

Table 16-11 Valve information example

Valve	TVD	Port ID	Av/Ab	1-R	TRO@60°F
1	2367	0.1875	0.0380	0.9620	898
2	3947	0.1875	0.0380	0.9620	858
3	5027	0.1875	0.0380	0.9620	845
4	5712	0.1875	0.0380	0.9620	856

injection pressure will reach in the annulus. The deeper of these two choices is used. Using the kickoff pressure approach, the calculation would be to divide the injection kickoff pressure, p_{ko}, minus the separator pressure, p_{sep}, by the gradient of the dead liquid in the tubing, dp/dz. This would be $L_1 = (p_{ko} - p_{sep})/(dp/dz)$. If the first valve is set on the fluid level, the definition of the depth of the first valve, expressed as a formula, is $L_1 = \text{TVD} - p_r/(dp/dz)$, where TVD is the depth to the perforations, and p_r is the shut-in pressure of the reservoir. In this case, the first valve is set using the available casing injection pressure and is not set on the fluid level, which is at a shallower depth.

This procedure is repeated using injection-pressure operated valves [Figure 16-31 (bottom)]. The plot shows that a four-valve design will reach near the bottom of the well with the objective tubing gradient and the injection pressure of 1000 psi. Note that each valve is spaced slightly more to the left as depth is increased to ensure that upper valves are not opened as lower valves are reached. Also, using injection pressure valves means that the injection pressure is dropped by the valve settings with depth to ensure that the upper valves stay closed as the lower valves are reached. Because the bottom valve is not exactly at the well bottom, the design is such that the expected production, given the well PI, will be slightly less than the desired 1000 STB/d if the well data and the objective gradient turn out to be accurate when the well is unloaded.

Table 16-11 lists valve information for the four valves that were selected for this particular design. A_v/A_b is the ratio of the port diameter to the bellows diameter. The next column is one minus this ratio. The port IDs are selected to pass enough gas during the unloading process, and the bottom valve is selected to pass enough gas to reach the objective tubing gradient when the desired amount of gas is injected down the casing and through the bottom valve. The "TRO" is the test rack opening pressure of the valves, which is the pressure needed to open the valve in the shop when pressure is applied from the casing ports (for these injection-pressure operated valves) with atmospheric pressure only on

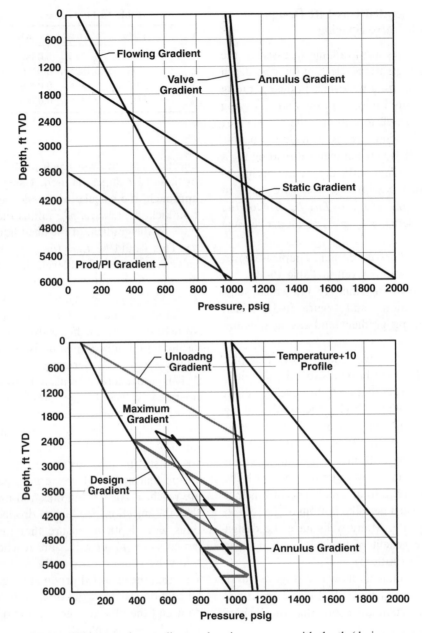

Figure 16-31 Tubing spacing gradient and casing pressure with depth (design example)

the port. The TRO is the shop pressure set so that the installed valves will have the correct nitrogen-charged dome pressure in the well to facilitate the unloading process.

The first valve can be placed on the fluid level during shut-in, or if the fluid level is unknown, the first valve can be placed no deeper than the available kickoff casing pressure can reach if the annulus is considered full of fluid to begin with.

Gas-lift installations provide flexibility throughout the life of the field because it is easy to redesign the completion for changing well conditions, and the same downhole components may be used again in the recompleted well. As the bottomhole pressure declines, at some point the well may not be able to be lifted economically by continuous lift. At this point, the well may be more economically and productively lifted by switching to intermittent lift.

16-4.5 Intermittent Lift

In intermittent lift, the well is produced and is periodically shut in until sufficient pressure builds in the casing to lift liquids and to produce the well for a brief time before shutting the well in again. The same downhole equipment that was used for continuous lift may be adapted to perform equally well intermittently.

The optimum transfer point where wells on continuous lift should be converted to intermittent lift is depicted in Table 16-12.

Efficient intermittent lift is achieved by using the largest possible orifice in the operating valve. When the valve opens, there should be no restriction to the flow of the injection gas, ensuring that the slug of liquid is transferred to the surface with the minimum amount of lift gas.

To enhance the production from wells on intermittent lift, a plunger is often inserted into the tubing string, forming a physical barrier between the lift gas and the fluid to reduce the fallback associated with the intermittent lift operation. In fact, a plunger is suggested in any intermittent lift operation unless it is operated at such a high frequency that there is no time for the plunger to fall during shut-in cycles.

Plungers are also used in many gas wells to unload fluids that tend to load the tubing, thereby reducing the inflow of gas. A typical plunger lift installation is shown in Figure 16-32. Plungers are operated with the gas from the well pressuring the annulus during a shut-in period. The well is opened, the plunger and accumulated liquids are allowed to surface, and gas is permitted to flow. If the well is low in the ratio of gas to produced liquids, the well may require gas to be injected down the annulus, or it may enter some unloading intermittent gas-lift valves located in the tubing. Plungers can pass gas-lift mandrels if the plunger has sufficient length to span the opening. The selection of the mandrel depends on the specific well conditions and servicing needs of gas-lift wells.

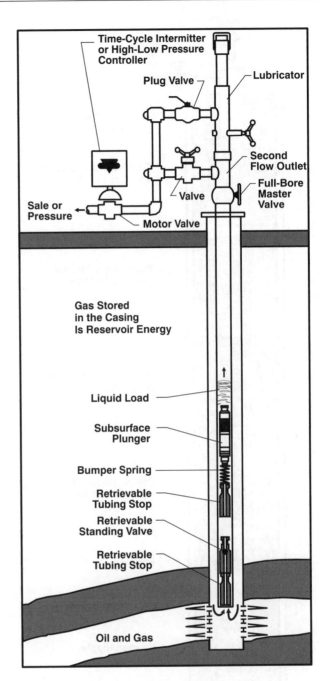

Figure 16-32 Plunger installation

Table 16-12 Maximum flow rates for intermittent gas lift

Tubing size	Flow rate
$2\frac{3}{8}$ in	150 STB/d
$2\frac{7}{8}$ in	250 STB/d
$3\frac{1}{2}$ in	300 STB/d
$4\frac{1}{2}$ in	Not advised

16-4.6 Gas-Lift Completions

The lowest-cost completion design uses conventional mandrels. This is shown in Figure 16-33. One of the benefits of using conventional mandrels is that it is very easy to integrate plunger lift with conventional mandrels, which is not true of installations with SPMs. The internal pocket of an SPM presents a problem for plun-

extensions have only worked for limited depths. Some typical gas-lift installations follow.

Figure 16-34 depicts a standard completion with retrievable valves in place. This is the most generic installation schematic for gas-lift operations. It presupposes that the casing string is able to withstand the injection and/or unloading pressures, which is not the case in many

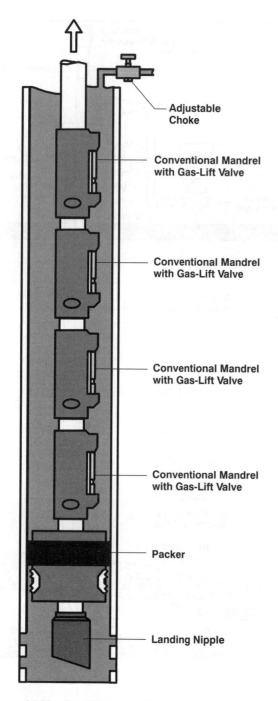

Figure 16-33 Gas-lift completion—conventional mandrels (injection-pressure operated valves)

ger operations because as the plunger assembly enters the SPM, the eccentric pocket will allow bypass of the lift gas, typically resulting in a loss of velocity, and in some cases may make it difficult for the plunger to reach the surface. Some operators have successfully used extensions to the plunger to straddle the pocket, but these

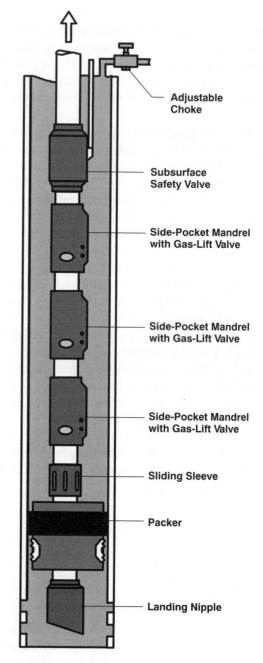

Figure 16-34 Gas-lift completion—SPMs and retrievable valves

older wells, where a separate tubing string is used to contain the lift gas.

In wells with questionable casing integrity, and where the high injection gas pressure cannot be allowed to contact the casing, a side-pocket mandrel is used. The valves are retrievable and the gas enters the pocket from a smaller side-mounted tubing string rather than from a gas-charged annulus, as shown in Figure 16-35. Gas-lift operations are efficient even in high-rate wells.

When the flow rates exceed the flow potential of the tubing, it is possible to reverse the production and inject down the tubing and flow up the annulus between the tubing and casing, as shown in Figure 16-36.

Wells that cross multiple producing intervals may use a dual-zone gas-lift scheme as shown in Figure 16-37. The annulus between the tubing strings and casing is used as a common supply for the lift gas. When this producing scheme is used in intermittent-lift wells, careful attention must be paid to the valve operating conditions to prevent the well from operating in an unstable fashion; one zone may operate as a thief zone, using an inordinate amount of the available lift gas, and making the other zone produce less than it is capable of and thereby driving both strings into a zone of unstable flow.

When the completion configuration prevents the point of injection from achieving the desired depth, or when the volume of gas in an intermittent-lift installation is less than acceptable, a chamber lift design is used. The basic concept is to create a chamber of sufficient volume to achieve the desired result, as is shown in Figure 16-38.

The number of horizontal wells has increased recently, leading a number of operators to consider gas lift in the horizontal section of the well. This practice is not advisable because in most near-horizontal pipelines, increasing the GLR will cause additional friction pressure in the horizontal section. The preferred completion configuration for a horizontal well on gas lift is to use the gas-lift mandrels only in the portion of the wellbore where the deviation from vertical is less than 70°. This configuration also allows the valves to be replaced by wireline if necessary.

The efforts to reduce the cost in all oilfield operations have led a number of suppliers to offer a gas-lift mandrel that is deployed on coiled tubing. This mandrel provides a quicker installation time, reduces cost, and may offer the advantage of reducing the flow area to make the overall gas-lift operation more efficient. While this is still a relatively new concept, it may become more popular for use in future completions. A typical mandrel is shown in Fig.16-39.

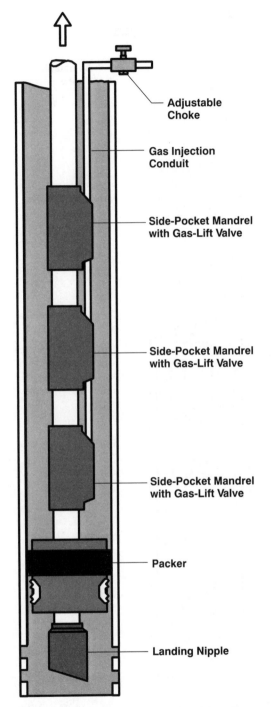

Figure 16-35   Gas-lift completion—retrievable valves, standpipe injection

In summary, gas lift provides an efficient method to produce wells that have a high GLR, where it is inadvisable to operate other pumping methods. Gas lift is a density-reduction technique rather than a pressure-boosting technique, as is the case with many pump meth-

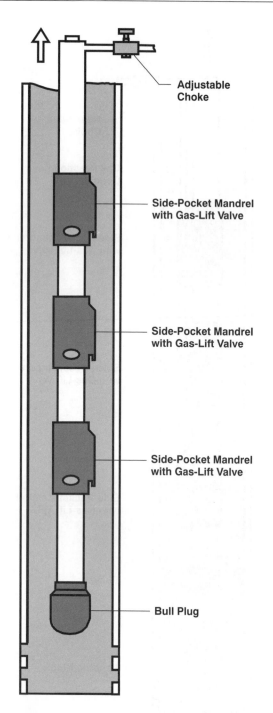

Figure 16-36 Gas-lift completion—retrievable valves, annular flow

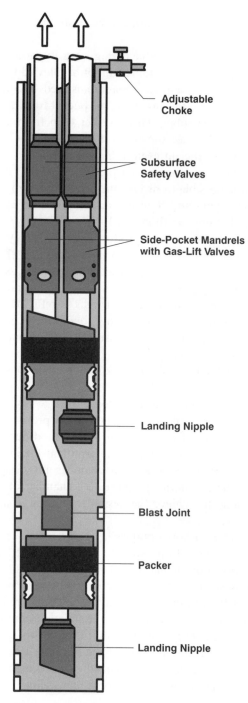

Figure 16-37 Gas-lift completion—retrievable valves, dual zone

ods. The compatibility of the injected gas and the produced fluids and the general cleanliness of the lift gas are critical considerations. Dirty lift gas may cause premature fouling of the gas-lift valves, increasing operating expenses considerably.

16-5 HYDRAULIC PUMP INSTALLATIONS

Hydraulic pumps provide another efficient artificial-lift option. Well candidates that lack the bottomhole pressure to sustain acceptable flowrates and have low GLRs, but which still have acceptable productivity capacity, are

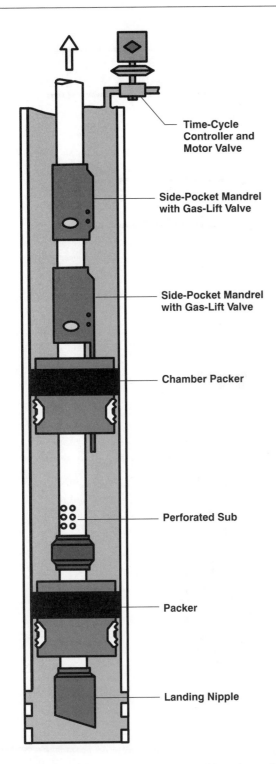

Figure 16-38 Gas-lift completion—retrievable valves, chamber installation

cies, and are serviceable with slickline well servicing operations.

Hydraulic pumps use a source of high-pressure power fluid—typically degassed and filtered production oil—to drive a downhole, positive-displacement piston pump, or to create a suction at the outlet of a venturi nozzle. While hydraulic jet pump drawdown performance does not match that of an ESP, hydraulic pumps are still capable of flowing as much as 15,000 STB/d or more. In many cases, one barrel of power fluid is required to lift one barrel of produced fluid, although this ratio can be much higher. Often the economics of providing the required volume of power fluid is the factor which determines whether a hydraulic pump should be used. For horizontal applications, the hydraulic pump may find increased acceptance because power cables or rod strings lack power. All that is required is a suitably sized flow conduit for the power fluid.

Figure 16-40 depicts the surface power unit and related accessories. This same type of power unit can be used for either a jet pump or a piston pump. In most installations, the power fluid, is commingled with the produced fluids, and a separator is required to ensure the removal of water or gas from the power fluid. The life of the downhole pump is affected by the cleanness of the power fluid. In the jet pump, the power fluid is always commingled with the production, whereas in some piston pumps (closed systems), the fluids may be segregated into separate flow conduits. The benefit in separating the power fluid is that less processing of the power fluid is required.

When the casing string is incapable of withstanding the pressure from the power fluid and/or when a closed system is used, the operator will install a separate tubing string for injecting the power fluid. A simple landing nipple for the pump assembly is all that is required downhole in most installations. No special packers are needed unless there are special circumstances for isolating an upper zone.

The piston pump is a positive displacement device (Figure 16-41). The volume of the produced fluid depends on the pump diameter and the number of strokes per minute, and the boost pressure depends on the pressure of the power fluid. The operation of the jet pump is significantly different. Jet pumps have no moving parts (Figure 16-42). The power fluid is directed through a converging nozzle into a plenum where the exit pressure creates a zone of low pressure. This drawdown is used to enhance the flow of fluids from the formation, and the two fluids are mixed in the plenum. A diverging nozzle is used to restore the pressure of the

ideal candidates for hydraulic pumps. Two basic pump types are piston pumps and jet pumps. Both pump types require similar surface equipment, offer good lift efficien-

The following labels appear in Figure 16-38:

- Time-Cycle Controller and Motor Valve
- Side-Pocket Mandrel with Gas-Lift Valve
- Side-Pocket Mandrel with Gas-Lift Valve
- Chamber Packer
- Perforated Sub
- Packer
- Landing Nipple

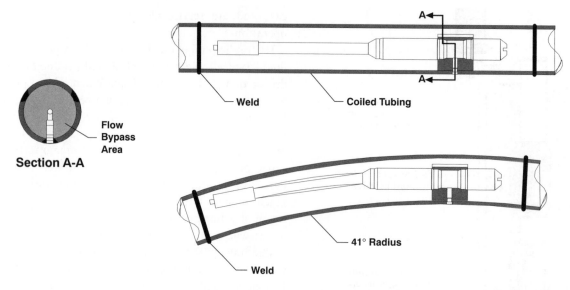

Figure 16-39   Spoolable gas-lift valve

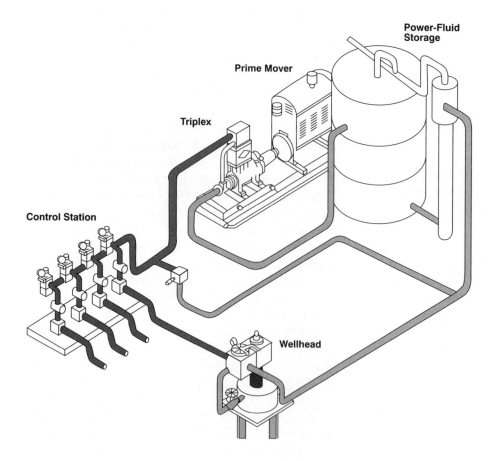

Figure 16-40   Hydraulic lift—central battery installation

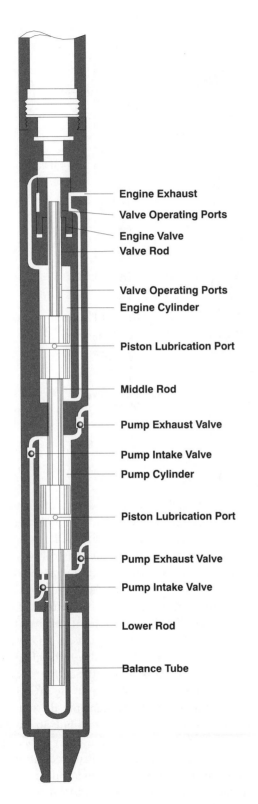

Engine Exhaust
Valve Operating Ports
Engine Valve
Valve Rod

Valve Operating Ports
Engine Cylinder

Piston Lubrication Port

Middle Rod

Pump Exhaust Valve

Pump Intake Valve
Pump Cylinder

Piston Lubrication Port

Pump Exhaust Valve

Pump Intake Valve

Lower Rod

Balance Tube

Figure 16-41 Hydraulic lift—downhole piston pump

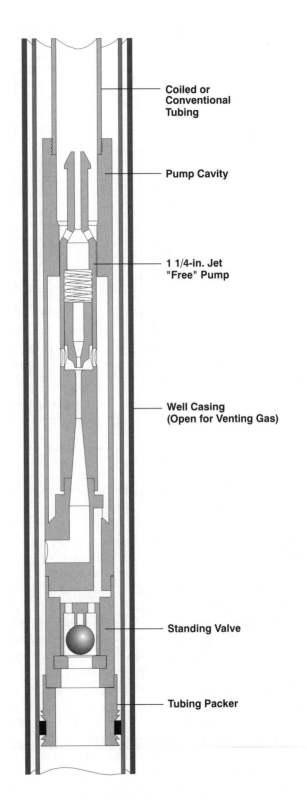

Coiled or
Conventional
Tubing

Pump Cavity

1 1/4-in. Jet
"Free" Pump

Well Casing
(Open for Venting Gas)

Standing Valve

Tubing Packer

Figure 16-42 Jet pump

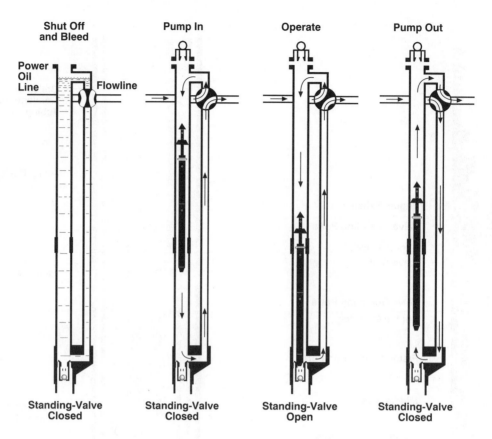

Figure 16-43 Hydraulic lift—free pump operation

commingled fluid so that it may be moved to the surface. Both types of pumps are easily serviceable.

Hydraulic pumps are designed to be easily removable. The power fluid may be used to circulate the pump to the surface, or it may be retrieved with slickline service operations; no costly service work is required. Figure 16-43 shows a "free" pump being circulated to the surface. This makes the operating economics very attractive. The downhole components (the pump receiver assembly) that are installable on the tubing string have no moving parts and require minimal service. Of course, the surface equipment is readily accessible at all times.

A typical jet pump with the pump installed inside a sliding side-door circulating device is very common and is the preferred design because it allows isolation of the casing and tubing whenever service work is performed on the well. Newer applications include small-diameter jet pumps run on small coiled tubing to unload gas wells.

Petrie *et al.* (1983), Petrie *et al.* (1983, 1984), and the petroleum production handbook (Frick, 1962) provide a more detailed explanation of the design technique used for hydraulic pumps. A field example of offshore applications is given by Boothby *et al.* (1988).

16-6 SUMMARY

Many challenges still exist in the design of artificial-lift systems. Modern well geometries differ significantly from the vertical wellbores where artificial lift has historically been applied. Horizontal wells, multilateral wells, high-temperature wells, wells with sand-laden production, wells with corrosive fluids, and other conditions all contribute to problems which must be considered when using artificial lift. The lift requirements may be for production of several thousands of barrels per day or may be for a very few barrels of water per day to be lifted from a gas well so that gas can produce efficiently.

Unique challenges also exist in applying lift systems in subsea wells. Subsea well requirements do not significantly change the artificial-lift system but do emphasize the need for reliability more than in other applications.

Even more complexity can be introduced if artificial-lift systems are considered in combination as presented, for example, by Divine *et al.* (1990) where ESPs in combination with gas lift are discussed as an alternative to increasing compression capabilities for the gas-lift system.

These and other design issues will provide the future advances in the application of artificial-lift systems in well completions.

REFERENCES

API 11S4, Recommended Practice for Sizing and Selection of Electric Submersible Pump Installations, First Edition (1993).

API Gas Lift Book 6, Am. Pet. Inst., Dallas.

Bennett, P.: "Artificial Lift Concepts & Timing," *Pet. Eng. Int.* (May 1980) 144–162.

Blais *et al.*: Pennwell Artificial Lift Methods Poster, 1986.

Boothby, L.K., Garred, M.A., and Woods, J.P.: "Application of Hydraulic Jet Pump Technology on an Offshore Production Facility," paper SPE 18236, 1988.

Brown, K.E. (ed.): *Technology of Artificial Lift Methods*, Pennwell Books, Tulsa, OK (1980).

Brown, K.E.: "Overview of Artificial Lift Systems," *JPT* (Oct. 1982) 2384–2394.

Clegg, J.D.: "High-Rate Artificial Lift," *JPT* (March 1988) 277–282.

Clegg, J.D.: "Another Look at Gas Anchors," *Proc., 1989 Annual Meeting of the SWPSC*, Lubbock, TX.

Clegg, J.D., Bucaram, S.M., and Hein, N.W.: "Recommendations and Comparisons for Selecting Artificial-Lift Methods," *JPT* (Dec. 1993) 1128–1167.

Divine, D.L., Eads, P.T., Lea, J.F., and Winkler, H.W.: "Combination Gas Lift and Electrical Submersible Pump Lift System," 1990 SPE Electric Submersible Pump Workshop, Houston, April 23–25.

Dublanko, G., Borgan, P., and Dwiggins, J.: "The Utilization of a Cable Deployed Pumping System," 1995 Annual SPE-ESP Workshop, Houston.

Frick, T. and Taylor, R.W. (eds.) *Petroleum Production Handbook, Vol. 1, Mathematics and Production Equipment SPE*, Richardson, TX (1962).

Gibbs, S.G.: "A General Method for Predicting Rod Pumping System," paper SPE 6850, 1977.

Gilbert, W.E.: "Flowing and Gas-Lift Well Performance," *Drill. and Prod. Prac.*, API (1954) 143.

Johnson, L.D.: "Here are Guidelines for Picking Artificial Lift Method," *O&GJ* (1968) 110–114.

Lea, J.F., and Kol, H.: "Economic Comparison of ESP vs. Hydraulic Lift in the Priobskoye Field in Siberia," 1995 Annual SPE-ESP Workshop, Houston.

Neely, A.B.: "A.B. Neely Discusses Artificial Lift Techniques, Uses, and Developments," *JPT* (Sept. 1980) 1546–1549.

Nieberding, M.A., Schmidt, Z., Blais, R.N., and Doty, D.R.: "Normalization of Nitrogen-Loaded Gas-Lift Valve Performance Data," *SPEPF* (Aug. 1993) 203–210.

Petrie, H.L., Wilson, P.M., and Smart, E.E.: "The Theory, Hardware, and Application of the Current Generation of Oil Well Jet Pumps," 1983 Southwestern Petroleum Short Course Asso., April 27–28.

Petrie, H.L., Wilson, P.M., and Smart, E.E.: "Design Theory, Hardware Options and Application Considerations," *World Oil* (Nov. 1983–Jan. 1984).

Podio, A.L., McCoy, J.N., and Woods, M.D.: "Decentralized, Continuous-Flow Gas Anchor," paper SPE 29537, 1995.

Robison, C.E., and Cox, D.C.: "Alternate Methods for Installing ESP's," *Proc., 24th Annual SPE Offshore Technology Conference*, Houston (1992) **4**, 483–488.

Takacs, G.: *Modern Sucker Rod Pumping*, Pennwell Books, Tulsa, OK (1993).

Tovar, J., and Head, P.: "First Field Installation of a Coiled Tubing Powered Electrical Submersible Pump Completion," paper SPE 28914, 1994.

Wilson, B.L.: "Experiences with ESP's in Horizontal Wells," 1995 SPE-ESP Roundtable, Houston.

Winkler, H.W., and Smith, S.S.: *Camco Gas Lift Manual*, Camco Inc., Houston (1983).

17 Well Stimulation

Peter Valkó
Texas A&M University

Lewis Norman
Halliburton Energy Services

Ali A. Daneshy
Halliburton Energy Services

17-1 INTRODUCTION

Well stimulation ranks second only to reservoir description and evaluation as a topic of research or publication within the well construction process. The reason for this intense focus is simple: this operation increases the production of petroleum from the reservoir. Thus, this facet of the construction process actively and positively affects a reservoir's productivity, whereas most of the other operations in this process are aimed at minimizing reservoir damage or eliminating production problems.

Additionally, the scope of this topic is sufficient to warrant several entire books with in-depth theory and analysis. Several authors have already done this for the industry (Gidley *et al.,* eds. 1989; Economides and Nolte, eds., 1989). It is not the intention of this discussion of well stimulation to produce another of these works. Rather, this chapter contains an insightful presentation of basic stimulation theory along with a discussion of several operational techniques not normally presented in technical literature. This information, woven into a well-referenced outline of stimulation functions, should prepare the reader to understand the basics and operation of well stimulation.

17-2 WELL STIMULATION AS A MEANS OF INCREASING THE PRODUCTIVITY INDEX

The primary goal of well stimulation is to increase the productivity of a well by removing damage in the vicinity of the wellbore or by superimposing a highly conductive structure onto the formation. Commonly used stimulation techniques include hydraulic fracturing, *fracpack*, carbonate and sandstone matrix stimulation (primarily acidizing), and fracture acidizing. Each of these stimulation techniques is intended to provide a net increase in the productivity index, which can be used either to increase the production rate or to decrease the drawdown pressure differential. A decrease in drawdown can help prevent sand production and water coning, and/or shift the phase equilibrium in the near-well zone toward smaller fractions of condensate. Injection wells also benefit from stimulation in a similar manner.

17-2.1 Productivity Index Before Stimulation

In discussing the productivity of a specific well, we think of a linear relation between production rate and driving force (drawdown):

$$q = J\Delta p \tag{17-1}$$

where the proportionality coefficient J is called the productivity index (PI). During its lifespan, a well is subject to several changes of flow conditions, but the two most important idealizations are constant production rate

$$\Delta p = \frac{Bq\mu}{2\pi kh}p_D \tag{17-2}$$

and constant drawdown pressure

$$q = \frac{2\pi kh\Delta p}{B\mu}q_D \tag{17-3}$$

The p_D and q_D functions are characteristic of the formation and are not independent from each other. For long-term behavior, the approximation $p_D \approx 1/q_D$ provides sufficient accuracy. It follows that

$$J = \frac{2\pi kh}{B\mu}q_D \approx \frac{2\pi kh}{B\mu p_D} \qquad (17\text{-}4)$$

therefore, it is easy to calculate the PI for an undamaged vertical well using the relations well known to petroleum engineers (Table 17-1).

Because of the radial nature of flow, most of the pressure drop occurs near the wellbore, and any damage in this region significantly increases the pressure loss. The impact of damage near the well can be represented by the skin effect s, added to the dimensionless pressure in the expression of the PI:

$$J = \frac{2\pi kh}{B\mu(p_D + s)} \qquad (17\text{-}5)$$

The skin is an idealization, capturing the most important aspect of near-wellbore damage; the additional pressure loss caused by the damage is proportional to the production rate.

bigger impact on the production of an "average" oil well than reducing it from 25 to 20. In other words, the skin effect defines an intuitively "nonlinear" scale to represent the quality of the well.

17-2.2 Economic Impact of Well Stimulation

Selection of the optimum size of a stimulation treatment is based primarily on economics. The most commonly used measure of economic effectiveness is the net present value (NPV). The NPV is the difference between the present value of all receipts and costs, both current and future, generated as a result of the stimulation treatment. Future receipts and costs are converted into present value using a discount rate and taking into account the year in which they will appear. Another measure of the economic effectiveness is the payout period; that is, the time it takes for the cumulative present value of the net well revenue to equal the treatment costs. The NPV (as other equivalent indicators) is sensitive to the discount rate and to the predicted future hydrocarbon prices.

As with almost any other engineering activities, costs increase almost linearly with the size of the stimulation treatment but (after a certain point) the revenues increase only marginally or may even decrease. Therefore, there is

Table 17-1   Flow into an undamaged vertical well

Flow regime	Δp	$p_D \ (\approx 1/q_D)$
Transient (Infinite-acting reservoir)	$p_i - p_{wf}$	$p_D = -\frac{1}{2}Ei\left(-\frac{1}{4t_D}\right)$ where $t_D = \dfrac{kt}{\phi\mu c_t r_w^2}$
Steady state	$p_e - p_{wf}$	$p_D = \ln(r_e/r_w)$
Pseudosteady state	$\bar{p} - p_{wf}$	$p_D = \ln(0.472 r_e/r_w)$

Even with the best drilling and completion practices, some kind of near-well damage is present in most wells. In a way, the skin can be considered as the measure of the quality of a well. It is reasonable to look at any type of stimulation as an operation reducing the skin. With the generalization of negative skin factor, even those stimulation treatments that not only remove damage but also superimpose some new or improved conductivity paths with respect to the originally undamaged formation can be put into this framework. In the latter case, it is more correct to refer to the pseudoskin factor, indicating that stimulation causes some changes in the streamline structure as well.

An important aspect of the skin factor is illustrated in Figure 17-1. Reducing the skin from 1 to zero has a

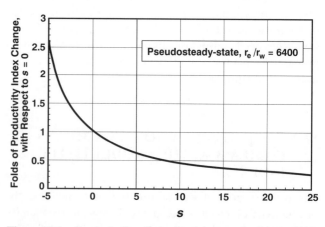

Figure 17-1   Typical oil-well productivity as a function of skin effect

an optimum size of the treatment that will maximize the NPV (Balen *et al.*, 1988). When using this general sizing technique, it is understood that for any given treatment size, we find the technically optimal way to create the fracture, and that is exactly when reservoir engineering knowledge becomes important.

17-3 BASIC PRINCIPLES OF HYDRAULIC FRACTURING

Hydraulic fracturing consists of injecting fluid into the formation with such pressure that it induces the parting of the formation. Proppants are used in hydraulic fracturing to prop or hold open the created fracture after the hydraulic pressure used to generate the fracture has been relieved. The fracture filled with proppant creates a narrow but very conductive path towards the wellbore. In almost all cases, the overwhelming part of the production comes into the wellbore through the fracture; therefore, the originally present near-wellbore damage is "bypassed," and the pretreatment positive skin does not affect the performance of the fractured well.

Perhaps the best single variable to characterize the size of a fracturing treatment is the amount of proppant placed into the formation. Obviously, more propped fracture volume increases the performance better than less, if placed in the right location. In accordance with the general sizing approach outlined above, the final decision on the size of the fracturing treatment should be made based on the NPV analysis. The question we have to answer here is more technical: how do we place a given amount of proppant in the best possible way?

17-3.1 Productivity Index Increase Because of Fracturing

Prats (1961) showed that except for the fracture extent, all the fracture variables affect the well performance through the combination

$$C_{fD} = \frac{k_f w}{k x_f} \qquad (17\text{-}6)$$

called dimensionless fracture conductivity, where k is the reservoir permeability, x_f is the half-length of the propped fracture, k_f is the permeability of the proppant pack, and w is the average fracture width. The product $k_f \times w$ itself is often referred to as the (dimensioned) fracture conductivity. When the dimensionless fracture conductivity is high (greater than 100, which is possible

to reach in very low-permeability formations with a massive hydraulic fracturing treatment), the behavior is similar to that of an infinite conductivity fracture studied by Gringarten and Ramey (1974). Interestingly enough, infinite conductivity behavior does not mean that we have selected the optimum way to place the given amount of proppant to the formation.

To characterize the impact of a finite-conductivity vertical fracture on the performance of a vertical well, it is indicative to present the pseudoskin factor as a function of the dimensionless fracture conductivity (Cinco-Ley and Samaniego, 1981). In other words, the PI for not-very-early times is given by

$$J = \frac{2\pi k h}{B\mu(p_D + s_f)} \qquad (17\text{-}7)$$

where the pseudoskin effect created by the propped fracture, s_f, can be read from the graph in Figure 17-2. The ordinate of the plot includes $\ln(r_w/x_f)$ because the skin effect is defined with respect to radial flow, where the inner boundary is the wellbore wall, while the performance of the fractured well is independent of the actual well radius. Hence, r_w is present only because of the definition of the skin effect, but drops out when added to the similar term in p_D. For large values of C_{fD}, the expression $\ln(x_f/r_w) + s_f$ approaches $\ln 2$, indicating that the production from an infinite conductivity fracture is $\pi/2$ times more than the production from the same surface arranged cylindrically (like the wall of a huge wellbore). In calculations it is convenient to use an explicit expression of the form

$$s_f = \ln \frac{r_w}{x_f} + \frac{1.65 - 0.328u + 0.116u^2}{1 + 0.18u + 0.064u^2 + 0.005u^3} \qquad (17\text{-}8)$$

where $u = \ln C_{fD}$ which reproduces the Cinco-Ley and Samaniego (1981) pseudoskin factor with sufficient accuracy. (An alternative concept to pseudoskin is the equivalent wellbore radius. In this chapter, we use only the pseudoskin concept because there is no reason to present the same results twice.)

17-3.2 Optimal Fracture Conductivity

In this context, a strictly technical optimization problem can be formulated: how to select the length and width if the propped volume of one fracture wing, $V_f = w \times h_f \times x_f$, is given as a constraint, and we wish to maximize the PI in the pseudosteady state flow regime.

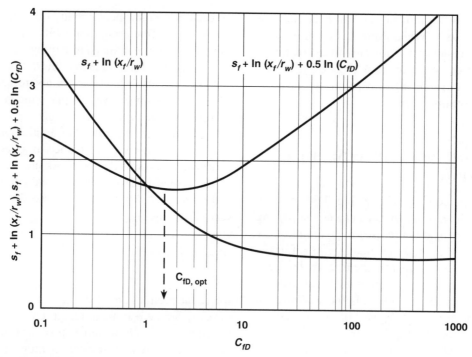

Figure 17-2   Pseudoskin factor for a finite-conductivity vertical fracture (after Cinco-Ley, Samaniego, and Dominguez, 1978)

It is assumed that the formation thickness, drainage radius, formation, and proppant pack permeabilities are known and the fracture is vertically fully penetrating; i.e., $h_f = h$. Selecting C_{fD} as the decision variable, the length is expressed as

$$x_f = \left(\frac{V_f k_f}{C_{fD} h k}\right)^{1/2} \qquad (17\text{-}9)$$

thus, the PI becomes

$$J = \frac{2\pi k h}{B\mu}$$

$$\times \frac{1}{\ln 0.472 r_e + 0.5\ln\dfrac{hk}{V_f k_f} + \left(0.5\ln C_{fD} + s_f + \ln\dfrac{x_f}{r_w}\right)} \qquad (17\text{-}10)$$

where the only unknown variable is C_{fD}. Since the drainage radius, formation thickness, two permeabilities, and propped volume are fixed, the maximum PI occurs when the quantity in parentheses

$$\left(0.51\ln C_{fD} + s_f + \ln\frac{x_f}{r_w}\right) \qquad (17\text{-}11)$$

becomes minimum. That quantity is also shown in Figure 17-2. Since the expression above depends only on C_{fD}, the optimum, $C_{fD,opt} = 1.6$, is a given constant for *any* reservoir, well, and proppant.

This result provides a deeper insight into the real meaning of dimensionless fracture conductivity. The reservoir and the fracture can be considered as a system working in series. The reservoir can deliver more hydrocarbon if the fracture is longer, but (since the volume is fixed) this means a narrower fracture. In a narrow fracture, the resistance to flow may be significant. The optimum dimensionless fracture conductivity corresponds to the best compromise between the requirements of the two subsystems. The optimum fracture half-length can be calculated as

$$x_f = \left(\frac{V_f k_f}{1.6 h k}\right)^{1/2} \qquad (17\text{-}12)$$

and, consequently, the optimum propped average width should be

$$w = \left(\frac{1.6 V_f k}{h k_f}\right)^{1/2} \qquad (17\text{-}13)$$

The most important implication of these results is that there is no theoretical difference between low- and high-permeability fracturing. In both cases, there exists a technically optimal fracture, and, in both cases, it should have a C_{fD} near unity. While in a low-permeability formation, this requirement results in a long and narrow fracture; in a high-permeability formation, a short and wide fracture provides the same dimensionless conductivity.

To illustrate this point, we will consider two examples. In the first example, we assume that the reservoir permeability is low, $k = 0.5 \times 10^{-15} \, m^2$ (0.5 md). If the reservoir drainage radius is 640 m (2100 ft), the formation thickness is 19.8 m (65 ft), the permeability of the created proppant bed is $60 \times 10^{-12} \, m^2$ (60,000 md), and the available proppant for one wing of the fracture will fill $6.13 \, m^3$ ($216.5 \, ft^3$); then, the optimum fracture half-length will be $x_f = 152.4$ m (500 ft) and the corresponding width will be 2.0 mm (0.08 in). If we assume a wellbore radius $r_w = 0.1$ m (0.328 ft) and a pretreatment skin effect $s = 5$, the optimum folds of increase in PI, calculated with Equations 17-5 and 17-7, will be 6.3. Figure 17-3 shows how the productivity index deteriorates with either too small or too large a dimensionless fracture conductivity, once the amount of proppant is fixed.

In the second example, we consider the same input data, but assume a high reservoir permeability, $k = 200 \times 10^{-15} \, m^2$ (200 md). Now the same amount of proppant should be placed into a 7.6-m (25-ft) long fracture with a width of 40.6 mm (1.6 in.). The optimum folds of increase will be "only" 2.57, but in production rate, this should be very significant.

Of course, the indicated "optimal fracture dimensions" may not be technically or economically feasible. In low-permeability formations, the indicated fracture length may be too large and the small width may be such that the assumed constant proppant permeability is not maintained any more. In high-permeability formations, the indicated large width might be impossible to create. In more detailed calculations, all the constraints have to be taken into account, but, in any case, a dimensionless fracture conductivity far away from the optimum indicates that either the fracture is a relative "bottleneck" ($C_{fD} < 1.6$) or it is unnecessarily packed with proppant instead of providing further reach into the formation ($C_{fD} > 1.6$).

An important issue to discuss is the validity of the pseudoskin concept for different reservoir types and flow regimes. In general, the pseudoskin created by a given fracture is constant only at late times. A fracture designed for optimal late-time performance may not be optimal at shorter times. In addition, for deeply penetrating fractures (that is, if the fracture half-length is already commensurate with the drainage radius), the pseudoskin

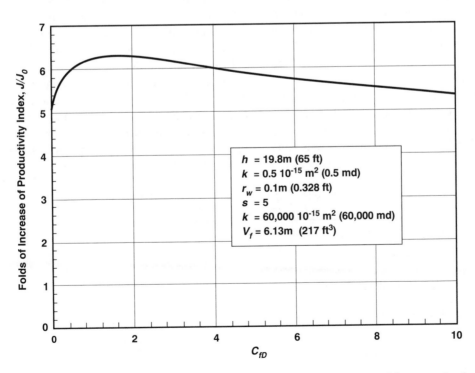

Figure 17-3   Example: folds of increase in productivity index because of fracture stimulation

function might deviate somewhat from the one shown in Figure 17-2. In fact, the whole concept of pseudoskin implies that we use one number to characterize the quality of a well and that number is the same for short and long times, for bounded or infinite-acting reservoirs. While not rigorous, it is a useful concept to understand the effect of dimensionless fracture conductivity on well performance.

A complete design process should include operational considerations such as tubing integrity, allowable surface pressure, and location place constraints. Geology also may place constraints on the treatment. For example, job size may be limited to keep the risk of fracturing into water zones reasonably low.

Once reservoir engineering considerations have dictated the fracture dimensions to create, the next issue is how to achieve that goal. The design of fracturing treatment involves the use of mechanical properties of the formation and the applied materials.

17-3.3 Mechanics of Fracturing: Linear Elasticity and Fracture Mechanics

Elasticity implies reversible changes. The appearance and propagation of a fracture means that the material has responded in an inherently non-elastic way and an irreversible change has occurred. Nevertheless, linear elasticity is a useful tool when studying fractures, because both the stresses and strains (except for the vicinity of the fracture face and especially the tip) may still be well described by elasticity theory.

A linear elastic material is characterized by elastic constants that can be determined in static or dynamic loading experiments. For an isotropic material, where the properties are independent of direction, two constants are sufficient to describe the behavior.

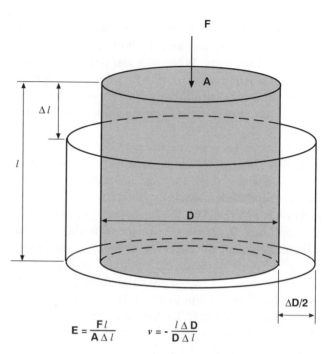

$$E = \frac{Fl}{A \Delta l} \qquad v = -\frac{l \Delta D}{D \Delta l}$$

Figure 17-4   Uniaxial loading test to obtain linear elasticity parameters

Figure 17-4 is a schematic representation of a static experiment with uniaxial loading. The two parameters obtained from such an experiment are the Young's modulus and the Poisson ratio. Table 17-2 shows the interrelation of those constants most often used in hydraulic fracturing. The plane strain modulus is the only elastic constant really needed in fracturing equations.

In linear elastic theory, the concept of plane strain is often used to reduce the dimensionality of a problem. It is assumed that the body is infinite at least in one direction, and external forces (if any) are applied parallel to this direction, "infinitely repeated" in every cross section. In such a case, it is intuitively obvious that the state of strain also "repeats itself" infinitely.

Table 17-2   Interrelations of various elastic constants of an isotropic material

Required/Known	E, v	G, v	E, G
Shear modulus, G	$\dfrac{E}{2(1+v)}$	G	G
Young's modulus, E	E	$2G(1+v)$	E
Poisson ratio, v	v	v	$\dfrac{E-2G}{2G}$
Plane strain modulus, E'	$\dfrac{E}{1-v^2}$	$\dfrac{2G}{1-v}$	$\dfrac{4G^2}{4G-E}$

Plane strain is a reasonable approximation in a simplified description of hydraulic fracturing. The main question is how to select the plane. Two possibilities have arisen, and they have generated two different approaches of fracture modeling. The state of plane strain was assumed horizontal by Khristianovitch and Zheltov (1955) and by Geertsma and de Klerk (1969) while plane strain in the vertical plane (normal to the direction of fracture propagation) was assumed by Perkins and Kern (1961) and Nordgren (1972). Often in the hydraulic fracturing literature, the term "KGD" geometry is used interchangeably with the horizontal plane-strain assumption, and "PKN" geometry is used as a substitute for postulating plane strain in the vertical plane.

Exact mathematical solutions are available for the problem of a pressurized crack in the state of plane strain. In particular, it is well known that the pressurized line crack has an elliptical width distribution (Sneddon, 1973):

$$w(x) = \frac{4p_0}{E'} \sqrt{c^2 - x^2} \qquad (17\text{-}14)$$

where x is the distance from the center of the crack, c is the half-length (the distance of the tip from the center), and p_o is the constant pressure exerted on the crack faces from inside. From Equation 17-14 the maximum width at the center is

$$w_0 = \frac{4cp_0}{E'} \qquad (17\text{-}15)$$

indicating that a linear relationship is maintained between the crack opening induced and the pressure exerted. When the concept of pressurized line crack is applied for a real situation, we substitute the p_0 with the net pressure, p_n, defined as the difference of the inner pressure and the minimum principal stress (also called far-field stress) acting from outside, trying to close the fracture (Hubbert and Willis, 1957; Haimson and Fairhurst, 1967).

Fracture mechanics has emerged from the observation that any existing discontinuity in a solid deteriorates its ability to carry loads. A hole (possibly small) may give rise to high local stresses compared to the ones being present without the hole. The high stresses, even if they are limited to a small area, may lead to the rupture of the material. It is often convenient to look at material discontinuities as stress concentrators that locally increase the otherwise present stresses.

Two main cases have to be distinguished. If the form of discontinuity is smooth (a circular borehole in a for-mation), then the maximum stress around the discontinuity is higher than the virgin stress by a finite factor, which depends on the geometry. For example, for a circular borehole, the stress concentration factor is 3.

The situation is different in the case of sharp edges (such as at the tip of a fracture). Then the maximum stress at the tip becomes infinite. In fracture mechanics, we have to deal with singularities. Two different loadings (pressure distributions) of a line crack result in two different stress distributions. Both cases may yield infinite stresses at the tip, but the "level of infinity" is different. We need a quantity to characterize this difference. Fortunately, all stress distributions near the tip of any fracture are similar in that they decrease with $r^{-1/2}$, where r is the distance from the tip. The quantity used to characterize the "level of infinity" is the stress intensity factor, K_I, defined as the multiplier to the $r^{-1/2}$ function. For the idealization of a pressurized line crack with half-length, c, and constant pressure, p_0, the stress intensity factor is given by

$$K_I = p_0 c^{1/2} \qquad (17\text{-}16)$$

In other words, the stress intensity factor at the tip is linearly proportional to the constant pressure opening up the crack and to the square root of the characteristic dimension.

According to the key postulate of Linear Elastic Fracture Mechanics (LEFM), for a given material, there is a critical value of the stress intensity factor, K_{IC}, called fracture toughness. If the stress intensity factor at the crack tip is above the critical value, the crack will propagate; otherwise, it will not. Fracture toughness is a useful quantity for safety calculations when the engineer's only concern is to avoid fracturing. In well stimulation, where the engineer's primary goal is to create and propagate a fracture, the concept has been found somewhat controversial because it predicts that less and less effort is necessary to propagate a fracture with increasing extent. In the large scale, however, the opposite is true. Fracture toughness discussion also appears in Chapter 6 along with a discussion of fracture propagation direction and effects of near-wellbore variations such as perforation orientation or wellbore deviation.

17-3.4 Fracturing Fluid Mechanics

Fluid materials deform (flow) continuously without rupture when subjected to a constant stress. Solids generally will assume a static equilibrium deformation under the same stresses. Crosslinked fracturing fluids usually behave as viscoelastic fluids. Their stress-strain material functions fall between those of pure fluids and solids.

From our point of view, the most important property of fluids is their resistance to flow. The local intensity of flow is characterized by the shear rate, $\dot{\gamma}$, measured in 1/s. It can be considered as the rate of change of velocity with the distance between sliding layers. The stress emerging between the layers is the shear stress, τ. Its dimension is force per area, in SI it is measured in Pa. The material function relating shear stress and shear rate is the rheological curve. Knowing the rheological curve is necessary to calculate the pressure drop (in fact the energy dissipation) in a given flow situation such as flow in pipe or flow between parallel plates.

Apparent viscosity is defined as the ratio of stress-to-shear rate. In general, the apparent viscosity varies with shear rate, but, for a very specific fluid, the Newtonian fluid, the viscosity is a constant. The rheological curve and the apparent viscosity curve contain the same information and are used interchangeably. Figure 17-5 shows typical rheological curves, and Table 17-3 lists some commonly used rheological constitutive equations. The model parameters vary with chemical composition, temperature and, to a lesser extent, with many other factors, including shear history. In the case of foams, the volumetric ratio between the gas and liquid phases plays an important role (Reidenbach, 1986; Winkler *et al.*, 1994).

Most fracturing gels exhibit significant shear thinning, such as loss of viscosity, with increasing shear rate. A constitutive equation capturing that most important

aspect of their flow behavior is the power law model. The flow behavior index, n, usually ranges from 0.3 to 0.6.

All fluids will exhibit some finite, high-shear, limiting viscosity. The build up of very high apparent viscosity at low shear might be approximated by the inclusion of a yield stress for certain fluids. Many fluids demonstrate what appear to be Newtonian low-shear plateaus. Much of the current rheology research in the oil field focuses on building more realistic apparent viscosity models that effectively incorporate each of the previously mentioned characteristics. Researchers also are exerting additional effort toward modeling the nonlinear, time-dependent, viscoelastic effects of crosslinked gels.

The main purpose for using a rheological model is to obtain the pressure-loss gradient expressed by the average flow velocity. The equations of motion have been solved for flow in circular tubes, annuli, and thin-gap parallel plates for most of the simple models. The solution is often presented as a relation between average linear velocity (flow rate per unit area) and pressure drop. In calculations, it is convenient to use the equivalent Newtonian viscosity, μ_e, which is the viscosity we should use in the equation of the Newtonian fluid to obtain the same pressure drop under the same flow conditions. While apparent viscosity (at a given local shear rate) is the property of the fluid, equivalent viscosity depends also on the flow geometry and carries the same information as the pressure drop solution. For more complex rheological models, there is no closed-form solution (neither for the pressure drop, nor for the equivalent Newtonian viscosity), and the calculations involve numerical root finding.

Of particular interest to hydraulic fracturing is the laminar flow in two limiting geometries. Slot flow occurs in a channel of rectangular cross section with an extremely large ratio of the longer side to the shorter side. Limiting ellipsoid flow occurs in an elliptic cross section with extremely large aspect ratio. The former corresponds to the KGD geometry and the latter to the PKN geometry. Collected in Table 17-4 are the solutions commonly used in hydraulic fracturing calculations.

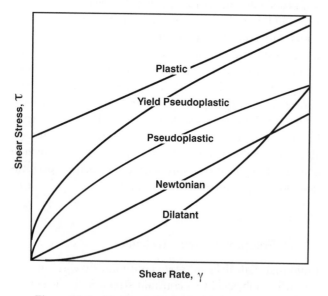

Figure 17-5 Idealized rheological behavior of fluids

Table 17-3 Commonly used rheological constitutive equations

$\tau = \mu\dot{\gamma}$	Newtonian
$\tau = K\dot{\gamma}^n$	Power law
$\tau = \tau_y + \mu_p\dot{\gamma}$	Bingham plastic
$\tau = \tau_y + K\dot{\gamma}^n$	Yield power law

Table 17-4 Pressure drop and equivalent newtonian viscosity

	Newtonian $\tau = \mu\dot\gamma$	Power law $\tau = K\dot\gamma^n$
Slot flow	$\dfrac{\Delta p}{L} = \dfrac{12\mu u_{\text{avg}}}{w^2}$	$\dfrac{\Delta p}{L} = 2^{n+1}\left(\dfrac{1+2n}{n}\right)^n Kw^{-n-1}u_{\text{avg}}$ $\mu_e = \dfrac{2^{n+1}}{3}\left(\dfrac{1+2n}{n}\right)^n Kw^{1-n}u_{\text{avg}}^{n-1}$
Flow in limiting elliptical cross section	$\dfrac{\Delta p}{L} = \dfrac{16\mu u_{\text{avg}}}{w_0^2}$	$\dfrac{\Delta p}{L} = \dfrac{2^{3+n}}{\pi}\left[\dfrac{1+(\pi-1)n}{n}\right]^n Ku_{\text{avg}}^n w_0^{-n-1}$ $\mu_e = \dfrac{2^{n-1}}{\pi}\left[\dfrac{1+(\pi-1)n}{n}\right]^n Kw^{1-n}u_{\text{avg}}^{n-1}$

Note that the equation for the laminar flow of a power law fluid in limiting ellipsoid geometry has not been derived. The solution presented here can be obtained by analogy considerations; for details, see Valkó and Economides (1995).

In calculating the friction pressure through the tubular goods, the well-known turbulent flow correlations are less useful for fracturing fluids, and special relations have to be applied because of the drag reduction phenomena caused by the long polymer chains. Rheological behavior also plays an important role in the proppant carrying capacity of the fluid (Roodhart, 1985; Acharya, 1986.)

17-3.5 Leakoff and Fracture Volume Balance

The polymer content of the fracturing fluid is partly intended to impede the loss of fluid into the reservoir. The phenomenon is envisioned as a continuous build up of a thin layer (the filter cake) that manifests an ever-increasing resistance to flow through the fracture face. In the formation, the actual leakoff is determined by a coupled system, of which the filter cake is only one element. A fruitful approximation dating back to Carter, 1957 (Appendix to Howard and Fast, 1957), considers the combined effect of the different phenomena as a material property. According to this concept, the leakoff velocity v_L, is given by the Carter I equation:

$$v_L = \frac{C_L}{\sqrt{t}} \tag{17-17}$$

where C_L is the leakoff coefficient (length/time$^{1/2}$) and t is the time elapsed since the start of the leakoff process. The integrated form of the Carter equation is

$$\frac{V_{\text{Lost}}}{A_L} = 2C_L\sqrt{t} + S_p \tag{17-18}$$

where V_{Lost} is the fluid volume that passes through the surface A_L during the time period from time zero to time t. The integration constant, S_p, is called the spurt loss coefficient. It can be considered as the width of the fluid body passing through the surface instantaneously at the very beginning of the leakoff process. Correspondingly, the term $2C_L\sqrt{t}$ can be considered the leakoff width. (Note that the factor 2 appears because of integration. It has nothing to do with "two wings" and/or "two faces" introduced later.) The two coefficients, C_L and S_p, can be determined from laboratory tests and/or from evaluation of a fracture calibration test.

17-3.5.1 Formal Material Balance: The Opening-Time Distribution Factor

Consider a fracturing treatment shown schematically in Figure 17-6. The volume V_i, injected into one wing during the injection time t_e, consists of two parts: the volume of one fracture wing at the end of pumping (V_e) and the volume lost. We use the subscript e if we wish to emphasize that a given quantity refers to the end of pumping. Note that all the variables are defined with respect to one wing. The area A_e denotes the surface of one face of one wing. Fluid efficiency is defined as the fraction of the fluid remaining in the fracture: $\eta_e = V_e/V_i$. The average width, $\overline{w}$, is defined by the relation $V = A\overline{w}$.

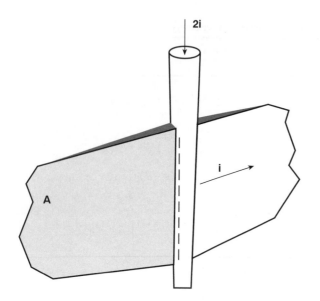

Figure 17-6   Material balance variables

A hydraulic fracturing operation may last from tens of minutes to several hours. Points of the fracture face near the well are opened at the beginning of pumping while the points at the fracture tip are younger. Application of Equation 17-18 necessitates the tracking of the opening-time of the different fracture face elements. If only the overall material balance is considered, it is natural to rewrite the injected volume as the sum of fracture volume, leakoff volume, and spurt volume using the formalism

$$V_i = V_e + K_L\left(2A_e C_L \sqrt{t_e}\right) + 2A_e S_p \qquad (17\text{-}19)$$

where the variable K_L is the opening-time distribution factor. It reflects the history of the evolution of the fracture surface, or rather the distribution of the opening time, hence the elegant name. In particular, if all the surface is opened at the beginning of the injection, then K_L reaches its absolute maximum, $K_L = 2$. The fluid efficiency is the ratio of the created to the injected volume. Dividing both volumes by the final area gives the efficiency as the ratio of the created width to the would-be width, where the would-be width is defined as the sum of the created and lost widths. Therefore, another form of Equation 17-18 is

$$\eta_e = \frac{\overline{w}_e}{\overline{w}_e + 2K_L C_L \sqrt{t_e} + 2S_p} \qquad (17\text{-}20)$$

showing that the term $2K_L C_L \sqrt{t_e}$ can be considered as the "leakoff width," while $2S_p$ is the spurt width.

Equation 17-20 can be rearranged to obtain the opening-time distribution factor in terms of fluid efficiency:

$$K_L = -\frac{S_p}{C_L \sqrt{t_e}} - \frac{\overline{w}_e}{2C_L \sqrt{t_e}} + \frac{\overline{w}_e}{2\eta_e C_L \sqrt{t_e}} \qquad (17\text{-}21)$$

Note that these relations are independent of the actual shape of the fracture face or the history of its evolution.

17-3.5.2 Constant Width Approximation (Carter Equation II)

To obtain an analytical solution for constant injection rate, Carter considered a hypothetical case when the fracture width remains constant during the fracture propagation (the width "jumps" to its final value in the first instant of pumping). Then, a closed-form expression can be given for the fluid efficiency in terms of the two leakoff parameters and the width:

$$\eta_e = \frac{\overline{w}_e\left(\overline{w}_e + 2S_p\right)}{4\pi C_L^2 t_e}\left[\exp(\beta^2)\mathrm{erfc}(\beta) + \frac{2\beta}{\sqrt{\pi}} - 1\right]$$

where

$$\beta = \frac{2C_L \sqrt{\pi t_e}}{\overline{w}_e + 2S_p} \qquad (17\text{-}22)$$

If calculated from Equation 17-21, the opening-time distribution factor remains between 4/3 (1.33) and $\pi/2$ (1.57) for any combination of the variables. This result is a consequence of the "fixed propagation width" assumption of the Carter II equation, but does not follow from the Carter I equation itself.

17-3.5.3 Power Law Approximation to Surface Growth

A basic assumption postulated by Nolte (1979, 1986) led to a remarkably simple form of the material balance. He assumed that the fracture surface evolves according to a power law

$$A_D = t_D^\alpha \qquad (17\text{-}23)$$

where $A_D = A/A_e$ and $t_D = t/t_e$ and the exponent α remains constant during the whole injection period. Nolte realized that in that case the factor is a function only of α. He denoted the dependence of the opening-time distribution factor on the exponent of surface growth by $g_0(\alpha)$ and presented g_0 for selected values of α. A simple expression first obtained by Meyer and Hagel (1989) can be used to obtain the value of the opening-time distribution factor for any α:

$$g_0(\alpha) - \sqrt{\pi}\alpha\Gamma(\alpha)/\Gamma(\alpha + 2/3) \qquad (17\text{-}24)$$

where $\Gamma(\alpha)$ is the Euler gamma function. In calculations, the following approximation to the g_0 function might be easier to use:

$$g_0(\alpha) =$$

$$\frac{2 + 2.06798\alpha + 0.541262\alpha^2 + 0.0301598\alpha^3}{1 + 1.6477\alpha + 0.738452\alpha^2 + 0.0919097\alpha^3 + 0.00149497\alpha^4}$$

$$(17\text{-}25)$$

Nolte assumed that the exponent remains between 0.5 and 1. With this assumption, the factor K_L lies between 4/3 (1.33) and $\pi/2$ (1.57), indicating that for two extremely different surface growth histories, the opening-time distribution factor varies less than 20% and, in general, the simple approximation $K_L = 1.5$ should provide enough accuracy for design purposes.

The exponent α has been related to the type of width equation and/or to the efficiency at the end of pumping and/or to the rheological behavior of the fluid. While none of these relations can be considered as proven theoretically, they are reasonable engineering approximations, especially because the effect of the exponent α on the final results of a model is limited. Our recommendation is to use $\alpha = 4/5$ for the PKN, $\alpha = 2/3$ for the KGD, and $\alpha = 8/9$ for the radial model, because these exponents can be derived from the no-leakoff equations shown later in Table 17-5. Numerically, the original constant-width approximation of Carter and the power-law surface-growth assumption of Nolte give very similar results when used for design purposes. The g_0-function approach has, however, technical advantages when applied to the analysis of calibration treatments.

17-3.5.4 Detailed Leakoff Models

The bulk leakoff model is not the only possible interpretation of the leakoff process. Several mechanistic models have been suggested in the past (Williams, 1970; Settari, 1985; Ehlig-Economides *et al.*, 1994; Yi and Peden, 1994; Mayerhofer *et al.*, 1995). The total pressure difference between the inside of the created fracture and a far point in the reservoir is written as a sum

$$\Delta p(t) = \Delta p_{\text{face}}(t) + \Delta p_{\text{piz}}(t) + \Delta p_{\text{res}}(t) \qquad (17\text{-}26)$$

Table 17-5 No-leakoff behavior of the basic width equations

Model	Perkins and Kern	Geertsma and deKlerk	Radial
Fracture extent	$x_f = c_1 t^{4/5}$	$x_f = c_1 t^{2/3}$	$R_f = c_1 t^{4/9}$
	$c_1 = c_1'\left(\dfrac{i^3 E'}{\mu h_f^4}\right)^{1/5}$	$c_1 = c_1'\left(\dfrac{i^3 E'}{\mu h_f^3}\right)^{1/6}$	$c_1 = c_1'\left(\dfrac{i^3 E'}{\mu}\right)^{1/9}$
	$c_1' = \left(\dfrac{625}{512\pi^3}\right)^{1/5} = 0.524$	$c_1' = \left(\dfrac{16}{21\pi^3}\right)^{1/6} = 0.539$	$c_1' = 0.572$
Width	$w_{w,0} = c_2 t^{1/5}$	$w_w = c_2 t^{1/3}$	$w_{w,0} = c_2 t^{1/9}$
	$c_2 = c_2'\left(\dfrac{i^2 \mu}{E' h_f}\right)^{1/5}$	$c_2 = c_2'\left(\dfrac{i^3 \mu}{E' h_f^3}\right)^{1/6}$	$c_2 = c_2'\left(\dfrac{i^3 \mu^2}{E'^2}\right)^{1/9}$
	$c_2' = \left(\dfrac{2560}{\pi^2}\right)^{1/5} = 3.04$	$c_2' = \left(\dfrac{5376}{\pi^3}\right)^{1/6} = 2.36$	$c_2' = 3.65$
Net pressure	$\overline{w} = \gamma w_{w,0}$ $\gamma = 0.628$ $p_{n,w} = c_3 t^{1/5}$	$\overline{w} = \gamma w_w$ $\gamma = 0.785$ $p_{n,w} = c_3 t^{-1/3}$	$\overline{w} = \gamma w_{w,0}$ $\gamma = 0.533$ $p_{n,w} = c_3 t^{-1/3}$
	$c_3 = c_3'\left(\dfrac{E'^4 \mu i^2}{h_f^6}\right)^{1/5}$	$c_3 = c_3'\left(E'^2 \mu\right)^{1/3}$	$c_3 = c_3'\left(E'^2 \mu\right)^{1/3}$ $c_3' = 2.51$
	$c_3' = \left(\dfrac{80}{\pi^2}\right)^{1/4} = 1.52$	$c_3' = \left(\dfrac{21}{16}\right)^{1/3} = 1.09$	

where Δp_{face} is the pressure drop across the fracture face dominated by the filtercake, Δp_{piz} is the pressure drop across a polymer-invaded zone, and Δp_{res} is the pressure drop in the reservoir. Depending on their significance under the given conditions, one or two terms may be neglected. While the first two terms are connected to the leakoff rate valid at the given time instant, the reservoir pressure drop is "transient" in the sense that it depends on the whole history of the leakoff process, not only on its instant intensity. Advantages of the detailed leakoff models include explicit pressure dependence of the leakoff volume and inclusion of physically more sound parameters, such as permeability and filtercake resistance. The application of these models is limited by the complexity of the mathematics involved and by the extra input they need.

17-3.6 Basic Fracture Geometries

Engineering models for the propagation of a hydraulically induced fracture combine elasticity, fluid flow, material balance, and (in some cases) an additional propagation criterion. Given a fluid injection history, a model should predict the evolution with time of the fracture dimensions and the wellbore pressure. For design purposes, a rough description of the geometry might be sufficient, and therefore, simple models predicting length and average width at the end of pumping are very useful. Models that predict these two dimensions, while the third one—fracture height—is fixed are referred to as two-dimensional (2D) models. In addition, if the fracture is postulated to have a circular surface (radial propagation), the model is still considered 2D. A further simplification occurs if we can relate length and width and at first neglect the details of leakoff. This idea is the basic concept of the early width equations. It is assumed that the fracture evolves in two identical wings, perpendicular to the minimum principal stress of the formation. Since the minimum principal stress is usually horizontal (except for very shallow formations), the fracture will be vertical.

17-3.6.1 Perkins-Kern Width Equation

The model assumes that the condition of plane strain holds in every vertical plane normal to the direction of propagation, but, unlike in a rigorous plane strain situation, the stress and strain state is not exactly the same in subsequent planes. In other words, the model applies a quasi-plane-strain assumption and the reference plane is vertical, normal to the propagation direction. Neglecting the variation of pressure along the vertical coordinate, the net pressure, p_n, is considered as a function of the lateral coordinate x. The vertically constant pressure at a given lateral location gives rise to an elliptical cross section. Straightforward application of Equation 17-15 provides the maximum width of the ellipse as

$$w_0 = (2h_f/E')p_n \qquad (17\text{-}27)$$

where h_f is the constant fracture height and E' is the plane strain modulus. The maximum width, w_0, is a function of the lateral coordinate. At the wellbore, it is denoted by $w_{w,0}$. The pressure drop of a Newtonian fluid in a limiting elliptical cross section is given in Table 17-4.

Perkins and Kern (1961) postulated that the net pressure is zero at the tip and they approximated the average linear velocity of the fluid at any location with the one-wing injection rate (i) divided by the cross-sectional area ($w_0 \times h_f \times \pi/4$):

$$\frac{dp_n}{dx} = -\frac{4\mu i}{\pi w_0^3 h_f} \qquad (17\text{-}28)$$

Combining Equations 17-27 and 17-28 and integrating with the zero net pressure condition at the tip results in the width profile

$$w_0(x) = w_{w,0}\left(1 - x/x_f\right)^{1/4} \qquad (17\text{-}29)$$

where the maximum width of the ellipse at the wellbore is given by

$$w_{w,0} = \left(\frac{512}{\pi}\right)^{1/4}\left(\frac{\mu i x_f}{E'}\right)^{1/4} = 3.57\left(\frac{\mu i x_f}{E'}\right)^{1/4} \qquad (17\text{-}30)$$

In reality, the flow rate in the fracture is less than the injection rate, not only because part of the fluid leaks off, but also because the increase of width with time "consumes" another part of the injected fluid. In fact, what is more or less constant along the lateral coordinate at a given time instant is not the flow rate, but rather the flow velocity (flow rate divided by cross section). The interesting fact is, however, that repeating the Perkins-Kern derivation with a constant-flow velocity assumption hardly affects the final results.

Equation 17-30 is the Perkins-Kern width equation. It shows the effect of the injection rate, viscosity, and shear modulus on the width, once a given fracture length is achieved. Knowing the maximum width at the wellbore, we can calculate the average width, multiplying it by a constant shape factor, γ,

$$\bar{w} = \gamma w_{w,0} \quad \text{where} \quad \gamma = \frac{\pi}{4}\frac{4}{5} = \frac{\pi}{5} = 0.628 \quad (17\text{-}31)$$

The shape factor contains $\pi/4$ because the vertical shape is an ellipse. It also contains another factor, 4/5, which accounts for the lateral variation of the maximum width of the ellipse.

In the petroleum industry, the modification of Equation 17-30 with a slightly different constant is used more often and is referred to as the Perkins-Kern-Nordgen (PKN) width equation (Nordgren, 1972):

$$w_{w,0} = 3.27 \left(\frac{\mu i x_f}{E'}\right)^{1/4} \quad (17\text{-}32)$$

17-3.6.2 Khristianovich-Zheltov-Geertsma-deKlerk-Daneshy Width Equation

The first model of hydraulic fracturing, elaborated by Khristianovich and Zheltov (1955) envisioned a fracture with the same width at any vertical coordinate within the fixed height, h_f. The underlying physical hypothesis is that the fracture faces slide freely at the top and bottom of the layer. The resulting fracture cross section is a rectangle. The width is considered as a function of the coordinate x. It is determined from the plane strain assumption, now applied in (every) horizontal plane. The Khristianovich and Zheltov model contained another interesting assumption: the existence of a non-wetted zone near the tip. Geertsma and deKlerk (1969) accepted the main assumptions of Khristianovich and Zheltov and reduced the model into an explicit width formula.

Recalling the plane strain solution, the fracture width at the wellbore (now not varying along the vertical coordinate) is

$$w_w = \frac{4 x_f \bar{p}_n}{E'} \quad (17\text{-}33)$$

where $\bar{p}_n$ is some equivalent lateral average of the net pressure. Geertsma and deKlerk assumed a rectangular cross-sectional area. The pressure gradient at some location x, where the width is w and the average linear velocity is known, can be obtained from the corresponding slot-flow entry of Table 17-4. The average linear velocity of the fluid at any location was approximated with the one-wing injection rate (i) divided by the cross-sectional area ($w \times h_f$):

$$\frac{dp_n}{dx} = -\frac{12\mu i}{w^4 h_f^2} \quad (17\text{-}34)$$

Combining Equations 17-33 and 17-34 is not straightforward, but Geertsma and deKlerk estimated $\bar{p}_n$ and arrived at the explicit formula for the width at the wellbore:

$$w_w = \left(\frac{336}{\pi}\right)^{1/4}\left(\frac{\mu i x_f^2}{E' h_f}\right)^{1/4} = 3.22 \left(\frac{\mu i x_f^2}{E' h_f}\right)^{1/4} \quad (17\text{-}35)$$

In this case, the shape factor, relating the average width to the wellbore width, has no vertical component and, because of the elliptical shape, we obtain

$$\bar{w} = \gamma w_w \quad \text{where} \quad \gamma = \frac{\pi}{4} = 0.785 \quad (17\text{-}36)$$

It is convenient to refer to Equation 17-35 as the GdK width equation. If, however, the concept of horizontal plane strain approximation is to be emphasized, it is usual to refer to KGD geometry or the KGD view of the fracture.

Daneshy's (1978) model considers a non-constant pressure distribution along the fracture length, and a non-Newtonian fracturing fluid with properties that can change with time and temperature. The numerical computations calculate the specific leakoff, increase in width, and flow rate at points along the fracture length during fracture extension. The computations are further refined by ensuring compatibility between the fluid and fracture mechanics aspects of the problem.

We can calculate the ratio of the Geertsma-deKlerk average width to the Perkins-Kern average width:

$$\frac{\bar{w}_{GdK}}{\bar{w}_{PK}} = \left(\frac{21 \times 625}{32 \times 512}\right)^{1/4}\left(\frac{2 x_f}{h_f}\right)^{1/4} = 0.95 \left(\frac{2 x_f}{h_f}\right)^{1/4} \quad (17\text{-}37)$$

For short fractures, where $2x_f < h_f$, the horizontal plane strain assumption (KGD geometry) is more appropriate, and for $2x_f > h_f$, the vertical plane strain assumption (PKN geometry) is physically more sound. The obtained ratio of the average widths shows that the "transition" between the two commonly used width equations is essentially "smooth," since the two width equations give nearly the same average width for a fracture with equal vertical and horizontal dimension ($2x_f = h_f$).

17-3.6.3 Radial (Penny-Shaped) Width Equation

This situation corresponds to horizontal fractures from vertical wells, vertical fractures extending from horizontal wells, or, when fracturing relatively thick homogeneous formations, all from a limited perforated interval. While the computations of fracture width are sensitive to how the fluid enters the fracture (a "true" point source would give rise to infinite pressure), a reasonable model can be postulated by analogy, which results in the same average width as the PK equation when $R_f = x_f = h_f/2$. The result is

$$\bar{w} = 2.24 \left(\frac{\mu i R_f}{E'} \right)^{1/4} \qquad (17\text{-}38)$$

(Depending on the particular author's preference in applying analogies, various constants are used in the literature.)

Considering the effect of the input data on the evolving fracture, the real significance of the simple models is the insight they provide. Additional insight can be gained by comparing the behavior of the models in the case when leakoff can be neglected. Table 17-5 shows the no-leakoff behavior of the basic models.

The last row of the table deserves particular attention. The net pressure increases with time for the PK width equation but decreases with time for the other two equations. This result is a well-known result that raises some questions because, in massive hydraulic fracturing, the net treating pressure most often increases with time. As a consequence, the calculated net pressure from the GdK and radial width equations is of limited practical value. Even more startling is the other (less well-known) consequence of the GdK and radial width equations: the net pressure as a function of time does not depend on the injection rate. The reason is that the KGD (and radial) view of the fracture implies that the larger the fracture extent, the lower the net pressure needed to maintain a certain width. While it is a consequence of linear elasticity theory and the way the plane strain assumption is applied, it leads to absurd results in the large scale, characteristic of hydraulic fracturing. This poses a very serious obstacle to using the KGD and radial width equations for pressure related investigations.

17-3.7 Fracture Design

At this point, we have sufficient information to present a simple design procedure. It is assumed that h_f, E', i, μ, C_L, and S_p are known, and a target length, x_f is specified. The problem is to determine the pumping time, t_e using the combination of a width equation such as PKN and material balance. The first part of a typical design procedure is shown in Table 17-6. Refinement techniques are delineated in Tables 17-7 to 17-9.

If the permeable height, h_p is less than the fracture height, it is convenient to use exactly the same method, but with apparent leakoff and spurt loss coefficients. The apparent leakoff coefficient is the "true" leakoff coefficient (the value with respect to the permeable layer) multiplied by the factor r_p defined as the ratio of permeable to fracture surface (see Figures 17-7 and 17-8). For the PKN and KGD geometries, it is the ratio of permeable to fracture height

---

Table 17-6 Pumping time determination

1 Specify a certain target length: x_f.

2 Calculate the wellbore width at the end of pumping from the PKN (or any other) width equation:

$$w_{w,0} = 3.27 \left(\frac{\mu i x_f}{E'} \right)^{1/4}$$

3 Convert wellbore width into average width
$$\bar{w}_e = 0.628 w_{w,0}$$

4 Assume a $K_L = 1.5$.

5 Solve the following material balance equation for t (note that selecting $\sqrt{t}$ as the new unknown means a simple quadratic equation has to be solved):

$$\bar{w}_e + 2S_p + 2K_L C_L \sqrt{t} = \frac{i}{h_f x_f} t$$

Denote the larger root by t_e.

6 At this point, one may stop or refine K_L.

Table 17-7 Refinement of K_L using the Carter II equation

Calculate an improved estimate of K_L from

$$K_L = -\frac{S_p}{C_L \sqrt{t_e}} - \frac{\bar{w}_e}{2 C_L \sqrt{t_e}} + \frac{\bar{w}_e}{2 \eta_e C_L \sqrt{t_e}}$$

where

$$\eta_e = \frac{\bar{w}_e (\bar{w}_e + 2S_p)}{4 \pi C_L^2 t_e} \left[\exp(\beta^2) \mathrm{erfc}(\beta) + \frac{2\beta}{\sqrt{\pi}} - 1 \right]$$

and $$\beta = \frac{2 C_L \sqrt{\pi t_e}}{\bar{w}_e + 2 S_p}$$

If K_L is near enough to the previous guess, stop; otherwise, continue the iteration with the next estimate of K_L.

Table 17-8 Refinement of K_L by linear interpolation according to Nolte

Estimate the next K_L from

$$K_L = 1.33\eta_e + 1.57(1 - \eta_e)$$

where

$$\eta_e = \frac{\overline{w}_e x_f h_f}{it_e}$$

If K_L is near enough to the first guess, then stop; otherwise, continue the iteration with the new K_L.

Table 17-9 K_L from the α method

Assume a power law exponent α and calculate $K_L = g_0(\alpha)$ using Equation 17-25. Use the obtained K_L instead of 1.5 in the material balance. (Note that this is not an iterative process.)

$$r_p = \frac{h_p}{h_f} \qquad (17\text{-}39)$$

while for the radial model, it is given by

$$r_p = \frac{2}{\pi}\left\{ \frac{h_p}{2R_f}\left[\frac{h_p}{2R_f}\left(1 - \frac{h_p}{2R_f} \right) \right]^{1/2} + \arcsin\frac{h_p}{2R_f} \right\} \quad (17\text{-}40)$$

There are several ways to incorporate non-Newtonian behavior into the width equations. A convenient procedure is to add one additional equation connecting the equivalent Newtonian viscosity with the flow rate. Assuming power law behavior of the fluid, the equivalent Newtonian viscosity can be calculated for the average cross section using the appropriate entry from Table 17-4. After substituting the equivalent Newtonian viscosity into the PKN width equation, we obtain

$$w_{w,0} = 9.15^{\frac{1}{2n+2}} \times 3.98^{\frac{n}{2n+2}}\left[\frac{1+2.14n}{n} \right]^{\frac{n}{2n+2}} K^{\frac{1}{2n+2}}\left(\frac{i^n h_f^{1-n} x_f}{E'} \right)^{\frac{1}{2n+2}} \tag{17-41}$$

As an example, consider a formation with Young's modulus $E = 30.3\,\text{GPa}$ ($4.4 \times 10^6\,\text{psi}$), Poisson ratio $\nu = 0.2$, permeable height $12.8\,\text{m}$ ($42\,\text{ft}$). Assume we wish to create a $x_f = 152.4\,\text{m}$ ($500\,\text{ft}$) fracture, pumping a power law fluid with flow behavior index $n = 0.7$ and consistency $K = 0.393\,\text{Pa.s}^{0.7}$ ($0.0082\,\text{lb}_f/\text{ft}^2$). The slurry injection rate for one wing is $i = 0.0212\,\text{m}^3/\text{s}$ ($16\,\text{bbl/min}$ for two wings.) The leakoff coefficient is $C_{L,p} = 0.13 \times 10^{-3}\,\text{m/s}^{1/2}$ ($3.3 \times 10^{-3}\,\text{ft/min}^{1/2}$) and the spurt loss coefficient is $S_{p,p} = 1.14\,\text{mm}$ ($0.028\,\text{gal/ft}^2$), both with respect to the permeable layer. Assuming PKN geometry with fracture height $h_f = 19.8\,\text{m}$ ($65\,\text{ft}$), calculate the necessary pumping time. The first step is to convert the leakoff parameters into apparent values with respect to the fracture area. Since $r_p = h_p/h_f = 0.646$, the apparent leakoff coefficient is $C_L = 0.0838 \times 10^{-3}\,\text{m/s}^{1/2}$ ($2.13 \times 10^{-3}\,\text{ft/min}^{1/2}$) and the apparent spurt loss coefficient is $S_p = 0.737\,\text{mm}$ (0.0181 gal/ft^2). The maximum width at the wellbore from Equation 17-40 is $w_{w,0} = 5.3\,\text{mm}$ ($0.209\,\text{in.}$) and the

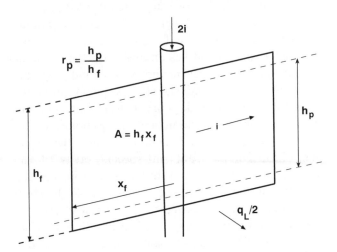

Figure 17-7 Ratio of permeable to total surface area, KGD and PKN geometry

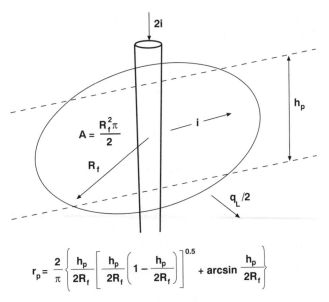

Figure 17-8 Ratio of permeable to total surface area, radial geometry

average width is $w_e = 3.33$ mm (0.131 in.) Using the rough approximation $K_L = 1.5$, the positive solution to the quadratic equation in Table 17-7 is $2461 s$; therefore, the necessary pumping time is 41 min. The fluid efficiency is 19.2 %. The calculated pumping time is 40.5 min using the Carter II equation method and 38.5 min using the α-method. In fact, the refinement is not significant.

The design above is based on the assumption of constant height and non-retarded fracture propagation. If evidence indicates that any of these assumptions is too restrictive, a more sophisticated design should be based on the methods discussed in Section 17-6.

A typical design process then continues with the determination of the proppant schedule. The fluid injected at the beginning of the job without proppant is called *pad*. It opens up the fracture and mostly leaks off into the formation. The selection of the pad volume is the key factor in avoiding screenout, which is the premature

---

Table 17-10   Proppant Schedule

1 Calculate the injected slurry volume
$$V_i = it_e$$

2 Calculate the exponent of the proppant concentration curve
$$\varepsilon = \frac{1 - \eta_e}{1 + \eta_e}$$

3 Calculate the pad volume and the time needed to pump it:
$$V_{\text{pad}} = \varepsilon V_i$$
$$t_{\text{pad}} = \varepsilon t_e$$

4 The required proppant concentration (mass/unit injected slurry volume) curve is given by
$$c = c_e \left(\frac{t - t_{\text{pad}}}{t_e - t_{\text{pad}}} \right)^\varepsilon$$

where c_e is the maximum proppant concentration in the injected slurry.

5 Calculate the mass of proppant placed into one wing
$$M = \eta_e c_e V_i$$

6 Calculate the propped width:
$$w_p = \frac{M}{(1 - \phi_p)\rho_p x_f h_f}$$

where ϕ_p is the porosity of the proppant bed and ρ_p is the true density of the proppant material

7 Calculate the dimensionless fracture conductivity:
$$C_{fD} = \frac{k_f w_p}{k x_f}$$

where k_f is the proppant bed permeability.

---

bridging of proppant leading to sudden increase of injection pressure. A too-large pad volume, however, can decrease the final propped width, especially in the target layer near the wellbore. Once the pad is pumped, the proppant concentration of the injected slurry is elevated step-by-step until the maximum possible value is reached. While many sophisticated methods are available, a simple design technique given in Table 17-10 using material balance and a prescribed functional form (power law) (Nolte, 1986) may be satisfactory to calculate the proppant schedule.

Ideally, the proppant schedule above results in a uniform proppant concentration in the fracture at the end of pumping, with the value of the concentration equal to c_e. The schedule is derived from the requirement that (1) the whole length created should be propped; (2) at the end of pumping, the proppant distribution in the fracture should be uniform; and (3) the schedule curve should be in the form of a delayed power law with the exponent and fraction of pad (ε) being equal. More complex proppant scheduling calculations may take into account the movement of the proppant both in the lateral and the vertical directions, variations of the viscosity of the slurry with time and location (resulting from temperature, shear rate and solid content changes), width requirement for free proppant movement, and other phenomena (Babcock *et al.*, 1967; Daneshy, 1978; Shah, 1982).

Note that in the schedule above, the injection rate i refers to the slurry (not clean fluid) injected into one wing. The obtained proppant mass M refers to one wing. The concentrations are given in mass per unit volume of slurry, and other types of concentration (added mass to unit volume of neat fluid) have to be converted first. The last two steps are intended to obtain the C_{fD}. A reasonable design should yield C_{fD} around 1.6. If the resulting C_{fD} is far from the technical optimum, alternates should be considered for the following decision variables: target fracture length, fluid properties (including rheology and leakoff), proppant properties (including proppant pack permeability), maximum possible proppant concentration, and/or injection rate.

Continuing our previous example, assume that maximum 12.5 lb light proppant (with specific gravity 2.1) can be added to 1 gal neat fluid. In terms of proppant concentration in the slurry, this becomes $c_e = 7.3$ lb/gal of slurry volume, which means the maximum proppant concentration of the slurry is $c_e = 875$ kg/m^3. Since the fluid efficiency is 19.3 %, the proppant exponent and the fraction of pad volume is $\varepsilon = 0.677$. Therefore, the pad injection time is 27.8 min, and, after the pad, the proppant concentration of the slurry should be continuously elevated according to the schedule, $c = 875\left(\dfrac{t - 1666}{794}\right)^{0.677}$ where c is in kg/m^3 and t is in seconds, or $c = 7.3\left(\dfrac{t - 27.8}{13.3}\right)^{0.677}$ where c is in lb/gal of slurry volume and t is in min. The obtained proppant curve is shown in Figure 17-9.

Since, at the end of pumping, the proppant concentration is equal to c_e everywhere in the fracture, the mass of proppant placed into one wing is $M = V_e \times c_e = \eta_e \times V_i \times c_e$; in our case, $M = 8760$ kg (19,400 lb). The average propped width after closure on proppant can be determined if we know the porosity of the proppant bed. Assuming $\phi_p = 0.3$, the propped volume is $V_p = M/[(1 - \phi_p)\rho_p]$, in our case, 6.0 m^3. The average propped width is $w_p = V_p/(x_f h_f)$, or 2 mm (0.078 in.). A quick check of the dimensionless fracture conductivity substituting the propped width shows that $C_{fD} = (60 \times 10^{-12} \times 0.002)/(5 \times 10^{-16} \times 152) = 1.6$; therefore, the design is optimal.

17-4 FRACTURING MATERIALS

Now, after the discussion of fracturing theory, we will turn attention to the materials used to create propped hydraulic fractures. Materials used in the fracturing process can be categorized into fracturing fluids, additives, and proppants. The fluid and additives act jointly to produce the hydraulic fracture when pumped, transport the proppant into the fracture, then flow back to allow the generated propped fracture to produce. Stimulation costs for materials and pumping are estimated at 46% for pumping, 25% for proppants, 19% for fracturing chemicals, and 10% for acid. Materials and proppants used in hydraulic fracturing have undergone changes since the first commercial fracturing treatment was performed in 1949 with a few sacks of coarse sand and gelled gasoline.

17-4.1 Fracturing Fluids

The functionality of the fracturing fluid was outlined above. Factors to consider when selecting the fluid include availability, safety, ease of mixing and use, compatibility with the formation, ability to be recovered from the fracture, and cost.

Fracturing fluids can be categorized as (1) oil- or water-based, (2) mixtures of oil and water called emulsions, and (3) oil- and water-based systems containing nitrogen or carbon dioxide gas. Fluid use has evolved from exclusively oil-based in the 1950s to more than 90% crosslinked, water-based in the 1990s. Nitrogen and carbon dioxide systems in water-based fluids are used in about 25% of all fracture stimulation jobs.

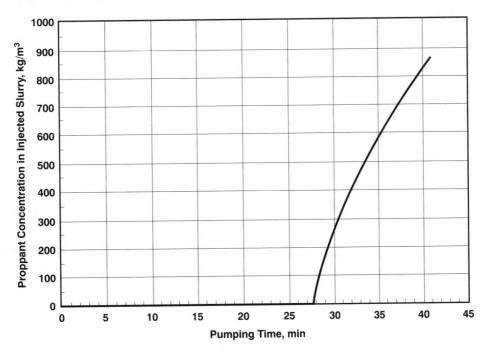

Figure 17-9 Example: Proppant concentration in the injected slurry

Table 17-11   Crosslinked fluid types

Crosslinker	Gelling Agent	pH range	Application Temp. °F
B, non-delayed	guar, HPG	8-12	70–300
B, delayed	guar, HPG	8–12	70–300
Zr, delayed	guar	7–10	150–300
Zr, delayed	guar	5–8	70–250
Zr, delayed	CMHPG, HPG	9–11	200–400
Zr-a, delayed	CMHPG	3–6	70–275
Ti, non-delayed	guar, HPG, CMHPG	7–9	100–325
Ti, delayed	guar, HPG, CMHPG	7–9	100–325
Al, delayed	CMHPG	4–6	70–175
Sb, non-delayed	guar, HPG	3–6	60–120

a–compatible with carbon dioxide

Table 17-11 lists commonly used fracturing fluids in order of current usage. The choice of which crosslinking method to use is based on the capability of a fluid to yield high viscosity, while meeting cost and other performance requirements.

Viscosity is one of the most important qualities associated with a fracturing fluid. The most efficient viscosity-producing gum is guar, produced from the guar plant. Guar derivatives called hydroxypropyl guar (HPG) and carboxymethylhydroxypropyl guar (CMHPG) are also used because they provide lower residue, faster hydration, and rheological advantages because less gelling agent is required if the guar is crosslinked. The base guar or guar-derivative is reacted with a metal that couples multiple strands of gelling polymer. Crosslinking effectively increases the size of the base guar polymer, increasing the viscosity in the range of shear rates important for fracturing from 5- to 100-fold. Boron (B) often is

used as the crosslinking metal, followed by zirconium (Zr) and, to a smaller extent, titanium (Ti), antimony (Sb), and aluminum (Al). Foams (Chambers, 1994) are especially advantageous in water-sensitive formations. Their application minimizes fracture-face damage and eases the cleanup of the wellbore after the treatment.

17-4.2 Fluid Additives

Gelling agent, crosslinker, and pH control (buffer) materials define the specific fluid type and are not considered to be additives. Fluid additives are materials used to produce a specific effect independent of the fluid type. Table 17-12 lists commonly used additives.

Biocides effectively control bacterial contamination. Most waters used to prepare fracturing gels contain bacteria originating from contamination of either the source water or the storage tanks on location. The bacteria pro-

Table 17-12   Fracturing fluid additives

Additive	Concentration gal or lb added to 1000 gal clean fluid	Purpose
biocide	0.1–1.0 gal	prevent guar polymer decomposition by bacteria
fluid loss	10–50 lb	decrease leakoff of fluid during fracturing
breakers	0.1–10 lb	provide controlled fluid viscosity reduction
friction reducers	0.1–1.0 gal	reduce wellbore friction pressure loss while pumping
surfactants	0.05–10	reduce surface tension, prevent emulsions, and provide wetting
foaming agents	1–10 gal	provide stable foam with nitrogen and carbon dioxide
clay control	—	provide temporary or permanent clay-water compatibility

Table 17-13 Fracturing fluid breakers

Breaker	Application temp, °F	Comments
Enzyme	60–200	Efficient breaker. Limit to below pH 10.
Encapsulated Enzyme	60–200	Allows higher concentrations for faster breaks.
Persulfates (Sodium, Ammonium)	120–200	Economical. Very fast at higher temp.
Activated persulfates	70–120	Low temperature and high pH
Encapsulated persulfates	120–200	Allow higher concentrations for faster breaks.
High temperature oxidizers	200–325	Used where persulfates are too quick.

duce enzymes that can destroy viscosity rapidly. Raising the pH to >12, adding bleach, or using a broad-spectrum biocide effectively controls bacteria.

Fluid loss control materials provide spurt-loss control. The materials consist of finely ground particles ranging from 0.1 to 50 microns. The lowest cost, most effective material is ground silica sand. Starches, gums, resins, and soaps also provide some degree of cleanup from the formation because of their water solubility. The guar polymer eventually controls leakoff once a filter cake is established.

Breakers reduce viscosity by reducing the size of the guar polymer. Much of the current fluid development efforts center around breaker testing and development. Table 17-13 summarizes several breaker types and application temperatures.

Surfactants provide water wetting, prevent emulsions, and lower surface tension. Reduction of surface tension allows improved fluid recovery. Surfactants are available in cationic, nonionic, and anionic forms and usually are included in most fracturing treatments. Some specialty surfactants provide improved wetting and fluid recovery.

Foaming agents provide the surface-active stabilization required to maintain finely divided gas dispersion in foam fluids. These ionics also act as surfactants and emulsifiers. Stable foam cannot be prepared without a surfactant for stabilization.

Clay control additives produce temporary compatibility in water-swelling clays using various salts. Organic chemical substitutes used at lower concentrations now are also available.

17-4.3 Fluid Testing

A large amount of laboratory testing and development is needed in the qualification of any fracturing fluid. Fluid rheology and frictional pressure, viscosity reduction (called breaking), fluid leakoff, compatibility with common additives, temperature stability, proppant transport characteristics, fracture conductivity, and possible damage from gel residue, ease of handling and mixing, special equipment requirements, availability, and environmental concerns are some of the dominant areas of testing. Testing methodology attempts to model the actual formation conditions where the fluid system will be applied. Following are some general considerations addressed by fluid testing:

- **Compatibility** Precipitation of solids may cause fracture face damage, reducing the overall performance. Therefore, the fluid should be chemically compatible with the formation at in-situ temperatures and pressures.

- **Rheology** Most data for crosslinked fluids are obtained from a concentric cylindrical geometry or Couette-type instrument. The rheometer provides power law fluid data (n and K-values) for use in fracturing calculations. While use and suitability of these data are sometimes questioned, the industry has a large base of experience with the results of such data and can use it for comparative purposes. A considerable amount of fluid conditioning precedes the passage of fluids through these instruments to subject them to the same shear history as they are likely to experience during the actual applications. Geometric similarity, large scale, and advanced measurement techniques are advantageous for fluid characterization (Mears *et al.*, 1993).

- **Fluid loss** Accurate prediction of fluid leakoff is important in simulating and designing hydraulic fractures. Current leakoff testing is quite comprehensive, with elaborate fluid preparation exposing it to the same shear and temperature as in the fracture, and a slot-testing geometry. Using formation core on both sides of this slot allows fluid leakoff measurements that allow for better modeling of actual conditions.

- **Breaking (viscosity reduction)** While a fracturing fluid must retain viscosity during the treatment, ultimately, the viscous fluid must be removed from the fracture before producing the well. Laboratory testing is done to establish the details and ease of this process. These tests consist of heating the fluid under formation conditions and observing the build-up and the reduction of its viscosity with time. During the early part of the test, the measured viscosities correspond to those experienced by the fluid during the extension of the fracture. A premature breaking of the fluid in this stage is considered undesirable. Later, the fluid should lose most of its viscosity soon enough that it does not significantly impede production of the reservoir fluid.

- **Proppant-carrying capacity** Current industrial practices include measurements of viscosity in the appropriate shear rate and observation of proppant settlement in static fluid columns. Several other experimental approaches are looking into future techniques for refinement of the testing process. There are at least two types of proppant movement inside the fracture that affect proppant transport. The first is single-particle settling, where the individual particles fall through the fluid because of their (usually) higher density. The size and density of the particles as well as the effective viscosity of the fluid affect single-particle settling. The concentration of particles also affects the rate of settling because the particles tend to clump and fall faster than expected at low concentrations (Shah, 1982). At higher concentrations, particles interfere with each other and fall slower than expected, producing hindered settling. In most cases, however, single particle settling is an insignificant parameter in the final propped fracture geometry where the fracturing fluid still maintains an apparent viscosity of 50 to 100 cp at the fracture tip and at the end of the job. This condition covers most treatments that use viscous gels. Only non-gelled water treatments, which intentionally aim for a banking-type design, experience significant single-particle settling.

The other proppant transport mechanism is convection. The word refers to the movement of proppant along with the flowing fluid (a desirable phenomenon) and also to the downward movement of later stages of the proppant laden fluid with respect to the earlier entered stages (a less understood and uninvited phenomenon). In either particle settling or convection, rigorous tracking of proppant placement requires sophisticated modeling techniques. Most of the current development in fracturing simulators is connected with improved proppant transport description (Cleary and Fonseca, 1992; Barree and Conway, 1995).

- **Residue in the proppant pack** The resulting permeability of the proppant pack will be contaminated because of the residue of the gelled fracturing fluid occupying part of the pore space. Breakers are intended to reduce the viscosity of the fracturing fluid to an extent that it flows back, freeing the porous space for hydrocarbon flow.

- **Filter-cake residue** Ultimately, the productivity of a fracture depends on the pressure-drop components in subsequent locations on the way of the produced hydrocarbon fluids towards the wellbore: in the reservoir, through the filter cake at the fracture face—which may contain 10 to 20 times the polymer content of the original fluid—in the fracture, and, finally, through the perforations. Tests to model the build-up, breaking, and residues of the filter cake have advanced significantly during recent times. These tests have shown that foams, borates, and uncrosslinked gels provide the best overall performance of the fracture.

17-4.4 Proppant Selection

Since proppants are used to hold open the fracture after the hydraulic pressure used to generate the fracture has been relieved, their material strength is of crucial importance. The propping material has to be strong enough to bear the closure stress; otherwise, the conductivity of the crushed proppant bed will be considerably less than the design value (both the width and the permeability of the proppant bed decrease). Other factors considered in the selection process are size, shape, and composition. There are two main categories of proppants: naturally occurring sands and manmade ceramic and bauxite proppants. Sands are used for lower-stress applications in formations at approximately 6000 ft and, preferably, considerably less. Man-made proppants are used for high-stress situations in formations generally deeper than 8000 ft. Between these two values, the magnitude of the stress is the deciding factor.

Three ways to increase the fracture conductivity are to (1) increase the proppant concentration to produce a wider fracture, (2) use larger proppant size to produce a more permeable fracture, or (3) change the proppant type to gain more strength. Figures 17-10, 17-11, and 17-12 illustrate the three methods of increasing conductivity through proppant choice. Since closure stress is a main concern when selecting the right proppant for a given job, Figure 17-13 provides a selection guide based on closure stress.

17-5 TREATMENT EXECUTION

17-5.1 Pump Schedule

An accurate pump schedule will generate the designed propped-fracture geometry. The first stage of the treatment is the pad. The pad protects the treatment from fluid loss and generates fracture length and width. Typically, 30 to 60% of the fluid pumped during a treatment leaks off into the formation while pumping, and the pad provides most of this necessary extra fluid. The pad also generates sufficient fracture length and fracture width to allow proppant placement. Too little pad results in shorter fracture lengths. Too much pad results in too much fracture length and height growth, and the final propped width is too narrow. Even if fluid loss were zero, a minimum pad volume would be required to open sufficient fracture width to allow proppant admittance. Generally, a fracture width of three times the proppant diameter is considered as necessary to avoid bridging. Following the pad stage, the remaining stages of the treatment, which are the proppant-carrying stages, are pumped. Figure 17-14A illustrates the proppant distribution in the fracture after the first proppant-carrying stage. Most fluid loss occurs near the fracture tip, where the pad is located; however, some fluid loss occurs along the fracture. Additional fluid loss dehydrates and con-

centrates the proppant-laden stages. Figure 17-14B shows the concentrated initial proppant-carrying stage. The concentration can fall to 0.5 to 3 lb proppant per gal of clean fluid (lb/gal) As the treatment continues, the proppant concentration increases. In the later stages, high proppant concentrations in the pumped fluid are more acceptable because the fluid is exposed to less fluid loss in the remaining period of time until the end of pumping.

Figure 17-14C shows a typical treatment in which the pad is totally depleted just as pumping ends, and the first proppant-carrying stage has concentrated to a final designed concentration. The second proppant stage has undergone less dehydration but also has concentrated to the same, final designed concentration. At shutdown, ideally the entire fracture—wellbore to tip—is filled with a uniform concentration of proppant slurry.

During the job execution, a situation may arise when the proppant gets stuck in the fracture. Such bridging of the proppant blocks additional proppant from entering the fracture. Soon, the treating pressure rises above the technical limit, and/or the well is filled with sand. This phenomenon, called *screenout* should be avoided, because it may damage the equipment and/or cause a premature end of the job without fulfilling the design target. The critical process controlling the screenout is fluid leakoff; therefore, considerable effort is spent on determining the leakoff coefficient. This goal is the main objective of the so-called "minifrac" injection test.

In highly permeable and soft formations, however, a screenout might be desirable. A well-timed bridging of the proppant near the fracture tip may arrest fracture propagation, resulting in the subsequent inflation of the fracture width. A shorter but wider fracture then results

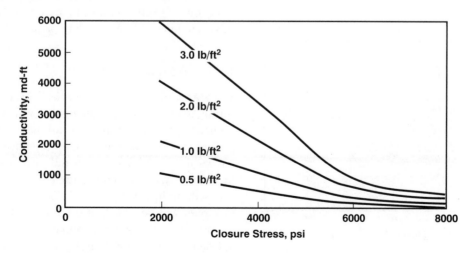

Figure 17-10   Fracture conductivity for various areal proppant concentrations

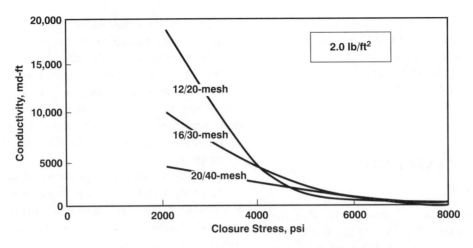

Figure 17-11   Fracture conductivity for various mesh sizes

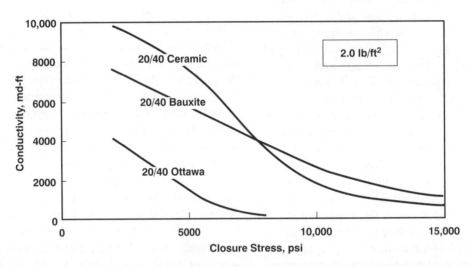

Figure 17-12   Fracture conductivity for proppants of various types

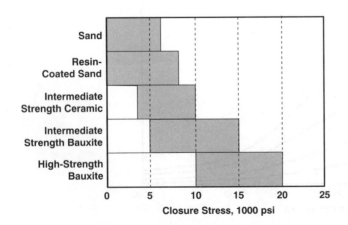

Figure 17-13   Proppant selection guide

in a more desirable C_{fD}, whereas the C_{fD} without the "tip screenout" (TSO) would be much lower. More details of this technique will be given in Chapter 19.

17-5.2 Treatment Flowback

The standard procedure at the end of a hydraulic fracturing treatment is to shut in the well for a matter of hours, overnight, or for several days, particularly in nonenergized treatments. The extended shut-in time allows the fracture to close and allows any viscosified fluids to break completely back to water. The procedure is based on the principle that if the fracture is not closed, proppant will flow back into the wellbore, and any fluid viscosity would enhance proppant removal.

Fractures, particularly in tight reservoirs, may require long periods to close, and, during this time, excessive

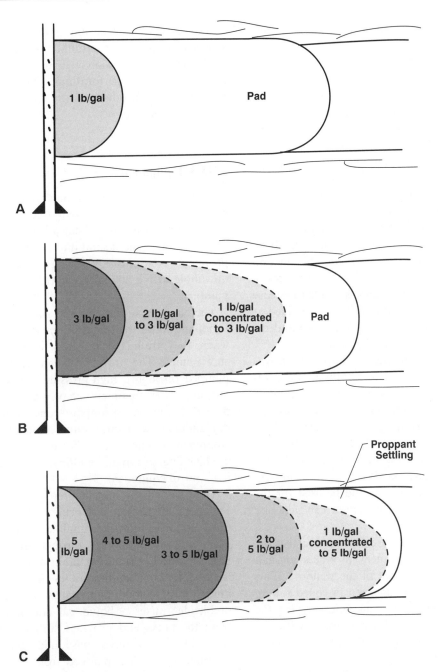

Figure 17-14 Evolution of proppant distribution during pumping (A) At time the first proppant stage is injected; (B) At intermediate time; (C) At end of pumping

near-wellbore proppant settling may occur. If the near-wellbore area loses conductivity, the hydraulic fracturing treatment may fail. Any pinching effect in the near-wellbore area or decrease in conductivity in the proppant pack may counteract fracturing fluid cleanup. Therefore, a technique called *forced closure* is applied. It consists of flowing back the well quickly and thereby slowing the rate of proppant settling in the near-wellbore area. Forced closure does not necessarily cause rapid

fracture closure and might even involve reverse gravel-packing of proppant at the perforations. The flowback must begin within seconds of shutdown. One of the major benefits of immediate flowback is the use of a supercharge of pressure that was built up during the fracture treatment. Conventional flowback methods allow this pressure to dissipate before flowing the fluids back when shutting in the well for an extended period.

This pressure buildup helps produce more flowback, even in underpressured wells.

Aggressively flowing a well with viscous fluids will not carry proppant out through the perforations, but low flowback rates might not be sufficient to avoid proppant settling near the wellbore. Overbreaking of the fracturing gel can be detrimental to the job because of rapid settling of the proppant, even where immediate flowback is used. The well should produce some amount of unbroken gel to achieve a successful fracture treatment. A major mechanism in stopping proppant production is the creation of a proppant pack opposite the perforations. The flowback procedure allows better grain-to-grain contact on resin-coated sand than in conventional flowback operations and prevents later proppant production caused by consolidation when high velocity occurs. While aggressive flowback may cause a sudden increase of the load on the proppant, crushing can be avoided as long as the well maintains a hydrostatic column and the proppant strength has been selected properly.

When flowing back a well where a short section was stimulated, a flowback rate of 2 to 3 bbl/min is considered advantageous. Fewer and smaller perforations allow for better bridging of proppant. Typical casing gun perforations are the most advantageous. Smaller perforations sometimes can cause problems in proppant placement, and larger perforations are difficult to bridge and may yield sand production. When perforating sections, 2 to 3 bbl/min per section should be used for the reverse gravel-packing mode. This action can generate large flow rates.

The buildup of artificial pressure produced in the formation by the fracturing treatment produces enough energy to clean the proppant out of the wellbore. This procedure eliminates the costs incurred in coiled tubing cleanout and sand bailing. A viscous fluid carrying the sand acts as a buffer, and less abrasion occurs with viscous fluid than with low-viscosity fluids or fluids containing gas during flowback.

Foam or energized treatments must be flushed either wholly or partially with the base fluid to provide control over the flushed volumes. The volume must not be overflushed. Using liquids for flushing also gives the operator some control of flow rate before the gas hits the surface. Flow rates can be controlled using pressure and choke tables. Energized and foam treatments should be flowed back quickly and aggressively to take advantage of the energized gas. Shutting in a fracture treatment that uses CO_2 or N_2 for any period is counterproductive. Reservoirs with any permeability quickly absorb the energizing gas.

If a well that has been shut in for an extended period is suspected of proppant settling, it is recommended to flow back the well more slowly to prevent sand production through the open perforations. Once there is no more danger of sand production, the flowback rate should be increased gradually.

17-5.3 Equipment

Stimulation equipment has undergone extensive changes since the first commercial hydraulic fracturing treatment was performed in 1949. That job involved hand-mixing five sacks of 20-mesh sand into 20 bbl of fluid (0.6-lb/gal proppant concentration). The mixture was pumped downhole with a 300-hp cementing and acidizing triplex pump.

The basics have not changed since the first treatment. Proppant and a treating fluid are delivered to a blender where they are mixed and transferred to the high-pressure pumps. The proppant-laden treating fluid is then pumped through a high-pressure manifold to the well. However, the treatments have grown in complexity, flow rate, pressure, proppant concentration, liquid and dry additives, and total volumes of fluid. The types of equipment required to perform stimulation treatments are blending, proppant handling, pumping, and monitoring/control equipment.

The blending equipment prepares the treatment fluid by mixing the correct proportions of liquid and dry chemical additives into the stimulation fluid. Mixing can be done either continuously throughout the treatment or batch-mixed in fracturing tanks before the stimulation treatment. For continuous mixing of the treatment fluid, the base fluid is prepared by a preblender. The preblender mixes a liquid gel concentrate with water from the tanks and provides sufficient hydration time to yield the required base-fluid gel viscosity. The hydrated gel is then pumped from the hydration tank to the blender. In addition to the additives, the blender mixes proppant with the treatment fluid.

Quality control of the mixing process is now computer controlled. Set points for the mixture concentrations are entered into a computer. The computer maintains the required concentrations, regardless of the mixing flow rate. Operational parameters of the blender such as tub level, mixer agitation, and pressures also have been placed under automatic control, thereby minimizing the potential for human error.

Proppants are stored, moved, and delivered to the blender using several methods. The sacked proppant can be handled manually or with dump trucks/trailers and pneumatic systems. Ever-increasing quantities of proppant in treatments have necessitated the use of field storage bins. When the required proppant amount exceeds field storage bin capacity, multiple bins are located around a gathering conveyor that transfers the proppant to the blender. Because of the distances that the proppant often must travel from the farthest storage bin to the blender, automatic control systems have been added to the storage bins and gathering conveyors to allow uninterrupted delivery of proppant.

Pumping equipment has grown from 300-hp pumps in 1949 to today's 2000+ hp produced from a single crankshaft pump. The pressure requirements for treatments also have grown from 2000 psi to 20,000 psi. Full-power shift transmissions under computer control synchronize the engine speed with the gear shift such that the pump rates before and after the shift are the same. Computer-controlled pumping equipment also allows automatic pressure and/or rate control.

Monitoring stimulation treatments has progressed from pressure gauges, stopwatches, and chart recorders to full computer monitoring and control. More than a thousand different parameters can be monitored and recorded during a stimulation treatment. These parameters are not limited to treatment fluids, proppant, and additives, but also to the equipment performing the treatment. Monitoring the treatment fluids is an essential part of the quality control. Primary fluid parameters monitored and recorded during a stimulation treatment include, but are not limited to, pressures, temperatures, flow rates, proppant and additive concentrations, pH, and viscosity. Any of the parameters can be plotted or displayed during the job along with real-time calculation of downhole parameters. Parameters monitored and recorded on the equipment provide data for future design improvements and diagnosis of equipment problems.

17-5.4 Microfracture Tests

The microfracture stress test (microfrac) determines the magnitude of the minimum principal in-situ stress of a formation. The test usually is performed by injecting pressurized fluid into a small, isolated zone (4 to 15 ft, 1.2 to 4.6 m) at low rates (1 to 25 gal/min, 0.010 to 0.095 m^3/min). The minimum principal in-situ stress can be determined from the pressure decline after shut-in or the pressure increase at the beginning of an injection cycle. The fracture closure pressure and fracture

reopening pressure are good approximations to the minimum principal in-situ stress. Additional discussion on the microfrac technique is presented in Chapter 6.

17-5.5 Minifracs

The most important test on the location before the main treatment is known as a minifrac, or a fracture calibration test. The minifrac is actually a pump-in/shut-in test with fluid volumes typically around thousands of gallons. The information gathered from a minifrac includes the closure pressure, p_c, net pressure, and entry (perforations and near-wellbore) conditions. The fall-off part of the pressure curve is used to obtain the leakoff coefficient for a given fracture geometry. Figure 17-15 illustrates the strategic locations on a typical pressure response curve registered during the calibration activities.

A minifrac design should be performed with the initial treatment design. The design goal for the minifrac is to be as representative as possible of the main treatment. To achieve this objective, sufficient geometry should be created to reflect the fracture geometry of the main treatment and to obtain an observable closure pressure from the pressure decline curve. The most representative minifrac would have injection rate and fluid volume equal to the main treatment, but this is not often a practical solution. In practice, several conflicting design criteria must be balanced, including minifrac volume, created fracture geometry, damage to the formation, a reasonable closure time, and cost of materials and personnel.

The fracture closure typically is selected by examining several different plots of the pressure decline curve and, if available, integrating prior knowledge obtained, such as from microfrac tests. An added complication is that temperature and compressibility effects may cause pressure deviations. However, with the aid of an analysis program, a temperature-corrected decline curve can be generated, and all normal interpretations of the different plot types still hold (Soliman, 1986).

The original concept of pressure decline analysis is based on the observation that during the closure process, the rate of pressure decline contains useful information on the intensity of the leakoff process (Nolte, 1979; Soliman and Daneshy, 1991) in contrast to the pumping period, when the pressure is affected by many other factors.

Assuming that the fracture area has evolved with a constant exponent α and remains constant after the pumps are stopped, at time $(t_e + \Delta t)$ the volume of the fracture is given by

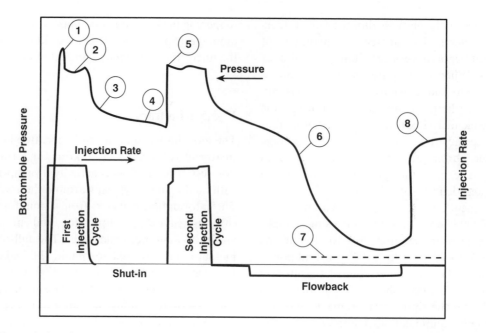

Figure 17-15 Strategic locations on a pressure response curve
(1) Formation breakdown; (2) Propagation; (3) Instantaneous shut-in; (4) Closure pressure from fall-off; (5) Reopening; (6) Closure pressure from flowback; (7) Asymptotic reservoir pressure; (8) Closure pressure from rebound

$$V_{t_e+\Delta t} = V_i - 2A_e S_p - 2A_e g(\Delta t_D, \alpha)C_L\sqrt{t_e} \quad (17\text{-}42)$$

where the dimensionless delta time is defined as

$$\Delta t_D = \Delta t/t_e \quad (17\text{-}43)$$

and the two-variable function $g(\Delta t_D, \alpha)$ can be obtained by integration. Its general form is given by (Valkó and Economides, 1995)

$$g(\Delta t_D, \alpha)$$
$$= \frac{4\alpha\sqrt{\Delta t_D} + 2\sqrt{1 + \Delta t_D} \times F\left[\frac{1}{2}, \alpha; 1 + \alpha; (1 + \Delta t_D)^{-1}\right]}{1 + 2\alpha}$$
$$(17\text{-}44)$$

The function $F[a, b; c; z]$ is the *Hypergeometric function* available in the form of tables or computing algorithms. For computational purposes, the approximations in Table 17-14 might be useful.

Dividing Equation 17-42 by the area, the fracture width at time Δt after the end of pumping is given by

$$\overline{w}_{t_e+\Delta t} = \frac{V_i}{A_e} - 2S_p - 2C_L\sqrt{t_e}g(\Delta t_D, \alpha) \quad (17\text{-}45)$$

Hence, the *time variation* of the width is determined by the $g(\Delta t_D, \alpha)$ function, the length of the injection period and the leakoff coefficient, but is not affected by the *fracture area*.

Table 17-14 Approximation of the g-function for various exponents α ($d = \Delta t_D$)

$$g\left(d, \frac{4}{5}\right) = \frac{1.41495 + 79.4125\,d + 632.457d^2 + 1293.07\,d^3 + 763.19d^4 + 94.0367\,d^5}{1. + 54.8534\,d^2 + 383.11\,d^3 + 540.342\,d^4 + 167.741\,d^5 + 6.49129\,d^6}$$

$$g\left(d, \frac{2}{3}\right) = \frac{1.47835 + 81.9445\,d + 635.354\,d^2 + 1251.53\,d^3 + 717.71\,d^4 + 86.843\,d^5}{1. + 54.2865\,d + 372.4\,d^2 + 512.374\,d^3 + 156.031\,d^4 + 5.5955\,d^5 - 0.0696905\,d^6}$$

$$g\left(d, \frac{8}{9}\right) = \frac{1.37689 + 77.8604\,d + 630.24\,d^2 + 1317.36\,d^3 + 790.7\,d^4 + 98.4497\,d^5}{1. + 55.1925\,d + 389.537\,d^2 + 557.22\,d^3 + 174.89\,d^4 + 6.8188\,d^5 - 0.0808317\,d^6}$$

The decrease of average width cannot be observed directly, but the net pressure during closure is already directly proportional to the average width according to

$$p_n = S_f \overline{w} \tag{17-46}$$

simply because the formation is described by *linear elasticity theory*, Equation 17-15. The coefficient S_f is the fracture stiffness, expressed in Pa/m (psi/ft). Its inverse, $1/S_f$, is called the fracture compliance. For the basic fracture geometries, expressions of the fracture stiffness are given in Table 17-15.

The combination of Equations 17-45 and 17-46 yields (Nolte, 1979)

$$p = \left(p_C + \frac{S_f V_i}{A_e} - 2S_f S_p \right) - \left(2S_f C_L \sqrt{t_e} \right) \times g(\Delta t_D, \alpha) \tag{17-47}$$

Equation 17-47 shows that the pressure fall-off in the shut-in period will follow a straight line trend

$$p = b_N - m_N \times g(\Delta t_\Delta, \alpha) \tag{17-48}$$

if plotted against the *g*-function (transformed time, Castillo 1987). The *g*-function values should be generated with the exponent, α, considered valid for the given model. The slope of the straight line, m_N is related to the unknown leakoff coefficient by

Table 17-15 Proportionality constant, S_f and suggested α for basic fracture geometries

	PKN	KGD	Radial
S_f	$\dfrac{2E'}{\pi h_f}$	$\dfrac{E'}{\pi x_f}$	$\dfrac{3\pi E'}{16 R_f}$
α	4/5	2/3	8/9

$$C_L = \frac{-m_N}{2\sqrt{t_e} s_f} \tag{17-49}$$

Substituting the relevant expression for the fracture stiffness, the leakoff coefficient can be estimated as given in Table 17-16.

This table shows that for the PKN geometry, the estimated leakoff coefficient does not depend on unknown quantities since the pumping time, fracture height, and plain-strain modulus are assumed to be known. For the other two geometries considered, the procedure results in an estimate of the leakoff coefficient which is strongly dependent on the fracture extent (x_f or R_f).

From Equation 17-47, we see that the effect of the spurt loss is concentrated in the intercept of the straight-line with the $g = 0$ axis:

$$S_p = \frac{V_i}{2A_e} - \frac{b_N - p_C}{2S_f} \tag{17-50}$$

As suggested by Shlyapobersky *et al.* (1988), Equation 17-50 can be used to obtain the unknown fracture extent if we assume there is no spurt loss. The second row of Table 17-16 shows the estimated fracture extent for the three basic models. Note that the no-spurt loss assumption results in the estimate of the fracture length also for the PKN geometry, but this value is not used for obtaining the leakoff coefficient. For the other two models, the fracture extent is obtained first, and then the value is used in interpreting the slope. Once the fracture extent and the leakoff coefficient are known, the lost width at the end of pumping can be easily obtained from

$$w_{Le} = 2g_0(\alpha) C_L \sqrt{t_e} \tag{17-51}$$

and the fracture width from

$$\overline{w}_e = \frac{V_i}{x_f h_f} - w_{Le} \tag{17-52}$$

Table 17-16 Leakoff Coefficient, C_L, and no-spurt fracture extent for the different fracture geometries

	PKN	KGD	Radial
Leakoff coefficient, C_L	$\dfrac{\pi h_f}{4\sqrt{t_e}E'}(-m_N)$	$\dfrac{\pi x_f}{2\sqrt{t_e}E'}(-m_N)$	$\dfrac{8R_f}{3\pi\sqrt{t_e}E'}(-m_N)$
Fracture extent	$x_f = \dfrac{2E'V_i}{\pi h_f^2(b_N - p_C)}$	$x_f = \sqrt{\dfrac{E'V_i}{\pi h_f(b_N - p_C)}}$	$R_f = 3\sqrt{\dfrac{3E'V_i}{8(b_N - p_C)}}$

for the two rectangular models, and

$$\overline{w}_e = \frac{V_i}{R_f^2 \pi/2} - w_{Le} \qquad (17\text{-}53)$$

for the radial model.

Often, the fluid efficiency is also determined

$$\eta_e = \frac{\overline{w}_e}{\overline{w}_e + w_{Le}} \qquad (17\text{-}54)$$

Note that neither the fracture extent nor the efficiency are model parameters. Rather, they are state variables and, hence, will have different values in the minifrac and the main treatment. The only parameter that is transferable is the leakoff coefficient itself, but some caution is needed in its interpretation. The bulk leakoff coefficient determined from the method above is apparent with respect to the fracture area. If we have information on the permeable height, h_p, and it indicates that only part of the fracture area falls into the permeable layer, the apparent leakoff coefficient should be converted into a "true value" with respect to the permeable area only. This is done simply by dividing the apparent value by r_p (see Equation 17-39).

While adequate for many low permeability treatments, the outlined procedure might be misleading for higher permeability reservoirs. The conventional minifrac interpretation determines a single effective fluid-loss coefficient, which usually slightly overestimates the fluid loss when extrapolated to the full job volume (Figure 17-16). This overestimation typically provides an extra factor of safety in low-permeability formations to prevent screen-

out. However, when this same technique is applied in high-permeability or high-differential pressure between the fracture and the formation, it significantly overestimates the fluid loss for wall-building fluids if extrapolated to the full job volume (Dusterhoft *et al.*, 1995). Figure 17-17 illustrates the overestimation of fluid loss which might be detrimental in high-permeability formations where the objective often is to achieve a tip screen-out. Modeling both the spurt loss and the combined fluid-loss coefficient by performing a net pressure match in a 3D simulator is an alternative to classical fall-off analysis. This approach is illustrated in Figure 17-18.

It is important to note that the incorporation of more than one leakoff parameter (and other adjustable variables) increases the number of degrees of freedom. As a result, a better match of the observed pressure can be achieved, but the match may be not unique in the sense that other values of the same parameters may provide a similar fit.

17-6 TREATMENT ANALYSIS

17-6.1 Treating Pressure Analysis

Fracturing pressure is often the only available direct information on the evolution of the fracture during the treatment. Thus, fracturing pressure interpretation and subsequent decisions (depending on the result of interpretation) are some of the primary responsibilities of the fracturing engineer. A log-log plot of bottomhole treating pressure vs. time suggested by Nolte and Smith (1981) is used for this purpose. First, the occurrence of

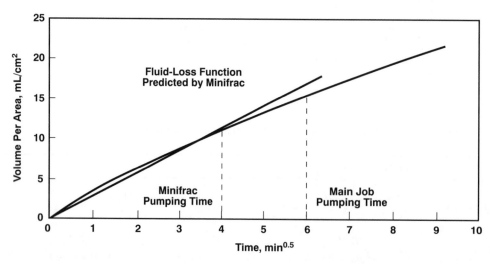

Figure 17-16 Fluid leakoff extrapolated to full job volume, low permeability (after Dusterhoft *et al.*, 1995)

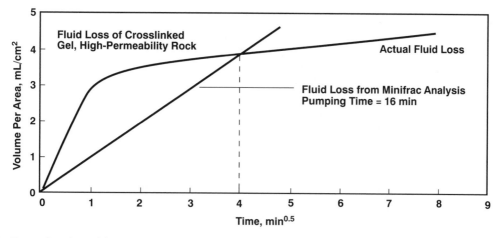

Figure 17-17   Overestimation of fluid leakoff extrapolated to full job volume, high permeability (after Dusterhoft *et al.*, 1995)

the main events (change in injection rate, fluid quality, proppant concentration) should be indicated as in Figure 17-19. Then, a qualitative interpretation follows. A steady positive slope of order 0.25 is interpreted as unrestricted (normal) fracture propagation. An abrupt increase in the fracture surface (caused by fast height growth into another layer) is assumed if the slope changes to a negative value. An increasing slope approaching the value of unity is considered a sign of restricted tip propagation, and it is often followed by an even larger slope indicating the fast fill-up of the fracture with proppant (screenout). The quantitative interpretation needs a more rigorous description of the leakoff rate and some significant assumptions on fracture geometry.

17-6.2 Well Testing

Often, a test before the fracturing treatment is not possible at all in low-permeability formations, and we have limited information on the permeability. In such cases, the buildup test of the fractured well is intended to obtain the permeability and the fracture extent simultaneously. Unfortunately, this is an ill-posed problem in the sense that many different combinations of the unknown parameters give a good match. In high-permeability formations, where the permeability is usually known, the primary goal of a post-treatment test is to evaluate the created fracture.

For well-testing purposes, an infinite-acting reservoir can be considered. The transient behavior of a vertical

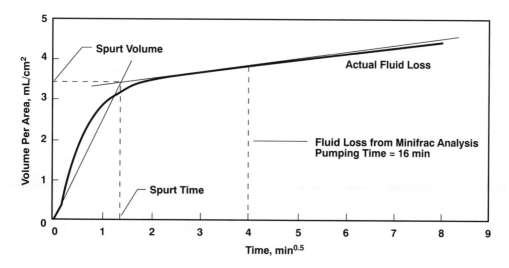

Figure 17-18   Realistic estimation of leakoff using parameters obtained from performing a net pressure match in a 3D simulator (after Dusterhoft *et al.*, 1995)

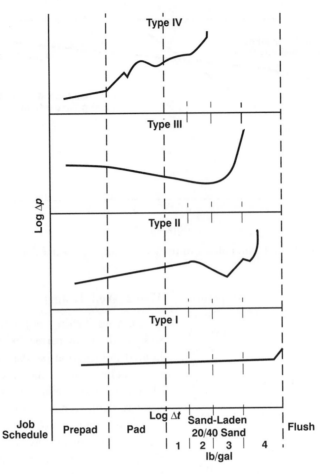

Figure 17-19   Treatment pressure response types in order of increasing danger of screenout, Nolte-Smith Plot (after Conway *et al.*, 1985)

well intersected by a finite conductivity fracture is well known from the works of Cinco-Ley and coworkers (1978, 1981). Figure 17-20 shows the log-log plot of the dimensionless pressure and the "time-log derivative" parameterized by the dimensionless fracture conductivity. In the so-called bilinear flow regime, in which the flow is determined by both the reservoir and fracture properties, the plot shows a quarter slope because in this flow regime, the dimensionless pressure can be expressed as

$$p_D \approx \frac{\pi}{\Gamma(5/4)\sqrt{2C_{fD}}} t_{Dxf}^{1/4} \qquad (17\text{-}55)$$

where t_{Dxf} is the dimensionless time based on the fracture half-length as characteristic dimension. Once such a regime is identified, a specialized plot of the pressure vs. the quarter root of time can be constructed. The

slope m_{bf} of the straight line fitted is a combination of the reservoir and fracture properties:

$$
\begin{aligned}
m_{bf} &= \frac{\pi}{2\sqrt{2}\Gamma(5/4)\sqrt{C_{fD}}} \frac{B\mu q}{\pi k h} \left(\frac{k}{\phi \mu c_t x_f^2} \right)^{1/4} \\
&= \left(0.390 \frac{B\mu^{3/4}q}{h\phi^{1/4}c_t^{1/4}} \right) \frac{1}{k^{1/4}k_f^{1/2}w^{1/2}}
\end{aligned}
\qquad (17\text{-}56)
$$

It can be used to obtain one or the other quantity or their combination, depending on the available information. As is obvious from the equation above, the formation permeability and the fracture conductivity cannot be determined simultaneously from this regime. Knowing the formation permeability, the fracture conductivity ($k_f \times w$) can be determined from the slope, but the fracture extent cannot. Our suggestion is to assume $C_{fD} = 1.6$, determine an equivalent fracture conductivity

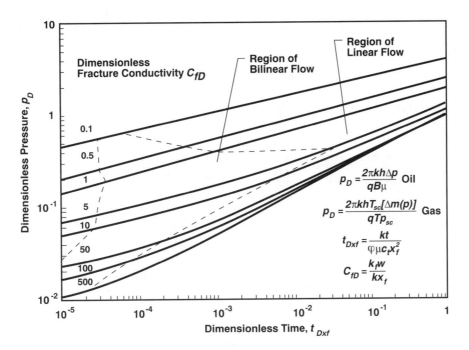

Figure 17-20 Dimensionless pressure and pressure-log-derivative of a vertically fractured well (after Cinco-Ley and Samaniego, 1981)

from Equation 17-56, and calculate an equivalent fracture length:

$$x_{f_{eq}} = \frac{(k_f w)_{eq}}{1.6k} \qquad (17\text{-}57)$$

Comparing the equivalent fracture length to the design length may provide valuable information on the success of the fracturing job.

The actual fracture extent might also be determined from an earlier flow regime called formation linear flow or from the late-time pseudoradial flow. Unfortunately, the formation flow regime might be limited to unrealistically small times, and therefore, masked by wellbore phenomena, and, at least for low permeability formations, the pseudoradial flow regime might not be available during the limited period of time of a standard well test.

In the literature, several other effects are considered, including the influence of boundaries, reservoir shape, and well location, commingled reservoirs, partial vertical penetration, non-Darcy flow in the fracture and/or in the formation, permeability anisotropy, double porosity, phase changes, fracture face damage, and spatially varying fracture conductivity. Several of these issues are addressed also in Chapter 15.

17-6.3 Logging Methods and Tracers

Once a reservoir interval has been fracture stimulated, several logging methods are available to image the created fracture and evaluate the treatment performance. The most widely used methods include gamma ray, spectral gamma ray, temperature, production, full wave-form sonic, and oriented gamma ray logging. Spectral gamma ray images use multi-isotope tracers to identify fracturing outcomes such as (1) propped vs. hydraulic fracture height at the wellbore, (2) proppant distribution at the wellbore, (3) amount of understimulated interval or fracture height, and (4) fracture conductivity as a function of fracture width/proppant concentration.

Staging efficiency can be measured with separate tracers in each multiple stage used to fracture or acidize the well. Larger stress contrast or pore pressure contrast between reservoir layers may exist than was anticipated, causing inefficient coverage from a single-stage treatment. Tracers also may determine that multiple stages are unnecessary, and fewer stages can be used. Diversion effectiveness can be determined with different tracers in individual stages between diverter stages (such as gel, foam, or mechanical). Tracers also can establish that ball sealers were effective in distributing the fracturing or acidizing treatment over the entire interval.

Proppant redistribution may be indicated by using "zero wash" tracers or by running more than one post-treatment image. Zero wash tracers determine if proppant settling or the return of early or late proppant to the wellbore has occurred.

It is recommended to use tracers when one or more of the following conditions exist:

- Thick intervals of reservoir to be stimulated (>45 ft).

- Stress contrast between the zone of interest and adjacent barriers is less than 700 psi.

- Limited entry stimulation technique is planned.

- Specialty proppants (different size or type) will be used or tailed-in at the end of the treatment.

- Fluid leakoff is expected to be higher than usual or is unknown.

Temperature logging can determine post-treatment hydraulic fracture height and fluid distribution at the wellbore but is not indicative of proppant placement or distribution. Cold fluids (ambient surface temperature) injected into the formation can be detected readily by a change in the temperature profile within a wellbore. A series of logging passes usually is sufficient to determine the total treated height. Intervals that received a large volume of injected fluids and/or proppant will require a much longer time to return to thermal equilibrium.

17-7 ADVANCED MODELING OF FRACTURE PROPAGATION

17-7.1 Height Containment

The vertical propagation of the fracture is subjected to the same mechanical laws as the lateral propagation, but the minimum horizontal stress may vary significantly with depth and that variation may constrain vertical growth. The equilibrium height concept of Simonson *et al.* (1978) provides a simple and reasonable method of calculating the height of the fracture if there is a sharp stress contrast between the target layer and the over- and under-burden strata. If the minimum horizontal stress is considerably (several hundred psi) larger in the over- and underburden layers, we may assume that the fracture height is determined by the requirement of reaching the critical stress intensity factor at both the top and bottom tips. As the pressure at the reference point (at the center of perforations) increases, the equilibrium penetrations into the upper (Δh_u) and lower (Δh_d) layers increase.

The requirement of equilibrium poses two constraints (one at the top, one at the bottom), and the two penetrations can be obtained by solving a system of two equations. If the hydrostatic pressure component is neglected, the solution is unique up to a certain pressure called the run-away pressure. Above the run-away pressure there is no equilibrium state. This concept does not mean necessarily that an unlimited height growth occurs, but there is no reason to assume that the vertical growth will be more constrained than the lateral one. As a consequence, we can assume a radially propagating fracture. In case of extremely large negative stress contrast, unlimited height growth may actually happen, leading to the loss of the well.

The equilibrium height concept can be applied in an averaged manner, determining a constant fracture height with an average treatment pressure (Rahim and Holditch, 1993). If the concept is applied for every time instant at every lateral location, we arrive at the so-called pseudo-3D (P3D) models discussed below.

17-7.2 Simulators

One inherent drawback of the algebraic 2D models is that they cannot account for the variation in flow rate with the lateral coordinate caused by inflation of the width and leakoff. Figure 17-21 shows the basic notation of Nordgren's (1972) model, which is based on the continuity equation

$$\frac{\partial q}{\partial x} + \frac{2 h_f C_L}{\sqrt{t - \tau(x)}} + \frac{\partial A_c}{\partial t} = 0 \qquad (17\text{-}58)$$

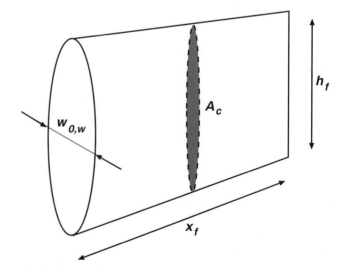

Figure 17-21 Basic notation for Nordgren's differential model

where q is the flow rate, $A_c = (\pi/4)w_0 h_f$ is the cross-sectional area (both varying with location and time) and $\tau(x)$ is the opening time corresponding to a given location x. After substituting the relations between width and pressure for vertical plane strain and between pressure and flow rate for limiting elliptical flow geometry, this equation takes the form

$$\frac{E'}{128\mu}\left(\frac{\partial^2 w_0^4}{\partial x^2}\right) = \frac{8h_f C_L}{\pi\sqrt{t-\tau}} + h_f\left(\frac{\partial w_0}{\partial t}\right) \quad (17\text{-}59)$$

where w_0 is the maximum width of the ellipse at location x.

The initial conditions are and $w_0 = 0$ and $x_f = 0$. The wellbore boundary condition is given by

$$\text{at } x = 0: \frac{\partial w_0^4}{\partial x} = -\frac{512\mu}{\pi E'}i \quad (17\text{-}60)$$

where the injection rate, i may vary with time.

The moving boundary (Stefan's) condition valid for the tip was rigorously formulated by Kemp (1990):

$$\text{at } x = x_f: -\frac{\pi E'}{4\times 128\mu}\left(\frac{\partial w_0^4}{\partial x}\right) = \frac{\pi}{4}u_f h_f w_0 + 2u_f h_f S_p$$
$$(17\text{-}61)$$

where the tip propagation velocity, u_f, is the growth-rate of the fracture length, i.e. $u_f = dx_f/dt$. This equation shows that the flow rate at the tip (left-hand side) equals the sum of volume growth rate at the tip (first term on the right-hand side) plus the rate of spurt caused by creating new surface (second term).

It should be emphasized that the propagation velocity is an additional variable. To obtain a closed system, we need another boundary condition that determines the propagation velocity, either directly or indirectly.

Nordgren considered the special case when (1) there is no spurt loss and (2) the net pressure is zero at $x = x_f$, and, hence, the width at the moving fluid front is zero. If these two assumptions are accepted, the two boundary conditions at the tip reduce to

$$\text{at } x = x_f: w_0 = 0 \text{ and } u_f = -\frac{E'w_0^2}{32\mu h_f}\left(\frac{\partial w_0}{\partial x}\right) \quad (17\text{-}62)$$

The model can be written in dimensionless form where the solution is unique (at least for constant injection rate). In practical terms, this means that it is enough to solve the system once (by an appropriate numerical method). Most of the P3D models are generalizations of Nordgren's equation for non-constant height (Palmer and Caroll, 1983; Settari and Cleary, 1986; Morales and Abou-Sayed, 1989). They relax the assumption of fixed height and introduce some kind of coupled description of the height evolution (Figure 17-22).

The even more useful impact of Nordgren's equation (or rather of its moving boundary condition) is that it reveals the basic source of controversy in hydraulic fracture modeling. At first, it may look natural to assume a zero net pressure at the tip. However, the zero-net-pressure assumption implies that the energy dissipated during the creation of new surface can be neglected. Evidence of "abnormally high" fracturing pressures (Medlin and Fitch, 1988; Palmer and Veatch, 1990) suggests that the zero-net-pressure assumption cannot be valid in general. Our present understanding is that in most cases the fracture propagation is retarded, which means higher than zero net pressure at the tip, because there is an intensive energy dissipation in the near-tip area. Several attempts have been made to incorporate this tip phenomenon into fracture propagation models. One reasonable approach is to introduce an apparent fracture toughness which increases with the size of the fracture (Shlyapobersky *et al.*, 1988). Other approaches include a controlling relationship for the propagation velocity, u_f, incorporating some additional mechanical property of the formation.

Additional modeling activities, including more rigorous description of the leakoff process, using more advanced techniques to solve the embedded problem of linear elasticity, incorporating temperature variations, dealing with non-homogenous rock properties, and taking into account poroelasticity effects, temperature variations, and improving the fluid flow description have provided further insight (Clifton and Abou-Sayed, 1979; Advani, 1982; Barree, 1983; Thiercelin *et al.*, 1985; Boutéca, 1988; Detournay and Cheng, 1988).

17-8 MATRIX STIMULATION

Matrix stimulation can remove the damage in a localized region around the wellbore, and, in some cases, it can even create a high-permeability network to bypass the damage.

Acidizing treatments, the most common matrix stimulation treatments, are often categorized by formation rock type (sandstone or carbonate) and acid type. Any acidizing treatment, regardless of formation type or acid, is performed through the naturally existing flow

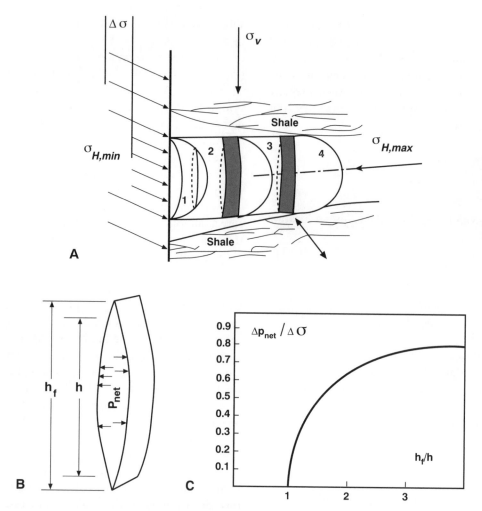

Figure 17-22   Fracture geometry and height growth

channels in the rock. Acid injection into the formation is carried out so that the pressures achieved during the treatment do not exceed the fracture pressure of the formation.

Acidizing is usually very economically attractive, because relatively small treatments may improve the well performance considerably. The potential improvement is, however, often limited compared to proppant fracturing. In addition, the risks associated with this kind of treatment cannot be neglected. Matrix acidizing involves complex chemical and transport phenomena that, while effective in removing one kind of damage, may create another one. Acid placement and damage removal from laminated formations where some perforations penetrate very high-permeability layers is especially problematic. Diversion techniques, discussed later, are intended to force acid flow into the damaged layers instead of selecting the clean zones.

17-8.1 Carbonate Reservoirs

Carbonate acidizing is the reaction of hydrochloric acid on calcium carbonate (calcite) or calcium magnesium carbonate (dolomite) formation to form water (H_2O), carbon dioxide (CO_2), and a calcium or magnesium salt. As the carbonate is dissolved, wormholes form in the formation (Nierode and Kruk, 1973). Wormholes are highly conductive, open channels radiating stochastically from the point of acid injection (Figure 17-23). Wormholes form because the porous media are not homogeneous. As permeability increases through the enlarged pores or pore throats, flow increases into the wellbore. The length, direction, and number of wormholes generated by the treatment depend on the reactivity of the formation and the rate at which acid leaks off to the matrix. These channels may be only a few inches in length or as long as several feet, depending on the char-

Figure 17-23   Wormholes Created by Matrix Acidizing

acteristics of the rock. The efficiency of the stimulation is determined by the structure of the wormholes. Once formed, the wormholes carry virtually all the flow.

Matrix acidizing treatments in limestone reservoirs commonly use a 15% concentration of hydrochloric acid, although concentrations from 7.5 to 28% by weight can be used. Additives such as polymers or surfactants may help increase the effective length of the generated wormholes. Organic acids, such as acetic or formic acid, also can be used if extreme corrosion, special metals, long contact times, or crude oil sludging are significant problems. Acid volumes for matrix acidizing range from 15 to 200 gal/ft of carbonate interval.

17-8.2 Sandstone Reservoirs

Matrix acidizing in sandstone reservoirs removes near-wellbore skin damage caused by drilling, completions, fines migration, or clay swelling. Small concentrations of HCl-HF acid react within 1 to 6 in. of the wellbore to dissolve damaging materials within the rock matrix.

The reaction of HF acid with sand dissolves small amounts of rock. The reaction of HF acid with alumino-silicates also produces a tertiary reaction. During the primary reaction, HF dissolves alumino-silicates (M-Al-Si-O), forming a mixture of silicon and aluminum fluoride complexes. Any other metal ions associated with clay also are released into solution. The silicon and aluminum fluorides formed in the primary reaction continue to react with the alumino-silicates, consuming additional acid, which can lead to potential precipitation.

17-8.3 Diversion

Diversion methods help place the acid effectively in the formation, while reducing injection into clean zones. Both chemical and mechanical diversion tools are available to direct fluid flow at the perforations.

Chemical diverters include slugs of solids, viscous pills, or foams. Commonly used solids include oil-soluble resins, benzoic acid flakes, rock salt, or calcium carbonate. The solids are held in suspension with a viscous carrier fluid. Viscous pills contain high loadings of gelling agent with delayed crosslinkers. Occasionally, solids are added to help establish fluid-loss control before crosslinking takes place. Viscous pills are most frequently used for zonal isolation where the bottom zone requires protection. Foam is used commonly in matrix acidizing of sandstones; however, research indicates that foam also can divert acid in limestone formations.

Mechanical diverting methods include ball sealers and injection-type packers. Ball sealers are dropped during acidizing to plug perforations taking fluid. Injection-type packers isolate an interval before pumping an acid sequence to provide positive fluid placement. However, injection-type packers are more costly than other diversion methods.

17-9 FRACTURE ACIDIZING

In fracture acidizing, acid is injected at a rate high enough to generate the pressure required to fracture the formation. Differential etching occurs as the acid chemically reacts with the formation face. Areas where the rock has been removed are highly conductive to hydrocarbon flow after the fracture closes. Fracture acidizing differs from hydraulic fracturing in that hydraulic fracturing fluids usually are not chemically reactive, and a proppant is placed in the fracture to keep the fracture open and provide conductivity.

As a general guideline, fracture acidizing is used on formations with >80% hydrochloric acid solubility. Low-permeability carbonates (>20 md) are the best candidates for these treatments. Fluid loss to the matrix and natural fractures can also be better controlled in lower-permeability formations.

A conventional fracture acidizing treatment involves pumping an acid system after fracturing. It may be preceded by a nonacid preflush and usually is overflushed with a nonacid fluid.

Several acid fracturing methods have evolved by controlling fluid loss during the treatment, pumping a high-viscosity preflush ahead of the acid solution, or controlling the densities of the preflush, acid, and overflush fluids used in the treatment. One technique uses nonacid phases containing fluid-loss control additives pumped at intervals during the treatment to re-establish fluid-loss control. Another technique uses a high-viscosity preflush ahead of the acid solution. The acid fingers through the preflush, forcing differential etching. A smaller percentage of the total fracture is exposed to acid, and less acid is required to obtain the required penetration distance. The density-controlled acidizing technique controls the density of the preflush, acid, and overflush fluids allowing gravity separation to direct the flow of acid into a vertical fracture.

The closed-fracture acidizing (CFA) technique enhances fracture conductivity. The technique is used in conjunction with existing fractures in the formation. The fractures can be natural, previously created fractures, or fractures hydraulically induced just before the CFA treatment. The CFA treatment involves pumping acid at low rates below fracturing pressure into a fractured well. The acid preferentially flows into areas of higher conductivity (fractures) at low rates for extended contact times, resulting in enhanced flow capacity. Larger flow channels tend to remain open with good flow capacity under severe closure conditions.

Proppants usually are required in wells with closure stresses greater than 5,000 psi. Differential etching caused by fracture acidizing cannot support such high stresses. As the drawdown increases and additional closure pressure is applied, the etched regions tend to crush or collapse, severely restricting flow in the wellbore area. A tail-in stage of proppant following a fracture acidizing treatment allows the formation to provide conductivity as formation stresses increase.

REFERENCES

Acharya, A.: "Particle Transport in Viscous and Viscoelastic Fracturing Fluids," *SPEPE* (March 1986) 104–110.

Advani, S.H.: "Finite Element Model Simulations Associated With Hydraulic Fracturing," *SPEJ* (April 1982) 209–18.

Babcock, R.E., Prokop, C.L., and Kehle, R.O.: "Distribution of Propping Agents in Vertical Fractures," *Producers Monthly* (Nov. 1967) 11–18.

Balen, R.M., Meng, H.Z., and Economides, M.J.: "Application of the Net Present Value (NPV) in the Optimization of Hydraulic Fractures," paper SPE 18541, 1988.

Barree, R.D.: "A Practical Numerical Simulator for Three Dimensional Fracture Propagation in Heterogeneous Media," paper SPE 12273, 1983.

Barree, R.D., and Conway, M.W.: "Experimental and Numerical Modeling of Convective Proppant Transport," *JPT* (Mar. 1995) 216.

Boutéca, M.J.: "Hydraulic Fracturing Model Based on a Three-Dimensional Closed Form: Tests and Analysis of Fracture Geometry and Containment," *SPEPE* (Nov. 1988) 445–454, *Trans.*, AIME, **285.**

Castillo, J.L.: "Modified Pressure Decline Analysis Including Pressure-Dependent Leakoff," paper SPE 16417, 1987.

Chambers, D.J.: "Foams for Well Stimulation" in "Foams: Fundamentals and Applications in the Petroleum Industry," *ACS Advances in Chem. Ser.* (1994) No. 242, 355–404.

Cinco-Ley, H., Samaniego, F., and Dominguez, F.: "Transient Pressure Behavior for a Well With Finite-Conductivity Vertical Fracture," *SPEJ* (1978) 253–264.

Cinco-Ley, H., and Samaniego, F.: "Transient Pressure Analysis for Fractured Wells," *JPT* (1981) 1749–1766.

Cleary, M.P., and Fonseca Jr., A.: "Proppant Convection and Encapsulation in Hydraulic Fracturing: Practical Implications of Computer and Laboratory Simulations," paper SPE 24825, 1992.

Clifton, R. J., and Abou-Sayed, A. S.: "On the Computation of the Three-Dimensional Geometry of Hydraulic Fractures," paper SPE 7943, 1979.

Conway, M.W., McGowen, J.M., Gunderson, D.W., and King, D.: "Prediction of Formation Response From Fracture Pressure Behavior," paper SPE 14263, 1985.

Daneshy, A.A.: "Numerical Solution of Sand Transport in Hydraulic Fracturing," *JPT* (Nov. 1978) 132–140.

Detournay, E., and Cheng, A.H-D.: "Poroelastic Response of a Borehole in a Non-Hydrostatic Stress Field," *Int. J. Rock Mech., Min. Sci. and Geomech. Abstr.* (1988) **25,** No. 3, 171–182.

Dusterhoft, R., Vitthal, S., McMechan, D., and Walters, H.: "Improved Minifrac Analysis Technique in High-Permeability Formations," paper SPE 30103, 1995.

Economides, M.J., and Nolte, K. (eds.), *Reservoir Stimulation* (2nd ed.), Prentice Hall, Englewood Cliffs, NJ (1989).

Ehlig-Economides, C.A., Fan Y., and Economides, M.J.: "Interpretation Model for Fracture Calibration Tests in Naturally Fractured Reservoirs," paper SPE 28690, 1994.

Geertsma, J. and de Klerk, F.: "A Rapid Method of Predicting Width and Extent of Hydraulically Induced Fractures," *JPT* (Dec. 1969) 1571–81.

Gidley, J.L., Holditch, S.A., Nierode, D.E., and Veatch, R.W., Jr. (eds.), *Recent Advances in Hydraulic Fracturing*, Richardson, TX (1989), SPE Monograph 12.

Gringarten, A.C., and Ramey Jr., H.J.: "Unsteady State Pressure Distributions Created by a Well with a Single-Infinite Conductivity Vertical Fracture," *SPEJ* (Aug. 1974) 347–360.

Haimson, B.C., and Fairhurst, C.: "Initiation and Extension of Hydraulic Fractures in Rocks," *SPEJ* (Sept. 1967) 310–318.

Howard, G.C., and Fast, C.R.: "Optimum Fluid Characteristics for Fracture Extension," *Drill. and Prod. Prac.*, API (1957), 261–270 (Appendix by E.D. Carter.)

Hubbert, M.K., and Willis, D.G.: "Mechanics of Hydraulic Fracturing," *Trans.* AIME, **210**, 153–166 (1957).

Kemp, L. F.: "Study of Nordgren's Equation of Hydraulic Fracturing," paper SPE 18959, 1990.

Khristianovitch, S. A., and Zheltov, Y. P.: "Formation of Vertical Fractures by Means of Highly Viscous Fluids," *Proc.,* World Pet. Cong., Rome (1955) **2**, 579–586,.

Mayerhofer, M.J., Economides, M.J., and Ehlig-Economides, C.A.: "Pressure-Transient Analysis of Fracture-Calibration Tests," *JPT* (March 1995) 1–6.

Mears, R.B., Sluss Jr., J.J., Fagan, J.E., and Menon, R.K.: "The Use of Laser Doppler Velocimetry (LDV) for the Measurement of Fracturing Fluid Flow in the FFCF Simulator," paper SPE 26619, 1993.

Medlin, W.L., and Fitch, J.L.: "Abnormal Treating Pressures in Massive Hydraulic Fracturing Treatments," *JPT* (1988) 633–642.

Meyer, B.R., and Hagel, M.W.: "Simulated Mini-Fracs Analysis," *JCPT* (1989) **28**, No. 5, 63–73.

Morales, R.H. and Abou-Sayed, A.S.: "Microcomputer Analysis of Hydraulic Fracture Behavior with a Pseudo-Three Dimensional Simulator," paper SPE 15305, 1989.

Nierode, D.E., and Kruk, K.F.: "An Evaluation of Acid Fluid Loss Additives Retarded Acids, and Acidized Fracture Conductivity," paper SPE 4549, 1973.

Nolte, K.G.: "Determination of Proppant and Fluid Schedules from Fracturing Pressure Decline," *SPEPE* (July 1986) 255–265 (originally SPE 8341, 1979).

Nolte, K.G., and Smith, M.B.: "Interpretation of Fracturing Pressures," *JPT* (Sept. 1981) 1767–1775.

Nordgren, R.P.: "Propagation of a Vertical Hydraulic Fracture," *SPEJ* (Aug. 1972) 306–14; *Trans.,* AIME, **253**.

Palmer, I.D., and Carroll, H.B., Jr.: "Three-Dimensional Hydraulic Fracture Propagation in the Presence of Stress Variations," paper SPE 10849, 1983.

Palmer, I.D., and Veatch Jr., R.W.: "Abnormally High Fracturing Pressures in Step-Rate Tests," *SPEPE* (Aug 1990) 315–23, *Trans.,* AIME, **289**.

Perkins, T.K., and Kern, L.R. : "Width of Hydraulic Fractures," *JPT* (Sept 1961) 937–49, *Trans.,* AIME, **222**.

Prats, M.: "Effect of Vertical Fractures on Reservoir Behavior - Incompressible Fluid Case," *SPEJ* (June1961) 105–118, *Trans.,* AIME, **222**.

Rahim, Z., and Holditch, S.A.: "Using a Three-Dimensional Concept in a Two-Dimensional Model to Predict Accurate Hydraulic Fracture Dimensions," paper SPE 26926, 1993.

Reidenbach, V.G., Harris, P.C., Lee, Y.N., and Lord, D.L.: "Rheological Study of Foam Fracturing Fluids Using Nitrogen and Carbon Dioxide," *SPEPE* (Jan. 1986) **1**, 39–41.

Roodhart, L.P.: "Proppant Settling in Non-Newtonian Fracturing Fluids," paper SPE 13905, 1985.

Settari, A.: "A New General Model of Fluid Loss in Hydraulic Fracturing," *SPEJ* (1985) 491–501.

Settari, A., and Cleary, M. P.: "Development and Testing of a Pseudo-Three-Dimensional Model of Hydraulic Fracture Geometry," *SPEPE* (Nov 1986) 449-66; *Trans.,* AIME, **283**.

Shah, S.N: "Proppant Settling Correlations for Non-Newtonian Fluids under Static and Dynamic Conditions," *SPEJ* (1982) 164–170.

Shlyapobersky, J., Walhaug, W.W., Sheffield. R.E., and Huckabee, P.T.: "Field Determination of Fracturing Parameters for Overpressure Calibrated Design of Hydraulic Fracturing," paper SPE 18195, 1988.

Simonson, E.R., Abou-Sayed, A.S., and Clifton, R.J.: "Containment of Massive Hydraulic Fractures," *SPEJ* (1978) 27–32.

Sneddon, I.N.: "Integral Transform Methods," in *Mechanics of Fracture I, Methods of Analysis and Solutions of Crack Problems,* G. C. Sih (ed.), Nordhoff International, Leyden (1973).

Soliman, M.Y.: "Technique for Considering Fluid Compressibility Temperature Changes in Mini-Frac Analysis," paper SPE 15370, 1986.

Soliman, M.Y., and Daneshy, A.A.: "Determination of Fracture Volume and Closure Pressure From Pump-In/Flowback Tests," paper SPE 21400, 1991.

Thiercelin, M.J., Ben-Naceur, K., and Lemanczyk, Z.R.: "Simulation of Three-Dimensional Propagation of a Vertical Hydraulic Fracture," paper SPE 13861, 1985.

Valkó, P., and Economides, M.J.: *Hydraulic Fracture Mechanics,* Wiley, Chichester, England (1995).

Williams B.B: "Fluid Loss from Hydraulically Induced Fractures," *JPT* (1970) 882-888; *Trans.,* AIME, **249**.

Winkler, W., Valkó, P., and Economides, M.J.: "Laminar and Drug Reduced Polymeric Foam Flow," *J. of Rheology* (1994) **38**, 111–127

Yi, T., and Peden, J.M.: "A Comprehensive Model of Fluid Loss in Hydraulic Fracturing," *SPEPF* (Nov. 1994) 267–272.

18 Sand Stabilization and Exclusion

R. Clay Cole
Halliburton Energy Services

Colby Ross
Halliburton Energy Services

18-1 INTRODUCTION

The production of formation sand with oil and/or gas from sandstone formations creates a number of potentially dangerous and costly problems (Coberly and Wagner, 1938; Suman *et al.*, 1983; Decker and Carnes, 1977). Losses in production can occur as the result of sand partially filling up inside the wellbore. If the flow velocities of the well cannot transport the produced sand to the surface, this accumulation of sand may shut off production entirely. If shutoff occurs, the well must be circulated, or the sand in the casing must be bailed out before production can resume.

Once produced sand is at the surface and no longer threatens to erode pipe or reduce productivity, the problem of disposal remains. Sand disposal can be extremely costly, particularly on offshore locations where environmental regulations require that the produced sand must be free of oil contaminants before disposal.

Subsurface safety valves can become inoperable, leading to large economic loss and personal hazards, particularly at offshore and remote locations. Erosion-damaged surface and subsurface equipment (Figure 18-1) is expensive to replace, and valuable time is lost during replacement and repair.

Formation damage is another problem associated with wells that produce sand unchecked. The possible creation of void spaces behind the casing can leave the casing and any shaly streaks in the reservoir unsupported. Specifically, the casing can be subjected to excessive compressive loading, causing collapse or buckling. The much less permeable shaly streaks that remain can collapse around the perforated casing, causing severe and irreparable restrictions to production. Failure to prevent formation sand production in its early stages can therefore be very expensive in terms of eventual lost revenue and additional operating costs. In addition, it can create potentially hazardous conditions at the wellsite.

18-2 PREDICTING SAND PRODUCTION

Several methods are used for predicting the likelihood of whether a sandstone reservoir will produce sand. Prediction methods vary in complexity, and no method is completely reliable. All methods depend on various reservoir characteristics, such as drilling rate, density, modulus of elasticity, drillstem test data, logs, bulk compressibility, and other characteristics. Conditions such as drawdown pressures, surge pressures, and treating fluid types may vary from well to well, limiting the reliability of these prediction methods (Morita *et al.*, 1989; Mullins *et al.*, 1974). Nevertheless, the following information sources are suggested as guidelines for predicting sand production: (1) experience in the area, (2) drilling data, (3) core-sample evaluation, (4) drillstem tests, (5) logs, and (6) production data.

18-3 SAND PRODUCTION MECHANISMS

Before sand exclusion measures can be optimized, the following sand production mechanisms must be understood:

- **Grain-by-Grain Movement** Sand movement away from the formation face probably accounts for most formation failures. A well with this type of formation rock may eventually require some type of sand con-

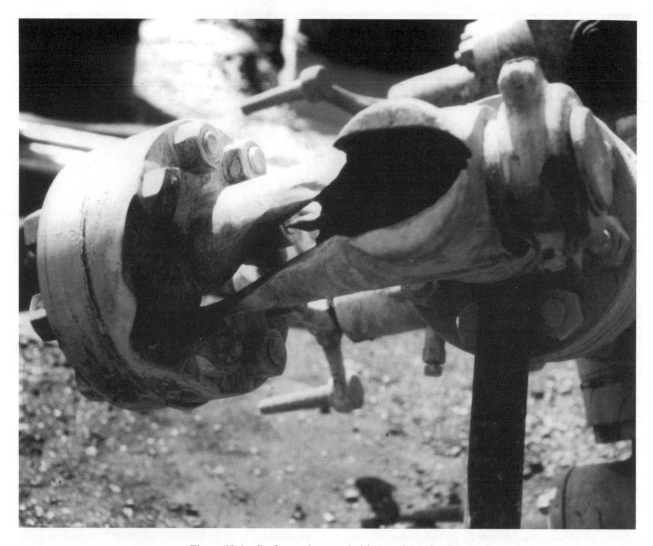

Figure 18-1 Surface valve, eroded by sand production

trol. If the well is allowed to produce sand for an extended period, the choice of sand control method may become more limited.

- **Movement of Small Masses** Under some circumstances, formation rock can break away, resulting in rapid failure. Generally, this type of failure will result in a sanded wellbore that will not produce after sand covers the perforations.

- **Massive Fluidization** The massive amount of sand produced either prevents production or causes erosion; sand disposal problems become too great for production to continue.

Most authorities (Suman *et al.*, 1983; Decker and Carnes, 1977) recommend that sand control techniques be applied immediately upon indication that a formation will produce sand. This practice will allow the highest success rate and the lowest production loss possible after sand control is applied.

Laboratory studies have shown that once an unconsolidated sand is disturbed, the sand cannot be packed back to its original permeability (Decker and Carnes, 1977). Therefore, sand control should be applied before the reservoir rock is seriously disturbed by sand production. Regaining the original density and porosity is not difficult, but the permeability will always be much less than the original permeability. Therefore, if a well is allowed to produce enough sand to seriously disturb the reservoir rock around the wellbore, restressing the reservoir during sand control application can result in a lowered permeability zone around the wellbore. Chapter 6 describes the physics of the mechanisms of sand production and formation deconsolidation.

Wells that have no sanding tendencies can be completed in a number of conventional ways. These options are discussed in Chapter 14. If no initial indication of sand production exists, but it could become a problem in the future, retrievable packer equipment should be used during the completion so that workovers can be performed easily if sand control treatments become necessary.

18-4 METHODS FOR SAND EXCLUSION

Four broad groups of sand-exclusion methods are available: (1) production-restriction methods, (2) mechanical methods, (3) in-situ chemical consolidation methods, and (4) combination methods. With the exception of production restriction methods, the remaining methods provide some means of mechanical support for the formation and help prevent formation movement during stresses that result from fluid flow or pressure drop in the reservoir.

18-4.1 Production Restriction

In a sandstone formation, stresses resulting from fluid production or pressure drop act on the minerals that bond the sand grains together, resulting in sand production. One means of reducing this sand production is to restrict the production rate. This method has the lowest initial cost and, in some cases, may be a successful alternative to other available methods. However, in most cases, it is not a durable or economical solution to formation sand production.

Case histories of horizontal wells have verified that production rate restriction can initially reduce sand production. When a long interval of formation is exposed, equivalent flow volumes can be produced with much lower fluid velocities in the formation. Thus, controlling production in horizontal wells has provided a viable sand-exclusion technique that can be used in formations that were previously completed with vertical wells.

However, in many instances, the reduced production rate that excludes formation production sand is not profitable. In addition, the production rate is not always the only factor contributing to sand production. The degree of consolidation of the formation, the type and amount of cementitious material present, and the amount of water being produced are also significant contributing factors to sand production. These other factors may allow a well to produce sand even after the production rate has been severely restricted.

18-4.2 Mechanical Methods

Mechanical methods are the most common well treatments for excluding sand production. Mechanical sand-control methods are diverse, but they always include some type of device installed downhole that causes the sand to bridge or filters the produced fluids or gases. These devices include a wide array of slotted liners, wire-wrapped screens, prepacked screens, and metal filter screens which are generally used with gravel-pack procedures.

Screens and slotted liners encompass a broad range of downhole-filtration devices. These devices (1) filter out the formation sand, (2) retain any particulate materials, graded sand, or other proppants placed against the formation to support it, and (3) strain out the naturally loose component grains. Variations in screen designs, which will be discussed later, are affected by cost, durability, and flow-through characteristics.

Gravel-packing is a mechanical bridging technique that involves placing and tightly packing a large volume of sized proppant between the formation face or perforation tunnels. A filter device is used that is fine enough to retain the proppant in direct contact with the formation sand, thus preventing its movement toward the wellbore. The packed proppant is supported in the wellbore by a slotted liner or screening device. Advanced gravel-pack designs serve the dual purpose of stopping sand movement and providing production enhancement.

18-4.3 In-Situ Chemical Consolidation Methods

Sand control by chemical consolidation involves the process of injecting plastics or plastic-forming chemicals into the naturally unconsolidated formation, which provides grain-to-grain cementation. The objective of formation sand consolidation is to cement sand grains together at the contact points, maintaining maximum permeability. Consolidation systems are covered in Section 18-7.

18-4.4 Combination Methods

Methods in this category combine technologies of both chemical consolidation and mechanical sand-control; a sand filter bed is one example. In these processes, a gravel-pack is performed with a resin-coated pack-sand as the "gravel." Instead of being held in place with a screen or liner, the pack-sand is held in place by the cured resin, which provides a compressive strength of several hundred to several thousand pounds per square inch. Although procedures vary, the objective is to secure the pack in place while leaving the casing unobstructed. Combination methods include the use of products such as semicured, resin-coated proppants and liquid, resin-coated gravel-pack sands.

18-5 SELECTING THE APPROPRIATE SAND-EXCLUSION METHOD

Once a well is identified as requiring sand exclusion, the appropriate exclusion method must be determined based on the following criteria:

- **Economics**—the initial cost of the treatment and its effect on production

- **Historical success**
- **Applicability**—degree of difficulty to perform treatment

- **Length of service**—estimations of sand-free production and need/frequency rates for repetition of the treatment

Choosing the appropriate technique for sand exclusion requires an in-depth understanding of each sand-exclusion method and its many modifications and variations. Table 18-1 shows the merits and limitations of the many types of mechanical and chemical consolidation methods covered in this chapter.

18-6 MECHANICAL METHODS

Gravel-packing techniques were developed for the water well industry in the early 1900s (Coberly and Wagner, 1938; Decker and Carnes, 1977). The early gravel-pack techniques adopted by the petroleum industry consisted of running a slotted liner to depth, then pouring gravel down the annulus from the surface. This technique was sufficient on shallow, straight wells, but on deeper wells, the gravel would either not reach bottom or it would bridge in the wellbore.

For improved gravel placement, specialized tools were developed. Tools such as set shoes, washdown shoes, clutch assemblies, and stuffing boxes allowed the downhole placement of gravel through such techniques as the *washdown* or *reverse-circulation* methods.

As gravel-placement problems persisted, tool systems were developed that allowed gravel to be pumped down the workstring at greater velocities, which helps prevent premature bridging (Suman *et al.*, 1983). The new tool systems also helped minimize contamination of the

gravel with debris from the annulus, the premature bridging of sand in the annulus, and stuck pipe. Tools developed for this system include the cup-packer crossover, circulating crossovers, and hookwall packer systems.

During the 1960s and 1970s, systems were developed in which the gravel-pack screen, (Suman *et al.*, 1983; Decker and Carnes, 1977) production packer, and circulating service tools were run in the well in one trip. Workstring reciprocation was used to operate the mechanically set packer in the squeeze, circulating, and reverse positions. When the gravel-pack was complete, the crossover service tool assembly was retrieved and replaced by the production seal assembly.

During the 1980s, tools were further developed, and major advancements in carrier fluids were made. Filtered gelled polymers, such as hydroxyethyl-cellulose (HEC) were used for suspending the sand during pumping. Slurry sand concentrations could be increased to over 15 lb/gal, which both reduced pumping time and improved gravel placement into the formation tunnels for a more effective pack that did not contain commingled pack and formation sands.

In the 1990s, industry experts introduced the concept of increasing the pack-sand volume placed outside the casing (Chapter 19). These *fracpacks* required even more specialized tool designs to withstand the high rates and volumes that were being pumped at high pressure. Synthetic proppants became more frequently used since they were more resistant to crushing and had higher permeability under high confining stress. However, because synthetic proppants are significantly more erosive than sands, they pose additional design problems for tool designers. Sand control continues to evolve. New gravel-pack systems, fluids, and chemicals are continually being developed for improved sand placement and pack performance.

Four main components of a gravel-pack must be considered: pack-sand, screen type, carrier fluid, and tools. Each of these components is discussed in the following paragraphs.

18-6.1 Mechanical Components

18-6.1.1 Pack-Sands

For a successful gravel-pack, the pack proppant must be carefully selected. The American Petroleum Institute (API) has issued document RP-58 (1995), which presents recommended practices concerning the evaluation of pack-sand. Accepted by most oil and service companies, this document sets quality standards regarding sand size

Table 18-1   Merits and limitations of sand exclusion methods

Formation Characteristics	Mechanical Methods, Gravel-Packs, and Screens	Chemical and Combination Methods, Consolidation with Resins and Resin-Coated Sands
Formation Strength	Will not change formation strength	Adds considerable formation strength, with exception of the resin-coated sand
Permeability	Applicable. Certain techniques may reduce permeability	Applicable. Certain techniques may reduce permeability
Poorly Sorted Grain Sizing	Applicable using special job designs	Applicable with few restrictions
< 10% Fines and Clays	Very Applicable. Good anticipated job life	Very Applicable. Good anticipated job life
> 10% Fines and Clays	Applicable using special job techniques	Marginally applicable. Good resin injection and coverage is difficult
> 10% Acid Solubility	Applicable with restricted acid pretreatments	Not applicable with acid-hardened type resins
< 10% Acid Solubility	Very Applicable. Good anticipated job life	Very Applicable. Good anticipated job life
< 50° Hole Angle	Very Applicable. Good anticipated job life	Very Applicable. Good anticipated job life
> 50° Hole Angle	Applicable using special tools, screens and techniques	Not applicable. Poor job success history, uniform coverage problems
Open Hole	Applicable using special job techniques	Not applicable. Poor job success history, uniform coverage problems
Cased Hole	Very Applicable. Good anticipated job life	Usually very applicable. Good anticipated job life
Slim Casing	Marginally applicable. Severe tool and screen restrictions	Very Applicable. Good anticipated job life
Single Zone	Very Applicable. Good anticipated job life	Very Applicable. Good anticipated job life
Multiple Zones	Applicable using special tools, screens and techniques, leaving screen in wellbore	Very Applicable. Leaves clear wellbore, should be done as an initial measure
< 30 ft Interval Length	Very Applicable. Good anticipated job life	In most instances. Very Applicable. Good anticipated job life
> 30 ft Interval Length	Very Applicable. May require special tools, screens, and designs	Not applicable. High costs and uneven resin coverage
High Water Producer	Applicable. May require additional chemical fines control	Very Applicable. Good anticipated job life
Gas Producer	Very Applicable. Good anticipated job life	Applicable. Some resin systems clean up better with good anticipated job life
Oil Producer	Very Applicable. Job life depends on screen and proppant quality.	Very Applicable. Anticipated job life of 3 to 8 years
Low BHST, < 120°F	Very Applicable. Good anticipated job life	Provisional. Difficult curing conditions for some resins.
Medium BHST	Very Applicable. Good anticipated job life	Very Applicable. Good anticipated job life
High BHST, > 250°F	Provisional. May require special proppant and tool designs	Provisional. Limited placement time for some resins; durability reduced
Steam Injection	Provisional. Likely will require special proppant and tool designs	Marginally applicable. Some resins are more resistant than others

gradations, shape (sphericity and roundness), strength, and allowable percentages of foreign material in sand that will be used for gravel-packing.

Sand size should be quality-controlled by sieving so that a proper formation sand-to-pack gravel size ratio is maintained and absolute permeability is not reduced. Grain roundness is a measure of the relative sharpness of grain corners, or of grain curvature (how nearly the edges of the grains approach that of a circle). Roundness is critical because angular sand has a greater tendency to form premature bridges than rounded sands.

Angular sand is also more likely to chip and fragment during placement, which reduces the permeability of the pack and can clog slots and screens.

18-6.1.2 Liners and Screens

Slotted liners and screens are downhole filters that provide differing mechanisms and levels of sand retention or pack-sand support. Before a liner or screen is chosen, the well must be carefully evaluated so that the most applicable product can be selected (Cole *et al.*, 1992). Screen construction and shape can influence (1) how well sand becomes packed in the annulus, (2) the flow capacity of the covered zone, and (3) how long the composite pack might last.

The simplest slotted liner is made of oilfield pipe that has been slotted with a precision saw or mill. These slots must be cut longitudinally, otherwise the tubing would become weak under tension. The individual slots can be as small as 0.016 in. or as large as required for the gravel size that will be used.

Mill-slotted pipe provides strength and economical service; it is particularly well-suited for water wells. Generally, slotted liners are used for oilfield applications when wire-wrapped screens cannot be used economically. For example, slotted liners are typically used in wells that have long completion intervals or low productivity. Although slotted liners are much less expensive than wire-wrapped screens, they do not have high-inlet-flow areas and thus are not as useful for high-rate wells.

Wire-wrapped screens consist of keystone-shaped, stainless-steel or corrosion-resistant wire wrapped around a drilled or slotted mandrel made of standard oilfield tubular goods or special alloys. Spacing or stand-off between the mandrel and wire wrap allows for maximum flow through the screen. In some screens, this spacing is created by grooves cut into the pipe; in most screens, ribs are affixed under the wire wrap to the pipe (Figure 18-2).

In a prepacked screen, the annulus between the outer jacket and the pipe base is packed with gravel-pack sand. This sand may or may not be resin-coated. Three types of prepacked screens are available: a dual-wrapped prepack, a casing external prepack, and low-profile screen.

The oldest screen type is a dual-wrapped pipe-based screen prepack. This screen provides built-in sand control when a gravel-pack fails or is not feasible (Figure 18-3).

The casing-external prepack screen has an inner wire-wrapped pipe base centered inside a large piece of perforated pipe. Primarily used for horizontal wells, this type of prepack screen is packed with a graded resin-coated pack-sand that is sized to bridge formation sand. This resin-coated pack-sand, cured at 350°F, bonds the sand grains together to prevent the sand from coming out the perforations of the outer perforated pipe (Figure 18-4).

The third type of prepack screen is the low-profile screen. This screen is similar to the dual-screen prepack, but it consists of an inner microscreen and a regular outer screen jacket. The prepacked sand layer is very

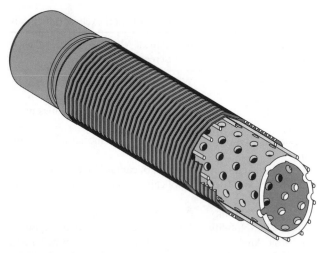

Figure 18-2   All-welded, pipe-base wire-wrapped oilwell screen

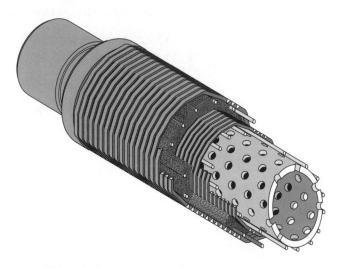

Figure 18-3   Dual-wrapped prepacked well screen

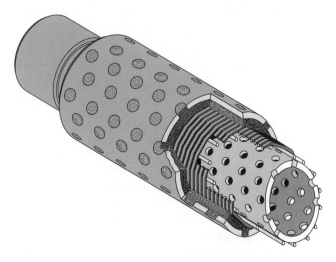

Figure 18-4   Casing-external prepacked gravel-pack screen

thin compared to regular prepacked screens. Low-profile prepacked screens are recommended for wells that will be gravel-packed but in which casing sizes and tubing requirements restrict the outside diameter of the screen. The thin layer of gravel between the outer and inner screen jackets provides insurance in case a void develops in the gravel pack.

Alternative screen designs are continually being developed that provide more durability, compensate for voids left in the gravel pack, or allow better sand-packing while preventing premature sandouts. Most of these screen products have limited use, and their performance in selected jobs is documented in the literature (Cole *et al.*, 1992; Ali and Dearing, 1996).

The sintered metal screen was designed as an improvement over low-profile prepacks. Instead of the prepacked sand layer around the center pipe, a jacket of seamless, porous, sintered stainless steel is used. The uniform porosity of the metal allows the use of any grade of gravel-pack sand (Cole, 1992).

The woven metal-wrapped screen was introduced to fill a similar need as the sintered metal screen, and it can also be used with any grade of gravel-pack sand. This screen consists of layers of various grades of woven metal that are wrapped around a perforated center pipe, which allows the packed zone to be post-acidized without damaging any prepacked sand layer in the screen.

The auger-head screen design has been used as a conventional screen with a washdown shoe. The auger end allows the screen to be screwed into the sand-filled casing or hole. As the hole angle becomes more horizontal, this design becomes more effective than conventional washdown configurations.

Another alternative screen design, the shunt tool, may reduce premature sandouts during the packing of very long intervals. This screen design provides an extra shunt path for gel-sand slurry, which reduces the possibility of slurry dehydration when high-permeability streaks are packed over the composite interval.

18-6.1.3 Carrier Fluids

A fluid used for gravel-packing has three primary functions. First, it must transport the sand or packing medium to the location in the well where the gravel pack will be established. During a horizontal gravel-pack treatment, the packing grains will change direction several times. The velocity or viscosity of the fluid must be able to influence the direction of the grains. Without a transport medium, the grains will always travel down-

ward. If sufficient lift is not provided throughout placement, premature sandout could occur. Sandout is caused when an agglomeration of grains develops in critical pathways, blocking the progress of the remaining grains necessary to form the desired pack.

The second function of carrier fluids is to separate themselves from the grains to allow the close contact desired. The degree to which this fluid loss occurs is critical to gravel-packing success: if the fluid separates too early or too much, a premature sandout may occur; if the fluid separates too late or incompletely, a void may be left when the pack grains settle out. As the grains arrive at the packing location in the well, they must be deposited compactly and sequentially against the formation. The fluid may either exit through the screen and washpipe and return to the surface or, more conventionally, exit to the adjacent formation.

The important third function of a gravel-pack fluid is to return from the formation without reducing permeability within the zone treated. When the gravel pack is finally established, the operator will want to put the well on production. The gravel-pack fluid lost to the formation must now change direction and flow back into the wellbore. Often, a fluid property that is desirable for grain transport may prevent the fluid from rapidly returning from the formation. A high-viscosity fluid is such an example. To rid a fluid of its viscosity before recovery from the formation, a gel breaker must be used. Once in the formation, the fluids have the opportunity to react with the formation grains. These reactions can increase the difficulty of recovering the packing fluid as well as attaining suitable hydrocarbon flow from the packed zone (Almond and Bland, 1984). For example, clay swelling can be caused when clays are exposed to low-salinity water. If the fluid contains surfactants, their influence on the water and hydrocarbon flow from the formation must also be considered. Certain packing fluids that include polymers to create viscosity will leave a chemical residue in the formation when the gel is broken by chemical breakers. The amounts of gel residue associated with the different polymers varies greatly. Before a job is performed, the proposed polymer and mixing brine should be carefully analyzed. When the connate water in some oil-producing formations contacts the dense brines (zinc bromide) used in the gravel pack, a pH change could occur; specifically, zinc hydroxide could form, plugging the pores of the formation.

In addition to various brines that have been successfully used in gravel-packing, several polymers have been added to these brines to enhance the degree of sand transport and packing efficiency in gravel-packing operations. Some of the common polymers and their properties are listed in Table 18-2 (Scheuerman, 1986; Torrest, 1982; Sparlin, 1969).

Gelled fluids made from the polymers in Table 18-3 and ungelled brines each have their champions in the industry. Since good, long-lasting gravel-pack jobs have been reported with each of the contending fluids, it appears that technique is as important as fluid properties in the outcome of the gravel pack (Maly, 1979). Most research now shows that the best perforation packing can be achieved with a viscosified fluid (Schroeder, 1987). However, the same study shows that the best

annulus packing is attained with ungelled brine. Staged gravel-pack treatments allow the use of both packing methods. First, a number of gelled brine-sand slurry stages are pumped to fill the perforations, then the treatment is completed with brine and low sand concentrations that pack the annulus.

18-6.1.4 Tools and Accessories

The tools and accessories for gravel-pack treatments either (1) remain in the well or (2) are run, placed, and retrieved before the production tubing is run.

Table 18-2   Common polymers used in gravel-packing

Gelling Agent	Chemical Category	Common Trade Names
HEC	Hydroxyethyl cellulose	HEC-10, HYDROPAC, AQUAPAC
SGC	Succinoglycan, biopolymer	ShellFlo S, FLO-PAC
Xanthan	Xanthan, clarified biopolymer	Xanvis, Bi-O-PAC
Surfactant Gel	Ammonium quat	PermPAC AV

Table 18-3   Gelled fluids and their characteristics

Polymer	HEC	SGC	Xanthan	Surfactant Gel
Usage Level	60 to 80 lbm/gal	20 to 50 gal/gal	15 to 35 lbm/gal	20 to 40 gal/gal
Cost Factor[a]	1.00	1.07	1.50	2.52
Physical Form	Dry powder	Jelly-like liquid[b]	Dry powder	Pourable liquid[c]
Preparation Steps	Adjust to pH = 3 to 4, add polymer and raise pH to 7	Add over-the-top into high shear area; No pH adjustments.	Use Fe sequestering agent; Add polymer; add salts last.	Batch or continuous mix of surfactant with chloride brines
Requirements for Optimum Performance	Preshearing and filtration to 5 to 10 μm	Complete dispersion of the polymer	Extensive and specific preshearing then filtration	Complete dispersion of surfactant
Restrictions	Does not yield in three-salt brines, mostly 15 to 18 lbm/gal	Slow to yield in brines with limited free water	Does not tolerate divalent brines, i.e. $CaCl_2$	Does not tolerate saturated brines or ones containing $ZnBr_2$
Maximum Viscosity at 75°F (cp)	160	23	21	240
Sand settling	Slight	Almost none	Almost none	Almost none
Minimum Viscosity at 75°F (cp)	45	14 to 16	14 to 16	66
Sand Settling	Moderate	Slight to Moderate	Slight to Moderate	Almost none
Temp. Limit	190 to 230°F, depending on the mix brine	160°F to 210°F, Very dependent on mix brine.	190°F to 230°F, Depending on the mix brine	200°F
Common classes of Breakers Used	Acids Oxidizers Specific Enzymes	Oxidizers Oxidizers with Enhancers	Oxidizers Oxidizers with Enhancers	No internal breakers. Broken upon dilution by formation fluids or flushes

[a]Least cost of material factor = 1.0
[b]8.48 lbm/gal
[c]8.52 lbm/gal

Completion Tools

Gravel-Pack Packer—The gravel-pack packer is one of several packer types and designs. Generally, a sealbore-retrievable or permanent packer is used. If a retrievable packer contains slips above the element package, sand could settle on top of the packer, causing retrieval problems.

Flow Sub—A flow sub can either be a ported sub or a sleeve-closure device combined with a sealbore and lower pup joint. The flow sub provides a location for the service-tool gravel exit ports and a path that directs flow to the outside of the screen. Closing sleeves are primarily used (1) if sufficient lengths of blank are not available to restrict flow up the annulus instead of through the screen, and (2) if production seals cannot be run below the ported sub to prevent sand production.

Mechanical Fluid-Loss Device—A mechanical fluid-loss device stops uncontrolled fluid losses to the formation while the service tools are being tripped out of the well and production tubing is being run. These tools can be any one of a number of flapper, ball, or plug-type devices that can later be broken, dissolved, or expended.

Safety Joint—A safety joint allows operators to detach the screen from the packer during retrieval operations. It also provides a safeguard against extreme loads being placed on the gravel-pack packer as formation compaction or gravel-pack settling occurs. This tool should maintain the integrity between the inner and outer mandrels until approximately 12 in. of movement occurs.

Blank Pipe—Blank pipe above the screen provides a reservoir of proppant that can fill in from the top of the screen as settling occurs. Generally, at least 60 ft (two joints of pipe) should be used on high-rate water-packs (HRWPs) and fracpacks. Gravel packs using high-density slurry should include enough blank pipe above the screen to accommodate the entire slurry volume in the annular space below the blank. If less than 60 ft of blank pipe is used, the following issues must be considered:

- Gravel-pack quality (compacted, void-free pack)

- Screen damage from high sandout pressures or direct impingement from adjacent perforations

- Carrier-fluid viscosity (sandout can proceed up to and into the workstring)

Production screens are installed next in the string and are described in Section 18-6.2.

Tell-Tale Screen—A tell-tale screen is a short, upper or lower section of screen that indicates when the proppant pack has achieved a certain height during pumping. It is separated from the rest of the screen assembly by a seal that opens or closes, circulating flow as the gravel-pack service tool is manipulated. In new applications, tell-tale screens have been virtually replaced by large-diameter washpipe.

Seal Assembly—The seal assembly provides a seal between the bottom of the screen assembly and the sump packer. It is typically run with a slotted cylindrical collet, which has a raised diameter that interferes with the inside of the sump packer sealbore. This collet indicates the presence of drag when the seal assembly passes through the sealbore of the sump packer.

Sump Packer—The sump packer is the lowermost packer in the sand-control completion assembly. A variety of packer designs can function as a sump packer, which must provide a solid bottom that can contain the proppant at the lower end of the gravel-pack interval. Without this packer, any proppant settling in the rathole could result in voids in the proppant pack around the screen. Packers normally used in this application are permanent or retrievable sealbore packers (Chapter 14) that are run on electric line or hydraulic setting tools, generally before perforating. This packer also provides a location where weight can be applied to position the gravel-pack assembly in the wellbore.

Service Tools

Crossover Service Tool—The crossover service tool provides channels for the circulation of proppant slurry to the outside of the screen and returns circulation of fluid through the screen and up the washpipe. Most tools also include the hydraulic setting tool that sets the gravel-pack packer.

Reverse-Ball Check-Valve—A reverse-ball check-valve can be as simple as a ball sitting on a restricted-diameter crossover in the service tool or as complex as a multiple-actuating ball valve. Regardless of the valve design used, the valve should not create a hydraulic lock or result in excessive differentials to the formation.

Swivel Joint—The swivel joint allows operators to assemble concentric tubing strings by freely rotating the joint as threads are made up on the outer strings of pipe.

Washpipe—The washpipe is attached to the gravel-pack service tool and run inside the screen. The washpipe serves two functions. First, it provides a return fluid circulation path that can be spaced out at the very end of the screen interval. This path forces the proppant slurry to flow to the lowermost screen before bridging in the screen-casing annulus.

The second function of the washpipe is to prevent the proppant carrier fluid from flowing through the outside of the screen. Loss of fluid from a proppant slurry can cause premature and rapid bridging to occur, especially in high-density proppant concentrations. A number of studies have verified that the wash pipe diameter should be at least 80% of the inside diameter of the screen base pipe.

Shifting Tools—Shifting tools position the closing sleeves or close the flapper valves as the service tool is pulled from the well.

Tool Selection—Once the well objectives are understood, a completion technique and the well completion tools can be selected. The service provider and the tool supplier must know the casing size and weight, zone depths, bottomhole pressure, expected pressure differential, and the well fluids (both completion fluids and produced fluids). Generally, a packer is selected first on the basis of the size, weight, and grade of the casing. The bore size of the packer is typically influenced by the tubing size used, which affects the size of many accessory items. Because well fluids are occasionally corrosive, the packer and completion tools may need to be made from corrosion-resistant materials.

18-6.2 Mechanical Techniques and Procedures

A complete, void-free gravel pack is one of the most effective sand-control measures available. The annular portion of the pack alone cannot sustain high-rate well productivity over a long period. The external gravel pack (the area either in a perforation tunnel or fracture that extends past any near-wellbore damage) is a key to prolonged trouble-free production. Since the perforation tunnels are the only communication from the formation to the wellbore, they must remain unclogged by formation sand and fines. If gravel-packing is used for controlling sand production, the pack-sand should fill the perforation tunnels and void spaces behind the casing (Decker and Carnes, 1977; van Poolen *et al.*, 1958; Penberthy, 1988). Sometimes, the formation can even be restressed, although restressing is seldom achieved except through pressure gravel-packing.

18-6.2.1 Gravity Pack

A gravity pack is one of the most primitive gravel-packing techniques. With water in the hole, a screening device is run into the hole on tubing with a backoff joint. Sand is then slowly poured down the annulus while fluid is running into the hole. This technique is one of the least expensive approaches, and it provides very little control of sand placement. In fact, it is unlikely that any sand is forced through the perforations; sand may bridge around the tubing collars during sand placement. Consequently, when this method is used, flow into the wellbore may quickly become restricted by formation fines invading the pack.

18-6.2.2 Washdown Method

The washdown method consists of depositing gravel to a predetermined height above perforations, then running the screen and liner assembly with a washpipe and a circulating shoe. The screen is washed down through the gravel. When the shoe is on the bottom, gravel is allowed to settle back around the screen and liner. The most basic washdown tool system has minimal provisions for compacting the gravel in the screen-casing annulus, and no means of squeezing gravel through the perforations. However, perforations are often pressure-packed before the screen is washed down. In more complex tool systems, the tools can perform the washdown technique followed by a circulating, squeeze, or fracpack after the packer has been set. During the washdown procedure, the gravel sizes tend to segregate. In addition, the tools could stick or stop short of the bottom. If viscous fluids are used as a means of suspending the sand, they may damage the zone.

18-6.2.3 Circulation Packs

Circulation packing, sometimes called conventional gravel-packing, normally involves the placement of gravel that is suspended in a low-viscosity transport fluid pumped at low gravel concentrations. Circulation packing is usually conducted in an open hole or in a cased hole after the perforations have been prepacked with gravel. Typically, the transport fluid is filtered brine with gravel added at a concentration of 0.5 to 1.0 lb/gal. The gravel is commonly mixed into the fluid through a gravel injector while fluid is pumped at approximately 0.5 to 3.0 bbl/min. Gravel is transported into the annulus between the screen and casing (or the screen and the open hole), where it is packed into position from the bottom of the completion interval upward. The transport fluid then returns to the annulus through the washpipe inside the screen that is connected to the workstring. Circulation packing is compatible with essentially all the subsequent placement techniques discussed in this chapter.

18-6.2.4 Reverse-Circulation Pack

As the term implies, the *reverse-circulation method* involves the reverse circulation of water-sand slurry or gelled water-sand slurry. For the placement of some sand through the perforations, returns can be shut off and pumping can be continued until a pressure increase indicates that sand has covered the tell-tale screen. If no positive pressure is exerted, only sand-packing inside the wellbore is likely to be achieved. Two or three joints of unslotted tubing are generally run above the screen. These tubing joints provide a space for sand to fill the annulus above the screen, which normally prevents fluid and sand movement up the annulus and provides a reserve of pack-sand if the pack should settle.

18-6.2.5 Bullhead Pressure Packs

This inexpensive procedure requires a packer and a releasing crossover tool; a tell-tale screen or washpipe is not required, since no returns are taken during the procedure. The water-gel sand slurry is pumped down the tubing behind a gel prepad. Slurry is then directed from the tubing into the casing-screen annulus through the releasing crossover tool below the packer. Pumping pressure increases as more perforations are covered with pack-sand. After final sandout, the packer is released from the screen and blank pipe at the crossover tool.

18-6.2.6 Circulating-Pressure Packs

A *circulating pressure pack* requires a packer and crossover tool as well as a screen with an internal washpipe. This technique allows slurry to be injected down the tubing, which prevents the scouring of drilling fluid, rust, pipe dope, and scale from the tubing-casing annulus and reduces the potential of pack permeability damage. The slurry crosses over into the annulus below the packer. The carrier fluid deposits the sand and enters the screen and washpipe. Fluid is then conveyed to the annulus above the packer by means of the crossover and then returned to the surface.

For sand to be placed through the perforations, the surface returns should be stopped at some point, and slurry should be squeezed against the formation before the perforations are tightly packed with sand.

18-6.2.7 Slurry Packs

Slurry packing can be performed with a variety of tools and pumping techniques. The procedure consists of pumping gravel at high concentrations in a viscous transport fluid. The slurry is usually batch-mixed in a blender or paddle tank before it is pumped. Although the fluid can be either water- or oil-based, hydroxyethyl cellulose (HEC)-viscosified brine is the usual choice (Decker and Carnes, 1977; Penberthy, 1988; Rensvold, 1978). The typical polymer loading is 60 to 80 lb/Mgal. In some situations, the gelled fluid is crosslinked so that it creates very high viscosities. The concentrations of gravel usually pumped with slurry-pack fluids is about 10 lb/gal. In certain situations, however, concentrations have ranged from much less than 10 to more than 18 lb/gal. When these high-density fluids are pumped, the fluid and gravel tend to move as a mass. Compared with the low-density conventional gravel-pack fluid, these slurry systems have significantly greater gravel-suspending capabilities. Depending on well conditions, pump rates normally range from $\frac{1}{2}$ to 4 or 5 bbl/min when gravel is placed around the screen.

When high-viscosity fluids transport gravel into the completion interval, a reserve volume must be specified. To allow for subsequent gravel settling, this volume is usually higher than the reserve volume required for low-viscosity fluids. Well conditions affect the gravel reserve volume when viscous fluids are used. Before the gravel is pumped, the service provider should calculate the theoretical amount of gravel required to pack the annulus around the screen and the amount of gravel reserve. Placement should be approximately 100% under ideal conditions. Depending on the additional gravel placed through the perforations, possible bridging in the tubulars, or pumping conditions, the amount of gravel placed could differ from the theoretical volume. If the amount of gravel is considerably less, premature bridging has probably occurred. Therefore, before the well is placed on production, gravel must be settled around the screen and additional gravel must be pumped.

When high-viscosity fluids circulate gravel around the screen, the pump rate should be low enough so that the viscous drag forces do not exceed the gravitational forces once the slurry has reached the completion interval. If the viscous forces become dominant, the gravel will be drawn into the screen rather than settling to the bottom of the well. As a result, the gravel will pack radially outward from the screen, possibly resulting in premature indications of a completed gravel pack. This type of packing geometry and sequence is not desirable; ideally, the gravel should dehydrate from the bottom of the completion interval upward. The potential for nonuniform packing is one of the disadvantages of slurry packs. Therefore, the actual volume of gravel pumped should be carefully compared to the theoretical volume. If the pumped-gravel volume is much less than the theoretical volume when high-viscosity fluids are used, the gravel should be allowed to settle before additional gravel is pumped for completing the pack.

An upper tell-tale screen should be avoided when high-viscosity fluids are used for circulating gravel because gravel will probably screen off on the upper tell-tale. This screenoff could prevent gravel placement over the main completion interval. When no upper tell-tale is used, the screen or slotted liner should be extended above the completion interval before the blank tubing is added so that a gravel reserve is available. Because of gravel settling, part of the gravel reserve volume will be filled if well deviations do not exceed approximately 60°. Sometimes, a lower tell-tale screen is used with slurry packing. If a lower or upper tell-tale screen is not used with this system, the washpipe extends to the bottom of the screen section.

18-6.2.8 Staged Prepacks and Acid Prepacks

Productivity can usually be increased when perforations are prepacked with gravel immediately before the annulus gravel pack is pumped. The prepack provides a more complete gravel fill of the perforation and as a result, it increases perforation permeability. The gravel pack prevents perforations from collapsing of filling with formation sand (Hall and Pace, 1990).

Hall and Pace reported that the best results were achieved when alternating stages of acid treatment and gravel slurry were displaced in volumes sufficient to treat 20 to 30 ft of the well interval. Therefore, if the treatment interval is 100 ft, then four or five phases of acid and slurry will be needed to provide enough gravel-pack material for the entire interval.

18-6.2.9 Water-Packs and High-Rate Water-Packs

As its name suggests, a *water-pack* uses water as a carrier fluid for gravel-pack sand. Because the polymer residue

from slurry packing could potentially damage formation permeability, water-packs have become a popular alternative to slurry packing in recent years. Water-packs can help form very tightly packed annular packs, but they have a high leakoff rate in high-permeability zones, which can result in bridging in the screen/casing annulus. This bridging can then cause premature screenout of the treatment.

High-rate water-packs were developed to overcome the high leakoff problems encountered with standard water-packs in high-permeability formations (Rensvold, 1978; Ledlow and Johnson, 1993; Penberthy and Shaughnessy, 1992). High-rate water-packs are usually preceded by an acid prepack, and they place far more sand (up to 700 lb/ ft of perforations) since they exceed the formation-parting pressure. Other successful high-rate water-pack treatments have been reported from geopressured reservoirs where very little differential has existed between static formation pressure and formation parting pressure (Penberthy and Shaughnessy, 1992).

Although the term *water-pack* suggests that only water is used for proppant transport, a lightly gelled slurry (25 lb HEC/Mgal) is frequently pumped (Penberthy, 1988). The danger that zonal isolation may be compromised is possible with high-rate water-packs, just as it is with fracpack completion services. Tracer log data from some water-packed wells indicate that incomplete entry of the tracer over the entire interval height has occurred, leaving only the high-permeability area of the zone packed. Therefore, only wells with formations that can sufficiently resist fracture height growth are candidates for high-rate water-packing. Since the pumping rate, fluid volume, and pressure in this type of completion are large, the tools used must withstand the more severe well conditions. If tools are not properly selected, downhole equipment failures could be caused by collapse and flow erosion. Both the fines in the fluid stream and the return flow rate must be controlled so that the possibility of screen erosion during the packing operation is limited.

18-6.2.10 Fracpacks

Fracpack treatments are a special type of gravel-packing in which the volume of sand placed outside the perforations far exceeds the volume of the voids behind the pipe. The theory and methods used to optimize this type of gravel-pack treatment are discussed in Chapter 19.

18-6.2.11 Summary

Choice of given technique is driven by economics. A number of factors have a relationship to completion cost. These factors include

- Design simplicity

- Minimal rig time through the elimination of trips

- Minimal rig floor assembly time

- Well depth

- Well control problems (fluid loss, high-pressure zones)

- Zone spacing and the number of zones to be completed

- Bottomhole pressure, temperature, and fluids

- Availability of surface pumping and blending equipment

- Required well completion life

- Workover costs

- Safety considerations

Each of these items contributes to the system selection process. These decisions also vary geographically.

18-6.3 Mechanical Job Designs

Effective mechanical sand exclusion requires good job execution and a good gravel-pack design. The primary objective of the application should be to control the production of formation sand without excessively reducing well productivity. During the job design process, the parties involved must choose a pack-sand grade, a screening device, a carrier fluid, chemical pretreatments, and placement techniques. The first step of the job design is to evaluate the formation that requires the gravel pack.

18-6.3.1 Formation Characteristics

Formation permeability influences the type of carrier fluid that will optimize leakoff during pack placement. Sieve analysis can reveal the formation's average structural grain size, allowing job designers to determine the correct grain size of pack sand. Knowledge of the formation mineralogy, usually measured by X-ray diffraction instrumentation, helps job designers identify troublesome feldspars and clay minerals that are prone to swell and/or migrate as fluids are produced through the pack.

The relationship between the particle size distribution of a given formation sand and the critical size required for gravel-pack sand is significant. Therefore, job designers must distinguish the size of the load-bearing solids from the size of formation fines. Fines are very small particles of loose solid materials in the pore spaces of nearly all sandstone reservoirs. Produced fines likely originate at the interface between the gravel pack and the near-well formation, rather than from distant points in the reservoir. The higher flow velocities near the well probably contribute to increased fines mobility in this region.

The integrity of the formation analysis depends on the quality of the sample used in the analysis (Himes, 1986; Maly and Krueger, 1971). The most accurate and preferred sample type is a full core obtained during drilling across the expected interval that will be completed. However, this type of sample is expensive to obtain, especially if the exact completion interval is unknown, and long intervals of formation must be cored.

Bailed or produced samples are generally unacceptable because sample segregation can occur. Produced samples will include finer material and bailed samples are composed of the coarser fractions. These samples tend to represent an average of the particle sizes across the interval, and the representative average is usually larger than the finest sands in the formation.

Obtaining samples of the formation with a wireline sidewall coring tool or gun is the most popular method. The gun is run into the open hole before it is cased, and a hollow core barrel is fired through the filter cake into the formation of interest. Therefore, the formation material obtained is of the immediate wellbore region. This formation material is usually flushed with drilling fluid filtrate that penetrates the formation before filter-cake buildup; therefore, cores that are taken subsequently are usually contaminated with drilling fluid clays. Contamination with these fines can lead to erroneous results during sample analysis. The amount of artificially introduced fines can be as high as 30% by weight. Since most of this material is 20 μm or smaller, a sieve analysis of these samples will be skewed toward a smaller size and result in a D_{50f} value that is much smaller than the true value; consequently, a finer pack-sand is recommended, and lower productivity may result.

Two acceptable methods are used for determining the particle size of formation sand. The first method is a sieve analysis, in which a sample is cleaned of oils and non-formation constituents, and then dried and passed through a specific set of sieves. For all comparative analyses, the industry has adopted the *US mesh series*, which consists of a standard series of 12 sieves and a bottom pan. In this series, each sieve opening has twice the cross-sectional area as the sieve below it in the series. Table 18-4 lists the sieve numbers and their opening sizes.

The sieve analysis is performed according to ASTM Procedure C136-84A, which requires that the portion of the sample retained on each sieve is weighed, and that a weight percentage is calculated. When graphically displayed, the sieve analysis data can reflect the sizes of the component grains and their comparative contribution as a percentage of weight. Figure 18-5 shows how the D_{50f} point of an analysis is determined graphically.

Table 18-4 Standard series of 12 sieves for a sample analysis

US Sieve Number	Sieve Opening (in.)
10	0.0787
20	0.0331
30	0.0232
40	0.0165
60	0.0098
80	0.0070
100	0.0059
120	0.0049
140	0.0041
170	0.0035

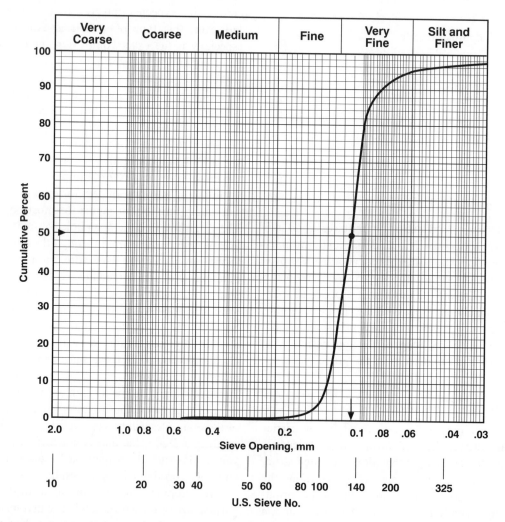

Figure 18-5 A typical sieve analysis plot showing the value of the D_{50f} point (in.), found by extending a perpendicular line down from the D_{50f} point on the curve

This point represents the median grain diameter of the formation tested. Although other significant points along the sieve curve have been used for calculating a pack-sand size, especially where nonuniform sands are involved, the D_{50f} point is nearly universally accepted for this purpose.

A second method is to use an electronic particle-counter. Suitable instruments for this task use either the principle of light blockage or laser beam technology to sort the constituent grains of the sample according to size. Some of these instruments have variable size ranges that can be examined. Most instrumentation uses selected size channels (usually designated in microns) that roughly correspond to the sizes of the US mesh series of sieves discussed earlier. These instruments can calculate the D_{50} point and other points of the sample, and they can also produce the typical S-shaped curve shown in Figure 18-5, from which the D_{50f} point and grain size distribution of the sample can be observed. Instrumental sizing methods generally require smaller sample volumes, reduce the likelihood of human error, and allow for more complete particle separation than measurements taken through the use of dry sieving (Figure 18-5).

18-6.3.2 Pack-Sand Selection Criteria

One of the most important parameters in a gravel-pack design is the ratio of the gravel grain size, D_{50p}, to the formation sand grain size, D_{50f}. When the D_{50p}/D_{50f} ratio is high, the oversized pack-sand will allow the invasion of formation sand, which reduces the overall permeability of the packed zone (often to less than the native reservoir's permeability). Conversely, if undersized gravel is used, it will provide excellent sand control, but it may jeopardize productivity in certain situations.

Figure 18-6 shows that maximum permeability occurs when the D_{50p}/D_{50f} ratios are less than or greater than 10. Specifically, a ratio of approximately 6 provides the maximum gravel-pack permeability with good sand control. At a ratio of 15, the pack permeability is good, but sand control is poor because the formation sand tends to move into the pack-sand. At a ratio of 10, the formation sand can move into the gravel pack, but it will have difficulty moving through it, causing a severe loss in overall productivity (Suman *et al.*, 1983; Saucier, 1974; Hill, 1941).

Many years of field-proven experience have shown that a D_{50p}/D_{50f} ratio of 5 to 6 helps compensate for possible sampling errors or a lack of samples from the entire zone in question. A suggested procedure for properly determining a workable grade of pack-sand is as follows (Figure 18-7):

1. Determine the D_{50f} value from a representative formation sample.

2. Multiply that value by 5, which results in the D_{50p}.

3. Compare the calculated D_{50p} required to the values in Table 18-5 (for API grades of pack-sand).

4. Select the grade of API-approved pack-sand from Table 18-5 that is nearest to the calculated value.

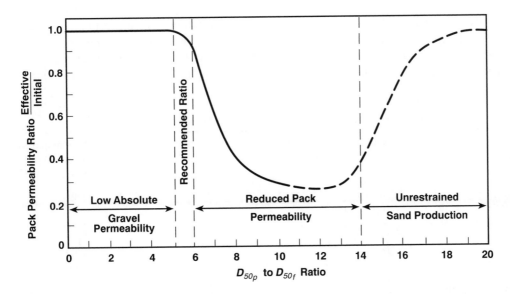

Figure 18-6   Effect of D_{50p}/D_{50f} ratios on sand control and pack permeability, after Saucier (1974)

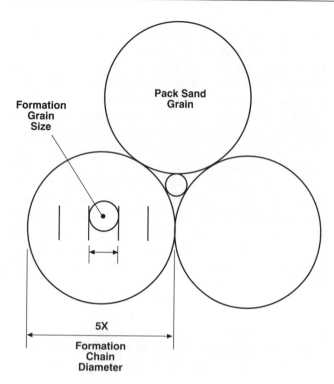

Figure 18-7 Relationship of pack to formation grains at $D_{50p}/D_{50f} = 5$

18-6.3.3 Screen Selection Criteria

The screen is not the primary source of sand stoppage in a gravel-pack completion. Instead, the gravel-pack sand is primarily responsible for stopping the formation sand from moving toward the wellbore. Therefore, the screen must be 100% effective in retaining the gravel-pack sand in place.

Screen quality covers a variety of properties: flow capacity, tensile strength, collapse strength, and corrosion resistance (Figure 18-8). Most high-quality oilwell screens are constructed of alloys that have maximum strength and resistance to the fluids in the well. Suggested alloys include 304 stainless steel, 316D stainless steel, and Incoloy 825 for high-temperature use. These screens include a densely perforated center pipe for support against collapse and gauge distortion. The keystone-shaped screen wire minimizes plugging from particles that are small enough to enter from the exterior. This keystone-shaped wire is wrapped with the smaller dimension toward the mandrel, which reduces plugging since only a minimal contact area exists between the wire and particle. The screen outer wire is joined by a weld to each vertical support rod (Figure 18-9). The gauge of the screen should be consistent throughout the tool, with a tolerance range from ± 0.001–0.002 in.

In addition to the above features, prepack screen quality can be assessed on the basis of the uniformity of its sand-pack bed; the bed should have no voids. Any movement of the prepack sand bed within the screen indicates a poorly packed product.

The length of the production screen depends on the length of the perforated interval. The general rule in high-density slurry packs is to use a 5-ft overlap below the bottom of the perforated interval and a 5- to 15-ft overlap above the top of the perforated interval. This overlap maximizes productivity and possibly compensates for any depth measurement errors involving the location of the sump packer in relation to the perforations. If the well deviation is increased to more than 40° to 50° from vertical, the overlap above the top of the perforated interval should be increased. Experience in a particular field or formation will also influence the amount of top overlap required.

For proper gravel-pack installation and operation, an annular clearance of $\frac{3}{4}$ in. to 1 in. between the OD of the screen and the ID of the casing is necessary. The greater the annular clearance, the greater the pack placement and function efficiency. Pack placement problems and

Table 18-5   Recommendations for pack-sands and screen devices

D_{50f} Times Factor 5 to 5.5 (in.)	Recommended Grade of Pack-Sand	Recommended Slot Width (in.)	Recommended Screen Wire Spacing (in.)
0 to 0.0125	50 to 70-mesh	NA	0.005
0.0125 to 0.017	40 to 60-mesh	NA	0.006
0.017 to 0.023	30 to 50-mesh	NA	0.008
0.023 to 0.030	20 to 40-mesh	NA	0.012
0.030 to 0.0455	16 to 30-mesh	0.016	0.016
0.0455 to 0.0595	12 to 20-mesh	0.025	0.025
0.0595 to 0.0715	10 to 16-mesh	0.035	0.035
0.0715 and Larger	8 to 12-mesh	0.05	0.05

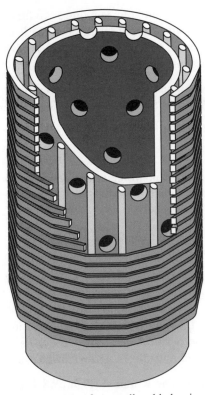

Figure 18-8 Cut-away of an all-welded pipe-base wire-wrapped screen

premature sand bridging in the annulus most frequently occur when the annular clearance is less than $\frac{1}{2}$ to $\frac{3}{4}$ in. Table 18-6 shows suggested screen OD measurements in relation to various casing sizes.

The annular clearance should allow the screen to be washed over easily during any subsequent workovers that might be required. Anything less than a $\frac{3}{4}$-in. clearance is not recommended because of the extreme difficulty in washing over the screen.

Manufacturers of wire-wrapped screens express the space between the wires in units of 0.001 in., which is referred to as the *gauge* of the screen. The correct screen gauge is chosen according to the grade of pack-sand that the screen will have to retain. Since 100% retention of all of the pack-sand is essential during all phases of well life, the decision cannot be safely based on simple sand bridging. A method for determining the safest wire spacing is described in the following paragraphs.

The smallest pack-sand grains are represented by the highest mesh number in the grade designation. For example, 60-mesh (0.0097-in. diameter) is the smallest grain size in a 60/40 grade of pack-sand. To safely retain 60/40-mesh pack-sand, the screen gauge should be 0.5 to 0.9 times 0.0097 in., or 0.0049 in. to 0.0088 in. Therefore, a 6-gauge screen should be selected. The spacing between

Table 18-6 Recommended screen diameters for adequate gravel-pack annulus

Casing OD (in.)	$4\frac{1}{2}$		5		$5\frac{1}{2}$	
	Max. ID	Min. ID	Max. ID	Min. ID	Max. ID	Min. ID
	4.09	3.826	4.56	4.00	5.044	4.548
Base pipe OD (in.)	1.66	1.315	1.9	1.66	2.375	1.90
Screen Wire-Wrap OD (in.)	2.26	1.94	2.55	2.26	2.97	2.55
Clearance (in.)	0.92	0.94	1.01	0.87	1.04	1.00

Casing OD (in.)	6		$6\frac{5}{8}$		7	
	Max. ID	Min. ID	Max. ID	Min. ID	Max. ID	Min. ID
	5.524	5.132	6.135	5.675	6.538	5.92
Base pipe OD (in.)	2.875	2.375	3.50	2.875	4.00	3.50
Screen Wire-Wrap OD (in.)	3.48	2.97	4.13	3.48	4.5	4.13
Clearance (in.)	1.02	1.08	1.00	1.10	1.02	0.90

Casing OD (in.)	$7\frac{5}{8}$		$8\frac{5}{8}$		$9\frac{5}{8}$	
	Max. ID	Min. ID	Max. ID	Min. ID	Max. ID	Min. ID
	7.125	6.435	8.097	7.511	9.063	8.125
Base pipe OD (in.)	4.50	4.00	5.50	5.00	6.625	5.50
Screen Wire-Wrap OD (in.)	5.12	4.50	6.19	5.64	7.23	6.19
Clearance (in.)	1.00	0.97	0.96	0.95	0.92	0.97

Note: The above dimensions may vary between different manufacturers

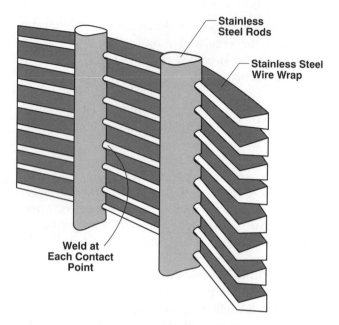

Figure 18-9   Detail of well screen wires relative to the vertical rods

the wire should be 0.5 to 0.9 times the diameter of the smallest pack-sand grains (Figure 18-9).

18-6.3.4 Gravel-Pack Job Calculations

Pack-Sand Volume Required

The next stage in the design of the gravel pack is determining the amount of pack-sand required for the treatment. Generally, a worksheet containing information about the geometry of the well is completed (Table 18-7). This worksheet focuses specifically on the spaces that will be occupied by pack-sand. The standard supplier issue for gravel-pack sand is in 100-lb units (one sack), and the standard sack is considered to occupy 1 ft^3. Therefore, all of the worksheet information must be converted to cubic feet, as shown in the following equations.

Step 1—Sand Volume Required to Fill Casing/Screen Annulus (Figure 18-10)

$$V_{sa} = (L_s + h) \times V_a \qquad (18\text{-}1)$$

where V_{sa} is the sand volume required to fill the annulus, L_s is the combined screen length, h is the height of sand above the screen (with a volume surplus of 40 to 60 ft), and V_a is a volume factor (ft^3/ft).

Step 2—Additional Sand Volume Required to Fill the Rathole (When no Sump Packer is Used) (Figure 18-11)

$$V_{sr} = L_r \times V_c \qquad (18\text{-}2)$$

where V_{sr} is the sand volume required to fill the rathole, L_r is the rathole length, and V_c is the volume factor for the casing of hole.

Step 3—Sand Volume Required for Outside the Wellbore, Perforation Tunnels and the Formation (Figure 18-12)

$$V_{sf} = L_p \times F \qquad (18\text{-}3)$$

where V_{sf} is the sand volume for the formation, L_p is the length of the perforated interval, and F is the volume factor for sand outside the perforations (usually 1 ft^3/ft).

Step 4—Total Sand Required (Figure 18-13)

$$V_{st} = V_{sa} + V_{sr} + V_{sf} \qquad (18\text{-}4)$$

where V_{st} is total sand volume. For the calculation of total sand weight,

$$W_{st} = V_{st}\rho \qquad (18\text{-}5)$$

where W_{st} is total sand weight and ρ is the density factor (100 lb/ft^3).

Carrier-Fluid Volume

The volume of carrier fluid required to place 10 ft^3 (1000 lb) of sand has been calculated for various sand concentrations per gallon of fluid. Ten cubic feet of sand occupies a real volume of 6.5 ft^3 (48.62 gal) based on a specific gravity of 2.63 (for true all-silica sand) and an absolute volume of 0.0456 gal/lb. Table 18-8 shows the carrier-fluid requirements and slurry volumes for multiples and partial increments of 10 ft^3 of sand.

Table 18-7   Job calculation worksheet (Required Information)

1) depth of the top of the sump packer	ft	5) Inside diameter of casing		in.
2) Length of O-ring sub	ft	6) Outside diameter of screen		in.
3) Length of Tell-tale screen	ft	7) Length of rathole		ft
4) Length of sump packer seal	ft	8) Void volume outside of perforations		ft^3

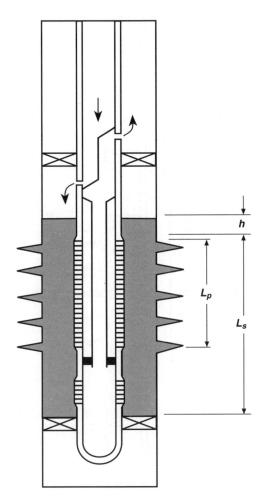

Figure 18-10 Well diagram showing the position of sand packed in the annulus

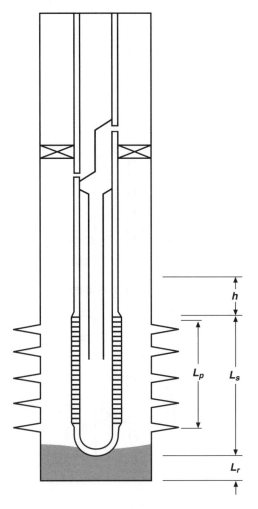

Figure 18-11 Well diagram showing the position of sand packed in the rathole

When carrier-fluid volumes are determined for a job, additional fluid is usually set aside for use as a prepad ahead of the sand slurry and as push pad following the slurry. The recommended volume of additional fluid varies highly throughout the industry, but 6 to 10 bbl is common for most slurry-packing operations.

18-6.3.5 Predicting Job Outcome by Computer Modeling

Sand-control engineers can use a variety of versatile computer programs to evaluate different gravel-pack designs. These programs allow engineers to simulate a gravel-pack treatment and track the treatment's progress from initial placement through screenout (Figure 18-14).

Most numerical simulators model the flow of slurry in the wellbore during a gravel pack. Each fluid stage, from the surface to downhole, is assigned its own fluid rheo-logical properties and solids loading. The pumping can be in the upper circulation, lower circulation, or squeeze mode.

Model input screens prompt the user to include detailed parameters from the reservoir, the fluids/slurry anticipated, the packer and tools, and well geometry. These research parameters must be thoroughly researched and properly entered before the model can perform accurate and realistic job predictions.

18-7 CHEMICAL CONSOLIDATION TECHNIQUES

One of the most compelling reasons to consider chemical consolidation rather than other types of sand exclusion is that chemical consolidation allows the wellbore to be free of tools, screens, and pack-sand. In certain instances,

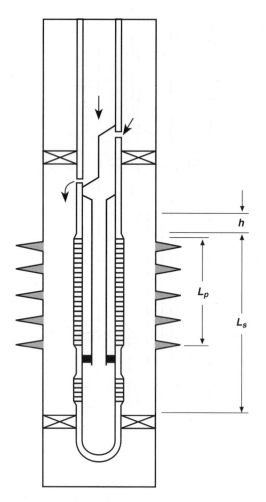

Figure 18-12 Well diagram showing the position of sand packed in the perforation tunnels

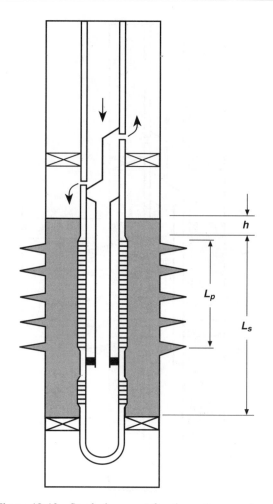

Figure 18-13 Sand placement for the entire gravel pack

chemical consolidations can also be performed on the hole without a rig.

Sand exclusion by chemical consolidation involves the process of injecting chemicals into naturally unconsolidated formations as a means of providing grain-to-grain cementation. For formation sand consolidation to occur with just the contact points of the grain cemented (which creates a continuous flow matrix), excess resin material must be displaced from pore spaces by an overflush fluid at some time in the sequential treatment process. Techniques for successfully accomplishing chemical consolidation are some of the most sophisticated in completion work.

Chemical sand consolidation treatments are formulated to coat individual particles of sand and lock them in place without significantly sacrificing permeability. With the sand consolidated into a hard, permeable mass around the perforations, sand production is mini-

mized and hydrocarbon production is possible for many years (Figure 18-15).

Before a chemical consolidation can be seriously considered, the treatment zone must meet all necessary criteria that will allow the systems to function successfully (Table 18-1). Although exceptions exist, the following criteria must be met (Suman *et al.*, 1983; Murphey *et al.*, 1974):

- Zone length must generally be no more than 25 ft so that resin and hardener chemicals can be accurately directed to the targeted area.

- Zone temperature should not exceed 280°F so that chemicals can be properly placed.

- Formation permeability should be equal to or greater than 100 md or higher, with less than 15% presence of clays and feldspars.

Table 18-8 Carrier fluid requirement and slurry volume to place 10 ft³ of sand

Sand per Gallon of Fluid (lb)	Total Slurry Volume (gal)	Carrier Fluid Volume (gal)
1.00	1122.91	1074.29
2.00	585.77	537.14
3.00	406.72	358.10
4.00	317.20	268.57
5.00	263.48	214.87
6.00	227.67	179.05
7.00	202.09	153.47
8.00	182.91	134.29
9.00	167.99	119.36
10.00	156.05	107.43
11.00	146.29	97.66
12.00	138.15	89.52
13.00	131.26	82.64
14.00	125.36	76.74
15.00	120.24	71.62

- The formation must contain less than 5% calcareous material.

- The zone must be cased, correctly cemented, and perforated.

The long-term durability of the resin treatment for a chemical consolidation varies. If the previously listed conditions are met to an optimum degree, sand-free service has been documented for as long as 10 years; in other situations, however, sand production may return in only a few months.

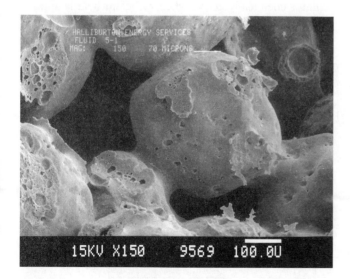

Figure 18-15 Sand grains locked together by in-situ resin consolidation, leaving pore spaces open to flow

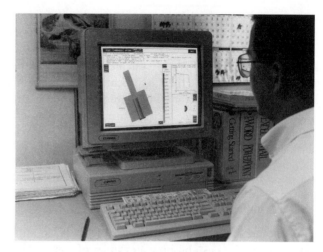

Figure 18-14 Use of computer software to generate an optimum gravel-pack design

The treatment's degree of success depends on the ultimate well conditions imposed and the exact method of consolidation selected (Rensvold, 1982). Because of the inherent difficulties associated with chemical consolidation methods, they have a lower success rate than gravel packs. Most failures (either short life, or incomplete sand stoppage) can be traced to conditions that exceed the limits of the chemical system used.

18-7.1 Internally Activated Systems

Internally activated systems consist of an epoxy resin that contains a hardener and accelerator. Most of these systems are very time-dependent and react rapidly at high temperatures, resulting in a limited amount of working time. Hardening of the resin begins at the surface. These systems have limited success in very hot wells or in zones that have high clay content. A major attraction for internally activated system is that they ensures that all resin placed is mixed with hardener. Under most conditions, these systems demonstrate better-than-average durability.

18-7.2 Externally Activated Systems

Externally activated or overflush systems, as they are sometimes called, use a high-yield furan resin solution. Permeability is established when a specific volume of spacer is pumped into the formation. This spacer displaces all but a residual resin coating at the grain-to-grain contact points. Afterward, an overflush hardener solution containing an extremely reactive acid component is pumped. A surfactant, which helps the resin adhere to the sand while extracting the hardener solution's reactive component, also causes the resin to polymerize (Figure 18-16).

Externally activated epoxy systems are also available. Overflushes can be either hydrocarbon or aqueous. Normally, overflushes contain a hardener or accelerator chemical, and some are viscous to improve sweep efficiency. In cases where a two-step overflush is used, one flush restores permeability and the second introduces a cure activator.

18-7.3 Application

The amount of resin solution required for a given application is based on establishing a cylinder of consolidated formation that is 4 ft in diameter. Theoretically, formation inconsistencies nearly always interfere with uniform penetration of the resin solution. For a cylinder of consolidated formation in typical Miocene sands, 80 to 90 gal of consolidating fluid is required per foot of formation, based on 20% to 30% porosity. This cylindrical matrix provides some casing support and minimizes particle migration. The greater the cylinder's theoretical diameter, the lower the flow velocity at the extremity. Therefore, particles are less likely to be transported at lower velocities.

If a consolidation technique is selected for use on an older well that has produced an appreciable quantity of sand, the well should be packed with clean sand before consolidation. When the liquid resin is pumped into the formation through a cavity or loose zone, uniform distribution of liquid resin is difficult if not impossible. Restressing the formation to some degree is also desirable.

Even if a well or zone meets all the qualifying parameters, other possible objections to performing chemical consolidation include (1) environmental considerations, (2) logistic and storage problems, (3) lack of experience, and (4) the increased cost of resins and treating chemicals.

18-8 COMBINATION METHODS

Combination methods are sand exclusion methods that incorporate chemical technology, usually resins, to enhance conventional gravel-packing, with or without screens. Two types of resin applications are available for combination treatments: (1) semicured resin coatings can be used on seemingly dry sands with proven latent reactivity, and (2) highly reactive, liquid resins can be mixed with hardeners and applied to the pack grains just before the treatment is pumped into the well.

18-8.1 Semicured Resin-Coated Pack Gravels

Semicured, resin-coated gravel products are characteristically high-purity, round, crystalline silica sand or bauxite. The base mineral grains are coated with a heat-reactive phenolic resin formulation that contains additives that promote uniform behavior under temperature, fluid flow, and closure stress. The resin coating accounts for 2 to 4% of the products' weight. The resin is formulated to securely bond to the sand-grain surface. A typical 20/40-mesh round sand has a resin layer less than 0.001 in. thick. The product is dry to the touch and remains chemically active when sufficient heat is applied (Sinclair and Graham, 1977). This condition is referred

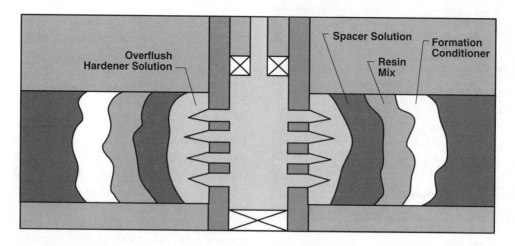

Figure 18-16 Application of in-situ resin consolidation steps for externally hardened system

to as a "B" stage cure. The ultimate and final cure occurs when temperatures above 130°F are applied for sustained periods. When bottomhole temperatures are low, heated oil or water can be circulated to further warm the resin-coated particles and help them bond together. If heated fluids are not used, bonding chemicals should be used. These chemicals consist of alcohols and surfactants that soften the resin coating and promote self-adhesion.

In the deliverable form, most curable resin-coated gravel products meet API RP58 specifications and are delivered to the wellsite in 100-lb bags. A variety of mesh sizes are offered.

The resin coating is tough and slightly deformable, which helps prevent crushing, embedding, and flowback after treatments. These resin-coated particles can bond together behind screens or casing with heat or chemical aids. After placement, the bonded, resin-coated gravel is locked into place for long periods. The phenolic resin base used in semicured, resin-coated gravel materials provides excellent resistance to HCl and HF attack. Future acid cleanout treatments should have little effect on the coating's integrity or strength. The coating helps protect the sand from hot water dissolution to temperatures as high as 600°F.

Curable, resin-coated gravel products are versatile and can be added as regular gravel to the water or gelled fluid; they can be placed with conventional equipment in a linerless gravel pack or behind a screen or slotted liner. However, since little closure stress is on the resin-coated gravel in these situations, maximum possible strength is seldom attained. Extended shut-in times must often be used so that bond strength can increase.

When correctly placed behind a screen or liner, resin-coated gravel can form a permanent downhole filter, preventing the pack from shifting or mixing with formation fines when the well is opened or shut in. To a limited extent, resin-coated gravel is used outside cased gravel packs in high-rate wells and/or for long intervals.

Under some conditions, RCS has successfully patched existing gravel-pack screens and liners. During a successful repair, resin-coated particles are injected into the holes of a failed screen or liner. After the particles bond, those remaining in the screen are removed, leaving the hole repaired.

These semicured, resin-coated products are commonly used in the manufacture of prepacked screens and liners. Once tight grain-to-grain contact is attained in the tool, optimum heat curing is applied under controlled conditions. As a result, instead of being loose and mobile, the resin-coated particles within the screen lock together to provide extra protection against particle abrasion and/or erosion.

18-8.2 Liquid Resin-Coated Pack Gravel

Procedures to resin-coat pack-sands and pack gravels with liquid resins have been successfully applied and commercially available for several years. These processes differ significantly from ones that feature semicured resin coatings. The liquid resin solution is job-mixed and added to the sand-carrier fluid slurry. Variations of these processes use either an internal or an external resin hardener system. Through processes of chemical attraction, liquid resin is preferentially drawn to the silica sand surfaces instead of well tubular goods.

For the internally hardened variations, hardeners, adhesion promoters, and coating aids are added to the liquid resin at the surface. The base resin for these systems is predominantly an epoxy. If required, excess resin can be added to the process, which will permit consolidation of the pack-sand as well as some of the adjacent formation.

Externally hardened systems also begin with liquid resin mixtures added to the sand-fluid slurries. Preferential sand coating with resin also occurs in these processes. When the well no longer accepts the resin-coated sand, the sand slurry remaining in the wellbore is circulated out, leaving the hole clean. This cleanout is followed by a spacer and catalyst solution that harden the resin-coated sand packed into the perforation and beyond.

Performing these treatments is similar to performing a gravel pack, except the pack-sand ultimately becomes resin-consolidated, and, theoretically, eliminates the need for screening devices. Several variations of placement procedures are available; some can be continuously blended while others must be batch-mixed.

A common procedure, used in perforated-cased holes requires the use of a viscosified carrier fluid and high concentrations of pack-sand. The sand concentration may be as high as 15 lb/gal. The internally hardened resin concentration is usually 1 gal of resin per sack of pack-sand used. During mixing, the resin coats the pack-sand in the slurry. No provision is made for excess resin to coat any of the formation face, except at the interface. The sand, which has a tacky resin coating, remains dispersed and suspended in the carrier fluid. The mixture is pumped to the sanding zone in the well (Figure 18-17). Because of the high sand concentration, little carrier fluid is lost to the formation. Since the system is internally

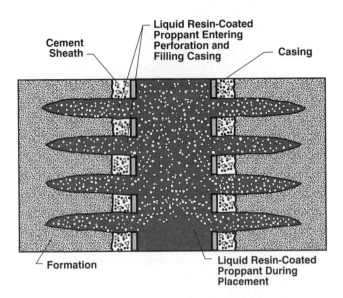

Figure 18-17   Application of resin-coated sand slurry

catalyzed, flushing out the tacky sand remaining in the wellbore is difficult. Therefore, the entire mass is allowed to set up, and the sand remaining in the casing is drilled out before production (Figure 18-18).

A second procedure used in perforated-cased completions is to disperse resin with low concentrations of sand in a thin carrier fluid in a mixing tank. The sand-to-carrier fluids ratio is normally $\frac{1}{2}$ lb/gal. The externally hardened resin-to-sand ratio is from 3 to 5 gal/sk. During mixing, some of the resin automatically coats the pack-

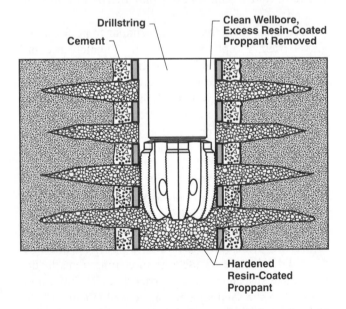

Figure 18-18   After the sand grains are locked together, the sand in the casing is drilled out, leaving the wellbore clear.

sand, and the excess resin remains dispersed in the carrier fluid. As the mixture is pumped, the resin-coated sand screens off against the formation and in the perforation tunnels. As the carrier fluid moves into the formation, tight grain-to-grain contact is established in the pack-sand. The excess resin in the carrier fluid can coat some of the formation sand. When the well no longer takes any more coated sand, the solids left in the wellbore can be circulated out. The cleanout, in this situation, is followed by a spacer and catalyst solution. This treatment results in a consolidation of both the pack-sand and the formation sand.

18-9 HORIZONTAL GRAVEL-PACKING

Openhole horizontal completions may be extremely difficult to achieve in unconsolidated or weakly consolidated sandstone formations because the horizontal hole is more likely to collapse and fill. Unless certain safeguards are implemented during drilling, the hole is likely to collapse before a gravel pack can be performed.

Gravel-packing of high-angle cased wells with long perforated intervals can now be successfully completed sand-free with minimal production loss. Much of the technology used for gravel-packing these highly deviated wells will also likely be used for the successful gravel-packing of true horizontal completions (Elson *et al.*, 1984).

Some carrier fluid technology used for vertical wells cannot be successfully used for gravel-packing in high-angle holes. For example, as a gelled carrier fluid breaks back in a vertical wellbore, excess sand in the annular reservoir above the production screen tends to settle and fill any voids in the pack. In a true horizontal gravel pack, the same gravity that corrects a poor gravel pack in a vertical wellbore actually leaves a void across the entire upper side of the annulus. In addition, gravity permits sand that has packed into the upper perforations to settle back into a loosely packed annulus (Schroeder, 1987; Elson *et al.*, 1984).

18-9.1 Variables that Affect Sand Delivery

According to test results, for a sand-pack without voids, optimum grain-to-grain contact must occur the instant that fluid motion is stopped in a horizontal gravel-packing operation. Therefore, any fluid that does not provide for either complete leakoff or complete sand settling could potentially result in a poorly packed annulus.

For gravel-packing horizontal wells, the wellbore deviation is not a variable, nor for all practical purposes are the sand size and density. Therefore, to achieve efficient sand-packing, we are restricted to the variables associated with (1) either the carrier fluid viscosity and density, (2) the sand concentration used, and (3) the fluid velocity.

On the basis of field experience and model testing, the following conclusions can be made concerning fluid choice:

- Gelled fluids with 100% sand suspension ability are available, such as xanthan gum gels. However, these gels are generally not good horizontal gravel-pack fluids because of their very low fluid-loss rate to the formation. To ensure a tight pack, the carrier fluid should have extremely high fluid loss.

- Hydroxyethyl cellulose (HEC) carrier fluids have good sand suspension and better fluid-loss capabilities than xanthan, but they tend to leave voids in the pack because of slow gel-breaking and the retarded development of a tight sand-pack in the annulus.

18-9.2 Pump Rate and Fluid Velocity

The term *pump rate* is often misused as related to gravel-pack operations. Actually, under these circumstances, pump rate is really referring to the fluid velocity in the annulus or through individual perforations.

The design parameter of pump rate (fluid velocity) directly affects packing efficiency. Tests have shown that in highly deviated wells, a rate increase from 0.6 to 1.0 bbl/min increased the percentage of packed volume from approximately 60% to approximately 93%. At 1.4 bbl/min, the packed volume increased to approximately 96%. An increase in placement rate virtually eliminates bridges formed in the annulus between the blank pipe and casing above the perforations (Peden *et al.*, 1984).

If we assume that the gel carrier fluid does not have 100% sand suspension capabilities because it was selected to maintain a good fluid loss and allow for quick sand settling, we must determine a compensating pump rate (velocity). The velocity of the carrier fluid through a perforation is affected by two limitations:

- The *maximum velocity* through a perforation must not exceed a rate that would cause a jetting action on the formation, which would intermix the formation sand and pack-sand.

- The *minimum velocity* through a perforation must be high enough to ensure that the pack-sand will flow into the perforation and not past it.

Some studies have shown that viscous carrier fluids provide more efficient perforation packing than brine carrier fluids. This same study also showed that at lower flow rates, the viscous systems tended to pack irregularly, which indicates that higher rates will be beneficial in horizontal gravel packs.

18-9.3 Alpha and Beta Wave Progression Through the Annulus

The basic method for horizontal sand-packing with brine is a two-step procedure, which includes an alpha wave sand deposition in one direction and a beta wave sand deposition in the opposite direction (Dickerson and Anderson, 1987). Water-based sand slurry is pumped down the vertical workstring out the horizontal portion of the screen-casing annulus. The slurry flow rate is controlled so that the slurry velocity is above 7 ft/sec within that tubing. The 7-ft/sec velocity is generally required for the pack-sand to become a slurry.

At the crossover exit into the annulus, the velocity of the slurry rapidly drops to about $\frac{1}{2}$ ft/sec (0.2 m/s) in a normal 4-in. (10-cm) borehole in the formation. The sand immediately falls out of slurry suspension, and a sand dune rapidly builds up in the borehole, both in the forward direction (away from the vertical wellbore) and in the reverse direction (back toward the vertical wellbore). The sand dune fills the horizontal borehole annulus to about 70 to over 90% fill. This deposition is known as the alpha wave sand deposition.

The wedge-shaped form of the leading edge of the deposited dune of sand proceeds at about 1 ft/min (0.5 cm/s) back toward the vertical wellbore. The carrier water fluid splits into two paths within the horizontal borehole. After depositing its sand burden, the main portion of the water progresses through the washpipe and back to the wellbore. The other portion of the brine enters the formation, depending upon its permeability.

The brine velocity in the ullage flow space appears to stabilize automatically. If the water velocity is greater than 1 to 3 ft/sec (30 to 91 cm/s), the sand is washed ahead to open a larger ullage flow space. If the water velocity is less than 1 to 3 ft/sec (30 to 91 cm/s), the sand deposits more rapidly to fill in and reduce the cross section of the ullage flow space. Based upon many experiments with widely varying conditions of flow, sand concentration, and sand size and shape, this sand deposition process appears to be self-regulating.

The actual movement of sand in the ullage space is not caused by the development of slurry. Instead, the sand moves through the hopping (saltation) of individual sand grains along the top of the sand dune, much like the sedimentation process in a river bed.

The leading edge of the sand dune progresses toward the toe of the wellbore until it reaches the end of the screen. At this point, the beta wave deposition of sand in the horizontal borehole begins. In this beta wave, the sand flows out automatically and simultaneously from the toe of the wellbore back toward the heel end of the horizontal borehole along the ullage flow space.

The sand movement during in the beta deposition occurs in successive waves, during which the sand moves in a wedge-shaped front along the flat top of the sand dune, which had been placed in the alpha wave. After the water propels the sand along the dune top (by saltation), the water then either moves into the formation or is returned by the crossover tool to the surface. The final result is a fully packed horizontal borehole with a screen in place.

18-9.4 Sand Concentration

The design parameter of sand concentration affects the quality of the gravel pack. Gruesbeck *et al.* (1978) illustrated that "packing efficiency in deviated wellbores increases with (1) lower gravel concentration, (2) decreased particle diameter, (3) decreasing particle density, (4) higher fluid density, (5) higher flow rates, and (6) increasing resistance to fluid flow in the tailpipe annulus."

In most horizontal completions, the leakoff rate should be relatively uniform, since the permeabilities should be relatively uniform. Compensation for some nonuniformity can be accomplished through the use of fluid-loss additives and larger pad volumes ahead of the gravel-pack slurry. These fluid-loss additives and larger pad volumes leaked off to the formation can temporarily modify the injection profile, allowing a more uniform slurry loss to the formation while reducing the potential for a premature sand bridge across a higher-permeability interval. Additional compensation can also be accomplished by small reductions in sand concentration and increased pump rates.

18-9.5 Placement Procedure and Tool Configuration

A number of decisions must be made during the design of a horizontal completion system. These decisions affect both the tools that will be run and the operational aspects of running the completion and completing the well.

The job design must be simple to improve reliability and should be integrated so that all components work together efficiently. If the job will be performed offshore, the service provider should preassemble as many tools as possible to accelerate rig floor assembly. If possible, the tools should be run in a lower-cost, less harsh environment so that compatibility can be ensured. Before anything is finalized, the completion tools that extend to the surface should be specified and checked for compatibility. A completion design that does not allow access to tools lower in the completion or a design that restricts flow should be corrected before the design of the horizontal tool system is completed.

Service company personnel should carefully review and be able to complete all aspects of the completion procedures that are developed. These procedures should provide extensive details of all operations so that the job will be performed smoothly without the need for problem-solving and decision-making. Checkpoints, operational warnings, contingency plans for difficult operations, and setting depths should all be carefully considered. Fluid-loss control and well control considerations should also be included as part of both the primary procedure and contingency plans so if decisions must be made quickly, they will not severely impact cost or productivity. For offshore operations, rig-floor assembly should also be discussed as a means of ensuring thorough planning and efficient operation. Future access to the formation to remove fluid-loss devices, bring the well on line, and perform service operations must also be considered. All of these preparations provide a means of measuring success and improving future well completions.

More complex horizontal tool completion systems allow fluids to be circulated within the wellbore. The primary circulation path is down the workstring and out the shoe at the end of the string. Reverse-circulating is also possible if the system has a shoe that allows circulation in both directions. This shoe must be closed later to prevent the influx of sand. After the packer is set, the circulating paths are commonly alternated from the *heel* (the area below the packer) to the *toe* (the wash shoe at the very end of the completion string). This circulation pattern can be used for displacing fluid-loss materials and drilling solids, and/or for stimulating to remove filter cake. It can be followed by a discrete washing of the screen section with cup-packer washing tools.

Once the well has been stimulated and the workstring is removed, a number of mechanical and chemical fluid-loss options exist that can control the well while the uphole completion is run and the wellhead is landed. These options include ceramic flappers, pressure-operated reverse flappers, and carbonate fluid-loss materials.

Once the well is controlled, the workstring can be pulled and completion equipment can be run for the uphole completion. Often, the completion equipment is operated just as it would be in a conventional vertical well completion. Equipment should be chosen based on the need for future access to the horizontal section and the possibility that coiled tubing may have to be run through flow-control devices.

18-9.6 Liner/Tailpipe Ratio

Numerous studies (Schroeder, 1987; Elson *et al.*, 1984; Peden *et al.*, 1984; Dickerson and Anderson, 1987) have concluded that the liner/tailpipe ratio is probably the most critical aspect of gravel-pack design. For horizontal well completions, this ratio is even more critical. These studies indicate that a ratio of 0.8 or greater is mandatory for the successful gravel-packing of horizontal well completions. The magnitude of the problem and limitations of this chapter do not allow for a more detailed discussion of this very important aspect. However, extensive studies reporting on this critical aspect of gravel-pack design are listed in the reference section.

18-9.7 Screen/Casing Clearance

Studies indicate that the screen/casing clearance directly affects pack efficiency. If the clearance is reduced, velocity will increase, which improves sand transport efficiency. Under these circumstances, however, a greater tendency for premature sand bridging in the screen/casing annulus will occur for two reasons:

- A reduced screen/casing annulus will increase the resistance to flow, which increases flow in the screen-tailpipe annulus.

- Reduced volume in the screen/casing annulus increases the effect of slurry dehydration as fluid leaks off to the formation.

Therefore, in horizontal wells, the most desirable means of reducing the tendency for premature sandoff in the screen/casing annulus is to increase the size of the annulus and decrease the sand concentration. Based on the above conclusions, the radial screen/casing clearance for gravel-packing horizontal completions should range from 1.0 to 1.5 in.

18-9.8 Perforation Phasing

The industry has two seemingly conflicting views on perforation phasing in horizontally cased wells. Some experts recommend perforating only the lower half of the casing or some portion of the bottom half of casing. Others recommend perforating radially around the total casing circumference. Extensive literature is available regarding perforation size, shot density, and perforation penetration as they relate to gravel-packing and sand control, but very few references are available regarding perforation phasing as it relates to sand control or gravel-packing. Current practices indicate that horizontal wells can be perforated either 360° or the bottom 120 to 180°.

REFERENCES

Ali, S. A., and Dearing, H. L.: "Sand Control Screens Exhibit Degrees of Plugging," *Pet. Eng. Intl.*, (July 1996) 36–41.

Almond, S. W., and Bland, W. E.: "The Effect of Break Mechanism on Gelling Agent Residue and Flow Impairment in 20/40 Mesh Sand," paper SPE 12485, 1984.

API RP-58, second edition, Am. Pet. Inst., Dallas,1995.

Coberly, C.J., and Wagner, E.M.: "Some Consideration in the Selection and Installation of Gravel-Packs for Oil Wells," *JPT* (Aug. 1938) 1–20.

Cole, R. C.: "A Study of the Properties, Installation, and Performance of Sintered Metal Gravel-Pack Screens," paper OTC 7012, 1992.

Decker, L.R., and Carnes, J.D.: "Current Sand Control Practices," paper IPA, Jarkarta, Indonesia, 1977.

Dickerson, W. R., and Anderson, R. R.: "Gravel-packing Horizontal Wells," paper SPE 16931, 1987.

Elson, T.D., Darlington, R.H., and Mantooth, M.A.: "High-Angle Gravel-Pack Completion Studies," *JPT* (Jan. 1984) 69–78.

Gruesbeck, C., Salathiel, W.M., and Echols, E.E.: "Design of Gravel-pack in Deviated Wells," paper SPE 6805, 1978.

Hill, K.E.: "Factors Affecting the Use of Gravel in Oil Wells," *Drill. & Prod. Prac.*, API, (1941) 134–43.

Himes, R. E.: "New Sidewall Core Analysis Technique to Improve Gravel-pack Designs," paper SPE 14813, 1986.

Ledlow, L.B., and Johnson, M. H.: "High Pressure Packing with Water: An Alternative Approach to Conventional Gravel-packing," paper SPE 26543, 1993.

Maly, G.P., and Krueger, R.F.: "Improper Formation Sampling Leads to Improper Selection of Gravel Size," *JPT* (Dec. 1971) 1403–1408.

Maly, G. P.: "Close Attention to the Smallest Job Details Vital for Minimizing Formation Damage," paper SPE 5702, 1979.

Morita, N. *et al.*: "Realistic Sand-Production Prediction: Numerical Approach," *SPEPE* (Feb. 1989) 14–24; *Trans.*, AIME. **287**.

Mullins, L.D., Baldwin, W.F., and Berry, P.M.: "Surface Flowline Sand Detection," paper SPE 5152, 1974.

Murphey, J.R., Bila, V.J., and Totty, K.: "Sand Consolidation Systems Placed with Water," paper SPE 5031, 1974.

Peden, J.M., Russel, J., and Oyeneyin, M.B.: "The Design and Optimization of Gravel-packing Operations in Deviated Wells," paper SPE 12997, 1984.

Penberthy, W. L., and Shaughnessy, C. M. "Sand Control," SPE Series on Special Topics, Vol. 1, SPE 1992.

Penberthy, W.L. Jr.: "Gravel Placement Through Perforations and Perforation Cleaning for Gravel-packing," *JPT* (Feb. 1988) 229-236; *Trans.*, AIME, **285**.

Rensvold, R. F.: "Full Scale Gravel-pack Model Studies," EUR 39, London, England, Oct. 1978.

Rensvold, R. F., "Sand Consolidation Resins, Their Stability in Hot Brines," paper SPE 10653, 1982.

Saucier, R.J.: "Considerations in Gravel-pack Design," *JPT* (Feb. 1974) 205-212; *Trans.*, AIME, **257**.

Scheuerman, R.F.: "A New Look at Gravel-Pack Carrier Fluid," *SPEPE* (Jan. 1986) 9–16.

Schroeder, D.E. Jr.: "Gravel-pack Studies in a Full-Scale, High-Pressure Wellbore Model," paper SPE 16890, 1987.

Sinclair, A.R., and Graham, J.W.: "Super Sand Control—With Resin Coated Gravel," *Oil & Gas J.* (April 18, 1977) 58–60.

Sparlin, D.D.: "Fight Sand with Sanda Realistic Approach to Gravel-packing," paper SPE 2649, 1969.

Suman, G.O., Ellis, R.C., and Snyder, R.E.: *Sand Control Handbook*, second edition, Gulf Publishing Co., Houston (1983) 67–68.

Torrest, R.S.: "The Flow of Viscous Polymer Solutions for Gravel-packing Through Porous Media," paper SPE 11010, 1982.

van Poolen H.J., Tinsley, J.M., and Saunders, C.D.: "Hydraulic Fracturing—Fracture Flow Capacity vs. Well Productivity," *Trans.*, AIME (1958) **213**, 91–95.

19 High-Permeability Fracturing

Ronald E. Oligney
Texas A&M University

Peter Valkó
Texas A&M University

Michael J. Economides
Texas A&M University

Sanjay Vitthal
Halliburton Energy Services

19-1 INTRODUCTION

Since it was first used to improve production from marginal wells in Kansas in 1946, and its rapid, widespread acceptance in the early 1950s, massive hydraulic fracturing (MHF) has become the dominant completion technique in the U.S. In 1993, 40% of new, completed oil wells and 70% of gas wells were fracture-treated (Figure 19-1). With improved modern fracturing capabilities and the advent of high-permeability fracturing (HPF), also referred to as a *fracpack* or other variants, the industry is increasingly recognizing the tremendous advantages of fracturing most wells. Even near water or gas contacts, prospects considered the bane of fracturing, HPF is being applied, because it offers controlled fracture extent and limits drawdown (Mullen *et al.*, 1996; Martins *et al.*, 1992).

The rapid ascent of high-permeability fracturing from a few isolated treatments before 1993 (Martins *et al.*, 1992; Grubert, 1990; Ayoub *et al.*, 1992) to some 300 treatments per year in the U.S. by 1996 (Tiner *et al.*, 1996) suggests that HPF is becoming a dominating optimization tool for integrated well completion and production and one of the major recent developments in petroleum production (Table 19-1).

As recently as 1993, hydraulic fracturing was considered simply a means of production enhancement, and was used almost exclusively for low-permeability reservoirs. The large fluid leakoff and unconsolidated sands associated with high-permeability formations would ostensibly prevent the initiation and extension of a single, planar fracture with sufficient width to accept a meaningful proppant volume. Moreover, such fracture morphology, even if successfully created and propped, would be incompatible with the defined needs of moderate- to high-permeability reservoirs, which require large conductivity (width).

The key feature in high-permeability fracturing is the tip-screenout (TSO) technique, which arrests lateral fracture growth and allows for subsequent fracture inflation and packing. The result is short but exceptionally wide fractures. While in traditional, unrestricted fracture growth, an average fracture width of 0.25 in. would be considered normal, in TSO treatments, widths of 1 in. or even larger are commonly discussed.

Fundamental modeling and field evidence have suggested that HPF treatments are primarily effective because they bypass near-well damage (DeBonis *et al.*, 1994; Grubert, 1990; Hannah *et al.*, 1993; Hunt *et al.*, 1994; Martins *et al.*, 1992; Montagna *et al.*, 1995; Monus *et al.*, 1992; Mullen *et al.*, 1994; Patel *et al.*, 1994; Reimers and Clausen, 1991; Smith *et al.*, 1987; Stewart *et al.*, 1995a and 1995b; Wong *et al.*, 1993). This benefit is both the controlling and the necessary mechanism for appreciable production enhancements from HPF jobs.

Fundamentally, high-permeability fracturing would always result in a negative skin effect, although it would be of much smaller absolute value than the skins

The authors wish to thank the Gas Research Institute for their support of many research developments that appear in this chapter.

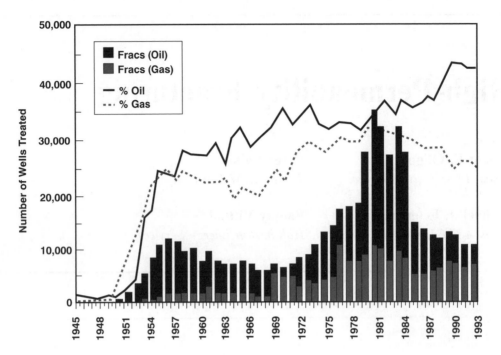

Figure 19-1   The importance of hydraulic fracturing

obtained in low-permeability reservoirs. However, post-treatment positive skins have been measured in many wells, and these must be attributed to the connectivity (choke) between the well and the fracture. This connectivity is related, for instance, to the number and condition of the perforations, and the fracture tortuosity from its initiation direction to its ultimate propagation direction. These issues are addressed in Section 19-6.3 because, although they affect all fractures, they have a particularly serious effect on the low conductivity expected in high-permeability treatments. Thus, although from an "accounting" point of view, a pretreatment skin effect of 10 compared to a post-treatment skin of 5 would imply a skin reduction equal to 5, in reality, the pretreatment skin of 10 is totally eliminated and supplanted by a new skin equal to 5. This issue is important because it suggests the direction of indicated improvements in these types of stimulation treatments.

While production enhancement is of primary importance, there are actually a number of reasons to consider fracturing a high-permeability formation:

- Bypassing formation damage

- Controlling sand deconsolidation

- Reducing fines migration and asphaltene production

- Reducing bottom water coning

- Improving communication between reservoir and wellbore

- Stimulating wells

In contrast to virtually all conventional hydraulic fracturing, positive post-treatment skin effects are possible after HPF treatments. This effect is commonly attributed to fracture-face damage that results from excessive fluid leakoff, but non-Darcy flow in the formation and especially in the fracture may also be a reasonable hypothesis.

It is interesting that HPF (fracpack) treatments did not necessarily originate as an extension of hydraulic fracturing—although they borrowed heavily from established techniques—but rather as a means of sand-production control. In controlling the amount of surface sand production, two distinctly different activities can be performed downhole: *sand exclusion* and *sand*

Table 19-1   Fracturing role expanded

Permeability	Gas	Oil
Low	$k < 0.5$ md	$k < 5$ md
Moderate	$0.5 < k < 5$ md	$5 < k < 50$ md
High	$k > 5$ md	$k > 50$ md

deconsolidation control. Sand exclusion refers to all filtering devices such as screens and gravel packs. Gravel-packing, the historically preferred well completion method to remedy sand production, is one such technique. These techniques do not prevent sand migration in the reservoir, so fines migrate and lodge in the gravel pack and screen, causing large damage skin effects. Well performance progressively deteriorates and is often not reversible with matrix stimulation treatments. Attempts to stem the loss in well performance by increasing the pressure drawdown often aggravates the problem further and may potentially lead to wellbore collapse.

A more robust approach is the control of sand deconsolidation, the prevention of fines migration at the source. It is widely perceived that the use of HPF accomplishes this by mating with the formation in its (relative) undisturbed state and reducing fluid velocities or "flux" at the formation face.

Actually, three factors contribute to sand deconsolidation: (1) *pressure drawdown* and the resulting flux of the fluid, (2) the *strength of the rock* and integrity of the natural cementation, and (3) the *state of stress.* Of these three, the only factor that can be readily altered is the distribution of flow and pressure drawdown. With the introduction of formation fluids to the well along a more elongated path, such as a hydraulic fracture or horizontal well, it is entirely possible to reduce the fluid flux and, in turn, control sand production.

Of course, little can be done to affect the state of stress. The magnitude of earth stresses depends primarily on reservoir depth and to some extent, pressure, and the situation becomes more complicated at depths of 3000 ft or less. Pressure maintenance with gas or water flooding may be counterproductive unless maintenance of reservoir pressure allows economic production at a smaller drawdown. While various innovations have been suggested to remedy the incompetent formations or improve on natural cementation—for example, by introducing complex well configurations (Section 19-8.2) or various exotic chemical treatments—relatively little can be done to control this factor, either. Rock mechanics issues were discussed more broadly in Chapter 6.

In light of the discussion above, it should not be surprising that HPF is rapidly replacing gravel packs in many petroleum provinces that are susceptible to sand production. As with any stimulation technique that results in a productivity index improvement (defined as the production rate divided by the pressure drawdown), the operator is responsible for allocating this new productivity index either to a *larger rate* or a *lower drawdown,* or any combination of the two.

The present trend in HPF indicates a marked departure from the heritage of gravel-packing, incorporating more and more from hydraulic-fracture technology. This trend can be seen, for instance, in the fluids and proppants applied. While the original fracpack treatments involved sand sizes and "clean" fluids common in gravel-packing, now the typical proppant size for hydraulic fracturing (20/40-mesh) seems to be dominant. The increasing application of crosslinked fracturing fluids also supports the trend.

For this reason, the term *high-permeability fracturing* (HPF) seems more appropriate than *fracpack,* and this term will be used throughout the chapter.

In the following section, HPF is compared in a semi-quantitative way to competing technologies. This comparison is followed by a discussion of the key issues in high-permeability fracturing including design, execution, and evaluation.

19-2 HPF VS. COMPETING TECHNOLOGIES

19-2.1 Gravel Pack

The term *gravel pack* refers to the placement of gravel (actually, carefully selected and sized sand) between the formation and the well as a means of filtering out (retaining) reservoir particles that migrate through the porous medium. A "screen" is used to hold the gravel pack itself in place. This manner of excluding reservoir fines from flowing into the well causes an accumulation of fines in the near-well zone and an attendant reduction in gravel-pack permeability (i.e. damage).

The progressive deterioration of gravel-pack permeability (increased skin effect) leads, in turn, to a decline in well production. Increasing the pressure drawdown to counteract production losses can result in accelerated pore-level deconsolidation and additional sand production.

Any productivity index relationship, e.g. the steady-state expression for oil, can be used to demonstrate this point:

$$J = \frac{q}{p_e - p_{wf}} = \frac{kh}{141.2B\mu \left(\ln \dfrac{r_e}{r_w} + s \right)} \tag{19-1}$$

If we assume that $k = 50$ md, $h = 100$ ft, $B = 1.1$ res bbl/STB, $\mu = 0.75$ cp and $\ln r_e/r_w = 8.5$, the productivity indexes for an ideal (undamaged), a relatively damaged ($s = 10$), and a typical gravel-packed well ($s = 30$) would be 5, 2.3 and 1.1 STB/d/psi, respectively. For a drawdown of 1,000 psi, these productivity indexes would result in production rates of 5000, 2300 and 1100 STB/d, respectively. Clearly, the difference in production rates between the ideal and gravel-packed wells can be considerable and very undesirable.

High-permeability fracturing under the same scenario would combine the advantages of propped fracturing to bypass the near-wellbore damage and gravel-packing to provide effective sand control. Figure 19-2 is the classic presentation of fracture-equivalent skin effect (Cinco-Ley *et al.*, 1978) in terms of dimensionless fracture conductivity, $C_{fD}(= k_f w/k x_f)$, and the fracture half-length, x_f.

As shown in Figure 19-2, even with a lackluster hydraulic fracture ($C_{fD} = 0.5$) and short fracture length ($x_f = 50$ ft), the skin effect, s_f (using again $r_w = 0.328$ ft), would be equal to -3.

A negative skin effect equal to -3 applied to Equation 19-1 yields a productivity index of 7.7 STB/d/psi, more than a 50% increase over the ideal PI, and seven times the magnitude of a damaged gravel-packed well. Even with a damaged fracture (leakoff-induced damage as described by Mathur *et al.*, 1995) and a skin equal to -1, the productivity index would be 5.6 STB/d/psi, a five-fold increase over a damaged gravel-packed well.

This calculation brings forward a simple, yet frequently overlooked issue. Small negative skin values have a much greater impact on well performance than comparable magnitudes (absolute value) of positive skin. Furthermore, in the example calculation, a five-fold increase in the productivity index suggests that the production rate would increase by the same amount if the drawdown is held constant. Under an equally possible scenario, the production rate could be held constant and the drawdown reduced to $\frac{1}{5}$ its original value. Any other combination between these two limits can be envisioned. The utility of high-permeability fracturing is, thus, compelling.

19-2.2 High-Rate Water Packs

As shown in Table 19-2, distilled from recent empirical data collected and reported by Tiner *et al.* (1996), high-

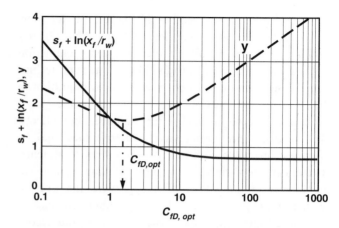

Figure 19-2   Pseudoskin factor of a vertical well intersected by a finite-conductivity vertical fracture (after Cinco-Ley *et al.*, 1978)

rate water packs seem to have an advantage over gravel packs, but they do not provide the productivity improvement of HPF. This improvement over gravel packs is reasonable because of the additional proppant placed in the perforation tunnels.

While not shown in the table, the performance of these completions over time is also of interest. It is commonly reported that production from high-rate water packs (as in the case of gravel packs) deteriorates with time. By contrast, Stewart *et al.* (1995), Mathur *et al.* (1995), and Ning *et al.* (1995) all report that production may progressively improve (skin values may decrease) during the first several months after an HPF treatment.

19-2.3 Performance of Fractured Horizontal Wells in High-Permeability Formations

Two of the most important recent developments in petroleum production are horizontal wells and high-permeability fracturing. Considerable potential is possible when the two technologies are combined. Horizontal wells can be drilled either transversely or longitudinal to the fracture azimuth. The transverse configuration is appropriate for low-permeability formations and has been widely used and documented in the literature. The longitudinally fractured horizontal well warrants further attention, specifically in the case of high-permeability

Table 19-2   Skin values reported by Tiner *et al.* (1996)

Gravel-pack	High-Rate Water Pack	HPF
+5 to +10, excellent	+2 to +5, reported	0 to +2, normally
+40 and higher are reported		

Table 19-3 Discounted revenue in $ millions US

Configuration	$k = 1\,md$	$k = 10\,md$	$k = 100\,md$
Vertical well	0.73	6.4	57.7
Horizontal well	3.48	14.2	78.8
Fractured vertical well, $C_{fD} = 1.2$	2.59	13.4	89.6
Fractured horizontal well, $C_{fD} = 1.2$	3.88	16.3	95.8
Infinite-conductivity fracture (upper bound for both horizontal and vertical well cases)	3.91	16.3	103.3

formations. HPF often results in low dimensionless-conductivity hydraulic fractures, yet such fractures installed longitudinally in horizontal wells in high-permeability formations can have the net effect of installing a (relative) high-conductivity streak in an otherwise limited-conductivity flow conduit. Using a generic set of input data, Valkó and Economides (1996) showed discounted revenues for 15 cases that demonstrate this point.

Table 19-3 shows that for a given permeability, the potential for the longitudinally fractured horizontal well is always higher than that of a vertical well, and that the horizontal well may approach the theoretical potential of an infinite-conductivity fracture when realistic fracture widths are considered.

Furthermore, Valkó and Economides showed that the horizontal well fractured with 10-fold less proppant ($C_{fD} = 0.12$) still outperforms the fractured vertical well for $k = 1$ and 10 md, and that the horizontal well is competitive at 100 md. In fact, for the range of 1 to 10 md, even a 100-fold reduction in fracture width ($C_{fD} = 0.012$) is more than enough. Thus, with the longitudinal configuration, orders-of-magnitude less fracture width (than that suggested for a fractured vertical well) might be sufficient to achieve a certain production goal.

(and assuming the pump rate is larger than the rate of leakoff to the formation), continued pumping will inflate the fracture (increase fracture width). This TSO and fracture inflation should be accompanied by an increase in net fracture pressure. Thus, the treatment can be conceptualized in two distinct stages: fracture creation (equivalent to conventional designs) and fracture inflation/packing (after tip-screenout).

Figure 19-4 (after Roodhart *et al.*, 1993) compares the two-stage HPF process with the conventional single-stage fracturing process. Creation of the fracture and the arrest of its growth (tip-screenout) is accomplished by injecting a relatively small pad and a 1 to 4-lb/gal sand slurry. Once fracture growth has been arrested, further injection builds fracture width and allows injection of high-concentration (10 to 16-lb/gal) slurry. Final areal proppant concentrations of 20 lb/ft^2 are not uncommon. The figure also illustrates the common practice of retarding injection rate near the end of the treatment (coincidental with opening the annulus to flow) to dehydrate/pack the near wellbore and screen. Rate reductions may also be used to force tip-screenout in cases where no TSO event is observed on the downhole pressure record.

19-3 KEY ISSUES IN HPF

19-3.1 Tip-Screenouts

The critical elements of high-permeability fracturing treatment design, execution, and interpretation are substantially different than for conventional fracturing treatments. In particular, HPF relies on a carefully timed *tip-screenout (TSO)* to limit fracture growth and allow for fracture inflation and packing (Figure 19-3). The TSO occurs when sufficient proppant has concentrated at the leading edge of the fracture to prevent further fracture extension. Once fracture growth has been arrested

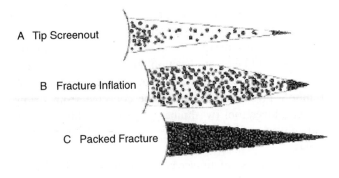

Figure 19-3 Width inflation with the tip-screenout technique

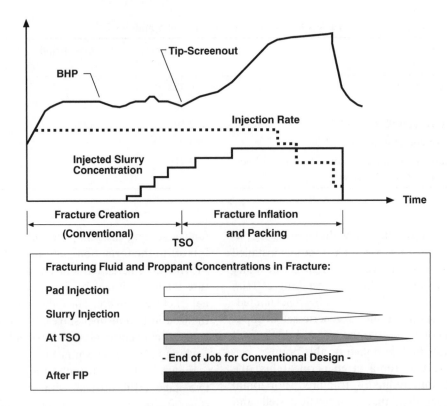

Figure 19-4   Comparison of conventional and HPF design concepts (after Roodhart *et al.*, 1993)

Industry experience suggests that the tip-screenout can be difficult to model, affect, or even detect. The many reasons for this difficulty include a tendency toward overly conservative design models (resulting in no TSO), partial or multiple tip-screenout events, and inadequate pressure monitoring practices.

It is now well accepted that accurate bottomhole measurements are imperative for meaningful treatment evaluation. Calculated bottomhole pressures are unreliable because of the dramatic friction pressure effects associated with pumping high sand concentrations through reduced-ID tubulars and service-tool crossovers. Surface data may indicate that a TSO event has occurred when the bottomhole data shows no evidence, or the opposite may be true. Even in the case of downhole pressure data, there has been some discussion regarding where measurements should be taken. Friction and turbulence concerns have caused at least one operator to conclude that bottomhole pressure data should be collected from below the crossover tool (washpipe gauges) in addition to the data collected from the service tool bundle (Mullen *et al.*, 1994).

The detection of tip-screenout is discussed further in Section 19-7 along with the introduction of a simple screening tool to evaluate bottomhole data.

19-3.2 Net Pressure and Fluid Leakoff Considerations

The entire HPF process is dominated by net pressure and fluid leakoff considerations, first because high-permeability formations are typically soft and exhibit low elastic modulus values, and second, because the fluid volumes are relatively small and leakoff rates are high (high permeability, compressible reservoir fluids, and non-wall-building fracturing fluids). As described previously, the tip-screenout design itself also affects net pressure. While traditional practices applicable to design, execution, and evaluation in MHF continue to be used in HPF, these are frequently not sufficient.

19-3.2.1 Net Pressure, Closure Pressure, and Width in Soft Formations

Net pressure is the difference between the pressure at any point in the fracture and the fracture closure pressure. This definition involves the existence of a unique closure pressure. Whether the closure pressure is a constant property of the formation, or it depends heavily on the pore pressure (or rather on the disturbance of the pore pressure relative to the long-term steady value) is an open question.

In high-permeability, soft formations, it is difficult (if not impossible) to suggest a simple recipe to determine the closure pressure as classically derived from shut-in pressure decline curves (Section 19-4.3.2). Furthermore, because of the low elastic modulus values, even small, induced uncertainties in the net pressure are amplified into large uncertainties in the calculated fracture width.

19-3.2.2 Fracture Propagation

Fracture propagation, the availability of sophisticated 3D models notwithstanding, is not yet a well-described phenomenon. Recent studies (Chudnovsky *et al.*, 1996) emphasize the stochastic character of this propagation in competent hard-rock formations. No serious attempt has been made to describe the physics of fracture propagation in soft rock, but it is reasonably expected to involve incremental energy dissipation and more severe tip effects (with the effect of increasing net pressures). Again, because of the low modulus values, an inability to predict net-pressure behavior may lead to significant differences between predicted and actual treatment performance. Ultimately, the classic models may not reflect even the main features of the propagation process.

Currently, fracture propagation and net-pressure features are "predicted" through the use of a computer fracture-simulator. This trend of substituting clear models and physical assumptions with "knobs" such as (1) arbitrary stress barriers, (2) friction changes (attributed to erosion, if decreasing, and sand resistance, if increasing) and (3) poorly understood properties of the formation expressed as dimensionless "factors," does not help clarify the issue.

19-3.2.3 Leakoff in the High-Permeability Environment

Considerable effort has been expended on laboratory investigation of the fluid leakoff process for high-permeability cores. A comprehensive report can be found in both Vitthal and McGowen (1996) and McGowen and Vitthal (1996). The results raise some questions about how effectively fluid leakoff can be limited by filter-cake formation.

In all cases, but especially in high-permeability formations, the quality of the fracturing fluid is only one of the factors that influence leakoff, and it is often not the determining one. Transient fluid flow in the formation

might have an equal or even larger impact. Transient flow cannot be understood by simply fitting an empirical equation to laboratory data; the use of models based on solutions to the fluid flow in porous media is an unavoidable step.

19-3.3 Fundamentals of Leakoff in HPF

In the following, three models are considered that describe leakoff in the high-permeability environment. Use of the traditional Carter leakoff model requires some modification for use in HPF as shown. (Note: While this model continues to be used almost exclusively across the industry, it is not entirely sufficient for the HPF application.) An alternate, filter cake-based leakoff model has been developed based on the work by Mayerhofer *et al.* (1993). The most appropriate but not yet widespread leakoff model for high-permeability formations may be that of Fan and Economides (1995), which considers the series resistance caused by (1) the filter cake, (2) the polymer-invaded zone, and (3) the reservoir. While the Carter model is the most common in current use, the models of Mayerhofer *et al.* and Fan and Economides represent important building blocks and provide a conceptual framework for understanding the critical issue of leakoff in HPF.

19-3.3.1 Fluid Leakoff and Spurt Loss as Material Properties: The Carter Leakoff Model with Nolte's Power Law Assumption

To make use of material balance, the term V_L, the lost volume, must be described. For rigorous theoretical development, V_L is the volume of liquid entering the formation through the two created fracture surfaces of one wing. There are two main philosophies concerning leakoff. The first considers the phenomenon as a *material property* of the fluid/rock system. The basic relation (called the integrated Carter equation, also provided in Chapter 17) is given in consistent units as

$$\frac{V_L}{A_L} = 2C_L\sqrt{t} + S_p \qquad (17\text{-}18)$$

where A_L is the area and V_L is the total volume lost during the period from time zero to time t. The integration constant, S_p, is called the spurt-loss coefficient, which is measured in meters. It can be considered as the *width* of the fluid body passing through the surface instantaneously at the very beginning of the leakoff process, while $2C_L\sqrt{t}$ is the width of the fluid body following the first slug. The two coefficients, C_L, and S_p can be determined from laboratory tests.

As discussed in more detail in Chapter 17, Equation 17-18 can be visualized assuming that the given surface element "remembers" when it has been opened to fluid loss and has its own "zero" time, which might be different from location to location on a fracture surface. Points of the fracture face near to the well are opened at the beginning of pumping while the points at the fracture tip are "younger." Application of Equation 17-18 or of its differential form necessitates the tracking of the opening time of the different fracture-face elements, as discussed in Chapter 17.

The second philosophy considers leakoff as a consequence of flow mechanisms into the porous medium and uses a corresponding mathematical description.

19-3.3.2 Filter Cake Based Leakoff Model According to Mayerhofer et al.

The method of Mayerhofer *et al.* (1993) describes the leakoff rate using two parameters that are physically more realistic than the leakoff coefficient: (1) filter-cake resistance at a reference time and (2) reservoir permeability. It is assumed that these parameters (R_0, the reference resistance at a reference time t_0, and k_r, the reservoir permeability) have been identified from a minifrac diagnostic test. In addition, reservoir pressure, reservoir fluid viscosity, porosity, and total compressibility are assumed to be known.

Total pressure gradient from inside a created fracture out into the reservoir, Δp, at any time during the injection, can be written as

$$\Delta p(t) = \Delta p_{\text{face}}(t) + \Delta p_{\text{piz}}(t) + \Delta p_{\text{res}}(t) \qquad (17\text{-}26)$$

where Δp_{face} is the pressure drop across the fracture face dominated by the filter cake, Δp_{piz}, is the pressure drop across a polymer-invaded zone, and Δp_{res} is the pressure drop in the reservoir. This concept is shown in Figure 19-5.

In a series of experimental works using typical hydraulic fracturing fluids (e.g. borate and zirconate-crosslinked fluids) and cores of permeability less than 5 md, no appreciable polymer-invaded zone was detected. This simplifying assumption is not valid for linear gels such as HEC (which do not form a filter cake), and the assumption may break down for crosslinked fluids at higher permeabilities, (50 + md). Thus, at least for crosslinked fluids, the second term in the right side of Equation 17-26 can reasonably be ignored, yielding

$$\Delta p(t) = \Delta p_{\text{face}}(t) + \Delta p_{\text{res}}(t) \qquad (19\text{-}2)$$

The filter-cake pressure term is proportional to R_0, the characteristic resistance of the filter cake. The transient pressure drop in the reservoir can be re-expressed as a series expansion of p_D, the dimensionless pressure function describing the behavior of the reservoir (unit response); t_D is the dimensionless time calculated with the maximum fracture length reached at time t_n, and r_p is the ratio of permeable height to the total height (h_p/h_f). With rigorous introduction of these variables and considerable rearrangement (not shown), an expression for leakoff can be written that is useful for both hydraulic fracture propagation and fracture-closure modeling:

$$q_n = \frac{\Delta p(t_n) - \dfrac{\mu_r}{\pi k_r r_p h_f} \left[-q_{n-1} p_D(t_{Dn} - t_{Dn-1}) + \sum_{j=1}^{n-1} (q_j - q_{j-1}) p_D(t_{Dn} - t_{Dj-1}) \right]}{\dfrac{R_0}{2 r_p A_n} \sqrt{\dfrac{t_n}{t_e}} + \dfrac{\mu_r p_D(t_{Dn} - t_{Dn-1})}{\pi k_r r_p h_f}}$$

$$(19\text{-}3)$$

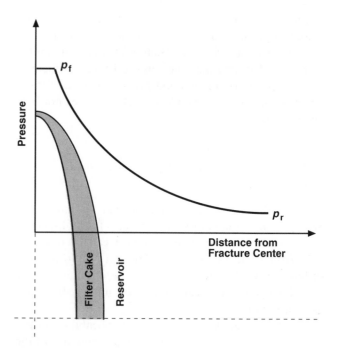

Figure 19-5 Filter cake plus reservoir pressure drop in the Mayerhofer *et al.* (1993) model

This expression allows for determination of the leakoff rate at time instant t_n, if the total pressure difference between the fracture and the reservoir is known, as well as the *history* of the leakoff process. The dimensionless pressure solution, $p_D(t_{Dn} - t_{Dj-1})$, has to be determined with respect to a dimensionless time that considers the *actual* fracture length at t_n.

The model can be used to analyze the pressure fall-off subsequent to a fracture injection (minifrac) test, as described by Mayerhofer *et al.* (1995). The method needs more input data than the similar analysis based on the Carter leakoff approach, but it offers the distinct advantage of differentiating between the two major factors of the leakoff process, filter-cake resistance and reservoir permeability.

19-3.3.3 Polymer-Invaded Zone-Based Leakoff Model of Fan and Economides

The leakoff model of Fan and Economides (1995) concentrates on the additional resistance created by the polymer-invaded zone. The total driving force behind fluid leakoff is the pressure difference between the fracture face and the reservoir, $p_{frac} - p_i$, which is equivalent to the sum of the following three separate pressure drops taken across the filter cake, in the polymer-invaded zone, and in the reservoir:

$$p_{frac} - p_i = \Delta p_{cake} + \Delta p_{inv} + \Delta p_{res} \qquad (19\text{-}4)$$

The fracture treating pressure is equivalent to the net pressure plus fracture closure pressure (minimum horizontal stress). When a non-filter-cake fluid is used, the pressure drop across the filter cake is negligible, which is the case for many HPF treatments. The physical model of this situation, (i.e., fluid leakoff controlled by polymer invasion and transient reservoir flow), is depicted in Figure 19-6. The polymer invasion is labeled in the figure as Region 1, while the region of reservoir fluid compression (transient flow) is denoted as 2.

By employing conservation of mass, a fluid-flow equation, and an appropriate equation of state, a mathematical description of this fluid leakoff scenario can be written. As a starting point, Equation 19-5 describes power-law fluid behavior in the porous medium:

$$\frac{\partial^2 p}{\partial x^2} = \frac{n\phi\mu_{eff}c_t}{k}\left(\frac{1}{u}\right)^{1-n}\frac{\partial p}{\partial t} \qquad (19\text{-}5)$$

where c_t is the system compressibility, k is the formation permeability, u is the superficial flow rate, n is the power law fluid-flow behavior index, ϕ is the formation porosity, and $\mu_{eff} = \frac{K'}{12}\left(9 + \frac{3}{n}\right)^n (150k\phi)^{\frac{1-n}{2}}$ is the fluid effective viscosity where K' is the power-law fluid consistency index.

Combining the description of the polymer-invaded zone and the reservoir, the total pressure drop is given by Fan and Economides (1995) as

$$p_{frac} - p_r = \frac{\sqrt{\pi}}{2}\frac{\phi\eta}{k}\left\{\mu_{app}\sqrt{\alpha_1}e^{\left(\frac{\eta}{\sqrt{4\alpha_1}}\right)^2}\mathrm{erf}\left(\frac{\eta}{\sqrt{4\alpha_1}}\right)\right.$$
$$\left. + \mu_r\sqrt{\alpha_2}e^{\left(\frac{\eta}{\sqrt{4\alpha_2}}\right)^2}\mathrm{erfc}\left(\frac{\eta}{\sqrt{4\alpha_2}}\right)\right\} \qquad (19\text{-}6)$$

where

$$a_1 = \frac{k}{n\phi\mu_{eff}\left(\frac{1}{u}\right)^{1-n}c_t} \quad \text{and} \quad a_2 = \frac{k}{\phi\mu c_t}$$

At given conditions, Equation 19-6 can be solved iteratively for the variable η. Once the value of η is found for a specified total pressure drop, the leakoff rate is calculated from

$$q_L = A\left(\frac{\eta}{2\phi}\right)\frac{1}{\sqrt{t}} \qquad (19\text{-}7)$$

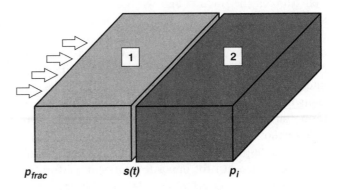

Figure 19-6 Fluid leakoff model with polymer invasion and transient reservoir flow

In other words, the factor $\eta/(2\phi)$ can be considered a pressure-dependent apparent leakoff coefficient.

19-4 TREATMENT DESIGN AND EXECUTION

In contrast to the preceding section, which was quite theoretical (appropriately so) and dealt fundamentally with key issues involved in high-permeability fracturing, especially leakoff, this section will provide practical detail as to the current best practice being applied in HPF design and execution.

Most HPF treatments are done with mechanical sand control equipment in place. While this is not always the case, and while there are many potential variations, a generalized job sequence follows:

1. Perforate the formation.
2. Run the gravel-pack screen assembly.
3. Spot/soak acid to clean up perforations.
4. Perform and interpret pretreatment diagnostic tests.
5. Design the TSO pumping schedule based on design variables from diagnostic tests.
6. Pump the TSO treatment until screenout or until the volume needed to form an annulus pack remains in workstring.
7. Slow the pump rate to 1 to 2 bbl/min and open the annulus valve to circulate in and dehydrate an annular pack.
8. Shut down the pumps when tubing pressure reaches its safe upper limit.
9. Prepare the well for production.

19-4.1 Perforations

It is widely agreed that establishing a conductive connection between the fracture and wellbore is critical to the success of HPF, but no consensus or study has emerged that gives definitive direction. In the context of high permeability and maximizing conductivity and fluid flow rate, a common response is to shoot the entire target interval with high shot-density and large holes (12 shots/ft with "big hole" charges). Concerns with clean formation breakdown (single-fracture initiation), near-well tortuosity, and perforations that are not packed with sand (especially in screenless HPFs) cause some operators to use just the opposite treatment: perforating the middle of the target zone only (possibly modifying the treatment up or down based on stress contrast) with a limited number of 0° or 180° phased perforations.

Arguments are made for and against underbalanced vs. overbalanced perforating: underbalanced perforating may cause formation failure and cause the guns to "stick," while overbalanced perforating eliminates a cleanup trip but may negatively impact the completion efficiency.

Solvent or other scouring pills are commonly circulated to the bottom of the workstring and then reversed out to remove scale, pipe dope, or other contaminants before they are pumped into the formation. Several hundred gallons (10 to 25 gal/ft) of 10 to 15% HCl acid will then typically be circulated or bullheaded down to the perforations and be allowed to soak, (to improve communication with the reservoir by cleaning up the perforations and dissolving debris in the perforation tunnel). Some operators are beginning to forego the solvent and acid cleanup (obviously to reduce rig time and associated costs) from the perspective that, in HPF, the damaging material is pumped deep into the formation and will not seriously impact well performance.

19-4.2 Mechanical Considerations

The vast majority of HPF treatments have been performed with the mechanical sand-control equipment in place. However, in some early jobs, the tip-screenout and gravel pack were done in two steps separated by a clean-out trip. Concerns with fluid loss/damage to the fracture and a desire to eliminate all unnecessary expense eventually discouraged this two-step approach. More recently, there is a trend toward screenless HPFs as described in Section 19-8.1.

Early treatments were plagued by rate and erosion-resistance limitations of the gravel-pack tools. Enlarged crossover ports have now been incorporated in the gravel-pack tools of all the major service companies, which minimize friction and erosion problems and allow for very aggressive treatment designs. The aggressive pumping schedules, in turn, have given rise to another problem: Tiner *et al.* (1996) report several instances where the blank liner above the screen has been collapsed at screenout. They suggest that the pressure outside the blank rises quicker than the internal pressure, resulting in a collapse of this "weak link." The suggested remedy is the use of P-110 grade pipe for the blank.

Limitations were also evident in the surface equipment used on early treatments. The tendency was to approach these treatments (especially offshore) as an oversized gravel-pack operation. While HPF volumes are relatively small for a fracture treatment, the high rates (20 bbl/min is common) and high proppant concentrations (up to 16 or 18 lb/gal) require high horsepower. Undersized gravel-pack units were often used in early jobs; otherwise, miscellaneous onshore fracturing units were hobbled together and placed on barges. This practice resulted in many failed treatments. Today, dedicated skid-mount units with fixed manifolds are widely available and provide adequate horsepower (including standby) within stringent space and weight limitations. Reliable mixing and blending equipment is now available to achieve the various fluid and additive specifications of HPF, including very-low to very-high proppant concentrations and slurry rates. Other than these considerations, the surface equipment is common to that used in conventional MHF operations.

19-4.3 Pretreatment Diagnostic Tests

The objective of pretreatment diagnostic tests (referred to as fracture calibration tests, minifracs, datafracs, etc.) is to determine within engineering bounds, the value of various parameters that govern the fracturing process. Fracture closure pressure (considered in most cases as equivalent to the minimum horizontal in-situ stress) and the fluid leakoff coefficient (used to describe bulk leakoff behavior) are the most common targets and are especially important in HPF as discussed previously. However, other information may also be sought or inferred, such as (1) fracture extension or propagation pressure (often referred to as formation parting pressure or FPP), (2) potential perforation or near-wellbore friction, (3) evidence of fracture-height containment, and (4) reservoir permeability.

Several features unique to high-permeability fracturing make well-specific design strategies highly desirable if not essential: (1) fracture design in soft formations is very sensitive to leakoff and net pressure, (2) the controlled nature of the sequential tip-screenout/fracture inflation and packing/gravel-packing process demands relatively precise execution strategies, and (3) the treatments are very small and typically "one-shot" opportunities. Furthermore, methods used in hard-rock fracturing for determining critical fracture parameters *a priori* (geologic models, log and core data or Poisson's ratio computational models based on poroelasticity) are of limited value or not yet adapted to the unconsolidated, soft, high-permeability formations.

The preceding discussion of advanced leakoff models and their applicability to pressure falloff analysis not-withstanding, three tests (with variations) form the current basis of pretreatment testing in high-permeability formations: step-rate tests, minifrac tests, and pressure falloff tests.

19-4.3.1 Step-Rate Tests

The step-rate test (SRT), as implied by its name, involves injecting clean gel at several stabilized rates, beginning at matrix rates and progressing to rates above fracture extension pressure. In a high-permeability environment, a test may be conducted at rate steps of 0.5, 1, 2, 4, 8, 10, and 12 bbl/min, and then at the maximum attainable rate. The injection is held steady at each rate step for a uniform time interval, typically 2 or 3 min at each step.

In principle, SRTs are intended to identify the fracture extension pressure and rate. The stabilized pressure (ideally bottomhole pressure) at each step is classically plotted on a Cartesian graph vs. injection rate. The point at which a straight line drawn through those points that are obviously below the fracture extension pressure (dramatic increase in bottomhole pressure with increasing rate) intersects with the straight line drawn through those points above the fracture extension pressure (minimal increase in pressure with increasing rate) is interpreted as the fracture extension pressure. The dashed lines on Figure 19-7 illustrate this classic approach.

While the conventional SRT is operationally simple and inexpensive, it is not necessarily accurate. A Cartesian plot of bottomhole pressure versus injection rate, in fact, does not generally form a straight line for radial flow in an unfractured well. Simple pressure transient analysis of SRT data through the use of de-superposition techniques shows that *with no fracturing,* the pressure vs. rate curve should exhibit upward concavity.

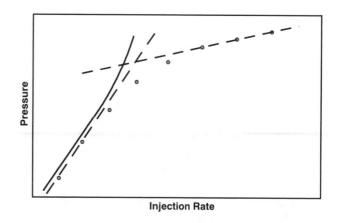

Figure 19-7 Ideal SRT—radial flow with no fracturing

Thus, the departure of the real data from ideal behavior may occur at a pressure and rate well below that indicated by the classic intersection of the straight lines (Figure 19-7).

The two-SRT procedure of Singh and Agarwal (1988) is more fundamentally sound. However, given the relatively crude objectives of the SRT in high-permeability fracturing, the conventional test procedure and analysis may be sufficient.

The classic test does indicate several things:

- Upper limit for fracture-closure pressure (useful in analysis of minifrac pressure falloff data)

- Surface treating pressure that must be sustained during fracturing (or whether sustained fracturing is even possible with a given fluid)

- Reduced rates that will ensure no additional fracture extension and (aided by fluid leakoff) packing of the fracture and near-wellbore with proppant

- Perforation and/or near-wellbore friction (indicated by bottomhole pressures that continuously increase with increasing rate, seldom a problem in soft formations with large perforations and high shot-densities)

- Expected casing pressure if the treatment is pumped with the service tool in the circulating position

A step-down option to the normal SRT is sometimes used specifically to identify near-wellbore restrictions (tortuosity or perforation friction). This test is usually done immediately following a minifrac pump-in stage. By observing how bottomhole pressure varies with decreasing rate, near-wellbore restrictions can be immediately detected; for example, bottomhole pressures that change only gradually during steps down in injection rate would indicate *no restriction*.

19-4.3.2 Minifrac Tests

Following the SRT, which establishes the fracture extension pressure and places an upper bound on fracture closure pressure, a minifrac is typically performed to tailor or redesign the HPF treatment with well-specific information. This test is the critical pretreatment diagnostic test. The minifrac analysis and treatment redesign is now commonly done on site in less than an hour, or 2 to 3 hours at the most.

Concurrent with the rise of HPF, minifrac tests and especially the use of bottomhole pressure information have become much more common. Otherwise, the classic minifrac procedure and primary outputs as described in Chapter 17 (fracture closure pressure and a single leakoff coefficient) are widely applied to HPF—this in spite of some rather obvious shortcomings.

The first step in analyzing a minifrac is determining fracture closure pressure, which is typically done by plotting the pressure decline after shut-in vs. some function of time. The main plots used to identify fracture closure are

- $p_{\text{shut-in}}$ vs. t

- $p_{\text{shut-in}}$ vs. $\sqrt{t}$

- $p_{\text{shut-in}}$ vs. g-function (and variations)

- $\log(p_{\text{ISIP}} - p_{\text{shut-in}})$

The selection of closure pressure using these plots, a difficult enough task in hard-rock fracturing, has proved to be arbitrary or nearly impossible in high-permeability, high fluid-loss formations. In some cases, the duration of the closure period is so limited (1 minute or less) that the pressure signal is masked by transient phenomena. Deviated wellbores and laminated formations (common in offshore U.S. Gulf Coast completions), multiple fracture closures, and other complex features are often evident during the pressure falloff. The softness (low elastic modulus) of these formations results in very subtle fracture closure signatures on the pressure decline curve. Flowbacks are not used to accent closure features because of the high leakoff and concerns with production of unconsolidated formation sand.

New guidelines and diagnostic plots for determining closure pressure in high-permeability formations are being pursued by various practitioners, and this information will eventually emerge to complement or replace the standard analysis and plots.

The shortcomings of classic minifrac analysis are further exposed when they are used (commonly) to select a single effective fluid-loss coefficient for the treatment. As described in Chapter 17, in low-permeability formations, this approach results in a slight overestimation of fluid loss and actually provides a factor of safety to prevent screenout. In high-permeability formations, the classic approach can dramatically underestimate spurt loss (zero spurt-loss assumption) and overestimate total fluid loss (Dusterhoft *et al.*, 1995). This uncertainty in leakoff behavior makes the controlled timing of a tip-screenout very difficult. Dusterhoft *et al.* outlined various procedures to correct for spurt loss and leakoff behavior that is not proportional to the square root of time; however, entirely new procedures based on sound fundamentals of leakoff in HPF (as outlined previously in Section 19-3.3) are ultimately needed. The traditional practice of accounting for leakoff with a bulk leakoff coefficient is simply not sufficient for this application.

19-4.3.3 Pressure Falloff Tests

A third class of pretreatment diagnostics for HPF has emerged that is *not* common to MHF: pressure falloff tests. Because of the high formation permeability, common availability of high-quality bottomhole pressure data and multiple pumping and shut-in cycles, matrix formation properties including *kh* and skin can be determined from short-duration pressure falloff tests with the appropriate transient flow equation. Chapman *et al.* (1996) and Barree *et al.* (1996) propose prefrac or matrix injection/falloff tests that involve injecting completion fluid below fracturing rates for a given period, and then analyzing the pressure decline with a Horner plot. The test is performed with standard pumping equipment, and it poses little interruption to normal operations. A test can normally be completed within 1 hour or may even make use of data from unplanned injection/shut-in cycles. The resulting permeability certainly relates to fluid leakoff as described earlier in this chapter (Section 19-3.3.2), and it allows the engineer to better anticipate fluid requirements. An initial skin value is useful in "benchmarking" the HPF treatment and for comparison with post-treatment pressure transient analysis.

19-4.3.4 Bottomhole Pressure Measurements

A discussion of pretreatment diagnostic tests requires a discussion of the source of pressures used in the analysis. Implicit to the discussion is that the only meaningful pressures are those adjacent to the fracture face, whether measured directly or translated to that point. At least four different types of bottomhole pressure data are available, depending on the location at which the real data were taken:

- Calculated bottomhole pressure (bottomhole pressure calculated from surface pumping pressure)

- Deadstring pressure (open annulus and bottomhole pressure determined based on the density of fluid in annulus; tubing may also be used as a deadstring when the treatment is pumped down the casing)

- Bundle carriers in the workstring (measured downhole, but above the service tool crossover)

- Washpipe data (from sensors attached to washpipe below the service-tool crossover)

Washpipe pressure data is the most desirable for HPF design and analysis because of its location adjacent to the fracture and downstream of all significant flowing pressure drops. Workstring bundle carrier data can introduce serious error in many cases because of fluid friction generated both through the crossover tool and in the casing/screen annulus. Without detailed friction-pressure corrections that account for specific tool dimensions and annular clearance, significant differences may exist between washpipe and workstring bundle carrier pressures (Mullen *et al.*, 1994). Deadstring pressures are widely used and considered acceptable by most practitioners; others suggest that redundant washpipe pressure data has shown that the deadstring can miss subtle features of the treatment. The use of bottomhole transducers with real-time surface readouts is suggested in cases where a deadstring is not feasible or when such well conditions as transients may obscure important information. The calculation of bottomhole pressures from surface pumping pressure is not recommended in HPF. The combination of heavy sand-laden fluids, constantly changing proppant concentrations, very high pump rates, and short pump times makes the estimation of friction pressures nearly impossible.

19-4.4 Tip-Screenout Design

The so-called tip-screenout or TSO design clearly differentiates high-permeability fracturing (HPF) from conventional massive hydraulic fracturing (MHF). While HPF introduces other identifiable differences, such as higher permeability, softer rock, smaller proppant volumes, etc., it is the tip-screenout which makes these fracturing treatments unique. Conventional treatments are designed to achieve TSO at the end of pumping. In high-permeability fracturing, the *fracture creation* stage that precedes TSO is followed by a *fracture inflation and packing* stage; this two-stage treatment gives rise to the vernacular *fracpack*. These conventional and HPF design concepts are illustrated and compared in Figure 19-4.

Because of the rapid ascent of high-permeability fracturing, many engineers did not have (and still do not have) computer models that accommodate the TSO design. By definition (Nolte, 1986), conventional fracture design systems were formulated with TSO as the endpoint. A no-growth fracture inflation and packing stage had not been envisioned, never mind entering the necessary design algorithms into a computer model. Recently, however, several of the commercially available simulators have been modified to accept the TSO designs. The in-house simulators of many producing companies and oilfield service companies have also been modified.

Given the near-crippling dependence of the modern petroleum engineer on "black-box" solutions, one is compelled to ask how engineers effected a TSO design before the modified computer programs were available. What is the key? An experienced engineer would recognize that after TSO (assuming complete arrest of fracture growth), the problem is reduced to a simple one of material balance.

Wong *et al.* (1995) offer the following algorithm that can be used with any conventional 2D simulator to develop a fundamentally sound tip-screenout design:

1. It is assumed that the following fracture parameters are known at the end of the TSO stage (from the simulator):

A_0	= fracture area at TSO
t_0	= total time to TSO
M_{ts0}	= total proppant mass
$p(t_0)$	= net pressure at TSO
$V_{F_{(t0)}}$	= fracture volume at TSO

2. For every i^{th} stage of the fracture inflation and packing (FIP) pumping schedule, the clean fluid volume (V_{ci}) and the pumping time for the ith stage (t_i) are given in terms of known slurry volume (V_i), proppant concentration (c_i), pump rate (q_i), and proppant specific density (ρ_p):

$$V_{ci} = V_i \rho_p / (\rho_p + c_i) \qquad (19\text{-}8)$$

and

$$t_i = V_i / q_i \qquad (19\text{-}9)$$

3. Cumulative time from TSO to the ith stage is simply

$$T_i = \sum t_i \qquad (19\text{-}10)$$

4. Assuming that the fracture area ceases to propagate after TSO, the fluid leakoff rate (q_l) and leakoff volume (V_l) at any time T_i are given (for low-efficiency conditions) as

$$q_l(T_i) = 2C_L A_0 \left(1/\sqrt{t_o}\right) \arcsin\left(1/\sqrt{\tau_i}\right) \qquad (19\text{-}11)$$

and

$$V_i(T_i) = 2C_L A_0 \sqrt{t_0} \left[\tau_i \arcsin(1/\tau_i) + \sqrt{(\tau_i - 1)} \right] \qquad (19\text{-}12)$$

where

$$\tau_i = (t_0 + T_i)/t_0 \qquad (19\text{-}13)$$

and C_L is the fluid leakoff coefficient.

5. The following material balance relations can be easily implemented in a spreadsheet program and used to calculate fracture parameters at any time T_i:

$$V_f(T_i) = \sum V_{ci} - V_i(T_i) \qquad (19\text{-}14)$$

$$V_f(T_i) = V_F(t_0) + \sum V_i - V_l(T_i) \qquad (19\text{-}15)$$

$$M_{fip}(T_i) = M_{ts0} + \sum (c_i V_{ci}) \qquad (19\text{-}16)$$

$$c_m(T_i) = M_{fip}(T_i)/V_f(T_i) \qquad (19\text{-}17)$$

$$APC(T_i) = M_{fip}(T_i)/A_0 \qquad (19\text{-}18)$$

and

$$\Delta p(T_i) = \Delta p(t_0) \frac{V_F(T_i)}{V_F(T_0)} \qquad (19\text{-}19)$$

where V_f is the total (two-wing) fluid volume, V_F is the total fracture volume, M_{fip} is the total amount of proppant, c_m is the average proppant concentration loading, APC is the average areal proppant concentration, and Δp is the net pressure.

Using the relations above, a TSO design is developed that specifies pump rate, slurry volume, and proppant loading during fracture inflation and packing in as many stages as deemed appropriate. Design objectives include (1) achieving a desired fracture width (from areal proppant concentration) and (2) ensuring that the proppant does not dehydrate prematurely ($c_m < 28 \, \text{lb}/\text{gal}$).

Early TSO treatment designs commonly called for 50% pad (similar to conventional fracturing) and a fairly aggressive proppant ramping schedule; however, it is now increasingly common to reduce the pad to 10 to

15% of the treatment and extend the 0.5 to 2 lb/gal stages (which combined, may comprise 50% of total slurry volume, for example). This practice is intended to "create width" for the higher concentration proppant addition (12 to 14 lb/gal).

19-4.5 Pumping a TSO Treatment

Anecdotal observations related to real-time HPF experiences are abundant in the literature and are not the focus of this text. However, some observations related to treatment execution are necessary:

- Most treatments are pumped with a gravel-pack service tool in the "circulate" position with the annulus valve closed at the surface. This practice allows for live annulus monitoring of bottomhole pressure (annulus pressure + annulus hydrostatic head) and real-time monitoring of the progress of the treatment.

- When no evidence exists of the planned TSO on the real-time pressure record, the late treatment stages can be pumped at a reduced rate to effect a tip-screenout. Obviously, this practice requires reliable bottomhole pressure data and direct communication with the frac unit operator.

- Near the end of the treatment, the pump rate can be slowed to gravel-packing rates, and the annulus valve can be opened to begin circulating a gravel pack. The reduced pump rate is maintained until tubing pressure reaches a safe upper limit, signaling that the screen/casing annulus is packed.

- Because very high proppant concentrations are used, the sand-laden slurry used to pack the screen/casing annulus must be displaced from the surface with clean gel well before the end of pumping. Thus, proppant addition and slurry volumes must be metered carefully to ensure that there is sufficient proppant left in the tubing to place the gravel pack (to avoid "overdisplacing" proppant into the fracture).

- Conversely, if an HPF treatment sands out prematurely (with proppant in the tubing), the service tool can be moved into the "reverse" position and the excess proppant can be circulated out.

- Movement of the service tool from the squeeze/circulating position to the reverse position can create a sharp instantaneous drawdown effect, and it should be done carefully to avoid swabbing unstabilized formation material into the perforation tunnels and annulus.

19-4.5.1 Swab Effect Example

The following simple equation, given by Mullen *et al.* (1994) can be used to convert swab volumes into oilfield-unit flow rates.

$$q_s = 2057 \frac{V_s}{t_m} \qquad (19-20)$$

where q_s is the instantaneous swab rate in bbl/d, V_s is the swabbed volume of fluid in gal, t_m is the time of tool movement in seconds, 2057 is the conversion factor for gal/sec to bbl/d.

The volume of swabbed fluid is calculated from the service-tool diameter and the length of stroke during which the sealed service tool does not allow fluid bypass. The average swab volume of a 2.68-in. service tool is 2.8 gal when the service tool is moved from the squeeze position to the reverse-circulation position. Assuming a rather normal movement time of 5 seconds, this represents an instantaneous production rate of 1103 bbl/d.

19-5 FRACTURE CONDUCTIVITY AND MATERIALS SELECTION

19-5.1 Optimum Fracture Dimensions

Much has been published recently concerning optimum fracture dimensions in HPF. While there are debates regarding the optimum dimensions, fracture conductivity is largely regarded as more important than fracture length. Of course, this intuitive statement only recognizes the first principle of fracture optimization: Higher permeability formations require higher fracture conductivity to maintain an acceptable value of the dimensionless fracture conductivity, C_{fD}.

So how long should the fracture be? A "rule of thumb" is that fracture length should be equal to half the perforation height (thickness of producing interval). Alternatively, Hunt *et al.* (1994) showed that cumulative recovery from a well in a 100-md reservoir with a 10-ft damage radius is optimized by extending a fixed 8000-md-ft conductivity fracture to any appreciable distance beyond the damaged zone. This result implies that there is little benefit to a 50-ft fracture length compared to a 10-ft fracture length. Two observations are in order: first, the Hunt *et al.* evaluation is based on cumulative recovery; second, their assumption of a fixed fracture conductivity implies a decreasing dimensionless fracture conductivity with increasing fracture length (less than optimal placement of the proppant).

It is generally true that if an acceptable C_{fD} is maintained, additional length will provide additional production. (An acceptable C_{fD} may require an increase in areal proppant concentration from $1.5\,\text{lb/ft}^2$, which is common in hard-rock fracturing, to $20\,\text{lb/ft}^2$ or more.) Ultimately, the decision becomes one of economics and/or optimal placement of a finite proppant volume (as in offshore environments where total fluid and proppant volumes may be physically limited).

These issues are discussed more rigorously in the following sections.

19-5.1.1 Fracture Width as a Design Variable

In practice, fracture extent and width have been difficult to influence separately. Once a fracturing fluid and injection rate are selected, the fracture width evolves with increasing length according to strict relations (at least in the well-known PKN and KGD design models). Therefore, the key decision variable has been the fracture extent. Once a fracture extent is selected, the width is calculated as a consequence of technical limitations, (maximum realizable proppant concentration). Knowledge of the leakoff process helps to determine the necessary pumping time and pad volume.

The tip-screenout (TSO) technique has brought a significant change to this design philosophy. Through TSO, fracture width can be increased without increasing the fracture extent. In this context, a strictly technical optimization problem can be formulated: How does one independently select the optimum fracture length and width under a given proppant volume constraint? The problem is one of maximizing the PI in the pseudosteady-state flow regime. The answer is of primary importance in understanding HPF, but is also necessary for understanding hydraulic fracturing in general.

The same propped volume can be used to create a narrow, elongated fracture or a wide, short fracture. It is convenient to select C_{fD} as the decision variable, and then the fracture half-length can be expressed using the propped volume of one wing, V_f, as

$$x_f = \left(\frac{V_f k_f}{C_{fD} h k}\right)^{1/2} \tag{17-9}$$

The productivity index (Equation 19-1) after the creation of a fracture of half-length, x_f, can be written in oilfield units as

$$J = \frac{kh}{141.2 B \mu \left[\ln \dfrac{0.472 r_e}{r_w} + \ln \dfrac{r_w}{x_f} + \left(\ln \dfrac{x_f}{r_w} + s_f\right)\right]} \tag{19-21}$$

where s_f is the Cinco-Ley *et al.* pseudoskin appearing because of the fracture. The quantity $\ln x_f/r_w + s_f$ can be obtained from the dimensionless fracture conductivity, C_{fD}, (Equation 17-8). The wellbore radius drops out and the fracture half-length is substituted from Equation 17-9. The resulting productivity index is

$$J = \frac{kh}{141.2 B \mu \left[\ln 0.472 r_e + 0.5 \ln \dfrac{hk}{V_f k_f} + 0.51 \ln C_{fD} + \left(\ln \dfrac{x_f}{r_w} + s_f\right)\right]} \tag{19-22}$$

where the only unknown is C_{fD}. Since the drainage radius, formation thickness, two permeabilities, and the propped volume are fixed, the maximum PI occurs when the quantity

$$y = 0.51 \ln C_{fD} + \ln \frac{x_f}{r_w} + s_f \tag{19-23}$$

becomes a minimum. The quantity y is also shown on Figure 19-2. Since it depends only on C_{fD}, the optimum C_{fD}, opt $= 1.6$ is a *given constant* for any reservoir, well, and proppant. (Note: Depending on the accuracy of the calculations and the graphical representation, some authors have suggested the value 1.2.) As explained in Chapter 17, the optimum dimensionless fracture conductivity corresponds to the best compromise between the capacity of the fracture to conduct and the capacity of the reservoir to deliver hydrocarbon.

19-5.1.2 Technical Optimization

Once the volume of proppant that can be placed into one wing of the fracture, V_f, is known, the optimum fracture half-length can be calculated as

$$x_f = \left(\frac{V_f k_f}{1.6 h k}\right)^{1/2} \tag{19-24}$$

and consequently, the optimum propped average width should be

$$w = \left(\frac{1.6 V_f k}{h k_f}\right)^{1/2} \qquad (19\text{-}25)$$

These results have several implications. Most important, there is no theoretical difference between low- and high-permeability fracturing. In both cases, a *technically* optimal fracture exists, and it should have a dimensionless fracture conductivity of order unity. In a low-permeability formation, this requirement results in a long and narrow fracture. In high-permeability formations, a short and wide fracture may provide the same dimensionless conductivity. In practice, not all proppant will be placed into the permeable layer, so in the relation above, the *effective* volume should be used, subtracting the proppant placed in the nonproductive layers. It is also important to recognize that the indicated "optimal fracture" may not always be feasible. In high-permeability formations, departure from the optimum dimensionless fracture conductivity might be justified by several factors (e.g. the indicated large width may be impossible to create): a minimum length may be dictated by the damage radius, severe non-Darcy effects in the fracture may strongly reduce the apparent permeability of the proppant pack, and considerable fracture width can be lost because of proppant embedment into the soft formation.

19-5.1.3 Economic Optimization

Having settled the optimization of fracture length vs. width for a fixed proppant volume, the remaining task is to optimize proppant volume. Obviously, this is an economic optimization issue rather than a technical one. The more proppant that is placed in the formation (otherwise optimally), the better the performance of the well. At this point, economic considerations must dominate. The additional revenue at some point becomes marginal compared to the linearly (or even more strongly) increasing costs. This situation is properly treated by applying net present value (NPV) analysis (Balen *et al.*, 1988). Though a NPV analysis always provides an "optimum design," it should not replace the understanding of the underlying technical optimization issues.

19-5.2 Proppant Selection

The primary and unique issue relating to proppant selection for high-permeability fracturing is *proppant sizing*. Proppant strength, shape, composition, and other factors are included in a more general discussion of proppant

selection in Section 17-3.4. Resin-coated proppants are discussed briefly as an emerging HPF technology in Section 19-8.1. While specialty proppants (intermediate-strength and resin-coated proppants) have certainly been used in HPF, most treatments are pumped with standard graded-mesh sand.

When selecting a proppant size for HPF, the engineer faces competing priorities: sizing the proppant to address concerns with sand exclusion, or using maximum proppant size to ensure adequate fracture conductivity.

As with equipment choices and fluids selection, the gravel-packing roots of fracpack are also evident when proppant selection is considered. Engineers initially focused on sand exclusion and a gravel pack derived sizing criteria such as that proposed by Saucier (1974). Saucier recommends that the mean gravel size (D_{g50}) be five to six times the mean formation grain size (D_{f50}). The so-called "4-by-8 rule" represents minimum and maximum grain-size diameters that are distributed around Saucier's criteria, i.e. $D_{g,\min} = 4 D_{g50}$ and $D_{g,\max} = 8 D_{g50}$, respectively. Thus, many early treatments were pumped with standard 40/60-mesh or even 50/70-mesh sand. The somewhat limited conductivity of these gravel-pack mesh sizes under in-situ formation stresses may not be adequate in many cases. Irrespective of sand mesh size, fracpacks tend to reduce concerns with fines migration by reducing fluid flux at the formation face.

The current trend in proppant selection is to use fracturing-size sand. A typical HPF treatment now uses 20/40 proppant (sand). Maximizing the fracture conductivity can itself help prevent sand production by reducing drawdown. Results with the larger proppant have been encouraging, both in terms of productivity and limiting or eliminating sand production (Hannah *et al.*, 1993).

It is interesting to note that the topics of formation competence and sanding tendency, major issues in the realm of gravel-pack technology, have not been widely studied in the context of HPF. In many cases, HPF is providing a viable solution to completion failures *despite* the industry's limited understanding of (soft) rock mechanics.

This move away from gravel-pack practices toward fracturing practices is common to many aspects of HPF with the exception (so far) of downhole tools, and it seems to justify changing our terminology from *fracpack* to *high-permeability fracturing*. The following discussion of fluid selection is also consistent with this perspective.

19-5.3 Fluid Selection

Fluid selection for HPF has always been driven by concerns with damaging the high-permeability formation, either by filter-cake buildup or (especially) polymer invasion. Most early treatments were performed with HEC, the classic gravel-pack fluid, because it was perceived to be less damaging than guar-based fracturing fluids. While the debate continues and many operators continue to use HEC fluids, the fluid of choice is increasingly borate-crosslinked HPG.

Based on a synthesis of reported findings from several practitioners, Aggour and Economides (1996) provide a well-reasoned rationale to guide fluid selection in HPF. Their findings suggest that if the extent of fracturing fluid invasion is minimized, the degree of damage (permeability impairment caused by filter-cake or polymer invasion) is of secondary importance. They use the effective skin representation of Mathur *et al.* (1995) to show that if fluid leakoff penetration is small, even severe permeability impairments can be tolerated without exhibiting positive skin effects. In this case, the obvious recommendation in HPF is to use high-polymer concentration, crosslinked fracturing fluids with fluid-loss additives, and an aggressive breaker schedule. The polymer, crosslinker, and fluid-loss additives limit polymer invasion, and the breaker ensures maximum fracture conductivity, a critical factor which cannot be overlooked. Experimental work corroborates these contentions.

Linear gels have been known to penetrate cores of very low permeability (1 md or less) whereas crosslinked polymers are likely to build filter cakes at permeabilities two orders of magnitude higher (Roodhart, 1985; Mayerhofer *et al.*, 1991). Filter cakes, while they may damage the fracture face, clearly reduce the extent of polymer penetration into the reservoir that is normal to the fracture face. At extremely high permeabilities, even crosslinked polymer solutions may invade the formation.

Cinco-Ley and Samaniego (1981) and Cinco-Ley *et al.* (1978) described the performance of finite-conductivity fractures and delineated the following three major types of damage affecting this performance:

- *Reduction of proppant-pack permeability* resulting from either proppant crushing or (especially) unbroken polymer chains, leads to fracture conductivity impairment. This condition can be particularly problematic in moderate- to high-permeability reservoirs. Extensive progress in breaker technology has dramatically reduced concerns with this type of damage.

- *Choke damage* refers to the near-well zone of the fracture that can be accounted for by a skin effect. This damage can result from either overdisplacement at the end of a treatment or by fines migration (native or proppant) during production and the accumulation of fines near the well but within the fracture.

- *Fracture-face damage* implies permeability reduction normal to the fracture face and includes permeability impairments caused by the filter cake, polymer-invaded zone, and filter cake-invaded zone.

19-5.3.1 Composite Skin Effect

Mathur *et al.* (1995) provide the following representation for effective skin resulting from radial wellbore damage and fracture-face damage:

$$s_d = \frac{\pi}{2}\left[\frac{b_2 k_r}{b_1 k_3 + (x_f - b_1)k_2} + \frac{(b_1 - b_2)k_r}{b_1 k_1 + (x_f - b_1)k_r} - \frac{b_1}{x_f}\right]$$

(19-26)

Figure 19-8 depicts the two types of damage accounted for in s_d, (fracture-face and radial wellbore damage). The b- and k- terms are defined graphically in Figure 19-9 and represent the dimensions and permeabilities of various zones included in the finite-conductivity fracture model of Mathur *et al.*

The equivalent damage skin from Equation 19-26 can be added directly to the undamaged fracture skin effect to obtain the total skin:

$$s_t = s_d + s_f$$

(19-27)

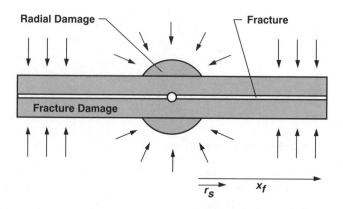

Figure 19-8   Fracture-face damage

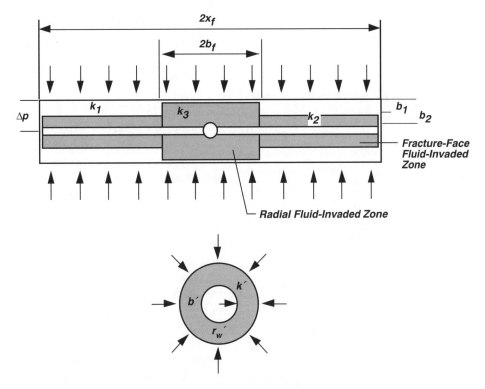

Figure 19-9   Fluid-invaded zones

19-5.3.2 Parametric Studies

Aggour and Economides (1996) used the Mathur *et al.* model (with no radial wellbore damage) to evaluate total skin and investigate the relative effects of different variables. Their results related the total skin in a number of discrete cases to (1) the depth of fluid invasion normal to the fracture face and (2) the degree of permeability reduction in the fluid-invaded zone. A sample of their results (for $x_f = 25$ ft, $C_{fD} = 0.1$, and $k_f = 10$ md), expressed initially in terms of damage penetration ratios, b_2/x_f, and permeability impairment ratios, k_2/k_r, are re-expressed in real units on Table 19-4. Under each of these conditions, the total skin is equal to zero.

These results suggest that for a (nearly impossible) 2.5-ft penetration of damage, a positive skin is obtained only if the permeability impairment in the invaded zone is more than 90%. For 1.25-ft damage penetration, the permeability impairment would have to be over 95% to achieve positive skins. If the penetration of damage can be limited to 0.25 ft, even a 99% permeability reduction in the invaded zone would not result in positive skins. At a higher dimensionless conductivity equal to unity, even higher permeability impairments can be tolerated without suffering positive skins.

At least one obvious critical conclusion can be made from this work: The extent of damage normal to the fracture face is more important than the degree of damage. If fluid invasion can be minimized, even 99% damage can be tolerated. The importance of maximizing C_{fD} is also illustrated; certainly, a good proppant-pack should not be sacrificed in an attempt to minimize the fracture-face damage.

This conclusion emphasizes the selection of appropriate fracturing fluids:

- Linear gels because of their considerable leakoff penetration, are *not* recommended

- Crosslinked polymer fluids with high gel loadings appear to be much more appropriate

- Aggressive breaker schedules are imperative

Table 19-4   Total skin

Depth of Fluid Invasion	Permeability Reduction
2.5 ft	90%
1.25 ft	95%
0.25 ft	99%

After Aggour and Economides (1996)

- Filter-cake building additives may also be considered to minimize the spurt loss and total leakoff

19-5.3.3 Predicted Well Performance with Cleanup

Fracture-face damage should not significantly alter long-term HPF performance. Previous work has confirmed this conclusion. Mathur *et al.* (1995) used a case study from the Gulf Coast and Ning *et al.* (1995) studied gas wells in Alberta, Canada. The Mathur *et al.* study assumed a linear cleanup of the fracture and observed an early-time improvement of the production rate. The Ning *et al.* study showed that the effect of fracture conductivity on the long-term production rate was the significant factor, whereas the effects of fracture-fluid invasion damage were minimal.

19-5.3.4 Experiments in Fracturing Fluid Penetration

McGowen *et al.* (1993) presented a series of experiments showing the extent of fracturing fluid penetration in cores of various permeabilities. Fracturing fluids used were 70-lb/Mgal HEC and 30- or 40-lb/Mgal borate-crosslinked HPG. Filtrate volumes were measured in mL/cm^2 of leakoff area (centimeters of penetration) for a 10-md limestone and 200- and 1,000-md sandstones at 120°F and 180°F.

Several conclusions are drawn from this work:

- Crosslinked fracturing fluids are far superior to linear gels in controlling fluid leakoff. For example, 40-lb/Mgal borate-crosslinked HPG greatly outperforms 70-lb/Mgal HEC in a 200-md core at 180°F.

- Linear gel performs satisfactorily in 10-md rock but fails dramatically at 200 md. Even aggressive use of fluid-loss additives (40-lb/Mgal silica flour) does not appreciably alter the leakoff performance of HEC in a 200-md core.

- Increasing crosslinked gel concentrations from 30- to 40-lb/Mgal has a major impact on reducing leakoff in 200-md core. Crosslinked borate maintains excellent fluid-loss control in 200-md sandstone and performs satisfactorily even at 1000 md.

This experimental work strongly corroborates the modeling results of Aggour and Economides (1996) and suggests the use of higher-concentration crosslinked polymer

fluids with, of course, an appropriately designed breaker system.

19-5.3.5 Viscoelastic Carrier Fluids

HEC and borate-crosslinked HPG fluids are the dominant fluids currently used in HPF; however, a third class of fluid deserves to be mentioned, so-called viscoelastic surfactant (VES) fluids. There is little debate that these fluids exhibit excellent rheological properties and are nondamaging, even in high-permeability formations. The advantage of VES fluids is that they do not require the use of chemical breaker additives; the viscosity of this fluid conveniently breaks (leaving considerably less residue than polymer-based fluids) either when it contacts formation oil or condensate, or when its salt concentration is reduced. Brown *et al.* (1996) present typical VES fluid performance data and case histories.

The vulnerability of VES fluids is in their temperature limitations. The maximum application temperature for VES fluids has only recently been extended from 130°F to 200°F.

19-5.4 Quality Control

Many early HPF treatments failed because of equipment problems and a lack of quality control on fluids and proppants. In general, the intense quality control that is standard for onshore massive hydraulic fracture treatments was not immediately adopted on small offshore fracpack treatments. This invited skepticism of the merits of the process and somewhat slowed the introduction of HPF technology. In addition to quality-control procedures that have been instituted by all major service companies, it is now common for producing companies to supply a consultant or in-house specialist to oversee the quality control on most HPF treatments.

A number of control checks should be performed before each HPF treatment to verify the performance of all fluids and proppants. The treatment itself should also be closely monitored so that (1) to the extent possible, real-time modifications can be made that will improve the outcome of the treatment and (2) unavoidable deficiencies in the treatment execution can be appropriately evaluated *post-mortem*. The reader is referred to the *Stimulation Engineering Handbook* (Ely, 1994) for a detailed explanation of fracturing quality-control procedures.

19-6 EVALUATION OF HPF TREATMENTS: UNIFIED APPROACH

19-6.1 Production Results

The evaluation of HPF treatments can be viewed on several different levels. Economic justification (production results) is the first level on which HPF technology was (and continues to be) evaluated. Simply put, HPF has gained widespread acceptance because it allows operators to produce more oil at less cost.

McLarty and DeBonis (1995) report that fracpack treatments typically result in production increases 200 to 250% times that of comparable gravel packs, and offer the example cases in Table 19-5.

Similar reports of production increase are scattered throughout the body of HPF literature. Stewart *et al.* (1995) present a relatively comprehensive economic justification for HPF that considers (in addition to productivity improvements) the incremental cost of HPF treatments and the associated payouts, operating expenses, relative decline rates, and reserve recovery acceleration issues.

19-6.2 Evaluation of Real-Time Treatment Data

There is increasing recognition of the value of real-time HPF treatment data. Complete treatment records and digital treatment datasets are now routinely collected and evaluated as part of post-treatment analysis. In fact, a fracpack cooperative has been established at Texas A&M University to facilitate databanking of these real-time engineering datasets.

Treatment reconstruction and post-mortem diagnosis hold tremendous potential to improve HPF design and execution, but the usefulness of many ongoing efforts in this regard is limited. With the proliferation of user-friendly, "black box software," many engineers embrace and increasingly confuse technology with computer software and simulations. A popular approach to evaluation of real-time datasets (pretreatment and main treatment) is net-pressure history-matching, although this approach is not advocated.

The incorporation of multiple leakoff, stress, friction, and other variables in a 3D simulator, while it may (and invariably does) lead to an excellent "match," unfortunately sacrifices the uniqueness (usefulness) of the evaluation by introducing multiple degrees of freedom. These activities may provide operators with qualitative direction on a case-by-case basis, but they also conceal the real issues and retard fundamental development of the technology.

In contrast to this approach, consider the step-wise approach for the evaluation of bottomhole treating pressures outlined by Valkó *et al.* (1996):

- A leakoff coefficient is determined from an evaluation of minifrac data using a minimum number of assumptions, minimum input data, and minimum user interaction. Radial fracture geometry and a combined Nolte-Shlyapobersky method are suggested.

- When the obtained leakoff coefficient is used, an almost automatic procedure is suggested to estimate the created fracture dimensions and the areal proppant concentration from the bottomhole-pressure curve monitored during the execution of the HPF treatment. This procedure (termed "slopes analysis") is further developed and expanded in Section 19-7 as a fundamental and potentially important building block for the evaluation of real-time HPF data.

- The obtained fracture dimensions and areal proppant concentration can be converted into an equivalent fracture extent and conductivity. The actual performance of the well is analyzed on the basis of well-test procedures, and these results are compared to the results of the slopes analysis.

Conducting the procedure above for a large number of treatments originating from various operators will result in a data bank that ultimately improves the predictability and outcome of HPF treatments.

At present, there seems to be a trend in the industry to support joint efforts and assist mutual exchange of information. The procedure above provides a coherent (though not exclusive) framework to compare HPF data from various sources through the use of a common, cost-effective evaluation methodology.

Table 19-5 Example Production Results

Job Type	Before	After
New Well Comparison	460 bopd	1,216 bopd
Recompletion (oil)	1,300 bopd	2,200 bopd
Recompletion (gas)	3.8 MMcf/d	13.2 MMcf/d
Sand Failure	200 bopd	800 bopd

After McLarty and DeBonis (1995)

19-6.3 Post-Treatment Pressure-Transient Analysis

For post-treatment evaluation, temperature logs and various fracture-mapping techniques, such as triaxial borehole seismics and radioactive tracer mapping, have gained increasing importance. However, from the basis of future production, by far the most important evaluation is pressure transient analysis. While avoiding an exhaustive treatment of the subject, it is appropriate at this juncture to discuss several issues related to pressure-transient analysis in HPF wells, especially positive skin factors, which pose the largest challenge to treatment evaluation.

The performance of a vertically fractured well under pseudosteady-state flow conditions was investigated by McGuire and Sikora (1960) through the use of a physical analog (electric current). A similar study for gas wells was conducted by van Poolen *et al.* (1958). For the "unsteady-state" case, a whole series of works was initiated by Gringarten and Ramey (1974), and continued by Cinco-Ley *et al.* (1978) They clarified concepts of the infinite-conductivity fracture, uniform-flux fracture, and finite-conductivity fracture. From the formation perspective, double-porosity reservoirs, multi-layered reservoirs, and several different boundary geometries have been considered. The typical flow regimes (fracture linear, bilinear, pseudoradial) have been well documented in the literature. Deviations from ideality (non-Darcy effects) have also been considered.

Post-treatment pressure transient analysis for HPF wells starts with a log-log diagnostic plot that includes the pressure derivative. Once the different flow regimes are identified, specialized plots can be used to obtain the characteristics of the created fracture. In principle, fracture length and/or conductivity can be determined using the prior knowledge of permeability. For HPF, however, the relatively large arsenal of pressure-transient diagnostics and analysis for fractured wells has proven somewhat ineffective. Often, it is difficult to reveal the marked characteristics of an existing fracture on the diagnostic plot. In fact, the well often behaves similar to a slightly damaged, unstimulated well. An HPF treatment is often considered successful if a large positive skin of order +10 or more is decreased to the range of +1 to +4. These (still) positive skin factors create the largest challenge of treatment evaluation.

The obvious discrepancy between theory and practice has been attributed to several factors, some of which are well documented and understood and some others of which are still in the form of hypotheses:

- **Factors causing decrease of apparent permeability in the fracture** The most familiar factor that decreases the apparent permeability of the proppant pack, and therefore fracture conductivity, is *proppant-pack damage*. The reduction of permeability because of the presence of residue from the gelled fluid and failure of proppant because of closure stress are well understood. Since those phenomena exist in any fracture, they cannot be the general cause of the discrepancy in high-permeability fracturing. *Non-Darcy flow in the fracture* is also reasonably well understood. Separation of rate-independent skin from the variable-rate component by multiple-rate well testing is a standard practice. The effect of *phase change in the fracture* is less straightforward to quantify.

- **Factors decreasing the apparent width** Embedment of the proppant in a soft formation is now well documented in the literature (e.g. Lacy *et al.*, 1996).

- **Fracture-face skin effect** The two sources of this phenomenon are filter-cake residue and the polymer-invaded zone. Sometimes the long-term cleanup (decrease of the skin effect) of a stimulated well is considered as indirect proof of such damage. It is assumed that linear polymer fluids invade more deeply into the formation and therefore, cause more fracture-face damage, as discussed by Mathur *et al.* (1995).

- **Permeability anisotropy** While the anisotropy of permeability has only a limited effect on pseudoradial flow, the early-time transient flow regime of a stimulated well is very sensitive to anisotropy. This fact is often neglected when the well is characterized with one single skin effect.

- **Concept of skin** It has to be emphasized that the concept of negative skin as the only measure of the "quality" of a well might be a source of the discrepancy itself.

19-6.3.1 Validity of the Skin Concept in HPF

There is, in fact, no clear theoretical base for obtaining negative skin from short-time well-test data distorted by wellbore storage if the well has been stimulated. The use of infinite-acting reservoir + wellbore storage + skin type-curves in this case is not based on sound physical principles and might cause unrealistic conclusions.

In addition, the validity of the pseudoskin concept during the transient production period is an important issue. In general, the pseudoskin concept is valid only at late times. Thus, a fracture designed for optimal late-time performance may be not optimal at shorter times. One may ask how much performance is lost in selecting fracture dimensions that are optimal for a late time. This question has not been investigated, but it is reasonable to assume that the loss in performance is negligible for high-permeability reservoirs where the dimensionless times corresponding to a month or year are much higher than for low-permeability reservoirs.

19-6.3.2 Effect of Non-Darcy Flow in the Fracture

Non-Darcy flow is another important issue that deserves specific consideration in the context of HPF. Non-Darcy flow in gas reservoirs causes a reduction of the productivity index by at least two mechanisms. First, the apparent permeability of the formation may be reduced (Wattenbarger and Ramey, 1969) and second, the non-Darcy flow may decrease the conductivity of the fracture (Guppy *et al.*, 1982).

Consider a closed gas reservoir producing under pseudosteady-state conditions, and apply the concept of pseudoskin effect determined by dimensionless fracture conductivity.

Definitions and Assumptions

Gas production is calculated from the pseudosteady-state deliverability equation:

$$q = \frac{\pi k h T_{sc}[m(\bar{p}) - m(p_{wf})]}{p_{sc} T}$$
$$\times \frac{k_{r,\text{app}}}{k_r\left[f_1(C_{fD,\text{app}}) + \ln\left(\frac{0.472 r_e}{x_f}\right)\right]} \quad (19\text{-}28)$$

where $m(p)$ is the pseudopressure function, $k_{f,\text{app}}$ is the apparent permeability of the proppant in the fracture, and $k_{r,\text{app}}$ is the apparent permeability of the formation. (All the equations in this subsection are given for a consistent system of units, such as SI.) The function f_1 was introduced by Cinco-Ley and Samaniego (1981); it was presented in Chapter 17 as

$$f_1(C_{fD}) = s_f + \ln\frac{x_f}{r_w} = \frac{1.65 - 0.328u + 0.116u^2}{1 + 0.18\ln u + 0.064u^2 + 0.005u^3}$$
$$(17\text{-}8)$$

where $u = \ln C_{fD}$.

The apparent dimensionless fracture conductivity is defined by

$$C_{fD,\text{app}} = \frac{k_{f,\text{app}} w}{k_{r,\text{app}} x_f} \quad (19\text{-}29)$$

The apparent permeabilities are flow-rate dependent; therefore, the deliverability equation becomes implicit in the production rate.

Proceeding further requires a model of non-Darcy flow. Almost exclusively, the Forcheimer equation is used in the industry:

$$-\frac{dp}{dx} = \frac{u}{k}v + \beta\rho|v|v \quad (19\text{-}30)$$

where $v = q_a/A$ is the Darcy velocity and β is a property of the porous medium.

A popular correlation was presented by Firoozabadi and Katz (1979) as

$$\beta = \frac{c}{k^{1.2}} \quad (19\text{-}31)$$

where $c = 8.4 \times 10^{-8}\ \text{m}^{1.4}\ (= 2.6 \times 10^{10}\ \text{ft}^{-1}\ \text{md}^{1.2})$.

To apply the Firoozabadi and Katz correlation, we write

$$-\frac{dp}{dx} = \mu v\frac{1}{k}\left(1 + \frac{\beta k\rho|v|}{\mu}\right) = \mu v\frac{1}{k}\left(1 + \frac{c\rho|v|}{k^{0.2}\mu}\right) \quad (19\text{-}32)$$

showing that

$$\frac{k_{\text{app}}}{k} = \frac{1}{1 + \dfrac{c\rho|v|}{k^{0.2}\mu}} \quad (19\text{-}33)$$

The equation above can be used both for the reservoir and for the fracture if correct representative linear velocity is substituted. In the following, it is assumed that $h = h_f$.

A representative linear velocity for the reservoir can be given in terms of the gas-production rate as

$$v = \frac{q_a}{4hx_f} \quad (19\text{-}34)$$

where q_a is the in-situ (actual) volumetric flow rate; therefore, for the reservoir non-Darcy effect

$$\left(\frac{c\rho v}{k^{0.2}\mu}\right)_r = \left(\frac{c\rho q_a}{2h\mu}\right)\frac{1}{2x_f k_r^{0.2}} \quad (19\text{-}35)$$

A representative linear velocity in the fracture can be given in terms of the gas-production rate as

$$v = \frac{q_a}{2hw} \tag{19-36}$$

Thus, for the non-Darcy effect in the fracture, one can use

$$\left(\frac{c\rho v}{k^{0.2}\mu}\right)_f = \left(\frac{c\rho q_a}{2h\mu}\right)\frac{1}{wk_f^{0.2}} \tag{19-37}$$

The term ρq_a is the mass flow rate, and it is the same in the reservoir and in the fracture; $c\rho q_a$ is expressed in terms of the gas-production rate as

$$\frac{c\rho q_a}{2h\mu} = \frac{c\rho_a\gamma_g}{2h\mu}q = c_0 q \tag{19-38}$$

where q is the gas-production rate in standard volume per time, γ_g is the specific gravity of gas with respect to air, and ρ_a is the density of air at standard conditions. The factor c_0 is constant for a given reservoir-fracture system.

The final form of the apparent permeability dependence on production rate is

$$\left(\frac{k_{\text{app}}}{k}\right)_r = \frac{1}{1 + \dfrac{c_0 q}{2x_f k_r^{0.2}}} \tag{19-39}$$

for the reservoir and

$$\left(\frac{k_{\text{app}}}{k}\right)_f = \frac{1}{1 + \dfrac{c_0 q}{wk_f^{0.2}}} \tag{19-40}$$

for the fracture. As a consequence, the deliverability equation becomes

$$q = \frac{\pi k h T_{sc}\left[m(\bar{p}) - m(p_{wf})\right]}{p_{sc}T}$$
$$\times \frac{1}{\left(1 + \dfrac{c_0 q}{2x_f k_r^{0.2}}\right)\left[f_1\left(C_{fD,\text{app}}\right) + \ln\left(\dfrac{0.472r_e}{x_f}\right)\right]} \tag{19-41}$$

where

$$c_{fD,\text{app}} = \frac{k_f w}{k_r x_f} \frac{1 + \dfrac{c_0}{wk_f^{0.2}}q}{1 + \dfrac{c_0}{2x_f k_r^{0.2}}q} \tag{19-42}$$

The *additional* skin effect, s_{ND}, appearing because of non-Darcy flow, can be expressed as

$$s_{ND} = \left(1 + \frac{c_0 q}{2x_f k_r^{0.2}}\right)\left[f_1\left(C_{fD,\text{app}}\right) + \ln\left(\frac{0.472r_e}{x_f}\right)\right]$$
$$- \left[f_1\left(C_{fD}\right) + \ln\left(\frac{0.472r_e}{x_f}\right)\right] \tag{19-43}$$

The additional non-Darcy skin effect is always positive and depends on the production rate in a nonlinear manner.

Equations 19-29 and 19-31 are of primary importance for interpreting post-fracture well-testing data and to forecast production. If the mechanism responsible for the post-treatment skin effect is not understood clearly, the evaluation of the treatment and the production forecast might be severely erroneous.

19-6.3.3 Case Study in Effect of Non-Darcy Flow

As discussed above, non-Darcy flow in a gas reservoir reduces the productivity index by at least two mechanisms. First, the apparent permeability of the formation may be reduced, and second, the non-Darcy flow may decrease the fracture conductivity. In this case study, the effect of non-Darcy flow on production rates and observed skin effects is investigated. Reservoir and fracture properties are given in Table 19-6.

A simplified form of Equation 19-41 in field units is

$$q = \frac{\bar{p}^2 - p_{wf}^2}{\dfrac{1424\mu ZT}{k_r h}}$$
$$\times \frac{1}{\left(1 + \dfrac{c_0 q}{2x_f k_r^{0.2}}\right)\left[f_1\left(F_{CD,\text{app}}\right) + \ln\left(\dfrac{0.472r_e}{x_f}\right)\right]} \tag{19-44}$$

where $c_0 = \dfrac{c\rho_a\gamma_g}{2h\mu}$ has to be expressed in ft-md$^{0.2}$ MMscf/d. In the given example, $c_0 = 73$ ft-md$^{0.2}$ MMscf/d and

Table 19-6. Data for fractured well in gas reservoir

r_e	ft	1 500
μ	cp	0.02
Z	N/A	0.95
T	^{0}R	640
k_r	md	10
h	ft	80
h_f	ft	80
k_f	md	10,000
x_f	ft	30
w	inch	0.5
γ_g	N/A	0.65
$\bar{p}$	psi	4 000
r_w	ft	0.328

$$C_{fD,\text{app}} = \left(\frac{k_f w}{k_r x_f}\right) \frac{1 + \dfrac{c_0}{2 x_f k_r^{0.2}} q}{1 + \dfrac{c_0}{w k_f^{0.2}} q} = \left(\frac{k_f w}{k_r x_f}\right) \frac{1 + c_{0r} q}{1 + c_{0f} q}$$

(19-45)

where

$$c_{0r} = 2.34 \times 10^{-3} \, \text{m}^3/\text{s} = 7.67 \times 10^{-2} \, (\text{Mscf/d})^{-1}$$

$$c_{0f} = 6.14 \times 10^{-1} \, \text{m}^3/\text{s} = 2.78 \times 10^2 \, (\text{Mscf/d})^{-1}$$

Therefore, in field units

$$C_{fD,\text{app}} = 1.39 \frac{1 + 0.76q}{1 + 280q} \qquad (19\text{-}46)$$

and

$$q = \frac{4000^2 - p_{wf}^2}{21.645} \times \frac{1}{(1 + 0.76q)[f_1(F_{CD,\text{app}}) + 3.16]}$$

(19-47)

The non-Darcy component of the skin effect can be calculated as

$$s_{ND} = (1 + 0.00076q)[f_1(F_{CD,\text{app}}) + 3.16] - 4.619$$

(19-48)

The results are shown graphically in Figs. 19-10 through 19-12. It is apparent that the effect of the fracture (negative skin on the order of -3) is hidden by the positive skin effect induced by non-Darcy flow. The zero or posi-

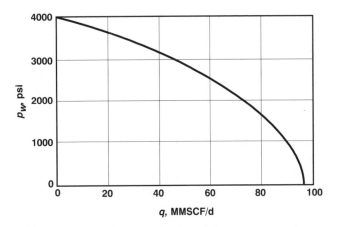

Figure 19-10 Inflow performance of the fractured gas reservoir, non-Darcy effect from the Firoozabadi - Katz correlation

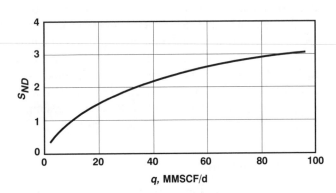

Figure 19-11 Additional skin effect from non-Darcy flow in the fracture

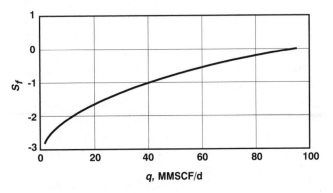

Figure 19-12 Observable pseudoskin, the resulting effect of the fracture with non-Darcy flow effect

tive observable skin effect, while directly attributable to the inevitable effect of non-Darcy flow, might be interpreted as an unsuccessful HPF job.

19-7 SLOPES ANALYSIS

Complete tip-screenout is expected to produce a distinct behavior in the treating pressure; that is, the treating pressure should markedly increase with time. However, HPF treatments often exhibit numerous increasing-pressure intervals that are interrupted by anomalous pressure decreases, most probably because fracture extension can still occasionally occur (in many cases, a single complete tip-screenout is not achieved).

This work (Valkó *et al.*, 1996) provides a simple tool for examining such behavior. Treating-pressure curves are analyzed to gain insight to the evolution of fracture extent and a plausible end-of-job proppant distribution.

In developing the tool, several design parameters were intentionally imposed: the method should require minimum user input beyond the real treatment data, it should be relatively independent of the fracture propagation model used, and it should not be a history-matching procedure. In accordance with the basic requirement of model independence, the slopes analysis method is a screening tool based on simple equations and a well-defined (reconstructible) algorithm. Based on its simplicity, the tool lends itself to real-time use as well.

19-7.1 Assumptions

During tip-screenout, the fracture width is inflated while the area of the fracture faces remains theoretically constant. This phenomenon should appear as a marked increase in the treating pressure. In practice, the increasing pressure intervals may be interrupted by an anomalous pressure decrease because fracture extension can still occur occasionally. Based on this rationale, the HPF treatment is considered a series of (regular) arrested extension/width growth intervals interrupted by (irregular) fracture-area extension intervals.

In this case, the treatment can be decomposed into sequential periods of constant fracture area separated by periods (possibly several) of fracture extension. The time periods are located by a simple processing of the treatment-pressure curve.

If this vision of the treatment is accepted, then the slope of the increasing-pressure curve during a width-inflation period may be interpreted to obtain the "pack-

ing radius" of the fracture at that point during the treatment, which is characteristic for the given period. Putting together a sequence of packing-radii estimates gives a scenario which—combined with additional information on the proppant injection history—yields the final proppant distribution.

In transforming the idea to a working algorithm, several assumptions must be made, both regarding fracture geometry and the character of the leakoff process. The following assumptions are made:

- The created fracture is vertical with a radial geometry.
- Fluid leakoff can be described by the Carter leakoff model (Howard and Fast, 1957) in conjunction with the power-law type area growth used by Nolte (1979), or by one of the detailed leakoff models discussed in Section 19-3.3.
- Fracture-packing radius may increase or decrease with time.
- Hydraulic-fracture radius (which defines leakoff area) cannot decrease; it is the maximum of the packing radii that have occurred up to the given time.
- During regular width-inflation periods, the pressure slope is defined by linear, elastic rock behavior and fluid-material balance with friction effects being negligible.
- Injected proppant is distributed evenly along the actual packing area during each incremental period of arrested extension/width growth.

The suggested method consists of several steps. First, those portions of the bottomhole pressure curve are selected that show positive slope. The slope is then interpreted assuming that the pressure increase is caused by width inflation. The interpretation results in a packing radius that corresponds to a given time point. A step-by-step processing of the entire curve gives a history of the packing radius, though it still does not provide information regarding those intervals when the slope is negative. The history is made complete by interpolating between the known values.

Based on this history of packing-radius evolution, the final proppant distribution is easily determined by superimposing real-time proppant injection data. Final proppant distribution (which implies fracture length and width) is the practical result of the proposed slopes analysis.

19-7.2 Restricted-Growth Theory

Tip-screenout can be considered to be inflating the fracture width while the area of the fracture face does not increase. If the average width is denoted by w and the fracture-face area (one wing, one face) is denoted by A, then

$$\frac{dw}{dt} = \frac{1}{A}(i - q_L) \qquad (19\text{-}49)$$

where i is the injection rate (per one wing) and q_L is the fluid-loss rate (from one wing).

The basic notation is shown in Figure 19-13. Assuming that the fracture is radial with radius R, then

$$A = \frac{\pi R^2}{2} \qquad (19\text{-}50)$$

As a first approximation, assume that the pressure in the inflating fracture does not depend on location (it is homogeneous). The net pressure (the excess pressure above the minimum principal stress) is directly proportional to the average width:

$$p_n = \frac{3\pi E'}{16R} w \qquad (19\text{-}51)$$

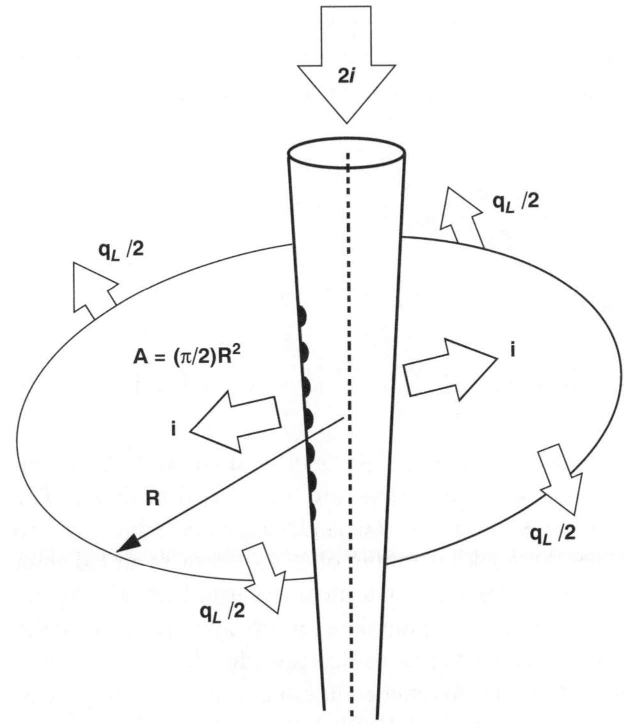

Figure 19-13   Schematic of fracpack, radial-fracture geometry

where E' is the plane-strain modulus (Chapter 17).

Substituting Equations 19-50 and 19-51 into Equation 19-49, the time derivative of net pressure is obtained as

$$\frac{dp}{dt} = \left(\frac{3\pi E'}{16R}\right)\left(\frac{2}{\pi R^2}\right)(i - q_L) \qquad (19\text{-}52)$$

where the subscript for net pressure is dropped because the derivative of bottomhole pressure and that of net pressure are equal.

Recording the bottomhole pressure and injection rate provides the possibility of using Equation 19-52 to determine R. For this purpose, an estimate of q_L is needed.

Details of the Carter leakoff model are given in Chapter 17. Assuming that the fracture has extended up to the given time t according to Nolte's power-law assumption, and that it is arrested at the given time instant t, the leakoff rate $q_{L,t}$ immediately after the arrest is given by

$$q_{L,t} = 2AC_L \frac{1}{\sqrt{t}} \left(\frac{\partial g(\Delta t_D, \alpha)}{\partial \Delta t_D}\right)_{\Delta t_D = 0} \qquad (19\text{-}53)$$

where A is the current fracture area and α is the power-law exponent of the areal growth. The two-variable g-function was discussed in Chapter 17.

For a radial fracture created by injection of a Newtonian fluid, the exponent is taken as $\alpha = \frac{8}{9}$, and the derivative of the g-function is

$$\left[\frac{\partial g(\Delta t_D, \frac{8}{9})}{\partial d\Delta t_D}\right]_{\Delta t_D = 0} = 1.91 \qquad (19\text{-}54)$$

Therefore, the estimate of leakoff rate is obtained as

$$q_{L,t} = 2AC_L \frac{1}{\sqrt{t}} 1.91 \qquad (19\text{-}55)$$

Equations 19-52 and 19-55 were developed explicitly in the text, and they form the core basis for the slopes analysis method. The use of these relations is demonstrated in the following section.

19-7.3 Slopes Analysis Algorithms

The restricted-growth theory is combined with simple material-balance computations to form the slopes analysis method as demonstrated below using a sample set of HPF data provided by Shell E&P Technology Co.

19-7.3.1 Selecting Intervals of Width Inflation

Figure 19-14 is the bottomhole pressure recorded during a HPF treatment. While it may look "not typical," most of the data sets available (without the natural self-censoring of publishing authors) are "not typical" in one or more respects. The recommended approach of avoiding premature assumptions about the form of the pressure curve is based exactly on this fact. The slopes analysis approach can be better described as a signal processing operation than one of fitting a given model to the data.

The suggested method consists of selecting those portions of the bottomhole pressure curve that show positive slope. Straight lines are fitted to the points corresponding to each such interval.

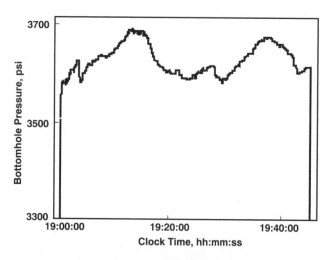

Figure 19-14   Bottomhole treating pressure from fracpack treatment

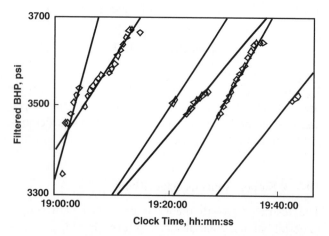

Figure 19-15   Bottomhole pressure points corresponding to width inflation intervals and corresponding "straight lines"

Using a simple algorithm, one can select points satisfying the criterion of restricted fracture growth. Straight lines are fitted to the individual series to arrive at the plot shown in Figure 19-15.

The slope of the straight line gives an average pressure derivative corresponding to the given time interval of restricted growth. In view of the stated assumptions, these slopes contain information that defines the actual packing radius corresponding to discrete moments during the HPF treatment.

*19-7.3.2 Determining the Packing Radius
        Corresponding to a Width-Inflation Period*

Substituting the obtained expression for the leakoff rate, Equation 19-52 can be rewritten as

$$m = \left(\frac{3\pi E'}{16R}\right)\left(\frac{2}{\pi R^2}\right)\left[i - 2\left(\frac{\pi R^2}{2}\right)C_L\frac{1}{\sqrt{t}}1.91\right] \quad (19\text{-}56)$$

Rearranging Equation 19-56, we obtain

$$R^3 + R^2\left(\frac{2.25E'C_L}{m\sqrt{t}}\right) - \left(\frac{0.375E'i}{m}\right) = 0 \quad (19\text{-}57)$$

Once a restricted-growth interval is selected, knowing the slope, m, and the injection rate, i at a given time, t, Equation 19-57 can be solved for R. Since the equation is cubic, an explicit solution can be given, which (in consistent units) is given by

$$R_p = \frac{0.7501}{m\sqrt{t}}\left[a - C_L \times E' + \frac{(C_L \times E')^2}{a}\right] \quad (19\text{-}58)$$

where

$$a = \left[0.4443 \times E'i \times m^2 \times t^{3/2} - (C_L \times E')^3\right]^{1/3} \quad (19\text{-}59)$$

Equations 19-58 and 19-59 can be used with the actual one-wing slurry injection rate, i, recorded at time t. The obtained solution is the *packing radius*. Figure 19-16 shows the packing radius obtained from recorded data of the example HPF treatment. As seen from the figure, after a certain pumping time (approximately 25 min), the packing radius begins to decrease. In other words, near the end of the treatment, only the near-wellbore part of the fracture was "packed." This condition is consistent with the treatment objectives, and it was achieved by

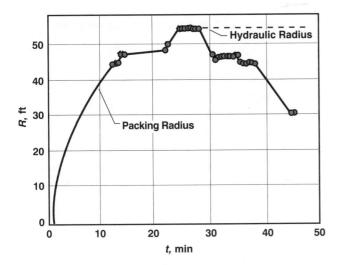

Figure 19-16 Estimated packing radius with interpolation to fill in the "gaps"

gradually decreasing the injection rate at the final stages of the treatment.

19-7.3.3 Interpolation Between Known Values of the Packing Radius

Since the packing radius is obtained only in those selected intervals where width inflation can be assumed, a simple tool is needed to fill in the "gaps." A simple logarithmic interpolation is used to estimate the packing radius in between the known values.

In addition, one can estimate the "hydraulic" fracture radius at time t as the maximum of the packing radii up to that point (dashed line in Figure 19-16). While proppant is placed within the actual packing radius, leakoff occurs along the area determined by the hydraulic-fracture extent. Knowledge of the hydraulic-fracture extent is useful for further material balance considerations.

19-7.3.4 Determining the Final Areal Proppant Concentration

Final proppant concentration (proppant distribution) in the fracture can be derived in a relatively straightforward fashion from the packing radius curve and knowledge of the bottomhole proppant concentration as a function of time. (The standard job record typically includes this information.)

Calculation of the final areal proppant concentration in the fracture follows the simple scheme:

1. For every time interval, Δt, determine the mass of proppant entering the fracture.
2. Assume this mass to be uniformly distributed inside the packing radius corresponding to the given time step.
3. Obtain the mass of proppant in a "ring" between radius R_1 and R_2 by summing up (accumulating) the mass of proppant placed during the whole treatment.
4. Repeat Step 3 for all rings to obtain the areal proppant concentration as a function of radial location R.

Application of the scheme above to the example data results in the areal proppant concentration as a function of the radial distance from the center of the perforations, R. The areal proppant concentration distribution for the example dataset is shown in Figure 19-17.

The proposed method for evaluating pressure behavior of HPF treatments is not based on specific fracture mechanics and/or proppant transport models. Rather, it takes the pressure curve "as is" and processes it using minimum additional data. The usual data records of a job (slurry injection rate, bottomhole proppant concentration, and bottomhole pressure) can be used for estimating fracture extent and the distribution of proppant in the fracture. The only other additional input parameters necessary for the analysis are plane-strain modulus and leakoff coefficient.

Success of the procedure depends on the validity of the key assumption that positive slopes observed in the bottomhole pressure curve are caused by restricted fracture extension/width growth. If there is no time interval satisfying the criterion of restricted extension or if no other phenomena involved mask the effect, such as (1) pressure

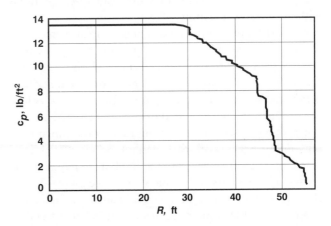

Figure 19-17 Final areal proppant concentration as a function of radial distance from the center of the perforations

transients caused by sharp changes of the injection rate or (2) dramatic changes in friction pressure resulting from proppant concentration changes, the estimated packing radius might be in considerable error. Nevertheless, the suggested procedure is considered a substantive first step in the analysis of HPF treatment pressure data.

19-8 EMERGING HPF TECHNOLOGIES

19-8.1 Screenless and Rigless HPF Completions

On the basis of a recent industry survey, Tiner *et al.* (1996) report that the most common HPF technology advance being sought by producing companies is one that will allow removal or simplification of gravel-pack screens and tools, which are still used in most HPF completions. The most likely alternative is to eliminate the screen completely and use conventional fracturing methods, with a "twist": the final proppant stage should be tailed-in with resin-coated sand to control proppant flowback. A number of these screenless HPF treatments have been completed, apparently with considerable success (Kirby *et al.*, 1995).

Screenless HPFs have the potential of dramatically reducing treatment costs and simplifying treatment execution; however, some questions remain: Can the resin-coated proppant in fact be placed as needed to prevent proppant flowback and ensure a high-conductivity connection between the fracture and the wellbore? What about formation sand production from those perforations that are not connected to the fracture? If successful, screenless HPFs would also allow the development of multiple-zone HPF completions and through-tubing HPF recompletions. The major benefit of through-tubing completions, of course, is that they can often be done without a rig on location.

New HPF operations and equipment are also emerging to allow rigless coiled tubing completions in wells that are completed with gravel-pack screens (Ebinger, 1996). Depending on the particular configuration, the treatment is pumped through a fracturing port/sleeve located below the production packer and above the screen. The port is opened and closed with a shifting tool on the coiled tubing. Because a gravel pack cannot be circulated into place, prepacked screens are required. This requirement seems to be the largest drawback to the technique. While the rigless HPFs may be uniquely suited to dual-zone completions, the primary influence behind this trend is cost reduction by eliminating rig costs and inefficiencies associated with rig timing.

19-8.2 Complex Well-Fracture Configurations

Vertical wells are not the only candidates for hydraulic fracturing. Figure 19-18 shows some basic configurations for single-fractured wells. Horizontal wells using HPF with the well drilled in the expected fracture azimuth (thereby ensuring a longitudinal fracture) appear to be (at least conceptually) a very promising prospect as discussed in Section 19-2.3. However, a horizontal well intended for a longitudinal fracture configuration would have to be drilled along the maximum horizontal stress. This requirement, in addition to well-understood drilling problems, may contribute to long-term stability problems.

Figure 19-19 illustrates two multiple-fracture configurations. A rather sophisticated conceptual configuration would involve the combination of HPF with multiple-fractured vertical branches emanating from a horizontal parent well drilled above the producing formation. Of course, horizontal wells, being normal to the vertical stress, are generally more prone to wellbore stability problems. Such a configuration would allow for placement of the horizontal borehole in a competent, nonproducing interval. Besides, there are advantages to fracture-treating a vertical section over a highly deviated or horizontal section: (1) multiple starter fractures, fracture turning, and tortuosity problems are avoided, (2) convergence-flow skins ("choke" effects) are much less of a concern, and (3) the perforating strategy is simplified.

19-8.3 Technology Demands: Where Do We Go From Here?

19-8.3.1 Candidate Selection

Wellbore stability is viewed in a holistic approach with horizontal wells and hydraulic fracture treatments. Proactive well completion strategies are critical in well-

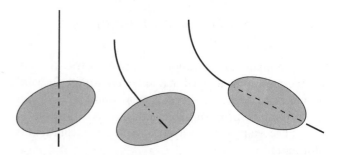

Figure 19-18 Single-fracture configurations for vertical and horizontal wells

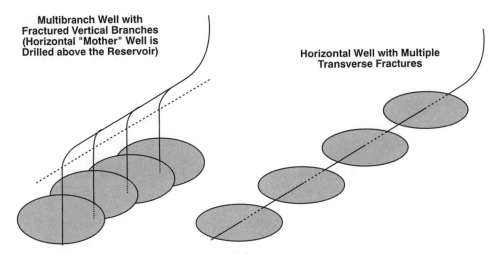

Figure 19-19   Multibranched, multiple-fracture configurations for horizontal wells

bore stability and sand-production control to reduce drawdown while obtaining economically attractive rates. Reservoir candidate recognition for the correct well configurations is the critical element. Necessary steps in candidate selection include (1) appropriate reservoir engineering, (2) formation characterization using modern techniques, (3) wellbore-stability calculations, and (4) the substantive combination of production forecast with an assessment of sand-production potential. The mixed origin of the HPF (fracpack) community (gravel-packing and fracturing) still exists. For gravel-packers, it is still difficult to consider correctly the whole reservoir and not merely the near-wellbore region. On the other hand, for practitioners of fracturing, it is still difficult to understand the mechanisms involved in sanding and its control.

19-8.3.2 Completion Hardware

Completion hardware should be improved. As discussed above, there is a need to simplify, eliminate, or otherwise advance beyond the modified gravel-pack hardware currently being used in HPF. There is also a need for improved zonal isolation hardware, (for the execution of hydraulic fracturing in complex well-fracture configurations). In fact, lack of appropriate drilling, completion, and stimulation hardware is often the limiting factor in the indicated new completion configurations. Clearly, all fracture treatments must be conducted separately, in stages. Thus, inexpensive and robust zonal isolation schemes are necessary. Certain zonal-isolation hardware is available, but it is expensive and often logis-

tically difficult to use. Other currently available techniques such as polymer or sand plugs, are prone to failure.

19-8.3.3 The Fracture-Well Connection

New fracture-and-well interfaces should be developed, which could include the next generation of screens and replacement technologies. Hydraulic fractures are prone to sand production, both from the reservoir and the proppant-pack itself. This situation is particularly important in high-rate wells where, although the reservoir problem may be resolved, the near-well fracture portion may be susceptible to sand production. The current solution (gravel-pack screens) should be abandoned. Although they are reasonably effective, these screens can cause a serious choke effect at the fracture-well interface. New consolidation techniques or perhaps, oriented long perforations and alternative "sieves" can be envisioned.

19-8.3.4 Next-Generation Completion Fluids

The next generation of drilling and well completion fluids should also be developed. The envisioned "smart" fluids would consider application and formation-specific issues affecting wellbore stability and damage. New gels, polymers, and leakoff-control additives are urgently needed for the drilling of complicated wellpaths through difficult formations. Nondamaging fluids will be critical to the future of production engineering. Stimulation is often expensive, cumbersome, and at times, unsuccessful. Production-induced problems such as paraffin and asphaltene deposition, and especially sand production, can be avoided if the well is undamaged, and no large drawdown is necessary for appropriate production.

19-8.3.5 Treatment-Pressure Analysis

Treatment-pressure analysis is based on understanding the leakoff process and on the concept of net pressure. Soft formations and their rock mechanics are still not well understood, and an understanding of which mechanisms control fracture length and width is more evasive than ever. Many of the debates could be settled by determining the closure pressure (and hence the net pressure) with more confidence. While there is a lot of activity in this area, new results are very limited. Reiterating old ideas is common—(flowback or not, two minifracs or three, crosslinked fluid or not, step-rate or not, where to draw the straight line, etc.)—but these ideas are revealing little new information. There is definitely a need for innovative thinking here. For example, an independent and possibly direct instrumental determination of closure pressure would be a significant step ahead in the engineering of hydraulic fracturing in general, and especially for the hydraulic fracturing of soft formations.

19-8.3.6 Improved Well Test Interpretation for HPF

From the point of view of evaluating high-permeability fracture treatments, it is imperative to improve the well-test interpretation procedure and understand the phenomenon of positive apparent skins. The lack of desired (negative) post-treatment skins is still haunting the industry. Well productivity is obviously being improved by HPF, but it is not clear whether the treatments are optimum or whether they could be significantly improved through the use of more aggressive treatment parameters. Would more aggressive schedules provide additional benefit, or are the possibilities limited by choke effect at the perforations? These issues are not clearly distinguished by current pressure-transient methods.

19-8.3.7 Global Databases

There is an emerging consensus as to the importance of global databases of formation and other properties, at least for a specific geographic area and specific activity. The main difficulty is not to find funding for such data-bases, but to provide incentive and methods to encourage continuous input of data and the tailoring of data formats toward common standards. Once a database is established, simple evaluation tools (such as the slopes-analysis method presented) can be used to evaluate a large number of treatments efficiently. Estimated fracture dimensions and conductivities could then be compared with results from pressure-transient analysis and production results. Discrepancies should be resolved using a large number of data sets from various independent sources.

REFERENCES

Aggour T.M., and Economides, M.J.: "Impact of Fluid Selection on High-Permeability Fracturing," paper SPE 36902, 1996.

Ayoub, J.A., Kirksey, J.M., Malone, B.P., and Norman, W.D.: "Hydraulic Fracturing of Soft Formations in the Gulf Coast," paper SPE 23805, 1992.

Balen, R.M., Meng, H.-Z. and Economides, M.J.: "Application of the Net Present Value (NPV) in the Optimization of Hydraulic Fractures," paper SPE 18541, 1988.

Barree, R.D., Rogers, B.A., and Chu, W.C.: "Use of Frac-Pac Pressure Data to Determine Breakdown conditions and Reservoir Properties," paper SPE 36423, 1996.

Brown, J.E., King. L.R., Nelson, E.B., and Ali, S.A.: "Use of a Viscoelastic Carrier Fluid in Frac-Pack Applications," paper SPE 31114, 1996.

Chapman, B.J., Vitthal, S., and Hill, L.M.: "Prefacturing Pump-In Testing for High-Permeability Formations," paper SPE 31150, 1996.

Chudnovsky, A., Fan J., Shulkin, Y., Dudley, J.W., Shlyapobersky, J, and Schraufnagel, R.: "A New Hydraulic Fracture Tip Mechanism in a Statistically Homogeneous Medium," paper SPE 36442, 1996.

Cinco-Ley, H., and Samaniego, V.F.: "Transient Pressure Analysis: Finite Conductivity Fracture Case Versus Damage Fracture Case," paper SPE 10179, 1981.

Cinco-Ley, H., Samaniego, V.F., and Dominquez, N.: "Transient Pressure Behavior for a Well With a Finite Conductivity Vertical Fracture," *SPEJ* (Aug. 1978) 253–264.

Cinco-Ley, H., and Samaniego, V.F.: "Transient Pressure Analysis for Fractured Wells," *JPT* (Sept. 1981) 1749-1766.

DeBonis, V.M., Rudolph, D.A., and Kennedy, R.D.: "Experiences Gained in the Use of Frac-Packs in Ultra-Low BHP Wells, U.S. Gulf of Mexico," paper SPE 27379, 1994.

Dusterhoft, R., Vitthal, S., McMechan, D., and Walters, H.: "Improved Minifrac Analysis Technique in High-Permeability Formations," paper SPE 30103, 1995.

Ebinger, C.D.: "New Frac-Pack Procedures Reduce Completion Costs," *World Oil* (April 1996).

Ely, J.W.: *Stimulation Engineering Handbook*, Pennwell, Houston (1994).

Fan, Y., and Economides, M.J.: "Fracturing Fluid Leakoff and Net Pressure Pressure Behavior in Frac&Pack Stimulation," paper SPE 29988, 1995.

Firoozabadi, A., and Katz, D.L.: "An Analysis of High-Velocity Gas Flow Through Porous Media," *JPT* (Feb. 1979) 211–216.

Gringarten, A.C., and Ramey, A.J., Jr.: "Unsteady State Pressure Distributions Created by a Well With a Single-Infinite Conductivity Vertical Fracture," *SPEJ* (Aug. 1974) 347–360.

Grubert, D.M.: "Evolution of a Hybrid Frac-Gravel Pack Completion: Monopod Platform, Trading Bay Field, Cook Inlet, Alaska," paper SPE 19401, 1990.

Guppy, K.H, Cinco-Ley, H., Ramey Jr., H.J., and Samaniego-V., F.: "Non-Darcy Flow in Wells With Finite-Conductivity Vertical Fractures," *SPEJ* (Apr. 1982) 681–698; **Trans.,** AIME **273**.

Hannah, R.R., Park, E.I., Walsh, R.E., Porter, D.A., Black, J.W., and Waters, F.: "A Field Study of a Combination Fracturing/Gravel Packing Completion Technique on the Amberjack, Mississippi Canyon 109 Field," paper SPE 26562, 1993.

Howard G.C., and Fast, C.R.: "Optimum Fluid Characteristics for Fracture Extension," *Drill. and Prod. Prac.*, API (1957) 261–270.

Hunt, J.L., Chen, C.-C., and Soliman, M.Y.: "Performance of Hydraulic Fractures in High-Permeability Formations," paper SPE 28530, 1994.

Kirby, R.L., Clement, C.C., Asbill, S.W., Shirley, R.M., and Ely, J.W.: "Screenless Frac Pack Completitions Utilizing Resin Coated Sand in the Gulf of Mexico," paper SPE 30467, 1995.

Lacy, L.L., Rickards, A., and Bilden, D.M.: "Fracture Width and Embedment Testing in Soft Reservoir Sandstone," paper SPE 36421, 1996.

Martins, J.P., Collins, P.J., and Rylance, M.: "Small Highly Conductive Fractures Near Reservoir Fluid Contacts: Application to Prudhoe Bay" paper SPE 24856, 1992.

Mathur, A.K., Ning, X., Marcinew, R.B., Ehlig-Economides, C.A., and Economides, M.J.: "Hydraulic Fracture Stimulation of High-Permeability Formations: The Effect of Critical Fracture Parameters on Oilwell Production and Pressure," paper SPE 30652, 1995.

Mayerhofer, M.J., Economides, M.J., and Nolte, K.G.: "An Experimental and Fundamental Interpretation of Fracturing Filter-Cake Fluid Loss," paper SPE 22873, 1991.

Mayerhofer, M.J., Economides, M.J., and Ehlig-Economides, C.A.: "Pressure Transient Analysis of Fracture Calibration Tests," paper SPE 26527, 1993.

McLarty, J.M., and DeBonis, V.: "Gulf Coast Section SPE Production Operations Study Group - Technical Highlights from a Series of Frac Pack Treatment," paper SPE 30471, 1995.

McGuire W.J., and Sikora, V.J.: "The Effect of Vertical Fractures on Well Productivity," *JPT* (Oct. 1960) 72.

McGowen, J.M., Vitthal, S., Parker, M.A., Rahimi, A., and Martch Jr., W.E.: "Fluid Selection for Fracturing High-Permeability Formations," paper SPE 26559, 1993.

McGowen J.M., and Vitthal S.: "Fracturing Fluid Leakoff Under Dynamic Conditions Part 1: Development of a Realistic Laboratory Testing Procedure," paper SPE 36492, 1996.

Montagna, J N., Saucier, R.J., and Kelly, P.: "An Innovative Technique for Damage By-Pass in Gravel Packed Completions Using Tip Screenout Fracture Prepacks," paper SPE 30102, 1995.

Monus, F.L., Broussard, F.W., Ayoub, J.A., and Norman, W.D.: "Fracturing Unconsolidated Sand Formations Offshore Gulf Mexico," paper SPE 22844, 1992.

Mullen, M.E., Norman, W.D., and Granger, J.C.: "Productivity Comparison of Sand Control Techniques Used for Completions in the Vermilion 331 Field," paper SPE 27361, 1994.

Mullen, M.E., Norman, W.D., Wine, J.D., and Stewart, B.R.: "Investigation of Height Growth in Frac Pack Completions," paper SPE 36458, 1996.

Mullen, M.E., Stewart, B.R., and Norman, W.D.: "Evaluation of Bottomhole Pressures in 40 Soft Rock Frac-Pack Completions in the Gulf of Mexico," paper SPE 28532, 1994.

Nolte, K.G.: "Determination of Proppant and Fluid Schedules from Fracturing Pressure Decline," *SPEPE* (July 1986) 255–265; also paper SPE 8341, 1979.

Ning, X., Marcinew, R.P., and Olsen, T.N.: "The Impact of Fracturing Fluid Cleanup and Fracture-Face Damage on Gas Production," paper CIM 95-43, 1995.

Patel, Y.K., Troncoso, J.C., Saucier, R.J., and Credeur, D.J.: "High-Rate Pre-Packing Using Non-Viscous Carrier Fluid Results in Higher Production Rates in South Pass Block 61 Field," paper SPE 28531, 1994.

Reimers, D.R., and Clausen, R.A.: "High-Permeability Fracturing at Prudhoe Bay, Alaska," paper SPE 22835, 1991.

Roodhart, L.P.: "Fracturing Fluids: Fluid-Loss Measurements Under Dynamic Conditions," *SPEJ* (Oct. 1985) 629–636.

Roodhart, L.P., Fokker, P.A., Davies, D.R., Shlyapobersky, J., and Wong, G.K.: "Frac and Pack Stimulation: Application, Design, and Field Experience From the Gulf of Mexico to Borneo," paper SPE 26564, 1993.

Saucier, R.J.: "Considerations in Gravel Pack Design," *JPT* (Feb. 1974) 205212.

Singh, P.K., and Agarwal, R.G.: "Two-Step Rate Test: A New Procedure for Detremining Formation Parting Pressure," paper SPE 18141, 1988.

Smith, M.B., Miller II, W.K., and Haga, J.: "Tip Screenout Fracturing: A Technique for Soft, Unstable Formation," *SPEPE* (May 1987) 95–103.

Stewart, B.R., Mullen, M.E., Ellis, R.C., Norman, W.D., and Miller, W.K.: "Economic Justification for Fracturing Moderate to High-Permeability Formations in Sand Control Environments," paper SPE 30470, 1995a.

Stewart, B.R., Mullen, M.E., Howard, W.J., and Norman, W.D.: "Use of a Solids-free Viscous Carrying Fluid in Fracturing Applications: An Economic and Productivity Comparison in Shallow Completions," paper SPE 30114, 1995b.

Tiner, R.L., Ely, J.W., and Schraufnagel, R.: "Frac Packs — State of the Art," paper SPE 36456, 1996.

Valkó, P., Oligney, R.E., and Schraufnagel, R.A.: "Slopes Analysis of Frac & Pack Bottomhole Treating Pressures," paper SPE 31116, 1996.

Valkó, P., and Economides, M.J.: "Performance of Fractured Horizontal Wells in High-Permeability Reservoirs," paper SPE 31149, 1996.

van Poolen, H.K., Tinsley, J.M., and Saunders, C.D.: "Hydraulic Fracturing—Fracture Flow Capacity vs. Well Productivity" *Trans.* AIME (1958) **213,** 91.

Vitthal S., and McGowen J.M.: "Fracturing Fluid Leakoff Under Dynamic Conditions Part 2: Effect of Shear Rate, Permeability and Pressure," paper SPE 36493, 1996.

Wattenburger, R.A., and Ramey Jr., H.J.: "Well-Test Interpretation of Vertically Fractured Gas Wells," paper SPE 2115, 1969.

Wong, G.K., Fors, R.R., Casassa, J.S., Hite, R.H., Wong, G.K., and Shlyapobersky, J.: "Design, Execution and Evaluation of Frac and Pack (F&P) Treatments in Unconsolidated Sand Formations in the Gulf of Mexico," paper SPE 26564, 1993.

20 Water Control

Mary Hardy
Halliburton Energy Services

Thomas Lockhart
Eniricerche

20-1 IS WATER PRODUCTION A PROBLEM?

In petroleum production, a certain amount of water production is expected and sometimes even necessary in the initial phases of the life of the reservoir or well. A petroleum engineer will have to be able to decide when water control solutions should be applied. If the costs associated with a water production rate still allow for an acceptable operating profit from produced oil or gas, that water production rate is considered acceptable. If the costs associated with a water production rate are too high to allow for an acceptable operating profit margin, the water rate is considered excessive.

Excessive water production can be caused by the natural depletion of a reservoir where an active water drive (either natural or artificial) has simply swept away most of the oil that the reservoir can produce, and there is little left to produce but water. The best completions and production practices can delay, but not stop this water production. Most cases where water-production rates have become a problem could have been avoided or delayed. Understanding reservoir behavior provides a basis for determining whether excessive water production is a concern and to determine if current water production is excessive. The following issues should be considered when estimating optimum water production rates:

- Current and projected oil prices
- Relative cost of high-capacity water handling facilities
- Cost per volume to dispose of produced water (treatment, transporting,
- reinjecting, etc.)

- Relative expense of completing wells to maintain low water production rates
- Water production needed to produce sufficient oil rates
- Surface or downhole facilities limited by fluids rate
- Water production rate effect on bypassed oil
- Reservoir maturity
- Water production rate effect on corrosion rates
- Water production rate effect on sand production
- Water production rate effect on scale formation

This chapter describes some of the currently available methods to predict, prevent, delay, and reduce excessive water production. These methods include methods for calculating potential water rates, options for redesigning well completions, and mechanical and chemical methods for minimizing water production.

20-1.1 Oil and Water Production Rates and Ratios

Operators who own a small percentage of the wells in a field often want to produce the wells as quickly as possible, without regard to total reservoir drainage effects. However, fluid production rates need to be controlled because excessive production rates can result in lower ultimate recoveries on a reservoir scale or in a shorter economic lifespan of an individual well.

If a water source exists in the reservoir, oil production rates influence current water production rates and the rate of water production increase. Material mass balance equations, Darcy's law, or simulators are commonly used tools to predict fluid production ratios and rates for reservoirs and individual wells.

20-1.1.1 Material Mass Balance

Material mass balance calculations will help estimate total production of the reservoir fluids (first published by Schilthuis, 1936). These calculations combine the classic concepts of Newton's conservation of mass, the ideal gas law, liquid compression, and material solubilities. The derivation and application of this type of technique is described in many reservoir engineering textbooks (a good example can be found in Dake, 1994, and references therein). The Material Balance Equation was once described (van Everdingen, 1953) as

"(Cumulative oil produced and its original dissolved gas + Cumulative free gas produced + Cumulative water produced) − (Cumulative expansion of oil and dissolved gas originally in reservoir − Cumulative expansion of free gas originally in reservoir) = (Cumulative water entering original oil and water reservoir)."

The mass balance equation can also be used to predict fluid flow from a well. The cumulative produced fluids would be redefined as well production history, and the stock-tank oil initially in place would change to well-drainage radius.

20-1.1.2 Darcy's Law

Darcy's law relates permeability and pressure drop to fluid flow rate. Flow into the wellbore is often considered radial for wells completed in non-fractured zones. Petroleum engineers often perform flooding experiments with reservoir core samples and produced fluids to illustrate the relationship between fluid saturation and the relative permeability of oil and water. This relationship is illustrated in Figure 20-1. Initial fluid saturations measured after drilling and water-oil rock relative permeability curves (like those in Figure 20-1) can be used to estimate k_{ro} and k_{rw}. Then Equations 20-1 and 20-2 can be used to predict relative flow rates of oil and water from a zone.

$$q_w = \frac{2\pi k k_{rw} h(p_e - p_w)}{\mu_w \ln r_e/r_w} \qquad (20\text{-}1)$$

$$q_o = \frac{2\pi k k_{ro} h(p_e - p_w)}{\mu_o \ln r_e/r_w} \qquad (20\text{-}2)$$

In these equations, conditions that have a significant effect on flow rates, well completion type, tubing size, and skin damage have been effectively accumulated into the p_w term, rather than considered separately.

The calculations can only provide predictions of relative fluid flow rates and are only useful for the assumed conditions. Saturations, pressures, fluid viscosities, relative permeabilities, and skin factors can vary significantly over the lifespan of a well or reservoir; therefore, the relative flow rates calculated with the equations above will also vary over time.

20-1.1.3 Productivity Index

The productivity index (*PI*) is the ratio of liquid production rate to the pressure drop at the center of the completed interval. *PI* is a measure of a well's potential and can be extrapolated to estimate field potentials. Conditions such as relative permeabilities, skin factors, reservoir pressure, and oil viscosity can change throughout the well or reservoir life and can change the *PI* (Beggs, 1991).

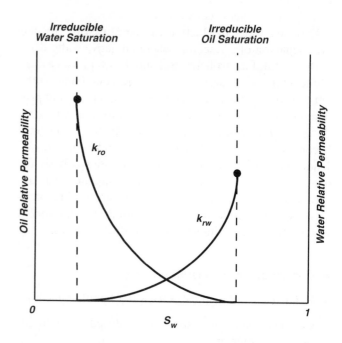

Figure 20-1 Relative permeability curves for water and oil, after Lake (1989)

20-1.1.4 Simulators

Reservoir simulators are tools that use the ability of computers to incorporate concepts such as material mass balance and Darcy's law to predict total reservoir performance. The engineer must be aware of the abilities and limitations of the various types of simulators. Many simulators take into account the varying conditions in a reservoir; however, not all take into account the same variables or assume the same change profiles for these variables. Simulators offer a more convenient way to incorporate reservoir changes when predicting fluid production. Simulator predictions are history-matched with reality to update predictions (Coats, 1987).

Simulators have characterized the production of several reservoirs. Simulations of the Gullfaks and Forties fields in the North Sea have been used to plan future well locations, water production, secondary recovery programs, and future production volumes (Tollefsen *et al.*, 1994; Maldal *et al.*, 1996; Brand *et al.*, 1996). The South Tano Field in Ghana was simulated to estimate reserves and the risk of field development (Hassall *et al.*, 1995). Since the simulation results were marginal, development of this field has been suspended.

20-1.2 Rate-Limited Facilities

Downhole equipment and surface facilities have a maximum rate at which they can handle fluids. For the downhole equipment, maximum rates generally depend on tubing and orifice sizes, pressure drawdowns (also a function of fluid density), and fluid viscosities.

Factors such as imposed regulations, surface equipment, and transport rates can also limit the maximum production rate of a well or group of wells to less than their potential. In the following cases, water production can seriously reduce the oil production rate.

Total water production rates must not exceed the maximum disposal rate. Maximum disposal rates are defined by allowable water discharge volumes, limited separator rates (Georgie *et al.*, 1992), amount of water that can be transported efficiently from the facility, total water that a pipeline operator allows to flow through an available pipeline, and rate at which water may be reinjected.

An increased water cut can significantly increase the hydrostatic head in the wellbore. This phenomenon will decrease the drawdown pressure at the wellbore and consequently decrease the maximum fluid production rate or stop production entirely.

If the well produces at the maximum rate and water rates increase, oil rates will suffer. In these cases, the oil production decline can be as simple as a one-to-one exchange (one fewer barrel of oil produced for one more barrel of water). However, the change in oil production will be a function of friction pressure and viscosity variations with different water-oil ratios. Bourgoyne *et al.* (1986) present a good discussion of these effects.

20-1.3 Water Production Effect on Bypassed Oil

When water influx is the result of deeper, reservoir-related water production mechanisms, unchecked water production can result in a significant decrease in the total volumes of accessible, mobile oil (Kortekaas, 1985). Higher water production rates from a zone implies that both the relative permeability to water and the water saturation in that zone is increasing. The higher these parameters are allowed to climb, the more difficult it will be to produce oil from that zone again. For example, where excessive water coning has been allowed to occur, pockets of unswept oil can be left.

20-1.4 Reservoir Maturity

When an operator initiates production in a field, that reservoir may be a recent find or one that has been produced for several years. Conditions in the reservoir change while it is being produced. These changes include the following:

- The remaining movable oil in place declines
- Principle recovery mechanisms often shift from primary to secondary.
- Reservoir pressures can drop
- Oil viscosities may increase if pressures decline below the bubblepoint
- Connecting aquifer depths change

20-1.5 Water Production Rate Effect on Corrosion Rates

Water production rates can significantly affect corrosion rates of downhole and surface equipment. Corrosion rates can be connected to kinetic and erosive/corrosive effects. The rate at which corrosion will occur depends on the concentration of corrosive materials (oxygen, H_2S, CO_2, salts, etc.). The sooner fresh volumes of water containing these corrosive materials come into contact with the metal surfaces of either the downhole or surface equipment, the sooner corrosion can take place (this is not necessarily to imply a linear relationship). Some corrosion products can act as a protective coating against further corrosion (such as low-solubility iron oxides). However, if the flow rate becomes high enough, it can erode the coating from the tubing surface and expose the fresh metal surface to the corrosive materials.

20-1.6 Water Production Rate Effect on Scale Deposition Rates

Water production affects scale deposition rates in a number of ways. Just as water rates can affect corrosion, if the produced water tends to cause scaling, the faster the water is produced, the faster the scaling deposition. Erosion again affects this process. Extreme friction can help erode scale deposits from the tubing. When a waterflooding program is in place, another consideration is the injection water composition (Patton, 1974). If scaling is increased when the injection and formation water mix, scaling can be dramatically increased when injection water breaks through (Thomas, 1987).

Several of the North Sea fields inject seawater in their waterflooding programs. The formation water in many of these fields contains barium and strontium, and the seawater contains sulfate. The intermixing of these chemicals will increase the probability of scaling. The water cuts for the fields may be as low as 2% to 3%. However, there are cases when additional seawater is produced. Wells that begin to produce the seawater in conjunction with the bottom water are often plugged with scale in a matter of weeks. It is therefore important to know not only how much water can be handled economically, but what the potential water source and water production mechanisms will be.

20-1.7 Water Production Rate Effect on Sand Production

Water can weaken cementitious materials that hold the formation in place, allowing sand production (Coulter *et al.*, 1987; Muecke 1979). Zones that produce water may therefore have a lower maximum pressure drop at which sand-free production exists. A broad discussion of sand control concepts is found in Chapter 18 of this book.

20-2 WATER PRODUCTION MECHANISMS

Predicting, preparing for, and treating for water production involves knowing how that water may be produced. Factors that help determine water production mechanisms include the reservoir drive mechanism, production rates (reservoir and well), connate water and irreducible oil saturations, permeabilities (vertical and horizontal), porosity, permeability anisotropy and heterogeneity, relative permeability/mobility to water and oil, location and continuity of impermeable barriers, reservoir dip, original water-oil contact, portion of productive interval completed, completion type (*e.g.* perforated, open hole, etc.) location, and quality of primary cement job. The commonly observed water production mechanisms and expected water production rates are described below. Figure 20-2 illustrates water-production histories often associated with these mechanisms.

20-2.1 Completions-Related Mechanisms

Casing leaks Casing leaks can occur in tubing and collars and are often caused by poor completions practices: improper tightening of joints (too loose causing no seal, or too tight causing excess strain), or tubing incompatibilities with downhole conditions (temperature, corrosive materials, pressure, etc.). Casing leaks are often observed by a sudden, rapid increase in the water cut (Curve 2 in Figure 20-2).

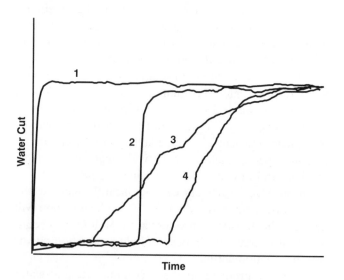

Figure 20-2 Example water production history curves.

Channel behind casing Poor cement/casing or cement/formation bonds often lead to channels in the casing-formation annulus. These channels can occur at any time in the life of the well, but are usually observed by a rapid increase in water production immediately after a stimulation treatment, or an unexpectedly high water cut immediately after completion, as in Curves 1 and 2 in Figure 20-2. Channels behind the casing are much more common than casing leaks.

Completion into water This phenomenon generally occurs when the available data (core data, driller's reports, and openhole logs) is either misinterpreted, of poor quality, or unavailable. As in some casing leaks, a symptom of completion into water is an unexpectedly high water-cut immediately after production begins, as in Curve 1 in Figure 20-2.

20-2.2 Reservoir-Related Mechanisms

Bottomwater This mechanism is the only commonly occurring water production mechanism that is unavoidable. When a reservoir has an active aquifer driving oil production, as the reservoir depletes, water slowly displaces the oil, and the wells in the field slowly begin to produce water. The water production history of wells producing water caused by this mechanism even rise if the water table is similar to Curve 3 in Figure 20-2.

Barrier breakdown Natural low-permeability barriers, such as dense shale layers, sometimes separate the oil zone from an aquifer. This barrier can break down for various reasons. If drawdown pressure during production exceeds what the barrier can withstand, it will fail, allowing water to break through and produce. The barrier can also either fracture or dissolve as a result of hydraulic fracturing or matrix acidizing treatments, respectively. A rapid increase in the water production rate can also be an indication of this mechanism. If the barrier is broken during completion (either while drilling or stimulating), a water production history similar to that depicted in Curve 1 of Figure 20-2 would be more representative. If it is caused by pressure depletion or stimulation treatments later in the well life, water production may look more like Curve 2.

Coning and cresting In a waterdrive reservoir, the drawdown pressure at the wellbore will tend to pull water up into the wellbore. When extreme drawdowns exist in a vertical well, the resulting shape of the near-wellbore water-oil contact is conical; in a horizontal well (Figure 20-3A), the shape is more like a crest of a wave (Figure 20-3B).

Coning and cresting can be avoided if the well is produced below its critical rate, which is the maximum water-free production rate. Critical rates have been studied extensively. Probably the earliest documentation of critical rate studies was presented in Muskat and Wyckoff (1935). Studies since then have invoked a wide variety of considerations (or conditions) into evaluating critical rates: unsteady states, pseudo-steady states, permeability heterogeneities, horizontal wells, and three-phase flow. Most critical rate calculations assume that the rate and cone or crest shape are affected by the ratio of vertical to horizontal permeability, oil zone thickness, ratio of gravity and viscous forces, well penetration, and mobility ratios.

The primary differences in the calculation methods are in the assumptions made to implement simplifications. Muskat and Wyckoff assumed linear flow, whereas Meyer and Garder (1954) invoked radial flow. Chierici, *et al.* (1964) assumed no influence in cone shape, whereas Wheatly (1985) applied cone-shape calculations. The best method for determining critical rates (or many other values) is the method which most accurately assumes the well conditions at hand. For further review of coning and cresting, see Yang and Wattenbarger (1991), Chaperone (1986), Guo and Lee (1993), and the references previously mentioned in this section. The water production history of a well with a coning problem may look something like Curve 3 in Figure 20-2.

Channeling through high permeability In an ideal, homogeneous, waterdrive reservoir, the oil is uniformly displaced by the water. However, "ideal" and "homogeneous" are rarely applicable to reservoirs; there are often layers of varying permeability within a producing interval. As quantified by Darcy's law, the flow rate is faster through higher-permeability layers. The result is a high water production rate through these layers before water has swept oil from the surrounding layers. In fields where a waterflooding program is in place, this can result in injection and immediate

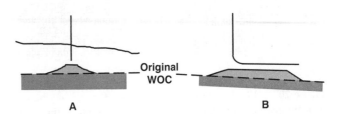

Figure 20-3   (a) Water coning, (b) Water cresting

production of injected water with no significant displacement of additional oil. A side view of a 2D simulation of water flowing from an injector to a producer is shown in Figure 20-4 (simulator described and developed in van Batenburg, 1991). In these simulations, 50% of the total pore volume between the injector and the producer was swept with water. Portions (grid blocks) of the rock that were not swept are represented as outlined blocks (•); those that were swept with water have no outline. The bold lines are streamlines representing isoflow.

In the cases where a permeability contrast exists (Figure 20-4B), a significant amount of oil is by-passed by the injection water, compared to the case where no permeability contrast exists (Figure 20-4A). A high-permeability layer can result in a rapid rise in the water cut after breakthrough, as shown by Curve 4 in Figure 20-2.

Fracture communication between injector and producer Natural fractures can provide a direct link between an injector and a producer, allowing the water to flow primarily through these high-permeability channels, and bypass oil within the adjacent rock matrix. For highly communicating fracture networks, the water production history may look more like Curve 2 in Figure 20-2, where the abrupt onset of water occurs within a couple of days (or even hours) after injection begins. If the fracture network does not give as direct a path between the injector and producer, the water production history may look more like Curve 4 in Figure 20-2.

Stimulation out of zone In a production well, this occurs when an aquifer is stimulated during a fracturing or matrix acid treatment. In an injection well, this would encompass stimulation treatments that result in decreased sweep efficiency. The effect that this type of problem would have on the water production history is illustrated

by Curve 2 in Figure 20-2, where the onset of water would coincide closely with the stimulation treatment.

20-3 PREVENTING EXCESSIVE WATER PRODUCTION

Excessive water production can be treated either in the completion stage or after it becomes a problem. Historically, water production was ignored until it was a real problem—wells often producing in excess of 90% water. Postponing treatments can seriously jeopardize the total well/reservoir productivity, particularly if the water production mechanism is reservoir-related. Medical doctors often promote prevention as the best cure; the same can be said for excessive water production. Simulators can be extremely helpful in deciding which prevention option will be the most successful, both from an incremental oil and from an initial investment point of view. Simulators can also help determine if it is worth preventing excessive water production or treating for it later. The methods below focus on prevention. Some of the options use chemical treatments. Treatment selection and design are discussed in Section 20-4.

20-3.1 Preventing Casing Leaks

Casing leaks can be prevented if tubing is selected that will withstand long-term exposure to the projected pressure, temperature, and chemical environments that may exist downhole. Temperature, pressure, and chemical resistivities of tubing materials and tubing selection considerations are found in Gutzeit *et al.* (1987) and Craig *et al.* (1992). The literature describes several field cases where tubing selection has prevented casing leaks (Turki, 1985; Cox and Babitzke, 1989; Chitwood and Coyle, 1994).

20-3.2 Preventing Channels Behind Casing

A good primary cement job will usually prevent channels behind the casing. Methods to achieve a good bond between the reservoir and the casing have been established; however, poor primary cement jobs are relatively common. Chapters 8, 10, and 11 discuss techniques to achieve successful primary cementing.

20-3.3 Preventing Coning and Cresting

Because coning and cresting result from a low pressure at the wellbore pulling up the water-oil contact (WOC),

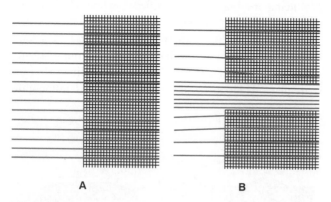

A   **B**

Figure 20-4 Flow pattern of a waterflood through (a) homogeneous permeability and (b) a zone with a high-permeability streak

techniques to prevent coning and cresting involve ways to minimize the drawdown on the WOC. As described in Section 20-2.2, holding production rates under the critical rate was the original technique implemented in coning prevention. However, limiting production rates to minimize coning also limits revenue. Other methods to prevent coning involve maximizing the critical rate.

20-3.3.1 Perforating

In the completion stage, the location of perforations has been used to help prevent coning. Muskat and Wyckoff (1935) recognized that the farther away the perforations are from the WOC, the lower the tendency for coning will be. The effect of perforation location on the critical rate calculations is illustrated in Equation 20-3 (Chaperone, 1986).

$$q_{oc} = \frac{4.888 \times 10^{-4} k_H h_o^2 \Delta \rho q_c}{\mu_o B_o} \qquad (20\text{-}3)$$

Pirson (1977) and Meyer and Garder (1954) calculated the height of the bottom of the optimum well completion using Equation 20-4. Guo and Lee (1993) presented work that indicates that the completion interval "...should be less than one-third of the total thickness of the oil zone, depending on oil-zone thickness, wellbore radius, and drainage area radius"

$$h_{cb} = h_o - (h_o - h_p)\frac{\Delta \rho_{og}}{\Delta \rho_{wg}} \qquad (20\text{-}4)$$

The concept of maximizing the distance between the WOC and the perforations applies to both vertical and horizontal wells. The equations above are designed for application in vertical wells. Calculations for horizontal wells are presented in Chaperone (1986) and Yang and Wattenbarger (1991). This technique is limited to the height of the oil zone. If the zone is thin, the distance between the WOC and the perforations will be small, limiting the completion interval. Another drawback is that the upper interval of an oil zone may have a very low permeability.

In completing horizontal wells in the Troll field, Norway, shot density was also optimized (Brekk and Lien, 1994). The primary objective was to minimize the pressure loss in the wellbore by minimizing the shot densities to reduce the friction factor across the perforated interval.

20-3.3.2 Fracturing

Fracturing a perforated interval to prevent coning has not been very successful. If designed properly, an hydraulic fracture in a vertical well can help dissipate the wellbore drawdown in a way similar to that of a horizontal well (the water is pulled by a "line" of pressure, rather than a "point" of pressure). Generally, vertical permeabilities in hydraulic fractures are very high compared to the matrix horizontal permeability, so extreme care must be taken to ensure that fracture growth does not propagate too close to the WOC. For specifics on design parameters for hydraulic fracturing, see Chapter 17 and Gidley *et al.* (1989).

Fracturing horizontal wells has been shown to maximize well PIs. There are several horizontal wells in the Dan field, Denmark. A feasibility study indicated that horizontal wells would produce the same amount as fractured vertical wells, but would be more expensive to complete. The study also indicated that horizontal wells would not significantly affect the coning breakthrough time (Andersen *et al.*, 1988). The success of this field has been attributed to horizontal wells with multiple fractures. Initial results showed that of the 41 wells in the field, 25% of the total oil production from the Dan field came from the three wells completed in this manner.

20-3.3.3 Artificial Barriers

Placing an artificial impermeable barrier between the WOC and the completed interval will greatly reduce coning tendencies and increase a well's maximum water-free production rate (Richardson *et al.*, 1987). Artificial gel barriers have often been used to decrease coning, but the vast majority have not been placed until after water breakthrough has occurred. Problems in waiting come from many sources. In the end, it is very difficult to treat a well for coning and not to plug the entire completed interval. The problem is aggravated by short completion intervals due to thin zones, high k_v/k_h, or both.

Success rates have been very high in cases where artificial barriers were placed between the WOC and the completion interval before the wells were placed on production. The general procedure is to (1) drill to the water zone and perforate, (2) place an artificial barrier of a radius sufficient to prevent (or significantly delay) coning, and (3) move up the wellbore and perforate into the oil zone. [Equations that can help determine the required barrier radius can be found in Meyer and Garder (1954).]

A successful example of this procedure is in the Miocene sands in central Louisiana, where severe coning problems plagued operators. Water breakthrough from coning can occur within weeks (or even days) of well completion. Water breakthrough was delayed for several months in wells that were completed with artificial barriers separating the water from the oil.

20-3.3.4 Dual Completions

Wells can be produced from the water zone and the oil zone to prevent coning. When producing both zones, two production strings are used to keep the oil and water separated (Othman, 1987)—hence the process is referred to as "dual completions." Dual completions are successful at reducing water coning potentials by decreasing the pressure in the water zone in the wellbore area. Dual completions also reduce the surface handling costs of the produced water by reducing the need for separating the water from the oil.

20-3.3.5 Horizontal Wells to Prevent Coning

Horizontal wells have been shown to be quite useful in minimizing coning effects. However, cresting still can occur in horizontal wells. Horizontal wells offer the flexibility to disperse the drawdown pressure at the wellbore. In a vertical well, the water is pulled into the wellbore by pressure concentrated in one spot (the radius of the wellbore), whereas in a horizontal well, the pressure is spread out along the length of the wellbore. This is not meant to imply that the pressure is uniformly dispersed down the wellbore; generally it is found that the drawdown at the heel of a horizontal section is higher than at the toe. In thin oil zones, vertical completion lengths are short, which minimizes the well drainage radius and the amount of oil that can be pulled into that wellbore.

Despite these advantages of horizontal wells, the cost of drilling them is still significantly higher than for a conventional well. (This disparity is declining as the industry improves horizontal well drilling technology.) The popularity of horizontal wells has increased over the years because of their increasing capability to prove their worth. Ahmed (1991) summarizes the experiences with horizontal wells drilled to minimize coning in fields in the U.S., Canada, and Asia. Shell has implemented horizontal well completions to reduce coning of both water and gas in the Rabi field, Gabon (Pelgrom *et al.*, 1994), and horizontal wells drilled in the Bombay Offshore basin have also been very productive (Srinivasam *et al.*, 1996). Mukherjee and Economides (1991) presented a study that helps engineers evaluate

the horizontal-vs.-vertical option. Gilman *et al.* (1994) describe the application of short-radius horizontal wells in the Yates field unit in west Texas. This work also shows an example of how production was maximized by orienting the wellbores optimally, as was predicted from a dual-permeability simulator.

The fundamental concept applied to preventing coning in vertical wells will also apply in preventing cresting in horizontal wells: the drawdown the WOC "sees" should be minimized. Chaperone (1986) calculated cresting critical rates, and Yand and Wattenbarger (1991) predicted breakthrough times. As in a vertical well, water breakthrough times and critical rates are maximized when the lateral portion of the well is placed at the top of the oil zone.

In general, frictional pressure losses cause the pressure drop between the wellbore and the reservoir to be higher at the heel of a lateral than at the toe. This condition is exaggerated in wells that are highly productive, are completed with small tubing, or are extended reach, and becomes significant if the flowing pressure gradient in the lateral is similar to the producing drawdown pressure (Dikken, 1990). Because of this, water breakthrough from cresting most often occurs at the heel of a lateral. If the drawdown pressure can be better dispersed down the lateral, many cresting problems can be either minimized or delayed. Keeping this in mind, some general statements can be made about completing horizontal wells with significant pressure drop along the lateral: (1) Shot densities should be higher at the toe than at the heel (Landman and Goldthorp, 1991), (2) stimulation treatments should concentrate more at the toe than at the heel, and (3) Impermeable barriers can be placed at the heel of the lateral to decrease cresting tendencies.

20-3.4 Preventing Channeling through High Permeability

High-permeability streaks can connect a producer either to an underlying aquifer or to an injection well. Minimizing flow through a high-permeability channel is often referred to as profile modification. Preventing flow through high-permeability streaks can involve partial perforating, stimulating, or partial blocking. In general, this mechanism can best be prevented if the water flowing through the high-permeability layer is somewhat confined (crossflow minimal) so that water does not flow around blocking treatments or flow into stimulation treatments. Crossflow (illustrated in Figure 20-5) is usually minimized by either a very low-permeability

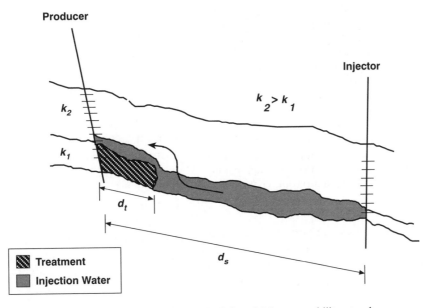

Figure 20-5   Crossflow bypassing a block in a high-permeability streak

layer adjacent to the high-permeability streak, or a low vertical-to-horizontal permeability ratio.

Preventive measures are also successful when the water source and the producer are relatively close: For best results, maximize d_t/d_s (d_t = treatment radius, d_s = distance from well to water source). The most effective way to prevent channeling through a high-permeability streak is to reduce the permeability of the entire streak. However, this usually requires such large treatments that it is economically unjustifiable.

20-3.4.1 Perforating

This option is relatively straightforward: Perforating into the high-permeability streak is minimized (either by density and/or depth) or avoided. This option is particularly successful when crossflow is minimal.

20-3.4.2 Stimulation Techniques

Stimulation treatments can help improve the *PI* of a well and disperse the drawdown over the entire perforated interval, thereby decreasing the pressure drop across the high-permeability layer. Because the effect of stimulation techniques on preventing channeling through high permeability will be more confined to the near wellbore area, they would be best suited when the water source is relatively close. Possible options include

- **Stimulating past drilling/completions fluids damage by near wellbore fracturing (Chapter 19) or selective HF acidizing (Chapter 5)** This would be most applicable when the fluids flowing through the high-permeability layer are confined by a low k_v/k_h and when the skin factors in the low-permeability intervals are high enough so that near wellbore stimulation treatments will have a significant effect. (This means the low-permeability layers have sufficient permeability to allow leakoff during drilling and completion.)

- **Large-scale hydraulic fracturing treatments (Chapter 17) to increase the effective drainage area of the low-permeability layers** This will be most applicable in cases where the high-permeability layer is sufficiently separated from the interval targeted for fracturing to prevent the fracture growth from entering the high-permeability layer. Statoil has reported success in using this technique in the North Sea (Bale *et al.*, 1994).

20-3.4.3 Permeability Reduction

These treatments are injected into the higher-permeability layer(s) to reduce their permeability so that it will be roughly equivalent to the lower-permeability intervals. Treatments of this sort should be only partially, rather than completely, plugging so that the oil in the high-permeability layer is not lost.

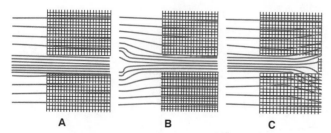

Figure 20-6 (a) flow pattern of a waterflood through a zone with a high-permeability streak; (b) high-permeability streak treated before watering out (injector treated); (c) high-permeability streak treated before watering out (producer-treated).

If the high-permeability streak can potentially connect an injector to a producer, an engineer must decide which well to treat, if both should be treated, or if treatment should be delayed until breakthrough occurs. If the water flow through the high-permeability layer is confined, options such as a near wellbore stimulation in the producer and avoiding perforating into the high-permeability streak at the injector may be suitable. When there is little chance of preventing the water from finding the high-permeability channel, treating both the injector and the producer may be necessary. The following example illustrates how a simulator (van Batenburg, 1992) might help determine which option would be the most beneficial. In these simulations, 50% of the total pore volume between the injector and the producer was swept with water. Portions (grid blocks) of the rock that were not swept are represented as outlined blocks (•); those that have been swept with water have no outline. The bold lines are streamlines representing isoflow.

This simulation compares the difference in sweep patterns if the injection profile is allowed to go unchecked (Figure 20-6A), the injector is treated with a partial blocking treatment before injection is started (Figure 20-6B), or the producer is treated with the same treatment before injection begins (Figure 20-6C). Preventing breakthrough by treating the injector will leave the least amount of unswept oil, according to these calculations.

20-3.5 Preventing Fracture Communication between Injector and Producer

Again, this water production mechanism is most often treated after water breaks through a fracture network by plugging the fracture at either the injector or the producer with a chemical gellant. This technique can also be used before breakthrough. Problems with this technique arise when the permeability of the surrounding rock matrix is too low to maintain acceptable production or

injection rates after treatment. Other problems associated with this technique have involved poor treatment design. If the agent used to plug the fracture is not strong enough to withstand the drawdown at the wellbore, it may be produced back (further discussion appears in Section 20-4.2.3).

Another technique focuses on well placement. If an operator acquires an older field or a series of wells where it is not economically feasible to drill new wells, this option will be limited. Choosing the location of the injectors relative to the producers is the prime factor, whether in drilling new wells or converting existing wells from producers to injectors. An aerial view of an example well pattern in a reservoir with the indicated fracture orientation is illustrated in Figure 20-7. In this case, the only well that could be considered as an injector is well A-01. A location of a future injector might be where F-01 is indicated.

If waterflooding programs already exist, reorienting the direction of the waterflood front is also possible. The McElroy field in the Permian Basin is an example where this was successful (Nolen-Hoeksema *et al.*, 1994). In this case, the reservoir was naturally and hydraulically fractured. New injection wells were drilled to maximize the injection efficiency, the placements of which were determined by considering the existing wells and the relative fracture orientation.

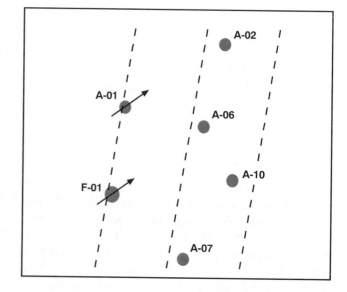

Figure 20-7 Example well pattern: choosing injectors

20-3.6 Completing to Accommodate Future Water Production Rates

20-3.6.1 Future Zonal Isolation

Completing wells with the capabilities to selectively shut off production in a zone (or zones) is a relatively common way to prepare for increased water rates. There are three general ways to achieve this. One method is to build mechanical isolation equipment into the well when it is completed. This procedure is primarily done when the excessive water production is expected to enter the wellbore at an interval upstream from the end of the well. Sliding sleeves in combination with packers are commonly used to achieve this effect. As illustrated in Figure 20-8, the sleeves are built into the tubing and packers are placed between the sleeves. If a higher-permeability layer begins to produce water at high rates, the sleeve through which that water is produced can be closed. Case histories are presented in Harrison *et al.* (1994) and Moradi (1988).

Mechanical zonal isolation is also used to shut off water production from the end of a well. Settable packers such as bridge plugs or slickline-settable plugs are tools of choice for this. This procedure is applicable when excessive water rates caused by bottomwater are expected. Water would enter the wellbore first at the bottom set of perforations. A slickline-settable packer can be placed to shut off the bottom set of perforations and incrementally moved up the hole as the coning progresses. Placing plugs such as these is often difficult if the nipple and/or tubing restrictions are small. In the Everest and Lomond fields in the North Sea, a slick monobore completion was developed that helped maximize these constrictions so that rigless mechanical zonal isolation was achievable (Laing *et al.*, 1993*)*.

Planning for selectively injecting chemical plugging agents also involves consideration of the tools required to isolate the target zone (e.g. single packers and straddle packers). Sliding sleeves can also be used when in-depth blocking or wellbore blocking is necessary. Before treatment, all sleeves are closed except that in which a treat-

ment will be injected. After the treatment, the closed sleeves are reopened to production and the sleeve that was opened for treatment is closed.

20-3.6.2 Creative Water Management

Even when the best engineering techniques for prevention have been implemented, there are cases where excessive water production will be imminent. In other cases, the definition of excessive water production is based on the costs associated with lifting and handling of the water. These include costs to replace corroded production strings, mill out upstream carbonate scale (deposition of which is aggravated by release of CO_2), run an ESP (electrical submersible pump) to offset increased hydrostatic head from an increased height of the water column, separate oil from water to environmentally safe levels (< 40 ppm for the North Sea), and to store and transport produced water.

An option to reduce water storage costs, to separate water for disposal, and to transport produced water is to reinject the water either as a part of a waterflooding program or into a non-communicating reservoir. The cost of drilling enough injection wells and/or converting producers to injectors will at least partially offset the benefits of a reinjection program. Careful planning must be done to accurately anticipate how much water can be reinjected and to avoid typical injection well problems such as scaling and corrosion. Several reinjection programs have been very effective. Examples are found in the Forties field, North Sea (Brand *et al.*, 1995) and in Yemen's Masila block (Wilkie *et al.*, 1996).

Another option that is currently under investigation is to reinject produced water without bringing it to the surface. This option potentially avoids all the costs associated with lifting and handling water. A well completed to handle this might look similar to the one illustrated in Figure 20-9, where multilateral technology is used. This hypothetical well is located in a reservoir that has a thin oil zone with an active aquifer; therefore, water cresting potentials are high. Lateral L1 is drilled into the payzone and L2 is drilled into the water zone. Oil and water will be produced from L1 and will be fed into a downhole separator (a device that works like a centrifuge). The oil is then directed to the surface from the separator, and the water is directed down to L2 for subsequent reinjection. The two laterals are directed away from each other to avoid aggravating cresting problems with locally high

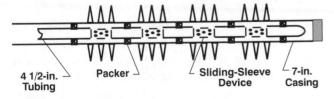

Figure 20-8 Sliding sleeves added to completion to allow future zonal isolation

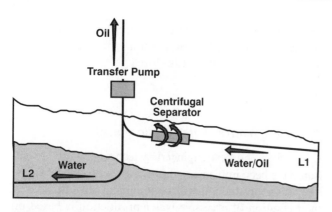

Figure 20-9 Multilateral well completed to accommodate reinjection rather than lifting of produced water

water pressures caused by injection. Advantages to a design such as this are listed as follows:

- Upstream corrosion and scaling are avoided

- Oil does not have to be separated as completely from the water as it would if the water were to be disposed of

- There will be no pressure reduction resulting from increased hydrostatic pressure because no water column is formed in the production string

- Because the water will be pushed down rather than up, the strain on the downhole pump will be minimized and its life span lengthened

20-4 TREATMENTS USED TO REDUCE EXCESSIVE WATER PRODUCTION

Excessive water production is most often treated rather than prevented. The budget available to treat the problem reduces as the water cut increases. This means that little money is available to determine where the water is coming from and why, to pay for the most technically viable solution, or to employ proper placement techniques. The keys to successfully shutting off or preventing excessive water are proper problem characterization, appropriate treatment design, and effective treatment placement.

20-4.1 Characterizing the Problem

There is no set method for characterizing water-production problems; each case presents a different set of avail-

able data for making design decisions. However, the following is a very effective generic procedure.

- Using all available data on a candidate, assign/calculate values to as many of the factors that help determine water-production mechanisms (listed in Section 20-2) as possible.

- List the unknowns and how they can be determined. For example, PLT or downhole video can help pinpoint where water is entering the wellbore (case histories wherein this technique helped prevent squeezing the wrong set of perforations is presented in Maddox *et al.*, 1995).

- Compare the risk associated with designing a treatment with the given knowns *vs.* the cost of collecting more data. For instance, if an engineer cannot distinguish between a channel behind the casing or a high-permeability streak without a cement bond log, the engineer must determine if it is worth the risk of treating for the wrong mechanism, rather than spending the money to run the log. Often treatment volumes for high-permeability streaks are higher (and therefore more expensive) than for channels behind the casing. If the engineer chooses to squeeze cement to fill a channel behind the casing when the problem was really a high-permeability streak, he or she may stop water production for a short time, but an additional treatment may be necessary shortly after the squeeze treatment. On the other hand, if the engineer chooses to treat for a high-permeability streak when it was really a channel behind the casing, he or she risks spending more for a larger treatment when it may not have been necessary. The cost to run the log must also be considered. If the candidate well is an offshore subsea completion well, the cost to run the log may be several times more than the cost of either treatment.

- Estimate post-treatment production potential.

20-4.2 Treatment Design

Whether preventing or shutting off excessive water, there are two primary questions that must be answered for proper solution designs: (1) What is the treatment expected to do? (2) What conditions must the treatment withstand? This section will focus on shutting off excessive water; however, many of the concepts are transferable to preventing excessive water production.

20-4.2.1 Expected Treatment Effect on Water Production

Success must be realistically defined for any operation to be successful. This sounds simplistic; however, there are many cases where an operator does not fully communicate what results are expected of a treatment, or recognize expectations that are unrealistic. It may not be possible to decrease the water cut to 0% and to increase oil production by a factor of 10. The same information used for identifying the water production mechanisms and to determine the maximum amount of water is also used to realistically define what a treatment is expected to do.

As an example, enough information was available to determine that Well A-01 is producing excessive water through the high-permeability interval, Z1, from an underlying aquifer and no water production is coming from zones Z2 or Z3 (Figure 20-10). The potential for vertical communication between Z1 and the adjacent zones is known to be minimal, but the water saturation of Z3 was 35% when the well was new (and at least that currently). If production from Z1 is completely shut off, water will likely begin to produce from Z3. Although this zone was not contributing to the water production prior to treatment, the producing pressure that Z3 sees would increase due to fewer perforations open and possibly a

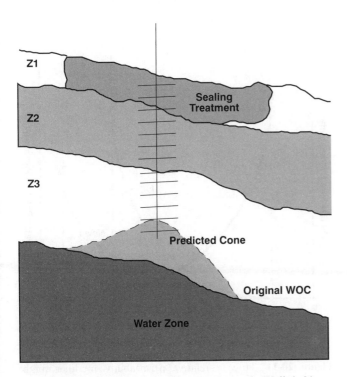

Figure 20-10 Treatment effect on Example Well A-01

shorter water column. Assuming Z1 is shut off, an operator can predict what the new production index and water rate will be. The operator can also calculate the new critical rate for the well to determine if it is worth keeping the production rate low enough to prevent additional water production due to coning.

In this example, the treatment was expected to completely shut off production in the target interval; however, complete seals are not always advantageous. Whenever water permeability is decreased, the water-drive mechanism displacing oil is decreased. Generally, treatments can only be effectively placed a short distance from the wellbore—rarely more than 50 ft, a very small area on a reservoir scale. Although water shut-off treatments may not significantly affect the waterdrive mechanism on a reservoir scale, completely cutting off a well from the water source can have serious detrimental effects on the well's productivity. Also, the target interval may still contain a significant volume of mobile oil.

20-4.2.2 Treatment Types

Zone Sealants

Definition When a zone sealant is properly applied, all production of the treated zone is shut off. There are two classes of zone sealants: wellbore and matrix sealants. Wellbore sealants can be either chemical or mechanical. The chemical sealants can either be strong gellants or cements and are designed not to appreciably penetrate the rock matrix. Mechanical sealants are generally packers or sliding sleeves that are either built into the completion or run into the well after excessive water production occurs. Matrix sealants are chemical treatments that are injected into the rock matrix of the target zone and subsequently reduce the absolute permeability of the treated rock to zero. The size of any wellbore sealant must be sufficient to cover the entire interval.

Mechanism A zone sealant must be capable of completely plugging all flow channels that connect the reservoir to the wellbore. Mechanical and chemical wellbore sealants rely on the strength of the bulk material and the bond between the material and the tubing; weakening of either can cause the seal to fail. Matrix sealants must also be resistant to downhole environments. Because the materials are injected farther into the rock matrix, if some degradation or syneresis of the treatments occurs, the seal can still remain intact (Bryant, 1996). The paths through which fluids flow through matrices comprise pore spaces connected by pore throats, rather than bundled capillary paths. Because of this, matrix zone sealants do not need to plug every pore space and throat, just enough of them so that there are no open flow channels connecting the reservoir to the wellbore.

Lifetime Generally, sealants are expected to last indefinitely, or at least until the economic benefit or the treatment sufficiently offsets its cost.

Permeability-Reducing Agents (PRAs)

Definition A PRA must be able to reduce the water production from the target interval. Most PRAs are matrix treatments, but sand plugs placed in the wellbore can also be designed to reduce, rather than plug, permeability. A PRA can also reduce the oil permeability of the target zone.

Mechanism A PRA that decreases the permeability 50% can work either by completely plugging half of the flow channels in a rock matrix, by partially blocking all the flow channels, or by completely plugging less than half of the flow channels and partially blocking others. Materials that are used for zone sealants may also be designed to partially reduce permeability by adjusting formulation, placement, volume, or a combination thereof (see Section 20-4.2.3).

Lifetime Generally, sealants are expected to last indefinitely, but the ability of a matrix sealant to maintain its ability to reduce water rates can be a function of the zone water saturation (see Section 20-4.2.2).

Relative Permeability Modifiers (RPMs)

Definition The true definition of an RPM has been the subject of a great deal of debate. In the purest sense, a material that reduces the relative permeability to water more than to oil is considered to be an RPM. Some argue that if a treatment will decrease the water-oil ratio (WOR) that is produced from the target zone, it should also be considered an RPM. Both definitions imply that when an RPM is applied, the relative water-oil producing rate will decrease. The first definition also requires an RPM to shift the relative permeability vs. saturation curves (Figure 20-11) so that when residual oil saturation is reached in the rock matrix (at maximum water saturation), the k_w/k_o is lower.

Lifetime The issue of RPM lifetime has also been debated. Chemically, RPM lifetimes are defined by the time required for the treatment to produce back. Engineering studies indicate that when a material reduces the WOR of an interval, eventually the water driving production will build up on the other side of the treated area. When the water saturation of the outside area rises to match the maximum water saturation of the original

matrix, the oil permeability will go to zero: no more oil will be produced from that interval. Simulations can predict how long a scenario of this type will take. When an RPM loses effectiveness in field applications, a retreatment can help determine if the original treatment failed because it was producing back slowly throughout its lifetime, if it was allowing water saturation buildup outside the treatment area, or if another reason caused it to fail (e.g. water breakthrough from a new water-producing mechanism). If the second treatment is successful, it is likely that the first treatment failed to remain in the rock matrix.

Mechanism A third issue that still has to be resolved is the mechanism or mechanisms by which RPMs reduce water permeability more than oil permeability. Selective pore throat plugging and selective rock surface effects are among the most postulated mechanisms (Liang *et al.*, 1994).

Selective pore-throat plugging implies that the disproportionate permeability effects rely on the ability of an hydrophilic RPM to selectively invade and subsequently plug more water-saturated pore spaces. Laboratory data indicate that some RPMs do, at least partially, rely on this mechanism. RPM formulations that are designed to form a weak gelatinous material after they are placed in the rock matrix will likely rely the most on this mechanism.

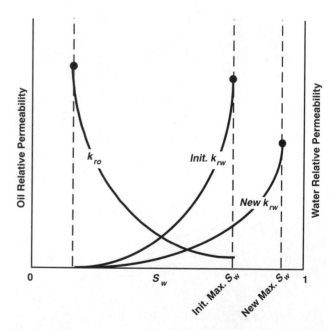

Figure 20-11 How a relative permeability modifier might adjust relative permeability curves

It is also postulated that some RPMs adsorb to the surface of the rock and selectively interact with the water. This selective interaction may include a lubrication or hydrophilic film effect (Zaitoun and Kohler, 1988; Sparlin and Hagen, 1984) where the RPM increases the relative oil mobility, or a gel shrinking and swelling effect (Sparlin and Hagen, 1984), where the polymers tend to shrink more in oil than in water, causing the pore sizes to be smaller when more water is present. An RPM system that is designed to be adsorbing, but will not necessarily thicken in the rock matrix, will probably rely on this mechanism.

Description of Previously Applied Treatments

The applications of currently available systems to either promote an effect or to treat a specific water production mechanism are listed in Table 20-1. Brief descriptions of these systems are provided below.

Mechanical plugs These include mechanical tubing packers and sliding sleeve devices. These can be built into the tubing and inflated or closed by wireline or slickline devices. Mechanical packers can also be run into the well and set after the well is completed. Mechanical water shut-off devices have been used successfully throughout the world, and are most useful when the there is little potential for the water to flow to another open section of the well (Brand *et al.*, 1995).

Mechanical plugs can be used to seal any section of the completed interval.

Sand plugs Sand can be placed in the wellbore to reduce or shut off production of the lower interval of a well. Sand sieve size and plug size can be adjusted to cause either a total seal or a partial plug. Often when a total seal is required, the last portion of the plug can be mixed with cement (or another chemical sealant) to reinforce the seal.

Water-based cement These include either standard or ultrafine cements slurried in water. The small particle size of the ultrafine cements can allow more complete penetration into micro-channels.

Hydrocarbon-based cements These also include both standard and ultrafine cements, but are slurried in oils (usually diesel). The slurries contain surface-active materials that allow them to absorb water from an external source. Oil-based cements are designed to be placed anywhere in a well, but only set if they come into contact with water, allowing a certain amount of selectivity (Crook *et al.*, 1994).

Externally activated silicates These systems generally comprise two stages. The first stage contains a material that will gel a silicate instantaneously upon contact (usually a water-thin $CaCl_2$ brine). The second stage contains the silicate source (viscosity varies from one similar to honey to that of water, depending on the concentra-

Table 20-1   Systems used to affect permeability and to treat specific water-production mechanisms

Desired Treatment Effect	Mechanical plugs	Sand plugs	Water-based cements	Hydrocarbon-based cements	Internally activated silicates	Externally activated silicates	Monomer systems	Crosslinked polymer systems	Wettability modifiers	Microbial systems	Foams
Zone isolation	x		x	x	x	x	x	x			
Permeability reduction					x		x	x		x	x
Relative permeability modification								x	x		
Water production problem											
Acidized into water	x				x		x	x			
Barrier breakdown			x		x	x	x	x			
Bottomwater	x	x	x		x	x	x	x			
Casing leaks			x	x	x	x	x				
Channel behind casing			x	x	x						
Coning or cresting					x		x	x	x	x	x
Fractured into water				x	x	x	x	x			
High permeability channel from injector			x	x	x		x	x	x	x	x
High permeability channel to producer					x		x	x	x	x	x
High permeability streak into aquifer	x		x	x	x		x	x			x

tion of silicate). These silicates are pumped with inert spacers between the stages to keep them separate until they reach the target area. Resulting gels are stiff, brittle solids. Case histories where these silicates have been used are found in Vidick *et al.* (1988) and Murphey *et al.* (1982).

Internally activated silicates (IAS) These systems are generally placed as water-thin freshwater based solutions: a silicate source and an activator designed to trigger gelation of the silicate at a predesignated time. The gel times of silicates depend on the system pH and temperature. Gel times of most currently applied IAS systems are controlled by pH, taking the downhole temperatures into account. The target pH is either achieved on the surface by strong or weak acids or in situ by materials that slowly degrade (either thermally or with time) to form acids. Resulting gels are stiff, brittle solids. Although their effective permeability reduction in a rock matrix can decrease with silicate concentration, they tend to reduce oil and water permeability equivalently. IAS systems have been very effective in field applications (Vinot *et al.*, 1989; Herring *et al.* 1984).

Monomer systems These systems are placed as water-thin solutions containing a low molecular weight material (monomers or oligomers) and an activator. After placement, the activator initiates the polymerization of the monomeric or oligomeric material and results in a solution with a much higher viscosity. Polymerizations are usually activated by adjusting the system to a pH that will allow polymerization at the required time at downhole temperatures (similar to the IAS systems), or by the slow decomposition (either thermal or with time) of the activator to form free radicals capable of initiating polymerization. Monomer systems that have been used commercially include (1) phenol and formaldehyde controlled with pH (solid gels); (2) resorcinol and formaldehyde controlled by pH (stiff fragile gels); (3) acrylamide and an optional bisacrylamide-crosslinker activated by the decomposition of a time-delayed oxidizer (molasses-like to rigid ringing gels); and (4) bifunctional aminoacrylate initiated by thermal decomposition of an oxidizer (lipping to rigid ringing gels). The ability of these systems to reduce permeability is relative to monomer concentration. They also seem to have some relative-permeability effects at low concentrations.

Crosslinked polymer systems These are polymer- and crosslinker-containing systems that are placed in the rock matrix at viscosities low enough to allow injectability (generally, 10 to 200 cp: injectability will depend upon formation permeability). After placement, the systems crosslink to form thick viscoelastic gels (from lipping

gels to rigid gels, while the concentration and crosslinking is increased). At high concentrations, these systems are used for zone isolation. At lower concentrations, they have been used for either permeability reduction or relative permeability modification.

The polymers used are normally water soluble: partially hydrolyzed polyacrylamides (PHPA), thermally stabilized copolymers of PHPA, non-hydrolyzed polyacrylamide (NHPA), cationic polyacrylamide, polyvinyl alcohol, guar, guar derivatives, xanthan, and scleroglucan. PHPA and its copolymers are most commonly used; biopolymers have rarely positive results. Most of the polymers start out crosslinkable, so they need to rely on the crosslinker chemistry for delay. NHPA has no place for the crosslinkers to attach, so its crosslinking is delayed by the slow hydrolysis of the polymer to form crosslink sites (Sydansk, 1993).

Metallic and organic crosslinkers have been used, both of which are generally pumped as "masked" materials that are unable to interact with the polymer until their masks are removed. Metallic "masks" are called ligands, which are strongly attracted to the metal ion by ionic forces. The stronger this attraction and larger the ligand, the longer it takes for the metal to release to crosslink the polymer. The rate at which the metal is released can be controlled by pH or by the ligand concentration in the system. Excess ligand can be added to the crosslinker or the polymer/crosslinker solution to delay metal release (Lockhart and Albonico, 1992), but this often results in weaker crosslinking interactions of the metal with the polymer. Metallic crosslinkers used commercially include chromium acetate (Sydansk, 1993), chromium propionate (Mumallah, 1988), zirconium lactate (Moffit *et al.*, 1996), and aluminum citrate (Stavland and Jonsbraten, 1996).

Organic crosslinkers work in one of two ways: (1) a weakly-attached organic group that is connected to the part of the crosslinker molecule that would crosslink slowly hydrolyzes off, leaving the crosslinker molecule free to react with the polymer (glyoxal: Zaiton *et al.*, 1991; glutaraldehyde: Matre, 1994); (2) components that can slowly form the crosslinker are added to the polymer solution, rather than a crosslinker (phenol/formaldehyde: Moradi-Araghi, 1994).

Surface-active RPMs These systems are generally pumped as low-viscosity solutions containing materials (generally polymers) that adsorb to the surface of the rock matrix. When in place, these materials primarily interact with the rock surface; no significant crosslinking or polymerization is expected (further explanation is given above in the RPM section). Materials that have been used for this include PHPA (Lake, 1989), aminoacrylate copolymers (Pace and Weaver, 1983), and surfactant-alcohol blends (Llave and Dobson, 1994).

Foams Foams have been used to reduce water production, but not often. Generally, these materials are placed as solutions with either dissolved gas that expands after placement, or with no gas so that it subsequently foams with contact with gas downhole. Foams are more commonly used for more in-depth mobility control but can be used as blocking agents as well. A description of the use of foams is found in Seright and Liang (1995) and references therein.

2-4.2.3 Treatment Lifetime

In general, treatments are expected to remain effective at reducing or shutting off water production under downhole conditions indefinitely, depending on the stability of the treatment and the predicted water-production rates in the months and years following the treatment. For treatments that completely shut off production of the target interval, the treatment should prevent further production from that interval indefinitely. For treatments that are designed only to reduce permeability of a zone (reducing either the absolute permeability, the ratio of oil and water permeabilities, or both), the ability of the treatment to maintain low water rates after placement will change as the relative saturation of water changes. In some cases, the relative water saturation will remain constant, but more often, water saturations of a given producing interval increase with the age of that interval. As the relative permeability curves in Figure 20-1 indicate, the relative water permeability of that interval will also rise if the treatment will allow it. Studies have predicted that if treatments force the relative water permeability of an interval to remain constant even as water saturations rise, a water block may result that shuts off all production from that interval.

For many treatments that are to be injected into the rock matrix, the ability of the treatment to reduce permeability relies on the material's resistance to degradation at downhole conditions. This is often tested in the laboratory by exposing bulk treatment samples to reservoir fluids and temperatures. Stability is then judged by a treatment's ability to maintain its initial bulk properties. Most chemical gellants are often only considered stable if the gel viscosity remains constant and if it does not synerese (shrink; leaving partially dehydrated gel balls in water that was originally a part of the bulk gel).

If a gel is exposed to reservoir fluids and temperatures and maintains its bulk viscosity with no apparent syneresis, it will probably survive in the rock matrix. The opposite cannot be assumed: if a bulk gel sample degrades or synereses when exposed to reservoir fluids or temperatures, it may still maintain all of its ability to reduce/stop water flow in the rock matrix (Bryant *et al.*, 1996). The reason for this behavior has not yet been defined. Although bulk testing is relatively convenient, more accurate stability determinations are performed by flooding core material with candidate treatments and monitoring the treatments' long-term effectiveness. It is not completely understood why there is this disparity. However, if the bulk treatment degrades but maintains effectiveness in a core, it is possible that gel in the rock matrix did not degrade because it was exposed to less oxygen (oxygen can be very effective at enhancing polymer degradation). If the bulk gel synereses when exposed to reservoir fluids and temperatures but maintains effectiveness in the rock matrix, it is possible that the gel does not synerese in the rock matrix, or that after syneresis, the remaining concentrated gel particles block pore throats in such a way that no open channels exist through which fluids can flow.

20-4.2.4 Selecting Treatment Composition and Volume

Choosing which treatment to use to either reduce or prevent excessive water production is an exercise in balancing technical aspects (strength, depth, and stability requirements) with economic aspects (volumes, concentrations, and treatment types). The best technical solution may also be twice the cost of the next best option, but may potentially result in three times the economic benefits. On the other hand, an inexpensive, technically inferior treatment may not reduce water production enough to offset even its minimal cost. Some general relationships can be drawn: strength is proportional to concentration, and depth and volume requirements decrease with increasing gel strength. Another important factor is environmental regulations. When a system is chosen, it must also meet the environmental and toxicity requirements of the area of application. Currently, these requirements are most stringent in the Norwegian sector of the North Sea.

When the expectations of a treatment are defined, and the placement and long-term downhole conditions are determined, available treatments are screened for their capability to fulfill these needs. This screening, along with matching environmental legislation, may result in one choice or several. It may be determined that there are a few candidate systems either based on different chemicals, or based on the same chemicals but varying in concentration or penetration requirements. For instance, a crosslinked polymer treatment made of the same chemical components will seal the target zone if Formulation A is used, and will only reduce the target zone's permeability if Formulations B or C are used (Table 20-2).

Formulation B is an example of a lower concentration system that will produce a gel that causes less damage to a given volume of rock than Formulation A; Formulation B will result in a weaker gel. Formulation C can be just as damaging to the rock in which it is injected, but there is less of it to create the effect. This example can only apply if Formulations A and C are of low enough concentration that their ability to create damage is relative to their penetration depth (some systems are so concentrated that they need no penetration to form a seal).

20-4.3 Placement

Bullheading This is a single injection method (one fluid pumped at a time) where the treatment is pumped down the existing tubing with no mechanical zone isolation to divert it into the target zone. This can be an effective method if the only open zone is the target zone, or if the treatment will not significantly damage zones adjacent to the target zone. Bullheading is often the least expensive and most simple placement technique.

Mechanical packer placement This is also a single injection technique that uses a packer to isolate the target zone. This includes isolating with straddle packers, single packers, sand plugs, or a combination.

Dual injection This technique is a hybrid of the mechanical packer technique. Mechanical isolation of the target zone is first achieved. To prevent invasion of the treatment into an adjacent zone, a second non-damaging fluid is pumped simultaneously. If the target zone is above the zone that needs protection, the treatment is placed down the tubing annulus and the second fluid is placed down the tubing; if the target zone is below the zone to be protected, the treatment is placed down the tubing and the second fluid is placed down the annulus. The injection pressure of both fluids is kept equivalent.

Isoflow This placement technique provides a method of directing the treatment to the target zone without using mechanical zone isolation. The end of a tubing string is placed at the top of the target interval. Two fluids are pumped simultaneously, one down the tubing and the other down the annulus, depending on the relative location of the target zone. A logging tool is hung at the end of the tubing. The fluid that is pumped down the annulus is doped with a material that can be detected by the logging tool. During placement, the location of the interface between the two fluids is monitored, and the pump rates are adjusted throughout the job so that the level of this interface remains at the end of the tubing. When the logging tool detects the material going down the annulus, the relative pump rate of the fluid going down the tubing is increased.

20-4.3.1 Viscosity Considerations

When zone sealing or absolute permeability reduction fluids are needed, zonal isolation to prevent damage of non-target intervals (or to prevent unnecessary waste of treatment) is recommended. In some cases, however, the target zone cannot be completely isolated or is internally laminated with varying permeability lenses so that it is not easy (or possible) to isolate the fraction of water producing perforations.

Systems that do not completely seal the zone are often chosen for this type of application. In this case, it might be most effective to treat with a sealing formulation that will preferentially penetrate in the higher-permeability, watered-out lens. Studies have indicated that materials that are placed as water-thin (or very low viscosity) tend to preferentially inject into the higher-permeability layers; higher viscosity fluids tend to self-divert (Sorbie and Seright, 1992). The risk of damaging adjacent lower-

Table 20-2 Comparison of treatments

Formulation A: Zone sealant	Formulation B: PRA	Formulation C: PRA
1.5% Polymer A 1000 ppm Crosslinker X Mixed in 2% KCl brine Treatment Volume: 100 m³	0.7% Polymer A 500 ppm Crosslinker X Mixed in 2% KCl brine Treatment Volume: 100 m³	1.5% Polymer A 1000 ppm Crosslinker X Mixed in 2% KCl

permeability lenses can also be reduced by careful job monitoring and/or variations in treatment formulation throughout the job.

20-4.3.2 Temperature Considerations

For systems that react after placement, treatment compositions also often determine set times. When wells are treated with systems that gel in the rock matrix, wells are shut in after placement until the treatment thickens to prevent placed treatments from either being produced back, flowing into a non-target zone by crossflow, or being pushed away from the wellbore by injection water. Historically, treatment gel times were designed based on the placement time required (pump-rate dependent) and the bottomhole static temperature (BHST). This assumes the treatment will be at BHST while it is setting, an appropriate assumption for low-volume treatments. However, when fluids are injected into a zone, the temperature of that zone cools down significantly: cooldowns are more dramatic at higher injection rates and volumes.

Surprisingly, simulations indicate that when wells are shut in after treatment, the time required for a zone to reheat to reservoir temperature can be days or even weeks (Figure 20-12). If a large treatment is required, cooldown simulations should be performed and treatments should be designed to react in a reasonable time frame. This can mean either adjusting the pumping rate, the treatment formulation, or both. The decision can also depend on the system chosen; there may be temperature/time limitations, or time/strength dependency. For instance, if an IAS system is to be placed, it might be better to inject the treatment at as high a rate as possible and adjust the activator concentration or composition so that the material placed last will set within a couple of hours at the predicted low temperature, rather than to place a slow-setting IAS formulation slowly to avoid cooldown. The former will result in a stronger gel at the wellbore, where needed, and a shorter placement time (shorter down-time of the well, less loss of revenue, and faster treatment pay out).

REFERENCES

Ahmed, U.: "Horizontal Well Completion Recommendations Through Optimized Formation Evaluation," paper SPE 22992, 1991.

Andersen, S.A., Hansen, S.A., and Fjeldgaard, K.: "Horizontal Drilling and Completion: Denmark," paper SPE 18349, 1988.

Bale, A., Smith, M.B., and Settari, A.: "Post-Frac Productivity Calculation for Complex Reservoir/Fracture Geometry," paper SPE 28919, 1994.

Beggs, H.D.: "Production Optimizing Using Nodal Analysis," Oil and Gas Consultants Int. Inc., OK (1991).

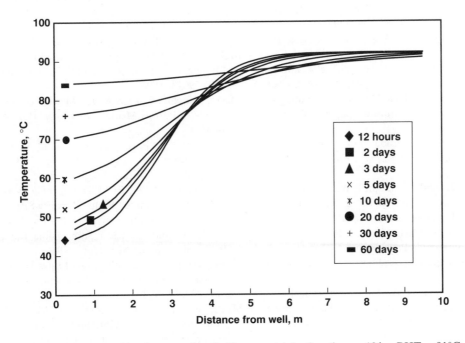

Figure 20-12   Predicted heat-up profile of a treated well: Treatment injection time = 12 hr, BHT = 91°C, treatment radius = 7

Bourgoyne, A.T. Jr., Millheim, K.M., Chenevert, M.E., and Young, F.S. Jr.: *Applied Drilling Engineering*, SPE, Richardson, TX (1986).

Brand, P.J., Clyne, P.A., Kirkwood, F.G., and Williams, P.W.: "The Forties Field: 20 Years Young," paper SPE 30440, 1996.

Brekk, K., and Lien, S.C.: "New, Simple Completion Methods for Horizontal Wells Improve Production Performance in High-Permeability Thin Oil Zones," paper SPE 24762, 1994.

Bryant, S.L., Rabaioli, M.R. and Lockhart, T.P.: "Influence of Syneresis on Permeability Reduction by Polymer Gels," paper SPE 35446, 1996.

Chaperone, I.: "Theoretical Study of Coning Toward Horizontal and Vertical Wells in Anisotropic Formations: Subcritical and Critical Rates," paper SPE 15377, 1986.

Chierici, G.L., Ciucci, G.M., and Pizzi, G.: "A Systematic Study of Gas and Water Coning by Potentiometric Models," *JPT* (Aug, 1964) p. 923–929.

Chitwood, G.B., and Coyle, W.R.: "New Options To Eliminate Corrosion of Completion Equipment in Water Injection Service," SPE paper 27853, 1994.

Coats, K.H.: "Reservoir Simulation," *Petroleum Engineering Handbook*, SPE, Richardson, TX (1987), Chapter 48.

Coulter, A.W. Jr., Martinez, S.J., and Fischer, K.F.: "Remedial Cleanup, Sand Control, & Other Stimulation Treatments," *Petroleum Engineering Handbook*, SPE, Richardson, TX (1987), Chapter 56.

Cox, J.B., and Babitzke, E.H.: "Completion Design for Wells with High H_2S and CO_2 Content," unsolicited, SPE 19513, 1989.

Craig, B.D., Straatmaan, J., and Petersen, G.J.: "Careful Planning Improves CRA Selection for Corrosive Environments," *Pet. Eng. Intl.* (July 1992), p. 26–30.

Crook, R., Hadar, S., and Hardy, M.: "Conformance Control Extends Well's Life," paper ADSPE 105, 1994.

Dake, L.P.: *The Practice of Reservoir Engineering*, Elsevier, The Netherlands (1994).

Dikken, B.J.: "Pressure Drop in Horizontal Wells and Its Effect on Production Performance," paper SPE 19824, 1990.

Georgie, W.J., Bryne, K.H., and Kjaerland, G.: "Handling of Produced Water: Looking to the Future," paper SPE 25040, 1992.

Gidley, J.L., Holditch, S.A., Nierode, D.E., and Veatch, R.W.Jr.: *Recent Advances in Hydraulic Fracturing*, SPE, Richardson, TX (1989) **12**.

Gilman, J.R., Bowzer, J.L., and Rothkopf, B.W.: "Application of Short-Radius Horizontal Boreholes in the Naturally Fractured Yates Field," paper SPE 28568, 1994.

Gutzeit, J., Merrick, R.D., and Scharfstein, L.R.: "Corrosion in Petroleum Refining and Petrochemical Operations," *Metals Handbook*, 9th Edition, ASM International (1987) **13**, Chapter 3.

Guo, B., and Lee, R.L-H.: "A Simple Approach to Optimization of Completion Interval in Oil/Water Coning Systems," paper SPE 23994, 1993.

Harrison, R.D., Restarick, H., and Grigsby, T.F.: "Case Histories: New Horizontal Completion Designs Facilitate Development and Increase Production Capabilities in Sandstone Reservoirs," paper SPE 27890, 1994.

Hassall, A., Lockwood, D., and Akpor, N.A.: "Reservoir Description of South Tano Field, Offshore Ghana: A Case History," AAPG Poster Session (Mar. 1995).

Herring, G.D., Milloway, J.T., and Wilson, W.N.: "Selective Gas Shut-Off Using Sodium Silicate in the Prudhoe Bay Field, AK," paper SPE 12473, 1984.

Kortekaas, T.F.M.: "Water/Oil Displacement Characteristics in Crossbedded Reservoir Zones," *SPEJ* (Dec. 1985) 917–926.

Laing, C.M., Ogier, M.J., and Hennington, E.R.: "Everest and Lamond Completion Design Innovations Lower Completion and Workover Costs," paper SPE 26743, 1993.

Lake, L.W.: *Enhanced Oil Recovery*, Prentice Hall, NJ (1989).

Landman, M.J., and Goldthorpe W.H.: "Optimization of Perforation Distribution for Horizontal Wells," paper SPE 23005, 1991.

Liang, J.T., Sun, H., and Seright, R.S.: "Why Do Gels Reduce Water Permeability More Than Oil Permeability?," paper SPE 27829, 1994.

Llave, F.M., and Dobson Sr., R.E.: "Field Application of Surfactant-Alcohol Blends for Conformance Control," paper SPE 28618, 1994.

Lockhart, T.P., and Albonico P.: "A New Gelation Technology for In-Depth Placement of Cr^{3+}/Polymer Gels in High-Temperature Reservoirs," paper SPE 24194, 1992.

Maddox, S., Gibling, G.R., and Dahl, D.: "Downhole Video Services Enhance Conformance Technology," paper SPE 30134, 1995.

Maldal, T., Gilje, E., Kristensen, R., Kårstad, T., Nordbotten, A., Schilling, B.E.R., and Vikane, O.: "Planning and Development of Polymer Assisted Surfactant Flooding for the Gullfaks Field, Norway," paper SPE 35378, 1996.

Matre, B., Boutelje, J., Tengberg-Hansen, H., and Tyvold, T.: "Evaluation of Gel Systems for High Temperature," paper SPE 29013, 1994.

Meyer, H.I., and Garder, A.O.: "Mechanics of Two Immiscible Fluids in Porous Media," *J. Applied Phys.* (Nov. 1954) **25**, 1400.

Muecke, T.W.: "Formation Fines and Factors Controlling Their Movements in Porous Media," *JPT* (1979), 144.

Moffit, P.D., Moradi-Araghi, A., and Ahmed, I.: "Development and Field Testing of a New Low Toxicity Polymer Crosslinking System," paper SPE 35173, 1996.

Moradi, S.C.: "An Innovative Single Completion Design With 'Y Block' and Electrical Submersible Pump for Multiple Reservoirs," paper SPE 17663, 1988.

Moradi-Araghi, A.: "Application of Low-Toxicity Crosslinking Systems in Production of Thermally Stable Gels," paper SPE 27826, 1994.

Mukerjee, H., and Economides, M.J.: "A Parametric Comparison of Horizontal and Vertical Well Performance," paper SPE 18303, 1991.

Mumallah, N.A,: "Chromium (III) Propionate: A Crosslinking Agent for Water-Soluble Polymers in Hard Oilfield Brines," *SPERE* (Feb. 1988) 243–250.

Murphey, J., Young, W., and Oberpriller, F.: "Treatment of Lost Circulation and Water Production Problems with a Powdered Silicate," paper CIM 82-33-46, 1982.

Muskat, M. and Wyckoff, R.D.: "An Approximate Theory of Water Coning in Oil Production," *Trans. AIME* (1935), **114**, 144–161.

Nolan-Hoeksema, R.C., Avasthi, J.M., Pape, W.C., and El Raba, A.W.M.: "Waterflood Improvement in the Permian Basin: Impact of In-Situ-Stress Evaluations," paper SPE 24873, 1994.

Othman, M.E.: "Review of Dual Completion Practice for Upper Zakum Field," paper SPE 15756, 1987.

Pace, J.R., and Weaver, J.D.: "The Use of Water-Oil-Ratio Control Agents To Improve Oil Production," paper SPE 11448, 1983.

Patton, C.C.: "Water Formed Scales," *Oilfield Water Systems*, Campbell Petroleum Series, Oklahoma City, OK (1974).

Pelgrom, J.J., Mann, J.P., Davidson, C.J., Mattjes, G.R., and Risseeuw, A.S.: "Improved Well Design Maximizes Oil-Rim Profitability," paper SPE 27737 (1994).

Pirson, S.J.:*Oil Reservoir Engineering*, R.E. Krieger Publishing Co., New York (1977).

Richardson, J.G., Sangree, J.B., and Sneider, R.M.: "Coning," paper SPE 15787, 1987.

Schilthuis, R.J.: "Active Oil and Reservoir Energy," *Trans.,* AIME 1936.

Seright, R.S., and Liang, J.: "A Comparison of Different Types of Blocking Agents," paper SPE 30120, 1995.

Sorbie, K.S., and Seright, R.S.: "Gel Placement in Heterogeneous Systems With Crossflow," paper SPE 24192, 1992.

Sparlin, D.D., and Hagen, R.W.: "Controlling Water in Producing Operations - Part 5," *World Oil.* (June, 1984), 137-142.

Srinivasan, S., Joshi, J.M., Narasiham, J.L., Rao, M., Barua, S., and Jha, K.N.: "Feasibility of Development of Marginal Fields Through Horizontal Well Technology," paper SPE 35439, 1996.

Stavland, A., and Jonsbraten, H.C.: "New Insight Into Aluminum Citrate/Polyacrylamide Gel for Fluid Control," paper SPE 35381, 1996.

Sydansk, R.D.: "Acrylamide-Polymer/Chromium (III) - Carboxylate Gels for Near Wellbore Matrix Treatments," paper SPE 20214, 1993.

Thomas, C.E., Mahoney, C.F., and Winter, G.W.: "Water-Injection Pressure Maintenance and Waterflood Processes," *Petroleum Engineering Handbook*, SPE, Richardson, TX (1987), Chapter 44.

Tollefsen, S., Graue, E., and Svinddal, S.: "Gullfaks Development Provides Challenges," *World Oil* (Apr. 1994), 45–54.

Turki, W.H.: "Drilling and Completion of Khuff Gas Wells, Saudi Arabia," paper SPE 13680, 1985.

van Batenburg, D.W.: "The Effect of Capillary Forces in Heterogeneous 'Flow-Units': A Streamline Approach," paper SPE 22588, 1991.

van Everdingen, A.F., Timmerman, E.H., and McMahon, J.J.: "Application of the Material Mass Balance Equation to a Partial Water-Drive Reservoir," *Trans.,* AIME (1953) **198**, 51–60.

Vidick, B., Yearwood, J.A., and Perthuis, H.: "How to Solve Lost Circulation Problems," paper SPE 17811, 1988.

Vinot, B., Schechter, R.S., and Lake, L.W.: "Formation of Water-Soluble Silicate Gels by the Hydrolysis of a Diester of Dicarboxylic Acid Solubilized as Microemulsions," paper SPE 14236, 1989.

Wheatly, M.J.: "An Approximate Theory of Oil Water Coning," paper SPE 14210, 1985.

Wilkie, D.I., Kennedy, W.L., and Tracy, K.F: "Produced Water Disposal—A Learning Curve in Yemen," paper SPE 35030, 1996.

Yang, W., and Wattenbarger, R.A.: "Water Coning Calculations for Vertical and Horizontal Wells," paper SPE 22931, 1991.

Zaitoun, A., and Kohler, N.: "Two-Phase Flow Through Porous Media: Effect of an Adsorbed Polymer Layer," paper SPE 18085, 1988.

21 Designing Well Completions for the Life of the Field

Shari Dunn-Norman
University of Missouri-Rolla

Clark Robison
Halliburton Energy Services

21-1 INTRODUCTION

21-1.1 Background

Ideally, every well completion design should optimize production vs. costs and, therefore, provide the most profitable operation of the well over its producing life. The incremental net present value (NPV) criterion and the comparative cost-to-benefit ratio are two means of quantifying the optimization exercise.

The completion design should also ensure certain aspects of the well's operation, such as safety, availability, and efficient use of equipment inventory. Unfortunately, not all well completions achieve this objective. In part, their failure to do so is a function of the manner in which they have been designed.

Most frequently, well completion designs have some historical basis. Specifically, they are based on previous designs that were proven acceptable in a different field or well and appear to satisfy requirements similar to the current design problem. The engineer uses this historical design and adapts it to meet specific criteria and accommodate local operating experience.

This approach to well completion design is problematic for several reasons. First, the conditions and rationale used to develop a design are not readily discernible by inspecting its completion schematic. When using this approach, engineers must have a high level of experience and operations knowledge to apply designs correctly.

When historical designs are adapted, a coherent and consistent design philosophy may or may not be applied. Rather, the design philosophies used for formulation of

the historical design, regardless of whether they are evident or not, are simply translated into the current design. With multiple engineers specifying completions in many different areas, this approach can lead to inexplicable geographical completion design practices.

The third, and perhaps most significant limitation is that the use of historical designs does not stimulate consideration of a wide range of design alternatives. If all design options are not explored, the selected design cannot be verified as being optimum. Although the design selected could be the best, no basis will be available for demonstrating it.

An alternative methodology for designing wellbore compilations is to formulate designs anew, by structuring and applying their design reasoning. Several possible mechanisms for this method include the use of checklists or design manuals. Ideally, such approaches should be formulated into design models.

21-1.2 Historical Modeling Attempts in Well Completion Design

Patton and Abbot (1985) introduced the first formalized approach to well completion design, in which they depicted it as a closed system consisting of components that interact with one another. Components within the completion system include any object described by attributes. Components outside the system are considered to be part of the environment, but these components may influence the system. Both types of components can act as constraints (limitations placed on the operation of the system) or as resources on which the system depends.

Patton and Abbot (1985) used the system description as the basis for the systems approach to well completion design. This approach is an integrated design procedure in which each component of the design is configured with respect to the other system elements. To formalize the systems approach, Patton and Abbott used flow diagrams that demonstrated the information required for design decisions.

Peden and Leadbetter (1986) extended Patton and Abbott's diagrammatic representation of well completion design. In their work, each component was treated individually (Figures 21-1 and 21-2). The design links between components were noted, although their exact means of interaction was not shown. This work is significant because it recognizes the decomposable nature of the completion design problem.

Dunn-Norman (1990) developed a prototype knowledge-based system in well completion design. This model included a representation of the design process applied to a North Sea completion. This work demonstrates the relationship between heuristic knowledge and engineering computations in well completion design.

None of the attempts to formalize the completion design process have resulted in an agreed methodology for completion design. However, the historical work has value because it identifies many of the elements (functional requirements, time-related changes, etc.) that must be considered when new completion designs are formulated.

21-2 THE PROCESS OF WELL COMPLETION DESIGN

The process of well completion design includes the reasoning and functions that must be performed to specify the components required to complete a well. This process encompasses production engineering, detailed in previous chapters of this book, as well as operational considerations.

Figure 21-3 depicts the process of completion design. As shown, no deterministic procedure exists for a well completion design. Rather, key topics such as perforating or tubular selection are considered after the gross objectives and functional requirements of the completion have been identified.

At the outset of the design process, some initial knowledge of the reservoir is known in addition to wellbore, fluid, and well test data. Based on an understanding of the reservoir and other parameters, general types of well completions (cased hole, gravel-pack, subsea) can be considered. These general types of completions act as a guide in formulating the final design. Identifying relevant types of completions could also aid in formulating functional requirements for the design. This initial design process is highlighted in Figure 21-4 and detailed in Figure 21-5.

21-2.1 Functional and Well Service Requirements

Functional requirements are the operational or functional specifications that a well completion must satisfy. These requirements act as a framework for the developing design.

Table 21-1 lists example functional requirements for a subsea well completion. These requirements can be either implicit or explicit. Implicit requirements are functions generally expected of all well completions in a certain geographical area, while explicit requirements are those functions that the designer uniquely specifies. In practice, experienced engineers may satisfy implicit functional requirements without consciously noting these in a list.

Some functional requirements may not exist at the outset of the design process; instead, they evolve as certain design decisions are made. For example, if a decision is made to run a permanent packer with a stab-in seal assembly, then this decision would generate a requirement to set the permanent packer (either with wireline or a workstring) before the main completion is run.

Functional requirements can also evolve in conjunction with time-dependent design parameters. For example, if reservoir pressure declines, then artificial lift may be required at some time.

By specifying the functional and well servicing requirements at the outset of the completion design process, engineers can simplify the selection of preliminary completion concepts (Figure 21-5), and can highlight the key tradeoffs between completion design alternatives.

Ideally, a complete and comprehensive list of functional requirements should be specified at the outset of the design process. This is not a trivial task. Even highly experienced and skilled engineers may have difficulty foreseeing all the operational consequences of a particular design, or the design provisions that must be made for decreases in reservoir pressure, changes in fluid saturation, and other time-dependent design criteria. Nevertheless, specifying functional requirements is a fundamental step in the completion design process.

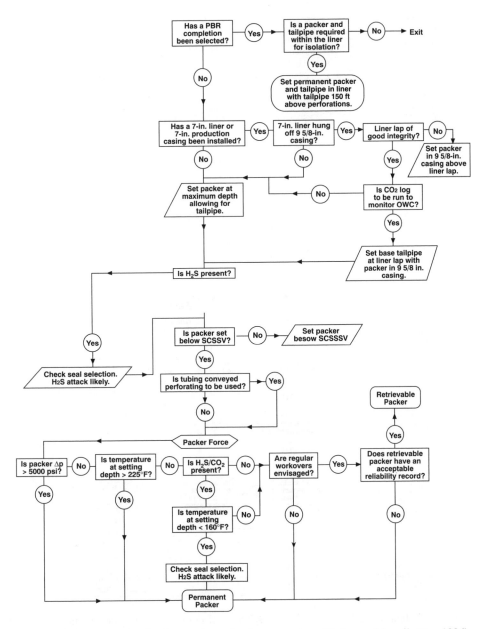

Figure 21-1   Production packer selection procedure (Peden and Leadbetter, 1986)

The following example illustrates how two different types of completions can be considered at the outset of the completion design process and how the completion alternatives can be compared on the basis of NPV.

21-2.2 Example Application for Well Completion Selection: A Production Engineering Approach

21-2.2.1 Background

Suppose that a 2000-ft horizontal well is drilled in a sandstone reservoir. The principal functional require-

ments of the completion for this reservoir (Figure 21-4) are to (1) provide an acceptable productivity index improvement over a vertical well and (2) control sand production at the surface.

Both cementing/perforating and a gravel pack with screen could potentially fulfill the second requirement. While the first requirement is met by both completions initially, the gravel-pack completions appear to deteriorate with time. Because sand particles are retained by the gravel pack, skin effects equal to 3, 5, and 10 respectively have been measured in the first three successive years. In contrast, cementing and perforating is more expensive

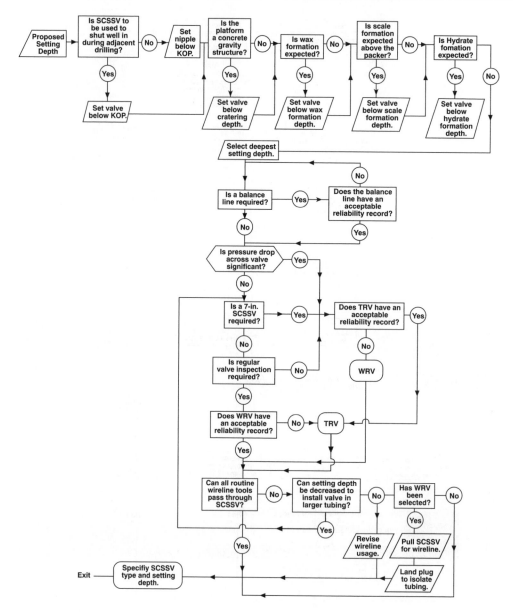

Figure 21-2   Surface-controlled subsurface safety valve (SCSSV) selection procedure (Peden and Leadbetter, 1986)

and far more cumbersome to install. Which of the two completion styles should be selected?

Other pertinent variables are $p_e = 4000\,\mathrm{psi}$, $B = 1.1\,\mathrm{resbbl/STB}$, $\mu = 1\,\mathrm{cp}$, $r_e = 1500\,\mathrm{ft}$, $k_H = 3\,\mathrm{md}$, $k_V = 0.5\,\mathrm{md}$, $r_w = 0.328\,\mathrm{ft}$.

Assume that the reservoir is under strong edgewater drive, and thus the production mechanism is essentially steady-state for the first few years.

21-2.2.2 Solution

Equation 15-27 and ancillary expressions for the index of anisotropy and the half-axis of the drainage ellipse a are sufficient for the solution to this problem. The damage skin is added to the second logarithmic term in the denominator of Equation 15-27, and it is multiplied by the scaled aspect ratio, $\beta h/L$. Thus, the index of anisotropy is $\beta = 2.4$, and the half-axis of the drainage ellipse is $a = 1675\,\mathrm{ft}$.

The production rates for $s = 0$, 3,5, and 10 are 885, 795, 745 and 644 STB/d, respectively. If the price of oil is \$20/bbl and the discount rate is 20%, then the 3-year NPVs of the production from the two wells are \$13.6 million and \$11.3 million, respectively. This results in an incremental discounted revenue of \$2.3 million, which should readily cover the incremental cost of

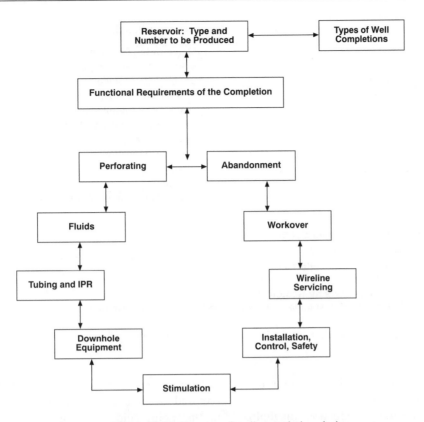

Figure 21-3 Process of well-completion design

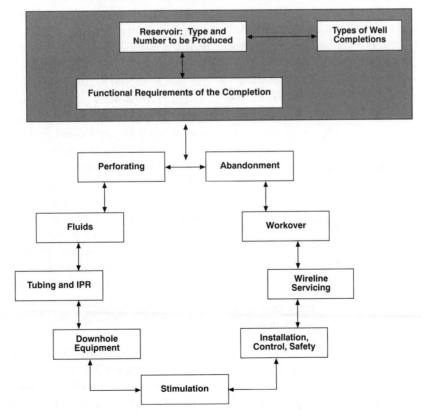

Figure 21-4 Highlight of initial process of well-completion design

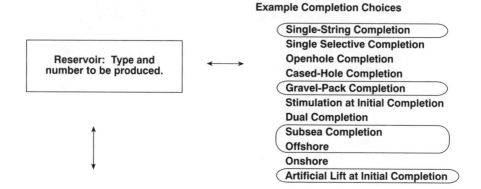

Example Completion Choices

Single-String Completion
Single Selective Completion
Openhole Completion
Cased-Hole Completion
Gravel-Pack Completion
Stimulation at Initial Completion
Dual Completion
Subsea Completion
Offshore
Onshore
Artificial Lift at Initial Completion

Example Functional Requirements of the Well Completion

- Commingle Reservoirs
- Provide Sand Control at Initial Completion
- Use SCSSVs for Offshore Well
- Equip with Dual SCSSVs for Redundancy in Subsea Well
- Provide SPMs in Design at Initial Completion, to Allow for Future Gas-Lift

Figure 21-5   Specifying functional requirements and types of completions

cementing and perforating the well. The remaining revenue would be incremental benefit.

(For the 3-year NPVs, the daily production was multiplied by 365 days and \$20; then, the figure was divided by $(1 + i)^n$, where i is the discount rate and n is the number of years. For the damaged well, the calculated rates for Years 1 through 3 were used. For the undamaged well, a constant rate equal to 885 STB/d was used.)

Once the cemented and perforated completion design is selected, other requirements of the completion are identified and a detailed design is formalized.

21-2.3 Detailing the Completion Design

Tubing specification is one of the earliest decisions made in a completion design, provided that the well has already been drilled and cased, or that the drilling/casing program has been established. Tubing specification includes the size, grade, and weight of the tubing, as well as the length of the string to be run.

The tubing is sized based on the reservoir's inflow capacity (Chapter 15), which will have been measured in well tests. Because this capacity is time-dependent, flow-rate changes must be predicted so that the selected tubing size can produce the expected range of rates.

The tubing grade and weight are based on the corrosivity of the produced fluids and the strength that the tubulars must provide (Chapter 7). The corrosivity of the environment will be known, although it might

become more severe with time (Chapter 14). The required strength of the tubulars might not be known at this point, and a tubing weight might need to be assumed on a tentative basis.

The reason that tubing design is one of the earliest decisions in the design process is that all other components in the string are sized with respect to the tubing. Unfortunately, this design process falsely suggests that the tubing should *always* be the first item specified for the design. Subsea wells can be an exception, since long lead times for subsea trees and wellheads may mean that these components are considered first, and then designated as constraints for the remaining design.

A perforating program (Chapter 13) might also need to be specified before the tubing size is determined, since the perforation density and length of interval must be known before the well's inflow capacity can be calculated. The perforating program will include the type of gun to be run, the charge type and size, the shot density, and phasing. Mechanical aspects of the perforating program (how the guns will be conveyed to depth) might depend on the specifications of other components in the completion, particularly if tubing-conveyed perforating is used.

The packer (Chapter 14) is another component that is specified early in the design process, because it must coordinate with a number of other devices, such as the tubing-to-packer connection, the seal assembly, and any components that compensate for tubing movement. The

Table 21-1   Example functional requirements–Subsea completions (IHRDC, 1987)

Completion Considerations		Importance of Need	Competion Design Implications
Rates	High	None	- Favors Two Small Tubing Strings with Chokes
	Moderate	High	
	Low	Possible	
	Variable	Critical	
Pressures	High	None	
	Low	Probable	- Artificial Lift Required
Producing Characteristics	Multiple Zones	Possible	- Stack Completions
	Minimize Costs	Moderate	- Review Costs
	Access Difficulty	High	- TFL/New Technology
	Uptime	High	- Minimize Difficulty to Future Workovers
	Rate Control	Critical	- Chokes Needed
	Rate Stability	Critical	- Wellhead Chokes Needed
	Long Life	Unlikely	- Carbon Steel Sufficient
	Other		
	● Density of Kill Fluid	Moderate	- Kickoff with Gas-Lift
	● Safety During Vessel Re-entry	Critical	- 2 SSSVs and Kill System
	● Wellhead Damage	Possible	- Annular SSSV
Monitoring	Test Frequency	High	- Critical Choke Beam or Dedicated Flowline
	Pressure Measurement	Moderate	- TFL Access for Downhole Tools
	Special BHP Surveys	Some Needed	- TFL Access for Downhole Tools
	Log Contacts	Critical	- Vertical Access Required
	Production Logs	Some Needed	- Vertical Access Required
	Tubing Investigation	High	- TFL Access and/or Vertical Access
Artificial Lift	Intermittent	High	- Gas-List is Optimal Method with Maintenance via TFL and Vertical Access
	Continuous	Possible	
	Increasing Gross Rate	High	
	Pressure Depletion	Possible	
Kick-Off	Initial Completion	Moderate	- Use Gas-Lift System
	Routine Operations	High	- Gas Compressor Supply Required
	Depleted Conditions	Possible	
	High Water-Cut	High	
	Critical Rate	High	- GLV Maintenance System
	Frequency	High	
	Gas Supply Volume	Moderate	- Gas Compressor Special Requirements
	Gas Supply Pressure	Design Variable	
Repairs	Cement	High	- Future Concurrent Production and Workover Operations; Easy Access; Robust Tubing Joints
	Gravel-Pack	Critical	
	SSSV	Probable	
	Tubulars	Low	
	New Interval	Possible	- Multizone Completion Design
Recompletions	Uphole	Moderate	- Large Casing Preferable
	Deepen	None	- Limit Depth of Rathole
	Sidetrack	Possible	- Maximize Casing Size
	Function Change	Moderate	- Large CSG Preferable
Well Kill	Frequency	High or Low	- Operations Procedure
	Difficulty	Mod - High	- Alternate Methods
Production Problems	Sand Control	Critical	- Gravel-Pack Required
	Paraffin	Possible	- TFL Access for Scraping
	Emulsions	Possible	- Chemical Injection Capability
	Water Cut	High	- Artificial Lift Required
	Scale	Possible	- TFL Access
	Corrosion	Moderate	- Carbon Steel and Downhole Chemical Inhibitor Injection
	Other		
	● Erosion	Low	
	● Fines	Probable	- Frequent Acid Jobs Required
	● GP Failure	Moderate	- TFL with Annular Kill Valve

packer and its design links to other components determine a significant part of the completion design.

Once the packer has been selected, there is less natural sequential ordering to the completion design problem, and only the interactions between completion components, fluids, and operations must be considered (Figure 21-6). For example, a subsurface safety valve (SSV) can be selected based on the considerations discussed in Chapter 14. Valve metallurgy will be determined based on the reservoir fluids to be produced and any anticipated future well treatments. Valve type and retrievability will depend on whether full tubing workovers can be justified, or whether wireline servicing is needed.

These points illustrate a few of the critical design considerations undertaken after the general completion type has been decided. Specific functional requirements such as servicing an SSV without tubing removal, tubing strength great enough to withstand 8000 psi pressure for future stimulation, or a provision for 12 ft of tubing shortening serve as drivers for the detailed design.

21-3 SPECIFYING THE LIFE OF A WELL COMPLETION

The life of a well completion is the time that a particular design will be in service without major modification. Engineers designing well completions must specify this life, or at least have a period of service in mind, when designing a well completion. Ideally, the completion should remain in the ground until the reservoir is abandoned. However, in many cases, this goal might not be feasible with current completion technology. In such cases, shorter periods of service life are specified, and these periods are determined in various ways.

In flowing wells, a common method of determining design life is to match the tubular performance for a particular tubing size to the inflow performance of the reservoir over a period. If the tubing size must be reduced as reservoir pressure declines, then the design life would be from the initial time until the time when the tubing must be changed. Other downhole equipment, fluids, and procedures would be specified in accordance with this period.

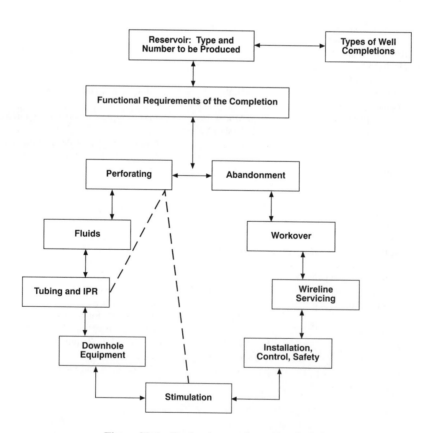

Figure 21-6   Design interactions (detailed design phase)

Well servicing can be another consideration in determining well life. For flowing wells, servicing may not be a major concern, unless tubing leaks or productivity declines are significant issues. However, in wells produced with artificial lift, or wells equipped with safety valves that must be periodically tested, the time to downhole/equipment failures will certainly affect the duration of a completion's design life. In practice, engineers apply their operating experience to make this determination.

Recompletion to another reservoir, or stimulation requirements may also determine the life of a completion. Unless the initial design provides a wireline or coiled tubing method of recompletion or stimulation, then the initial completion must be modified or removed to perform these tasks.

In general, any change in the reservoir or produced fluids that dictates a change in the completion that cannot be provided for at the outset of the completion design, or that cannot be achieved through surface or remote control, will require completion modification. The time of this modification is the design life for the completion.

21-4 PREDICTING AND ACCOMMODATING COMPLETION DESIGN CHANGES

To design completions that can operate for the life of the field or for some shorter period of service, engineers must determine the design parameters that will change as a function of *time*. The following discussions detail some typical reservoir and fluid parameters that may change with time.

21-4.1 Reservoir Parameters

21-4.1.1 Pressure Depletion and Material Balance

During primary recovery, a reservoir produces fluids without additional pressure maintenance. If the reservoir is initially above bubble point, then the wells may be choked so that reservoir pressure can be maintained at the bubble point, which prevents premature gas production. Eventually, however, the reservoir pressure will decline as a function of reservoir voidage (pseudosteady state), or an increased drawdown will be required to produce the well at a constant rate (steady state).

Material balance equations predict reservoir pressure decreases as a function of time. Dake (1978) presented the general material balance equation (ignoring water influx and production):

$$
\begin{aligned}
N_p\big[B_o + (R_p - R_s)B_g\big] &= NB_{oi} \\
&\times \left[\frac{(B_o - B_{oi}) + (R_{si} - R_s)B_g}{B_{oi}} + m\left(\frac{B_g}{B_{gi}} - 1\right)\right] \\
&+ (1 + m)\left(\frac{c_w S_{wc} + c_f}{1 - S_{wc}}\right)\Delta p\bigg]
\end{aligned} \tag{21-1}
$$

Equation 21-1 can be simplified further if the reservoir is assumed to have no initial gas cap but rapidly passes below the bubble-point pressure after production begins. These assumptions encompass a large number of reservoirs. This mode of recovery is known as solution gas drive, and the calculation method outlined below is referred to as Tarner's (1944) method. It was described further by Craft and Hawkins (1991).

$$
N_p\big[B_o + (R_p - R_s)B_g\big] = N\big[(B_o - B_{oi}) + (R_{si} - R_s)B_g\big] \tag{21-2}
$$

Since G_p (gas produced) $= N_p R_p$, then

$$
N_p(B_o - R_s B_g) + G_p B_g = N\big[(B_o - B_{oi}) + (R_{si} - R_s)B_g\big] \tag{21-3}
$$

and therefore

$$
N = \frac{N_p(B_o - R_s B_g) + G_p B_g}{(B_o - B_{oi}) + (R_{si} - R_s)B_g} \tag{21-4}
$$

Equation 21-4 has two distinct components: the oil cumulative production, N_p, and the gas cumulative production, G_p. These components are multiplied by groups of thermodynamic variables with the following definitions

$$
\Phi_n = \frac{B_o - R_s B_g}{(B_o - B_{oi}) + (R_{si} - R_s)B_g} \tag{21-5}
$$

and

$$
\Phi_g = \frac{B_g}{(B_o - B_{oi}) + (R_{si} - R_s)B_g} \tag{21-6}
$$

Equation 21-4 becomes

$$
N = N_p \Phi_n + G_p \Phi_g \tag{21-7}
$$

Let $R = GOR$, the instantaneous, producing gas-oil ratio. Then

$$R = \frac{\Delta G_p}{\Delta N_p} \qquad (21\text{-}8)$$

If $N = 1$ STB, Equation 21-7 in terms of increments of production becomes

$$1 = (N_p + \Delta N_p)\Phi_n + (G_p + \Delta G_p)\Phi_g$$
$$= (N_p + \Delta N_p)\Phi_n + (G_p + R\Delta N_p)\Phi_g \qquad (21\text{-}9)$$

The instantaneous value of R is the GOR within a production interval with incremental cumulative production ΔN_p. Therefore, in a stepwise fashion between intervals i and $i+1$

$$1 = \left(N_{pi} + \Delta N_{pi \to i+1}\right)\Phi_{n,av} + \left(G_{pi} + R_{av}\Delta N_{pi \to i+1}\right)\Phi_{g,av} \qquad (21\text{-}10)$$

and solving for $\Delta N_{pi \to i+1}$

$$\Delta N_{pi \to i+1} = \frac{1 - N_{pi}(\Phi_n)_{av} - G_{pi}(\Phi_g)_{av}}{(\Phi_n)_{av} + R_{av}(\Phi_g)_{av}} \qquad (21\text{-}11)$$

This finding is important. It can predict incremental production for a decline in the reservoir pressure. The following steps are needed:

1. Define Δp for calculation.
2. Calculate Φ_n and Φ_g for the two different pressure values. Obtain averages for the interval.
3. Assume a value of R_{av} in the interval.
4. Calculate $\Delta N_{pi \to i+1}$ from Equation 21-11.
5. Calculate $N_{pi \to i+1}$.
6. Calculate $G_{pi+1}\left(\Delta G_{pi \to i+1} = \Delta N_{pi \to i+1}R_{av}\right)$.
7. Calculate the oil saturation from

$$S_o = \left(1 - \frac{N_p}{N}\right)\frac{B_o}{B_{oi}}(1 - S_w) \qquad (21\text{-}12)$$

8. Obtain the relative permeability ratio k_g/k_o from a plot vs. S_o. Relative permeability curves are usually available.
9. Calculate R_{av} from

$$R_{av} = R_s + \frac{k_g \mu_o B_o}{k_o \mu_g B_g} \qquad (21\text{-}13)$$

10. Compare calculated R_{av} in the interval with the assumed value.

With the average pressure within the interval, the Vogel correlation can be used for calculating the well's inflow performance relationship (IPR). The IPR is combined with the vertical lift performance (VLP) to provide the well production rate. Finally, with the material balance outlined above, rate and cumulative production vs. time can be established.

21-4.1.2 Example: Forecast of Well Performance in a Two-Phase Reservoir

Background

A well is 8000 ft deep, completed with a 3-in. ID tubing and with 100 psi flowing wellhead pressure. If the drainage area is 40 acres with $h = 115\,\text{ft}$, $\phi = 0.21$, and $S_w = 0.3$, develop a forecast of oil rate and oil and gas cumulative production vs. time until the average reservoir pressure declines to 3350 psi ($\Delta \bar{p} = 1000\,\text{psi}$).

Solution

Since $p_i = p_b = 4350\,\text{psi}$, the solution gas drive material balance and Tarner's method outlined in Section 21-4.1.1 are appropriate. With the properties in Table 21-2, the variables Φ_n and Φ_g can be calculated for a range of reservoir pressures.

For the following sample calculation, assume a pressure interval between 4350 and 4150 ft. The average B_o, B_g, and R_s are 1.42 res bbl/STB, 7.1×10^{-4} res bbl/scf and 820 scf/STB, respectively. Noting that B_{oi} and R_{si} (at p_i) are 1.43 res bbl/STB and 840 scf/STB, Equation 21-5 yields

$$\Phi_n = \frac{1.42 - (820)(7.1 \times 10^{-4})}{(1.42 - 1.43) + (840 - 820)(7.1 \times 10^{-4})} = 199 \qquad (21\text{-}14)$$

From Equation 21-6

$$\Phi_g = \frac{7.1 \times 10^{-4}}{(1.42 - 1.43) + (840 - 820)(7.1 \times 10^{-4})} = 0.17 \qquad (21\text{-}15)$$

Similarly, these variables are calculated for four additional 200-psi intervals and are shown in Table 21-2. For the first interval, assume that the producing $R_{av} = 845$ scf/STB. Then, from Equation 21-11

Table 21-2 Physical Properties for Tarner's Calculation

$\bar{p}$ (psi)	B_o (res bbl/STB)	B_g (res bbl/scf)	R_s (scf/STB)	Φ_n	Φ_g
4350	1.43	6.9×10^{-4}	840		
	1.42	7.1×10^{-4}	820	199	0.17
4150					
	1.395	7.4×10^{-4}	770	49	0.044
3950					
	1.38	7.8×10^{-4}	730	22.6	0.022
3750					
	1.36	8.1×10^{-4}	680	13.6	0.014
3550					
	1.345	8.5×10^{-4}	640	9.42	0.010
3350					

$$\Delta N_{pi \to i+1} = \frac{1}{199 + (845)(0.17)} = 2.92 \times 10^{-3} \text{ STB} \tag{21-16}$$

which for the first interval, is the same as $N_{pi \to i+1}$. The incremental gas cumulative production is $(845)(2.92 \times 10^{-3}) = 2.48$ scf, which, for the first interval, also coincides with G_p.

From Equation 21-12

$$S_o = \left(1 - 2.92 \times 10^{-3}\right)\left(\frac{1.42}{1.43}\right)(1 - 0.3) = 0.693 \tag{21-17}$$

and from the relative permeability curves (not shown here), $k_g/k_o = 1.7 \times 10^{-4}$. From Equation 21-13, using $m_o = 1.7$ cp and $m_g = 0.023$ cp

$$R_{av} = 820 + 1.7 \times 10^{-4} \frac{(1.7)}{(0.023)} \frac{(1.42)}{(7.1 \times 10^{-4})}$$
$$= 845 \text{ scf/STB} \tag{21-18}$$

which agrees with the assumed value.

The calculation is repeated for all intervals and appears in summary from Table 21-3. For $A = 40$ acres, the initial oil-in-place is

$$N = \frac{(7758)(40)(115)(0.21)(1 - 0.3)}{1.43} = 3.7 \times 10^6 \text{ STB} \tag{21-19}$$

The IPR expression for this two-phase reservoir can be obtained from the Vogel correlation. With $k_o = 13$ md, $h = 115$ ft, $r_e = 1490$ ft, $r_w = 0.328$ ft, $m_o = 1.7$ cp, $s = 0$, and allowing B_o to vary with the pressure, the IPR becomes

$$q_o = 0.45 \frac{\bar{p}}{B_o}\left[1 - 0.2\frac{p_{wf}}{\bar{p}} - 0.8\left(\frac{p_{wf}}{\bar{p}}\right)^2\right] \tag{21-20}$$

The IPR curves for six average reservoir pressures (from 4350 to 3350 in increments of 200 psi) are shown in expanded form in Figure 21-7. At $q_o = 0$, each curve would intersect the p_{wf} axis at the corresponding $\bar{p}$.

Five VLP curves are also plotted for this well, each for the average producing gas-liquid ratio within the pressure interval. These values are listed in Table 21-3. The expected oil flow rates within each interval can be obtained from Figure 21-7.

All necessary variables are now available for calculating the duration of each interval. For example, the recovery ratio ($= \Delta N_p$ with $N = 1$ STB) in the interval between 4350 and 4150 psi is 2.92×10^{-3}, and since $N = 3.7 \times 10^6$ STB

$$\Delta N_p = \left(2.92 \times 10^{-3}\right) \times \left(3.7 \times 10^6\right) = 1.08 \times 10^4 \text{ STB} \tag{21-21}$$

Also

$$\Delta G_p = \left(1.08 \times 10^4\right)(845) = 9.1 \times 10^6 \text{ scf} \tag{21-22}$$

The average production rate within the interval is 1346 STB/d (from Figure 21-7), and therefore

$$t = \frac{1.08 \times 10^4}{1346} = 8 \text{ days} \tag{21-23}$$

The results for all intervals are shown in Table 21-4.

Table 21-3   Cumulative Production with Depleting Pressure for Well in Example 21-4.1.2 (N = 1 STB)

$\bar{p}$ (psi)	ΔN_p (STB)	N_p (STB)	R_{av} (scf/STB)	ΔG_p (scf)	G_p (scf)
4350					
	2.92×10^{-3}		845	2.48	
4150		2.92×10^{-3}			2.48
	8.41×10^{-3}		880	7.23	
3950		1.1×10^{-2}			9.71
	1.2×10^{-2}		1000	12	
3750		2.3×10^{-2}			21.71
	1.2×10^{-2}		1280	16.1	
3550		3.56×10^{-2}			37.81
	1.1×10^{-2}		1650	18.2	
3350		4.66×10^{-2}			56.01

This example illustrates how material balance methods can be used to determine pressure decline as a function of time. Reservoir pressure decline reduces the fluid flow to the surface and, ultimately, will cause a well to stop flowing naturally. When natural flow stops, the well must be produced by artificial lift. Preplanning an artificial lift system at the time of initial completion is critical in wells where significant reservoir pressure decline will occur.

Changes in reservoir pressure can also lead to sand influx. Sand production models can be used to predict when this could occur (Chapters 18 and 19), and in such cases, provisions for sand stabilization may be included at the time of initial completion.

Waterflooding and other enhanced production techniques typically arrest reservoir pressure declines. However, these techniques limit the injection pressure

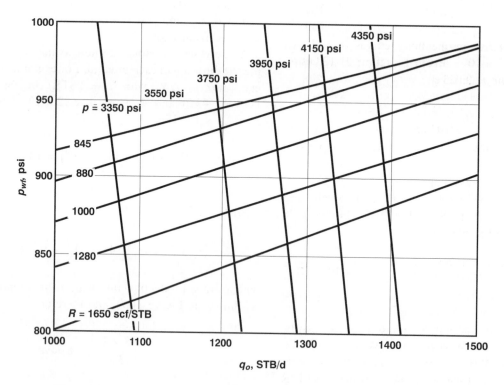

Figure 21-7   IPR and VLP curves for a solution-gas drive reservoir (Section 21-4.1.2)

Table 21-4 Production Rate and Oil and Gas Cumulative Recovery Forecast for Well in Example 21-4.1.2

$\bar{p}$ (psi)	q_o (STB/d)	ΔN_p (10^4 STB)	N_p (10^4 STB)	ΔG_p (10^6 scf)	G_p (10^6 scf)	Δt (d)	t (d)
4350							
	1346	1.08		9.1		8	
4150			1.08		9.1		8
	1288	3.11		27.4		24	
3950			4.19		36.5		32
	1234	4.44		44.4		36	
3750			8.63		80.9		68
	1176	4.66		59.6		40	
3550			13.29		140.5		108
	1120	4.07		67.2		36	
3350			17.36		207.7		144

the formation fracture gradient. In many cases, this pressure will be less than the initial reservoir pressure.

21-4.1.3 Fluid Saturation and Composition

Water coning, or water influx from a bottom-drive aquifer will increase the producing water-oil ratio over the life of many reservoirs. As a result, the fluid density increases as a function of time, which may accelerate the need for artificial lift. In more extreme cases, fluid composition may change as a function of time.

Sour gases, such as H_2S or CO_2, evolve during the life of many fields. The occurrence of sour gas production is believed to result from bacteria that is introduced into the reservoir through brine injection. At present, no models are available that can predict the occurrence of sour gas production. Nevertheless, the evolution of sour gases in a field will increase the need for corrosion protection of tubulars and surface equipment. Unless the casing, tubing, and other downhole equipment have been adequately protected during initial completion, a workover might be required at the time of corrosion failure.

Increased brine saturation might also lead to the production of barium sulfate and naturally occurring radioactive materials (NORM). NORM is believed to stem from naturally occurring radionuclides in produced water (Snavely, 1989). At present, no verified models are available that can predict the occurrence of NORM. If NORM is expected to occur during the life of the field then, during a workover, there must be a means of handling and disposing the material according to local or national regulations.

21-4.2 Well and System Performance

21-4.2.1 Inflow Performance Changes with Time

In steady-state well performance, a constant pressure boundary (natural aquifer or injector pressure maintenance) exists at the areal limit of the reservoir, and most of the pressure drop occurs near the wellbore. Steady-state inflow performance is constant as a function of time but can decrease with the evolution of skin with time as shown in Figure 21-8.

In pseudosteady-state well performance, the pressure at the outer boundary is not constant; it declines at a constant rate with time, as a function of drawdown and production. Typical time-related inflow performance changes for this producing regime are shown in Figure 21-9.

Chapter 15 discusses system performance curves. With time, a tubing intake curve will be shifted upward until it no longer intersects the IPR (Figure 21-10). At this point, the well requires artificial lift.

At the time of initial completion, engineers must either estimate when artificial lift is required, or they must make arbitrary provisions for some form of artificial lift. In offshore completions where other wells are already on gas lift, operators will commonly include one or more side-pocket mandrels equipped with dummy valves in a new completion. This arrangement allows for gas lift to be added to the new completion in the future with only wireline service.

Other methods of artificial lift, such as beam pumping and electrical submersible pumps (ESPs), require full tubing workovers for their installation. Hence, these artificial lift methods are not typically included at the time of initial completion in a flowing well.

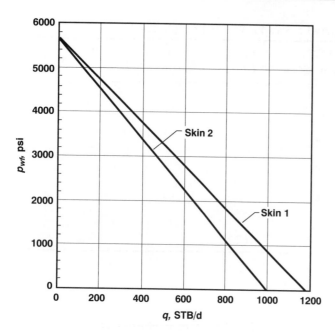

Figure 21-8   Steady-state IPR changes with skin

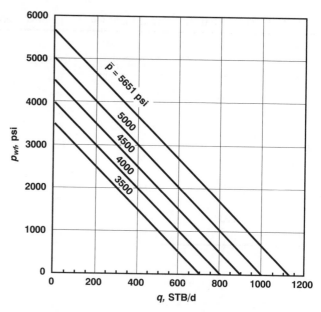

Figure 21-9   Pseudosteady-state IPR changes with time

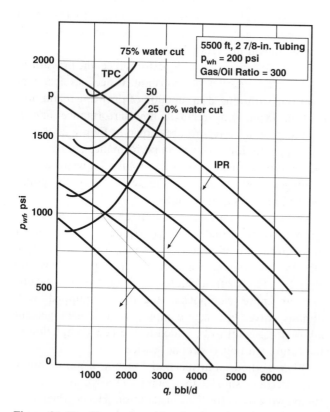

Figure 21-10   Change in tubing intake curve relative to IPR with time

21-4.2.2 Example: Gas-Lift Installation and Management

Background

For it to be economical, a well in an offshore reservoir producing at a WOR = 1 needs a minimum of 500 STB/d of production. The reservoir depth is at 8000 ft. Figure 21-11 describes the gradient curves for this well for $q = 500$ bbl/d and 50% oil, 50% water. The natural GLR is 300 scf/STB, and the IPR is given by

$$q = 0.39(\bar{p} - p_{wf}) \qquad (21\text{-}24)$$

Calculate the following:

1. At what average reservoir pressure should gas lift begin?
2. At which depth in the completion string should a gas-lift valve be initiated when the average reservoir pressure drops by 500 psi to maintain a production rate of 500 STB/d?

Solution

From Figure 21-11, at $H = 8000$ ft and GLR = 300 scf/STB (the fourth curve from the right), the required flowing bottomhole pressure to lift the fluid to the top should be 1750 psi.

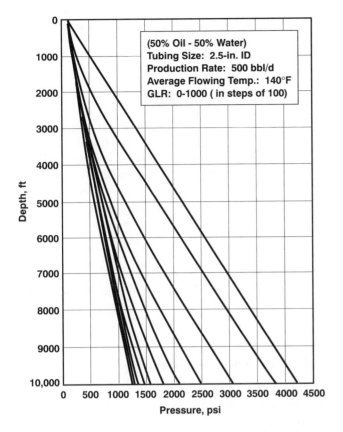

(50% Oil - 50% Water)
Tubing Size: 2.5-in. ID
Production Rate: 500 bbl/d
Average Flowing Temp.: 140°F
GLR: 0-1000 (in steps of 100)

Figure 21-11 Gradient curves for $q_1 = 500$ STB/d

From the IPR expression and rearrangement

$$\bar{p} = \frac{500}{0.39} + 1750 = 3050 \text{ psi} \qquad (21\text{-}25)$$

As long as the reservoir pressure is 3050 psi or greater, the well can produce at least 500 STB/d with natural lift. Once the pressure drops below this value, then gas lift may be required.

From the IPR expression and another rearrangement, the flowing bottomhole pressure must be (with new $\bar{p} = 2550$ psi)

$$p_{wf} = 2550 - \frac{50}{0.39} = 1270 \text{ psi} \qquad (21\text{-}26)$$

The flowing bottomhole pressure is related to the well-head pressure p_{tf} (here it is equal to 100 psi) by

$$p_{wf} = p_{tf} + H_{inj}\left(\frac{dp}{dz}\right)_a + (H - H_{inj})\left(\frac{dp}{dz}\right)_b \quad (21\text{-}27)$$

where H_{inj} is the injection depth, H is the well depth, $(dp/dz)_a$ is the flowing gradient above (at the artificial GLR), and $(dp/dz)_b$ is the flowing gradient below (at the natural GLR.)

The pressure gradient at the natural GLR at a depth between 5000 and 8000 ft is 0.33 psi/ft. (This result can be obtained graphically from Figure 21-11, but it can be obtained more accurately from any pressure-drop-in-pipes computer code). The search is by trial and error for finding the GLR curve with the gradient $(dp/dz)_a$, and a depth of H_{inj} would balance Equation 21-27.

A unique solution to the problem is available. For $H_{inj} = 7120$ ft and $(dp/dz)_a = 0.123$, corresponding to GLR = 750 scf/STB, Equation 21-27 and Figure 21-11 are satisfied. This solution also suggests the amount of gas that must be injected.

$$q_g = (750 - 300)500 = 2.25 \times 10^5 \text{ scf/d} \qquad (21\text{-}28)$$

From Figure 21-11, the pressure at the injection point $H_{inj} = 7120$ ft and GLR = 750 scf/STB) must be 980 psi.

From the classic approximate relationship

$$p_{inj} \approx p_{surf}\left(1 + \frac{H_{inj}}{40,000}\right) - \Delta p_{valve} \qquad (21\text{-}29)$$

and using $\Delta p_{valve} = 100$ psi, the surface injection pressure should be

$$p_{surf} = \frac{1080}{(1 + 7120/40,000)} = 915 \text{ psi} \qquad (21\text{-}30)$$

The calculation shows the position of the gas-lift valve, the type of valve needed, the volume of gas injected (the required handling capacity), and the expected compressor specifications.

This calculation, repeated for reservoir pressures between 3050 and 2550 psi, can allow gas-lift design for the life of the well within this interval.

21-4.2.2 Sources of Low Productivity (Scale, Perforation Plugging, etc.)

As water production increases, fines may migrate in the reservoir and scales may form (Chapter 5, Chapter 20). The perforations can become fully plugged or partially obstructed by such debris. Obstructions increase near-wellbore pressure loss or pressure loss across the completion and reduce flow capacity. As a result, the well's IPR may quickly decline, and the reduced productivity might require some form of reservoir stimulation.

A prudent approach in designing completions is to provide some means for stimulating the well without removing the completion. This approach is generally feasible for acidizing or acid fracturing without proppant. Hydraulic fracturing with proppant may or may not be possible through completion equipment with profiles and reduced internal diameters. If the completion includes an ESP or a beam pump, operators will not be able to stimulate the well without pulling the pump.

21-4.2.3 Water Coning

Water coning may occur as a result of poor perforation practices or excessive drawdown. Water coning can be predicted (Chapter 20). In reservoirs with bottom aquifers, or in wells located near an oil-water contact, such calculations should be made in conjunction with the perforating and production scheme. If water coning is indicated, then the perforated interval or drawdown may be reduced at the time of initial completion.

In some situations, it may be advantageous to use specialized completion schemes that limit downhole water production (Swisher and Wojtanowicz, 1995).

21-4.2.4 Recompletion Requirements

Many fields have multiple reservoirs, and not all reservoirs may be perforated and produced at the time of initial completion. If some zones cannot be commingled or produced separately at the time of initial completion, then they must be prepared for future production. Such preparations include cement squeezing, plugback, wireline operations, and reperforation.

If the reservoir for future production is located above the lowest producing zone, operators must perforate the upper zone at the time of initial completion and shut off production from this zone either with a sleeve or packers and tubing (assuming the completion will not be pulled). If the zone for future production is located below the lowest producing zone, operators may be able to perforate the zone using through-tubing guns without removing the existing completion.

Another consideration is whether it is necessary to recomplete in the uphole direction, or whether deeper zones can be completed effectively after shallower zones are depleted. Typically, shallow and depleted zones must be isolated from wellbore fluids during recompletion operations.

If a zone must be squeeze cemented in the recompletion process, then the existing completion will most likely be removed to accommodate the workover. Plugback and wireline operations may not require full tubing workovers.

21-4.2.5 Stimulation Requirements

With time, the IPRs may decline in wells that had no initial stimulation or wells that were initially acidized or hydraulically fractured. Stimulation during the initial completion is easier because the completion string typically has not been run in the wellbore. Stimulation can proceed through a workstring designed to tolerate the effects of proppant, acid, elevated pressures, and other aspects of the stimulation treatment.

Although many wells may be candidates for restimulation at some time, predicting that time is difficult. Nevertheless, planning for future stimulation treatments at the time of initial completion is prudent.

21-4.3 Mechanical Considerations

In addition to fluids and procedures, well completions also include equipment. Such equipment is subject to failure with time.

21-4.3.1 Equipment Failures

Typical failures include the inability to shift a sliding side-door, surface-controlled subsurface safety valve (SCSSV) failure (particularly following a required test), tubular collapse, tubular leaks, failure of a packer to hold pressure, and a myriad of failures associated with artificial lift equipment.

During initial completion, engineers should assume that each equipment item in the completion will fail, and they should formulate service plans for these events. Current completion technology requires full-tubing workovers for (1) tubing leaks or collapses, (2) failures of tubing-retrievable SCSSVs that do not have backup wireline valves included in the completion, and (3) most packer failures. Advances in technology may enable in-situ tubing and casing patches. This technology is being tested in the North Sea.

21-4.3.2 Equipment Redundancy and Risk Analysis

SINTEF (Molnes *et al.*, 1986) studied the reliability of SCSSVs and determined the typical service life for flapper and ball-type valves. This work eventually led to the development of dual SCSSVs and their application in subsea wells. Dual SCSSVs allow operators to lock open one valve and use a second valve, should the primary SCSSV fail. Providing such redundancy is important in extending the life of a subsea completion.

Risk analysis is another technique for predicting the frequency of expected equipment or system failures, and system safety. This type of analysis generally requires detailed examination of the producing system and a quantification of all the failures than can occur throughout the system. Such studies, while useful, are generally beyond the resource limitations of practicing engineers.

21-4.3.3 Workover Frequency

Well access, means of re-entry, and the number of trips to recover the completion (Johnson *et al.*, 1994) are all considerations in planning long-term well completions. For example, permanent packers with seal assemblies facilitate completion removal, but will require at least two trips: one trip to set the packer and one trip to run the completion. Retrievable packers must be unset before the completion is pulled.

Typical workover frequency should be identified at the time of initial completion. Design alternatives that could potentially reduce the workover frequency should also be considered. For example, if wells in a field are experiencing frequent tubing leaks, then a heavier-weight tubing, coated tubing, or special metallurgy should be considered in a new completion.

21-4.4 Design Optimization

The previous discussion detailed a few of the many considerations in designing well completions that can remain in service for an extended period.

If a single method of completion design were available, and if all possible design interactions between components and operations were known and could be quantified (Figure 21-6), a method might be developed for optimizing completion design. However, no such methodology currently exists.

Engineers have attempted to optimize completion designs. These "optimized designs" actually represent an evolution in the engineer's understanding of the functional requirements for the field and the reservoir. Alterations in the initial designs reflect the engineer's need to improve some limitation or enhance some aspect of the initial design (Kostol and Rasmussen, 1993; Mansour and Khalaf, 1989). The following example from the Statfjord Field illustrates this point.

21-4.4.1 Case Study—Statfjord Field

Kostol and Rasmussen (1993) presented a paper entitled "Optimized Well Completion Design in the Statfjord Field, North Sea." In this paper, the authors briefly describe the Statfjord Field reservoirs, discuss the geology and drainage strategy, and give an overview of the initial and optimized completion designs for the field.

The Statfjord wells were initially completed with 32-lb/ft, 7-in. L80 carbon steel production tubing (Figure 21-12). The stepdown nipple configurations and the limited internal diameters of the packer and expansion joints resulted in a 4.31-in. restriction in the tailpipe. For wells with 7-in. liner, a $5\frac{1}{2}$-in. tailpipe was run into the liner and a $9\frac{5}{8}$-in. production packer was set above the liner. In wells without liners, the packer was set about 50 m above the top of the reservoir (Kostol and Rasmussen, 1993).

A hydraulic packer was used with a 10-ft stroke expansion joint set above the packer. A landing nipple was included at 270 m for a wireline-retrievable downhole safety valve. The wireline-retrievable safety valve nipple had a minimum ID of 5.75 in., and the valve provided a 3.3-in. restriction (Kostol and Rasmussen, 1993).

The initial Statfjord completion was designed based on the assumption that the tubing would be retrieved for any workover or recompletion work. Since the CO_2 content of the produced fluids was low, corrosion problems were not anticipated. The large production rate dictated 7-in. tubing. Landing nipples were included to allow for temporary plugging of tubing. A 7-in. wireline-retrievable SCSSV was preferred over $5\frac{1}{2}$-in. tubing-retrievable SCSSV, so that restrictions could be avoided high up in the wellbore. The design was installed with a single trip in the hole. After the completion was installed, the wells were perforated with $2\frac{7}{8}$- or $3\frac{3}{8}$-in. wireline-conveyed through-tubing guns.

Operating experience with the Statfjord completions demonstrated the limitations of the initial design. Corrosion problems and elastomer leaks were experienced. The completion string design also limited the through-tubing perforating guns to a $3\frac{3}{8}$-in. size with 6 shots/ft. To perforate 12 shots/ft, two runs were required, and the shot distribution was constrained. These perforating limitations influenced both well productivity and sand production.

After a high number of workovers was performed, an optimized monobore completion was designed for the Statfjord Field (Figure 21-13). This completion optimized zone isolation requirements, the production strategy, well profiles, casing program, reservoir conditions, fluids, material selection, sand production and control,

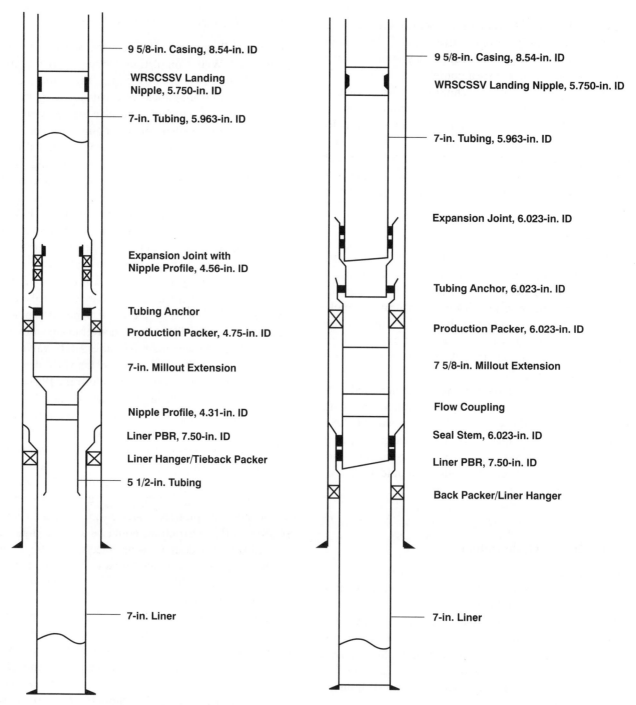

Figure 21-12   Stratfjord field—initial completion (Kostol and Rasmussen, 1993)

Figure 21-13   Stratfjord field—optimized monobore completion (Kostol and Rasmussen, 1993)

perforating techniques, logging and testing, safety, well control, and environmental considerations. Table 21-5 details some of the optimized parameters.

In most cases, the optimized monobore completion eliminated the need for recompletions that required tub-

ing retrieval. The optimized steel selection and the elastomers are expected to minimize corrosion leaks and failures. The completion allows for upward and downward recompletions, high shot-density, through-tubing perforating with $4\frac{1}{2}$-in. guns, and through-tubing

Table 21-5 Sample Optimized Parameters of the Statfjord Completions

Optimized Paramter	Limitation of Initial Completion	Optimized Monobore Design
Zone isolation and production strategy	No selectivity; recompletion required pulling tubing for workovers	Wells designed to allow for recompletions using wireline for running straddle packers and bridge plugs. Selectivity requires properly cemented casing and liners
Well Profile		Well profile optimized for minimum torque and drag in high angle wells. If possible, wells designed to stay within 65 degree limit for wireline service.
Casing Program	$9\frac{5}{8}$" casing, 7" liner	$9\frac{5}{8}$" casing 7" liner Use of 7" liner through the reservoir provides a flexible long term completion. Wells with $9\frac{5}{8}$" casing through part of the reservoir require heavy workover if recompleting downward from the upper zone.
Reservoir Conditions, Fluids, Material Selection	No corrosion provision	Most wells completed with 13% Cr tubing and liners. Carbon steel used for wells with short (2-3 year) periods of water production.
	No scale predicted	Scale inhibitor squeeze treatments planned and performed at first sign of water breakthrough
Sand Production	Wells were produced with limited drawdown. Four initial wells gravel packed.	Perforation techniques improved to allow use of larger size guns. Wells selectively perforated in the assumed strongest parts of the formation. Can gravel pack through tubing.
Perforation Technique	$2\frac{7}{8}$" or $3\frac{3}{8}$" through tubing guns 6 spf in single run	$4\frac{1}{2}$" through tubing wireline guns. use coiled tubing in deviation prevents wireline guns. Perforate underbalanced 30-45 bar. Gravel packed wells perforated 12 spf.
SSV	Wireline valve with 3.3" ID	7" nipples with valve restriction of 5.75". Dual control lines for redundancy in case of leaks.

gravel-packing for sand control. The large ID of the monobore completion enables maximum production.

The redesign of well completions in the Statfjord field is one example of the process that occurs when current completion technology is used to optimize well completions.

Many other completion designs are described in the literature. Examples include the Ekofisk Field, North Sea (Luppens, 1980), Everest and Lomond Field, North Sea (Laing *et al.*, 1993); Beta Field, offshore California (Bruist *et al.*, 1983), Kepiting Field, Indonesia (Coates and Kadi, 1987), Mobile Bay, Louisiana (Gordon *et al.*, 1994); Tuscaloosa trend (Huntoon, 1984); Thistle Field, North Sea (Moore and

Adair, 1988), and Arun, Indonesia (McKenna *et al.*, 1993).

The concept of "smart wells" is a burgeoning area in completion technology that may eventually allow engineers to design completions that can be completely manipulated, monitored, and serviced without tubing removal.

21-4.5 Future Considerations—Smart Wells

During the 1980s, many operators installed pressure and temperature instruments downhole to gather real-time production data. These data provided valuable information concerning how to produce the target zones more effectively. More recently, the industry has favored installations that have remotely operated controls placed in the well. Sensor requirements have now extended beyond just pressure and temperature to include flow rates, fluid composition, reservoir characteristics, etc. A well that has a system of downhole sensors and controls, and that includes a surface system to collect and transmit the production data to a remote facility for analysis, has been dubbed a "smart" well.

The most significant benefit that a "smart" well will provide is improved economics. The ability to sense the production processes and react to changing conditions to continuously optimize production will result in improved production and recoverable reserves. In addition, properly designed systems should reduce the operating expenses in a typical installation. Reduced capital expenses may also be realized when, for instance, a system is fielded that separates and reinjects the produced water downhole.

Imaginative solutions to commonplace production situations are being contemplated, and the technology to address these issues is under development. Some basic system designs will be necessary before any systems are used.

21-4.5.1 System Components

The subsystems that comprise a "smart" well include a telemetry system for conveying data to and from the surface, downhole sensors for collecting the desired parameters in the well, controls to reconfigure the downhole tools, and a surface subsystem. The surface subsystem includes a data collection terminal, software to analyze the data and make decisions based on the output, and some means of transmitting this data to a remote facility, if required. These systems are shown in Figure 21-14.

Telemetry

Most operators equate telemetry with a wireless system, but such systems may not be used in the earliest of "smart" wells. The technologies that are currently available are limited in various installations for specific reasons such as conductivity of the formation, variation in fluid properties, and noise either from surface facilities or from the flow in the well. Therefore, the proven hard-wired telemetry system will most probably be the first technology deployed in a "smart" well.

The telemetry system may need to transmit some considerable data streams depending on the needs of a particular well. For instance, if a well is equipped with 3D reservoir sensors that visualize the movement of oil, water, and gas near the wellbore, some significant data rates will be needed in the telemetry system. In the event of a multilateral or multizonal completion with multiple flow controls installed, the need for operational process data may exceed all other data. Since the well-control signals will be transmitted from the surface downhole, the telemetry system plays a very important role in the "smart" well.

Downhole Sensors

A "smart" well must control production processes. Therefore, sensors that provide information about the flowing conditions downhole will be important. Although the pressure and temperature instruments that were previously used are now more accurate and reliable, sensors will also be needed to determine multiphase flowing conditions and the rates and volumes of the flowstream constituent.

A major change in data retrieval has recently been introduced with the use of fiber-optic sensors in wells. These sensors use changes in the properties of the fiber to provide temperature and pressure profiles downhole. Optical energy is reflected from the point in the fiber where a temperature or pressure change has altered the optical properties of the fiber, giving the desired profile information (Phillips, 1997).

Microelectronic technology from the medical and defense industries now allows the oil industry to assess fluid viscosity, composition, and density reliably and continuously both inside the wellbore and in the reservoir adjacent to the wellbore (Phillips, 1997).

The "smart" well of the future will contain numerous sensors that will relay information about the reservoir, the fluids around the wellbore, and the produced fluids in the wellbore. Sensors will combine with well controls to mitigate or prevent adverse producing situations. For example, downhole sensors may detect the formation of scale; there may be controls built into the completion that will remove the scale and change the flowing conditions that caused it to form initially.

Current geophysical research is determining the interpretation problems associated with gathering between-well seismic data to provide a real-time 3D seismic within the reservoir. With such a capability, a reservoir management team could image reservoir features such as chan-

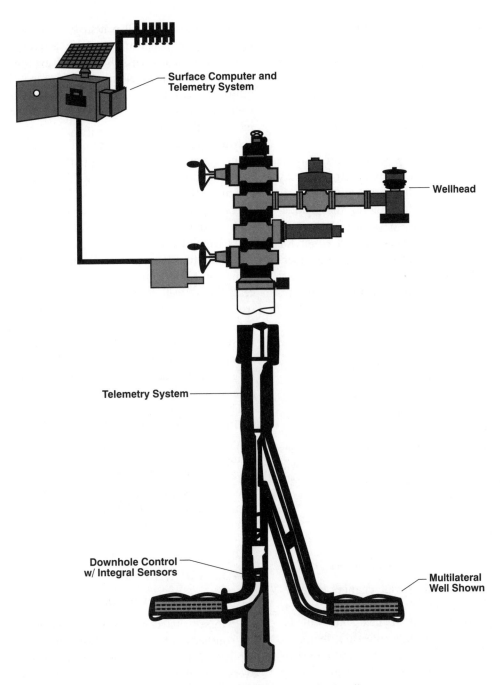

Figure 21-14 Typical "smart" well

nels, faults, and fractures; they could also "listen" to flood front movement within the reservoir (Phillips, 1997).

Controls

Before the data from wellbore sensors can be used to alter the producing scheme, remotely controlled downhole flow devices must be developed. A number of flow control devices exist that are now manipulated by wireline or hydraulic systems. The simplest of these is the downhole choke, which can be used between packers to regulate flow from a particular reservoir.

Used in the industry for more than 30 years, downhole chokes are the most direct means of controlling flow downhole. However, the existing chokes require intervention to extract and/or insert the flow control. The "smart" choke will be remotely operable, which represents a major improvement over existing completion methods.

Remote actuation of the controls is an entirely new subject (Botto *et al.*, 1996). Hydraulics have dominated this industry because of the high force capabilities and the robustness of the tools. However, recent advances in the areas of electromechanical actuation may change the technology used downhole.

More sophisticated downhole systems may also be necessary. For example, downhole separation is already being evaluated in field trials. When used with ESPs, this technology offers the prospect of avoiding the production of unwanted water or gas to surface with some method whereby the water or gas is injected either for reservoir pressure maintenance or into a disposal reservoir (Phillips, 1997). Electronically operated gas-lift valves are another example of developments being undertaken in this area.

Surface Systems

The surface system will typically include a computer that can collect and store production data, a software package that analyzes this data and helps users decide how the controls should be configured, and a telemetry system for transmitting this data to a remote terminal should it be required. Currently, a number of software packages are used for this task. In some cases, this software will function adequately; however it is currently considered a growth technology.

21-4.5.2 Reservoir Management with Smart Wells

Smart well technologies will provide changes in reservoir management. The variety and type of information available continuously throughout field life will be greatly enhanced. Both in-well data (pressure, temperature, viscosity, and compositional profiles) and between-well data (seismic, passive listening) will provide greater reservoir characterization. These enhancements will combine to advance reservoir management toward precise mapping of fluid fronts and reservoir properties throughout the reservoir (Phillips, 1997).

Most likely, reservoirs will be divided into discrete management intervals in the future. Systems and technologies will be developed to control that part of the reservoir with which they are in contact. Therefore, if more reservoir is contacted, greater control can be achieved, and potentially greater reserves can be accessed (Phillips, 1997).

Horizontal wells will benefit from smart well technology that can penetrate multiple fractures and isolate reservoir compartments. Reduced drawdown in horizontal wells may help eliminate unwanted water or gas influx. The capability to control production from such wells in a series of management intervals further increases their utility.

Multilaterals also offer significant reservoir management potential with smart well technology. Multilaterals enable a single wellbore to be used (1) for concurrently producing reserves from low-permeability areas of a heterogeneous reservoir, (2) to access multiple high-productivity fractures, (3) to access different reservoirs from the same wellbore, or (4) to better manage injection for pressure maintenance (Phillips, 1997). Flow control in multilateral wells is critical to their success. Smart well technology will maximize production from such completions (Robison and Clark, 1997).

Current projections for increased production and recoverable reserves that are directly attributable to "smart" systems is approximately 10 to 15%, but increases could be greater. In addition, this technology may make marginal field developments viable.

The facility to receive continuous in-well and between-well data, interpret these data, and remotely control or reconfigure the completion accordingly, is a fundamental change in current completion technology. These advances will enable new reservoir management schemes that can allow engineers to continuously characterize and monitor reservoir behavior, improve recovery, and enhance field economics.

21.4.5.3 Case Histories

Case History 1—Adriatic Sea

In 1996, a "smart" well was planned for completion in the southern Adriatic Sea (Botto, *et al.*, 1997). It was applied as a means of eliminating the need for workovers, thereby overcoming problems associated with deep-water (850 m), a remote field location, and limited recoverable reserves.

Two satellite subsea gas-lift wells (Aquila 2 and Aquila 3) are planned for "smart" completion. The proposed Aquila 3 completion is described in the following paragraphs; the proposed Aquila 2 completion is identical to the Aquila 3 above the upper packer, but the lower portion of the well is completed openhole.

The Aquila 3 is a medium-radius drainhole with a 3000-m vertical section and a 650-m horizontal section. The 650-m horizontal section intersects three zones with different permeabilities. From heel to toe, the permeability range of Zone A is 30 to 50 md; Zone B is 5 to 15 md; Zone C is 2 to 5 md. The completion will be designed to isolate Zone B, which is expected to conduct water (Botto, et al., 1997).

In Aquila 3, a $4\frac{1}{2}$-in. tubing string will be run through the 3000-m vertical section. This string will be equipped with a conventional SCSSV and an annular safety valve. Then, a $2\frac{7}{8}$-in. tubing will be run inside a slotted 5-in. liner through the 650-m horizontal section. The liner will be sectioned by two external casing packers (ECPs) and blanked to isolate Zone B. A seal receptacle in the blanked liner section and a seal assembly on the lower end of the inner tubing will direct fluid from Zones A and C to the liner-tubing annulus and to the inside of the inner string respectively (Botto *et al.*, 1997).

The first "smart" tool for the planned completion is a gas-lift interval control valve (ICV), which will be located immediately below the SCSSV. This four-position, remotely controlled sleeve has calibrated gas-lift orifices of different diameters (Botto *et al.*, 1997). This design will allow gas-lift performance to be fine-tuned throughout the life of the well.

The second proposed "smart" tool is a flow diverter, located below the chemical injection subassembly. The diverter is a three-position sleeve designed to control production from, or injection into, either Zone A alone, Zone C alone, or both zones commingled.

The planned control system for the well, referred to as the surface-controlled reservoir-analysis and management system (SCRAMS), was the result of a 2-year joint industry research and development project (Botto *et al.*, 1997). The system uses permanently installed electric cables (I-wires) and hydraulic lines that are protected within a reinforced "flat pack." Each flat pack contains one electric line and one hydraulic line. Multiple, redundant flat packs will be run in the well to ensure reliability. The lines will also be segmented so that signals can be diverted or steered into another viable line, should any failure occur in either the electric or hydraulic line.

The I-wires will power and communicate with each downhole sensor or tool. Each "smart" downhole tool will have actuator electronics modules (AEMs) containing solenoids. Electric signals passing through the AEMs will activate and control the hydraulic lines, which selectively manipulate each downhole tool (Botto *et al.*, 1997).

An independent hydraulic line will be provided for operating the SCSSV, and a single hydraulic line will connect the smart tools to the ESPs. This line will be isolated from the hydraulic lines in the flat pack(s),

which will allow the system to control, actuate, and set the production packer remotely through the SCRAMS network, without applying tubing pressure (Botto *et al.*, 1997).

The SCRAMS control system establishes a standard for multidrop downhole power distribution and communication referred to as Segnet (Botto *et al.*, 1997). This system uses an I-wire bus, which allows each downhole sensor or well tool to operate without interfering with other systems. The system also allows two-way digital communications for commands and data transfer.

Horizontal christmas trees with vertical access are another proposed feature of the Aquila completions. Normally, horizontal trees do not include vertical access. However, such access was believed necessary for quick detection of system damage or failure during installation. Botto *et al.*, (1997) describes the proposed design of these trees in detail.

Case History 2—Norway

Unlike the proposed Aquila well system, whose economic viability will depend on how effectively the system's gas-lift ICV and flow diverter can control production processes, a recently completed Norway well consists of a "smart" system (Fig. 21-15) that may be used for future subsea wells in the same area.

The Norway well is a single-zone producer that may cone water or gas. The "smart" well system used can restrict the influx of unwanted fluids by choking the flowstream. This completion included the following equipment: (1) a hard-wired telemetry system for transmitting data from the sensors to the surface and to pass control signals downhole, (2) sensors to continuously monitor the flowing pressure, temperature, gross flow rate, and fluid density, (3) flow controls to allow the fluid streams from the perforations to be commingled for optimum production, and (4) software to analyze the data and provide automatic control.

During installation, the control system set the production packer, which saved rig time. Additionally, the entire control system was monitored for faults, which would serve as a "blueprint" for future installations. The reliability and redundancy of the system is perhaps the most significant advancement this system provides over existing completions.

The control system uses proven solenoid technology from the aerospace industry to isolate the various hydraulic components. These solenoids are very reliable and are individually addressable through the communication system. Operating a particular device is simple;

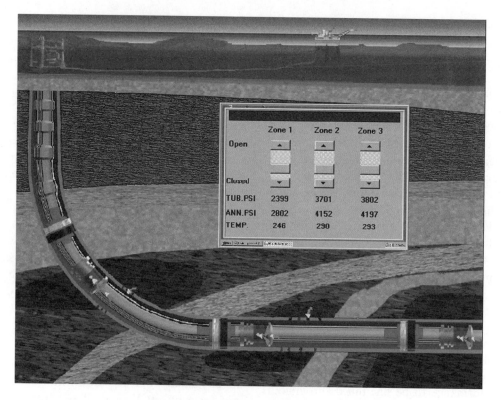

Figure 21-15   "Smart" monitoring system

the operator sends a coded message down the communication system, and the desired solenoid responds.

Redundant solenoids are provided at various positions in the completion so that the control signal can be diverted if a fault is detected in one section of the well. In this particular case, the operating current of one solenoid was gradually increasing. Although the solenoid never stopped operating, the decision was made to divert the signal path to bypass the troublesome component.

Because this system is the first of its kind to be used in this area, it is being carefully monitored. Since the well was completed, the inflow control valve has effectively controlled production from the producing interval. Depending on the long-term reliability of the system, it may be used in other subsea well completions in the same area.

REFERENCES

Botto, G., Maggioni, B. Guiliani, and Rubbo, D.: "Innovative Remote Control Completion for Aquila Deepwater Challenge," paper SPE 36948, 1996.

Bruist, E.H., Jeffris, R.G., and T.M. Botts. "Well Completion in the Beta Field, Offshore California," paper SPE 11696, 1983.

Coates, R.L., and Kadi, I. "Unique Subsea Completions in the Kepiting Field, Indonesia," paper SPE 16848, 1987.

Craft, B.C., and Hawkins, M. (Revised by Terry, R.E.): *Applied Petroleum Reservoir Engineering*, 2nd ed., Prentice Hall, Englewood Cliffs, N.J., 1991.

Dake, L.P. *Fundamentals of Reservoir Engineering*, Elsevier, Amsterdam, 1978.

Dunn-Norman, S.: "A Computational Model of Well Completion Design," PhD Thesis, Heriot-Watt University, Edinburgh, Scotland, 1990.

Gordon, J.R., Johnson, D.V., Herman, S.R., and Darby, J.B.: "Well Design and Equipment Installment for Mobile Bay Completions," paper OTC 7571, 1994.

Huntoon, G.G.: "Completion Practices in the Deep Sour Tuscaloosa Wells," SPE transactions, Vol. 277, January 1984, 79–95.

Johnson, K.J., Lorenzatto, R.A., Rittershaussen, J.H., Barreto, J.L. and Everaldo Lima Filho: "Single Trip Subsea Completions," paper OTC 7529, 1994.

Kostol, P., and Rasmussen, J.H.: "Optimized Well Completion Design in the Statfjord Field, North Sea," paper OTC 7327, 1993.

Laing, E.M., Ogier, M.J., and Hennington, E.R.: "Everest and Lomond Completion Design Innovations Lower Completion and Workover Costs," proceedings of Offshore European Conference, 1993.

Luppens, J.C.: "Completion Techniques for Albuskjell, Edda and Edlfisk Fields, Greater Ekofisk Area," paper OTC,1980.

Mansour, S., and Khalaf, F.: "Optimized Completion Significantly Reduced Cost and Time in Gulf of Suez Fields," paper SPE 17984, 1989.

McKenna, E., Sukup, R., and Biggs, R.A.: "Arun Indonesia: Big Bore Completion Tool Design, paper OTC 7328, 1993.

Molnes, E., Rausand, M., and Lindqvist, B.: "Reliability of Surface Controlled Subsurface Safety Valves," *SINTEF Phase II Main Report*, Trondheim, 1986, p5 ff.

Moore, P.C., and Adair, P.: "Dual Concentric Gas Lift Completion Design for the Thistle Field," paper SPE 18391, 1988.

Patton, L.D., and Abbott, W.A.: *Well Completions and Workovers: The Systems Approach*, Energy Publications, a Division of Harcourt Brace Jovanovich, Inc., Dallas, 1985.

Peden, J.M., and Leadbetter, A.: "Rationality in Completion Design and Equipment Selection in the North Sea," paper SPE 15887, 1986.

Phillips, I. C.: "Reservoir Management of the Future," Paper MS 383, presented at the Institute of Marine Engineers Innovative Technology for Challenging Environments Conference, April 8-9, London, England, 1997.

Robison, C. E.: "Overcoming the Challenges Associated With the Life Cycle Management of Multilateral Wells: Assessing Moves Towards the Intelligent Well," Paper OTC 8536, May 1997.

Snavely, Jr., E.: "Radionuclides in Produced Water - A Literature Review," study for the American Petroleum Institute, August, 1989.

Swisher, M.D., and Wojtanowicz, A.K.: "In Situ-Segregated Production of Oil and Water - A Production Method with Environmental Merit: Field Application," paper SPE 29693, 1995.

Tarner, J.: "How Different Size Gas Caps and Pressure Maintenance Programs Affect Amount of Recoverable Oil," *Oil Weekly*, 144: 32-34, June 12, 1944.

Index